HANDBOOK OF DEVELOPMENTAL SOCIAL NEUROSCIENCE

Handbook of
Developmental
Social
Neuroscience

EDITED BY

MICHELLE DE HAAN
MEGAN R. GUNNAR

THE GUILFORD PRESS
New York   London

Library of Congress Cataloging-in-Publication Data

Handbook of developmental social neuroscience / edited by Michelle de Haan, Megan R. Gunnar.
    p. ; cm.
  Includes bibliographical references and indexes.
  ISBN 978-1-60623-117-3 (hardcover : alk. paper)
  1. Developmental neurobiology—Handbooks, manuals, etc. 2. Social psychology—Handbooks, manuals, etc. 3. Neurosciences—Handbooks, manuals, etc. I. De Haan, Michelle, 1969–
II. Gunnar, Megan R.
  [DNLM: 1. Behavior—physiology. 2. Human Development—physiology. 3. Brain—physiology. 4. Interpersonal Relations. 5. Neurosciences—methods. WS 105 H2359 2009]
  QP363.5.H366 2009
  612.8—dc22

                                                                    2009001423

About the Editors

Michelle de Haan, PhD, is Reader in Developmental Cognitive Neuroscience in the Institute of Child Health at University College London. Dr. de Haan's research applies neuroimaging and neuropsychological methods to examine the neural correlates of typical and atypical cognitive and social development. She has published over 70 articles, books, and book chapters in this area, and is Associate Editor of the journal *Developmental Science*.

Megan R. Gunnar, PhD, is Regents Professor of Child Development in the Institute of Child Development at the University of Minnesota. Dr. Gunnar's research focuses on stress biology and the role of early adversity in shaping stress, emotion, and cognitive functioning in the developing child. She has published over 150 articles, chapters, and edited volumes in this area.

Contributors

Leslie Atkinson, PhD, Department of Psychology, Ryerson University, Toronto, Ontario, Canada

Jocelyne Bachevalier, PhD, Yerkes National Primate Research Center and Department of Psychology, Emory University, Atlanta, Georgia

Karen L. Bales, PhD, Department of Psychology and California National Primate Research Center, University of California, Davis, Davis, California

Robert James Richard Blair, PhD, Mood and Anxiety Disorders Program, National Institute of Mental Health, Bethesda, Maryland

Sarah-Jayne Blakemore, PhD, Institute of Cognitive Neuroscience, University College London, London, United Kingdom

C. Sue Carter, PhD, Department of Psychiatry and Brain–Body Center, University of Illinois at Chicago, Chicago, Illinois

Leslie J. Carver, PhD, Departments of Psychology and Human Development, University of California, San Diego, La Jolla, California

Tony Charman, PhD, Behavioural and Brain Sciences Unit, Institute of Child Health, University College London, London, United Kingdom

Suparna Choudhury, PhD, Division of Social and Transcultural Psychiatry, McGill University, Montréal, Québec, Canada

Barbara T. Conboy, PhD, Institute for Learning and Brain Sciences, University of Washington, Seattle, Washington

Lauren Cornew, PhD, Department of Radiology, Children's Hospital of Philadelphia, Philadelphia, Pennsylvania

Eveline A. Crone, PhD, Department of Developmental Psychology, Leiden University, Leiden, The Netherlands

Geraldine Dawson, PhD, Autism Speaks and Department of Psychiatry, University of North Carolina at Chapel Hill, Chapel Hill, North Carolina

Jean Decety, PhD, Departments of Psychology and Psychiatry, University of Chicago,
Chicago, Illinois

Michelle de Haan, PhD, Institute of Child Health, University College London,
London, United Kingdom

Monique Ernst, MD, PhD, Neurodevelopment of Reward Systems, Mood and Anxiety Disorders
Program, National Institute of Mental Health, Bethesda, Maryland

Susan Faja, MS, Child Clinical Psychology Program, Department of Psychology, University
of Washington, Seattle, Washington

Teresa Farroni, PhD, Centre for Brain and Cognitive Development, School of Psychology, Birkbeck,
University of London, London, United Kingdom; Department of Developmental Psychology and
Socialization, University of Padua, Padua, Italy

Elizabeth Finger, MD, Department of Neurology, University of Western Ontario,
London, Ontario, Canada

Alison S. Fleming, PhD, Department of Psychology, University of Toronto, Toronto, Ontario, Canada

Nathan A. Fox, PhD, Child Development Laboratory, Department of Human Development,
University of Maryland, College Park, College Park, Maryland

Louise Gallagher, MD, PhD, Department of Psychiatry, Trinity Centre for Health Sciences,
St. James's Hospital, Dublin, Ireland

Andrea Gonzalez, PhD, Department of Psychology, University of Toronto, Toronto, Ontario, Canada

Tobias Grossmann, PhD, Centre for Brain and Cognitive Development, School of Psychology,
Birkbeck, University of London, London, United Kingdom

Megan R. Gunnar, PhD, Institute of Child Development, University of Minnesota,
Minneapolis, Minnesota

Michelle K. Jetha, PhD, Department of Psychology, Neuroscience and Behaviour, McMaster
University, Hamilton, Ontario, Canada

David J. Kelly, PhD, Department of Psychology, University of Glasgow, Glasgow, United Kingdom

C. W. Lejuez, PhD, Center for Addictions, Personality, and Emotion Research, University
of Maryland, College Park, College Park, Maryland; Yale Child Study Center, Yale University,
New Haven, Connecticut

Yan Liu, PhD, Department of Psychology and Program in Neuroscience, Florida State University,
Tallahassee, Florida

Jessica Magidson, BA, Center for Addictions, Personality, and Emotion Research, University
of Maryland, College Park, College Park, Maryland

Donatella Marazziti, MD, Department of Psychiatry, Neurobiology, Pharmacology,
and Biotechnologies, University of Pisa, Pisa, Italy

Abigail Marsh, PhD, Department of Psychology, Georgetown University, Washington, DC

Anna Matheson, BSc, Institute of Child Health, University College London,
London, United Kingdom

Linda C. Mayes, MD, Yale Child Study Center, Yale University, New Haven, Connecticut

Meghan Meyer, MS, Department of Psychology, University of Chicago, Chicago, Illinois

Debra Mills, PhD, School of Psychology, Bangor University, Wales, United Kingdom

Masako Myowa-Yamakoshi, PhD, Graduate School of Education, Kyoto University, Kyoto, Japan

Charles A. Nelson, PhD, Laboratory of Cognitive Neuroscience, Developmental Medicine Center, Children's Hospital Boston, Harvard Medical School, Boston, Massachusetts

Sarah S. Nicholls, PhD, Yale Child Study Center, Yale University, New Haven, Connecticut

Olivier Pascalis, PhD, Department of Psychology, University of Sheffield, Sheffield, United Kingdom

Christa Payne, MS, Yerkes National Primate Research Center and Department of Psychology, Emory University, Atlanta, Georgia

Daniel S. Pine, MD, Child and Adolescent Research, Mood and Anxiety Disorders Program, National Institute of Mental Health Intramural Research Program, Bethesda, Maryland

Seth D. Pollak, PhD, Departments of Psychology, Anthropology, Pediatrics, and Public Affairs, and the Waisman Center for Human Development, University of Wisconsin at Madison, Madison, Wisconsin

Bethany C. Reeb, PhD, Child Development Laboratory, Department of Human Development, University of Maryland, College Park, College Park, Maryland

M. Mar Sanchez, PhD, Yerkes National Primate Research Center and Department of Psychiatry and Behavioral Sciences, Emory University, Atlanta, Georgia

Louis A. Schmidt, PhD, Department of Psychology, Neuroscience and Behaviour, McMaster University, Hamilton, Ontario, Canada

Gudrun Schwarzer, PhD, Department of Psychology, University of Giessen, Giessen, Germany

David Skuse, MD, Institute of Child Health, University College London, London, United Kingdom

Linda Patia Spear, PhD, Department of Psychology and Center for Developmental Psychobiology, Binghamton University, Binghamton, New York

Lindsey Sterling, MS, Child Clinical Psychology Program, University of Washington, Seattle, Washington

Masaki Tomonaga, PhD, Language and Intelligence Section, Primate Research Institute, Kyoto University, Inuyama, Aichi, Japan

Zuoxin Wang, PhD, Department of Psychology and Program in Neuroscience, Florida State University, Tallahassee, Florida

P. Michiel Westenberg, PhD, Department of Developmental Psychology, Leiden University, Leiden, The Netherlands

Joel C. Wommack, PhD, Department of Psychology and Program in Neuroscience, Florida State University, Tallahassee, Florida

Charles H. Zeanah, MD, Department of Psychiatry and Neurology, Tulane University Health Sciences Center, New Orleans, Louisiana

Preface

Developmental social neuroscience encompasses the empirical study of the neural mechanisms underlying the development of social processes, ranging from the perception of social signals to the expression of complex forms of social behavior. A basic assumption of this approach is that a full understanding of social development requires a multilevel analysis, wherein both biological and social levels of analysis and their relations are considered. In the last several years, there has been an increased interest in this area; this interest has been driven by many factors, including advances in technologies for measuring structure and function in the developing brain, and success in applying these techniques to questions of cognitive development. The *Handbook of Developmental Social Neuroscience* aims to provide a comprehensive and up-to-date source of information covering methodological, theoretical, and empirical aspects of the topic. It is the first volume to provide an integration of the information generated by this rapidly growing field. As such, we hope it will be a central resource both to those working in the area and to those who simply wish to learn about this newly emerging field, as well as a source of inspiration for those contemplating its possibilities.

Part I of this book consists of an introductory chapter that provides a brief background on the forces that have led to the rising popularity of social neuroscience, discusses why studying development is important for a full understanding of the neuroscience of social processing, and places the chapters that follow within this context. Following this brief introduction are five parts, each dealing with a major theme in developmental social neuroscience: methodological and biological background; perception of and communication with others; relationships; motivation and emotion; and psychopathology.

Part II, "Methodological and Biological Background," includes two chapters that provide background information on the basic brain circuits involved in social processing and the methods used to study them. In Chapter 2, Gunnar and de Haan outline a wide range of techniques used to study developmental social neuroscience: (1) measures of brain structure, including postmortem and brain imaging methods; (2) measures of brain activation, including functional magnetic resonance imaging (fMRI) and electroencephalography (EEG); (3) measures of neurochemical activation, including measuring bodily fluids and receptors;

(4) measures of peripheral activation, including cardiac activity, electrodermal response, and muscle activity; (5) measures of genetic processes, including behavioral and molecular genetics; (6) measures of neuropsychological processes; and (7) animal studies. This overview provides the basic background needed to evaluate the methodologies of the research described in subsequent chapters, and also highlights the particular challenges of using and integrating these methods to study social processes during development. In Chapter 3, Payne and Bachevalier overview the neuroanatomy of the developing social brain. They outline specific brain regions that, in adults, have been identified as being involved in social processing (including the amygdala, orbital frontal cortex, temporopolar cortex, anterior cingulate cortex, and superior temporal sulcus), and then detail the as yet limited information about their development. Their analysis suggests that maturation of these brain structures does not occur in synchrony but at different rates, and that these structures become increasingly integrated over development from infancy through adolescence to support the progressive emergence of complex social abilities.

The eight chapters in Part III, "Perceiving and Communicating with Others," cover the neural bases of several topics related to the development of the ability to exchange social signals with others. The first three chapters deal with processing of social information in the face and voice—an ability that emerges early in infancy. In Chapter 4, Pascalis, Kelly, and Schwarzer address several hotly debated questions about face perception, such as which components of the face-processing system are present at birth, which develop first, and at what stage the system is mature. They echo the views of Payne and Bachevalier in concluding that there is progressive specialization for face perception that occurs during development at a psychological and neural level. Grossmann and Farroni carry on with the topic of face perception in Chapter 5, but they focus on the perception of a particular facial feature: the eyes and their direction of gaze. They outline evidence suggesting that newborns arrive in the world prepared to process information about direction of eye gaze; however, they also argue that experience plays a key role in allowing infants' behavior to become more flexible and accurate not only in the basic ability to detect a person's gaze direction, but also in linking this information to particular objects and events in the environment. In Chapter 6, the third chapter to address face processing, de Haan and Matheson discuss the development of perception of emotion in faces and in voices. Again, their review of the evidence suggests that the basic brain network involved in processing emotional information in faces and voices is in place in children, but that the extent and pattern of this network's activation undergo developmental change. All three of these chapters note the challenges to studying the neurodevelopment of face perception that arise because of the different methods typically used for children of different ages; they also discuss the limitations of the current literature, such as its focus on static, silent facial stimuli.

The next three chapters in Part III focus on how infants and children use information gathered from the faces and actions of others to guide their own behaviors and interpret the behaviors of others. Infants look to adults when they encounter novel events, and use adults' facial and vocal expressions to regulate their own behavior. In Chapter 7, Carver and Cornew address the neural underpinnings of these abilities and their underlying components, and argue that development relies on automatic behaviors' progressively coming under more sophisticated and voluntary control through experience. Decety and Meyer echo this theme in Chapter 8's discussion of the neural bases and interrelation of the development of imitation and empathy. They argue that motor mimicry, which is unconscious and automatic, plays a key role in the development of emotion sharing, which represents a foundational process in the development of empathy. In Chapter 9, Choudhury, Charman, and Blakemore continue with the topic of understanding others' actions and minds: They discuss the literature on the neural underpinnings of mentalizing, with a particular focus on the period of adolescence. They raise the intriguing paradox that some of the brain regions involved in social cognition undergo dramatic development into adolescence, when the functions mediated by these regions, such as mentalizing, appear to mature much earlier. They also echo the theme that the answer to this puzzle may be in the progressive refinement of the neurobehavioral networks involved.

The last two chapters in Part III deal with the development of social communication. Mills and Conboy address this topic in Chapter 10 from the perspective of human development, describing the neural underpinnings of the development of speech perception, word recognition, and the semantic processing of words and gestures. They echo the conclusions from the preceding chapters in arguing for the importance of social interactions in setting up specializations for these abilities. In Chapter 11, the final chapter of the section, Myowa-Yamakoshi and Tomonaga focus on the evolutionary origins of social communication by discussing the development of imitation in humans and chimpanzees. They show how comparative study reveals differences between the two that are informative for understanding the debate regarding automatic and more voluntary, cognitive processes in early imitation.

Part IV focuses on "Relationships." Gonzalez, Atkinson, and Fleming address the psychobiology of one key relationship, that between mother and child, in Chapter 12. They take the view that early mother–infant interactions provide the baby with initial social experiences that lay a foundation influencing later relationships, and point out that much of this research has focused on the infant and much less on the neural underpinnings of caregiving behavior. Their review of animal studies suggests that the cortex works with the limbic system in mediating hypothalamic control of mothering, but they point out that in humans little work has been done on either the cortical or subcortical brain mechanisms of parenting, and as a consequence little is known about the neurobiological underpinnings of differential parenting. Bales and Carter continue in Chapter 13 with the topic of infant–caregiver bonding, but focus in detail on neuroendocrine aspects. They show that the neurochemicals crucial to formation of the adult bond, including oxytocin (OT), arginine vasopressin (AVP), corticotropin-releasing factor, and opioids, may also be involved in infant–parent bonds. However, they point out that theories regarding how these hormones interact to form these attachments are still emerging. In Chapter 14, Marazziti focuses on the neurobiological and neuroendocrine bases of adult romantic bonds; she describes research on these same neurochemicals and others, covering both the process of falling in love and that of maintaining the pair bond. Wommack, Liu, and Wang provide a perspective on this topic in Chapter 15 from animal studies, describing work with the prairie vole on the neurochemical bases of monogamy, as well as comparative studies with related species that follow a promiscuous lifestyle. Evidence from such studies points to the importance of AVP and OT in pair bonding, and also highlights the importance of dopaminergic mechanisms in partner preference. Wommack et al. suggest that an important avenue for further investigation is to understand how the systems involved in controlling monogamous relationships overlap with those involved in nonsocial behavior—a point that echoes the call in several other chapters for a better understanding of the overlap between brain networks involved in social and nonsocial behavior.

Motivations and emotions can change dramatically over development; the changes in social preoccupations and behaviors often observed during adolescence are classic examples. Part V, "Regulatory Systems: Motivation and Emotion," includes four chapters focusing on the neural and/or hormonal changes related to motivation and emotion over development. Chapter 16, by Schmidt and Jetha, focuses on individual differences in temperament, considering the biological origins of such differences and how they may in turn bias our experience by influencing how we process emotional information. This work points to the importance of such structures as the amygdala for temperamental style and its influence on processing affective information, and provides an excellent example of how different methodologies (in this case, EEG and fMRI) can complement one another in studying social processes. In Chapter 17, Ernst and Spear focus on the neural bases and development of reward systems. Social stimuli are powerful sources of motivation, and this chapter points out the overlap in the neural circuitries involved in social and reward processing. Mayes, Magidson, Lejuez, and Nicholls also pick up on this point in Chapter 18, discussing the caregiver–infant attachment as a primary motivator not only of early infant behavior but also of adult behavior in relationships. They discuss maternal substance abuse and addiction as an example of a type of psychopathology that originates from alterations in the mesolimbic dopaminergic reward systems involved in assigning salience to features of the environment—including, they argue, relationships such as the mother–child relationship.

Chapter 19 considers regulation of behavior with respect to social decision making. Crone and Westenberg break down the construct of social decision making into the components of cognitive control, future orientation, and perspective taking, and consider the neural bases of these components. Echoing the ideas set out in Chapter 3 by Payne and Bachevalier regarding the different rates of maturation of different components of the neural network for social processing, they argue that during development there is an imbalance of the emotion-inducing and emotion-regulating components, due to differential rates of maturation of the underlying neural substrates. In particular, they highlight the importance of understanding the different states of development of different regions of the prefrontal cortex during childhood and adolescence for understanding developmental changes in decision making.

Finally, Part VI, "Perspectives on Psychopathology," includes six chapters that explicitly bring a developmental social neuroscience perspective to our understanding of psychopathology in childhood and adolescence. The chapters in this part draw on themes developed in Parts II–V regarding social perception, relationships, motivation, and emotion, and several chapters also consider the role of genetic polymorphisms in the development of neural systems underlying socioemotional behavior and in modifying vulnerability to adverse social experiences during development. The first four chapters focus on developmental socioemotional disorders with a genetic basis, considering their etiologies and the relations of symptom expression to development of neural circuits and social experiences. In Chapter 20, Pine focuses on major depressive disorder (MDD), an emotional disorder that becomes prominent during adolescence. The chapter describes an information-processing approach to understanding the neuroscience of the disorder. Various social processing abilities are disrupted in adolescent MDD, including face processing, threat processing, memory, reward processing, decision making, and cognitive control; this profile is consistent with the evidence reviewed concerning disruptions of the medial temporal lobe, basal ganglia, and prefrontal cortex that have been documented in MDD. In Chapter 21, Blair, Finger, and Marsh take a similar approach to understanding the development and neural bases of psychopathy. They outline the neurocognitive characteristics of psychopathy, an emotional disorder involving atypical processing of reinforcement that at a neural level involves disruptions to the amygdala and orbital frontal cortex. They highlight areas that remain controversial, such as the involvement of other cortical areas and whether punishment processing is also affected; they also consider the genetic and environmental factors that lead to the condition.

In Chapter 22, Dawson, Sterling, and Faja provide an overview of the social neuroscience of autism. They review the strong evidence for genetic factors in the development of autism, and they describe the early appearance of atypical social orienting, attention to emotional cues, joint attention, motor imitation, and face processing in this condition. They argue that atypical brain development in autism may be further amplified by the altered social experiences that occur as a result of abnormalities in these social skills; they also point out the potential for intervention that this view provides. In Chapter 23, Skuse and Gallagher provide an overview of social and genetic factors in Turner syndrome, Williams–Beuren syndrome, and fragile X syndrome. They first provide an overview of different mechanisms by which genetic factors can lead to atypical development, and then discuss the physical, behavioral, cognitive, and neural phenotypes of the three syndromes. They point out that features of one of these disorders can also occur in another, and that it may thus be more useful to examine how genetic variation influences these more general features than to look for relationships with a specific disorder.

The final two chapters of the volume focus on two dimensions of atypical experiences that may influence the development of social neural systems. In Chapter 24, Reeb, Fox, Nelson, and Zeanah consider the absence of typical social experiences under conditions of deprivation and neglect. They describe the atypical social development that can occur in such circumstances, such as increased occurrence of disinhibited attachment relationships, and speculate that the neural correlates may involve the limbic system and its connectivity to frontal and temporal neural systems. In the final chapter, Chapter 25, Sanchez and Pollak address the presence of atypical social experiences as exemplified by physical abuse. They discuss alterations in the perception of threat that can occur in these circumstances, and indicate how neural systems such as the prefrontal cortex, limbic system, and neuroendocrine systems can modulate this process.

We hope that this brief overview of the contents of this volume will both tempt and guide readers as they approach the volume and read the full contributions. Together, the chapters of this volume provide a fascinating overview of the current state of developmental social neuroscience; in particular, they draw out important themes regarding the neural–hormonal circuits underlying social processing, and discuss how these are shaped by genetic and environmental factors in typical and atypical development. The chapters also highlight that the field is just at its beginning, with most areas in need of further investigation to fill in the outline that is beginning to form. With the many talented researchers both in and joining the field, and the exciting developments in technology and possibilities for cross-disciplinary collaboration that the future holds, the future looks very promising for this young field.

MICHELLE DE HAAN
MEGAN R. GUNNAR

Contents

IV. Relationships

V. Regulatory Systems: Motivation and Emotion

VI. Perspectives on Psychopathology

PART I

INTRODUCTION

The Brain in a Social Environment

Why Study Development?

MICHELLE DE HAAN
MEGAN R. GUNNAR

Recent years have seen a dramatic increase in the number of studies aimed at understanding the neural mechanisms underlying social behavior. This has led to the recognition of a new discipline, social neuroscience, and to the creation in 2006 of two new journals devoted to the topic: *Social Neuroscience* and *Social Cognitive and Affective Neuroscience*. Social neuroscience encompasses the study of a broad range of social abilities, including social-emotional perception, cognition, and behavior, which for the purposes of this chapter we collectively call "social processing." This explosion of research has included investigation of how social and biological factors interact during development, resulting in the subfield of developmental social neuroscience. The aim of this chapter is to provide a brief background on the forces that have led to the rising popularity of social neuroscience, and to highlight why studying development is important for a full understanding of the neuroscience of social processing.

The Rise of Social Neuroscience

Psychological and biological explanations of social processing have historically been considered incompatible. Psychology has traditionally been seen as a "social science," with a wide gulf separating it from the "natural sciences," such as biology and chemistry. One important factor that has enabled researchers to establish a connection between the two levels of explanation is the emergence of tools allowing measurement of human brain activity. New tools, such as functional magnetic resonance imaging (fMRI), have allowed scientists to examine brain activity while healthy human participants actually engage in social processing. Before such tools became available, opportunities to study brain activation in humans were much more limited. For example, researchers might add tasks to clinical assessment of diseased or injured brains, such as recording responses to faces and objects directly from the cortical surface in patients undergoing such assessment for evaluation of epilepsy (Allison, Puce, Spencer, & McCarthy, 1999). Such studies are still valuable, but the availability of such tools as fMRI allows more widespread study of both injured and healthy brains, with scientists even making efforts to overcome the practical and technical difficulties of applying such tools to healthy young infants. For example, fMRI has been used to examine the infant brain's response to auditory social stimuli, such as language (Dehaene-Lambertz, Dehaene, & Hertz-Pannier, 2002) and emotional sounds (Sauter, 2008).

Advances in the application of more traditional tools have also been important. The advent of high-density recordings for electroencephalography (EEG) has been particularly useful for better understanding the spatial characteristics of brain activation in infants and young children, for whom alternate imaging modalities are either not appropriate or more difficult to use. For example, use of high-density event-related potentials (ERPs) together with source localization software has allowed scientists to consider the similarities and differences in patterns of brain activation in infants and adults in response to visual social stimuli, such as faces (Johnson et al., 2005). It is important to recognize, however, that brain imaging alone cannot fully reveal the biological mechanisms of social processing. Genetic, hormonal, biochemical, physiological, and anatomical studies, as well as comparative studies, all provide critical data, and advances in techniques for collecting these types of data have contributed importantly to the growing field of social neuroscience. Indeed, a general lesson emerging from the new discipline is that the relations between neural processes and social or cognitive processing must be studied at multiple levels of analysis.

Another force contributing to the emergence of social neuroscience has been the precedent set by the success of the related field of cognitive neuroscience. The growing body of research identifying the neural mechanisms underlying cognitive processes illustrates how it is possible to study the biological underpinnings of mental processes. The mind–brain complex need no longer be considered an unobservable "black box." Subjective experiences, such as moods and emotions, can be studied with scientific methods. Researchers in cognitive neuroscience have also increasingly acknowledged that social behavior and emotion influence cognitive processes and thus are themselves topics in need of investigation. This has led to an interest in understanding the neural bases of individual differences in these processes as they relate to such social-emotional constructs as personality. For example, recent studies have shown that children scoring lower on socialization show a reduction in an ERP component linked to the anterior cingulate cortex (Santesso, Segalowitz, & Schmidt, 2005), and that children scoring higher on trait anxiety show an augmented ERP response to novel, unexpected sounds (Hogan, Butterfield, Philips, & Hadwin, 2007).

These forces have contributed to shaping the growing field of developmental social neuroscience, and have allowed a preliminary picture of the developing social brain to emerge. The available data highlight similarities across age in the basic networks involved in social processing, but also pinpoint differences in the way in which such networks are activated. As outlined by Payne and Bachevalier (Chapter 3, this volume), a complex network of interconnected subcortical and cortical brain structures has been implicated as a substrate to social cognition in adults. This network includes the hypothalamus, amygdala, anterior temporal lobe, posterior superior temporal sulcus, orbital prefrontal cortex, and medial prefrontal cortex. The limited information available so far suggests that a similar network is involved in children, but that it develops over an extended period from infancy until young adulthood, and that some components of the network (such as the amygdala) come online much earlier in development than other components (such as cortical regions). Functional imaging studies suggest that both the extent and pattern of cortical activation related to social processing change with age. For example, studies of the development of the cortical "fusiform face area" suggest that it is more diffuse and less specifically activated by faces in children than in adults (see Pascalis, Kelly, & Schwarzer, Chapter 4, this volume). Moreover, studies of the neural correlates of thinking about intentions suggest that although the same neural network is active in adults and adolescents, the relative roles of the different areas change, with activity moving from anterior (medial prefrontal) regions to posterior (temporal) regions with age (see Choudhury, Charman, & Blakemore, Chapter 9, this volume).

Although such results clearly make an important contribution to social neuroscience, applying neuroimaging techniques to developing populations is not simply a matter of taking the tools used with adults and using them to test children. Some brain imaging techniques used with adults are not usually applied to typically developing children, for ethical (e.g., positron emission tomography) and/or practical (e.g., fMRI) reasons. Even when techniques can be used with infants or children, important issues surrounding the analysis and interpretation of data need special consideration (see Gunnar & de Haan, Chapter 2, this volume).

The Components of Social Processing

Social processing includes our behavior as we interact with others, the thoughts and emotions we experience in relation to others, and our perceptions of others' social cues and behaviors. As outlined above and described in more detail by Payne and Bachevalier (Chapter 3, this volume), in adults a complex network of interconnected subcortical and cortical brain structures is believed to underlie social behavior, cognition, and emotions. Although a reasonable amount is known about the function of this network in adults, very little is known about its development and how it supports the progressive emergence of complex social abilities. A challenge is that social processes include such complex constructs as empathy, motivation, and theory of mind, which are difficult to map directly onto neural systems. For this reason, complex social processes are often broken down into more specific component processes. There are numerous examples of this approach in this volume. Carver and Cornew (Chapter 7) outline how the skill of social referencing can be broken down into the components of sharing attention, which is important for seeking social information; emotion recognition and associative learning, which are important for relating emotional information provided by others to novel events; and emotion regulation, which is important for using emotional information provided by others to govern one's own behavior. Similarly, Crone and Westenberg (Chapter 19) show how the use of social information to regulate decision making can itself be divided into the components of cognitive control, which keeps relevant information in an active state and exerts goal-directed behavior; future orientation, which involves anticipating consequences on the basis of reward and punishment; and perspective taking, which involves considering the thoughts and perspectives of other people. Mills and Conboy (Chapter 10) describe how language development in the first 3 years of life has been broken down into a set of skills including changes in the perception of phonetic contrasts in infants' native versus non-native language, associating words with meanings, producing the first words, combining words into two- or three-word utterances, and speaking in full sentences. In a final example, Decety and Meyer (Chapter 8) argue that the complex construct of empathy can be deconstructed in a model that includes (1) bottom-up processing of shared motor representations, which are navigated by (2) parietal areas known to be crucial in differentiating the perceiver's own perspective from those of others; all of these can be regulated by (3) top-down executive processing, in which the perceiver's motivation, intentions, and self-regulation influence the extent of an empathic experience. This focus on studying subcomponents of more complex behaviors can be particularly useful from a developmental perspective, when it is often the case that only some components of or precursors to more complex behaviors are observable. Developmental studies can provide unique opportunities to see how the components of the system interact in ways not possible in adults, where all the components are fully mature and operational.

Perceiving and Communicating with Others

One basic component of human social behavior is the ability to perceive social signals from others, and to display social signals in turn. One very active area of research in this domain is the study of the development of face processing (see Pascalis et al., Chapter 4, and de Haan & Matheson, Chapter 6, this volume). Electrophysiological (ERP) studies have identified a face-sensitive ERP component, the N170, and have used this response to begin mapping how the cortical response to faces evolves from infancy to adulthood. These studies have generated intriguing results, demonstrating that the precursors of the N170 are present in infancy but are less face-specific at this stage, and that the neural processing of eyes matures earlier than processing of the whole facial configuration does (see Pascalis et al., Chapter 4). These studies also suggest that although the basic neural network responding to faces is similar in infants and adults, the relative contributions of the components within the network differ with age (Johnson et al., 2005)—a conclusion also supported by the limited data from studies using fMRI in children (see Pascalis et al., Chapter 4). Many questions still remain to be answered, as few developmental studies have focused on how the emotional content of a face influences the brain's response, and even fewer have focused on how the brain processes emotional signals in voices. The existing literature has focused mainly on the perception of static images of faces; these can be considered quite removed from the actual types of social signals normally available in the environment, which would involve multimodal, moving inputs.

Beyond perceiving and displaying emotions, the abilities to communicate and to share affective experiences are also important for participating in the social world. Language allows young children to communicate their thoughts and intentions directly, and the development of language skills appears closely linked to that of other complex social skills, such as joint attention, empathy, and mentalizing. Less is known about the underlying neural bases of such complex skills in development, but the research described in this volume illustrates how even such complex areas are being tackled by innovative researchers. It has been established that a network including the temporal poles, the posterior superior temporal sulcus/temporoparietal junction, and the medial prefrontal cortex is involved in mentalizing. However, a puzzle for developmentalists is that some of these brain regions undergo very dramatic development into adolescence, even though the function of mentalizing appears to mature much earlier (see Choudhury et al., Chapter 9, this volume).

One important theme in much of the research investigating the neural correlates of perceiving and communicating social information has been the quest to understand how experience in the social world itself shapes the development of such abilities. For example, how do the faces that we see and the language that we hear affect the development of face- and speech-processing systems in the brain? Evidence suggests that at least in these domains, experiences in infancy have an important and lasting impact on how such social-perceptual information is processed (see Pascalis et al., Chapter 4, and Mills & Conboy, Chapter 10, this volume, for further discussion). A better understanding of the brain systems involved in these skills, and of the role experience plays in shaping their development, is important not only for the study of normative development; it may also provide insight into why these abilities sometimes do not develop normally and can also provide tools for earlier detection of disorders (see, e.g., Dawson, Sperling, & Faja, Chapter 22, this volume).

Relationships

Interpersonal relationships provide the context in which children construct a sense of who they are as individuals and social partners. Not only are relationships important for human development, but infants and parents are strongly motivated to form long-lasting relationships, termed "attachments," to one another. The recognition that attachment is a species-typical human motivational system has had a profound impact on the field of social development. Understanding the neurobiology of the attachment motivational system has become a focus of research in developmental social neuroscience. The study of species in which adults form pair bonds is providing insights into the genetic and neurobiological processes that support the formation of long-lasting attachments (see Wommack, Liu, & Wang, Chapter 15, this volume). Not only has this work been applied to human adult romantic relationships (see Marazziti, Chapter 14, this volume); it has guided researchers interested in the formation of parent–child attachment relationships to focus on the roles of two neuropeptide hormones, oxytocin and vasopressin, in the formation and maintenance of these relationships (see Gonzalez, Atkinson, & Fleming, Chapter 12, and Bales & Carter, Chapter 13, this volume). The influence of these neuropeptides on dopaminergic activity in the nucleus accumbens appears to be important in supporting motivated approach to the object of attachment. However, animal studies have also revealed that the hypothalamic–pituitary–adrenocortical system and its neuroactive peptide, corticotropin-releasing factor, along with pain-alleviating opiate peptides, all play a role in regulating parent–offspring attachments, implicating regulation of the fear system as an important component of these relationships (see Bales & Carter, Chapter 13). An understanding of the neural systems that underlie the formation and maintenance of parent–offspring relationships also helps contribute to our understanding of why disruptions in these relationships can have such long-term consequences for the developing child, as reviewed in this volume by Reeb, Fox, Nelson, and Zeanah (Chapter 24) and by Sanchez and Pollak (Chapter 25).

Motivation and Emotion

Explicating the neurobiology of motivation and emotion is central to the study of social neuroscience and critical to our understanding of social-emotional development. Any observer of young children quickly realizes that there are marked individual differences among children in their response to novel, strange, or unfamiliar events and in their emotional exuberance. Some of the earliest work in developmental social neuroscience has been directed at understanding the neural bases for these differences. As outlined by Schmidt and Jetha (Chapter 16, this volume), studies examining biases in EEG activity over the left and right frontal cortex provide evidence that processes influencing these biases are relevant for understanding both extremely inhibited or socially reticent children and those who are extremely uninhibited and socially outgoing. Not surprisingly, given the work on parent–offspring attachment and motivational systems, research on the approach/reward and avoidance/withdrawal systems more generally implicates the same neural systems that are discussed in work on parent–offspring attachment (see Ernst & Spear, Chapter 17, and Mayes, Magidson, Lejuez, & Nicholls, Chapter 18, this volume). Within the broader context of social neuroscience, interest in these systems reflects interest in understanding the neural bases of addiction and anxiety. From a developmental neuroscience perspective, these questions are motivated by an interest in understanding (1) why disturbances to the parent–offspring relationship increase the risk of substance abuse and depression; and (2) how individual differences in these systems may predispose children to different affective and behavioral pathologies, given exposure to adverse patterns of parenting.

Psychopathology

Studies of the neural bases of social processing can make an important contribution to our understanding of developmental disorders in social processing. The development of brain systems involved in social processing is shaped both by genetic information and through interaction with the environment, and disruption of either of these forces can lead to psychopathology. Children with genetic disorders that affect brain development may have atypical social behavior as a direct result of the genetic influence. Skuse and Gallagher (Chapter 23, this volume) outline work linking genotypes and phenotypes in the context of genetic developmental disorders, and highlight the need for the collection of more detailed information on the cognitive and behavioral phenotypes of brain structural anomalies observed in such disorders. Disordered social behavior in children with genetic disorders can also arise indirectly, as their atypical brains lead them to experience the world differently. For example, Dawson et al. (Chapter 22, this volume) describe how early genetic and environmental risk factors for autism can lead to abnormalities in brain development, which in turn alter children's patterns of responses to and perception of the environment. This view highlights the importance of identifying such early risk factors, in order to minimize their negative influence on further development. Finally, disordered social behavior can also result from negative environmental influences on brain development that otherwise would have occurred normally. In this volume, Reeb et al. (Chapter 24) and Sanchez and Pollak (Chapter 25) both discuss how complex sets of neural circuitry are shaped and refined over development by children's social experience. These studies of children who have experienced extreme social environments involving abuse, neglect, or deprivation are important not only for clarifying the processes through which social development goes awry in those cases, but also for providing a more general understanding of the interplay between social experience and brain development.

Summary and Conclusions:
The Importance of Developmental Social Neuroscience

As noted earlier, a complex network of interconnected subcortical and cortical brain structures has been implicated as a substrate to social behavior. The limited information available to date suggests that the neural structures in this network develop over an extended period from infancy until young adulthood. Importantly, some structures (such as the amygdala), come online much earlier in development than other (e.g., cortical) structures, which become specialized for processing social stimuli over a more protracted period. An important goal for future studies is to provide a more comprehensive picture of which specific social skills functionally mature at which points in development, and to determine which parts of the social brain network are mediating these skills.

One key question in social neuroscience has been whether the general cognitive processes involved in perception, language, memory, and attention are sufficient to explain social competence, or whether over and above these general processes there are specific processes that are special to social interaction. Developmental studies can be useful in this regard, as they allow study of the earliest periods when infants are first experiencing social stimuli, and they potentially allow the study of dissociations among skills that are not observable in adults as different skills emerge at different ages. Although brain structures mature at different tempos, they become increasingly integrated during infancy and adolescence to support the progressive emergence of complex social abilities.

Further developments in technology, and researchers' innovations in using these tools, will also be important for developmental social neuroscience. For example, applying fMRI to young infants provides information about the spatial pattern of brain activation that is more detailed than the data available from more commonly used measures, such as EEG/ERP. However, currently this type of work is very challenging and involves assessment of responses in sleeping children; this excludes certain types of studies, such as research on activation during visual processing. Better integration of the multiple levels of research—genetic, neurophysiological, neurohormonal, and neuroanatomical levels—within developmental social neuroscience is also an exciting prospect for the future.

Finally, better understanding and assessment of social environments will prove important as well. For example, evaluating children's exposure to different emotions during development can prove very challenging for researchers, yet data on such exposure would be very valuable to obtain. Given evidence that experience can have a very rapid (Maurer, Lewis, Brent, & Levin, 1999) as well as a lasting (Maurer, Mondloch, & Lewis, 2007) impact on human development, advances in this area not only will be important for understanding typical and atypical neurosocial development, but also may be very valuable for establishing evidence-based interventions for preventing or ameliorating disorders.

References

Allison, T., Puce, A. L., Spencer, D. D., & McCarthy, G. (1999). Electrophysiological studies of human face perception: I. Potentials generated in occipitotemporal cortex by face and non-face stimuli. *Cerebral Cortex, 8,* 415–430.

Dehaene-Lambertz, G., Dehaene, S., & Hertz-Pannier, L. (2002). Functional neuroimaging of speech perception in infants. *Science, 298,* 2013–2015.

Hogan, A. M., Butterfield, E. L., Phillips, L., & Hadwin, J. A. (2007). Brain response to unexpected novel noises in children with low and high trait anxiety. *Journal of Cognitive Neuroscience, 19,* 25–31.

Johnson, M. H., Griffin, R., Csibra, G., Halit, H., Farroni, T., de Haan, M., et al. (2005). The emergence of the social brain network: Evidence from typical and atypical development. *Developmental Psychopathology, 17,* 599–619.

Maurer, D., Lewis, T. L., Brent, H. P., & Levin, A. V. (1999). Rapid improvement in the acuity of infants after visual input. *Science, 286,* 108–110.

Maurer, D., Mondloch, C. J., & Lewis, T. L. (2007). Sleeper effects. *Developmental Science, 10,* 40–47.

Santesso, D. L., Segalowitz, S. J., & Schmidt, L. A. (2005). ERP correlates of error monitoring in 10-year-olds related to socialization. *Biological Psychology, 70,* 79–87.

Sauter, D. (2008, May). *Brain imaging studies: Theoretical issues and methodological challenges.* Paper presented at the pre-IMFAR workshop Progress in the Study of Brain Functions in Infants At-Risk for Autism, London, UK.

PART II
METHODOLOGICAL AND BIOLOGICAL BACKGROUND

CHAPTER 2

Methods in Social Neuroscience
Issues in Studying Development

MEGAN R. GUNNAR
MICHELLE DE HAAN

Developmental social neuroscience involves studying the development and neural underpinnings of social functioning. A full understanding of this topic requires use of a diverse range of methods, from counting synapses in the developing monkey brain to measuring the hormonal changes that occur when people fall in love. This chapter aims to provide an overview of the different methods that are used to assess neurobiological development and link it to social-behavioral development. Numerous methods are used, and doubtless many more will be developed, given the imagination and ingenuity of researchers in the field. It is beyond the scope of this chapter to provide an exhaustive list of every method employed in animal and human studies. Instead, we focus mainly on the noninvasive methods used in human research. The use of animal models can itself be considered a tool, which can reveal details of neural functioning that are beyond the reach of noninvasive measures used in human studies. However, there are challenges in translating findings across species, which we briefly discuss in this chapter. On the whole, though, we focus here on noninvasive tools used to study the neural bases of social functioning, including discussions of their relative strengths and weaknesses. We focus mainly on the tools that are mentioned throughout the book, and provide briefer information about other available tools that receive less attention within the chapters of this book. This chapter is meant to provide an introduction to these methods for those who are not experts; references are provided for those needing more detailed descriptions of particular methods.

Brain Structure

One basic issue for developmental social neuroscientists is determining when the brain regions involved in social behavior mature structurally (see Payne & Bachevalier, Chapter 3, this volume, for an overview of the basic neuroanatomy of social behavior). Below we discuss two important approaches that have been used to address this question: (1) study of postmortem brain tissue, and (2) structural magnetic resonance imaging (MRI).

Study of Postmortem Brain Tissue

Postmortem studies provide crucial information because they allow a microscopic examination of neural tissue that can reveal details of brain structure not possible to observe with brain imaging methods used *in vivo*, such as structural MRI. Knowledge of these microscopic changes is informative in its own right, and can also be helpful in understanding the reasons for developmental or pathological changes detected with structural MRI.

Techniques for Processing Brain Tissue

After death, brain tissue must be processed so that it is kept in a useful state for further study. An important first step is preserving the brain, either by freezing or by using a fixative (such as formalin) to harden the tissue and kill microorganisms. The tissue may also be embedded in a substance such as paraffin if it is to be cut. Ideally, the tissue will be processed in such a way not only that it is preserved, but that the greatest number of research techniques can subsequently be applied to it.

Staining, tracing, or other techniques can be used to highlight particular characteristics of the tissue. Staining is used to identify different types of cells within the tissue. For example, Nissl stains can be used to reveal the cell body; myelin stains can be used to show the fatty sheath that surrounds nerve cells; and Golgi–Cox stains can be used to visualize the branching and connections of individual neurons. Tracing is used to study connections between neurons. "Anterograde," or "forward," tracing visualizes connections by using special proteins that are taken up by cell bodies and transported down axons to the terminus, whereas "retrograde," or "backward," tracing involves injecting dyes into the terminal buttons of axons that are then transported back to the cell body. Histochemical techniques are used to identify the locations of neurons that secrete particular neurochemicals. Slices of brain tissue are exposed to an antibody for a particular neurotransmitter or enzyme that produces it, and are then viewed under ultraviolet light.

Advantages and Disadvantages

Although postmortem study of brain tissue has the great advantage of allowing more detailed analysis of brain tissue than is possible with noninvasive methods, there are also challenges and potential disadvantages to this technique. Among these disadvantages are the limited availability of such tissue from humans, and the fact that the samples available are not randomly obtained and so may not give a true picture of typical development. Factors that can be difficult to control (e.g., the cause of death, the time between death and tissue fixing or freezing) can all affect the quality of the tissue. The human infant brain, especially the premature one, is soft, is difficult to handle, and can easily be fragmented during removal. Moreover, interpretation of the immature brain can be difficult, because the gross and microscopic features of the brain can change from week to week. This is especially important to keep in mind when one is studying abnormalities, as characteristics that may be normal at one stage can be abnormal at another.

Brain Imaging

Whereas study of postmortem tissue with microscopic methods can provide important, detailed information about brain structure, structural MRI is an important complementary tool, as it allows study of brain structure in living humans.

Basics of the Method

MRI is a noninvasive method that can provide high-spatial-resolution structural images of the brain. MRI primarily images the signal coming from hydrogen nuclei, as hydrogen is part of both water and fat, and water and fat make up the largest proportion of the human body. Hydrogen nuclei possess a property called "spin," which can be thought of as a small magnetic field that produces a measurable signal. When an ambient magnetic field is present, the spinning nuclei tend to become aligned along the main axis of that field. Images are made by applying a series of brief energy pulses (radiofrequency pulses) throughout the field. Each pulse acts to momentarily tilt the hydrogen nuclei in a particular plane. This causes their energy state to change, and measuring this change in energy stage allows an image of the tissue to be created. For example, a measure called "T1 relaxation" is based on the energy released as the pulsed nuclei relax back to their initial, aligned position. Because different tissue types show different relaxation times, this allows differentiation of the different types of tissue in the image, such as gray matter and white matter. Another measure called "T2 relaxation" is based on the decrease in energy as the nuclei gradually fall out of alignment and no longer show uniform motion. This measure is especially useful for identifying brain regions.

More recently, a variant of conventional MRI known as diffusion tensor imaging (DTI) has been developed to study brain white matter in more detail. DTI relies on detection of microscopic movement of water within tissue. In white matter, tissue water does not diffuse freely, but rather in an anistropic, or directional, manner along the axis of the fiber bundle. DTI uses this property to identify and characterize white matter pathways, typically by calculating the overall amount of diffusion and the direction of diffusion.

Analysis of the Magnetic Resonance Signal

Once images are obtained, they needed to be evaluated. Several different methods are available for analysis of MRI. In visual inspection, a trained neuroradiologist inspects the image by eye to identify regions of abnormality. This method allows a relatively quick classification of brain scans as normal or abnormal, but does not provide quantitative information and may be insensitive to subtle variations among brain scans. Volumetrics is a quantitative technique in which a region of interest (e.g., the hippocampus) is identified either by manually tracing around the structure on the slices in which it is visible, or by using an algorithm that automatically identifies the boundaries of the structure. Though this method does provide a quantitative measure of the volume of the target structure, the process of manually tracing can be time-consuming, and many structures in the brain that are of interest for study do not have easily identifiable boundaries.

Voxel-based morphemetry is a quantitative method that allows comparison of the local concentrations of gray and white matter at the smallest three-dimensional unit of MRI measure, the voxel (see Ashburner & Friston, 2000). Before the images can be statistically compared, several steps of preliminary processing are needed to handle differences between images in gross anatomy and positioning. This processing involves first spatially normalizing images from all subjects into a common template (ideally an average of a large number of individual structural images, though standard templates such as Talairach's are also used); then segmenting the tissue into gray matter, white matter, and cerebrospinal fluid; and then smoothing the images. After this processing, statistical tests are used to compare the two groups voxel by voxel with correction for multiple comparisons, resulting in the statistical parametric map. Voxel-based morphometry is useful in that it is a fairly straightforward method that is less time-consuming and requires less neuroanatomical expertise than visual inspection or volumetrics. In addition, it allows a detailed analysis of the whole brain, rather than just a target region of interest. However, this method is sensitive to such factors as image quality and errors in segmentation and normalization; these factors are important to consider in developmental studies, where images may be noisier, and in studies of atypical populations, where abnormal neuroanatomy may result in misregistration during the normalization process.

Advantages and Disadvantages

The main advantage of structural MRI is that it is a noninvasive method that can be used with children to acquire detailed images of the living human brain. A limitation of structural MRI in developmental studies is the nature of the environment in which the images are typically acquired: Participants must stay motionless in a small, noisy space for prolonged periods. These requirements make it difficult to study children younger than 7–8 years, unless they are asleep or sedated. Even for older children, the environment is unusual and can be frightening, and so familiarization with the scanner environment prior to data collection is considered important to maximize cooperation (see Kotsoni, Byrd, & Casey, 2006, for discussion). Additional challenges related to use of MRI in developmental studies are discussed below in the section on functional MRI (fMRI).

Brain Activation

Although knowing whether a brain structure looks mature at a macroscopic or microscopic level is one indication of its development, this information does not necessarily indicate whether the structure is functionally mature. That is, is the structure activated within the same neural network, under the same circumstances, and so on, as in adults? Several tools have been used to assess brain activation during development, in order to increase our understanding of the functional development of brain areas involved in social behavior. Below we describe the two most commonly used methods: (1) fMRI and (2) electroencephalography (EEG)/event-related potentials (ERPs). Brief descriptions of and references for other methods, including positron emission tomography (PET), magnetoencephalography (MEG), near-infrared spectroscopy (NIRS), and transcranial magnetic stimulation (TMS), can be found in Table 2.1.

TABLE 2.1. Techniques for Studying Brain Structure and Function

Technique	Description	More details of method
	Brain structure	
Postmortem analysis	Brains are retrieved and processed after death to allow analysis of macroscopic and microscopic features of brain tissue.	
Structural magnetic resonance imaging (MRI)	A noninvasive method that can provide high spatial resolution (~ 1 mm^3) structural images of the living brain.	Gadian (2002)
Diffusion tensor imaging (DTI)	A variant of structural MRI that relies on detection of microscopic movement of water within tissue to allow more detailed visualization of white matter.	Cascio, Gerig, & Piven (2007)
	Brain function	
Functional MRI (fMRI)	A noninvasive method that measures brain activation indirectly by measuring changes in blood oxygenation. Provides good spatial resolution but more limited temporal resolution.	Kotsoni, Byrd, & Casey (2006)
Electroencephalography (EEG)/event-related potentials (ERPs)	A noninvasive method that measures the electrical activity of the brain via electrodes placed on the scalp. Provides good temporal resolution but more limited spatial resolution.	deBoer, Scott, & Nelson (2007); Taylor & Baldeweg (2002)
Magnetoencephalography (MEG)	A noninvasive measure of the magnetic fields produced by electrical activity in the brain. Physically, it measures the same currents as does EEG/ERP. Provides good spatial and temporal resolution.	Paetau (2002)
Near-infrared spectroscopy (NIRS)	A noninvasive method that, like fMRI, measures brain activation indirectly by measuring changes in blood oxygenation. NIRS accomplishes this by taking advantage of the different absorption characteristics of oxy- and deoxyhemoglobin to light in the near infrared part of the spectrum.	Meek (2002)
Transcranial magnetic stimulation (TMS)	Involves very short application of strong magnetic fields, using magnetic coils to induce an electric current in circumscribed brain regions.	Moll, Heinrich, & Rothenberger (2002)
Positron emission tomography (PET)	An invasive method wherein a radioactive tracer—commonly a radioactive version of glucose—is injected into the body and travels to tissues that are using glucose, which are then visualized in a scanner. Resolution is typically on the order of 5–8 mm.	De Volder (2002)

Functional Magnetic Resonance Imaging

Basics of the Method

fMRI is an indirect measure of brain activity. It measures changes in blood oxygenation, called the "blood-oxygen-level-dependent" (BOLD) response. This is accomplished by capitalizing on magnetic differences between oxygenated and deoxygenated blood. Deoxygenated hemoglobin becomes strongly paramagnetic, so it can be used as a naturally occurring contrast agent, with highly oxygenated brain regions producing a larger magnetic resonance signal than less oxygenated areas. The method assumes that the local increase in blood oxygenation reflects an increase in neuronal activity—an assumption that has been supported by studies in nonhuman primates (e.g., Logothetis, Pauls, Augath, Trinath, & Oeltermann, 2001). Local differences in the BOLD response between experimental conditions or groups of subjects are often quantified by using the voxel-based morphometry method described above.

Advantages and Disadvantages

As described above, MRI studies in very young children are difficult because of the nature of the scanning environment. fMRI studies present an additional challenge, because they typically rely on some sort of active response from the participant. This raises issues of task compliance and movement, so most fMRI studies have not tested children younger than school age (but see, e.g., Dehaene-Lambertz, Dehaene, & Hertz-Pannier, 2002, for an exception). Even for older children, preparation prior to scanning, so that the children are accustomed to the task itself, is important. For these reasons, fMRI is limited in its capacity to provide a full profile across the age span of functional brain development.

In addition to this issue, there are numerous other challenges for developmental fMRI studies, including developmental differences in anatomy and physiology that can affect the BOLD response, developmental differences in ability to perform the target task, and issues related to processing and analysis of pediatric data—all of which can potentially confound results. These problems are often magnified in studies dealing with atypically developing populations, who may have atypical brains and/or be of lower ability level. For example, age-related structural differences could influence the ability to localize brain activity in a common stereotactic space as that used with adults. Studies that have examined this issue suggest that, at least in children 7–8 years of age and above, anatomical differences between children and adults are small relative to the resolution of fMRI; these findings suggest that direct comparison in a common stereotactic space is reasonable (Kang, Burgund, Lugar, Peresen, & Schlaggar, 2003). However, the same conclusion does not necessarily hold for younger children or for clinical populations with atypical brain structure. A fuller discussion of these issues can be found in Kotsoni et al. (2006).

Electroencephalography/Event-Related Potentials

fMRI is a tool with good spatial resolution for identifying developmental differences in brain activation. However, because fMRI is not easily applied to infants and very young children, it cannot provide a full picture of functional brain development. ERPs constitute an important additional tool, because they can be applied across a wide span of ages and ability levels.

Basics of the Method

The EEG is the ongoing electrical activity of the brain, whereas the ERP is the subset of this activity that reflects processing of specific stimuli. In both methods, the electrical activity is measured via recording electrodes attached to the scalp. For brain electrical signals to be measurable from the scalp, the activity of large numbers of neurons must be summed together. It is believed that excitatory and inhibitory postsynaptic potentials probably provide the current that is detected by ERPs, as their time course is most compatible with that needed to provide sufficient summation of the signal (Allison, Woods, & McCarthy, 1986). When many neurons are activated simultaneously, they summate, and the activity propagates to the surface. The activity recorded at the surface is thought to come primarily from pyramidal cells of the cortex, as these cells tend to be aligned parallel to one another (which helps with summation of activity) and oriented more or less perpendicular to the scalp (which helps with propagation of activity to the surface).

Analysis of EEG/ERPs

Because EEG signals are typically of much larger amplitude than ERPs, the ERP signal must be extracted from the EEG. Typically this is accomplished by averaging the ERP signal over repeated presentations of a particular stimulus or stimulus condition. In this way, the part of the EEG that is random with respect to the timing of the event (and presumably unrelated to event processing) averages out to zero, while the part of the EEG that is time-locked to the stimulus (and presumably specifically related to its processing) is retained.

The ERP waveform itself typically consists of a series of peaks and troughs called "components." Components are usually characterized by the simultaneous consideration of their eliciting conditions, polarity (positive or negative) at particular recording sites, timing (latency), and scalp distribution (topography). Typically components are labeled according to their polarity (P for positive, N for negative) and either their order of occurrence (e.g., P1, P2, etc.) or mean latency from stimulus onset (e.g., N170, P300), although there are exceptions (e.g., the ERN, or error-related negativity is not labeled by its order of occurrence or latency). Peaks in the waveform are traditionally analyzed by quantifying the peak amplitude in microvolts relative to baseline and the timing of this peak relative to stimulus onset, and comparing these measures between conditions or participants via such methods as analysis of variance. Sometimes the ERP waveform also contains broad, sustained deviations from baseline, called "slow waves." These are usually more likely to be observed in younger individuals and/or in situations with a more prolonged recording epoch. Slow waves that do not have broad peaks are typically quantified by calculating the mean amplitude over the time in which they occur, or the area under the curve within this time.

EEG is not time-locked to an event in a precise fashion, but is typically recorded in ongoing, more prolonged conditions. For example, EEG may be recorded while a participant is expressing an emotion or when a certain mood is induced. The resulting electrical activity is traditionally analyzed by decomposing the EEG signal into its constituent frequencies by a method such as Fournier analysis, and quantifying the power in particular frequency bandwidths. Different bandwidths have been related to different states of consciousness or types of brain function. For example, power in the alpha bandwidth (~8–12 Hz in adults) over posterior regions is typically interpreted as being inversely related to brain activation, with high power in this bandwidth linked to low activation (see Stroganova & Orekhova,

2007, for further discussion of development of different bandwidths and their functional correlates). More recently, investigators have been interested in studying event-related oscillations (EROs), which are somewhat intermediate between EEG and ERP measures. EROs are bursts of EEG at particular frequencies that are approximately time-locked to task or stimulus presentation events. There is a growing literature on EROs in adults, particularly with respect to high-frequency (gamma-band) bursts, but this approach has only just begun to be applied to infants (see Csibra & Johnson, 2007, for a review).

Another important type of analysis that can be applied to EEG and ERP data is source analysis, which attempts to identify the location of the sources in the brain generating the scalp-recorded activity. One challenge for EEG/ERP source analysis is the inherently low spatial resolution of these measurements. The other one is known as the "inverse problem": There is not one unique solution that gives rise to a particular pattern of brain activity, but rather many different solutions are possible. For this reason, converging data from studies using other techniques (e.g., fMRI) are important to help constrain solutions. Developmental studies face additional issues, as most commercially available software packages for source localization are based on adult head models and do not take into account differences (in size and shape, volume conduction of signal, etc.) that can occur with development (see Johnson et al., 2001, for further discussion of source analysis in developmental studies).

Advantages and Disadvantages

Among the key advantages of EEG/ERP measures for developmental study are that they can be recorded even in passive tasks, when participants are simply looking at or listening to the stimuli; taking the measurement itself is not noisy; and they are not as sensitive to movement artifact as MRI methods. For all these reasons, EEG and ERPs have been used across a wider range of ages and ability levels than fMRI has. Another advantage is that ERPs provide much better information about the timing of brain events than can fMRI, with the former providing temporal resolution on the order of milliseconds and the latter on the order of seconds. The main disadvantages of EEG and ERPs are their limited spatial resolution, which is inferior to that of fMRI and many other methods, and their relative insensitivity to subcortical sources.

Neurochemical Measures

Studies of neurochemical activity (neurotransmitters, neuroactive peptides, neurohormones, hormones) also provide important insight into the functional development of the brain. Neurochemicals are critical to communication in both the central and peripheral nervous systems. Neurotransmitters relay, amplify, and modulate electrical signals between neurons; the classic neurotransmitters being gamma-aminobutyric acid (GABA), glutamate, the monoamines (norepinephrine [NE], epinephrine [E], dopamine [D or DA]), and acetylcholine. To these chemical messengers, however, we must add more than 50 neuroactive peptides, including oxytocin, vasopressin, and corticotropin-releasing factor (CRF), which have particular importance for social neuroscience. We must also add hormones produced by endocrine glands (e.g., the sex steroids, cortisol, thyroid hormones) that have central effects. Many of these neurochemicals not only regulate neural activity throughout the lifespan, but also serve as trophic factors during brain development (e.g., serotonin; Persico, Di Pino, & Levitt, 2006).

Measurement in Bodily Fluids

Neuroactive chemicals differ in how locally they produce their effects. Paracrine signaling is a form of signaling in which the target cell is close to the signal-releasing cell. Neurotransmitters fall into this category. Increases in activity of a paracrine-signaling cell may be highly local, requiring local measures to assess its impact most accurately. Hormones, on the other hand, are released into the bloodstream and can affect cells and tissues far from their sites of production. All neurochemicals are found in bodily fluids and are ultimately metabolized and excreted in urine. Neurochemicals can thus be measured relatively noninvasively in bodily fluids, including blood, saliva, and urine. Some studies, especially in nonhuman primates, also include assessments in cerebrospinal fluid. The critical measurement issue is whether the levels measured in such fluids reflect activity of the neurochemical in the neural system under study. For example, oxytocin and vasopressin figure prominently in current research in social neuroscience. However, the activity of these neuropeptides in the brain is what is important in neuroscience. Both oxytocin and vasopressin are produced by cells in the hypothalamus and released by the pituitary gland into general circulation. But the oxytocin and vasopressin acting in the central nervous system are produced in the brain, and are not derived from those released by the pituitary. Thus researchers question whether measures of oxytocin and vasopressin in plasma or urine reflect concentrations of these neurochemicals acting in the brain. Since similar problems arise for all neurochemicals acting in the brain that are produced in the brain, measures of the neurochemicals in bodily fluids create challenges to interpretation.

Assays

Regardless of the bodily fluid used or the neurochemical under study, immunoassays are the most widely used tool for measurement (Diamandis & Christopoulos, 1996). The logic of many of these assays is the same. An antibody is produced that is highly specific to the molecule of interest (i.e., an antibody to human cortisol, E, serotonin, etc.). The antibody has binding sites for the neurochemical. If a known amount of the neurochemical is placed in solution with an unknown amount, the two will compete for binding sites on the antibody. If the known amount is tagged or labeled in some measurable way, then the less labeled substance recovered from the assay, the more unlabeled substance in the unknown or subject's sample. The label used may consist of a radioisotope (as in a radioimmunoassay [RIA]) or an enzyme (as in an enzyme immunoassay [EIA]). Assays using enzymes rather than radioisotopes are becoming increasingly preferred, because they do not require meeting codes for the handling and disposing of radioactive material, as RIAs do. An EIA may use a label that produces color, fluorescence, or light, the intensity of which is then analyzed.

Issues in Analyzing Bodily Fluids

A number of problems challenge the assaying of neurochemicals in bodily fluids. Many neurochemicals are very unstable, because they are rapidly metabolized. These chemical reactions must be stopped immediately after sampling, via freezing or adding a chemical that stops the reaction. Some chemicals that affect neural activity, however, are quite stable (e.g., steroid hormones), and samples of these can remain at room temperature for days or weeks. The method used to collect and store the sample can be very critical. Some neurochemicals stick to glass (e.g., adrenocorticotropic hormone [ACTH]); others stick to the absorbent material that may be used to collect the bodily fluid; and sometimes the collection matrix introduces substances that interfere with the chemical reaction used in the assay. For example, interference has been reported for substances (e.g., citric acid) used to stimulate saliva flow in studies examining the concentration of cortisol in saliva (Schwartz, Granger, Susman, Gunnar, & Laird, 1998). A full discussion of the problems that can arise in collecting, storing, and assaying neurochemicals in bodily fluids is beyond the scope of this brief chapter. Let it suffice to say that in some cases, lack of agreement between studies may reflect differences in collection and handling of samples.

Receptors

All neurochemicals operate on their target tissues through interaction with their specific receptors. Often there is a family of receptors that may mediate quite different effects of the same neurochemical. For example, CRF interacts with CRF1 and CRF2 receptors, which have different neural distributions and functions. CRF1 receptors mediate many of the anxiogenic effects of this neuropeptide, whereas CRF2 mediates effects on eating, sleeping, and other more vegetative functions (Zoumakis, Rice, Gold, & Chrousos, 2006). Assessing the functional activity of a neurochemical system thus includes measurement of the neurochemical's receptor system.

All research on the functional activity and distribution of specific receptors—be they the receptors for a neurotransmitter (e.g., D2 receptors), a neuropeptide (e.g., CRF1 receptors), or a hormone (e.g., glucocorticoid receptors)—requires identifying ligands that are specific to the receptor. Novel ligand discovery is critical to both basic research and treatment in social neuroscience (Pike, Halldin, & Wikstrom, 2001). PET and its close cousin single-photon emission computed tomography (SPECT) are proving increasingly useful in the study of receptors, transporters, and other processes related to neurochemical signaling (Winnard & Raman, 2003). This area is referred to as "molecular imaging." In general, these procedures involve radiolabeling a ligand and examining its interaction with its specific receptor *in vivo*. Although none of the chapters in this volume discuss molecular imaging, we mention it here because the development of molecular imaging may well play an important future role in research relating the neurochemical substrates found in animal studies to disorders of social interaction in human psychopathology (see Frankle & Laruelle, 2002).

Receptor Agonists and Antagonists

Once specific agonists and antagonists for a receptor have been identified, they can be used in animal research to identify the role of the receptor in social and emotional behavior. Several animal social neuroscience studies have taken advantage of this means of investigating the role of neurochemical systems in social behavior. For example, antalarmin is an antagonist of the CRF1 receptor. It occupies this receptor, blocking its activation by CRF. In rhesus monkeys, oral administration of antalarmin not only reduced ACTH and cortisol production, but also significantly reduced fear behavior and increased exploration and sexual behavior in animals stressed by being placed adjacent to two unfamiliar males (Habib et al., 2000). Such data not only provide evidence of the role of CRF and the CRF1 receptor in fear behavior, but also stimulate interest in potential clinical applications (Zoumakis et al., 2006).

Oxytocin and vasopressin and their receptor systems are playing prominent roles in social neuroscience research (see, e.g., Marazziti, Chapter 14, and Wommack, Liu, & Wang, Chapter 15, this volume). Because of the role of oxytocin in human and animal parturition, there is considerable clinical interest in developing drugs that antagonize the oxytocin receptor, with some already approved in Europe for the treatment of preterm labor (Gimpl, Postina, Fahrenholz, & Reinheimer, 2005). To our knowledge, these ligands—some of which are highly specific to the oxytocin receptor, and others of which interact also with the vasopressin 1a receptor—have not been used in animal studies of the role of these neurochemicals in social interaction. However, some ligands of receptors of interest in social neuroscience have been studied in human research. For example, spirolactone, a diuretic used in the treatment of hypertension, occupies the mineralocorticoid receptors, which in the brain is one of the two types of receptors with which cortisol interacts. Studies of spirolactone in humans indicate that it elevates basal, but not stress-induced, cortisol and impairs cognitive functioning (Otte et al., 2007).

Peripheral Measures

In addition to the techniques aimed at assessing the central nervous system described above, researchers have also benefited from measuring activity of the peripheral nervous system, as this system is inherently involved in our experience of emotion. Measures of peripheral activation, such as heart rate, muscle activity, and electrodermal responses, can be easier to obtain than many of the measures described above, yet still provide important insights into neural bases of social behavior. Cardiac and electrodermal measures have been used to assess activity of the autonomic nervous system (ANS). When measures of muscle activity are sought, these are often measures of involuntary responses (e.g., eyeblink startle) believed to reflect activation of specific pathways in the central nervous system.

Measures of the Autonomic Nervous System

The ANS both responds to information processed by the central nervous system and provides feedback that affects central activity (Porges, 1995). The two arms of the ANS, the sympathetic nervous system (SNS) and the parasympathetic nervous system (PNS), are regulated through different neural pathways and, while often opposing one another, function separately (Berntson, Cacioppo, Quigley, & Fabro, 1994b). Because many target organs are innervated by both arms of the ANS, simple measures of their activity do not allow the researcher to decompose SNS and PNS contributions. For example, although an increase in heart rate may be interpreted as a sympathetic response, this may be inaccurate: Heart rate increases can reflect increased sympathetic input, decreased parasympathetic input, or some combination. Different psychological states (e.g., fear vs. positive excitement) produce heart rate accelerations through different patterns of sympathetic activation and parasympathetic withdrawal (Berntson, Cacioppo, Binkley, Uchino, & Quigley, 1994a). Therefore, researchers have sought pure measures of each arm of the ANS. We briefly describe commonly used procedures, beginning with those used to assess sympathetic activity.

Sympathetic Activity

The SNS uses acetylcholine as its preganglionic neurotransmitter, and NE (predominantly in most tissues) and E (predominantly from the adrenal medulla) as its postganglionic neurotransmitters. E and NE produce their effects via alpha-adrenergic (higher affinity for NE than E) and beta-adrenergic (higher affinity for E than NE) receptors. Although the SNS is capable of mass discharge, as in intense fight–flight responses, the differential sensitivity of alpha- and beta-adrenergic receptors to E and NE, and the fact that most E is produced by the adrenal medulla rather than the nerve endings of most postganglionic sympathetic fibers, mean that assessing sympathetic input to different organs provides information about different aspects of sympathetic functioning. In short, sympathetic reactivity is often highly nuanced.

CARDIAC ACTIVITY: PRE-EJECTION PERIOD

The pre-ejection period (PEP) is being used increasingly to index sympathetic activity. It is the time from the onset of ventricular depolarization, as indexed by the QRS complex in an electrocardiogram, until the beginning of left ventricular ejection (Berntson, Quigley, & Lozano, 2007). Stated simply, it is the time when the left ventricle of the heart is contracting against the closed aortic valve, or the time between when the heart says "Pump" and the valve is opened to allow the left ventricle to pump the blood to the right ventricle. Beta-adrenergic receptors responsive to E (particularly) regulate PEP duration, with increases in beta-adrenergic activity *shortening* the PEP.

To measure PEP, the researcher needs to know both when the QRS complex occurs (i.e., when the heart says "Pump") and when the blood is released from the left to right ventricle (i.e., when the aortic valve opens). The QRS complex can be identified in the electrocardiogram, but to obtain information on when the aortic valve opens, impedance cardiography is used. Electrodes are placed on the neck and thorax, and a low-grade electrical current (2–4 mA) is sent through the heart. "Impedance" is resistance to the flow of an electrical current, and because blood is a better conductor than tissue, impedance is lower when the left ventricle is full of blood than when it is empty. The increase in impedance when the ventricle empties (and thus the aortic valve has opened) is indexed by a particular waveform (dZ/dt). Thus to measure PEP, one identifies the time difference between the QRS signal in the electrocardiogram and the dZ/dt complex in the impedance cardiogram (see also Lozano et al., 2007). PEP is measured with each heartbeat and typically averaged. Because sympathetic activation of the heart may be very brief in response to many psychological challenges imposed in the laboratory, the issue for the researcher is to identify the smallest number of heartbeats to average to capture a sympathetic response while still averaging out errors in measurement. Two-minute periods are often preferred (e.g., Demaree et al., 2006). Developmentally, although PEP has been measured in infants, children, and adults, there is evidence that this measure may be less responsive to stimulation in children (Quigley & Stifter, 2006).

ELECTRODERMAL MEASURES: GALVANIC SKIN RESPONSE

The galvanic skin response (GSR), also called the skin conductance response or electrodermal response, is a method of measuring the electrical resistance of the skin. This resistance is decreased with sweating, as electrical signals flow more readily through fluid than tissue. Sweat is produced by eccrine glands that are under sympathetic regulation. Thus GSR provides a measure of sympathetic activity. GSR is conducted by attaching electrodes to the skin, usually on the fingers. There are various automated systems for GSR collection.

SALIVARY ALPHA-AMYLASE

Amylase, an enzyme, is found in saliva where its function is to break down starches. The isoform, alpha-amylase, is the predominant form in mammals. Sympathetic innervation of the salivary glands (predominantly NE) stimulates viscous alpha-amylase secretion. Although there is still considerable disagreement over interpretation, researchers have recently begun measuring alpha-amylase concentrations in saliva as an indirect index of sympathetic activity (Granger, Kivlighan, el-Sheikh, Gordis, & Stroud, 2007).

Parasympathetic Activity

The parasympathetic arm of the ANS complements sympathetic activity in the organs and tissues they jointly innervate. Whereas sympathetic activity typically increases activity, parasympathetic activity typically provides the brakes (Berntson et al., 2007). The PNS uses only acetylcholine as its neurotransmitter, operating through muscarinic and nicotinic cholinergic receptors. Typically, parasympathetic ganglia are close to their areas of innervation, allowing the PNS to operate in a highly nuanced fashion. Most of the interest in social neuroscience has been directed toward the portion of the PNS associated with the 10th cranial or vagus nerve, arising from cell groups in the brainstem at the level of the inferior olivary complex. Two cell groups in this complex regulate different aspects of vagal activity: The nucleus ambiguus provides vagal innervation to the heart, while the motor nucleus primarily regulates activity at the level of the gut.

This differentiation of vagal regulation is the basis for Porges's (1995) polyvagal theory. Porges argues that cardiac measures of vagal activity are closely orchestrated with neural activity innervating facial muscles—and thus with the capacity of humans and other higher primates to engage in complex social interchanges that involve controlling emotional expressions, as well as speech, which requires coordination of breathing, tongue, and larynx. Because of vagal feedback to the nucleus tractus solitarius in the midbrain, which sends afferents to the hypothalamus, amygdala, and cingulate gyrus, activity of this segment of the PNS may also play a role in inhibiting or regulating central emotional reactivity. From an individual-differences standpoint, considerable attention has been paid to differences in vagal regulation as it relates to social and emotional competence (e.g., Bornstein & Suess, 2000; Calkins, 1997; Katz & Gottman, 1995).

RESPIRATORY SINUS ARRHYT MIA OR VAGAL TONE

Vagal input to the heart slows heart rate by reducing the conduction velocity of the sinoatrial node. During expiration (breathing out), vagal input increases, slowing the heart; during inspiration (breathing in), vagal input decreases, allowing the heart to beat faster. Variations in heart rate associated with breathing index the degree of vagal input to cardiac control—hence the terms respiratory sinus arrhythmia (RSA) or vagal tone (VT). Average heart rate variability provides a rough index of vagal regulation, but because only some beat-to-beat variability reflects breathing, heart rate variability is not a specific index of vagal activity. All techniques for measuring RSA or VT depend on isolating the beat-to-beat variability dependent on breathing. This can be done somewhat directly by simultaneously collecting respiration data and then isolating co-occurring variability in interbeat intervals, or through examining variation in the spectral frequency that is typically associated with breathing. Both methods provide fairly similar results in healthy children and adults under most laboratory conditions of psychosocial measurement (Berntson et al., 1997). RSA or VT tends to decrease when individuals actively process social information and when they speak. Developmentally, baseline RSA or VT increases over the first years of life (Porges, Doussard-Roosevelt, Portales, & Greenspan, 1996), although changes in RSA or VT to social and emotional stimuli are observed in both children and adults (Quigley & Stifter, 2006).

ELECTROMYOGRAPHY

Electromyography (EMG) detects the electrical potential generated by muscle cells—potentials that change when a muscle contracts. In addition to its use in detecting eye movement artifact during EEG recordings, EMG is used to detect startle reflexes.

Startle Response.   Startles, components of the defensive motivational system, are responses to sudden, unexpected stimuli (e.g., a flash of light or a loud noise). They include involuntary contractions of muscles in the legs, arms, and around the eye (i.e., the blink reflex). Startle responses figure in social and affective neuroscience because they are modulated by emotion-evoking stimuli that produce different motivational states (Lang, Bradley, & Cuthbert, 1998). Emotion modulation of the startle response involves pathways from the amygdala and bed nucleus to brainstem nuclei that generate startle responses (Davis, Walker, & Lee, 1997). It is common to measure startle by placing electrodes over the orbicularis oculi, a muscle group just below the eye, and then measuring the EMG intensity of blink in response to a startle (often acoustic) probe. The motivational or emotional states of participants can be manipulated by conditioning them to threat-predicting versus safety-predicting cues (Lissek et al., 2005), or by showing them static photos or movie clips rated to evoke different emotional/motivational states (for a review, see Bradley, 2007). Photos of people displaying different emotional expressions also modulate the startle reflex, with increases in response to threatening facial expressions and decreases to positive facial expressions; interestingly, though, there is some evidence that anger expressed by male faces produces larger startle increases than that expressed by female faces (Hess, Sabourin, & Kleck, 2007).

Postauricular Reflex.   Whereas the startle reflex increases under aversive and decreases under positive or pleasant motivational conditions, researchers are now beginning to examine a second reflex that is elicited by the same probes but bears the opposite relation to motivational states. This reflex is produced by the postauricular (PA) muscle that lies just behind each ear. Although its neural circuitry is not yet understood, its evolutionary origins probably lie in the motivation to increase auditory sensitivity in the presence of approach or appetitive stimuli, and to decrease such sensitivity and protect the ear in the face of threat. The PA reflex is assessed via EMG, with two electrodes—one on the surface of the pinna, and the other on the scalp over the PA muscle. It is becoming more common in startle studies to measure both the eyeblink startle to assess defensive motivation and the PA reflex to examine appetitive motivation (see Benning, Patrick, & Lang, 2004; Hess et al., 2007).

Genetics

With the sequencing of the human genome and rapid advances in genetic technologies, there are increasing attempts to bridge genetics with both social and cognitive neuroscience. The burning question is this: Can the function of specific genes be linked to specific social functions or dysfunctions? Attempts to answer this question have involved a number of different methods (see Table 2.2 for an overview). Here we briefly discuss two of these, behavioral genetics and molecular genetics.

Behavioral Genetics

Quantitative behavioral genetics involves indirect methods that do not assess actual differences in DNA sequences, but instead involve assessing how these differences influence the development of individual differences in behavior by examining families or special cases, such as twins or adoptees (see Pennington, 2002, or Plomin, DeFries, McClearn, & Rutter, 1997 for more information about these methods). Family studies allow assessment of the familiality of a trait, which is measured as the correlations among relatives for a continuous trait or the relative risk for a categorical trait. It is important to note that familiality does not prove genetic influence, as the environment can also mediate familiality. Twin and adoption studies are important for understanding whether familiality is in part due to genetic influence. This type of study allows investigators to estimate the proportion of variance in a trait that can be attributed to genes and to the environment. For example, one study examined 3000 pairs of 7-year-olds to understand the genetic basis of social and nonsocial (e.g., obsessive behaviors) impairments in autism spectrum disorders (Ronald, Happé, & Plomin, 2005). This study was able to show that although social and nonsocial behaviors are both highly heritable, their genetic overlap is modest, with most of the genetic influence being specific to either social or nonsocial behaviors. Segregation analysis is another behavioral genetic method that formally tests different models for the mode of transmission of a genetic trait in families. If a major locus effect is observed, molecular analyses are needed to identify the locus in the genome.

Molecular Genetics

Molecular genetic methods aim to relate variations, or polymorphisms, of a gene of interest across individuals to variations in their social function. This includes linkage studies, which aim to find the approximate location of a major locus influencing a trait, and association studies, which aim to identify the allele that influences a trait. For example, polymorphisms in the gene for the D4 receptor have been related to disorganized attachment in infants (Lakatos et al., 2002), novelty seeking in infants and adolescents (Laucht, Becker, & Schmidt, 2006) and attention-deficit/hyperactivity disorder (Frank, Pergolizzi, & Perilla, 2004).

Molecular geneticists use a variety of methods to obtain and manipulate the DNA needed for such studies. DNA must first be isolated, or removed from the cell so that it is separated from its other components. Typically DNA must also be amplified, or copied many times to have sufficient copies for analysis. A common technique for copying, or cloning, DNA is called the "polymerase chain reaction." The DNA must then be prepared for analysis. In this process, such techniques as gel electrophoresis are used to separate DNA fragments

TABLE 2.2. Commonly Used Approaches for Mapping Genotype to Cognitive Phenotype, with Some of Their Merits and Limitations

Approach	Studied populations and recent review articles	Question(s) addressed	Advantages	Limitations
Genetic manipulations in animal models of cognitive functioning	Clearly defined lines of experimental animals (e.g., rodents) for which established behavioral measures exist (Flint, 2003).	What are the effects of single-gene (or, more recently, multiple-gene) mutations (e.g., gene silencing) on cognitive and behavioral markers, compared to those in wild-type littermates?	Precise control over the number and locus of manipulated genes allows systematic studies of the effects of defective genes.	• Functional homology of genes across species is accepted for many, but not all, genes of interest. • Behavioral effects of genetic mutations are often found only in specific lines, highlighting the importance of genetic background in modulating gene-specific effects.
Molecular genetics	Typically or atypically developing individuals characterized by different polymorphisms of a gene of interest (Goldberg & Weinberger, 2004).	How does cognitive performance vary for individuals characterized by different polymorphisms?	It allows testing the effects of variability of genes coding for proteins whose functions are relatively well understood (e.g., dopamine receptor types).	• Selection of candidate genes is somewhat focused on well-understood proteins, perpetuating interest in a set of proteins (but note advances in using data-driven approaches—e.g., in proteomics).
Quantitative behavioral genetics	Large samples of: • Monozygotic and dizygotic twins (Plomin et al., 2001). • Subgroups of interest (e.g., poorly functioning on the constructs of interest) (Plomin & McGuffin, 2003; Bishop, 2001).	What proportion of the variance of a given cognitive measure is due to the genetic variation? What are the additive or multiplicative effects of genetic variation on any cognitive measure?	Large samples mean that multiple factors affecting variability (genetics, shared and nonshared environments, etc.) can be assessed.	• Limited inferences can be drawn from individual cases. • Amount of the variance on higher-level cognitive measures accounted for by individual genes is usually exceedingly small.
Cognitive functioning in developmental disorders	Individuals with: • Monogenic disorders (involved gene known—e.g., Lai et al., 2001). • Polygenic disorders (involved loci known—e.g., Bellugi et al., 1999; Donnai & Karmiloff-Smith, 2000). • Behaviorally defined syndromes for which genetic contributions are as yet unknown (although multiple candidate genes or loci may have been identified by linkage analysis).	How do individuals with a genetic disorder differ from comparison groups in terms of selected cognitive skills?	It allows pitting atypical groups against typical and atypical comparison groups either at discrete time points in development (cross-sectional designs) or over developmental time (longitudinal designs).	• Inferences that can be made from the comparison across groups depend on the correct choice of matching criteria, but the chosen matching criteria themselves affect the direction of group differences. • Relatively small samples of individuals preclude interpretation of null findings (low statistical power). • Restricted availability of sufficient individuals of various age groups limits drawing full developmental trajectories of performance.
Computational modeling	Neural networks designed to test distinct effects of various parameter settings on performance during and after training (Thomas & Karmiloff-Smith, 2002, 2003).	What changes in the parameters of a network alter its learning in a way that models cognitive functioning and its development in typical and atypical populations?	It allows clear definition of which computational properties or changes thereof best account for performance during and after training.	It is not always easy to operationalize neural variables into meaningful parameters of neurobiologically plausible models.

Note. From Scerif and Karmiloff-Smith (2005). Copyright 2005 by Elsevier. Reprinted by permission. Divisions across approaches are somewhat arbitrary, and many researchers in the field use more than one. Advantages and limitations of each approach are also not mutually exclusive, with some overlap across categories.

of different lengths, Blotting techniques are used to break the DNA into single strands (so that it is receptive to a "probe" that can bind to it) and to transfer it from the delicate gel to a sturdier membrane. Once DNA has been isolated and purified, it can be further analyzed in a variety of ways, such as to identify the presence or absence of specific sequences or to locate mutations within a specific sequence.

Advantages and Disadvantages

Genetic studies provide important information both about the relative contributions of genetic and environmental influences on social functioning, and about the biological mechanisms underlying social functioning. One challenge is to integrate this information with other approaches in developmental social neuroscience to document the path from genes to behavior. Answers to such complicated analyses require dialogue between researchers using different approaches, and are complicated because complex social behaviors probably do not rely on a single gene.

Neuropsychological Approaches

Neuropsychologists have traditionally studied individuals with brain lesions to understand how perceptual, cognitive, and social functions are organized and how they are instantiated in the brain. One important contribution of the neuropsychological approach is use of the "double dissociation," in which the performance of two patient groups on two different types of tasks are contrasted. If one group is impaired on one task and intact on another, and the second group shows the opposite pattern, it is concluded that the skills tapped by the different tasks are mediated by independent processes. For example, double dissociations in face and object processing have been used to argue that faces are processed separately from other types of objects in the brain.

Neuropsychological studies also provide an important contribution by helping pinpoint which brain regions are strictly necessary for a particular function. For example, although neuroimaging studies of healthy individuals provide a picture of which brain areas are typically activated by a task, they do not indicate which parts of the activated network are absolutely necessary for successful performance. When applied to development, neuropsychological studies of children with brain injury or disease can also provide insight into the plasticity of the developing brain and its social functions. In general, neuropsychological studies can take either a "forward approach," in which individuals with known brain injury are studied and the behavioral correlates are identified, or a "backward approach," in which individuals with an impairment in social function are studied and the neural correlates of these impairments are identified. In both approaches, neuropsychological tests that are believed to tap specific brain regions or networks are useful tools.

Neuropsychological Tests
Traditional Standardized Tests

Neuropsychological tests consist of tasks that are specifically designed to measure a psychological function known to be linked to a particular brain structure or pathway. Traditionally, neuropsychological tests are formal standardized tests administered in a very specific way, and an individual's results on these tests are compared with data obtained from a large normative sample to obtain standardized scores for the individual's performance relative to that of the normative population. For example, the Wisconsin Card Sorting Test and the Tower of London are tests typically thought to tap frontal lobe functioning. Advantages of standard tasks are that their widespread use facilitates comparison of results across studies, since administration procedures are standard, and the availability of normative data facilitates classification of performance at the individual level. Traditionally, standard neuropsychological assessment has been more limited for social functions than for other domains, such as cognition, perception, and motor skills. This situation is currently changing; for example, developmental batteries such as the NEPSY now include scales of social perception.

Experimental "Marker" Tasks

Experimental "marker" tasks have also been applied with a logic similar to that of traditional neuropsychological tasks: A task that has been shown to rely on a particular brain region is applied to a population to track the functioning of that system during development or disease. For example, the Iowa Gambling Task is a test of social decision making that, based on results from adult patient populations and neuroimaging studies, is believed to rely on ventromedial frontal cortex (see Crone & Westenberg, Chapter 19, this volume). This task has been applied to the study of typically developing children as a way of understanding the functional development of this brain region (e.g., Crone & van der Molen, 2004), as well as to children with developmental disorders of social functioning as a way of understanding the role of this brain region in the disorder (e.g., Blair, Colledge, & Mitchell, 2001). Unlike those of standard neuropsychological tests, the details and administration of experimental marker tasks often vary across studies, and data are typically compared to those for a control group rather than to a larger normative database. However, such tasks have two advantages: (1) They are quicker to develop and so may be closer to the "cutting edge" than standard tests, which can take years to develop and standardize; and (2) they often can target more specific or subtle questions than standard tests can, since the latter are often aimed at detecting significant impairments rather than more subtle variations in performance.

Studies of Persons with Brain Lesions

Studying individuals with brain injury is the main focus of neuropsychology. In a simple view, if an individual with damage to a particular brain region is impaired in a particular psychological function, then that function must depend on that brain structure. For example, the damage to the frontal lobes sustained by Phineas Gage caused a severe disruption of his social functioning, pointing to a key role of the frontal lobes in normal social behavior (see Macmillan, 2002, for the story of Phineas Gage). Although this concept is quite straightforward, application of this approach does have its challenges. For example, there can be challenges in interpretation: Does the damaged brain region actually normally mediate the impaired function, or does it instead provide input to another structure, which carries out the function? Has the injury actually affected the competence (level of skill for that function), or has it instead affected performance (ability to apply one's competence in a particular setting)? Another challenge is that neuropsychologists do not have control over the location and extent of a brain injury, and so even though patients may be grouped together, no two will have exactly the same pattern or size of lesion. Here MRI studies have been helpful in more precisely characterizing brain injuries and determining, for example, which areas of injury are common in a particular group of patients and which are idiosyncratic.

Studies of Persons with Atypical Social Behavior

Another approach to understanding the neural bases of social function is to study individuals with documented impairments in social function, to determine which brain areas are affected in such cases. Perhaps the most studied developmental examples are the neural bases of autism spectrum disorders (see Dawson, Sterling, & Faja, Chapter 22, and Decety & Meyer, Chapter 8, this volume). Identifying brain areas that are atypical in developmental disorders of social behavior provides an insight into the neural bases of such disorders, and also into the brain structures that normally mediate the affected abilities. The patterns of brain abnormality in these disorders may be more complex, diffuse, or subtle than those observed in acquired brain injury.

Issues in Applying Neuropsychological Approaches

In addition to the challenges already described, neuropsychological approaches have additional challenges when they are applied in developing populations. The classic approach, borrowed from the adult model, assumes that a given brain injury is static and that the pattern of functional performance is a direct consequence of that particular injury. However, the developing system is a dynamic one, with any acquired injury superimposed on a changing system, and any genetic disorder unfolding within this changing system. This makes establishing brain–behavior relations less straightforward. For example, the effects of a brain injury acquired in infancy may not even be apparent until years later. This is illustrated, for example, in cases of developmental amnesia, where injury to the medial temporal lobe sustained perinatally results in a memory impairment that is not apparent until school age (Gadian et al., 2000).

In addition, specific, highly selective impairments may be less likely in developmental studies, since even an impairment that begins selectively can have widespread effects on the development of other systems, resulting in a pattern of associated deficits rather than a single selective deficit. For example, a specific impairment in auditory processing can affect many aspects of the developing speech and language system. It also cannot be assumed that the consequences of an injury or genetic disorder are constant across the lifespan. This is nicely illustrated by the example of Williams syndrome. Whereas adults with the syndrome show a profile with relative strength in verbal communication skills and relative weakness in number processing, infants with Williams syndrome appear to show the opposite pattern of strength and weakness (Paterson, Brown, Gsodl, Johnson, & Karmiloff-Smith, 1999). These and other issues in applying neuropsychological frameworks to studying development are discussed by Bishop (1997).

Animal Studies

Preclinical or animal research is critical in neuroscience, as animal models allow the use of invasive procedures that cannot be employed in studies with humans. In many cases, invasive procedures are necessary for a full understanding of molecular and cellular mechanisms operating to support cognitive and social functioning. However, despite the fact that such research is critical to the study of social neuroscience, translating the findings from preclinical work to the human case poses challenges that must be considered. This is especially true in the case of developmental research, because the timing of neural development differs across species.

Given the common use of rats in animal neuroscience research, we must be particularly concerned about developmental timing differences between the rat pup and the human child. The brain is less mature at birth in the rodent than in the human, for example. Thus studies examining early postnatal development in the rat are likely to bear on issues of prenatal development in the human. In some cases, this is a distinct advantage. For example, sexual differentiation of the brain extends into early postnatal development in the rodent, allowing researchers to manipulate testosterone levels *postnatally* in the rat to test its effects on differentiation of sexual behavior (Harris & Levine, 1965). In humans, the androgens that operate to differentiate the brain do so prenatally, as is also the case in nonhuman primates, where comparable experimental manipulations require injecting the androgen into the pregnant female in order to affect levels in the central nervous system of her developing fetus (Thornton & Goy, 1986).

In other cases, however, the greater immaturity of the rodent brain than of the human brain at birth can create challenges in translating the implications of the rodent studies. For example, variations in maternal behavior that arise naturally or via experimental manipulations have been found to have profound effects on fear, anxiety, and stress reactions in the rat, and it is now clear that epigenetic mechanisms account for these effects (Meaney & Szyf, 2005). However, these effects occur during the first week of the rat's life—a period most comparable to the last trimester of human brain development. Furthermore, the aspect of maternal care in the rat that produces these epigenetic effects appears to be maternal licking and grooming of the pup. This might lead to the hypothesis that "touch" or "massage" of the human infant is an important mechanism in shaping the development of neural systems involved in stress and emotion regulation (Field, 1995). On close inspection of the rat findings, however, we find that licking of the anogenital region is what seems to be needed for regulation of the stress neuraxis in the rat pup (Suchecki, Rosenfeld, & Levine, 1993), raising questions about "direct" translation to the human case.

In addition to species differences in timing, species and subspecies differences in neuroanatomy exist on both the structural and molecular levels. In some cases, these differences can be exploited to provide vital insights into processes underlying social behavior. For example, researchers studying montane and prairie voles—subspecies that differ in their propensity for pair bonding—have now shown that these two types of voles do not differ substantially in their production of a critical neuropeptide, but that the distribution of their receptors for this peptide differ markedly. In the pair-bonding voles, these receptors are located in brain regions where the presence of the peptide supports positive, approach behavior to the partner; in the non-pair-bonding voles, this is not the case (Insel & Young, 2001). In other cases, differences in neuroanatomy, including receptor distributions, among species may make translation of findings across species more challenging. For example, there is now evidence that while in rodents mineralocorticoid receptors, which are important in mediating health-promoting effects of glucocorticoids (cortisol in humans, corticosterone in rats), are found predominantly in the hippocampus. In nonhuman primates (and possibly in humans), the frontal cortex is rich in these receptors, raising the possibility that elevations in cortisol may affect somewhat different neural systems in rats than in primates (Sanchez, Young, Plotsky, & Insel, 2000).

These and other challenges accompanying attempts to translate social neuroscience research with certain animals to studies with other species, including humans, do not negate the critical importance of animal model research in the social neuroscience. However, these challenges do indicate that one must exercise caution in translating findings. They also point to the importance of animal and human researchers' working closely together in order to facilitate the building of translational bridges.

Summary and Conclusions

In this chapter, we have overviewed the diverse methods that are used to study the neural bases of social development. The range of methods available is exciting, as it provides the opportunity to understand this topic at many different levels of analysis.

From a developmental perspective, one challenge is that it is not always possible to use the same techniques across development, from infancy to adolescence and beyond. For example, MRI techniques have been used mainly with children older than 7 years, at least for studies of normative development, because of issues of cooperation in the scanning environment. Similarly, the range of standardized neuropsychological tests for infants and young children is quite limited, and few would be usable across the span from infancy through adolescence. This can make it difficult to document a comprehensive developmental trajectory of structural and functional brain development. Such techniques as EEG/ERP measures, which can be applied to participants across a wide range of age and ability levels, can provide important information in this regard. However, few researchers to date in developmental social neuroscience have actually applied these techniques using the same paradigm across a wide age span.

Comparisons across age are themselves not always easy even when the same tool can be applied. For example, physical changes in respiratory systems or skull thickness can influence MRI or EEG measurements, respectively, and one must be informed so as not to misinterpret changes caused by non-neural factors as changes in brain function or structure. Behavioral tests are no less immune to difficulties. For example, a particular task may be more sensitive to a particular function at one age than at another; thus a change in performance over age can be misinterpreted as a true change in competence when it is, rather, a change in test sensitivity.

Ultimately, many of the challenges of the methods described in this chapter are related to the fact that studying development involves studying a developing, changing system. The static models typically applied in adulthood cannot automatically be applied.

References

Allison, T., Woods, C. C., & McCarthy, G. M. (1986). The central nervous system. In M. G. H. Coles, E. Donchin, & S. W. Porges (Eds.), *Psychophysiology: Systems, processes, and applications* (pp. 5–25). New York: Guilford Press.

Ashburner, J., & Friston, K. J. (2000). Voxel-based morphometry: The methods. *NeuroImage, 11,* 805–821.

Bellugi, U., Lichtenberger, L., Mills, D., Galabruda, A., & Korenberg, J. R. (1999). Bridging cognition, the brain and molecular genetics: Evidence from Williams syndrome. *Trends in Neurosciences, 22*, 197–207.

Benning, S. D., Patrick, C. J., & Lang, A. R. (2004). Emotional modulation of the post-auricular reflex. *Psychophysiology, 41*, 426–432.

Berntson, G. G., Bigger, J. T., Jr., Eckberg, D. L., Grossman, P., Kaufmann, P. G., Malik, M., et al. (1997). Heart rate variability: Origins, methods, and interpretive caveats. *Psychophysiology, 34*, 623–648.

Berntson, G. G., Cacioppo, J. T., Binkley, P. F., Uchino, B. N., & Quigley, K. S. (1994a). Autonomic cardiac control: III. Psychological stress and cardiac response in autonomic space as revealed by psychopharmological blockades. *Psychophysiology, 31*(6), 599–608.

Berntson, G. G., Cacioppo, J. T., Quigley, K. S., & Fabro, V. T. (1994b). Autonomic space and psychophysiological response. *Psychophysiology, 31*, 44–61.

Berntson, G. G., Quigley, K., & Lozano, D. (2007). Cardiovascular psychophysiology. In J. T. Cacioppo, L. G. Tassinary, & G. G. Berntson (Eds.), *Handbook of psychophysiology* (3rd ed., pp. 182–210). New York: Cambridge University Press.

Bishop, D. V. M. (1997). Cognitive neuropsychology and developmental disorders: Uncomfortable bedfellows. *Quarterly Journal of Experimental Psychology, 50A*, 899–923.

Bishop, D. V. M. (2001). Motor immaturity and specific speech and language impairment: Evidence for a common genetic basis. *American Journal of Medical Genetics, 114*, 56–63.

Blair, R. J., Colledge, E., & Mitchell, D. G. (2001). Somatic markers and response reversal: Is there orbitofrontal dysfunction in boys with psychopathic tendencies? *Journal of Abnormal Child Psychology, 29*, 499–511.

Bornstein, M. H., & Suess, P. E. (2000). Physiological self-regulation and information processing in infancy: Cardiac vagal tone and habituation. *Child Development, 71*(2), 273–287.

Bradley, S. J. (2007). Examining the eyeblink startle reflex as a measure of emotion and motivation to television programming. *Communication Methods and Measures, 1*, 7–30.

Calkins, S. (1997). Cardiac vagal tone indices of temperamental reactivity and behavioral regulation in young children. *Developmental Psychobiology, 31*(2), 125–135.

Cascio, C. J., Gerig, G., & Pevin, J. (2007). Diffusion tensor imaging: Application to the study of the developing brain. *Journal of the American Academy of Child and Adolescent Psychiatry, 46*, 213–223.

Crone, E. A., & van der Molen, M. W. (2004). Developmental changes in real life decision making: Performance on a gambling task previously shown to depend on the ventromedial prefrontal cortex. *Developmental Neuropsychology, 25*, 251–279.

Csibra, G., & Johnson, M. H. (2007). Investigating event-related oscillations in infancy. In M. de Haan (Ed.), *Infant EEG and event-related potentials* (pp. 289–304). Hove, UK: Psychology Press.

Davis, M., Walker, D. L., & Lee, Y. (1997). Roles of the amygdala and bed nucleus of the stria terminalis in fear and anxiety measured with the acoustic startle reflex. *Annals of the New York Academy of Sciences, 821*, 305–331.

deBoer, T., Scott, L. S., & Nelson, C. A. (2007). Methods for acquiring and analysing infant event-related potentials. In M. de Haan (Ed.), *Infant EEG and event-related potentials* (pp. 5–37). Hove, UK: Psychology Press.

Dehaene-Lambertz, G., Dehaene, S., & Hertz-Pannier, L. (2002). Functional neuroimaging of speech perception in infants. *Science, 298*, 2013–2015.

Demaree, H. A., Schmeichel, B. J., Robinson, J. L., Pu, J., Everhart, D. E., & Berntson, G. G. (2006). Up- and down-regulating facial disgust: Affective, vagal, sympathetic, and respiratory consequences. *Biological Psychiatry, 71*, 90–99.

De Volder, A. G. (2002). Functional brain imaging of childhood clinical disorders with PET and SPECT. *Developmental Science, 5*, 344–360.

Diamandis, E. P., & Christopoulos, T. K. (Eds.). (1996). *Immunoassay*. San Diego, CA: Academic Press.

Donnai, D., & Karmiloff-Smith, A. (2000). Williams syndrome: From genotype through to the cognitive phenotype. *American Journal of Medical Genetics, 97*, 164–171.

Field, T. (1995). Massage therapy for infants and children. *Journal of Developmental and Behavioral Pediatrics, 16*(2), 105–111.

Flint, J. (2003). Animal models of anxiety and their molecular dissection. *Seminars in Cell and Developmental Biology, 14*, 37–42.

Frank, Y., Pergolizzi, R. G., & Perilla, M. J. (2004). Dopamine D4 receptor gene and attention deficit hyperactivity disorder. *Pediatric Neurology, 31*, 345–348.

Frankle, W. G., & Laruelle, M. (2002). Neuroreceptor imaging in psychiatric disorders. *Annals of Nuclear Medicine, 16*, 437–446.

Gadian, D. G. (2002). Magnetic resonance approaches to the identification of focal pathophysiology in children with brain disease. *Developmental Science, 5*, 279–285.

Gadian, D. G., Aicardi, J., Watkins, K. E., Porter, D. A., Mishkin, M., & Vargha-Khadem, F. (2000). Developmental amnesia associated with early hypoxic-ischaemic injury. *Brain, 123*, 499–507.

Gimpl, G., Postina, R., Fahrenholz, F., & Reinheimer, T. (2005). Binding domains of the oxytocin receptor for the selective oxytocin receptor antagonist barusiban in comparison to the agonists oxytocin and carbetocin. *European Journal of Pharmacology, 510*, 9–16.

Goldberg, T. E., & Weinberger, D. R. (2004). Genes and the parsing of cognitive processes. *Trends in Cognitive Sciences, 8*, 325–335.

Granger, D. A., Kivlighan, K. T., el-Sheikh, M., Gordis, E. B., & Stroud, L. R. (2007). Salivary alpha-amylase in biobehavioral research: Recent developments and applications. *Annals of the New York Academy of Sciences, 1098*, 122–144.

Habib, K. E., Weld, K. P., Rice, K. C., Pushkas, J., Champoux, M., Listwak, S., et al. (2000). Oral administration of a corticotropin-releasing hormone receptor antagonist significantly attenuates behavioral, neuroendocrine, and autonomic responses to stress in primates. *Proceedings of the National Academy of Sciences USA, 97*(11), 6079–6084.

Harris, G. W., & Levine, S. (1965). Sexual differentiation of the brain and its experimental control. *Journal of Physiology, 181*, 379–400.

Hess, U., Sabourin, G., & Kleck, R. E. (2007). Postauricular and eyeblink startle responses to facial expressions. *Psychophysiology, 44*, 431–435.

Insel, T. R., & Young, L. J. (2001). The neurobiology of attachment. *Neuroscience, 2*, 129–136.

Johnson, M. H., de Haan, M., Oliver, A., Smith, W., Hatzakis, H., Tucker, L. A., et al. (2001). Recording and analyzing high-density event-related potentials with infants: Using the geodesic sensor net. *Developmental Neuropsychology, 19*, 295–323.

Kang, H. C., Burgund, E. D., Lugar, H. M., Petersen, S. E., & Schlaggar, B. L. (2003). Comparison of functional activation foci in children and adults in using a common stereotactic space. *NeuroImage, 19*, 16–28.

Katz, L. F., & Gottman, J. M. (1995). Vagal tone protects children from marital conflict. *Development and Psychology, 7*, 83–92.

Kotsoni, E., Byrd, D., & Casey, B. J. (2006). Special considerations for functional magnetic resonance imaging of pediatric populations. *Journal of Magnetic Resonance Imaging, 23*, 877–886.

Lai, C. S., Fisher, S. E., Hurst, J. A., Vargha-Khadem, F., & Monaco, A. P. (2001). A forkhead-domain gene is mutated in a severe speech and language disorder. *Nature, 413*, 519–523.

Lakatos, K., Nemoda, Z., Toth, I., Ronai, Z., Ney, K., Sasvari-Szekely, M., et al. (2002). Further evidence for the role of the dopamine D4 receptor (DRD4) gene in attachment disorganization. *Molecular Psychiatry, 7*, 27–31.

Lang, P. J., Bradley, M. M., & Cuthbert, B. N. (1998). Emotion, motivation, and anxiety: Brain mechanisms and psychophysiology. *Biological Psychiatry, 44*, 1248–1263.

Laucht, M., Becker, K., & Schmidt, M. H. (2006). Visual exploratory behaviour in infancy and novelty seeking in adolescence: Two developmentally specific phenotypes of DRD4? *Journal of Child Psychology and Psychiatry, 47,* 1143–1151.

Lissek, S., Baas, J. M., Pine, D. S., Orme, K., Dvir, S., Rosenberger, E., et al. (2005). Sensation seeking and the aversive motivational system. *Emotion, 5*(4), 396–407.

Logothetis, N. K., Pauls, J., Augath, M., Trinath, T., & Oeltermann, A. (2001). Neurophysiological investigation of the basis of the fMRI signal. *Nature, 412,* 150–157.

Lozano, D. L., Norman, G., Knox, D., Wood, B. L., Miller, B. D., Emery, C. F., et al. (2007). Where to B in d*Z*/d*t*. *Psychophysiology, 44,* 113–119.

Macmillan, M. (2002). *An odd kind of fame: Stories of Phineas Gage.* Cambridge, MA: MIT Press.

Meaney, M. J., & Szyf, M. (2005). Environmental programming of stress responses through DNA methylation: Life at the interface between a dynamic environment and a fixed genome. *Dialogues in Clinical Neuroscience, 7,* 103–123.

Meek, J. (2002). Basic principles of optical imaging and application to the study of infant development. *Developmental Science, 5,* 371–380.

Moll, G. H., Heinrich, J., & Rothenberger, A. (2002). Transcranial magnetic stimulation in child psychiatry: Disturbed motor system excitability in hypermotoric syndromes. *Developmental Science, 5,* 381–391.

Otte, C., Moritz, S., Yassouridis, A., Koop, M., Madrischewski, A. M., Wiedemann, K., et al. (2007). Blockade of the mineralocorticoid receptor in healthy men: Effects on experimentally induced panic symptoms, stress hormones, and cognition. *Neuropsychopharmacology, 32,* 232–238.

Paetau, R. (2002). Magnetoencephalography in pediatric imaging. *Developmental Science, 5,* 361–370.

Paterson, S. J., Brown, J. H., Gsodl, M. K., Johnson, M. H., & Karmiloff-Smith, A. (1999). Cognitive modularity and genetic disorders. *Science, 286,* 2355–2358.

Pennington, B. F. (2002). *The development of psychopathology.* London: Taylor & Francis.

Persico, A. M., Di Pino, G., & Levitt, P. (2006). Multiple receptors mediate the trophic effects of serotonin on ventroposterior thalamic neurons *in vitro. Brain Research, 1095,* 17–25.

Pike, V. W., Halldin, C., & Wikstrom, H. V. (2001). Radioligands for the study of brain 5-HT1A receptors *in vivo. Progress in Medicinal Chemistry, 38,* 189–247.

Plomin, R., DeFries, J. C., McClearn, G. E., & McGuffin, P. (2001). *Behavioural genetics* (4th ed.). New York: Worth.

Plomin, R., DeFries, J. C., McClearn, G. E., & Rutter, M. (1997). *Behavioral genetics* (3rd ed.). New York: W. H. Freeman.

Plomin, R., & McGuffin, P. (2003). Psychopathology in the postgenomic era. *Annual Review of Psychology, 54,* 205–228.

Porges, S. W. (1995). Orienting in a defensive world: Mammalian modifications of our evolutionary heritage. A polyvagal theory. *Psychophysiology, 32,* 301–318.

Porges, S. W., Doussard-Roosevelt, J. A., Portales, A. L., & Greenspan, S. I. (1996). Infant regulation of the vagal "brake" predicts child behavior problems: A psychobiological model of social behavior. *Developmental Psychobiology, 29*(8), 697–712.

Quigley, K. S., & Stifter, C. A. (2006). A comparative validation of sympathetic reactivity in children and adults. *Psychophysiology, 43,* 357–365.

Ronald, A., Happé, F., & Plomin, R. (2005). The genetic relationship between individual differences in social and nonsocial behaviors characteristic of autism. *Developmental Science, 8,* 444–458.

Sanchez, M. M., Young, L. J., Plotsky, P. M., & Insel, T. R. (2000). Distribution of corticosteroid receptors in the rhesus brain: Relative absence of glucocorticoid receptors in the hippocampal formation. *Journal of Neuroscience, 20,* 4657–4668.

Scerif, G., & Karmiloff-Smith, A. (2005). The dawn of cognitive genetics? *Trends in Cognitive Science, 9,* 126–135.

Schwartz, E. B., Granger, D. A., Susman, E. J., Gunnar, M. R., & Laird, B. (1998). Assessing salivary cortisol in studies of child development. *Child Development, 69*(6), 1503–1513.

Stroganova, T., & Orekhova, E. V. (2007). EEG and infant states. In M. de Haan (Ed.), *Infant EEG and event-related potentials* (pp. 251–287). Hove, UK: Psychology Press

Suchecki, D., Rosenfeld, P., & Levine, S. (1993). Maternal regulation of the hypothalamic–pituitary–adrenal axis in the rat: The roles of feeding and stroking. *Developmental Brain Research, 75*(2), 185–192.

Taylor, M. J., & Baldeweg, T. (2002). Application of EEG, ERP and intracranial recordings to the investigation of cognitive functions in children. *Developmental Science, 5,* 318–334.

Thomas, M. S. C., & Karmiloff-Smith, A. (2002). Are developmental disorders like cases of adult brain damage?: Implications from connectionist modelling. *Behavioral and Brain Sciences, 25,* 727–750.

Thomas, M. S. C., & Karmiloff-Smith, A. (2003). Modelling language acquisition in atypical phenotypes. *Psychological Review, 110,* 647–682.

Thornton, J. E., & Goy, R. W. (1986). Female-typical sexual behavior of rhesus and defeminization by androgens given prenatally. *Hormones and Behavior, 20,* 129–147.

Winnard, P., & Raman, V. (2003). Real time non-invasive imaging of receptor–ligand interactions *in vivo. Journal of Cellular Biochemistry, 90,* 454–463.

Zoumakis, E., Rice, K. C., Gold, P. W., & Chrousos, G. P. (2006). Potential uses of corticotropin-releasing hormone antagonists. *Annals of the New York Academy of Sciences, 1083,* 239–251.

CHAPTER 3

Neuroanatomy of the Developing Social Brain

CHRISTA PAYNE
JOCELYNE BACHEVALIER

In humans and nonhuman primates, survival and species proliferation depend upon navigation within their complex species-specific social environment. Successful navigation requires monitoring and regulating one's own socially relevant behaviors, given the immediate social situation. To do so appropriately involves not only the identification of others' social and emotional behaviors, but also an appreciation for the personal significance of those behaviors. The conduct of others can provide information that may help one decide which actions and responses are appropriate to a specific instance; it may also offer more general guides to behavior, such as realizing relationships between interacting members. Thus social cognition (broadly construed) is dependent upon the ability to detect and interpret information about other individuals that is relevant to regulating one's own behavior according to the current emotional and social context (Baron-Cohen et al., 1999; Loveland, 2005). Given the remarkable behavioral (Brothers, 1995; Byrne & Whiten, 1988; Cheney & Seyfarth, 1990; DeWaal, 1989) and neurobiological (Barton & Aggleton, 2000; Petrides, 1994) similarities between humans and nonhuman primates, this chapter focuses only on the literature from these two species.

A complex neural network of interconnected structures, which includes brain regions within the frontal and temporal lobes, has long been implicated as a substrate to social cognition (Papez, 1937; MacLean, 1949). In particular, the amygdala, buried inside the anterior portion of the temporal lobe, and the orbital frontal cortex, covering the ventral surface of the frontal lobe, have been repeatedly implicated in the formation and maintenance of social interactions and the regulation of emotional behavior in adult primates (Adolphs, 2001; Amaral, Price, Pitkänen, & Carmichael, 1992; Barbas, 1993, 1995, 2000; Cavada, Company,

Tejedor, Cruz-Rizzolo, & Reinoso-Suarez, 2000; Dombrowski, Hilgetag, & Barbas, 2001; Emery & Amaral, 1999; Kling & Steklis, 1976; Machado & Bachevalier, 2003; Murray, 1991; Petrides, 1994; Satpute & Lieberman, 2006). Specifically, the amygdala is thought to be involved in processing emotional and social cues, in modulating memory, and in learning to fear potentially dangerous stimuli, whereas the orbital frontal cortex is thought to contribute to the more "cognitive" aspects of social cognition by influencing decision making in the context of emotional information. However, it is highly unlikely that these two structures are the only ones underlying social cognition. Some evidence (although it is limited) suggests that closely related regions—such as the temporopolar region, the cortex covering the superior temporal sulcus in the temporal lobe, and the anterior cingulate cortex surrounding the corpus callosum in the frontal lobe—also contribute to this functional domain (for reviews, see Adolphs, 1999, 2001; Bachevalier & Meunier, 2005; Grady & Keightley, 2002). Although most current research has begun to provide insights into the collaborative contributions of these structures in social cognition in adults, there is still a paucity of data regarding the development of this neural network and how it supports the progressive emergence of complex social abilities.

This chapter summarizes the neurodevelopment of specific brain regions relevant to social cognition, such as the amygdala, orbital frontal cortex, temporopolar cortex, anterior cingulate cortex, and superior temporal sulcus. Although most of the data reviewed come from nonhuman primate studies, direct comparisons and distinctions are made with the development of the same brain structures in humans whenever possible. In general, the developmental sequence is first given for nonhuman primates, followed by a comparison with what is known in humans, and throughout the text, the reader is notified when the text refers specifically to human infants. In addition, although there exists no established comparison between the sequence of anatomical and functional development in nonhuman primates and humans, a ratio of 1:4 (i.e., 1 month in infant monkeys relates to 4 months in human infants) has commonly been accepted for the development of visual functions. Thus the reader should assume that the parallels in the developmental sequence of the two species are viewed in a relative rather than an absolute sense.

The Amygdala

The primate amygdala is a subcortical group of 13 interconnected nuclei located in the anterior portion of the medial temporal lobe (Figure 3.1). The amygdala is highly interconnected with multimodal sensory cortical areas, and it interacts with cortical and subcortical areas involved in the initiation of motor behavior and the regulation of autonomic and visceral functions. The amygdala also has substantial mutual connections with areas involved in higher-order cognition (Amaral et al., 1992).

The cytological characteristics and connectivity of the amygdala have been extensively studied in nonhuman primates and have revealed strong similarities with those in humans throughout development. The lateral nucleus receives a large array of highly processed sensory information, including visual information from faces and facial expressions, gaze direction, body postures, and movements, as well as auditory information from specific vocal sounds and intonations. The reciprocity of these projections back to the sensory areas via the basal nucleus provides a route by which the amygdala can modulate the cortical processing of sensory stimuli (McGaugh, Ferry, Vazdarjanova, & Roozendaal, 2000). Interestingly, these

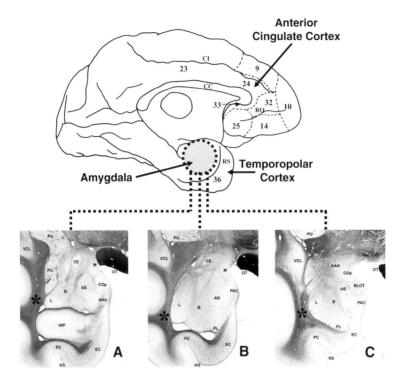

FIGURE 3.1. Medial view of the macaque brain (top), illustrating the architectonic areas of the medial prefrontal and medial temporal cortex, and depicting the anterior cingulate cortex (areas 24, 32, 33, and 25) and the temporopolar cortex (rostral to perirhinal area 36). The amygdala, a subcortical structure buried within the medial temporal lobe, is represented in gray in this medial view. Panels A–C (bottom) display myelin-stained coronal sections from the posterior, middle, and anterior thirds of the amygdala, respectively. AAA, anterior amygdaloid area; AB, accessory basal nucleus; AHA, amygdalohippocampal area; B, basal nucleus; CE, central nucleus; CI, cingulate sulcus; CO, cortical nucleus; Coa, anterior cortical nucleus; Cop, posterior cortical nucleus; EC, entorhinal cortex; HIP, hippocampus; L, lateral nucleus; M, medial nucleus; NLOT, nucleus of the lateral olfactory tract; OT, optic tract; PAC, periamygdaloid cortex; PC, perirhinal cortex; PL, paralamellar part of the basal nucleus; PU, putamen; RO, rostral sulcus; RS, rhinal sulcus; VCL, ventral claustrum. From Bachevalier and Meunier (2005). Copyright 2005 by Thomson Learning (now Cengage Learning). Reprinted by permission.

feedback projections to cortical sensory areas are widespread and reach not only the higher-order areas from which it receives major inputs, but also the primary sensory areas. As a result, the amygdala is able to influence sensory inputs at very early stages in their processing, such as the fusiform gyrus, known to be involved in face processing. Whereas the basal and lateral nuclei serve as an interface between sensory-specific cortical inputs, the central nucleus constitutes a relay to the brainstem and hypothalamus through which the amygdala putatively influences the autonomic and endocrine manifestations of emotion. Efferent projections emanate from the basal and accessory basal nuclei to the ventral striatum, providing a potential access to subcortical elements of the motor system. The connections in turn facilitate the influence of the amygdala on actions, such as the modulation of facial and vocal expressions, body postures, and movements. In addition, the amygdala is highly interconnected with the hippocampal formation, allowing for access to and modulation of stored information in cortical areas (such as past experience with an individual) (Amaral et al., 1992; Saunders & Rosene, 1988; Saunders, Rosene, & Van Hoesen, 1988).

Neurodevelopment

Embryological neurogenesis within the macaque amygdala occurs in a smooth medial-to-lateral wave across the amygdaloid nuclei. In contrast, because the temporal lobe rotates considerably during the latter stages of development, postnatal neurogenesis proceeds with a dorsal-to-ventral gradient (Kordower, Piecinski, & Rakic, 1992). Neurons first appear in the central, lateral, and basolateral nuclei and in the magnocellular and parvocellular divisions of the accessory basal nuclei at the end of the first month of gestation (gestation in macaques lasts approximately 165 days). During the following week, neuronal density continues to increase gradually in all nuclei, but by the middle of the second month of gestation a major reduction in neurogenesis, which originates in the central and medial nuclei and gradually spreads to include all nuclei, is observable. By the end of the second month of gestation, neurogenesis in the macaque amygdala has ceased.

Like neurogenesis, most of the amygdalocortical connections are already established by the time of birth, with the amygdala efferent projections largely resembling those seen in adult monkeys by 2 weeks of age (Amaral & Bennett, 2000). However, some immaturity in the pattern of afferent cortical projections is apparent. Specifically, as compared to the adult brains, visual inputs reaching the lateral nucleus of the amygdala originate from more widespread visual cortical areas in the infant brains (Rodman, 1994; Webster, Ungerleider, & Bachevalier, 1991), indicating that the amygdala receives less processed visual information during this neonatal period than in adulthood. In turn, this maturational refinement of corticoamygdala projections provides infant monkeys with increasingly detailed visual information as they mature.

The development of myelin either within the amygdala itself or within its afferent and efferent connections has yet to be thoroughly assessed. Gibson (1991) reports that early indications of myelination are observable in the macaque amygdala by the fourth postnatal week. Moreover, the stria terminalis (one of the major efferent projections of the amygdala) exhibits no sign of myelin until 4 weeks after birth and only a moderate level by 8 weeks. Accordingly, this fiber tract does not attain an adult level of myelination until more than 3 years postnatally (172 weeks), providing additional evidence that the influence of the amygdala on other neural systems gradually increases throughout the first years of life.

Whereas the neurochemistry of the adult amygdala has been characterized (Amaral et al., 1992), nothing is known about how these neurotransmitter and neuromodulatory systems actually develop. Nonetheless, several research groups have established that a portion of the amygdala's neuromodulatory qualities appear largely mature at birth. For example, the distribution of opiate receptors within the amygdala and other limbic-related structures (substantia innominata and cingulate cortex) does not differ between a 1-week-old and an adult macaque. The only observable distinction between age points is a slightly higher density of opiate receptors in the dorsal amygdala nuclei of infants, indicating some postnatal modification within this region (Bachevalier, Ungerleider, O'Neill, & Friedman, 1986). Furthermore, the overall pattern of serotonergic fibers in the amygdala of 2- and 5-week-old neonatal macaques is similar to that of adults. The only distinguishing characteristics are a slightly lower fiber density and somewhat less distinct areal distribution in the neonates (Prather & Amaral, 2000).

The observed gender differences in amygdala postnatal development as measured by structural magnetic resonance imaging (MRI) (Payne, Machado, Jackson, & Bachevalier, 2006) support a contribution of gonadal hormones to the development of the amygdala. Specifically, while levels of aromatase (an enzyme that converts testosterone into estradiol within tissues) in the amygdala of male fetuses are found at levels comparable to those of adult males (Clark, MacLusky, & Goldman-Rakic, 1988), those levels are significantly higher than those observed in female fetuses (Roselli & Resko, 1986). Interestingly, aromatase activity in female fetuses is notably higher than that observed in postnatal females (MacLusky, Naftolin, & Goldman-Rakic, 1986). In addition, the amygdala exhibits higher estradiol and testosterone levels than other neural areas, except the hypothalamus, throughout development (Abdelgadir, Roselli, Choate, & Resko, 1997; Bonsall & Michael, 1992; Michael, Zumpe, & Bonsall, 1992; Pomerantz & Sholl, 1987; Sholl, Goy, & Kim, 1989; Sholl & Kim, 1989).

In humans, amygdala neurogenesis begins earlier and lasts longer than in the macaque monkeys, but occurs in the same general medial-to-lateral pattern prenatally and dorsal-to-ventral pattern postnatally. An investigation using both cell body and myelin-specific histological staining techniques (Humphrey, 1968), revealed that at this age, the human amygdala can be distinguished by the fifth week of gestation, but no subdivisions are discernible. During the next 2 weeks, the cortical and medial nuclear groups begin to differentiate, followed by the emergence of a distinguishable undifferentiated deep nuclear complex, and finally by the differentiation of the central, basal, accessory basal, and lateral nuclei. Cytoarchitectonic development continues in all nuclei, especially the lateral nucleus, throughout the remainder of the first trimester and into the early stages of the second (Nikolić & Kostović, 1986). Although there exist no data on the connectional and neurochemical development of the amygdala in humans, mature levels of myelination in amygdalocortical connections are seen by 10 months of age (Brody, Kinney, Kloman, & Gilles, 1987; Kinney, Brody, Kloman, & Gilles, 1988). In addition, the sexual dimorphism in developmental patterns of the amygdala seen in nonhuman primates is also observed in humans. Thus boys have larger amygdalae than girls (Caviness, Kennedy, Makris, & Bates, 1995; Caviness, Kennedy, Richelme, Rademacher, & Filipek, 1996, Durston et al., 2001). However, there is conflicting evidence as to whether this sexual dimorphism is sustained in adulthood (Goldstein et al., 2001; Good et al., 2001; Brierly, Shaw, & David, 2002; Soinenen et al., 1994). It is worth noting that others have demonstrated that the apparent sex differences observed in amygdala volume disappear when ratios of amygdala volume to total cerebral volume are considered (Gur, Gunning-Dixon, Bilker, & Gur, 2002; Pruessner et al., 2002). However, between 4 and 18 years of age, boys, but not girls, exhibit a significant increase in the volume of the right amygdala (Giedd, Castellanos, Rajapakse, Vaituzis, & Rapoport, 1997; Giedd et al., 1996; Lenroot & Giedd, 2006). Unfortunately, no data exist concerning androgen levels and activity in the developing human brain.

Behavioral Considerations

Investigations of the maturation of social and emotional behavior over the first several years of a macaque's life suggest a convergence of anatomical and behavioral development (for a review, see Machado & Bachevalier, 2003). As noted above, the afferent and efferent connections of the amygdala at birth largely resemble the adult patterns of connectivity (Amaral & Bennett, 2000; Kordower et al., 1992; Webster et al., 1991). There also appear to be gender differences in the activity of gonadal hormones within the amygdala prior to birth (Roselli & Resko, 1986). Perinatally, the emotional repertoire of infant macaques is limited to the basic emotional states, such as relaxed, alert, or upset. They also appear to lack fear or defensive behaviors and are unable to comprehend the meaning of social signals (Mendelson, 1982a, 1982b; Suomi, 1984, 1990). However, between 1 week and 3 months of age, projections from higher-order visual cortices (areas TE and TEO) to the amygdala are significantly refined (Bachevalier, Hagger, & Mishkin, 1991; Rodman, 1994; Webster, Bachevalier, & Ungerleider, 1994), suggesting that during this maturational stage, the amygdala may receive increasingly accurate visual information regarding social signals. Interestingly, the initial refinements of these projections coincide with the emergence of appropriate responses to social signals by infant macaques. Moreover, as refinement of visual inputs into the amygdala nears completion during the third postnatal month (Bachevalier et al., 1991; Rodman, 1994), infant macaques begin to enter into social interactions, such as grooming and simple forms of play, for the first time (Hinde, Rowell, & Spencer-Booth, 1964; Suomi, 1990). Furthermore, gender differences in amygdala development appear to manifest themselves behaviorally during this time, in that male infants engage in play more than their female cohorts (Brown & Dixson, 2000; Itoigawa, 1973; Nakamichi, 1989; Seay, 1966). This correspondence of neuroanatomical and behavioral maturation suggests that amygdala development may play a critical role in the acquisition of associations between social signals and their appropriate responses during infancy.

A second example of behavioral development converging upon neuroanatomical maturation of the amygdala in macaques is seen in the myelination of the stria terminalis, one of the major extrinsic fiber tracts connecting the amygdala with brain structures involved in autonomic and endocrine manifestations of emotion, such as the hypothalamus, basal ganglia, basal forebrain, and areas of the brainstem (Amaral et al., 1992). As the projections between the amygdala and these structures are becoming fully myelinated, and presumably functionally mature, infant macaques show a pronounced increase in their fearful and defensive behaviors. Interestingly, a similar association is observed in humans: Children show stranger and separation fears at about 10 months of age, which coincides with the attainment of mature levels of myelination of amygdalocortical connections (Brody et al., 1987; Kinney et al., 1988).

The Orbital Frontal Cortex

The orbital region of the frontal cortex, a heterogeneous area that occupies the ventromedial surface of the frontal lobe, exhibits great homology among different primate species (Figure 3.2). This region has a complex organization that has been extensively discussed in the literature (for reviews, see Barbas, 2007; Cavada et al., 2000; Petrides & Pandya, 2002; Price, 2007; Semendeferi, Armstrong, Schleicher, Zilles, & Van Hoesen, 1998). There have been some exquisite lines of investigations that have used cytoarchitectonic characteristics and patterns of connectivity to parcel the ventromedial surface of the brain into over 20 distinct subdivisions organized into two neural networks: the orbital network and the medial network (for a review, see Price, 2007). However, as discussed below, the development of this region and its subdivisions have not been thoroughly characterized, nor have the functional characteristics of each subdivision. Therefore, for the purposes of this chapter, the "orbital frontal cortex" refers to areas 11, 13, and 14, the ventral portions of area 12, and the areas of agranular insular cortex (Ia; also referred to as areas OPA11 and OPro; Barbas, 2007) that extend onto the orbital surface.

As the amygdala does, the orbital frontal cortex receives highly processed information from all sensory modalities (visual, somatosensory, visceral, olfactory, and gustatory). This

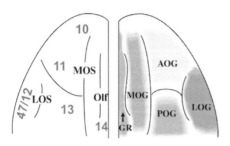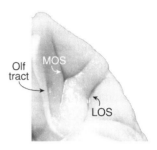

FIGURE 3.2. Ventral view of the macaque brain, illustrating the different cytoarchitectonic areas (10, 11, 12, 13, 14, and 47) composing the orbital frontal cortex. AOG, anterior orbital gyrus; GR, gyrus rectus; LOG, lateral orbital gyrus; MOG, medial orbital gyrus; MOS, medial orbital sulcus; Olf, olfactory sulcus; POG, posterior orbital gyrus. From Bachevalier and Meunier (2005). Copyright 2005 by Thomson Learning (now Cengage Learning). Reprinted by permission.

is particularly true posteriorly, in area Ia, which is distinctly multimodal, and robustly connected with the amygdala and temporopolar cortex (Ghashghaei & Barbas, 2002; Ghashghaei, Hilgetag, & Barbas, 2007; Kondo, Saleem, & Price, 2003). Medially, area 14 is extensively connected to the hippocampus and associated areas, such as the cingulate, retrosplenial, parahippocampal, and entorhinal cortices. In contrast, the more lateral regions are most heavily linked with premotor and sensory areas as well as with the amygdala. Areas 11 and rostral 12 are heavily interconnected with the insula, mediodorsal nucleus of the thalamus, inferior parietal lobule, and dorsolateral prefrontal cortex. The more caudal portions of area 12 and area 13 have significant connections with the amygdala, midline thalamus, and temporopolar cortex (Barbas, 1993, 1995; Carmichael & Price, 1994; Semendeferi et al., 1998).

As illustrated above, the orbital frontal area receives information about all aspects of the external and internal environment through its connections with thalamic nuclei involved in associative aspects of memory, as well as with the amygdala and temporopolar cortex, which are thought to be implicated in the regulation of emotion. Thus the orbital frontal cortex may serve as an integrative area using external and internal cues to facilitate the modulation and self-regulation of emotional behavior in relation to rapid changes in a social situation or context (as in dominance relationships and situational features). Finally, the orbital frontal cortex also sends efferents to brain regions critical to the hormonal modulation of emotions, such as the preoptic region of the lateral hypothalamus, as well as to areas critical to the motor control of emotions, such as the head of the caudate and the ventral tegmental area (Selemon & Goldman-Rakic, 1985); these efferent connections support the role of this structure in the regulation of emotions and goal-directed behaviors.

Neurodevelopment

Whereas numerous accounts utilizing a variety of methodologies have characterized the development of either the dorsolateral prefrontal cortex specifically (for a review, see Lewis, 1997) or the prefrontal cortex as a whole (for a review, see Benes, 2006), reports that specifically chronicle the maturation of the orbital frontal cortex in either macaque monkeys or humans are scarce. Unfortunately, the data are inconclusive as to whether synaptogenesis and synapse elimination occur concurrently in these two areas (Rakic, Bourgeois, Ecken-hoff, Zecevic, & Goldman-Rakic, 1986; for a counterargument, see Gibson, 1991). Thus the neurodevelopmental data generated from the dorsolateral prefrontal cortex may not be applicable to the orbital frontal cortex. To this extent, only data gathered directly from the orbital frontal cortex are discussed in this section.

There is no available information regarding the specific time course or pattern of the neurodevelopmental processes within the orbital frontal cortex during gestation in the monkeys. However, Schwartz and Goldman-Rakic (1991) have demonstrated that connections between the dorsolateral prefrontal and orbital frontal cortices emerge in the third trimester, and achieve an adult-like pattern approximately 1 month prior to birth. Notably, the corticocortical fibers connecting the left and right orbital frontal cortices become prominent and highly organized just prior to birth (Goldman & Nauta, 1977), and the afferent projections from the temporal cortical areas are secure in their adult patterns as early as 1 week of age (Webster et al., 1994). Conversely, the feedback projections from the orbital frontal cortex to the inferior temporal cortical regions exhibit protracted development and are not mature until 7 weeks after birth (Rodman & Consuelos, 1994).

Gibson (1991) reports an absence of myelination in the inferior frontal cortex at birth in monkeys. Myelination within the posterior portion of this cortical area begins in the first postnatal week and proceeds rapidly, reaching an adult stage by 1 year after birth (52 weeks). In contrast, the anterior portion of the inferior frontal cortex does not begin until the second postnatal week and follows a much more protracted pace of myelination, reaching an adult level by 2 years after birth (104 weeks). This time course differs from that of the primary sensory cortices, which achieve an adult level of myelination before 56 weeks of age. Such a protracted myelination of the frontal lobe, including the orbital frontal region, is also evidenced by recent MRI studies (Payne et al., 2006) showing that fibers in the frontal lobe are poorly myelinated until 40 weeks (10 months) of age in monkeys (see Figure 3.3).

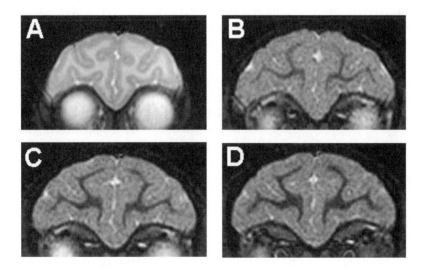

FIGURE 3.3. Coronal magnetic resonance images of the macaque frontal cortex through development. Images were acquired from a single male animal at 1 week (A), 27 weeks (B), 46 weeks (C), and 74 weeks (D), using a fast spin echo—inversion recovery sequence. Notice that the white matter becomes progressively darker, indicating an increase in myelination.

Dopamine innervation of the orbital frontal cortex attains an adult-like form by 3 days after birth (Berger, Febvret, Greengard, & Goldman-Rakic, 1990). However, maturational changes in the laminar distribution of dopamine innervation are evident until postnatal day 42. Neuropeptides may also play a prominent role in the maturation of the orbital frontal cortex. In particular, modification of neuropeptides in this cortical area, such as vasoactive intestinal polypeptide, somatostatin, and substance P, begins at approximately 4.5–5.5 months of gestation and continues into adulthood. Therefore, the modification of these neuropeptides may substantially contribute to late prenatal development of the macaque orbital frontal cortex (Hayashi & Oshima, 1986). Notably, the extensive postnatal neuronal pruning of synapses that utilize these three neuropeptides may minimize their influence after birth. Furthermore, neurons in the orbital frontal cortex containing the regulatory neuropeptide cholecystokinin (CCK) exhibit a rapid decrease after the first month of life, especially in areas 11 and 14, and achieve an adult pattern of expression by the fifth postnatal month (Oeth & Lewis, 1993). Even though CCK can either excite or inhibit neuronal firing, it is usually colocalized with the inhibitory neurotransmitter gamma-aminobutyric acid (GABA). As a result, the dramatic postnatal reduction in CCK-containing neurons may enhance the effect of GABA on the prefrontal cortex, which may in turn further facilitate its protracted functional development.

Although the concentration and activity of gonadal hormones have been shown to be sexually dimorphic in the macaque frontal cortex, no reports have specifically targeted the orbital frontal cortex. However, it is worth noting that gender differences in gonadal hormone action have been shown to hasten the functional maturation of the orbital frontal cortex in male macaques (Goldman, Crawford, Stokes, Galkin, & Rosvold, 1974; Clark & Goldman-Rakic, 1989; for a review, see Overman, Bachevalier, Schuhmann, & McDonough-Ryan, 1997). In addition, as in the amygdala, aromatase activity in the orbital frontal region of female fetal monkeys is considerably higher than in postnatal monkeys of either sex (MacLusky et al., 1986).

Little is known on the maturation of the orbital frontal cortex in humans, and only indirect evidence has been provided. Schore (1996) has indicated that the development of the human orbital frontal cortex occurs in two phases. The first phase occurs during the first year of life and is characterized by a structural maturation. The second developmental phase continues into the second year of life and includes the refinement of anatomical processes. However, it is important to note that the majority of the data regarding development of the orbital frontal cortex presented in that account were generated from nonprimate mammals, such as rats and mice, and thus may not be comparable to data from humans and other primates. Overman et al. (1997) have also reported gender differences in the functional maturation of the orbital frontal cortex in humans, which are similar to those described above in nonhuman primates.

Behavioral Considerations

As in the amygdala, investigations of behavioral maturation in rhesus macaques indicate a connection between neuroanatomical development of the orbital frontal cortex and the development of socioemotional behaviors (for a review, see Machado & Bachevalier, 2003). For example, the first year of a macaque's life is characterized by an increase in the frequency, duration, and complexity of social interactions with conspecifics. As this occurs, infant macaques must learn the various behavioral contingencies that accompany interactions with different individuals, each of which has its own set of motivations. Such contingencies are determined by a combination of factors, including dominance, gender, kinship, environment, and the immediate social context. Thus, in order to survive within their increasingly complex social environment, infant macaques must learn to accurately assess the emotional state and intentions of others and to self-regulate their responses appropriately. Interestingly, this learning roughly coincides with the completion of myelination in the posterior regions of the orbital frontal cortex (Gibson, 1991), and with a general refinement in the neurochemistry of this region (Oeth & Lewis, 1993). Such temporal coincidence suggests that a functional posterior orbital frontal cortex allows yearling macaques to integrate the multiple aspects of complex social interactions with memories of similar previous encounters to determine how they should react to a particular circumstance.

Correlations with behavior can also be seen during the maturation of the anterior orbital frontal cortex. Specifically, macaques begin to exhibit considerable gender differences in play, aggression, and sexual behavior, as well as some new gender differences in grooming and proximity, between 1 and 2 years of age (Bernstein, Judge, & Ruehlmann, 1993; Eaton, Johnson, Glick, & Worlein, 1986; Nakamichi, 1989)—the age at which the anterior orbital frontal cortex reaches an adult level of myelination (Gibson, 1991). Moreover, yearling macaques begin to acquire their own dominance ranking during this time (Bernstein & Williams, 1983). The correlation between neuroanatomical and behavioral development indicates that the anterior orbital frontal cortex may be crucially involved in learning dominance-rank-appropriate behaviors for both oneself and others.

The Temporopolar Cortex, Anterior Cingulate Cortex, and Superior Temporal Sulcus

Despite the paucity of information available regarding the individual characterizations of development in the temporopolar and anterior cingulate cortices and the superior temporal sulcus in nonhuman primates and humans, it is important to consider the available literature. Therefore, in this section we first describe the neuroanatomical organization of these cortical regions individually, and then review what is known about maturational processes of these regions as a group. When data are available, developmental features unique to each region are highlighted.

The temporopolar cortex is an area that encompasses the rostral tip of the temporal pole (Figure 3.1) and largely corresponds to area 38 (Chabardès, Kahane, Minotti, Hoffmann, & Benabid, 2002; Kondo et al., 2003; Markowitsch, Emmans, Irle, Streicher, & Preilowski, 1985). This cortical region is a site of convergence of extensive, highly processed sensory afferents originating from sensory cortical areas, as well as robust limbic afferents from the amygdala and orbital frontal cortex (Gloor, 1997; Moran, Mufson, & Mesulam, 1987). The temporopolar cortex has prominent reciprocal connections with the orbital frontal and medial frontal surfaces via the uncinate fasciculus. In contrast, it is only weakly interconnected with the anterior cingulate cortex. The efferent projections of the temporopolar cortex also include the lateral hypothalamus and the dorsolateral thalamic nucleus. Because of its diverse pattern of connectivity, the temporopolar cortex represents a putative site for the integration of internal and external inputs. Accordingly, this cortical region has also been associated with the regulation of autonomic functions and emotions (Chabardès et al., 2002), as well as with the maintenance of affiliative behaviors (Akert, Gruesen, Woolsey, & Meyer,

1961; Myers, 1958, 1975; Myers & Swett, 1970). However it remains unclear whether the temporopolar cortex uniquely contributes to this domain or whether it functions in concert with the amygdala and orbital frontal cortex (Bachevalier & Meunier, 2005).

The anterior cingulate cortex is located on the medial surface of the frontal lobe, surrounding the genu of the corpus callosum (Figure 3.1). It has been divided into two distinct functional subcomponents: the "cognitive" subdivision (area 32 and caudal area 24) and the "affective" subdivision (areas 25 and 33, and rostral area 24). The former is strongly interconnected with the dorsolateral prefrontal cortex, parietal cortex, and premotor and supplementary motor areas; it appears to mediate attention and executive functions, such as the detection of errors and response conflict. Conversely, the "affective" subdivision is interconnected with the limbic and paralimbic areas, such as the amygdala, nucleus accumbens, orbital frontal cortex, periaqueductal gray, anterior insula, and autonomic brainstem nuclei; it appears to be crucial in the assessment of the salience of emotional and motivational information, and may also regulate emotional responses during the monitoring of response conflict or errors (Dickman & Allen, 2000; Laurens, Ngan, Bates, Kiehl, & Liddle, 2003; Luu, Collins, & Tucker, 2000). The anterior cingulate cortex is interconnected with the basal and accessory basal nuclei of the amygdala, and receives dense afferents from the midline and intralaminar nuclei of the thalamus. In contrast to the dense subcortical inputs, projections from other cortical areas are relatively sparse. Specifically, only light projections are observable from the frontopolar, lateral prefrontal, temporopolar, parahippocampal, entorhinal, and posterior parietal cortices (Bachevalier, Meunier, Lu, & Ungerleider, 1997; Baleydier & Mauguière, 1980; Vogt, Rosene, & Pandya, 1979). Functionally, the anterior cingulate cortex has been implicated in the initiation of speech in humans (Barris & Schuman, 1953; Jürgens & von Cramon, 1982) and the production of vocalizations in monkeys (Jürgens & Ploog, 1970; Ploog, 1986; Robinson, 1967). To this extent, the interaction between the amygdala and the anterior cingulate cortex may serve as a pathway vital to the emotional modulation of speech and vocalizations. Moreover, the involvement of the anterior cingulate cortex in controlling visceromotor endocrine and skeletomotor outputs further implicates it in the control of emotional outputs for body postures and movements in general, in addition to internal emotional changes (Devinsky, Morrell, & Vogt, 1995; Vogt, Finch, & Olson, 1992). The anterior cingulate cortex is also thought to be involved in regulating exploratory behavior and attention to sensory stimuli, and therefore may be crucial in directing an individual's attention to events that are emotionally or motivationally significant.

The superior temporal sulcus is the longitudinal fissure located on the lateral aspect of the brain separating the superior and middle temporal gyri (Figure 3.4). Cells within this cortical area are sensitive to biological motion and gaze direction in both humans (Decety & Grezes, 1999; Hoffman & Haxby, 2000; Puce, Allison, Bentin, Gore, & McCarthy, 1998) and nonhuman primates (Perrett et al., 1985; Perrett, Hietanen, Oram, & Benson, 1992), and thus has been implicated in the analysis of intentions and observed actions of others (Mosconi, Mack, McCarthy, & Pelphrey, 2005). The polymodal cortical regions flanking the sulcus have been divided into several distinct cortical areas that differ in interconnectivity (Allison, Puce, & McCarthy, 2000; Seltzer & Pandya, 1978, 1989). Notably, there are discrete regions mutually connected to cortical areas representing a single sensory modality, as well as zones that are interconnected with regions representing multiple sensory modalities. For instance, area TAa, located anteriorly on the upper bank of the superior temporal sulcus, is mutually connected with the unimodal auditory cortex of the superior temporal gyrus; regions of the lower bank of the superior temporal cortex, such as area TEa, relay unimodal

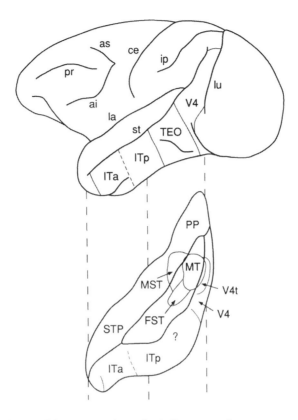

FIGURE 3.4. Lateral view of the macaque brain (top), illustrating the major cortical sulci of the macaque cortex and depicting the architectonic cortical areas of the inferior temporal (IT) cortex (ITa, ITp, TEO). The lower panel represents the standard flattened, or "opened," view of the superior temporal sulcus, which allows visualization of the cortical areas buried within the sulcus (FST, MST, MT, and STP). ai, inferior limb of arcuate sulcus; as, superior limb of arcuate sulcus; ce, central sulcus; FST, visual area of the fundus of the superior temporal sulcus; ip, intraparietal sulcus; ITa, anterior portion of IT visual area; ITp, posterior portion of IT visual area; la, lateral sulcus; lu, lunate sulcus; MT, middle temporal visual area; MST, medial superior temporal visual area; PP, posterior parietal cortex; pr, principal sulcus; st, superior temporal sulcus; STP, superior temporal polysensory area; TEO, visual area TEO; V4, visual area 4; V4t, transitional 4 visual area. Adapted from Rodman and Consuelos (1994). Copyright 1994 by Cambridge University Press. Adapted by permission.

visual information from both the ventral object recognition ("what") pathway and the dorsal spatial location–movement ("where") pathway to the temporopolar cortex. This convergence indicates that information about form and movement is integrated within this region (Beauchamp, Lee, Argall, & Martin, 2004), and provides further evidence for the role of the superior temporal sulcus in the detection of biological motion and gaze direction. Contrasting with the unimodal connectivity described above, projections to area TPO, located on the upper bank of the superior temporal sulcus and corresponding to the functionally defined area STP, originate from the superior temporal gyrus (auditory), inferior parietal lobule (somatosensory), and peristriate belt (visual). Consequently, the convergence of different sensory-related inputs to this cortical zone may foster the intermodal exchange of information. Interestingly, multisensory regions of the superior temporal sulcus also send diffuse projections to much of the frontal cortex, including the orbital frontal and dorsolateral prefrontal cortices, and have prominent reciprocal connections with the amygdala. In turn, the feedback projections from the amygdala may result in the "attentional amplification" (Posner & Dehaene, 1994) of activity within the superior temporal sulcus in response to socially salient stimuli.

Neurodevelopment

As illustrated in this section, there is an obvious dearth of information concerning the development of the temporopolar cortex, anterior cingulate cortex, and superior temporal cortex. Interestingly, in contrast to what is known about the development of the amygdala and orbital frontal cortex, many of the data on the maturational changes occurring within these regions were obtained from humans. Unfortunately, there have been no investigations assessing synaptogenesis within these regions specifically; however, in humans, the cingulate cortex is noticeable at about 18 gestational weeks (Monteagudo & Timor-Tritsch, 1997). Similarly, the superior temporal sulcus begins to be visible at approximately 23 gestational weeks (Chi, Dooling, & Gilles, 1977) in humans, and at about 12 gestational weeks (embryonic day 90) in macaque monkeys (Fukunishi et al., 2006).

As with the orbital frontal cortex, information concerning the specific time course and pattern of the neurodevelopmental processes within the temporopolar and anterior cingulate cortices or the superior temporal sulcus during gestation is not available. However, Rodman et al. characterized the postnatal development of the inferior temporal (IT) cortex of macaque monkeys (for a review, see Rodman, 1994). These investigations revealed a decrease in the density of projections from the temporopolar cortex to anterior IT cortex from 7 to 13–18 weeks after birth, at which point an adult pattern was attained. In addition, transient projections were identified projecting from the anterior cingulate cortex to anterior IT. Specifically, diffuse projections were observed at 7 weeks, with progressively fewer projections present at 13 and 18 weeks. Conversely, the projections from the superior temporal sulcus to anterior IT were secure in their adult pattern by 7 weeks.

Whereas Yakovlev and Lecours (1967) have contended that myelination within the cingulate cortex in humans is completed relatively early (long before 7 years), indirect measures of myelination suggests that this process continues throughout life within the anterior cingulate cortex. Thus, using MRI, Sowell et al. (2003) noted a 12% decrease in gray matter density in this cortical area from the youngest (7 years) to oldest (87 years) subjects. This observable decrease in gray matter density is thought to be attributable to an increase in myelination. Sowell et al. also reported a 12% decrease in gray matter density from 7 to 60 years and a 24% decrease from 40 to 87 years in the superior temporal sulcus, suggesting that the superior temporal sulcus also undergoes extended myelination. These later findings were corroborated in an investigation of cortical development from 4 to 21 years of age in humans, which indicated that the superior temporal sulcus was the last part of the temporal lobe to attain adult levels of myelination (Gogtay et al., 2004; for a review, see Lenroot & Giedd, 2006). Despite the lack of data on myelination of the temporopolar cortex, it is worth noting that in the telencephalon, myelination progresses from the central sulcus outward toward all the poles. As such, the temporal pole (and the cortex found in this region) is the last polar region to begin myelination, preceded by the occipital and frontal poles.

Dopaminergic innervation has also been assessed in the temporopolar and anterior cingulate cortices in macaque monkeys (Berger et al., 1990). As was observed in the orbital frontal cortex, dopaminergic innervation within the temporopolar cortex attained an adult-like pattern by 3 days after birth; however, the maturational changes in the laminar distribution observed in the orbital frontal cortex were not present in the temporopolar cortex. Conversely, subtle maturational changes were evident in the anterior cingulate cortex. That is, relative to adults, infant monkeys exhibited more pronounced labeling of dopaminergic processes, indicating that a considerable pruning of dopaminergic neurons occurs postnatally in the anterior cingulate cortex.

Behavioral Considerations

The paucity of information regarding the unique maturational processes of the temporopolar cortex, anterior cingulate cortex, and superior temporal sulcus severely limits the ability to draw parallels with socioemotional development. Mirroring the maturation of the visual inputs into the amygdala, projections from the anterior IT to the temporopolar and anterior cingulate cortices become increasingly refined during the fourth postnatal month in macaques. As previously noted, this coincides with the emergence of social interactions, such as grooming and simple forms of play, in infant monkeys. This concurrence of behavioral and neuroanatomical development further supports the putative roles of these structures in the maintenance of affiliative behaviors and emotions, and in the assessment of the behavioral significance of socioemotional information.

In contrast, the visual inputs into the superior temporal sulcus appear adult-like at birth; however, the myelination of the superior temporal sulcus occurs relatively late in development, suggesting a progressive maturation of abilities to integrate multisensory inputs mediating the perception of biological motion and gaze direction. Thus, in humans, recent event-related potential (ERP) data indicate that the functional contributions of the superior temporal sulcus to face processing may change with age. Notably, the pattern of ERP responses differs between adults and 3- to 12-month-old infants (see Pascalis, Kelly, & Schwarzer, Chapter 4, this volume). Specifically, several ERP studies have found that adults show an inversion effect, in that the "face-sensitive" negative deflection (N170) exhibits a larger amplitude and longer latency in the ERP generated in response to inverted compared to upright human faces (see Johnson et al., 2005). It is important to note that this response is selective for human faces, since there is no such effect elicited from upright versus inverted animal faces (de Haan, Pascalis, & Johnson, 2002). Infants as young as 6 months of age exhibit a comparable ERP component (N290) that has a smaller amplitude and longer latency (de Haan et al., 2002; Halit, de Haan, & Johnson, 2003); however, this component does not show an inversion effect. Until approximately 12 months of age, a second ERP component (P400) is affected by orientation, but this component is not specific to human faces (Halit et al., 2003). Recent source separation and localization analysis has revealed that generators within the right superior temporal sulcus discriminate between upright and inverted faces between 3 and 12 months of age, regardless of the morphology of the ERP response (i.e., N290 or P400; Johnson et al., 2005). Interestingly, the brain processes represented in these ERP components exhibit a protracted development: The inversion effect in the N290 is not evident in children until 8–11 years of age, and a fully adult-like response, which is characterized by a selective human inversion effect of the N170, is not seen until 13–14 years of age (Taylor, Batty, & Itier, 2004). Moreover, in human infants, the fusiform generators strongly contribute to the ERPs elicited by gaze direction. Conversely, the superior temporal sulcus generators are sensitive to gaze direction in adult subjects (Johnson et al., 2005; for a review, see Grossmann & Johnson, 2007; see also Grossmann & Farroni, Chapter 5, this volume). Taken together, these findings suggest that cortical processing of faces follows a developmental trajectory in humans; such that facial processing in infants and children is initially broad and poorly tuned, becoming gradually more specific to upright human faces with age. Thus the functional refinement of the superior temporal sulcus may result from the neuroanatomical maturation of the superior temporal sulcus and its interconnections with other relevant cortical structures, such as the fusiform area.

Conclusions

This review of the neurodevelopment of structures contributing to social cognition clearly demonstrates the enormous effort needed in this field of research to gain more substantive information on how the neuroanatomical social network organizes early in infancy to provide optimal functioning throughout life. Overall, the findings indicate that the neural structures implicated in social cognition develop over an extended period from infancy until young adulthood. Importantly, some structures, such as the amygdala, come online much earlier in ontogeny than other cortical structures, which become progressively specialized for processing social stimuli (Johnson, 2001). Thus, as researchers have begun to examine which specific social skills become available at different periods across development (for a review, see Grossmann & Johnson, 2007), future work will be necessary to determine which parts of the social brain are supporting these developing skills. Although the challenge is considerable, research is currently underway in several laboratories using noninvasive neuroimaging techniques to determine how the neural structures implicated in the adult social brain develop in early infancy. Such information on basic developmental processes is necessary not only to elucidate how the different components of social cognition are assembled together in the formation of the mature social brain network, but also to better understand the neural substrates of many developmental disorders associated with profound alterations in social skills, such as autism and Williams syndrome.

Acknowledgments

Preparation of this chapter was supported in part by grants from the National Institute of Mental Health (No. MH58846), the National Institute of Child and Human Development (No. HD35471), the Robert W. Woodruff Health Sciences Center Fund, Inc., Emory University, and the National Alliance for Autism Research Mentor-Based Predoctoral Fellowship; Yerkes Base Grant No. NIH RR00165; and the Center for Behavioral Neuroscience Grant No. NSF IBN-9876754. Christa Payne is supported by a Predoctoral Fellowship from the Integrated Training in Psychobiology and Psychopathology (No. NIMH T32-MH0732505).

References

Abdelgadir, S. E., Roselli, C. E., Choate, J. V., & Resko, J. A. (1997). Distribution of aromatase cytochrome P450 messenger ribonucleic acid in adult rhesus monkey brains. *Biology of Reproduction, 57,* 772–777.

Adolphs, R. (1999). Social cognition and the human brain. *Trends in Cognitive Sciences, 3,* 469–479.

Adolphs, R. (2001). The neurobiology of social cognition. *Current Opinion in Neurobiology, 11,* 231–239.

Akert, K., Gruesen, R. A., Woolsey, C. N., & Meyer, D. R. (1961). Klüver–Bucy syndrome in monkeys with neocortical ablations of the temporal lobe. *Brain, 84,* 480–498.

Allison, T., Puce, A., & McCarthy, G. (2000). Social perception from visual cues: role of the STS region. *Trends in Cognitive Sciences, 4,* 267–278.

Amaral, D. G., & Bennett, J. (2000). Development of amygdalo-cortical connections in the macaque monkey. *Society for Neuroscience Abstracts, 26,* 17–26.

Amaral, D. G., Price, J. L., Pitkänen, A., & Carmichael, S. T. (1992). Anatomical organization of the primate amygdaloid complex. In J. P. Aggleton (Ed.), *The amygdala: Neurobiological aspects of emotion, memory, and mental dysfunction* (pp. 1–66). New York: Wiley.

Bachevalier, J., Hagger, C., & Mishkin, M. (1991). Functional maturation of the occipitotemporal pathway in infant rhesus monkeys. In N. A. Lassen, D. H. Ingvar, M. E. Raichle, & I. Friberg (Eds.), *Alfred Benzon Symposium: Vol. 31. Brain work and mental activity* (pp. 231–240). Copenhagen: Munksgaard.

Bachevalier, J., & Meunier, M. (2005). The neurobiology of social-emotional cognition in nonhuman primates. In A. Easton & N. J. Emery (Eds.), *The cognitive neuroscience of social behaviour* (pp. 19–57). Hove, UK: Psychology Press.

Bachevalier, J., Meunier, M., Lu, M., & Ungerleider, L. G. (1997). Thalamic and temporal cortex input to medial prefrontal cortex in rhesus monkeys. *Experimental Brain Research, 115,* 430–444.

Bachevalier, J., Ungerleider, L. G., O'Neill, J. B., & Friedman, D. P. (1986). Regional distribution of [3H] naloxone binding in the brain of newborn rhesus monkey. *Brain Research, 390,* 302–308.

Baleydier, C., & Mauguière, F. (1980). The duality of the cingulate cortex in monkey: Neuroanatomical study and functional hypothesis. *Brain, 103,* 525–554.

Barbas, H. (1993). Organization of cortical afferent input to orbitofrontal areas in the rhesus monkey. *Neuroscience, 56,* 841–864.

Barbas, H. (1995). Anatomic basis of cognitive–emotional interactions in the primate prefrontal cortex. *Neuroscience and Biobehavioral Reviews, 19,* 499–510.

Barbas, H. (2000). Connections underlying the synthesis of cognition, memory and emotion in primate prefrontal cortices. *Brain Research Bulletin, 52,* 158–165.

Barbas, H. (2007). Specialized elements of orbitofrontal cortex in primates. *Annals of the New York Academy of Sciences, 1121,* 10–32.

Baron-Cohen, S., Ring, H. A., Wheelwright, S., Bullmore, E. T., Brammer, M. J., Simmons, A., et al. (1999). Social intelligence in the normal autistic brain: An fMRI study. *European Journal of Neuroscience, 11,* 1891–1898.

Barris, R. W., & Schuman, H. R. (1953). Bilateral anterior cingulate gyrus lesions: Syndrome of the anterior cingulate gyri. *Neurology, 3,* 44–52.

Barton, R. A., & Aggleton, J. P. (2000). Primate evolution and the amygdala. In J. P. Aggleton (Ed.), *The amygdala: A functional analysis* (2nd ed., pp. 479–508). Oxford: Oxford University Press.

Beauchamp, M. S., Lee, K. E., Argall, B. D., & Martin, A. (2004). Integration of auditory and visual information about objects in superior temporal sulcus. *Neuron, 41,* 809–823.

Benes, F. M. (2006). The development of the prefrontal cortex: The maturation of neurotransmitter systems and their interactions. In D. Cicchetti & D. J. Cohen (Eds.), *Developmental psychopathology: Vol. 2. Developmental neuroscience* (2nd ed., pp. 216–258). Hoboken, NJ: Wiley.

Berger, B., Febvert, A., Greengard, P., & Goldman-Rakic, P. S. (1990). DARPP-32, a phosphoprotein enreached in dopaminoceptive neurons bearing dopamine D1 receptors: Distribution in the cerebral cortex of the newborn and adult rhesus monkey. *Journal of Comparative Neurology, 299,* 327–348.

Bernstein, I. S., Judge, P. G., & Ruehlmann, T. E. (1993). Sex differences in adolescent rhesus monkey (*Macaca mulatta*) behavior. *American Journal of Primatology, 31,* 197–210.

Bernstein, I. S., & Williams, L. E. (1983). Ontogenetic changes and the stability of rhesus monkey dominance relationships. *Behavioural Processes, 8,* 379–392.

Bonsall, R. W., & Michael, R. P. (1992). Developmental changes in the uptake of testosterone in the primate brain. *Neuroendocrinology, 55,* 84–91.

Brierly, B., Shaw, P., & David, A. S. (2002). The human amygdala: A systematic review and meta-analysis of volumetric magnetic resonance imaging. *Brain Research Reviews, 39,* 84–105.

Brody, B. A., Kinney, H., Kloman, A., & Gilles, F. H. (1987). Sequence of central nervous system myelination in human infancy: 1. An autopsy study of myelination. *Journal of Neuropathology and Experimental Neurology, 46,* 283–301.

Brothers, L. (1995). Neurophysiology of the perception of intention by primates. In M. S. Gazzaniga (Ed.), *The cognitive neurosciences* (pp., 1107–1117). Cambridge, MA: MIT Press.

Brown, G. R., & Dixson, A. F. (2000). The development of behavioural sex differences in infant rhesus macaques (*Macaca mulatta*). *American Journal of Primatology, 41,* 63–77.

Byrne, R., & Whiten, A. (1988). *Machavellian intelligence: Social expertise and the evolution of intellect in monkeys, apes, and humans.* Oxford: Clarendon Press.

Carmichael, S. T., & Price, J. L. (1994). Architectonic subdivision of the orbital and medial prefrontal cortex in the macaque monkey. *Journal of Comparative Neurology, 15,* 366–402.

Cavada, C., Company, T., Tejedor, J., Cruz-Rizzolo, R. J., & Reinoso-Suarez, F. (2000). The anatomical connections of the macaque monkey orbitofrontal cortex: A review. *Cerebral Cortex, 10,* 220–242.

Caviness, V. S., Jr., Kennedy, D. N., Makris, N., & Bates, J. (1995). Advanced application of magnetic resonance imaging in human brain science. *Brain and Development, 17,* 399–408.

Caviness, V. S., Jr., Kennedy, D. N., Richelme, C., Rademacher, J., & Filipek, P. A. (1996). The human brain age 7–11 years: Volumetric analysis based on magnetic resonance images. *Cerebral Cortex, 6,* 726–736.

Chabardès, S., Kahane, P., Minotti, L., Hoffmann, D., & Benabid, A.-L. (2002). Anatomy of the temporal pole region. *Epileptic Disorders, 4*(Suppl. 1), 9–16.

Cheney, D. L., & Seyfarth, R. M. (1990). *How monkeys see the world.* Chicago: University of Chicago Press.

Chi, J. G., Dooling, E. C., & Gilles, F. H. (1977). Gyral development of the human brain. *Annals of Neurology, 1,* 86–93.

Clark, A. S., & Goldman-Rakic, P. S. (1989). Gonadal hormones influence the emergence of cortical function in nonhuman primates. *Behavioral Neuroscience, 103,* 1287–1295.

Clark, A. S., MacLusky, N. J., & Goldman-Rakic, P. S. (1988). Androgen binding and metabolism in the cerebral cortex of the developing rhesus monkey. *Endocrinology, 123,* 932–940.

Decety, J., & Grazes, J. (1999). Neural mechanisms subserving the perception of human actions. *Trends in Cognitive Sciences, 3,* 172–178.

de Haan, M., Pascalis, O., & Johnson, M. H. (2002). Specialization of neural mechanisms underlying face recognition in human infants. *Journal of Cognitive Neuroscience, 14,* 199–209.

Devinsky, O., Morrell, M. J., & Vogt, B. A. (1995). Contributions of the anterior cingulate cortex to behaviour. *Brain, 118,* 279–306.

DeWaal, F. (1989). *Peacemaking among primates.* Cambridge, MA: Harvard University Press.

Dickman, Z. V., & Allen, J. J. (2000). Error monitoring during reward and avoidance learning in high- and low-socialized individuals. *Psychophysiology, 37,* 43–54.

Dombrowski, S. M., Hilgetag, C. C., & Barbas, H. (2001). Quantitative architecture distinguishes prefrontal cortical systems in the rhesus monkey. *Cerebral Cortex, 11,* 975–988.

Durston, S., Hulshoff, H. E., Casey, B. J., Giedd, J. N., Buitelaar, J. K., & van Engeland, H. (2001). Anatomical MRI of the developing human brain: What have we learned? *Journal of the American Academy of Child and Adolescent Psychiatry, 40,* 1012–1020.

Eaton, G. G., Johnson, D. F., Glick, B. B., & Worlein, J. M. (1986). Japanese macaques' (*Macaca fuscata*) social development: Sex differences in juvenile behavior. *Primates, 27,* 141–150.

Emery, N. J., & Amaral, D. G. (1999). The role of the amygdala in primate social cognition. In R. D. Lane & L. Nadel (Eds.), *Cognitive neuroscience of emotion* (pp. 156–191). Oxford: Oxford University Press.

Fukunishi, K., Sawada, K., Kashima, M., Sakata-Haga, H., Fukuzaki, K., & Fukui, Y. (2006). Development of cerebral sulci and gyri in fetuses of cynomolgus monkeys (*Macaca fascicularis*). *Anatomy and Embryology, 211,* 757–764.

Ghashghaei, H. T., & Barbas, H. (2002). Pathways for emotion: Interactions of prefrontal and anterior temporal pathways in the amygdala of the rhesus monkeys. *Neuroscience, 115,* 1261–1279.

Ghashghaei, H. T., Hilgetag, C. C., & Barbas, H. (2007). Sequence of information processing for emotions based on the anatomic dialogue between prefrontal cortex and amygdala. *NeuroImage, 34,* 905–923.

Gibson, K. R. (1991). Myelination and behavioral development: A comparative perspective on questions of neoteny, altriciality and intelligence. In K. R. Gibson & A. C. Peterson (Eds.), *Brain maturation and cognitive development: Comparative and cross-cultural perspectives* (pp. 29–63). New York: Aldine de Gruyter.

Giedd, J. N., Castellanos, F. X., Rajapakse, J. C., Vaituzis, A. C., & Rapoport, J. L. (1997). Sexual dimorphism of the developing human brain. *Progress in Neuropsychopharmacology and Biological Psychiatry, 21,* 1185–1201.

Giedd, J. N., Vaituzis, A. C., Hamburger, S. D., Lang, N., Rajapakse, J. C., Kaysen D., et al. (1996). Quantitative MRI of the temporal lobe, amygdala, and hippocampus in normal human development: Ages 4–18 years. *Journal of Comparative Neurology, 366,* 223–230.

Gloor, P. (1997). *The temporal lobe and limbic system.* Oxford: Oxford University Press.

Gogtay, N., Giedd, J. N., Lusk, L., Hayashi, K. M., Greenstein, D., Vaituzis, A. C., et al. (2004). Dynamic mapping of human cortical development during childhood through early adulthood. *Proceedings of the National Academy of Sciences USA, 101,* 8174–8179.

Goldman, P. S., Crawford, H. T., Stokes, L. P., Galkin, T. W., & Rosvold, H. E. (1974). Sex-dependent behavioral effects of cerebral cortical lesions in the developing rhesus monkey. *Science, 186,* 540–542.

Goldman, P. S., & Nauta, W. J. (1977). Columnar distribution of cortico-cortical fibers in the frontal association, limbic, and motor cortex of the developing rhesus monkey. *Brain Research, 122,* 393–413.

Goldstein, J. M., Seidman, L. J., Horton, N. J., Makris, N., Kennedy, D. N., Caviness, V. S., Jr., et al. (2001). Normal sexual dimorphism of the adult human brain assessed by *in vivo* magnetic resonance imaging. *Cerebral Cortex, 11,* 490–497.

Good, C. D., Johnsrude, I. S., Ashbumer, J., Henson, R. N., Friston, K. J., & Frakowiak, R. S. (2001). A voxel-based morphometric study of ageing in 465 normal human brains assessed by *in vivo* magnetic resonance imaging. *NeuroImage, 14,* 21–36.

Grady, C. L., & Keightley, M. L. (2002). Studies of altered social cognition in neuropsychiatric disorders using functional neuroimaging. *Canadian Journal of Psychiatry, 47,* 327–336.

Grossmann, T., & Johnson, M. H. (2007). The development of the social brain in human infancy. *European Journal of Neuroscience, 25,* 909–919.

Gur, R. C., Gunning-Dixon, F., Bilker, W. B., & Gur, R. E. (2002). Sex differences in temporo-limbic and frontal brain volumes of healthy adults. *Cerebral Cortex, 12,* 998–1003.

Halit, H., de Haan, M., & Johnson, M. H. (2003). Cortical specialization for face processing: Face-sensitive event-related potential components in 3- and 12-month-old infants. *NeuroImage, 19,* 1180–1193.

Hayashi, M., & Oshima, K. (1986). Neuropeptides in cerebral cortex of macaque monkey (*Macaca fuscata*): Regional distribution and ontogeny. *Brain Research, 364,* 360–368.

Hinde, R. A., Rowell, T. E., & Spencer-Booth, Y. (1964). Behavior of socially living rhesus monkeys in their first six months. *Proceedings of the Zoological Society of London, 143,* 609–649.

Hoffman, E. A., & Haxby, J. V. (2000). Distinct representations of eye gaze and identity in the distributed human neural system for face perception. *Nature Neuroscience, 3,* 80–84.

Humphrey, T. (1968). The development of the human amygdala during early embryonic life. *Journal of Comparative Neurology, 132,* 135–165.

Itoigawa, N. (1973). Group organization of a natural troop of Japanese monkeys and mother–infant interactions. In C. R. Carpenter (Ed.), *Behavioral regulators of behavior in primates* (pp. 229–250). Lewisburg, PA: Bucknell University Press.

Johnson, M. H. (2001). Functional brain development in humans. *Nature Neuroscience, 2,* 475–483.

Johnson, M. H., Griffin, R., Csibra, G., Halit, H., Farroni, T., de Haan, M., et al. (2005). The emergence of the social brain network: Evidence from typical and atypical development. *Development and Psychopathology, 17,* 599–619.

Jürgens, U., & Ploog, D. (1970). Cerebral representation of vocalization in the squirrel monkey. *Experimental Brain Research, 10,* 532–554.

Jürgens, U., & von Cramon, D. (1982). On the role of the anterior cingulate cortex in phonation: A case report. *Brain Language, 15,* 234–248.

Kinney, H. C., Brody, B. A., Kloman, A., & Gilles, F. (1988). Sequence of CNS myelination in infancy. *Journal of Neuropathology and Experimental Neurology, 47,* 217–234.

Kling, A. S., & Steklis, H. D. (1976). A neural substrate for affiliative behavior in amygdala during social interactions in monkeys. *Experimental Neurology, 66,* 88–96.

Kondo, H., Saleem, K. S., & Price, J. L. (2003). Differential connections of the temporal pole with the orbital and medial prefrontal networks in macaque monkeys. *Journal of Comparative Neurology, 465,* 499–523.

Kordower, J. H., Piecinski, P., & Rakic, P. (1992). Neurogenesis of the amygdaloid nuclear complex in the rhesus monkey. *Developmental Brain Research, 68,* 9–15.

Laurens, K. R., Ngan, E. T. C., Bates, A. T., Kiehl, K. A., & Liddle, P. F. (2003). Rostral anterior cingulate cortex dysfunction during error processing in schizophrenia. *Brain, 126,* 610–622.

Lenroot, R. K., & Giedd, J. N. (2006). Brain development in children and adolescents: Insights from anatomical magnetic resonance imaging. *Neuroscience and Biobehavioral Reviews, 30,* 718–729.

Lewis, D. A. (1997). Development of the primate prefrontal cortex. In M. S. Keshavan & R. M. Murray (Eds.), *Neurodevelopment and adult psychopathology* (pp. 12–30). Cambridge, UK: Cambridge University Press.

Loveland, K. (2005). Social-emotional impairment and self-regulation in autism spectrum disorders. In J. Nadel & D. Muir (Eds.), *Typical and impaired emotional development* (3rd ed., pp. 365–382). Oxford: Oxford University Press.

Luu, P., Collins, P., & Tucker, D. M. (2000). Mood, personality, and self-monitoring: Negative affect and emotionality in relation to frontal lobe mechanisms of error monitoring. *Journal of Experimental and Psychological Genetics, 129,* 43–60.

Machado, C. J., & Bachevalier, J. (2003). Non-human primate models of childhood psychopathology: The promise and the limitations. *Journal of Child Psychology and Psychiatry, 44,* 64–87.

MacLean, P. D. (1949). Psychosomatic disease and the "visceral brain": Recent developments bearing on the Papez theory of emotion. *Psychosomatic Medicine, 11,* 338–353.

MacLusky, N. J., Naftolin, F., & Goldman-Rakic, P. S. (1986). Estrogen formation and binding in the cerebral cortex of the developing rhesus monkey. *Proceedings of the National Academy of Sciences USA, 83,* 513–516.

Markowitsch, H. J., Emmans, D., Irle, E., Streicher, M., & Preilowski, B. (1985). Cortical and subcortical afferent connections of the primate's temporal pole: A study of rhesus monkeys, squirrel monkeys, and marmosets. *Journal of Comparative Neurology, 242,* 425–458.

McGaugh, J. L., Ferry, B., Vazdarjanova, A., & Roozendaal, B. (2000). Amygdala: Role in modulation of memory storage. In J. P. Aggleton (Ed.), *The amygdala: A functional analysis* (2nd ed., pp. 391–424). Oxford: Oxford University Press.

Mendelson, M. J. (1982a). Visual and social responses in infant rhesus monkeys. *American Journal of Primatology, 3,* 333–340.

Mendelson, M. J. (1982b). Clinical examination of visual and social responses in infant rhesus monkeys. *Developmental Psychology, 18,* 658–664.

Michael, R. P., Zumpe, D., & Bonsall, R. W. (1992). The interaction of testosterone with the brain of the orchidectomized primate fetus. *Brain Research, 570,* 68–74.

Monteagudo, A., & Timor-Tritsch, I. E. (1997). Development of fetal gyri, sulci and fissures: A trans-vaginal sonographic study. *Ultrasound in Obstetrics and Gynecology, 9,* 222–228.

Moran, M. A., Mufson, E. J., & Mesulam, M. M. (1987). Neural inputs into the temporopolar cortex of the rhesus monkey. *Journal of Comparative Neurology, 256,* 88–103.

Mosconi, M. W., Mack, P. B., McCarthy, G., & Pelphrey, K. A. (2005). Taking an "intentional stance" on eye-gaze shifts: A functional neuroimaging study of social perception in children. *NeuroImage, 27,* 247–252.

Murray, E. A. (1991). Contributions of the amygdalar complex to behavior in macaque monkeys. *Progress in Brain Research, 87,* 167–180.

Myers, D. P. (1958). Some psychological determinants of sparing and loss following damage to the brain. In H. F. Harlow & C. N. Woolsey (Eds.), *Biological and biochemical bases of behavior* (pp. 173–192). Madison: University of Wisconsin Press.

Myers, R. E. (1975). Neurology of social behavior and affect in primates: A study of prefrontal and anterior temporal cortex. In K. J. Zulch, O. Creutzfeldt, & G. C. Galbraith (Eds.), *Cerebral localization* (pp. 161–170). New York: Springer–Verlag.

Myers, R. E., & Swett, C., Jr. (1970). Social behavior deficits of free-ranging monkeys after anterior temporal cortex removal: A preliminary report. *Brain Research, 18,* 551–556.

Nakamichi, M. (1989). Sex differences in social development during the first 4 years in a free-ranging group of Japanese monkeys, *Macaca fuscata. Animal Behaviour, 38,* 737–748.

Nikolić, I., & Kostović, I. (1986). Development of the lateral amygdaloid nucleus in the human fetus: Transient presence of discrete cytoarchitectonic units. *Anatomy and Embryology, 174,* 355–360.

Oeth, K. M., & Lewis, D. A. (1993). Postnatal development of the cholecystokinin innervation of monkey prefrontal cortex. *Journal of Comparative Neurology, 336,* 400–418.

Overman, W. H., Bachevalier, J., Schuhmann, E., & McDonough-Ryan, P. (1997). Sexually dimorphic brain–behavior development: A comparative perspective. In N. A. Krasnegor, G. R. Lyon, & P. S. Goldman-Rakic (Eds.), *Development of the prefrontal cortex: Evolution, neurobiology, and behavior* (pp. 17–58). New York: Elsevier Science.

Papez, J. W. (1937). A proposed mechanism of emotion. *Archives of Neurology and Psychiatry, 38,* 725–744.

Payne, C. D., Machado, C. J., Jackson, E. F., & Bachevalier, J. (2006). The maturation of the nonhuman primate amygdala: An MRI study. *Society for Neuroscience Abstracts, 32,* 718.9.

Perrett, D. I., Hietanen, J. K., Oram, M. W., & Benson, P. J. (1992). Organization and functions of cells responsive to faces in the temporal cortex. *Philosophical Transactions of the Royal Society of London, Series B, Biological Sciences, B355,* 23–30.

Perrett, D. I., Smith, P. A. J., Potter, D. D., Mistlin, A. J., Head, A. S., Milner, A. D., et al. (1985). Visual cells in the temporal cortex sensitive to face view and gaze direction, *Proceedings of the Royal Society of London, Series B, Biological Sciences, B223,* 293–317.

Petrides, M. (1994). Frontal lobes and behaviour. *Current Opinion in Neurobiology, 4,* 207–211.

Petrides, M., & Pandya, D. N. (2002). Comparative cytoarchitectonic analysis of the human and the macaque ventrolateral prefrontal cortex and corticocortical connection pattern in the monkey. *European Journal of Neuroscience, 16,* 291–310.

Ploog, D. (1986). Biological foundations of the vocal expressions of emotions. In R. Plutchik & H. Kellerman (Eds.), *Emotion: Theory, research, and experience* (Vol. 3, pp. 173–197). New York: Academic Press.

Pomerantz, S. M., & Sholl, S. A. (1987). Analysis of sex and regional differences in androgen receptors in fetal rhesus monkey brain. *Brain Research, 433,* 151–154.

Posner, M., & Dehaene, S. (1994). Attentional networks. *Trends in Neurosciences, 17,* 75–79.

Prather, M. D., & Amaral, D. G. (2000). The development and distribution of serotonergic fibers in the macaque monkey amygdala. *Society for Neuroscience Abstracts, 26,* 1727.

Price, J. L. (2007). Definition of the orbital cortex in relation to specific connections with limbic and visceral structures and other cortical regions. *Annals of the New York Academy of Sciences, 1121,* 54–71.

Pruessner, J. C., Kohler, S., Crane, J., Pruessner, M., Lord, C., Byrne, A., et al. (2002). Volumetry of the temporopolar, perirhinal, entorhinal and parahippocampal cortex from high-resolution MR images: Considering the variability of the collateral sulcus. *Cerebral Cortex, 12,* 1342–1353.

Puce, A., Allison, T., Bentin, S., Gore, J. C., & McCarthy, G. (1998). Temporal cortex activation in humans viewing eye and mouth movements. *Journal of Neuroscience, 18,* 2188–2199.

Rakic, P., Bourgeois, J. P., Eckenhoff, M. F., Zecevic, N., & Goldman-Rakic, P. S. (1986). Concurrent overproduction of synapses in diverse regions of the primate cerebral cortex. *Science, 232,* 232–235.

Robinson, B. W. (1967). Vocalization evoked from forebrain in *Macaca mulatta. Physiology and Behavior, 2,* 345–354.

Rodman, H. R. (1994). Development of inferior temporal cortex in the monkey. *Cerebral Cortex, 4,* 484–498.

Rodman, H. R., & Consuelos, M. J. (1994). Cortical projections to anterior inferior temporal cortex in infant macaque monkeys. *Visual Neuroscience, 11,* 119–133.

Roselli, C. E., & Resko, J. A. (1986). Effects of gonadectomy and androgen treatment on aromatase activity in the fetal monkey brain. *Biology of Reproduction, 35,* 106–112.

Satpute, A. B., & Lieberman, M. D. (2006). Integrating automatic and controlled processes into neurocognitive models of social cognition. *Brain Research, 1079,* 86–97.

Saunders, R. C., & Rosene, D. L. (1988). A comparison of the efferents of the amygdala and the hippocampal formation in the rhesus monkey: I. Convergence in the entorhinal, prorhinal, and perirhinal cortices. *Journal of Comparative Neurology, 271,* 153–184.

Saunders, R. C., Rosene, D. L., & Van Hoesen, G. W. (1988). Comparison of the efferents of the amygdala and the hippocampal formation in the rhesus monkey: II. Reciprocal and non-reciprocal connections. *Journal of Comparative Neurology, 271,* 185–207.

Schore, A. N. (1996). The experience-dependent maturation of a regulatory system in the orbital prefrontal cortex and the origin of developmental psychopathology. *Development and Psychopathology, 8,* 59–87.

Schwartz, M. L., & Goldman-Rakic, P. S. (1991). Prenatal specification of callosal connections in rhesus monkey. *Journal of Comparative Neurology, 307,* 144–162.

Seay, B. (1966). Maternal behavior in primiparous and multiparous rhesus monkeys. *Folia Primatologica, 4,* 146–168.

Selemon, L. D., & Goldman-Rakic, P. S. (1985). Longitudinal topography and interdigitation of corticostriatal projections in the rhesus monkey. *Journal of Neuroscience, 5,* 776–794.

Seltzer, B., & Pandya, D. N. (1978). Afferent cortical connections and architectonics of the superior temporal sulcus and surrounding cortex in the rhesus monkey. *Brain Research, 149,* 1–24.

Seltzer, B., & Pandya, D. N. (1989). Frontal lobe connections of the superior temporal sulcus in the rhesus monkey. *Journal of Comparative Neurology, 281,* 97–113.

Semendeferi, K., Armstrong, E., Schleicher, A., Zilles, K., & Van Hoesen, G. W. (1998). Limbic frontal cortex in hominoids: A comparative study of area 13. *American Journal of Physiological Anthropology, 106,* 129–155.

Sholl, S. A., Goy, R. W., & Kim, K. L. (1989). 5 alpha-reductase, aromatase, and androgen receptor levels in the monkey brain during fetal development. *Endocrinology, 124,* 627–634.

Sholl, S. A., & Kim, K. L. (1989). Estrogen receptors in the rhesus monkey brain during fetal development. *Developmental Brain Research, 50,* 189–196.

Soininen, H. S., Paranen, K., Pitkänen, A., Vainio, P., Hänninen, T., Hallikainen, M., et al. (1994). Volumetric MRI analysis of the amygdala and the hippocampus in subjects with age-associated memory impairment: Correlation to visual and verbal memory. *Neurology, 44,* 1660–1668.

Sowell, E. R., Peterson, B. S., Thompson, P. M., Welcome, S. E., Henkenius, A. L., & Toga, A. W. (2003). Mapping cortical change across the human life span. *Nature Neuroscience, 6,* 309–315.

Suomi, S. J. (1984). The development of affect in rhesus monkeys. In N. A. Fox & R. J. Davidson (Eds.), *The psychobiology of affective development* (pp. 119–159). Hillsdale, NJ: Erlbaum.

Suomi, S. J. (1990). The role of tactile contact in rhesus monkey social development. In K. E. Barbard & T. B. Brazelton (Eds.), *Touch: The foundation of experience* (pp. 129–164). Madison, CT: International Universities Press.

Taylor, M. J., Batty, M., & Itier, R. J. (2004). The faces of development: A review of early face processing over childhood. *Journal of Cognitive Neuroscience, 16,* 1462–1442.

Vogt, B. A., Finch, D. M., & Olson, C. R. (1992). Functional heterogeneity in cingulate cortex: The anterior executive and posterior evaluative regions. *Cerebral Cortex, 2,* 435–443.

Vogt, B. A., Rosene, D. L., & Pandya, D. N. (1979). Thalamic and cortical afferents differentiate anterior from posterior cingulate cortex in the monkey. *Science, 204,* 205–207.

Webster, M. J., Bachevalier, J., & Ungerleider, L. G. (1994). Connections of inferior temporal areas TEO and TE with parietal and frontal cortex in macaque monkeys. *Cerebral Cortex, 5,* 470–483.

Webster, M. J., Ungerleider, L. G., & Bachevalier, J. (1991). Connections of inferior temporal areas TE and TEO with medial temporal-lobe structures in infant and adult monkeys. *Journal of Neuroscience, 11,* 1095–1116.

Yakovlev, P. I., & Lecours, A.-R. (1967). The myelogenetic cycles of regional maturation of the brain. In A. Minkowski (Ed.), *Regional development of the brain in early life* (pp. 30–65). Oxford: Blackwell Scientific.

PART III

PERCEIVING AND COMMUNICATING WITH OTHERS

CHAPTER 4

Neural Bases of the Development of Face Processing

OLIVIER PASCALIS
DAVID J. KELLY
GUDRUN SCHWARZER

There is a general consensus that we are becoming face experts (Carey, 1992), and that the immature face-processing system present at birth develops with experience. Exactly which components of the face-processing system are present at birth, which develop first, and at what stage the system becomes adult-like are still hotly debated topics, however. The relation among brain development, neural network maturity, and the behavioral changes observed is even more difficult to grasp, due to the scarcity of data available. The aim of this chapter is to review current knowledge and understanding of the development of the face-processing system from birth, during infancy, and through childhood, until it becomes the sophisticated system observed in adults. We also review our understanding of the neural substrate of face processing during development.

Face Processing in Adults

Human adults are able to recognize hundreds of distinct faces (Bahrick, Bahrick, & Wittlinger, 1975). Faces constitute a category of stimuli that, unlike most other objects, are homogeneous in terms of the gross position of their elements (two eyes above the nose, nose above the mouth, etc.) and have to be discriminated on the basis of "relational information," such as the particular distance between the eyes or between lips and chin (see Figure 4.1) (Leder & Bruce, 2000). The ability to process relational information, called "configural processing," is posited to be the consequence of experience and thus can only be extended to

FIGURE 4.1. Examples of the "relational information" (see arrows) that is processed in configural face processing.

other categories that are discriminated on the basis of relational information and with which subjects are highly familiar (Diamond & Carey, 1986; Gauthier & Tarr, 1997; see Maurer, Le Grand, & Mondloch, 2002, for a definition). One of the most important indicators of expertise in adults is the "inversion effect"—the fact that faces are recognized more accurately and faster when presented in their canonical upright orientation than when presented upside down (e.g., Yin, 1969). Diamond and Carey (1986) suggested that the configural information required to accurately identify individual faces is disrupted by inversion, forcing a less accurate featural processing strategy. Therefore, an inversion effect with facial stimuli is evidence that the face-processing system has been engaged.

The adult face-processing system has been found to be species-specific and not flexible enough to process faces of other species at an individual level automatically (Dufour, Coleman, Campbell, Petit, & Pascalis, 2004). Adults are typically more proficient at recognizing faces from their own ethnic group than faces from other ethnic groups; this is commonly known as the "own-race bias" (ORB) (for reviews and discussion, see Bothwell, Brigham, & Malpass, 1989; Meissner & Brigham, 2001). One popular account of the ORB is known as the "contact hypothesis." This hypothesis simply asserts that the ORB is a consequence of primary exposure to faces from a single ethnic category. However, even with much support (e.g., Brigham, Maass, Snyder, & Spaulding, 1982; Chance, Goldstein, & McBride, 1975; Chiroro & Valentine, 1995; Cross, Cross, & Daly, 1971; Lindsay, Jack, & Christian, 1991), the contact hypothesis only offers an account of how the bias may arise; it fails to explain why recognition is poorer for faces from less familiar ethnic groups. Currently it seems that despite the availability of numerous explanations for the ORB (see Brigham & Malpass, 1985, for an overview), the exact nature of the bias remains elusive.

Valentine (1991) has proposed a "norm-based coding" model, in which faces are encoded as vectors according to their deviation from a prototypical average or origin. The origin of the space represents the average of all faces experienced by an individual throughout his or her lifetime. More typical faces are situated near the origin, and more distinctive faces further away. All new faces a person encounters are encoded in terms of their deviation from this norm face. This model can partly account for the ORB. Each individual's face space will be tuned toward the category of face most frequently seen in the environment. Consequently, the face space is less effective when a person is processing faces from a novel or infrequently encountered racial group.

Early Preference for Faces (over Objects)

Faces are ubiquitous in the infant's environment, and almost certainly are the most frequently encountered visual stimuli during the first few days of life. It is reasonable to assume that faces may consequently be processed preferentially, compared to other categories of visual stimuli. Fantz (1961) found that infants as young as 1 month of age showed a consistent spontaneous preference for face-like stimuli over non-face-like patterns. Morton and Johnson (1991) hypothesized that movement is an important component of the face preference shown during the first week of life. Indeed, some early studies (Goren, Sarty, & Wu, 1975; Johnson, Dziurawiec, Ellis, & Morton, 1991) showed that moving face-like patterns yielded greater tracking behavior than non-face-like patterns in newborns tested just a few minutes after birth. Preference has also been observed for static schematic face-like stimuli over equally complex visual stimuli (Kleiner, 1987; Macchi Cassia, Simion, & Umiltà, 2001; Mondloch et al., 1999; Simion, Valenza, Umiltà, & Dalla Barba, 1998; Valenza, Simion, Macchi Cassia, & Umiltà, 1996).

More recently, a preference has been demonstrated for using photographs of real faces when paired with inverted faces (Macchi Cassia, Turati, & Simion, 2004). Macchi Cassia et al.'s (2004) inverted stimuli were photographs of faces depicted from the crown of the head to the neck, which were "featurally inverted" (i.e., the mouth was positioned above the nose, and the eyes below it). This contrasts with classical inversion, where the entire face is simply rotated 180°. However, it can be argued that the preference for the upright face may be due to the discrepancy between the neck and the internal feature position. Collectively, these results show that newborns demonstrate a strong visual bias for human faces or human face-like stimuli. This preference is consistently found, regardless of whether stimuli are static or moving.

In a longitudinal study, Johnson et al. (1991) observed preferential tracking of faces over the first 5 months of life. They showed that 1-month-old infants tracked a schematic face-like pattern further than they did stimuli that possessed facial features in an atypical configuration or nonfacial features (three squares) in a facial configuration. They also demonstrated that this preferential tracking appeared to decline between 4 and 6 weeks after birth. The visual preference task was not well adapted for older infants, who habituated very quickly to the picture presented and were not willing to stay still on their parents' laps for long periods, so a preference for faces was not documented for older age groups.

Face Categorization

Categorization is "a mental process that underlies one's ability to respond equivalently to a set of discriminably different entities as instances of the same class" (Quinn & Slater, 2003). It is the construction of a prototypical average of a class of stimuli. Subsequent individual stimuli will be encoded according to their deviation of the prototype, as in Valentine's (1991) model (see above). The spontaneous preference for faces or face-like patterns, reviewed above, is indirect evidence of the existence of a face categorization process early in life. Any face is part of the same category of stimuli that are more attractive than other visual stimuli. Face processing develops quickly from birth, with infants receiving more and more experience of only a handful of face exemplars, mainly their parents. de Haan, Johnson, Maurer, and Perrett (2001) demonstrated that infants begin to show evidence of face prototype formation at 3 months of age. Before this age, face recognition seems to be exemplar-based: Each individual face is separately encoded. Quinn, Yahr, Kuhn, Slater, and Pascalis (2002) hypothesized that the face representation of 3-month-olds may be biased toward female faces, as their primary caregivers are usually their mothers. They supported their hypothesis by demonstrating that 3-month-old infants raised by their mothers preferred to look at female faces when these were paired with male faces. Quinn et al. also identified and tested a small population of 3-month-olds who had been raised by their fathers. In this instance, a preference for male faces was observed.

Kelly et al. (2005) further highlighted the role of experience in early infancy, demonstrating that 3-month-old Caucasian infants prefer to look at faces from their own racial group when paired with faces from other racial groups in a visual preference task. Caucasian newborns who were tested in an identical manner demonstrated no preference for faces from either their own or other racial groups. Kelly et al. concluded that the preference observed in 3-month-old infants is a direct consequence of predominant exposure to faces from their own race early in infancy. Two more recent studies have shown that this early preference is not confined to Caucasian infants. Kelly et al. (2007) replicated their findings with 3-month-old Chinese infants tested in China, and Bar-Haim, Ziv, Lamy, and Hodes (2006) found a preference for own-race faces in 3-month-old Israeli and Ethiopian infants, also tested in their native countries. Bar-Haim et al. also tested a population of Israeli-born, Ethiopian-descent infants who had received exposure to both African and Caucasian faces. These infants showed no preference for faces from either racial group. Collectively, these results clearly demonstrate how the early face processing system is influenced by the faces observed within infants' visual environment.

Recognition of Faces

In addition to face preference and categorization, a third and crucial aspect of the face-processing system is recognition—realizing that a specific face has been encountered before, and assessing its familiarity. The ability to learn and recognize individual faces is paramount for the development of attachments and social interaction. Recognition of the mother is particularly important for the development of attachment and emotional bonding between mother and child (Bowlby, 1969).

Psychologists have attempted to determine when and how the mother's face is first recognized by the infant, using both the real face and pictures of the mother's face. Field, Cohen, Garcia, and Greenberg (1984) first established that 4-day-olds looked longer at their mothers' live faces than at a stranger's face, even without voice cues. However, the possibility that olfactory cues may have facilitated discrimination of the mothers' faces cannot be discounted. To address this issue, Bushnell, Sai, and Mullin (1989) masked potential olfactory cues by spraying perfume in the testing area and ensured that the mothers and strangers were matched for hair color and length. Using a visual preference paradigm, they still found that newborns averaging 49 hours in age showed a preference for their mothers' faces. This result was replicated by Pascalis, de Schonen, Morton, Deruelle, and Fabre-Grenet (1995), who also recorded the delay between the time the mother was last seen and the recognition test. They showed that neonates could recognize their mothers even after a delay of 3 minutes. Bushnell (2001) established that a 15-minute delay between the last exposure to the mother's face and preference testing did not affect the strength of preference, suggesting that memory for the mother's face is robust within a few days of birth.

Pascalis et al. (1995) investigated which part of the face was important in this recognition and found that when external facial features were masked, newborns did not show a visual preference for their mothers' faces. Walton, Bower, and Bower (1992) used static images of the mother's face rather than real live faces. They found that newborns produced significantly more sucking responses to seeing images of their mothers' faces on the screen as opposed to a stranger's face. Sai and Bushnell (1988) showed that 1-month-olds could discriminate between their mothers' faces when these were paired with a stranger's face, and Barrera and Maurer (1981) showed that 3-month-old infants took longer to habituate to their mothers' faces than to a stranger's face.

A recent study by Sai (2005) elegantly demonstrated that newborns only recognized their mothers' faces if postnatal exposure to the mothers' voice–face combinations were available. It appears that a mother's face is in fact learned in conjunction with the mother's voice, which has been heard during gestation. If an infant is denied the auditory input of the mother's voice after birth, recognition of the mother's face is not demonstrated at this stage.

The mother's face is, however, a special case, as it is positively reinforced and is interrelated with attachment. Can newborns learn a second face during the first week of life? Two studies have shown that newborns can learn and discriminate schematic face-like stimuli, which differ by their inner elements (Simion, Farroni, Macchi Cassia, Turati, & Dalla Barba, 2002; Turati & Simion, 2002). Only a handful of studies have investigated recognition of strangers' faces in newborns. It has been shown that 4-day-old infants habituated to a photograph of a stranger's face can recognize it immediately and after a retention interval of 2 minutes (Pascalis & de Schonen, 2004; Turati, Cassia, Simion, & Leo, 2006). Neonates are able to learn and recognize another face as early as 4 days of age, even if the face is not presented with the full multimodal aspect. It is important to note that the pictures of the faces used in these experiments were identical in the familiarization task and during the recognition test. This raises an important question: Are newborns recognizing a face per se or just a picture? Certainly faces are matched with care, and recognition cannot be based on gross differences such as those in hair style. However, the fact remains that these are still essentially "picture-matching" tasks, and a degree of caution is required in interpreting such results. Long-term recognition has been investigated only with older infants, and these studies have shown that from 3 months of age faces can be learned and recognized after long delays (Fagan, 1973; Pascalis, de Haan, Nelson, & de Schonen, 1998). It is also from 3 months of age that infants are able to learn a face from a certain point of view (e.g., full face) and recognize it from a new point of view (e.g., three-quarters profile) (Pascalis et al., 1998). It is thus well established that infants are able to learn and recognize pictures of faces during the first months of life.

A strong belief that children are "terrible at face recognition" until 10 years of age (Carey, 1992, p. 12) was based on literature from the 1970s. A close inspection of more recent literature shows that children's performance levels can be increased by using tasks designed specifically for their age groups (see Want, Pascalis, Coleman, & Blades, 2003, for a review). Children from 4 to 5 years of age can achieve 80% accuracy when recognizing faces in a forced-choice task (Bruce et al., 2000). In a study with picture books, Brace et al. (2001) found that 5- to 6-year-olds performed almost at ceiling, achieving approximately 93% accuracy, whereas 2- to 4-year-olds achieved roughly 73% accuracy. Want et al. (2003) have shown that the same level of accuracy can be found in 5- and 10-year-olds if the familiarization period is adjusted appropriately for the age group. The face recognition system is not adult-like until late adolescence, however. In fact, a developmental "dip" is observed in some studies at about 12 years of age, disrupting the linear trend observed during childhood (see Leder, Schwarzer, & Langton, 2003, for a review). The "dip" may reflect either a change in the encoding strategy or biological changes occurring during puberty.

It is therefore clear that recognition per se is very efficient from an early age during childhood. However, the way in which faces are recognized may differ across development.

Configural Processing

The most influential explanation for the development of face processing has been proposed by Carey and colleagues (Carey & Diamond, 1977, 1994; Carey, Diamond, & Woods, 1980; Diamond & Carey, 1977). They proposed that children begin recognizing individuals largely on the basis of individual facial features, such as distinctive eyes or noses—a phenomenon called "featural processing." Adults, on the other hand, recognize individuals by processing the spatial relationships between the individual features of the face—a phenomenon called "configural processing,"as noted earlier. Only late in childhood, after sufficient exposure to faces, do children begin to recognize individuals through configural processing and achieve good levels of performance. This claim was originally based on evidence that children's recognition of faces is less affected by inversion than that of adults is. This view influenced the field of face-processing development for many years, and it is still widely accepted outside the field; however, a number of studies conducted with infants and children during the past decade have produced results that strongly challenge this account.

The Inversion Effect

The inversion effect is one of the most widely known and commonly studied effects in face recognition. As described earlier, refers to the fact that faces are recognized more accurately and quickly when presented in their canonical orientation than when presented upside down (Yin, 1969; Leder & Bruce, 2000; Maurer et al., 2002). As an explanation of the inversion effect, it has been proposed that in addition to coding of single facial features, upright face recognition relies on the configuration of facial features. It is this configural processing that may be disrupted when faces are turned upside down. Therefore, the inversion effect is typically accepted as an indirect measure of configural processing.

Several preferences have been found to be abolished by inverting face stimuli. Slater et al. (2000) found that newborns' preference for attractive faces was disrupted when the faces were inverted. Quinn et al. (2002) found that 3-month-old infants' preference for female faces over male faces was not found if the faces were presented inverted. Turati, Sangrigoli, Ruel, and de Schonen (2004) demonstrated 4-month-old infants' ability to recognize upright and inverted faces when familiarization and the test were done with the same pictures, but no recognition was found for inverted faces when the stimuli were learned in various poses.

Cohen and Cashon (2001) found signs of the inversion effect from 7 months of age. After being habituated to two adult female faces, infants were tested with a composite face constructed from the internal features of one face pasted onto the outer features of the other face. Results showed that in the upright condition, infants looked longer at the composite "switched" face than at the familiar face. This indicates that infants must have processed at least some configural properties of the face. However, infants could have combined all or only a number of internal and external features. In the inverted condition, infants did not look longer at the composite face than at the familiar face, showing no evidence of being able to process the configuration of inverted faces. The 7-month-olds showed an adult-like pattern of response; they processed the upright internal and external facial sections holistically and the inverted sections featurally.

The "switch" design used by Cohen and Cashon allows determination of a configural processing mode even when the faces are only presented in an upright orientation. The very fact that infants respond to composite faces made up of familiar features demonstrates their sensitivity to configural face information. Zauner and Schwarzer (2003) used such an approach and studied 4- to 8-month-olds' processing of schematically drawn faces. They showed that 4-month-olds processed the stimuli by features, whereas 6- and 8-month-olds used a more configural approach. In a follow-up study (Figure 4.2), Schwarzer, Zauner, and Jovanovic (2007) used real faces to determine which part of the face creates a switch effect. They compared small changes in either the eyes or the mouth with bigger changes in the eyes and mouth. They found that 10-month-old infants processed faces configurally for both small and big changes, whereas 6-month-olds processed eyes as features but the mouth as part of the whole face. Four-month-olds processed both the eyes and mouth as features. These mixed results are important, as it demonstrates that limited configural processing can be observed in infants, but is still very different from the configural system observed in adults. Carey and Diamond (1977) demonstrated a strengthening of the inversion effect between 6 and 10 years of age. In children, many studies have demonstrated the existence of an inver-

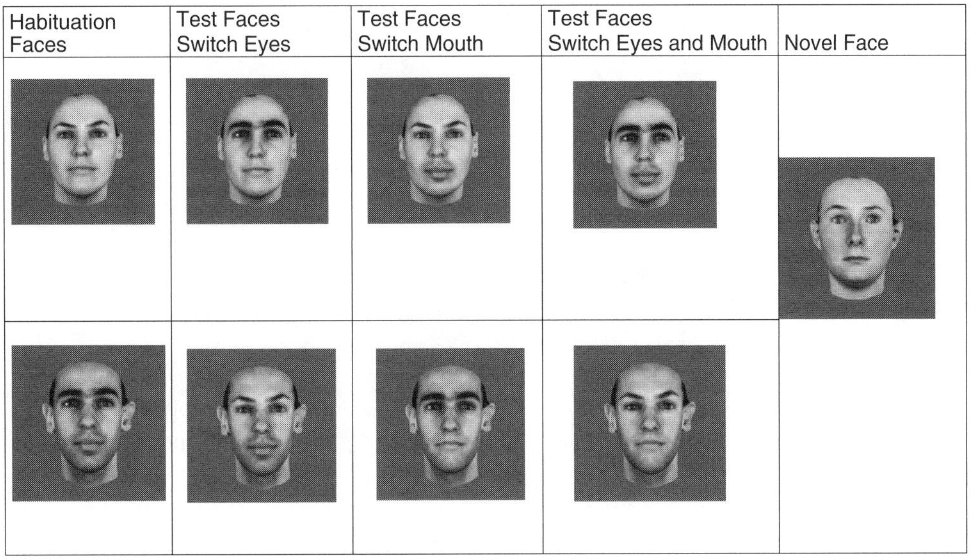

Habituation Faces	Test Faces Switch Eyes	Test Faces Switch Mouth	Test Faces Switch Eyes and Mouth	Novel Face

FIGURE 4.2. Faces used in the follow-up "switch" experiments. From Schwarzer, Zauner, and Jovanovic (2007). Copyright 2007 by Blackwell Publishing Ltd. Reprinted by permission.

sion effect even at a young age (Brace et al., 2001; Flin, 1985; Pascalis, Demont, de Haan, & Campbell, 2001; Want et al., 2003).

Holistic Processing and Sensitivity to Second-Order Relations in Children

Children's configural processing, in terms of holistic processing as "gluing together" facial features into a gestalt (see Maurer et al., 2002), has been demonstrated with the so-called "part–whole" face recognition task. In this task, children learn to identify a set of faces and are then asked to identify single features separately as well as in the context of the whole face. In one study, children between 6 and 10 years of age identified the facial features more easily and correctly in the whole-face condition than separately, whereas this advantage disappeared with inverted faces (Tanaka, Kay, Grinnel, Stansfield, & Szechter, 1998). Pellicano and Rhodes (2003) also found such sensitivity to holistic relations in even younger children, ages 4 and 5 years. Holistic processing in these young age groups has also been shown by the "composite-face effect" (e.g., Young, Hellaway, & Hay, 1987), in which the top half of a face is recognized correctly when presented apart, but recognition is slower when the top half was combined with the bottom half of a different face (see Figure 4.3). Recently, de Heering, Houthuys, and Rossion (2007) demonstrated that this effect is present from 4 years of age.

However, when children between 2 and 5 years of age were asked to categorize faces, they spontaneously categorized them analytically by focusing on single facial features instead of categorizing them holistically (Schwarzer, 2002). A shift toward holistic categorizations

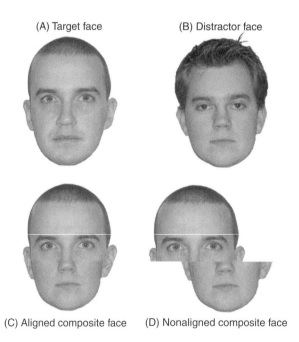

(A) Target face   (B) Distractor face

(C) Aligned composite face   (D) Nonaligned composite face

FIGURE 4.3. Examples of composite and noncomposite faces. *Note:* Faces C and D are composed of the top half of face A and the bottom half of face B.

was not observed until between 6 and 10 years of age (Schwarzer, 2000). The different modes of categorizing faces were observable not only in the different categorizations, but also in specific patterns of gaze behavior (Schwarzer, Huber, & Dümmler, 2005). Thus the explicit usage of holistic relations in a face—for example, for categorization—seems to develop later than the mere sensitivity to changes in holistic relations within a face.

Another type of configural face processing is the sensitivity to "second-order relations" in faces (i.e., information about the spacing among the internal features of a face). Mondloch et al. (2002) have provided evidence that this kind of configural processing also develops relatively late. They modified a single face to create new versions that differed in the shape of the internal features, the shape of the external contour, and the spacing among the internal features. Their results revealed that all children in the study (6- to 10-year-olds) made most errors on the spacing versions, whereas they were almost as accurate as adults on the other versions. These results are in line with those of Freire and Lee (2001), demonstrating that children ages 4–7 years performed only slightly better than chance when asked to detect a target face among distractors that differed only in the spacing of the features; their accuracy was higher when the target differed from the distractors in the shape of the features. Likewise, Mondloch, Leis, and Maurer (2006) could not find evidence of the sensitivity to the spacing of internal facial features in 4-year-olds, despite using a very child-friendly experimental setting. They therefore concluded that the ability to use such second-order cues to recognize facial identity emerges between 4 and 6 years.

Other studies, however, have shown 4-year-olds to be able to respond to changes of second-order relational information of faces under specific conditions. Pellicano, Rhodes, and Peters (2006) tested 4- and 5-year-olds with the part–whole paradigm in which facial features had to be recognized separately, in the context of the whole face with the same configuration, and (most interestingly) in the context of the whole face including space changes in the internal features. Children recognized facial features better with the same configuration than corresponding facial features with new spacing, indicating sensitivity to changes in the spacing of internal facial features. Precursors to sensitivity to second-order relational information can be seen even in infancy. Bhatt, Bertin, Hayden, and Reed (2005) habituated 5-month-old infants to a face in which the eyes were moved up and the mouth down. After habituation, the infants looked longer at the unaltered face and failed to discriminate the faces when they were inverted. Similarly, Thompson, Madrid, Westbrook, and Johnston (2001) provide evidence of 7-month-olds' preference for unaltered faces over faces with space changes of internal features (created by either shortening or lengthening the faces).

The Relationship between Processing Facial Identity and Obtaining Social Information from a Face

Faces convey not only information about the identity of a person, but also a variety of social information, such as emotional expression (see de Haan & Matheson, Chapter 6, this volume), speech support by lip reading, or gaze direction (see Grossmann & Farroni, Chapter 5, this volume). It is therefore crucial to analyze how the face-processing system masters this complexity. For example, it is possible to analyze whether and to what extent different types of information operate independently or in interaction with each other. Early research on adults' face recognition started from the assumption that processing of facial identity and emotional expression operate independently of each other (e.g., Bruce & Young, 1986; Etcoff, 1984), but more recent studies have demonstrated interactions of these dimensions. For example, judgments of facial expression were influenced by variations in identity (Schweinberger, Burton, & Kelly, 1999), and well-known faces were recognized better when showing an emotional expression (Gallegos & Tranel, 2005; Kaufmann & Schweinberger, 2004). This question has been discussed in depth from a neurological perspective (e.g., Calder & Young, 2005; Ganel, Valyear, Goshen-Gottstein, & Goodale, 2005, see Batty & Taylor, 2006, for an overview).

There is some evidence that even in infancy, face recognition is influenced when faces express social information. For example, studies by Johnson and Farroni (2003) show that faces with a direct gaze elicit a different electrophysiological reaction than faces with an averted gaze in 4-month-olds. The authors concluded that direct-gaze contact enhanced infants' processing of the faces. More recently, Farroni, Massaccesi, Menon, and Johnson (2007) have demonstrated that faces learned with direct gaze are recognized by 4-month-olds, whereas faces with averted gaze are not well recognized. The emotional expression of a face also provides information that has an impact on face processing. Groß and Schwarzer (2008) habituated 7-month-old infants to a full-frontal or three-quarter pose of a face (without hairline) with a neutral facial expression or with a smiling or angry expression. In the test phase, the habituation face was presented in a novel pose paired with a novel face in the same pose. Whereas the infants did not recognize the habituation face presented in the novel pose when the faces showed a neutral expression, they were able to do so when the faces expressed a positive or negative emotion. On the one hand, these results are in line with those of Cohen and Strauss (1979) indicating that infants were unable to recognize a face with a neutral expression from a novel pose when it was presented in only one other pose during habituation. On the other hand, the results parallel those by Pascalis et al. (1998) and Turati et al. (2004) demonstrating infants' ability to recognize a face from a novel position after the infants were habituated to different poses and emotional expressions of the face.

In a study with older children, Mondloch, Geldart, Maurer, and Le Grand (2003) analyzed the developmental course of 6- to 10-year-olds' performance on various recognition tasks: recognition of identity in spite of changes in pose and emotional expression; and recognition of facial expression, lip reading, and gaze direction regardless of changes in facial identity. The results indicated the slowest development for matching identity in spite of changes in the pose of the face. The performance of even the 6-year-olds was already nearing adult levels for the tasks involving matching facial expression and lip reading in spite of facial identity changes, followed by their performance on tasks for matching gaze direction and for matching identity despite changes in facial expression. These results are consistent with previous studies showing poor performance by young children on tasks that require recognizing faces' identity regardless of changes in head orientation, lighting, or facial expression (Benton & Van Allen, 1973, cited in Carey et al., 1980; Bruce et al., 2000; Ellis, 1992). In addition, Bruce et al. (2000) reported inconsistent correlations between tasks that involved matching faces for identity and matching them on the basis of expressive or other communicative dimensions. The authors interpreted the results as in line with Bruce and Young's (1986) early assumption of independence of these face-processing abilities in adults.

Recent studies, however, provide data necessitating a differentiation of this assumption. Spangler, Schwarzer, Korell, and Maier-Karius (2008) asked children between 5 and 10 years of age to sort faces according to facial identity while disregarding emotional expression, facial speech, and gaze direction, and then to sort the faces by emotional expression, facial speech, and gaze direction while disregarding facial identity. The data indicated that children's sorting according to facial identity operated independently of emotional expression, facial speech, and gaze direction, whereas sorting by emotional expression and facial speech was influenced by facial identity; sorting corresponding to gaze direction was uninfluenced by facial identity only. These results agree with those of Schweinberger and Soukup (1998), which also demonstrated an asymmetric relationship of processing facial identity, emotional expression and facial speech in adults. One reason for such asymmetry may be seen in De Sonneville et al.'s (2002) results suggesting the processing of facial identity to be faster than that of emotional expression. Finally, it can be concluded that face processing in adulthood, and even in early childhood, occurs as differential interaction between the processing of facial identity and the social information in the face.

From our review, it emerges that some aspects of face processing are already present during the first week of life. It then develops slowly toward a more elaborate, adult-like system. The transition to an adult-like face-processing system occurs in several different steps during infancy and childhood. In adults, we know that face processing is controlled by specialized cortical circuits, and recent investigations using electrophysiological and neuroimaging techniques have allowed us to start understanding how these neural circuits may develop their specificity. Before we discuss research on the neural substrate of face processing, however, we provide a brief discussion of prosopagnosia.

Prosopagnosia

The term "prosopagnosia" was coined by Bodamer (1947) to describe a specific form of agnosia that renders a person unable to recognize the faces of other humans. Most reported cases are attributed to either right-sided lesions (Barton, Press, Keenan, & O'Connor, 2002; de Renzi, 1986; Landis, Cummings, Christen, Bogen, & Imbof, 1986; Sergent & Signoret, 1992; Sergent & Villemure, 1989; Schiltz et al., 2006; Uttner, Bliem, & Danek, 2002) or bilateral lesions (Damasio, Damasio, & Van Hoesen, 1982; Meadows, 1974) of the medial occipito-temporal cortex. Typically, sufferers are able to classify a face as such, but are incapable of differentiating between faces and consequently cannot recognize others or even themselves. Some patients with prosopagnosia demonstrate this remarkable deficit while maintaining intact recognition abilities with other visually homogeneous stimuli, such as sheep (McNeil & Warrington, 1993) and cars (Henke, Schweinberger, Grigo, Klos, & Sommer, 1998; Rossion et al., 2003; Sergent & Signoret, 1992). Furthermore, there are cases in which individuals demonstrate a severe deficit in object recognition, but intact face recognition, suggesting that dissociable neural bases may underlie these abilities (Feinberg, Schindler, Ochoa, Kwan, & Farah, 1994; Moscovitch, Wincour, & Behrmann, 1997).

Until recently, the majority of studies have tested individuals with acquired prosopagnosia. However, in the past few years, numerous cases of developmental prosopagnosia have been reported in the literature (e.g., Duchaine, 2000; Barton, Cherkasova, Press, Intriligator, & O'Connor, 2003). Such individuals display face recognition deficits comparable to those of individuals with acquired prosopagnosia, but have no obvious cortical damage and are typically unaware of any incident during development (e.g., brain trauma) that could be responsible for the onset of their impairment. The number of reported cases of developmental prosopagnosia is relatively low, but it is clear that a great deal of variability exists among known cases. For example, Duchaine and Nakayama (2005) tested developmental prosopagnosic subjects with a range of face and object recognition tasks. All the subjects performed poorly on face tasks, but their performance on object tasks was highly variable: Although some patients performed as well as controls, others did much worse.

A key question for neuropsychologists is whether persons with acquired and developmental prosopagnosia actually have the same disorder. Benton (1990) dismissively suggested that developmental prosopagnosia "bears the same tenuous relationship to acquired prosopagnosia that developmental dyslexia bears to acquired alexia." Indeed, other researchers have even speculated that developmental prosopagnosia itself may be a heterogeneous disorder (Duchaine & Nakayama, 2006). To complicate matters further, some researchers argue that a distinction should be made between developmental and congenital prosopagnosia (Behrmann & Avidan, 2005). According to this interpretation, a patient should be considered to have developmental prosopagnosia unless it is certain that face recognition abilities have never been normal in his or her lifetime and that no neurological damage has been sustained. The patient may only be considered to have congenital prosopagnosia if this strict criterion is reached.

Currently, the precise nature and specificity of developmental prosopagnosia remain unclear. However, it is important to acknowledge that the existence of developmental prosopagnosia was not recognized until relatively recently. Furthermore, unlike patients who acquire face recognition difficulties, patients with developmental prosopagnosia may be largely unaware that their abilities are impaired, as they have never known anything different. It is likely that people may develop alternative strategies for recognizing individuals, and that their deficit thus remains unnoticed by themselves and others. Interestingly, other developmental disorders, such as amusia and specific language impairment (SLI), are known to have a familial component and a prevalence of roughly 5–7% (amusia—Peretz & Hyde, 2003; SLI—Tomblin et al., 1997). Therefore, an intriguing possibility is that developmental prosopagnosia may also possess a genetic component and have a similar prevalence rate. Future research is undoubtedly required to enhance our understanding of developmental prosopagnosia and its relation to the acquired form of the disorder.

The Neural Substrate of Face Processing

Electrophysiology

Face-selective electrophysiological activity has been observed in event-related potential (ERP) studies with adults. Faces elicit a negative deflection around 170 msec after stimulus onset (N170), which is of larger amplitude and shorter latency than the one elicited by objects. Its scalp distribution suggests that it is generated in the occipitotemporal/fusiform area (Bentin, Allison, Puce, Perez, & McCarthy, 1996). It has been reported to be influenced by stimulus inversion. The N170 is found to be of larger amplitude and longer latency for inverted human faces than for upright human faces (de Haan, Pascalis, & Johnson, 2002; see Figure 4.4a). This effect is specific to human face stimuli and has not been observed for animal faces (de Haan et al., 2002) or objects (Rossion et al., 2000).

Only a few studies have investigated the electrophysiological response elicited by face presentation in infant and children. At 6 months of age, infants show a positivity 400 msec after a face is presented, this wave being different when the face is inverted (de Haan et al., 2002). de Haan et al. (2002) have found an "infant N170" in 6-month-olds elicited by faces at 290 msec that precedes this positivity (Figure 4.4b). The "infant N170" showed simi-

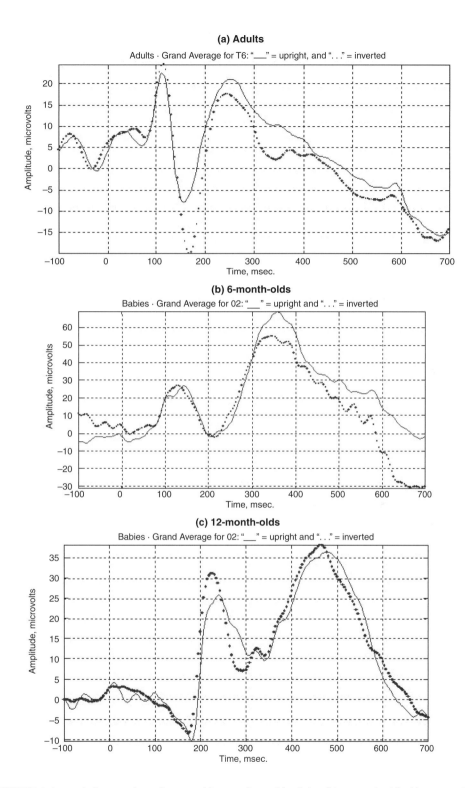

FIGURE 4.4. N170 for upright and inverted human faces. (a) Adults; (b) 6-month-olds; (c) 12-month-olds. Data in a and b from de Haan, Pascalis, and Johnson (2002). Data in c from Halit, de Haan, and Johnson (2003).

lar latency ranges for both upright and inverted faces. They also examined the influence of stimulus inversion for monkey faces and found that in adults such inversion did not affect the processing of monkey faces, whereas in 6-month-olds inversion affected the ERPs similarly for human and monkey faces. At about 12 months of age, the adult ERP pattern was observed (Figure 4.4c). The grand averages showed both the N290 and P400 components, and additionally showed a negativity that peaked slightly before 200 msec (Halit, de Haan, & Johnson, 2003). When 12-month-olds watched upright and inverted monkey faces, their waves were similar to those of adults, in that they were similar for both category of stimuli (Halit et al., 2003).

The N170 does, however, not become adult-like until late in childhood, as illustrated by a series of studies conducted by Taylor and collaborators (Itier & Taylor, 2004a, 2004b). Taylor, McCarthy, Saliba, and Degiovanni (1999) conducted ERP studies with 4- to 14-year-old children and adults in order to test whether qualitative or quantitative changes in face processing occur. Five categories (faces, cars, scrambled faces, scrambled cars, butterflies) were presented to the participants. In contrast to the other stimulus categories, the N170 was clearly found for faces across age groups. The latency of the N170, however, was much longer in young children and had not reached adult values by midadolescence. This demonstrates that the speed with which this processing occurs undergoes huge maturational changes between early childhood and adulthood. The authors interpreted the results in terms of quantitative changes in face-processing modes, in both featural and configural processing. Taylor, Edmonds, McCarthy, and Allison (2001) also showed these kinds of developmental changes in the N170 response to upright faces. Again, they found a slow maturation of configural processing. In contrast, the N170 in children was of shorter latency and greater amplitude in response to eyes alone than to whole faces, and was already mature by 11 years. This suggests that the
N170 responds to such specific features as the eyes as well as to configural information, and that this featural processing of the eyes is in place earlier than the configural processing.

Brain Imaging

The emergence of new technology in brain imaging has been and is still of great excitement, as it may help us to understand better how face processing differs across development before reaching full maturation. To date, functional magnetic resonance imaging (fMRI) studies have identified three cortical areas that appear to be at the core of face processing: the inferior occipital gyrus (the "occipital face area" or OFA), the middle fusiform gyrus (the "fusiform face area" or FFA), and the superior temporal sulcus (STS) (see Figure 4.5 and Plate 4.1). These areas are more activated in humans viewing images of human faces than in humans viewing images of other visual objects (Kanwisher, McDermott, & Chun, 1997; Puce, Allison, Asgari, Gore, & McCarthy, 1996; Haxby, Hoffman, & Gobbini, 2000). This system, which has received a great deal of attention in adults, is now well described and has even been the topic of a fierce discussion about the FFA's specificity for faces, which we do not discuss here (Kanwisher & Yovel, 2006). There is a general agreement that the neural circuit involved in face processing differs somehow during early infancy, but it remains controversial how much it differs, when it becomes adult-like, and whether additional neural substrates are involved in face processing during infancy and childhood. Figure 4.6 displays two alternative views of the development of the neural system of face processing. One favors a specialized system

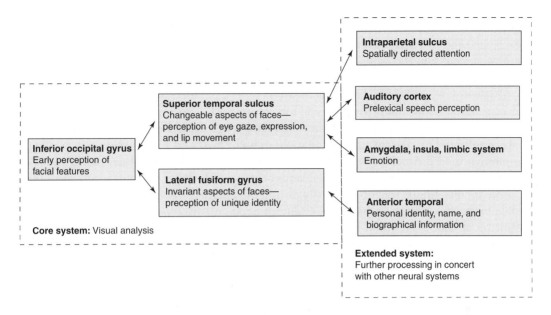

FIGURE 4.5. The three cortical areas identified by fMRI as basic to face processing, and their hypothesized effects on other parts of the brain.

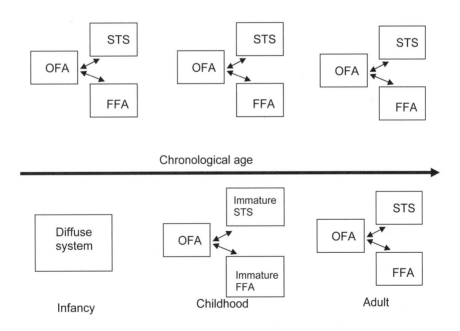

FIGURE 4.6. Two alternative views of the development of the neural system of face processing. Top: hypothesis 1; bottom: hypothesis 2.

from the beginning, which will of course develop with cortical maturation over the years. The second shows a scenario in which specialization does not appear before childhood and its development into the adult system is complete by only about 15 years of age. What evidence is there for these two hypotheses?

To date, only two studies have used neuroimaging techniques to examine face processing during infancy. Using a positron emission tomography (PET) technique, Tzourio-Mazoyer et al. (2002) found brain activation in 2-month-old infants presented with faces in brain regions that are activated by faces in adult populations in both fMRI and PET studies. It is important to note that the PET technique has poor spatial resolution and prevents firm conclusions on the similarity of the structures involved across the ages. However, faces also activated areas in these infants' brains that are typically devoted to language in adults, suggesting an early link between the visual and auditory systems or a more distributed network. More recently, Otsuka et al. (2007) recorded the hemodynamic response of the brain, using near-infrared spectroscopy, while 5- to 8-month-olds were watching upright or inverted faces and objects. The sensors were placed over the temporal and occipitotemporal region, which corresponds to the location of the STS in adults. They found that upright faces created a different signal from the one created by objects, but that inverted faces did not. This result is very similar to what has been recorded in the same region in adults with other imaging techniques (Bentin et al., 1996).

From infant research, it is thus possible to conclude that the neural substrate for face processing may be very similar early in life and in adulthood, which supports the behavioral data described earlier (hypothesis 1 in Figure 4.6). This conclusion is also supported by fMRI studies of language processing showing that the pattern of brain activation observed in infancy is very similar to that of adults, with extra areas recruited and a left-hemisphere advantage less pronounced (Dehaene-Lambertz, Hertz-Pannier, & Dubois, 2006). However, the conclusion that the neural substrate of face processing is very similar in infants and adults is weakened by the fact that the N170 in children differs dramatically from that of adults until 15 years of age.

Several fMRI studies have examined the development of face processing at different ages. Gathers, Bhatt, Corbly, Farley, and Joseph (2004) investigated the development of face perception in children ages 5–11 years who watched a sequence of faces or objects. They found that the area identified as the FFA was bilaterally activated in adults and in children from 9 to 11 years of age. No difference was observed in the FFA for faces and objects in the youngest age group. Moreover, the 9- to 11-year-olds presented an extended activation around the FFA. The 5- to 8-year-old group showed an even greater extension of bilateral activation in posterior and occipital regions. Does this result mean that the FFA is not implicated at all in face processing before 9 years of age and then suddenly starts to be involved? Almost certainly not. Conducting fMRI studies with children is highly problematic, and the null result observed could equally have been a consequence of generally lower levels of activation.

Passarotti et al. (2003) compared brain activation in 10- to 12-year-olds and adults in the FFA and in the middle temporal gyrus. They found that the overall amount of activation in the two regions did not differ across age, but that the children exhibited a more distributed activation than did the adults. Passarotti, Smith, DeLano, and Huang (2007) investigated the inversion effect in children (8–10 years of age), teenagers (13–17 years), and adults (20–30 years). They found that the right FFA showed a similar activation level for upright faces in the three age groups. The left FFA seemed to be more involved in the younger age group than in the others. The STS showed a very similar activation across ages. A difference was observed for inverted faces, with the youngest age group showing a greater activation for inverted than for upright faces.

Aylward et al. (2005) recorded brain activation in children ages 8–10 and 12–14 years, using a memory task in which the children had to recognize upright and inverted faces and houses. The older group showed an adult-like pattern of activation involving the fusiform gyrus and the occipital gyrus, whereas the younger group did not show activation of the fusiform area; instead, activity was recorded in the left inferior temporal lobe and the occipital lobe. The authors contrasted houses and faces, and found that the older age group processed the two types of stimuli differently, while the younger group did not. The authors suggested that even for the oldest group, the activity level of the fusiform area was still below the average in adults. They concluded that their results parallel the differences in face recognition across age observed in behavioral studies.

A more recent study by Golarai et al. (2007) provides a better understanding of the system. They compared recognition of objects and faces in children (7–11 years of age), teenagers, and adults. They found similar activation in the same areas for the three age groups for objects. For faces, they demonstrated that the FFA was three times larger in adults than in children, which may explain why some studies could not find it in young children.

The overall conclusion of these studies is that the FFA's activation develops during childhood toward a stronger and more localized pattern, as in adults. The results also support a specialization of the neural network from childhood. However, we have to be very cautious about interpreting the imaging studies, as these techniques have only recently been applied to children. Children's brain sizes vary more than those of adults, and most of those studies tested a wide range of ages (6–14 years) and collapsed these together as "children." We can then assume that the spatial resolution was affected, and this may partly explain the difficulty of identifying a clear area responding to faces. In the Golarai et al. (2007) study, the results of activation of the "FFA" varied, depending on which parameter was taken into account: global activation of the fusiform gyrus, or adult FFA coordinates applied to children's brains. After correction, the FFA in 7-year-olds presented in fact a very similar activation to that of the adult FFA. Moreover, the Golarai et al. (2007) study shows that other brain regions, such as the STS, are very much adult-like by 7 years of age. We can conclude that faces activate a neural network in the fusiform gyrus area before 10 years of age, but that the overall activation is more diffuse than that observed in adults. It is thus not possible to conclude that the FFA in children is adult-like until 10 years of age.

Conclusions

In the last 20 years, we have learned a lot about face-processing development and its underlying neural network. However, the differences in the tasks and techniques used to study different age groups make it rather difficult to draw firm conclusions about either the developmental course of face processing or its neural substrate. Most of our knowledge about the neural bases of infants' face processing comes from ERPs, whereas the children's studies have used both ERPs and fMRI.

From the fMRI studies done during childhood, it seems legitimate to conclude that the FFA undergoes significant changes during development; this is congruent with the results collected in both behavioral and electrophysiological studies. These results support hypothesis 2 in Figure 4.6. However, PET and ERP studies in infants have found selective responses to faces and nonfaces, supporting hypothesis 1. It is thus premature to draw any firm conclusions about the neural network involved in face processing during childhood. However, it is very unlikely that a brain region such as the FFA will suddenly become actively involved in face processing at 10 years of age, whereas it was not only a few months before. If we assume the existence of a semispecialized infant system, the face-processing system during childhood may be very immature in comparison with that in adulthood, but will be operational, with a moderate "working" FFA. This system will become adult-like during the late teenage years, after a lot of experience and interaction with faces. Experience in this way tunes the face-processing system at both a neural and psychological levels (Nelson, 2001). Such an interpretation is in line with both the neurophysiological and behavioral literature reviewed here.

References

Aylward, E. H., Park, J. E., Field, K. M., Parsons, A. C., Richards, T. L., Cramer, S. C., et al. (2005). Brain activation during face perception: Evidence of a developmental change. *Journal of Cognitive Neuroscience, 17,* 308–319.

Bahrick, H. P., Bahrick, P. O., & Wittlinger, R. P. (1975). Fifty years of memory for names and faces: A cross-sectional approach. *Journal of Experimental Psychology: General, 104,* 54–75.

Bar-Haim, Y., Ziv, T., Lamy, D., & Hodes, R. M. (2006). Nature and nurture in own-race face processing. *Psychological Science, 17*(2), 159–163.

Barrera, M. E., & Maurer, D. (1981). Recognition of mother's photographed face by the three-month-old infant. *Child Development, 52,* 714–716.

Barton, J. J. S., Cherkasova, M. V., Press, D. Z., Intriligator, J. M., & O'Connor, M. (2003). Developmental prosopagnosia: A study of three patients. *Brain and Cognition, 51,* 12–30.

Barton, J. J. S., Press, D. Z., Keenan, J. P., & O'Connor, M. (2002). Lesions of the fusiform face area impair perception of facial configuration in prosopagnosia. *Neurology, 58,* 71–78.

Batty, M., & Taylor, M. J. (2006). The development of emotional face processing during childhood. *Developmental Science, 9*(2), 207–220.

Behrmann, M., & Avidan, G. (2005). Congenital prosopagnosia: Face-blind from birth. *Trends in Cognitive Sciences, 9,* 180–187.

Bentin, S., Allison, T., Puce, A., Perez, E., & McCarthy, G. (1996). Electrophysiological studies of face perception in humans. *Journal of Cognitive Neuroscience, 8,* 551–565.

Benton, A. (1990). Facial recognition. *Cortex, 26,* 491–499.

Bhatt, R. S., Bertin, E., Hayden, A., & Reed, A. (2005). Face processing in infancy: Developmental changes. *Child Development, 76*(1), 169–181.

Bodamer, J. (1947). Die prosopagnosie. *Archiv für Psychiatrie und Nervenkrankheiten, 179,* 6–53.

Bothwell, R. K., Brigham, J. C., & Malpass, R. S. (1989). Cross-racial identification. *Personality and Social Psychology Bulletin, 15,* 19–25.

Bowlby, J. (1969). *Attachment and loss: Vol. 1. Attachment.* New York: Basic Books.

Brace, N. A., Hole, G. J., Kemp, R. I., Pike, G. E., Van Duuren, M., & Norgate, L. (2001). Developmental changes in the effect of inversion: Using a picture book to investigate face recognition. *Perception, 30,* 85–94.

Brigham, J. C., Maass, A., Snyder, L. D., & Spaulding, K. (1982). Accuracy of eyewitness identifications in a field setting. *Journal of Personality and Social Psychology, 42*(4), 673–681.

Brigham, J. C., & Malpass, R. S. (1985). The role of experience and contact in the recognition of faces of own- and other-race persons. *Journal of Social Issues, 41,* 415–424.

Bruce, V., Campbell, R. N., Doherty-Sneddon, G., Import, A., Langton, S., McAuley, S., et al. (2000). Testing face processing skills in children. British *Journal of Developmental Psychology, 18*, 319–333.

Bruce, V., & Young, A. (1986). Understanding face recognition. *British Journal of Psychology, 77*, 305–327.

Bushnell, I. W. R. (2001). Mother's face recognition in newborn infants: Learning and memory. *Infant and Child Development, 10*, 67–74.

Bushnell, I. W. R., Sai, F., & Mullin, J. T. (1989). Neonatal recognition of the mother's face. *British Journal of Developmental Psychology, 7*, 3–15.

Calder, A. J., & Young, A. W. (2005). Understanding the recognition of facial identity and facial expression. *Nature Reviews Neuroscience, 6*, 641–651.

Carey, S. (1992). Becoming a face expert. *Philosophical Transactions of the Royal Society of London, Series B, Biological Sciences, 335*, 95–103.

Carey, S., & Diamond, R. (1977). From piecemeal to configurational representation of faces. *Science, 195*, 312–313.

Carey, S., & Diamond, R. (1994). Are faces perceived as configurations more by adults than by children? *Visual Cognition, 1*, 253–274.

Carey, S., Diamond, R., & Woods, B. (1980). Development of face recognition: A maturational component? *Developmental Psychology, 16*, 257–269.

Chance, J., Goldstein, A. G., & McBride, L. (1975). Differential experience and recognition memory for faces. *Journal of Social Psychology, 97*, 243–253.

Chiroro, P., & Valentine, T. (1995). An investigation of the contact hypothesis of the own-race bias in face recognition. *Quarterly Journal Of Experimental Psychology, 48A*(4), 879–894.

Cohen, L. B., & Cashon, C. H. (2001). Do 7-month-old infants process independent features or facial configurations? *Infant and Child Development, 10*, 83–92.

Cohen, L. B., & Strauss, M. S. (1979). Concept acquisition in human infants. *Child Development, 50*, 419–424.

Cross, J. F., Cross, J., & Daly, J. (1971). Sex, race, age, and beauty as factors in recognition of faces. *Perception and Psychophysics, 10*(6), 393–396.

Damasio, A. R., Damasio, H., & Van Hoesen, G. W. (1982). Prosopagnosia: Anatomic basis and behavioural mechanisms. *Neurology, 32*, 331–341.

de Haan, M., Johnson, M. H., Maurer, D., & Perrett, D. I. (2001). Recognition of individual faces and average face prototypes by 1- and 3-month-old infants. *Cognitive Development, 16*, 659–678.

de Haan, M., Pascalis, O., & Johnson, M. H. (2002). Specialization of neural mechanisms underlying face recognition in human infants. *Journal of Cognitive Neuroscience, 14*, 199–209.

Dehaene-Lambertz, G., Hertz-Pannier, L., & Dubois, J. (2006). Nature and nurture in language acquisition: Anatomical and functional brain-imaging studies in infants. *Trends in Neurosciences, 29*(7), 367–373.

de Heering, A., Houthuys, S., & Rossion, B. (2007). Holistic face processing is mature at 4 years of age: Evidence from the composite effect. *Journal of Experimental Child Psychology, 96*, 57–70.

de Renzi, E. (1986). Prosopagnosia in two patients with CT scan evidence of damage confined to the right hemisphere. *Neuropsychologia, 24*(3), 385–389.

De Sonneville, L. M. J., Verschoor, C. A., Njiokiktjien, C., Op het Veld, V., Toorenaar, N., & Vranken, M. (2002). Facial identity and facial emotion: Speed, accuracy, and processing strategies in children and adults. *Journal of clinical and Experimental Neuropsychology, 24*(2), 200–213.

Diamond, R., & Carey, S. (1977). Developmental changes in the representation of faces. *Journal of Experimental Child Psychology, 23*, 1–22.

Diamond, R., & Carey, S. (1986). Why faces are and are not special. *Journal of Experimental Psychology General, 115*, 107–117.

Duchaine, B. C. (2000). Developmental prosopagnosia with normal configural processing. *Cognitive Neuroscience and Neuropsychology, 11*(1), 79–83.

Duchaine, B. C., & Nakayama, K. (2005). Dissociations of face and object recognition in developmental prosopagnosia. *Journal of Cognitive Neuroscience, 17*(2), 249–261.

Duchaine, B. C., & Nakayama, K. (2006). Developmental prosopagnosia: A window to content-specific face processing. *Current Opinion in Neurobiology, 16*, 166–173.

Dufour, V., Coleman, M., Campbell, R., Petit, O., & Pascalis, O. (2004). On the species-specificity of face recognition in human adults. *Current Psychology of Cognition, 22*(3), 315–333.

Ellis, H. D. (1992). The development of face processing skills. *Philosophical Transactions of the Royal Society of London, Series B, Biological Sciences, 335*, 105–111.

Etcoff, N. L. (1984). Selective attention to facial identity and facial emotion. *Neuropsychologia, 22*, 281–295.

Fagan, J. F. (1973). Infants' delayed recognition memory and forgetting. *Journal of Experimental Child Psychology, 16*, 424–450.

Fantz, R. L. (1961). The origin of form perception. *Scientific American, 204*, 66–72.

Farroni, T., Massaccesi, S., Menon, E., & Johnson, M. H. (2007). Direct gaze modulates face recognition in young infants. *Cognition, 102*, 396–404.

Feinberg, T. E., Schindler, R. J., Ochoa, E., Kwan, P. C., & Farah, M. J. (1994). Associative visual agnosia and alexia without prosopagnosia. *Cortex, 30*, 395–412.

Field, T. M., Cohen, D., Garcia, R., & Greenberg, R. (1984). Mother–stranger discrimination by the newborn. *Infant Behavior and Development, 7*, 19–25.

Flin, R. H. (1985). Development of face recognition: An encoding switch? *British Journal of Psychology, 76*, 123–134.

Freire, A., & Lee, K. (2001). Face recognition in 4- to 7-year-olds: Processing of configural, featural, and paraphernalia information. *Journal of Experimental Child Psychology, 80*, 347–371.

Gallegos, D. R., & Tranel, D. (2005). Positive facial affect facilitates the identification of famous faces. *Brain and Language, 93*(3), 338–348.

Ganel, T., Valyear, K. F., Goshen-Gottstein, Y., & Goodale, M. A. (2005). The involvement of the "fusiform face area" in processing facial expression. *Neuropsychologia, 43*, 1645–1654.

Gathers, A. D., Bhatt, R., Corbly, C. R., Farley, A. B., & Joseph, J. E. (2004). Developmental shifts in cortical loci for face and object recognition. *NeuroReport, 15*(10), 1549–1553.

Gauthier, I., & Tarr, M. J. (1997). Becoming a "Greeble" expert: Exploring the face recognition mechanism. *Vision Research, 37*, 1673–1682.

Golarai, G., Ghahremani, D. G., Whitfield-Gabrieli, S., Reiss, A., Eberhardt, J. L., Gabrieli, J. D. E., et al. (2007). Differential development of high-level cortex correlates with category-specific recognition memory. *Nature Neuroscience, 10*(4), 512–522.

Goren, C., Sarty, M., & Wu, P. Y. K. (1975). Visual following and pattern discrimination of face-like stimuli by newborn infants. *Pediatrics, 56*, 544–549.

Groß, C., & Schwarzer, G. (2008). *The generalization of facial identity over different viewpoints by seven- and nine-month-old infants: The role of facial expression.* Manuscript in preparation.

Halit, H., de Haan, M., & Johnson, M. H. (2003). Cortical specialisation for face processing: Face-sensitive event-related potential components in 3- and 12-month-old infants. *NeuroImage, 19*(3), 1180–1193.

Haxby, J. V., Hoffman, E. A., & Gobbini, M. I. (2000). The distributed human neural system for face perception. *Trends in Cognitive Sciences, 4*, 223–233.

Henke, K., Schweinberger, S. R., Grigo, A., Klos, T., & Sommer, W. (1998). Specificity of face recognition: Recognition of exemplars of non-face objects in prosopagnosia. *Cortex, 34*, 289–296.

Itier, R. J., & Taylor, M. J. (2004a). Face recognition memory and configural processing: A developmental ERP study using upright, inverted, and contrast-reversed faces. *Journal of Cognitive Neuroscience, 16*(3), 487–502.

Itier, R. J., & Taylor, M. J. (2004b). Face inversion and contrast-reversal effects across development: In contrast to the expertise theory. *Developmental Science, 7*(2), 246–260.

Johnson, M. H., Dziurawiec, S., Ellis, H., & Morton, J. (1991). Newborns' preferential tracking of face-like stimuli and its subsequent decline. *Cognition, 40*, 1–19.

Johnson, M. H., & Farroni, T. (2003). Perceiving and acting on the eyes: The development and neural basis of eye gaze perception. In O. Pascalis & A. Slater (Eds.), *The development of face processing in infancy and early childhood: Current perspectives* (pp. 155–167). New York: Nova Science.

Kanwisher, N. G., McDermott, J., & Chun, M. M. (1997). The fusiform face area: A module in human extrastriate cortex specialized for face perception. *Journal of Neuroscience, 17*, 4302–4311.

Kanwisher, N., & Yovel, G. (2006). The fusiform face area: A cortical region specialized for the perception of faces. *Philosophical Transactions of the Royal Society of London, Series B, Biological Sciences, 361*, 2109–2128.

Kaufmann, J. M., & Schweinberger, S. R. (2004). Expression influences the recognition of familiar faces. *Perception, 33*, 399–408.

Kelly, D. J., Liu, S., Ge, L., Quinn, P. C., Slater, A. M., Lee, K., et al. (2007). Cross-race preferences for same-race faces extend beyond the African versus Caucasian contrast. *Infancy, 11*(1), 87–95.

Kelly, D. J., Quinn, P. C., Slater, A. M., Lee, K., Gibson, A., Smith, M., et al. (2005). Three-month-olds, but not newborns, prefer own-race faces. *Developmental Science, 8*, F31–F36.

Kleiner, K. A. (1987). Amplitude and phase spectra as indices of infants' pattern preferences. *Infant Behavior and Development, 10*, 49–59.

Landis, T., Cummings, J. L., Christen, L., Bogen, J. E., & Imhof, H.-G. (1986). Are unilateral right posterior cerebral lesions sufficient to cause prosopagnosia?: Clinical and radiological findings in six additional patients. *Cortex, 22*, 243–252.

Leder, H., & Bruce, V. (2000). When inverted faces are recognized: The role of configural information in face recognition. *Quarterly Journal of Experimental Psychology, 53*, 513–536.

Leder, H., Schwarzer, G., & Langton, S. (2003). Face processing in early adolescence. In G. Schwarzer & H. Leder (Eds.), *The development of face processing.* Cambridge, MA: Hogrefe & Huber.

Lindsay, D. S., Jack, P. C., Jr., & Christian, M. A. (1991). Other-race face perception. *Journal of Applied Psychology, 76*(4), 587–589.

Macchi Cassia, V., Simion, F., & Umiltà, C. (2001). Face preference at birth: The role of an orienting mechanism. *Developmental Science, 4*(1), 101–108.

Macchi Cassia, V., Turati, C., & Simion, F. (2004). Can a nonspecific bias toward top-heavy patterns explain newborns' face preference? *Psychological Science, 15*(6), 379–383.

Maurer, D., Le Grand, R., & Mondloch, C. J. (2002). The many faces of configural processing. *Trends in Cognitive Sciences, 6*, 255–260.

McNeil, J. E., & Warrington, E. K. (1993). Prosopagnosia: A face specific disorder. *Quarterly Journal of Experimental Psychology, 46A*(1), 1–10.

Meadows, J. C. (1974). The anatomical basis of prosopagnosia. *Journal of Neurology, Neurosurgery, and Psychiatry, 37*, 489–501.

Meissner, C. A., & Brigham, J. C. (2001). Thirty years of investigating the own-race bias memory for faces: A meta-analytic review. *Psychology, Public Policy and Law, 7*, 3–35.

Mondloch, C. J., Geldart, S., Maurer, D., & Le Grand, R. (2003). Developmental changes in face processing skills. *Journal of Experimental Child Psychology, 86*, 67–84.

Mondloch, C. J., Le Grand, R., & Maurer, D. (2002). Configural face processing develops more slowly than featural face processing. *Perception, 31*, 553–566.

Mondloch, C. J., Leis, A., & Maurer, D. (2006). Recognizing the face of Jonny, Suzy, and me: Insensitivity to the spacing among features at 4 years of age. *Child Development, 77*(1), 234–243.

Mondloch, C. J., Lewis, T. L., Budreau, D. R., Maurer, D., Dannemiller, J. L., Stephens, B. R., et al. (1999). Face perception during early infancy. *Psychological Science, 10*(5), 419–422.

Morton, J., & Johnson, M. H. (1991). CONSPEC and CONLERN: A two-process theory of infant face recognition. *Psychological Review, 98*(2), 164–181.

Moscovitch, M., Wincour, G., & Behrmann, M. (1997). What is special about face recognition?: Nineteen experiments on a person with visual object agnosia and dyslexia but normal face recognition. *Journal of Cognitive Neuroscience, 9,* 555–604.

Nelson, C. A. (2001). The development and neural bases of face recognition. *Infant and Child Development, 10,* 3–18.

Otsuka, Y., Nakato, E., Kanazawa, S., Yamaguchi, M. K., Watanabe, S., & Kakigi, R. (2007). Neural activation to upright and inverted faces in infants measured by near infrared spectroscopy. *NeuroImage, 34*(1), 399–406.

Pascalis, O., de Haan, M., Nelson, C. A., & de Schonen, S. (1998). Long term recognition assessed by visual paired comparison in 3- and 6-month-old infants. *Journal of Experimental Psychology: Learning, Memory and Cognition, 24,* 249–260.

Pascalis, O., Demont, E., de Haan, M., & Campbell, R. (2001). Recognition of faces of different species: A developmental study between 5–8 years of age. *Infant and Child Development, 10,* 39–45.

Pascalis, O., & de Schonen, S. (1994). Recognition memory in 3–4-day-old human infants. *NeuroReport, 5,* 1721–1724.

Pascalis, O., de Schonen, S., Morton, J., Deruelle, C., & Fabre-Grenet, M. (1995). Mothers' face recognition by neonates: A replication and extension. *Infant Behavior and Development, 18,* 79–85.

Passarotti, A. M., Paul, B. M., Bussiere, J. R., Buxton, R. B., Wong, E., & Stiles, J. (2003). The development of face and location processing: An fMRI study. *Developmental Science, 6*(1), 100–117.

Passarotti, A. M., Smith, J., DeLano, M., & Huang, J. (2007). Developmental differences in the neural bases of the face inversion effect show progressive tuning of face-selective regions to the upright orientation. *NeuroImage, 34*(4), 1708–1722.

Pellicano, E., & Rhodes, G. (2003). Holistic processing of faces in preschool children and adults. *Psychological Science, 14*(6), 618–622.

Pellicano, E., Rhodes, G., & Peters, M. (2006). Are preschoolers sensitive to configural information in faces? *Developmental Science, 9*(3), 270–277.

Peretz, I., & Hyde, K. L. (2003). What is specific to music processing?: Insights from congenital amusia. *Trends in Cognitive Sciences, 7,* 362–367.

Puce, A., Allison, T., Asgari, M., Gore, J. C., & McCarthy, G. (1996). Differential sensitivity of human visual cortex to faces, letterstrings, and textures: A functional magnetic resonance imaging study. *Journal of Neuroscience, 16,* 5205–5215.

Quinn, P. C., & Slater, A. (2003). Face perception at birth and beyond. In O. Pascalis & A. Slater (Eds.), *The development of face processing in infancy and early childhood: Current perspectives* (pp. 3–12). New York: Nova Science.

Quinn, P. C., Yahr, J., Kuhn, A., Slater, A. M., & Pascalis, O. (2002). Representation of the gender of human faces by infants: A preference for female. *Perception, 31,* 1109–1121.

Rossion, B., Caldara, R., Seghier, M., Schuller, A. M., Lazeyras, F., & Mayer, E. (2003). A network of occipito-temporal face-sensitive areas besides the right middle fusiform gyrus is necessary for normal face processing. *Brain, 126,* 2381–2395.

Rossion, B., Gauthier, I., Tarr, M. J., Despland, P.-A., Linotte, S., Bruyer, R., et al. (2000). The N170 occipito-temporal component is enhanced and delayed to inverted faces but not to inverted objects: An electrophysiological account of face-specific processes in the human brain. *NeuroReport, 11,* 1–6.

Sai, F. Z. (2005). The role of the mother's voice in developing mother's face preference: Evidence for intermodal perception at birth. *Infant and Child Development, 14*(1), 29–50.

Sai, F. Z., & Bushnell, I. W. R. (1988). The perception of faces in different poses by 1-month-olds. *British Journal of Developmental Psychology, 6,* 35–41.

Schiltz, C., Sorger, B., Caldara, R., Ahmed, F., Mayer, E., Goebel, R., et al. (2006). Impaired face discrimination in acquired prosopagnosia is associated with abnormal response to individual faces in the right middle fusiform gyrus. *Cerebral Cortex, 16*, 574–586.

Schwarzer, G. (2000). Development of face processing: The effect of face inversion. *Child Development, 71*(2), 391–401.

Schwarzer, G. (2002). Processing of facial and non-facial visual stimuli in 2–5-year-old children. *Infant and Child Development, 11*, 253–269.

Schwarzer, G., Huber, S., & Dümmler, T. (2005). Gaze behavior in analytical and holistic face processing. *Memory and Cognition, 33*(2), 344–354.

Schwarzer, G., Zauner, N., & Jovanovic, B. (2007). Evidence of a shift from featural to configural face processing in infancy. *Developmental Science, 10*(4), 452–463.

Schweinberger, S. R., Burton, A. M., & Kelly, S. W. (1999). Asymmetric relationship between identity and emotion perception: Experiments with morphed faces. *Perception and Psychophysics, 61*, 1102–1115.

Schweinberger, S. R., & Soukup, G. R. (1998). Asymmetric relationship among perceptions of facial identity, emotion, and facial speech. *Journal of Experimental Psychology: Human Perception and Performance, 24*(6), 1748–1765.

Sergent, J., & Signoret, L. (1992). Varieties of functional deficits in prosopagnosia. *Cerebral Cortex, 2*, 375–388.

Sergent, J., & Villemure, J.-G. (1989). Prosopagnosia in a right hemispherectomized patient. *Brain, 112*, 975–995.

Simion, F., Farroni, T., Macchi Cassia, V., Turati, C., & Dalla Barba, B. (2002). Newborns' local processing in schematic facelike configurations. *British Journal of Developmental Psychology, 20*(4), 465–478.

Simion, F., Valenza, E., Umiltà, C., & Dalla Barba, B. (1998). Preferential orienting to faces in newborns: A temporal–nasal asymmetry. *Journal of Experimental Psychology: Human Perception and Performance, 24*(5), 1399–1405.

Slater, A., Bremner, G., Johnson, S. P., Sherwood, P., Hayes, R., & Brown, E. (2000). Newborn infants' preferences for attractive faces: The role of internal and external facial features. *Infancy, 1*(2), 265–274.

Spangler, S., Schwarzer, G., Korell, M., & Maier-Karius, J. (2008). *The relationship of processing facial identity, emotional expression, facial speech, and gaze direction in the course of development*. Manuscript submitted for publication.

Tanaka, W. J., Kay, J. B., Grinnel, E., Stansfield, B., & Szechter, L. (1998). Face recognition in young children: When the whole is greater than the sum of its parts. *Visual Cognition, 5*(4), 479–496.

Taylor, M. J., Edmonds, G. E., McCarthy, G., & Allison, T. (2001). Eyes first!: Eye processing develops before face processing in children. *NeuroReport, 12*, 1671–1676.

Taylor, M. J., McCarthy, G., Saliba, E., & Degiovanni, E. (1999). ERP evidence of developmental changes in processing of faces. *Clinical Neurophysiology, 110*, 910–915.

Thompson, L. A., Madrid, V., Westbrook, S., & Johnston, V. (2001). Infants attend to second-order relational properties of faces. *Psychonomic Bulletin and Review, 8*, 769–777.

Tomblin, J. B., Records, N. L., Buckwalter, P., Zhang, X., Smith, E., & O'Brien, M. (1997). Prevalence of specific language impairment in kindergarten children. *Journal of Speech, Language, and Hearing Research, 40*, 1245–1260.

Turati, C., Cassia, V. M., Simion, F., & Leo, I. (2006). Newborns' face recognition: Role of inner and outer facial features. *Child Development, 77*(2), 297–311.

Turati, C., Sangrigoli, S., Ruel, J., & de Schonen, S. (2004). Evidence of the face inversion effect in 4-month-old infants. *Infancy, 6*, 275–297.

Turati, C., & Simion, F. (2002). Newborns' recognition of changing and unchanging aspects of schematic faces. *Journal of Experimental Child Psychology, 83*, 239–261.

Tzourio-Mazoyer, N., de Schonen, S., Crivello, F., Reutter, B., Aujard, Y., & Mazoyer, B. (2002). Neural correlates of woman face processing by 2-month-old infants. *NeuroImage, 15*, 454–461.

Uttner, I., Bliem, H., & Danek, A. (2002). Prosopagnosia after unilateral right cerebral infarction. *Journal of Neurology, 249*, 933–935.

Valentine, T. (1991). A unified account of the effects of distinctiveness, inversion, and race in face recognition. *Quarterly Journal of Experimental Psychology, 43A*(2), 161–204.

Valenza, E., Simion, F., Macchi Cassia, V., & Umiltà, C. (1996). Face preference at birth. *Journal of Experimental Psychology: Human Perception and Performance, 22*(4), 892–903.

Walton, G. E., Bower, N. J. A., & Bower, T. G. R. (1992). Recognition of familiar faces by newborns. *Infant Behavior and Development, 15*, 265–269.

Want, S. C., Pascalis, O., Coleman, M., & Blades, M. (2003). Face facts: Is the development of face recognition in middle childhood really so special? In O. Pascalis & A. Slater (Eds.), *The development of face processing in infancy and early childhood: Current perspectives* (pp. 207–221). New York: Nova Science.

Yin, R. K. (1969). Looking at upside down faces. *Journal of Experimental Psychology, 81*(1), 141–145.

Young, A. W., Hellaway, D., & Hay, D. (1987). Configural information in face perception. *Perception, 16*, 747–759.

Zauner, N., & Schwarzer, G. (2003). Entwicklung der Gesichtsverarbeitung im ersten Lebensjahr. *Zeitschrift für Entwicklungspsychologie und Pädagogische Psychologie, 35*(4), 229–237.

CHAPTER 5

Decoding Social Signals in the Infant Brain

A Look at Eye Gaze Perception

TOBIAS GROSSMANN
TERESA FARRONI

An important social signal encoded in faces is eye gaze. The detection and monitoring of eye gaze direction are essential for effective social learning and communication among humans (Bloom, 2000; Csibra & Gergely, 2006). Eye gaze provides information about the target of another person's attention and expression, and it also conveys information about communicative intentions and future behavior (Baron-Cohen, 1995). It has been argued that an early sensitivity to eye gaze serves as a major foundation for later development of social skills (Baron-Cohen, 1995; Csibra & Gergely, 2006; Hood, Willen, & Driver, 1998).

Over the last two decades, a number of neuropsychological and neuroimaging studies have contributed to the identification of specialized cortical brain areas involved in gaze processing in adults (for an overview, see Table 5.1). Despite the progress in understanding adults' sophisticated gaze-reading capacities and their neural basis, the more basic questions about the development and precursors of these adult abilities have been virtually unaddressed. In order to answer these questions, it is of particular importance to look at the earliest stage of postnatal development (i.e., infancy). Therefore, the goal of this chapter is to review and integrate the accumulating behavioral and neuroimaging work on the emergence of the eye-gaze-processing system during infancy.

Before we turn to a review of the empirical work, it seems worthwhile to situate this work in the broader context of developmental theories of eye gaze perception. Let us therefore briefly consider the theoretical accounts that have been put forward. Theories in this area differ with respect to (1) the degree to which gaze processing depends on experience, and

TABLE 5.1. Overview of Brain Regions Implicated in Eye Gaze Processing in the Human Adult, and Their Probable Functional Roles

Brain regions	Functional properties	Studies
Right superior temporal sulcus (STS)	Posterior STS: No clear preference for specific gaze direction; sensitivity to others' (signalers') intentional states indicated by eye gaze	Calder et al. (2002); Pelphrey et al. (2003, 2004); Puce et al. (1998); Wicker et al. (1998)
	Anterior STS: Dissociable coding of different gaze directions (left vs. right gaze)	Calder et al. (2007)
Right inferior parietal lobule (IPL)/inferior parietal sulcus (IPS)	General role in orienting attention, which is initiated by viewing averted gaze	Calder et al. (2007); Hoffman & Haxby (2000)
Medial prefrontal cortex (MPFC)	General role in mentalizing, detection, and decoding of communicative/social intentions toward self (receiver), which can be triggered by eye gaze	Conty et al. (2007); Bristow et al. (2007); Kampe et al. (2003); Schilbach et al. (2006)
Amygdala	Extraction of affective significance; more responsive to direct gaze	George et al. (2001); Kawashima et al. (1999)
Fusiform face area (FFA)	Enhanced face encoding of direct-gaze faces; functional coupling with amygdala	George et al. (2001)

Note. This is not an exhaustive review of all neuroimaging studies available.

(2) the extent to which gaze processing is considered to be independent from other aspects of face processing. One influential theory proposed by Baron-Cohen (1994) stipulates that due to the adaptive importance of detecting eye gaze early in life, there is an innate module within the brain, the so-called "eye direction detector," which is specifically devoted to processing eye gaze. Furthermore, according to this account, the "eyes looking at me" simply pop out from a visual scene, which implies that this pop-out effect is independent from the face context. In contrast to this nativist and modular framework, other accounts argue that although there may be initial perceptual biases that guide infants' behavior to eye gaze, these biases may be linked to more general mechanisms for detecting faces (Farroni et al., 2005; Johnson, 2005; for a detailed discussion, see Farroni, Menon, & Johnson, 2006) and/ or communicative partners (Gliga & Csibra, 2007). These accounts further acknowledge that experience plays an important role in the creation of the gaze perception system (Farroni et al., 2006; Vecera & Johnson, 1995). This latter theoretical standpoint, which puts forward that initial biases present in newborns interact with experiential factors shaping the emergence of the gaze-processing system during infancy, is supported by the bulk of the empirical evidence presented in this chapter.

The structure of the chapter is as follows. First, we discuss behavioral data drawn from three areas that we consider most relevant for the discussion of eye gaze perception in infancy: (1) eye contact detection, (2) gaze following, and (3) joint attention. This section is followed by a thorough review of findings illuminating the neural basis of eye gaze perception in infancy. This represents the main focus of the chapter, and it builds and directly follows up on the behavioral phenomena outlined earlier. The data to be presented in this section are based on findings obtained by using electroencephalography (EEG)/event-related potential (ERP) methods, which are the neuroimaging methods most commonly used with infants. Note, however, that most of the neuroimaging work with adults relies on functional magnetic resonance imaging (fMRI), and the fact that such data are missing with infants partly limits the conclusions that can be drawn in terms of developmental comparisons between infants and adults, especially as far as brain structures are concerned. In closing, we integrate behavioral and neural findings, and look ahead to make suggestions for future work.

Infants' Behavior: Eye Contact Detection, Gaze Following, and Joint Attention

One critical aspect of gaze perception is the detection of eye contact, which enables mutual gaze with another person. Eye contact is considered the most powerful mode of establishing a communicative link between humans (Kampe, Frith, & Frith, 2003; Kleinke, 1986). Sensitivity to eye contact is evident early in human ontogeny, and is believed to provide a vital foundation for social-cognitive development (Baron-Cohen, 1995). From birth, infants have a bias to orient toward face-like stimuli (see Johnson, 2005) and prefer to look at faces with their eyes open (Batki, Baron-Cohen, Wheelwright, Connellan, & Ahluwalia, 2000). The most compelling evidence that human infants are born prepared to detect socially relevant information comes from work showing preferential attention in newborns to faces with direct gaze. In one study (Farroni, Csibra, Simion, & Johnson, 2002), human infants' eye movements were recorded while pairs of face stimuli were presented: One face in each pair directed eye gaze at the newborns, and the other directed eye gaze away (see Figure 5.1A). The analysis of

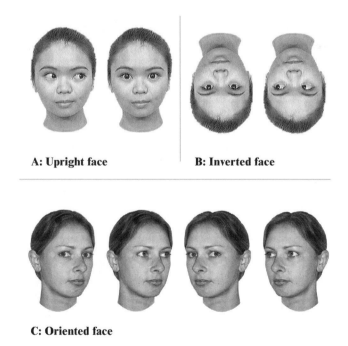

A: Upright face   **B: Inverted face**

C: Oriented face

FIGURE 5.1. Examples of direct-gaze and averted-gaze stimuli used in our behavioral and electrophysiological studies.

newborns' eye movements revealed that fixation times were significantly longer for the faces with direct gaze than for the faces with averted gaze. Furthermore, the number of orienting looks to the direct gaze was also higher than the number to the averted gaze.

Hitherto, the finding of an eye contact preference in newborns has been replicated in a study using a different set of photographic face stimuli (Farroni et al., 2006) and in another study using schematic faces (Farroni, Pividori, Simion, Massaccesi, & Johnson, 2004b). However, the preference for direct gaze in newborns is present only within the context of an upright face and a straight head (Farroni et al., 2006). In other words, when direct-gaze and averted-gaze faces are presented upside down or with the head turned to the side (see Figures 5.1B and 5.1C), then newborns fail to show a preference for eye contact. The finding that newborns do not show a preference when inverted faces are used rules out lower-level explanations of the original result (Farroni et al., 2002), such as a preference for symmetry (direct gaze) over asymmetry (averted gaze). Furthermore, the strong tendency of newborns to orient to faces with direct gaze only under conditions of an upright and straight-head face is consistent with the view that relatively simple perceptual biases and configuration-sensitive mechanisms may be driving this early eye contact preference (Farroni et al., 2002). This view was advanced by Farroni, Mansfield, Lai, and Johnson (2003), who tried to integrate the evidence of infants' perception of eye gaze with the earlier "two-process" theory of face recognition. More specifically, Johnson and Morton (1991) hypothesized that newborns have a primitive mechanism (termed "Conspec" for context-specific) that biases them to orient toward the basic configuration of faces, such as high-contrast areas corresponding to the eyes and mouth. Farroni et al. (2002, 2003) suggested that the same mechanism could also determine the preference of newborns for direct gaze. This is because a mechanism relying on darker elements on a lighter background may help the infant to find a face in the distance or the periphery, but could also support eye contact detection at close proximity (for a discussion, see Farroni et al., 2005). Such a basic mechanism may bias the input to the developing brain, ensuring that infants develop expertise for faces, including eye gaze processing, during later postnatal development. However, this idea remains to be tested.

Another important aspect of eye gaze processing is that averted gaze triggers a reflexive shift of an observer's visual attention (e.g., Driver et al., 1999). Numerous studies have investigated the effects of perceived gaze direction on spatial attention in adults (e.g., Friesen & Kingstone, 1998; Langton, Watt, & Bruce, 2000). The robust finding in these studies is that observers are faster to detect a target stimulus occurring in the peripheral visual field if it is preceded by a face looking in the same direction than when the preceding face is looking in the opposite direction. Newborns have also been found to be faster in making saccades to peripheral targets cued by the direction of eye movements of a schematic face, suggesting a rudimentary form of gaze following (Farroni et al., 2004b). By the age of 3 months, human infants are more likely to orient toward a target stimulus if it is preceded by a perceived gaze shift in the same direction when photographic images of a face are used (Hood et al., 1998).

Four-month-old infants were tested with a cueing paradigm adapted from Hood et al. (1998) to further examine the visual properties of the eyes that enable infants to follow the direction of someone's gaze (Farroni et al., 2003). This series of experiments revealed that infants, in contrast to adults, need to see eye movements to show a gaze-following effect; however, motion alone is insufficient to shift infants' attention, as gaze shifts in an inverted face does not elicit gaze following (this finding has been replicated with newborns; see Farroni et al., 2004b). Moreover, to find this effect in infants, the face has to be removed before the target object is presented—a finding that may be linked to young infants' difficulty in disengaging from attractive stimuli (Johnson & Morton, 1991)—and the gaze cueing has to be preceded by a period of mutual gaze in order for it to be effective. The latter finding supports what we have said earlier about the important role that eye contact plays in establishing social communication between humans. In summary, the critical features for eye gaze cueing in young infants are (1) an upright face, (2) a brief preceding period of eye contact, and (3) lateral motion of the eyes.

The youngest age at which infants have been found to follow the gaze of live partners is between 2 and 3 months (D'Entremont, Hains, & Muir, 1997; Scaife & Bruner, 1975). Again, the gaze-following response requires special triggering conditions, including constant infant-directed speech and target objects that are very close to the presenter's face. By about 6 months, infants follow gaze to more distant targets (Butterworth & Itakura, 2000; Butterworth & Jarrett, 1991), and gaze-following responses to a target become reliable between 7 and 9 months of age (Flom & Pick, 2005; Woodward, 2003). However, the precision of 9-month-olds' responses is still fragile when several potential targets are available (Flom, Deak, Phill, & Pick, 2004). This is because at about this age, infants usually direct their gaze to the first object on the correct side (Morales, Mundy, & Rojas, 1998). Furthermore, 9-month-old infants follow the head turn of someone whose eyes are closed, whereas only a month later they do not (Brooks & Meltzoff, 2005). Only by about 12 months do infants encode the psychological relationship between a person and the target of his or her gaze (Woodward, 2003). However, until the age of 14 months, infants follow blindfolded people's head turns (Brooks & Meltzoff, 2002); after this age, infants start to take into account whether the other has visual access to the target object (Caron, Keil, Dayton, & Butler, 2002; Dunphy-Lelii & Wellman, 2004) and to correctly integrate information from head and eye direction (Caron, Butler, & Brooks, 2002).

The development and improvement of the described gaze-following capacities in infants may be governed by different age-specific mechanisms (for a comprehensive theory, see Butterworth, 1991), and it is generally thought that this development serves the function of coordinating visual attention with other humans and thereby achieving joint attention with them (Tomasello, 1999). More specifically, it has been hypothesized to serve various functions, including (1) instrumental learning/obtaining rewards by catching sight of something interesting (Moore & Corkum, 1994), (2) identifying others' attentional or perceptual states (Baron-Cohen, 1991), and (3) finding out what the other person is communicating about (Csibra, 2008).

To summarize, newborns enter the world with initial perceptual and attentional biases that lead them to attend to eye gaze information encoded in a person's face. As outlined above, these biases seem to rely on certain triggering conditions and are specifically tuned to upright faces. With development, infants' behavior becomes more flexible and accurate not only in discerning a person's gaze direction, but also in linking another person's gaze and visual access to particular objects and events in the environment.

Although behavioral data remains an important source of information for developmental scientists, the successful application of neuroimaging methods to developmental populations has shed light on neural mechanisms that underlie infants' ability to read the language of the eyes. Therefore, in the next section we focus on the neural basis of eye gaze perception in infancy and try to integrate neural-level findings with the behavioral data presented in this section.

Neural Basis of Eye Gaze Perception in the Infant Brain

Gaze Direction Perception

The question we are trying to address in this section is this: What are the neural underpinnings of infants' behaviorally expressed preference for mutual gaze and the capacity to follow gaze? As noted at the start of this chapter, our review focuses on findings obtained by using EEG/ERP methods. Farroni et al. (2002) measured 4-month-old infants' ERPs, to examine neural processing of faces when accompanied with direct or averted gaze. In this study, an occipitotemporal ERP component (N170/N290) known to be sensitive to faces in adults (Bentin, Allison, Puce, Perez, & McCarthy, 1996) and infants (de Haan, Pascalis, & Johnson, 2002; Halit, de Haan, & Johnson, 2003; Halit, Csibra, Volein, & Johnson, 2004) was larger in amplitude in response to direct gaze than to averted gaze. This indicates that the presence of direct gaze enhances the neural processes in the infant brain that are associated with the earliest steps of face encoding (Farroni et al., 2002).

These findings have been replicated and extended by Farroni, Johnson, and Csibra (2004a), who found that in 4-month-old infants, an enhanced cortical processing of direct gaze was obtained even when the head was averted but direct mutual gaze was maintained. However, enhanced neural processing to faces with direct gaze was only found when eyes were presented in the context of an upright face, since in a second experiment, no differential cortical processing was observed when direct- and averted-gaze faces were presented upside down. It is interesting to note that these ERP experiments in 4-month-olds (Farroni et al., 2002, 2004a) followed the same logic (upright vs. inverted face, straight vs. oriented head) and used the same stimuli (see Figure 5.1) as the behavioral work with newborns, and therefore constitute an important extension of this work. This enables us to assess the developmental trajectory by comparing the behavioral data in newborns with the ERP data in 4-month-olds (see Table 5.2). Newborns show a strong preference for direct-gaze faces, and this preference requires the context of an upright human face (Farroni et al., 2006). Similarly, 4-month-olds show an enhanced processing of direct-gaze upright but not inverted faces (Farroni et al., 2002, 2004a). However, newborns do not exhibit a behavioral preference for direct-gaze faces when the head is averted, whereas direct gaze elicits an enhanced cortical response for both straight and averted head angles (see Table 5.2). This suggests that with development and experience, young infants' gaze perception abilities become more flexible, in the sense that they learn to extract information about mutual gaze independently of the head angle. A caveat to this conclusion is that different testing techniques were used with the different age groups; thus it remains possible that the differences observed were due to the measurement techniques rather than to the age groups tested.

Although the ERP studies have provided some insights into how infants process faces with direct and averted gazes, scalp-recorded ERPs do not yield direct information about the underlying brain sources. Johnson et al. (2005) applied independent-component analysis (ICA) to a previously published data set (Farroni et al., 2002), in order to uncover the brain sources sensitive to eye gaze. ICA is a statistical source separation technique (Makeig, Debener, Onton, & Delorme, 2004) that has been successfully employed to localize sources of infant electrophysiological recordings (Richards, 2004, 2005). Consistent with earlier ERP findings (Farroni et al., 2002), Johnson et al. (2005) identified brain sources in 4-month-old infants' occipital and temporal areas discriminating between direct and averted gaze. Whereas adults have been found to show specific activations associated with eye gaze perception in the superior temporal sulcus (STS) (Allison, Puce, & McCarthy, 2000), cortical

TABLE 5.2. Summary of Our Behavioral Findings in Newborns and Electrophysiological Findings in 4-Month-Olds

Face	Newborns	4-month-olds	
		ERPs	Gamma oscillations
Upright	Direct gaze: Eye contact preference	Direct gaze: Enhanced N290	Direct gaze: Enhanced occipital evoked gamma, prefrontal induced gamma
			Averted gaze: Temporoparietal induced gamma
Inverted	No preference	No ERP differences	No gamma activity differences
Oriented	No preference	Direct gaze: Enhanced N290	Direct gaze: Occipital evoked gamma, temporoparietal induced gamma

generators localized in the fusiform gyrus discriminated gaze direction best in infants. Furthermore, while it has been shown that the amplitude of the N290 in infants is modulated by eye gaze (Farroni et al., 2002, 2004a) and face orientation (de Haan et al., 2002; Halit et al., 2003), the amplitude of the adult N170 is only affected by face inversion not by direction of gaze (Grice et al., 2005; Taylor, Itier, Allison, & Edmonds, 2001). Taken together, these differences between infants and adults suggest that face and eye gaze share common patterns of cortical activation early in ontogeny, which later partially dissociate and become more specialized.

In addition, ICA revealed further sources that were sensitive to gaze direction, and a subsequent localization attempt estimated that these sources originated from the prefrontal cortex. Such an effect was not revealed in traditional ERP analyses, illustrating the power of statistical source separation methods (Makeig et al., 2004). These prefrontal sources are of particular interest, because fMRI studies show that prefrontal brain structures are activated by the detection of direct gaze and/or communicative intent in adults (Kampe et al., 2003; Schilbach et al., 2006).

Another technique that can reveal brain activation missed by averaging methods is the analysis of high-frequency oscillations in the gamma band (20–100 Hz). Such oscillations either are time-locked to eliciting stimuli (evoked gamma activity), or can be detected as induced gamma activity consisting of oscillatory bursts whose latency "jitters" from trial to trial and whose temporal relationship with the stimulus onset is fairly loose. Hence induced gamma activity is not revealed by classical averaging techniques, and specific methods based on time-varying spectral analysis of single trials are required to detect it (Tallon-Baudry & Bertrand, 1999). Gamma oscillations are of special interest, because they have been found to correlate with the blood-oxygen-level-dependent (BOLD) response used in fMRI (Foucher, Otzenberger, & Gounot, 2003; Fiebach, Gruber, & Supp, 2005), and it is thought that activity in the gamma range serves as a mechanism to integrate activity from various highly specialized brain areas (Gruber, Trujillo-Barreto, Giabbiconi, Valdés-Sosa, & Müeller, 2006; Tallon-Baudry & Bertrand, 1999; Rodriguez et al., 1999).

Gamma oscillatory activity in infants has primarily been studied only in the context of object processing (Csibra, Davis, Spratling, & Johnson, 2000; Kaufman, Csibra, & Johnson, 2003, 2005). However, in a recent study (Grossmann, Johnson, Farroni, & Csibra, 2007a), we examined gamma oscillations and its relationship to eye gaze perception in 4-month-old infants. This study was based on a time–frequency analysis performed on two previously published EEG data sets taken from Farroni et al.'s (2002, 2004a) ERP studies. Infants were presented with upright images of female faces directing their gaze toward them or to the side (see Figure 5.1A). We predicted a burst of gamma oscillation over prefrontal sites to direct gaze if gamma oscillations are indeed related to detecting eye contact/communicative intent, as suggested by adult fMRI work (Kampe et al., 2003; Schilbach et al., 2006). Averted gaze also serves an important function during communication by directing the perceiver's attention to certain locations or objects, and behavioral measures have shown that infants are sensitive to this aspect of eye gaze (Farroni et al., 2003; Hood et al., 1998). The right inferior parietal sulcus (IPS) and right STS, which have been identified as sensitive to averted gaze in the adult brain (Hoffman & Haxby, 2000), are potential candidates generating effects observable in infants. Therefore, we hypothesized that some activity over right posterior regions would be associated with the perception of averted gaze. In addition, another group of 4-month-old infants were presented with the same face stimuli upside down (see Figure 5.1B), which is thought to disrupt configural face processing (Rodriguez et al., 1999; Turati, Sangrigoli, Ruel, & de Schonen, 2004) and infants' preference for mutual gaze (Farroni et al., 2006). Thus we predicted that inverted faces would not induce activity in the gamma band that differed as a function of eye gaze.

The data revealed that evoked and induced gamma oscillations varied as a function of gaze direction in the context of an upright face, which extends previous ERP and source localization results (Farroni et al., 2002, 2004a; Johnson et al., 2005). In support of our hypotheses, specific effects with distinct spatial and temporal characteristics were observed, depending upon whether gaze was directed at or directed away from an infant. Direct gaze compared to averted gaze evoked early (100 msec) increased gamma activity (20–40 Hz) at occipital channels. Short-latency phase-locked oscillatory evoked gamma responses have been described in the visual modality in response to brief static stimuli in infant and adult EEG (Csibra et al., 2000; Tallon-Baudry & Bertrand, 1999). In adults, it has been shown that evoked gamma activity is significantly larger for items that match memory representations (Herrmann, Lenz, Junge, Busch, & Maess, 2003; Herrmann, Munk, & Engel, 2004). It is possible that for infants, a face with direct gaze represents a more prototypical ("better") face (Farroni, Massaccesi, Menon, & Johnson, 2007a), which is closer to what is represented in memory than a face with averted gaze, and therefore elicits an enhanced evoked oscillatory response. This interpretation is supported by, and may be linked to, findings showing an enhanced neural encoding (Farroni et al., 2002) and better recognition of upright faces with direct gaze in infants (Farroni et al., 2007a).

As predicted, direct gaze also elicited a late (300 msec) induced gamma burst over right prefrontal channels. In a previous analysis based on ICA, cortical sources sensitive to gaze direction had been identified in prefrontal regions (Johnson et al., 2005), which is consistent with the described finding. Directing eye gaze at someone (i.e., making eye contact) serves as an important ostensive signal in face-to-face interactions and helps in establishing a communicative link between two people. It has been argued that successful communication between two people depends crucially on the ability to detect the intention to communicate conveyed by signals directed at the self, such as making eye contact (Kampe et al., 2003). On a neural level, the dorsal part of the right dorsal medial prefrontal cortex (MPFC) has been found to be consistently activated when gaze is directed at, but not when gaze is averted away from, the self (Kampe et al., 2003; Schilbach et al., 2006). It is important to note that gamma oscillations measured with EEG have been found to correlate with the BOLD response used in fMRI (Fiebach et al., 2006; Foucher et al., 2003). It is thus possible that eye contact detection in 4-month-old infants recruits brain mechanisms very similar to those in adults. Alternatively, it has been found that emotional processing, regardless of valence, enhanced gamma band power at right frontal electrodes in adults (Müller, Keil, Gruber, & Elbert, 1999), and infants might have perceived the faces with direct gaze as more emotionally engaging, which resulted in similar gamma responses as in adults.

Averted gaze also serves an important function during social communication by directing the perceiver's attention to certain locations or objects, and there is behavioral evidence that 4-month-olds are sensitive to this aspect of eye gaze (Farroni et al., 2003; Hood et al., 1998). The right IPS and right STS have been identified as sensitive to averted gaze in the adult human brain (Haxby, Hoffman, & Gobbini, 2000; Hoffman & Haxby, 1999). It has been argued that activity in the IPS is specifically recruited when perceived eye gaze direction elicits a shift in spatial attention, whereas STS is more generally associated with eye and mouth movements (Haxby et al., 2000). Our finding of a late (300 msec) induced gamma burst in response to averted gaze over right occipitotemporal–parietal regions may reflect similar but perhaps more diffuse brain activations in infants.

In another study (Grossmann, Johnson, Farroni, & Csibra, 2008), we further examined how head orientation would influence infants' brain responses to eye gaze direction cues observed in the gamma band. This study was based on a time–frequency analysis performed on a previously published EEG data set taken from experiment 1 of Farroni et al.'s (2004a) ERP study. Infants were presented with upright images of female faces orienting their heads away from the infants, but directing their gaze either toward them or away from them (see Figure 5.1C). Corresponding with the findings reported for straight-head-angle faces, direct gaze compared to averted gaze elicited early (100 msec) increased gamma activity (20–40 Hz) at occipital channels. However, contrary to the previous findings (Grossmann et al., 2007a), no induced gamma burst was observed over prefrontal channels to direct gaze when the head was averted. This suggests that although infants at the age of 4 months can discriminate between direct and averted gaze in the context of averted heads as indicated by the increased evoked occipital gamma activity to direct gaze, in this context they do not yet recruit brain processes associated with detecting eye contact as a communicative signal. Hence it follows that by 4 months a frontal face is required to elicit activity in prefrontal brain structures involved in social communication. This finding stands in contrast to fMRI work showing that adults show specific activity in the prefrontal cortex in response to direct gaze, even when the head is averted from the perceiver (Kampe et al., 2003); it therefore suggests that development enabling the human brain to detect mutual gaze regardless of head orientation must occur after 4 months.

Furthermore, as in the previous study using faces with a frontal orientation (Grossmann et al., 2007a), we observed a late (300 msec) induced gamma burst over right temporoparietal regions, but whereas this burst was evoked by averted gaze when frontal faces were used, this activity occurred in response to direct gaze for oriented faces. These seemingly contradictory findings need explanation. Our suggestion is that the gamma activity observed over right temporoparietal channels is associated with neural computations integrating eye gaze direction information in relation to head angle. When the results are revisited from this perspective, it appears that this gamma burst is observed when eye direction is different from/incongruent with the head direction (i.e., averted gaze in a frontal face and direct gaze in an oriented face). This view is in accordance with adult fMRI data showing increased activity in the right STS to averted gaze in a frontal face (Hoffman & Haxby, 1999) and to direct gaze in an averted face (Pelphrey, Viola, & McCarthy, 2004).

More generally, the lateralization of the induced gamma band effects to gaze direction cues to the right hemisphere observed in these studies may be due to the fact that (1) the brain mechanisms underlying eye gaze perception show a high degree of specialization early in ontogeny, recruiting very similar brain areas in the right hemisphere as in adults; (2) the right-hemisphere dominance for the observed effects is simply due to the higher metabolic rate (Chiron et al., 1997) and earlier maturation of the right hemisphere observed during infancy (Thatcher, Walker, & Guidice, 1987); and/or (3) eye gaze perception triggers emotional processes in the infant, which have been shown to result in a lateralization of the gamma band effects to the right in adults (Müller et al., 1999).

The finding that inverted faces did not elicit gamma band responses that differed between direct and averted gaze is in line with, and adds further developmental evidence to, the notion that face inversion disrupts face processing (Rodriguez et al., 1999; Turati et al., 2004). This indicates that relatively early in development, cortical structures involved in face processing are already somewhat specialized to extract information about gaze direction from upright faces. It further shows that the gamma band effects observed in response to direct and averted gaze are not simply driven by "lower-level" perceptual parameters (e.g., symmetry [direct gaze] and asymmetry [averted gaze]), because then they should have occurred in the inverted condition as well. It is also important to note that these gamma band findings in infants show a high degree of correspondence in terms of timing and frequency content with previous findings in adults (Tallon-Baudry & Bertrand, 1999). This suggests continuity throughout development, and further emphasizes the functional importance of gamma band oscillations for social perception.

To summarize, the ERP, source localization, and gamma band findings reviewed here provide important insights into the neurodevelopmental origins of eye gaze direction perception in infancy (see Table 5.1). The systematic investigation of neural responses to direct and averted gaze in the context of upright versus inverted and frontal versus oriented faces reveal that (1) neural mechanisms are tuned to extract gaze information from upright faces; (2) with development young infants' gaze perception abilities become more flexible, in the sense that they learn to extract information about mutual gaze independent of the head angle; but (3) only frontal faces with direct gaze recruit brain processes, which may reflect the detection of eye contact as a communicative signal.

Neural Basis of Referential Gaze Perception and Joint Attention

As mentioned in previous sections of this chapter, one important communicative function of eye gaze is to direct attention to certain locations, events, and objects. Understanding the relations between eye gaze and target objects is particularly important for such aspects of development as word learning. Comprehending that another's gaze direction refers to a specific object allows a child to associate the object with a name or emotional expression (Baldwin & Moses, 1996). Adults' gaze has been found to facilitate object processing at a neural level in infants as young as 4 months (Reid, Striano, Kaufman, & Johnson, 2004). In this ERP study, objects that were previously cued by eye gaze elicited a diminished positive slow wave observed between 700 and 1000 msec over right frontotemporal channels. A diminished positive slow wave is thought to indicate deeper memory encoding (Nelson & Collins, 1991). This suggests that eye gaze as a social cue facilitates brain processes involved in memory encoding that may assist infants' learning.

In another ERP study, 9-month-old infants and adults watched a face whose gaze shifted either toward (object-congruent) or away from (object-incongruent) the location of a previously presented object (Senju, Johnson, & Csibra, 2006). This paradigm was based on that used in an earlier fMRI study (Pelphrey, Singerman, Allison, & McCarthy, 2003) and was designed to reveal the neural basis of "referential" gaze perception. When the ERPs elicited by object-incongruent gaze shifts were compared to the object-congruent gaze shifts, an enhanced negativity at about 300 msec over occipitotemporal electrodes was observed in both infants and adults. This suggests that infants encode referential information of gaze via neural mechanisms similar to those used by adults. However, only infants showed a fronto-central negative component that was larger in amplitude for object-congruent gaze shifts. It is thus possible that in the less specialized infant brain, the referential information of gaze is encoded in broader cortical circuits than in the more specialized adult brain. We return to this interesting finding on referential gaze in the following paragraphs, in which the neural basis of a very closely related social-cognitive phenomenon called "joint attention" or "shared attention" is discussed.

One of the major developmental changes in infants' engagement with others is the transition from participating in *dyadic* (face-to-face) interactions to developing a capacity to engage in *triadic* (infant–other—object) joint attention exchanges. Besides attending to an external object or event him- or herself, the ability to attend jointly with another person requires the infant to monitor (1) the other person's attention in relation to the self and (2) the other person's attention toward the same object or event. The establishment of joint attention and the monitoring of joint attention interchanges in preverbal infants rely heavily on eye gaze cues. Triadic relations between two minds and an object are thought to be uniquely human representations (Baron-Cohen, 1995; Tomasello, Carpenter, Call, Behne, & Moll, 2005), supporting shared attention and collaborative goals. The dorsal part of the MPFC has been identified as the neural substrate supporting these kinds of representations in the adult human brain (Frith & Frith, 2006; Saxe, 2006). It has been shown that already by 3 months of age, infants are able to behaviorally discriminate between dyadic and triadic joint attention interactions (Striano & Stahl, 2005). In this study, infants gazed and smiled more in the joint attention condition, in which an experimenter alternated visual attention between the infant and the object, than when the experimenter simply looked at the object without engaging or addressing the infant. Despite this early sensitivity to triadic interactions, a more robust understanding of joint attention is not in place until 9 months of age (Tomasello et al., 2005).

Striano, Reid, and Hoehl (2006b) used a novel interactive paradigm to examine the ERP correlates of joint attention in 9-month-old infants, in which an adult interacted live with each infant in two contexts. In the joint attention context, the adult looked at the infant and then at the computer screen displaying a novel object; in the nonjoint attention context, the adult only looked at the novel object presented on the screen. Compared to objects presented in the nonjoint attention context, objects presented in the joint attention context were found to elicit a greater negative component (Nc) peaking at about 500 msec, with a maximum over frontal and central channels. The Nc is generated in the prefrontal cortex and indicates the allocation of attention to a visual stimulus, with a greater amplitude indexing more allocation of attentional resources (Reynolds & Richards, 2005). Therefore, Striano et al. (2006b) suggested that infants benefit from the joint attention interaction and thus devote more attentional resources to those objects presented in this context. Another interesting finding from this study is that the amplitude of the Nc in both contexts appeared to be substantially larger than in previous ERP studies (de Haan & Nelson, 1997, 1999). It is possible that the change from viewing the stimuli passively on the screen (as in the previous work) to the application of a novel interactive paradigm increased the social significance for the infant, as reflected in a general increase in amplitude of the Nc.

It is worth noting that in Senju et al.'s (2006) study on referential gaze perception, discussed earlier, a frontocentral negativity very similar to that observed in Striano et al.'s (2006b) live joint attention study was found. This is not surprising, given the striking similarity between the two studies and the joint attentional nature of the situations presented to the infant in the two experiments. In Senju et al.'s (2006) study, the frontocentral negativity was obtained in 9-month-olds who watched a person on the screen make eye contact and then look at an object, and in Striano et al.'s (2006b) study, this brain response was obtained when a live experimenter established eye contact with the infant and then looked at an object on the screen. We therefore suggest that, as previously argued by Striano and colleagues, the joint attentional nature triggered mainly by eye gaze cues used in Senju et al.'s study can account for their finding of a frontocentral negativity to congruent gaze shifts.

In summary, there is evidence that joint triadic interaction with the infant indeed has an effect on brain structures, probably localized in the prefrontal cortex, that are associated with the allocation of attentional processing resources (Striano et al., 2006b). This provides support for the view that social information has an impact on and interacts with attentional processes (Grossmann, Striano, & Friederici, 2005, 2007b), which may point to an important mechanism of how social interaction influences information processing early in development. On the basis of behavioral findings, we have stated that one of the major developmental changes in infants' engagement with others is the transition from participating in *dyadic* (face-to-face) to *triadic* (infant–other–object) joint attention exchanges. This behaviorally evident transition has not been examined in the studies summarized here, since infants were only tested at ages at which they already engage successfully in triadic interactions. Thus, in order to elucidate the brain mechanisms underlying behavioral change, it seems worthwhile to test younger infants.

Conclusions

Various behavioral and neuroimaging methods (ERP, source localization, gamma band analysis) methods have been used successfully to study different aspects of eye gaze processing in infants. The available data on the behavioral and neural processes related to eye gaze direction processing, gaze following/referential gaze, and joint attention discussed in this chapter reveal insights into how the infant brain processes information about social gaze. Probably the most telling finding from all these studies is that the eyes provide a number of relevant social cues to which humans are sensitive from very early in development. Moreover, the infant's ability to decipher and interpret these various eye gaze cues in the face is associated with neural activity in a network of brain areas and cannot be assigned to a single module or mechanism. In summary, infants show specific modulations of cortical responses associated with eye gaze perception, which differ partly from those seen in adults. Nevertheless, the findings reviewed also suggest that infants successfully discriminate between direct and averted gaze, and can make use of gaze cues when encoding objects in memory, following the gaze of another person, or jointly attending with another person to an object. This suggests relatively early functioning of the brain structures involved in perceiving and acting on the eyes. However, this functioning may be more broadly tuned, and less specialized, than that seen in adults.

The findings on eye gaze perception reviewed here can be seen as one aspect of social brain development and should therefore be integrated into a broader theory of the emergence of the social brain network during infancy (see Grossmann & Johnson, 2007). Converging evidence indicates that even newborn humans are preferentially attentive to conspecifics, possibly due to subcortical routes for directing attention to social stimuli (Johnson, 2005). This preferential attention to conspecifics and their interactions will provide the appropriate input for developing cortical circuitry over the first few months. This input may ensure that a typical pattern of specialization into a cortical social brain network emerges (Johnson, 2005). According to this view, most parts of the social brain network (including the areas involved in gaze processing) can be activated in infants, though activation may also extend to other regions not activated under these circumstances in adults (Grossmann & Johnson, 2007). Furthermore, the social brain regions activated may have broader functions (i.e., may be less finely tuned) than in adults. Some of the evidence consistent with this view is that (1) compared to adults, infants activate additional cortical processes when processing eye gaze (Senju et al., 2006); and (2) the cortical mechanisms in infants for detecting eye contact as a communicative signal are less finely tuned than those in adults (Grossmann et al., 2007a, 2008). Furthermore, face perception and eye gaze perception have been shown to share common patterns of cortical activation early in ontogeny, which later partially dissociate and become more specialized (Farroni et al., 2002, 2004a; Grice et al., 2005; Johnson et al., 2005; Taylor et al., 2001). These findings support the view that structures in the social brain network initially have more homogeneous response properties, with common processing of many aspects of faces, eyes, bodies, and actions (Grossmann & Johnson, 2007). With experience, these structures may become more differentiated and specialized in their response properties, finally resulting in the specialized patterns of activation typically observed in adults. This view has implications for atypical development: Some developmental disorders that involve disruption to the social brain network, such as autism, may be characterized in terms of failures or delays of the specialization of structures in the cortical social brain network (see Johnson et al., 2005, for further discussion).

As part of integrating these findings into a broader theoretical framework of how the social brain develops, it also seems important to understand how eye gaze processing interacts with other domains attributed to the social brain. For example, very little is currently known about how eye contact influences the processing of emotional expressions in the infant brain. It has been previously argued that direct gaze is likely to be involved in the communication of increased emotional intensity, regardless of the particular emotion being expressed (Kimble & Olszewski, 1980; Kleinke, 1986). Despite the intuitive appeal of the long-standing notion that direct gaze may facilitate emotional expression processing, recent brain and behavioral research with adults has shown that the way in which gaze direction influences emotion perception depends on the emotion in question (Adams & Kleck, 2003, 2005; Adams, Gordon, Baird, Ambady, & Kleck, 2003). For example, averted gaze enhanced the perception of fear faces, whereas direct gaze enhanced the perception of angry faces (Adams & Kleck, 2005). The question of how eye gaze interacts with emotional expression has also been examined in young infants via ERP measures (Striano, Kopp, Grossmann, & Reid, 2006a). In this study, 4-month-old infants were presented with neutral, happy, and angry facial expressions when accompanied with direct and averted gaze. The results showed that neural processing of angry faces was enhanced in the context of direct gaze, whereas there was no effect of gaze on the processing of the other expressions. Although this neural sensitivity to the direction of angry gaze can be explained in different ways (for a discussion, see Grossmann & Johnson, 2007), it is nevertheless important to note that the neural processing of facial expression is influenced by the direction of eye gaze.

Whether or not this eye gaze interaction with emotional expression depends on experience has recently been investigated in newborns (Farroni, Menon, Rigato, & Johnson, 2007b). When a fearful expression was compared to a happy one, newborns looked significantly longer at the happy face. This preference for the happy face was not obtained when gaze was averted (Menon, Farroni, Rigato, & Johnson, 2008). One idea is that perception of facial expressions is acquired through experience, and that the perceptual dimensions relevant for different expressions are gradually discovered and used to differentiate perceptual inputs and associate them with different responses and consequences (Quinn & Johnson, 1997). It is possible that the reason why a happy face with direct gaze is preferred during the first few days of postnatal life is that this facial expression is likely to be the most common face stimulus in the newborns' experience with the visual world.

On a more general note, further progress in infant social-cognitive neuroscience depends crucially upon advances on three levels: (1) the development of experimental paradigms that better approximate real life, (2) improvement in neuroimaging methods that suit infants, and (3) integration of theories of functional brain development with theories of social-cognitive development.

It is experimentally challenging but worthwhile to continue to improve paradigms by modifying the tasks so that they are in accordance with an infant's world. For example, presenting multimodal stimuli allowed the use of an unusually high number of trials for analysis, suggesting an advantage of multimodal over unimodal stimuli in capturing infants' attention (Grossmann, Striano, & Friederici, 2006). This is of special interest when the generally low trial number of infant studies is considered. Moreover, in the recent study that employed a novel interactive paradigm to assess the neural correlates of joint attention in 9-month-old infants (Striano et al., 2006b), the amplitude of the elicited ERP component was substantially larger than that seen in previous literature, which have been due to the use of a live interaction. The use of multimodal stimuli and interactive paradigms has the advantages of greater social significance for the infants and higher ecological validity than the paradigms utilized in the past. Despite the advantages of using live interaction paradigms, we have to bear in mind that such paradigms also pose certain technical difficulties (e.g., complicated time control, increased motion artifacts).

Infant brain function has been predominantly investigated in studies using EEG and ERPs. Although these methods have helped to understand the neural underpinnings of infant social cognition, they suffer from relatively poor spatial resolution. Researchers have also begun to use fMRI to localize activity in the infant brain evoked by speech stimuli (Dehaene-Lambertz, Dehaene, & Hertz-Pannier, 2002), but fMRI requires rigid stabilization of the head and exposes infants to a noisy acoustic environment. These concerns make it unlikely that fMRI will be routinely used in paradigms with infants. A technique called near-infrared spectroscopy (NIRS), which permits the measurement of cerebral hemoglobin oxygenation in response to brain activation, has been successfully used to study infant visual, language, and memory abilities (for a review, see Aslin & Mehler, 2005). This novel technique does not suffer from the same problems in use with infants as fMRI does, and therefore NIRS may be a promising tool to study the infant social brain. Improving the tools used to localize functional activations in the infant brain is critical for our understanding of how the neural structures implicated in the adult social brain develop in early ontogeny.

Even once improved methods show us more clearly how neural events covary with infants' processing of social stimuli, the interpretation of such data will remain a major theoretical challenge. First attempts have been made to test theories of functional brain development (Johnson, 2001) by examining which parts of what is thought to be the adult social brain network are already active in infants (Johnson et al., 2005). The results of these studies provide preliminary support for the "interactive-specialization view" of human functional brain development, which assumes that developmental changes in the response properties of cortical regions occur as they interact and compete with each other to acquire their functional role (Grossmann & Johnson, 2007). This is in contrast to the "maturational view," in which the maturation of specific regions of the cortex is related to newly emerging cognitive and social functions, and also in contrast to the "skill-learning view," which purports that structures in the social brain are engaged by social stimuli due to their role in perceptual expertise. Future work with the imaging methods described above will be necessary to determine which of these viewpoints most accurately reflects the emerging social brain network during the first years.

Another important issue is that although dialogue between developmental psychologists and social-cognitive neuroscientists has begun on a theoretical level (Decety & Sommerville, 2004; Meltzoff & Decety, 2003), very little infant brain research is more directly informed and motivated by already existing theories of infant social-cognitive development (Csibra & Gergely, 2006; Meltzoff, 2002, 2005; Tomasello et al., 2005). The available theoretical frameworks are rich sources of hypotheses that are testable with neuroimaging tools. In this context, it will be of particular importance to identify the neural processes that underlie known social-behavioral and social-cognitive transitions. With respect to the topic of gaze perception, the proposed behavioral transition from simple gaze following to more sophisticated joint attention (Tomasello & Carpenter, 2007) is one area that, in our view, deserves closer examination on a neural level. Together, all the points listed here emphasize the truly interdisciplinary nature of the emerging field of social neuroscience of infancy, and present the field with the challenge of training scientists who are able to combine knowledge from cognitive neuroscience and developmental psychology in order to shed light on the developing social brain. It is our hope that this chapter will help to inform and guide future research.

Acknowledgments

Tobias Grossmann was partly supported by a Pathfinder Grant (CALACEI) from the European Commission, and by a Sir Henry Wellcome Postdoctoral Fellowship awarded by the Wellcome Trust (Grant No. 082659/Z/07/Z). Teresa Farroni was supported by the Wellcome Trust (Grant No. 073985/Z/03/Z). We would also like to thank Mark H. Johnson and Gergely Csibra for valuable contributions and helpful discussions.

References

Adams, R. B., Gordon, H. L., Baird, A. A., Ambady, N., & Kleck, R. E. (2003). Effects of gaze on amygdala sensitivity to anger and fear faces. *Science, 300,* 1536.

Adams, R. B., & Kleck, R. E. (2003). Perceived gaze direction and the processing of facial displays of emotion. *Psychological Science, 14,* 644–647.

Adams, R. B., & Kleck, R. E. (2005). The effects of direct and averted gaze on the perception of facially communicated emotion. *Emotion, 5,* 3–11.

Allison, T., Puce, A., & McCarthy, G. (2000). Social perception: Role of the STS region. *Trends in Cognitive Sciences, 4,* 267–278.

Aslin, R. N., & Mehler, J. (2005). Near-infrared spectroscopy for functional studies of brain activity in human infants: Promise, prospects, and challenges. *Journal of Biomedical Optics, 10,* 1–3.

Baldwin, D. A., & Moses, L. J. (1996). The ontogeny of social information gathering. *Child Development, 67,* 1915–1939.

Baron-Cohen, S. (1991). Precursors to a theory of mind: Understanding attention in others. In A. Whiten (Ed.), *Natural theories of mind: Evolution, development, and simulation of everyday mindreading* (pp. 233–251). Oxford: Blackwell.

Baron-Cohen, S. (1994). How to build a baby that can read minds: Cognitive mechanisms in mind-reading. *Cahiers de Psychologie Cognitive/Current Psychology of Cognition, 13,* 513–552.

Baron-Cohen, S. (1995). *Mindblindness: An essay on autism and theory of mind.* Cambridge, MA: MIT Press.

Batki, A., Baron-Cohen, S., Wheelwright, S., Connellan, J., & Ahluwalia, J. (2000). Is there an innate gaze module?: Evidence from human neonates. *Infant Behavior and Development, 23,* 223–229.

Bentin, S., Allison, T., Puce, A., Perez, A., & McCarthy, A. (1996). Electrophysiological studies of face perception in humans. *Journal of Cognitive Neuroscience, 8,* 551–565.

Bloom, P. (2000). *How children learn the meanings of words.* Cambridge, MA: MIT Press.

Bristow, D., Rees, G., & Frith, C. D. (2007). Social interaction modifies neural response to gaze shifts. *Social Cognitive and Affective Neuroscience, 2*(1), 52–61.

Brooks, R., & Meltzoff, A. N. (2002). The importance of eyes: How infants interpret adult looking behavior. *Developmental Science, 38,* 958–966.

Brooks, R., & Meltzoff, A. N. (2005). The development of gaze following and its relation to language. *Developmental Science, 8,* 535–543.

Butterworth, G. (1991). The ontogeny and phylogeny of joint visual joint attention. In A. Whiten (Ed.), *Natural theories of mind: Evolution, development, and simulation of everyday mindreading* (pp. 223–232). Oxford: Blackwell.

Butterworth, G., & Itakura, S. (2000). How the eyes, head and hand serve definite reference. *British Journal of Developmental Psychology, 18,* 25–50.

Butterworth, G., & Jarrett, N. (1991). What minds have in common in space: Spatial mechanisms serving joint visual attention in infancy. *British Journal of Developmental Psychology, 9,* 55–72.

Calder, A. J., Beaver, J. D., Winston, J. S., Dolan, R. J., Jenkins, R., Eger, E., et al. (2007). Separate coding of different gaze directions in the superior temporal sulcus and inferior parietal lobule. *Current Biology, 17,* 20–25.

Calder, A. J., Lawrence, A. D., Kaene, J., Scott, S. K., Owen, A. M., Christoffels, I., et al. (2002). Reading the mind from eye gaze. *Neuropsychologia, 40,* 1129–1138.

Caron, A. J., Butler, S. C., & Brooks, R. (2002). Gaze following at 12 and 14 months: Do eyes matter. *British Journal of Developmental Psychology, 20,* 225–239.

Caron, A. J., Keil, A. J., Dayton, M., & Butler, S. C. (2002). Comprehension of the referential intent of looking and pointing between 12 and 15 months. *Journal of Cognition and Development, 3,* 445–464.

Chiron, C., Jambaque, I., Nabbout, R., Lounes, R., Syrota, A., & Dulac, O. (1997). The right brain hemisphere is dominant in human infants. *Brain, 120,* 1057–1065.

Conty, L., N'Diaye, K., Tijus, C., & George, N. (2007). When eye creates the contact: ERP evidence for early dissociation between direct and averted gaze motion. *Neuropsychologia, 45,* 3024–3037.

Csibra, G. (2008). *Why human infants follow gaze: A communicative-referential account.* Manuscript submitted for publication.

Csibra, G., Davis, G., Spratling, M. W., & Johnson, M. H. (2000). Gamma oscillations and object processing in the infant brain. *Science, 290,* 1582–1585.

Csibra, G., & Gergely, G. (2006). Social learning and social cognition: The case for pedagogy. In Y. Munakata & M. H. Johnson (Eds.), *Attention and performance XXI: Processes of change in brain and cognitive development* (pp. 249–274). Oxford: Oxford University Press.

Decety, J., & Sommerville, J. A. (2003). Shared representations between self and others: A social cognitive neuroscience view. *Trends in Cognitive Sciences, 7,* 527–533.

de Haan, M., & Nelson, C. A. (1997). Recognition of the mother's face by six-month old infants: A neurobehavioral study. *Child Development, 68,* 187–210.

de Haan, M., & Nelson, C. A. (1999). Brain activity differentiates face and object processing by 6-month-old infants. *Developmental Psychology, 35,* 1114–1121.

de Haan, M., Pascalis, O., & Johnson, M. H. (2002). Specialization of neural mechanisms underlying face recognition in human infants. *Journal of Cognitive Neuroscience, 14,* 199–209.

Dehaene-Lambertz, G., Dehaene, S., & Hertz-Pannier, L. (2002). Functional neuroimaging of speech perception in infants. *Science, 298,* 2013–2015.

D'Entremont, B., Hains, S. M. J., & Muir, D. W. (1997). A demonstration of gaze following in 3- to 6-month-olds. *Infant Behavior and Development, 20,* 569–572.

Driver, J., Davis, G., Ricciardelli, P., Kidd, P., Maxwell, E., & Baron-Cohen, S. (1999). Gaze perception triggers reflexive visuospatial orienting. *Visual Cognition, 6,* 509–540.

Dunphy-Lelii, S., & Wellman, H. M. (2004). Infants' understanding of occlusion of others' line of sight: Implications for an emerging theory of mind. *European Journal of Developmental Psychology, 1,* 49–66.

Farroni, T., Csibra, G., Simion, F., & Johnson, M. H. (2002). Eye contact detection in humans from birth. *Proceedings of the National Academy of Sciences USA, 99,* 9602–9605.

Farroni, T., Johnson, M. H., & Csibra, G. (2004a). Mechanisms of eye gaze perception during infancy. *Journal of Cognitive Neuroscience, 16,* 1320–1326.

Farroni, T., Johnson, M. H., Menon, E., Zulian, L., Faraguna, D., & Csibra, G. (2005). Newborn's preference for face-relevant stimuli: Effects of contrast polarity. *Proceedings of the National Academy of Sciences USA, 102,* 17245–17250.

Farroni, T., Mansfield, E. M., Lai, C., & Johnson, M.H. (2003). Infants perceiving and acting on the eyes: Tests of an evolutionary hypothesis. *Journal of Experimental Child Psychology, 85,* 199–212.

Farroni, T., Massaccesi, S., Menon, E., & Johnson, M. H. (2007a). Direct gaze modulates face recognition in young infants. *Cognition, 102,* 396–404.

Farroni, T., Menon, E., & Johnson, M. H. (2006). Factors influencing newborns' preference for faces with eye contact. *Journal of Experimental Child Psychology, 95,* 298–308.

Farroni, T., Menon, E., Rigato, S., & Johnson, M. H. (2007b). The perception of facial expressions in newborns. *European Journal of Developmental Psychology, 4,* 2–13.

Farroni, T., Pividori, D., Simion, F., Massaccesi, S., & Johnson M. H. (2004b). Eye gaze cueing of attention in newborns. *Infancy, 5,* 39–60.

Fiebach, C. J., Gruber, T., & Supp, G. (2005). Neuronal mechanisms of repetition priming in occipitotemporal cortex: Spatiotemporal evidence from functional magnetic imaging and electroencephalography. *Journal of Neuroscience, 25,* 3414–3422.

Flom, R., Deak, G. O., Phill, C. G., & Pick, A. D. (2004). Nine-month-olds' shared visual attention as a function of gesture and object location. *Infant Behavior and Development, 27,* 181–194.

Flom, R., & Pick, A. D. (2005). Experimenter affective expression and gaze following in 7-month-olds. *Infancy, 7,* 207–218.

Foucher, J. R., Otzenberger, H., & Gounot, D. (2003). The BOLD response and the gamma oscillations respond differently than evoked potentials: An interleaved EEG–fMRI-study. *BMC Neuroscience, 4,* 22.

Friesen, C. K., & Kingstone, A. (1998). The eyes have it!: Reflexive orienting is triggered by nonpredictive gaze. *Psychonomic Bulletin and Review, 5,* 490–495.

Frith, C. D., & Frith, U. (2006). The neural basis of mentalizing. *Neuron, 50,* 531–534.

George, N., Driver, J., & Dolan, R. J. (2001). Seen gaze-direction modulates fusiform activity and its coupling with other brain areas during face processing. *NeuroImage, 13,* 1102–1112.

Gliga, T., & Csibra, G. (2007). Seeing the face through the eyes: A developmental perspective on face expertise. *Progress in Brain Research, 164,* 323–339.

Grice, S. J., Halit, H., Farroni, T., Baron-Cohen, S., Bolton, P., & Johnson, M. H. (2005). Neural correlates of eye-gaze detection in young children with autism. *Cortex, 41,* 342–353.

Grossmann, T., & Johnson, M. H. (2007). The development of the social brain in infancy. *European Journal of Neuroscience, 25,* 909–919.

Grossmann, T., Johnson, M. H., Farroni, T., & Csibra, G. (2007a). Social perception in the infant brain: Gamma oscillatory activity in response to eye gaze cues. *Social Cognitive and Affective Neuroscience, 2*(4), 284–291.

Grossmann, T., Johnson, M. H., Farroni, T., & Csibra, G. (2008). *The effect of head orientation on the neural correlates of eye gaze perception in 4-month-old infants.* Unpublished manuscript.

Grossmann, T., Striano, T., & Friederici, A. D. (2005). Infants' electric brain responses to emotional prosody. *NeuroReport, 16,* 1825–1828.

Grossmann, T., Striano, T., & Friederici, A. D. (2006). Crossmodal integration of emotional information from face and voice in the infant brain. *Developmental Science, 9,* 309–315.

Grossmann, T., Striano, T., & Friederici, A. D. (2007b). Developmental changes in infants' processing of happy and angry facial expressions: A neurobehavioral study. *Brain and Cognition, 64,* 30–41.

Gruber, T., Trujillo-Barreto, N. J., Giabbiconi, C. M., Valdés-Sosa, P. A., & Müller, M. (2006). Brain electrical tomography (BET) analysis of induced gamma band responses during a simple object recognition task. *NeuroImage, 29,* 888–900.

Halit, H., Csibra, G., Volein, A., & Johnson, M. H. (2004). Face-sensitive cortical processing in early infancy. *Journal of Child Psychology and Psychiatry, 45,* 1228–1234.

Halit, H., de Haan, M., & Johnson, M. H. (2003). Cortical specialization for face processing: Face-sensitive event-related potential components in 3- and 12-month-old infants. *NeuroImage, 19,* 1180–1193.

Haxby, J. V., Hoffman, E., & Gobbini, M. I. (2000). The distributed human neural system for face perception. *Trends in Cognitive Sciences, 4,* 223–233.

Herrmann, C. S., Lenz, D., Junge, S., Busch, N. A., & Maess, B. (2003). Memory-matches evoke human gamma-responses. *BMC Neuroscience, 5,* 13.

Herrmann, C. S., Munk, M. H., & Engel, A. K. (2004). Cognitive functions of gamma-band activity: Memory match and utilization. *Trends in Cognitive Sciences, 8,* 347–355.

Hoffman, E. A., & Haxby, J. V. (2000). Distinct representations of eye gaze and identity in the distributed human neural system for face perception. *Nature Neuroscience, 3,* 80–84.

Hood, B. M., Willen, J. D., & Driver, J. (1998). Adults' eyes trigger shifts of visual attention in human infants. *Psychological Science, 9,* 131–134.

Johnson, M. H. (2001). Functional brain development in humans. *Nature Reviews Neuroscience, 2,* 475–483.

Johnson, M. H. (2005). Subcortical face processing. *Nature Reviews Neuroscience, 6,* 766–774.

Johnson, M. H., Griffin, R., Csibra, G., Halit, H., Farroni, T., de Haan, M., et al. (2005). The emergence of the social brain network: Evidence from typical and atypical development. *Development and Psychopathology, 17,* 599–619.

Johnson, M. H., & Morton, J. (1991). *Biology and cognitive development: The case for face recognition.* Oxford: Blackwell.

Kampe, K., Frith, C. D., & Frith, U. (2003). "Hey John": Signals conveying communicative intention toward self activate brain regions associated with "mentalizing," regardless of modality. *Journal of Neuroscience, 12*, 5258–5263.

Kaufman, J., Csibra, G., & Johnson, M. H. (2003). Representing occluded objects in the human infant brain. *Proceedings of the Royal Society, Series B, Biological Sciences, 270*(Suppl. 2), 140–143.

Kaufman, J., Csibra, G., & Johnson, M. H. (2005). Oscillatory activity in the infant brain reflects object maintenance. *Proceedings of the National Academy of Sciences USA, 102*, 15271–15274.

Kawashima, R., Sugiura, M., Kato, T., Nakamura, A., Hatano, K., Ito, K., et al. (1999). The human amygdala plays an important role in gaze monitoring: A PET study. *Brain, 122*, 779–783.

Kimble, C. E., & Olszewski, D. A. (1980). Gaze and emotional expression: The effects of message positivity–negativity and emotional intensity. *Journal of Research in Personality, 14*, 60–69.

Kleinke, C. L. (1986). Gaze and eye contact: A research review. *Psychological Bulletin, 100*, 68–72.

Langton, S., Watt, R., & Bruce, V. (2000). Do the eyes really have it?: Cues to the direction of social attention. *Trends in Cognitive Sciences, 4*, 50–59.

Makeig, S., Debener, S., Onton, J., & Delrome, A. (2004). Mining event-related brain dynamics. *Trends in Cognitive Sciences, 18*, 204–210.

Meltzoff, A. N. (2002). Imitation as a mechanism of social cognition: Origins of empathy, theory of mind, and the representation of action. In U. Goswami (Ed.), *Handbook of childhood cognitive development* (pp. 6–25). Oxford: Blackwell.

Meltzoff, A. N. (2005). Imitation and other minds: The "like me" hypothesis. In S. Hurley & N. Chater (Eds.), *Perspectives on imitation: From cognitive neuroscience to social science* (pp. 55–77). Cambridge, MA: MIT Press

Meltzoff, A. N., & Decety, J. (2003). What imitation tells us about social cognition: A rapprochement between developmental psychology and cognitive neuroscience. *Philosophical Transactions of the Royal Society of London, Series B, Biological Sciences, 358*, 491–500.

Menon, E., Farroni, T., Rigato, S., & Johnson, M. H. (2008). *Constraints in the perception of facial expressions at birth.* Manuscript in preparation.

Moore, C., & Corkum, V. (1994). Social understanding at the end of the first year of life. *Developmental Review, 14*, 349–372.

Morales, M., Mundy, P., & Rojas, J. (1998). Following the direction of gaze and language development in 6-month-olds. *Infant Behavior and Development, 21*, 373–377.

Müller, M., Keil, A., Gruber, T., & Elbert, T. (1999). Processing of affective pictures modulates right-hemispheric gamma band activity. *Clinical Neurophysiology, 110*, 1913–1920.

Nelson, C. A., & Collins, P. F. (1991). Event-related potential and looking time analysis of infants' responses to familiar and novel events: Implications for visual recognition memory. *Developmental Psychology, 27*, 50–58.

Pelphrey, K. A., Singerman, J. D., Allison, T., & McCarthy, G. (2003). Brain activation evoked by perception of gaze shifts: The influence of context. *Neuropsychologia, 41*, 156–170.

Pelphrey, K. A., Viola, R. J., & McCarthy, G. (2004). When strangers pass: Processing of mutual and averted gaze in the superior temporal sulcus. *Psychological Science, 15*, 598–603.

Puce, A., Allison, T., Bentin, S., Gore, J. C., & McCarthy, G. (1998). Temporal cortex activation in humans viewing eye and mouth movements. *Journal of Neuroscience, 18*, 2188–2199.

Quinn, P. C., & Johnson, M. H. (1997). The emergence of category representations in infants: A connectionist analysis. *Journal of Experimental Child Psychology, 66*, 236–263.

Reid, V. M., Striano, T., Kaufman, J., & Johnson, M. (2004). Eye gaze cueing facilitates neural processing of objects in 4 month old infants. *NeuroReport, 15*, 2553–2556.

Reynolds, G. D., & Richards, J. E. (2005). Familiarization, attention, and recognition memory in infancy: An ERP and cortical source analysis study. *Developmental Psychology, 41*, 598–615.

Richards, J. E. (2004). Recovering dipole sources from scalp-recorded event-related-potentials using component analysis: Principal component analysis and independent component analysis. *International Journal of Psychophysiology, 54*, 201–220.

Richards, J. E. (2005). Localizing cortical sources of event-related potentials in infants' covert orienting. *Developmental Science, 8*, 255–278.

Rodriguez, E., George, N., Lachaux, J. P., Martinere, J., Renault, B., & Varela, F. (1999). Perception's shadow: Long-distance synchronization of human brain activity. *Nature, 397*, 430–433.

Saxe, R. (2006). Uniquely human social cognition. *Current Opinion in Neurobiology, 16*, 235–239.

Scaife, M., & Bruner, J. S. (1975). The capacity for joint visual attention in the infant. *Nature, 253*, 265–266.

Schilbach, L., Wohlschläger, A. M., Newen, A., Krämer, N., Shah, N. J., Fink, G. R., et al. (2006). Being with virtual others: Neural correlates of social interaction. *Neuropsychologia, 44*, 718–730.

Senju, A., Johnson, M. H., & Csibra, G. (2006). The development and neural basis of referential gaze perception. *Social Neuroscience, 1*, 220–234.

Striano, T., Kopp, F., Grossmann, T., & Reid, V. M. (2006a). Eye contact influences neural processing of emotional expressions in 4-month-old infants. *Social Cognitive and Affective Neuroscience, 1*, 87–95.

Striano, T., Reid, V. M., & Hoehl, S. (2006b). Neural mechanisms of joint attention in infancy. *European Journal of Neuroscience, 23*, 2819–2823.

Striano, T., & Stahl, D. (2005). Sensitivity to triadic attention in early infancy. *Developmental Science, 4*, 333–343.

Tallon-Baudry, C., & Bertrand, O. (1999). Oscillatory gamma activity in humans and its role in object representation. *Trends in Cognitive Sciences, 3*, 151–162.

Taylor, M. J., Itier, R. J., Allison, T., & Edmonds, G. E. (2001). Direction of gaze effects on early face processing: Eyes-only vs. full faces. *Cognitive Brain Research, 10*, 333–340.

Thatcher, R. W., Walker, R. A., & Giudice, S. (1987). Human cerebral hemispheres develop at different rates and ages. *Science, 236*, 1110–1113.

Tomasello, M. (1999). *The cultural origins of human cognition.* Cambridge, MA: Harvard University Press.

Tomasello, M., & Carpenter, M. (2007). Shared intentionality. *Developmental Science, 10*, 121–125.

Tomasello, M., Carpenter, M., Call, J., Behne, T., & Moll, H. (2005). Understanding and sharing intentions: The origins of cultural cognition. *Behavioral and Brain Sciences, 28*, 675–691.

Turati, C., Sangrigoli, S., Ruel, J., & de Schonen, S. (2004). Evidence of the face inversion effect in 4-month-old infants. *Infancy, 6*, 275–297.

Vecera, S. P., & Johnson, M. H. (1995). Eye gaze detection and cortical processing of face: Evidence from infants and adults. *Visual Cognition, 2*, 59–87.

Wicker, B., Michel, F., Henaff, A. M., & Decety, J. (1998). Brain regions involved in the perception of gaze: A PET study. *NeuroImage, 8*, 221–227.

Woodward, A. (2003). Infants' developing understanding of the link between looker and object. *Developmental Science, 6*, 297–311.

CHAPTER 6

The Development and Neural Bases of Processing Emotion in Faces and Voices

Michelle de Haan
Anna Matheson

Faces and voices reveal emotions, and thus are key components of everyday social interactions. The ability to correctly perceive and interpret these emotional cues is important across the lifespan—from infancy, when caregivers' emotional displays can be used to regulate behavior in novel situations (see Carver & Cornew, Chapter 7, this volume), to adolescence and adulthood, when emotional displays are used in more complex ways (see Choudhury, Charman, & Blakemore, Chapter 9, this volume). One issue that has interested researchers is understanding the extent to which the development of emotion perception is influenced by the maturation of neural circuits predisposed to mediate this ability, as well as the extent to which it is influenced by the social-emotional environments infants and children experience. In other words, to what extent do infants arrive in the world prepared to perceive emotions, and to what extent do they need experience and instruction in order to do so? This type of debate has implications for understanding not only how emotion processing typically develops, but how development may go awry in atypical circumstances. In this chapter, we describe the development and neural bases of the ability to perceive emotional information conveyed by faces and voices. We first review what is known about the neural bases of facial and vocal emotion perception in adults, and then focus on the development and neural bases of these abilities in infancy and childhood. We summarize and conclude by highlighting areas in need of further research.

Neural Bases of Facial and Vocal Emotion Perception in Adults

In order for a person to understand the meaning of facial or vocal displays of emotion, the relevant information in the sensory signal must be accurately perceived and appropriately discriminated. In everyday situations, these signals tend to be given by moving, vocalizing faces. In spite of this, most research on the neural bases of emotion perception and its development has involved static, silent faces. More recently, investigators have incorporated moving and multimodal stimuli into their experimental designs in order to understand how these attributes are integrated in the emotion-processing network. Below, we describe what is known in adults about the brain systems that mediate the perception of facial emotion (including motion), the perception of vocal emotion, and the integration of emotional information from these two modalities.

Faces: The Basic Brain Network

Recognition of facial emotion depends on both "perceptual processing," which involves distinguishing among the different facial configurations that denote given expressions, and "conceptual processing," which involves understanding the emotional meaning of a facial expression. In adults, these processes are mediated by a distributed neural network involving subcortical and cortical areas. Visual information about faces is initially passed along two neural pathways: (1) a subcortical system, which is involved in detecting faces and directing visual attention to them; and (2) a core cortical system, which is involved in the detailed visual-perceptual analysis of faces. Both of these components interact with (3) an extended cortical–subcortical system involved in further processing of faces, including the recognition of the emotional content of facial expressions (Gobbini & Haxby, 2007; Haxby, Hoffman, & Gobbini, 2000; Johnson, 2005).

In the subcortical pathway for face processing, information travels from the retina directly to the superior colliculus, then to the pulvinar, and on to the amygdala (de Gelder, Frissen, Barton, & Hadjikhani, 2003; Johnson, 2005). This route processes facial information quickly. For example, responses to faces in the amygdala are first observed at approximately 120 msec (Halgren et al., 1994), with a distinction seen between different emotional expressions at approximately 150 msec (Liu, Ioannides, & Streit, 1999). The subcortical pathway is believed to operate automatically, and to rely primarily on low-spatial-frequency information (reviewed in Johnson, 2005). Existence of such a pathway in humans is supported by studies showing that the emotional valence of facial expressions can be reliably discriminated even by patients with lesions to the primary visual cortex and without conscious visual experience (Morris, de Gelder, Weiskrantz, & Dolan, 2001; Morris, Öhman, & Dolan, 1999). This rapid pathway may allow some degree of face processing to be accomplished before slower, conscious cortical processing is completed, and thereby may mediate nonconscious responses to faces. This pathway may also modulate subsequent cortical processing of emotional signals (e.g., Tamietto & de Gelder, 2008). For example, projections from the amygdala to the occipital cortex may enhance visual processing of emotionally salient stimuli (Morris et al., 1998; Vuilleumier & Pourtois, 2007).

The core cortical system for visual analysis of faces receives input from the retina via the geniculostriate pathway, and includes the inferior occipital gyrus (encompassing the lateral occipital area, of which the "occipital face area" is a subregion), fusiform gyrus (including the "fusiform face area"), and posterior superior temporal sulcus (STS)/superior temporal gyrus. The inferior occipital gyrus mediates the early perception of faces and passes this information to two areas: (1) the fusiform gyrus and (2) the STS/superior temporal gyrus. There is evidence that the fusiform gyrus is primarily involved in the interpretation of the static components of facial expressions and identity (Kanwisher, McDermott, & Chun, 1997), whereas the superior temporal gyrus contributes to the recognition of the dynamic properties of facial expressions and eye gaze (Allison, Puce, & McCarthy, 2000; see below for further discussion).

The extended system receives input from, and in return communicates with, both the subcortical system and core cortical system. It encompasses a variety of regions involved in the further processing of these inputs to allow activities important in emotion processing, such as conscious emotional appraisal and interpretation of the intentions of others. The paralimbic and higher cortical areas (such as the medial prefrontal cortex, the somatosensory cortices, and the anterior cingulate) are involved in longer-latency processing of the conscious representations of emotional states, in controlling behavior in social situations, and in the planning of actions and goals.

Although this general network is activated by all emotional expressions, there is evidence that different emotions are associated with unique patterns of brain activation; this is particularly so for fear, anger, and disgust. For example, functional imaging studies have demonstrated that the amygdala is disproportionately activated by facial expressions of fear (reviewed in Vuilleumier & Pourtois, 2007). There is also evidence that the prefrontal cortex has a particular role in the recognition of anger, with activation in this region when anger is observed, compared to both happiness and sadness (Blair, Morris, Frith, Perrett, & Dolan, 1999; Harmer, Thilo, Rothwell, & Goodwin, 2001). The perception of facial expressions of disgust is regulated by both the insula and the basal ganglia. Patient studies have shown that damage to the left insula and basal ganglia prevents correct identification of the facial expression of disgust, and in fact causes an inability to experience the emotion itself (Adolphs, Tranel, & Damasio, 2003; Calder, Keanes, Manes, Antoun, & Young, 2000).

The Importance of Facial Motion

Humans are very sensitive to biological motion (Johansson, 1973). Head and facial motion provides information that is not present in the static image, with recognition accuracy for faces increasing with the addition of moving cues (Pike, Kemp, Towell, & Phillips, 1997). Humans may be especially sensitive to the temporal cues in facial expressions of emotion in the early stages of these expressions (Edwards, 1998).

The perception of biological motion in general, including implied motion, has been associated with a distributed network involving the posterior STS, amygdala, basal ganglia, intraparietal sulcus, and inferofrontal cortex (Grezes et al., 2001). Motion enhances activity in areas involved in emotion processing: Activation in regions of the right occipital and temporal cortices is enhanced by movement, with foci in the middle temporal gyrus and adjacent areas, including the STS and the amygdala (LaBar, Crupain, Voyvodic, & McCarthy, 2003; Sato, Kochiyama, Yoshikawa, Naito, & Matsumara, 2004). There is also a double dissociation between the brain networks involved in the processing of moving or changeable aspects of faces versus static, stable information. For example, single-unit recordings in macaques have shown that neurons within the inferior temporal gyri are involved in the perception of facial identity, whereas those within the STS are involved in the perception of facial movement and static images of changeable aspects of the face. In human studies, a double dissociation between the functional roles of the lateral fusiform gyrus and the STS has been observed: Selective attention to identity elicits a stronger response in the lateral fusiform gyrus than does selective attention to eye gaze, whereas the converse is true in the STS (Hoffman & Haxby, 2000; see also Steede, Tree, & Hole, 2007). The STS is in a prime position to integrate information about movement with information about object form, as it receives inputs from the anterior part of both the dorsal and ventral visual streams associated with processing of motion and form, respectively. Within the STS, the superior temporal polysensory area is particularly sensitive to biological motion (Oram & Perrett, 1994; Bruce, Desimone, & Gross, 1981), and is important for both emotion processing (Hasselmo, Rolls, & Baylis, 1989) and perception of the direction of eye gaze (Perrett et al., 1985).

Activation has also been recorded in the right ventral premotor cortex during the observation of both nonemotional mouth movements (Buccino et al., 2001) and dynamic facial expressions (Sato et al., 2004). This activity may reflect a "mirror system" where there is activity in the motor cortex in response to observing others' actions. Facial mimicry has been observed in both adults and infants who are viewing facial expressions (Dimberg, Thunberg, & Elmehed, 2000; Hess & Blairy, 2001).

Voices

The human voice conveys not only speech, but also paralinguistic information, such as emotional prosody. This type of information, together with nonspeech sounds (e.g., laughter), plays an important role in conveying emotional states.

Voices and nonspeech emotional sounds are both processed along the basic auditory pathway, wherein sound enters via the ear and passes first to the auditory midbrain, then to the auditory thalamus, and ultimately to the auditory cortex. As in the case of vision, a dual pathway for processing emotionally relevant auditory information has been proposed, with a rapid, subcortical thalamoamygdalar pathway and a slower, more detailed cortical pathway. The processing of emotional information by the rapid, subcortical auditory pathway has been studied most extensively in animals. For example, animals without an auditory cortex can learn an emotional association between a sound and receiving a shock, and will react in fear to the sound when it is presented alone (LeDoux, Sakaguchi, Iwata, & Reis, 1986). Studies have not examined in detail the dissociation of these pathways in humans with respect to processing of emotional information in the voice.

Neuroimaging studies do suggest that subcortical areas including the amygdala are important for processing nonspeech emotional sounds and for implicit processing of affective intonation (Wildgruber, Ackermann, Kreifelts, & Ethofer, 2006). By contrast, cortical areas including the right temporal lobe and bilateral orbital frontal cortex are activated when attention is given to emotional prosody (Wildgruber et al., 2006; reviewed in Belin, Fecteau, & Bedard, 2004).

Integration of Face and Voice Information

Although emotion recognition is typically better under bimodal conditions than under unimodal conditions (de Gelder & Vroomen, 2000; Kreifelts, Ethofer, Grodd, Erb, & Wildgruber, 2007), behavioral studies of adults' perception of facial and vocal emotion suggest that visual information is often more influential than auditory information (Hess, Kappas, &

Scherer, 1988). The role of auditory information increases in situations where the visual input is ambiguous or incongruent. Such studies with adults have led to a staged model of bimodal perception, wherein information from each modality is first evaluated separately, and then the data from both modalities are integrated (Massaro & Egan, 1996).

Neuroimaging studies suggest that the left middle temporal gyrus (Pourtois, de Gelder, Bol, & Crommelinck, 2005) and left posterior STS (Ethofer, Pourtois, & Wildgruber, 2006) and adjacent areas in the superior temporal gyrus (Kreifelts et al., 2007) are important for audiovisual integration of emotional information in the face and voice. In addition, there is evidence for early audiovisual integration within modality-specific cortices. There are two competing explanations for the latter finding—one that there is direct cross-talk between auditory and visual association cortices, and the other that there is feedback to these cortices from the STS. Lastly, studies have suggested interactions between lower- and higher-level processing, as unconscious processing of emotion in the face can affect conscious recognition of emotion in the voice (de Gelder, Pourtois, & Weiskrantz, 2002; Puce, Epling, Thompson, & Carrick, 2007).

Neural Bases of Adult Emotion Perception: Summary

Both facial and vocal information are thought to be processed by thalamoamygdalar and thalamocortical pathways, with the former providing both a mechanism for rapid response to emotional inputs and for modulating slower, more detailed cortical pathways. Though emotional faces and voices are not processed by identical networks, areas within the frontal and temporal lobes as well as the amygdala are important for processing unimodal information from emotion in both faces and voices, as well as for integrating information from the two modalities.

Development of Emotion Perception: Behavioral Studies

Processing of emotional expressions is one of the first elements of social skills to emerge. A clear understanding of the normative pattern of development is important, as it not only can provide insight into the functional development of the brain network described above, but may aid in early detection of such social-emotional disorders as autism.

Faces

Even though newborns' visual acuity at birth is between 10 and 30 times poorer than adults' (reviewed in Slater, 2001), they are able to process at least some of the internal features of the face. For example, within the first hours of life infants attend preferentially to face-like stimuli over other stimuli (Goren, Sarty, & Wu, 1975; Morton & Johnson, 1991), are able to discriminate among facial expressions (Field, Woodson, Greenberg, & Cohen, 1982), and are able to detect the direction of eye gaze in a face (Farroni, Csibra, Simion, & Johnson, 2002).

Results with older infants consistently show a good ability to discriminate happy facial expressions from most others: During the first few months of life, infants are able to discriminate happy from surprised, angry, and fearful expressions (Barrera & Maurer, 1981; Field et al., 1982; Kotsoni, de Haan, & Johnson, 2001; LaBarbera, Izard, Vietze, & Parisi, 1976; Nelson, Morse, & Leavitt, 1979; Nelson & Dolgin, 1985; Soken & Pick, 1999; Walker, 1982). One important exception is that infants do appear to have difficulty discriminating happy from sad expressions (Young-Browne, Rosenfield, & Horowitz, 1977; Oster & Ewy, 1980), even by 7 months of age (Soken & Pick, 1999; but see Caron, Caron, & MacLean, 1988; Walker, 1982). Interestingly, one study has demonstrated that infants as young as 3.5 months of age can discriminate between happy and sad expressions, but only when posed by their mothers and not when posed by a stranger (Kahana-Kalman & Walker-Andrews, 2001); these findings suggest that a familiar facial context may facilitate infants' abilities to tell expressions apart. Discrimination between other expressions has been less extensively studied.

Facial expression processing continues to improve into the preschool period, but fewer studies have examined the development beyond this time. This may be because school-age children often master the tasks typically used, such as labeling or matching facial emotions, and thus ceiling effects mask any further developmental change. In support of this idea, studies of older children and adolescents that have examined processing of more subtle aspects of facial emotion have found evidence of continued development (Durand, Gallay, Seigneuric, Robichon, & Baudouin, 2007; Thomas, DeBellis, Graham, & LaBar, 2007). There is evidence that emotion-processing skills continue to fluctuate even into adulthood, and that these skills may be related not only to age, but also to hormonal fluctuations such as those occurring during puberty (Wade, Lawrence, Mandy, & Skuse, 2006). In spite of the increasing interest in studying expression recognition in older children, results of existing studies are not entirely consistent on important questions, such as whether recognition of some expressions matures more quickly than that of others. One challenge for understanding the true developmental trajectory is that different methods are used at different ages. For example, there are conflicting findings in which infants have been reported to be better at recognizing expressions when they are displayed by familiar individuals (Kahana-Kalman & Walker-Andrews, 2001), but older children better when they are displayed by strangers (Herba et al., 2008). Whether this represents a true change in the way familiarity influences emotion processing or is an artifact of the different methods used to study the question at the different ages is unclear.

The research discussed so far has been based on use of static, silent facial images. There is much less information regarding how infants and children process more naturalistic, moving displays of facial emotion. This is not because infants cannot process motion; they are able to detect motion by 1 month of age, and can identify the direction of the motion over a wide range of velocities by 12 weeks (Wattam-Bell, 1996). There is also some evidence that infants are sensitive to motion information with emotional content (Soken & Pick, 1992), and that infants can both discriminate complex, subtle biological motion cues and detect invariants in such displays (Spencer, O'Brien, Johnston, & Hill, 2006). Infants are able to distinguish dynamic happy from angry and sad expressions (Walker, 1982), and to discriminate between other specific positive and negative dynamic emotional expressions (Soken & Pick, 1999). Motion may facilitate infants' processing of faces by providing specific movement patterns of the face that give multiple views of the face, which in turn aid perception of emotion (Nelson & Horowitz, 1983).

Available studies suggest that facial expression and facial motion are both salient dimensions for 2- to 3-month-olds, but do not clarify whether emotions are better perceived in static or in dynamic displays. This is because, while some studies have used dynamic stimuli to investigate infants' and children's emotion perception (e.g., Soken & Pick, 1999; Herba et al., 2008), they have not systematically compared infants' perception of static versus dynamic facial expressions. Since studies have shown that motion can benefit adults' processing of facial emotion, it is possible that prior studies with infants using static stimuli have underestimated their abilities, and that they would perform better with dynamic than with static stimuli in a direct comparison. On the other hand, studies suggesting that processing of visual form (ventral stream) develops earlier and more quickly than processing of visual motion (dorsal stream; e.g., Braddick, Atkinson, & Wattam-Bell, 2003; Mitchell & Neville, 2004) might suggest that processing of static faces would mature more quickly.

Voices

Few studies have examined the development of perception of vocal emotion. This is surprising, given the argument that infants show an auditory bias, in that they are more influenced by vocal than by visual cues to emotion (Caron et al., 1988; Vaish & Striano, 2004). One study showed that infants as young as 7 months of age differentiate angry, happy, and neutral prosody (Grossmann, Striano, & Friederici, 2005). A study with children ages 7–12 years suggests that the ability to recognize vocal expression of anger, sadness, and happiness is good by this age, as all children were over 80% accurate with the stimuli used (Shackman & Pollak, 2005).

Integration of Facial and Vocal Cues

There is some information regarding how the integration of facial and vocal displays of emotion develops. By 7 months of age, infants can correctly match vocal expressions of emotion to a dynamic facial display of emotion. Seven-month-olds look longer at the facial expression concordant with a vocal expression when tested with happy–angry (Soken & Pick, 1999; Walker, 1982), happy–neutral (Walker, 1982), angry–sad (Soken & Pick, 1999), interested–angry (Soken & Pick, 1999), and interested–happy (Soken & Pick, 1999) pairs, but they do not consistently do so for happy–sad pairs (see Soken & Pick, 1999, for a negative result, and Walker, 1982, for a positive result). Infants' matching appears to be based on information relevant to the emotion and not merely on temporal synchrony between the audio and visual portions of the stimuli, because these results hold even when the audio stimulus is played out of synchrony with the face (Soken & Pick, 1999; Walker, 1982).

Studies with adults have shown a greater reliance on visual than on auditory cues, and suggest a staged model in which information from each source is first evaluated separately prior to integration. By contrast, studies with infants suggest the opposite: As noted above, infants typically show an auditory bias, and are more influenced by vocal than by visual cues to emotion (Caron et al., 1988; Vaish & Striano, 2004). Young infants can use voice information to facilitate recognition of facial identity (Burnham, 1993), and they can sometimes selectively associate speech segments and faces on the basis of affective correspondence (Kaplan, Zarlengo-Strouse, Kirk, & Angel, 1997). By middle childhood, children show a bias for visual recognition of happy faces, but not angry or sad ones (Shackman & Pollak, 2005). Theories of multimodal perception suggest that, ontogenetically, individuals come to recognize the affective expressions of others through a perceptual differentiation process. In this view, recognition of affective expressions changes from a reliance on multimodally presented information to the recognition of vocal expressions and then of facial expressions alone (Walker-Andrews, 1997). Similarly, in the intersensory redundancy model of perception of multimodal stimuli, infants presented with a multimodal event first respond to the amodal information that is similar across modalities before they are able to focus on modality-specific features. For example, one study showed that in bimodal stimulation, discrimination of affect emerged by

4 months and remained stable across ages. However, in unimodal stimulation, detection of affect emerged gradually, with sensitivity to auditory stimulation emerging at 5 months and visual stimulation at 7 months (Flom & Bahrick, 2007).

Behavioral Studies of Emotion Perception Development: Summary

Infants are able to discriminate among static facial displays of emotion and to relate facial and vocal emotions within the first postnatal year of life. They are also able to discriminate among moving facial displays of emotion, but it is not clear whether motion facilitates infants' processing of emotion. Emotion-processing abilities continue to develop into childhood, but questions regarding differences in developmental profile for recognition of different emotions, the influence of person familiarity, and contributions of facial and vocal cues across age remain unclear, in part because of limited evidence and in part because of differences in methods used across available studies. Understanding the role of such factors as facial motion and audiovisual integration in the development of emotion recognition will be an important challenge for future studies.

Development of Emotion Perception: Neural Bases

Studying the neural bases of the development of emotion perception can provide insight into the mechanisms by which this ability develops, and can also provide clues to the possible neural bases of developmental disorders in which this skill is disrupted.

Faces

Studies with infants provide evidence that both the subcortical and core cortical pathways for face processing are operating to some extent within the first postnatal year of life. With respect to the subcortical pathway, there is evidence that newborns' preferential orienting to face-like patterns is mediated by a subcortical retinotectal pathway (Morton & Johnson, 1991). For example, newborns only show the preference under conditions to which the subcortical system is sensitive (when stimuli are moving and are in the peripheral visual field, but not when they are in the central visual field). Further experimental evidence in support of this view is that infants orient more toward face-like patterns than inverted face patterns in the temporal, but not the nasal, visual field (Simion, Valenza, Umiltà, & Dalla Barba, 1998). Since the retinotectal (subcortical) pathway is thought to have more input from the temporal hemifield and less input from the nasal hemifield than the geniculostriate (cortical) pathway has, this asymmetry in the preferential orienting to faces is consistent with subcortical, but not with cortical, involvement.

There is evidence that the amygdala also plays a role in processing facial expressions in infants and children. The eyeblink startle response is a reflex blink initiated involuntarily by sudden bursts of loud noise. In adults, these reflex blinks are augmented by viewing slides of unpleasant pictures and scenes, and they are inhibited by viewing slides of pleasant or arousing pictures and scenes (Lang, Bradley, & Cuthbert, 1990). From work in animals, it has been argued that fear potentiation of the startle response is mediated by the central nucleus of the amygdala, which in turn directly projects to the brainstem center, which mediates the startle and efferent blink reflex activity (Davis, 1989). One study applied this idea to 5-month-old infants (Balaban, 1995), showing them slides depicting happy, neutral, and angry expressions for 6 sec, followed 3 sec later by an acoustic startle probe. Consistent with the adult literature, infants' blinks were augmented when they viewed angry expressions and were reduced when they viewed happy expressions, relative to when they viewed neutral expressions. These results suggest that by 5 months of age, at least this aspect of the amygdala circuitry underlying the response to facial expressions is functional.

The amygdala may play an important role in the first years of life in establishing the functioning of the mature network. This idea is supported by studies showing that damage to the amygdala has a more pronounced effect on recognition of facial expression when it occurs early rather than later in life. For example, in one study of emotion recognition in patients who had undergone temporal lobectomy as treatment for intractable epilepsy, emotion recognition in patients with early right-sided mesial temporal sclerosis—but not those with left-sided damage or extratemporal damage—showed impairments on tests of recognition of facial expressions of emotion, but not on comparison tasks of face processing (Meletti, Benuzzi, Nichelli, & Tassinaria, 2003). This deficit was most pronounced for fearful expressions, and the degree of deficit was related to the age of first seizure and epilepsy onset. These results suggest that normal functioning of the amygdala in childhood is important for normal development of facial emotion processing.

There is evidence from event-related potential (ERP) studies that the core cortical system may be functional in infancy. ERP studies in adults have identified a face-sensitive component, the N170; this is believed to be generated by components of the core cortical system, such as the fusiform gyrus and STS. Studies in infants have identified an infant N170, with source analyses providing evidence for a source in the fusiform region (Johnson et al., 2005). Infant ERP studies do not show an influence of emotion on the infant N170 (Leppanen, Moulson, Vogel-Farley, & Nelson, 2007), but do show an influence of emotion at longer latencies. Brain potentials differ for happy compared to fearful faces within 400 msec after stimulus onset, although brain potentials for angry compared to fearful faces do not differ from one another at any latency up to 1700 msec after stimulus onset (Nelson & de Haan, 1996). Studies differ as to whether responses to fearful and neutral faces differ from one another at these latencies, with some studies finding a difference (Leppanen et al., 2007) and others failing to do so (Hoehl, Palumbo, Heinisch, & Striano, 2008). These infant ERP studies have not looked for rapid, pre-N170 influences of emotion that might reflect modulation by rapid subcortical pathways. The differences found at longer latencies could reflect activity generated in association cortex and possibly modulated by input from the amygdala, but this possibility has not been directly investigated. In adults, there is evidence that some modulations of posterior ERP activity by emotion reflect synchronization of these responses with more anterior frontotemporal activity (Sato, Kochiyama, Yoshikawa, & Matsumara, 2001).

Few studies have examined ERP responses to emotional expressions in children. Those existing suggest that emotion influences cortical processing from an early latency (Batty & Taylor, 2006; de Haan, Nelson, Gunnar, & Tout, 1998). There is a shift over the ages of 4–15 years: Facial expression influences the P1 in younger children, but this effect diminishes with age as an influence on the face-sensitive N170 component emerges (Batty & Taylor, 2006). The absence of influence of emotion on the N170 at the younger age is consistent with results of infant studies (Leppanen et al., 2007), though those infant studies did not look for earlier effects. These results may reflect a difference in the nature of facial emotion processing with age, with an initial focus on simpler perceptual differences between emotions and then a shift to a more "face-specific" effect. They may also reflect a difference in the balance between subcortical and cortical pathways in processing emotional signals. At least in 4- to 6-year-olds, the rapidity with which emotional information is processed may depend on the task: If there is also a familiarity dimension to the task, emotion processing may be delayed to a longer latency following familiarity processing (Todd, Lewis, Meusel, & Zelazo, 2008).

Imaging studies in children provide more direct evidence regarding development of brain regions involved in emotion processing. Facial expressions activate the amygdala in children and adolescents (Lobaugh, Gibson, & Taylor, 2006; Monk et al., 2003; Yang et al., 2007). One hypothesis is that amygdala activity during emotion perception may decrease with age as frontal circuits gain more control (Yurgelun-Todd & Killgore, 2006). There is not compelling evidence for a decrease in amygdala activation with age, at least between 8 and 15 years, though prefrontal activation to fearful faces does increase over this time (Yurgelun-Todd & Killgore, 2006). Studies with adults suggest that emotion-driven processing can be separated from attention-driven processing of emotional signals. For example, functional magnetic resonance imaging (fMRI) studies show that these increased responses in fusiform cortex to fearful faces are abolished by amygdala damage in the ipsilateral hemisphere, despite preserved effects of voluntary attention on fusiform cortex; by contrast, emotional increases can still arise despite deficits in attention or awareness following parietal damage (Vuilleumier & Pourtois, 2007). Developmental imaging studies suggest that emotion-driven and attention-driven processing also change with age: Compared to adults, adolescents exhibit greater modulation of prefrontal activation based on emotional content, and less modulation based on attentional demands (Monk et al., 2003). Imaging studies also suggest that different brain regions may be activated by different emotions (Lobaugh et al., 2006). However, there is limited information on this point, as many studies have focused on regions of interest (e.g., the amygdala) and facial expressions of interest (e.g., fear), rather than on whole-brain analysis or a wide range of facial emotions. Another limitation of existing imaging studies is that many group together children of a relatively wide age range (e.g., 8–17 years) and/or do not provide direct comparisons between age groups or between children and adults.

Voices, and Integration of Faces and Voices

Few developmental studies have examined the neural basis of emotion perception in voices or in audiovisual stimuli. Grossmann, Striano, and Friederici (2006) used ERPS to examine 7-month-olds' perception of congruent and incongruent face–voice pairings. Emotionally incongruent prosody elicited a larger negative component in infants' ERPs than did emotionally congruent prosody. Conversely, the amplitude of infants' positive component was larger to emotionally congruent than to incongruent prosody. The findings suggest that 7-month-olds can integrate emotional information across modalities and can recognize common affect in the face and voice. There is also some evidence that infants as young as 7 months of age show different ERPs to angry, happy, and neutral prosody. The pattern of waves obtained suggests that emotional prosody receives enhanced processing compared to neutral, and that angry prosody elicits greater attention than either happy or neutral (Grossmann et al., 2005).

In boys ages 7–17 years, medial prefrontal cortex and the right STS are activated when they attend to facial expressions and tone of voice in short scenarios involving irony (Wang, Lee, Sigman, & Dapretto, 2007).

Summary and Discussion

Perceiving emotion in faces and voices is an important part of everyday social life. Studies in adults suggest that both subcortical and cortical pathways play a role, with regions including the amygdala, ventral occiptotemporal cortex, STS, and prefrontal areas considered to play important roles. ERP and fMRI studies suggest that at least some of the regions implicated in emotion processing in adults are also activated in infants, children, and adolescents, but that the pattern and extent of activation change with age. It is important to reiterate that existing imaging studies are somewhat limited, in that many focus on a particular structure rather than whole-brain analysis; many group children of a wide age range (e.g., 8–17 years) into a single group; and many do not directly compare children to adults.

Developmental investigations have highlighted the competencies of infants for emotion perception, but have primarily involved studying the perception of static facial images. Investigating the perception of more realistic, moving displays is important, as motion is known to facilitate emotion processing in adults; thus studies limited to static stimuli may underestimate infants' and children's true abilities. Moving stimuli may also be more compelling, and thus may benefit studies where limits of attention constrain experimental designs, such as infant ERP studies.

There is as yet quite limited information on development of the neural bases of perception of emotion in the voice, or of how voice and face information interact. Given that infants can begin to learn information from the voice even prenatally, a better understanding of this topic is critical for a full understanding of how infants perceive emotion. For example, it is possible that emotional information from voices is processed prenatally, and that this knowledge then facilitates the perception of facial displays of emotion postnatally.

References

Adolphs, R., Tranel, D., & Damasio, A. R. (2003). Dissociable neural systems for recognizing emotions. *Brain and Cognition, 52,* 61–69.

Allison, T., Puce, A., & McCarthy, G. (2000). Social perception from visual cues: Role of the STS region. *Trends in Cognitive Sciences, 4,* 267–278.

Balaban, M. T. (1995). Affective influences on startle in five-month-old infants: Reactions to facial expressions of emotion. *Child Development, 66,* 28–36.

Barrera, M. E., & Maurer, D. (1981). The perception of facial expressions by the three-month-old. *Child Development, 52,* 293–206.

Batty, M., & Taylor, M. J. (2006). The development of emotional face processing during childhood. *Developmental Science, 9*(2), 207–220.

Belin, P., Fecteau, S., & Bedard, C. (2004). Thinking the voice: Neural correlates of voice perception. *Trends in Cognitive Sciences, 8,* 129–135.

Blair, R. J., Morris, J. S., Frith C. D., Perrett, D. I., & Dolan, R. J. (1999). Dissociable neural responses to facial expressions of sadness and anger. *Brain, 122,* 883–893.

Braddick, O., Atkinson, J., & Wattam-Bell, J. (2003). Normal and anomalous development of visual motion processing: Motion coherence and 'dorsal stream vulnerability.' *Neuropsychologia, 41,* 1769–1784.

Bruce, C., Desimone, R., & Gross, C. G. (1981). Visual properties of neurons in a polysensory area in superior temporal sulcus of the macaque. *Journal of Neurophysiology, 46,* 369–384.

Buccino, G., Binkofski, F., Fink, G. R., Fadiga, L., Fogassi, L., Gallese, V., et al. (2001). Action observation activates premotor and parietal areas in a somatotopic manner: An fMRI study. *European Journal of Neuroscience, 13,* 400–404.

Burnham, D. (1993). Visual recognition of mother by young infants: Facilitation by speech. *Perception, 22,* 1133–1153.

Calder, A. J., Keane, J., Manes, F., Antoun, N., & Young, A. W. (2000). Impaired recognition and experience of disgust following brain injury. *Nature Neuroscience, 3,* 1077–1078.

Caron, A. J., Caron, R. F., & MacLean, D. J. (1988). Infant discrimination of naturalistic emotional expressions: The role of face and voice. *Child Development, 59,* 604–616.

Davis, M. (1989). Neural systems involved in fear-potentiated startle. *Annals of the New York Academy of Sciences, 563,* 165–183.

de Gelder, B., Frissen, I., Barton, J., & Hadjikhani, N. (2003). A modulatory role for facial expressions in prosopagnosia. *Proceedings of the National Academy of Sciences USA, 100,* 13105–13110.

de Gelder, B., Pourtois, G., & Weiskrantz, L. (2002). Fear recognition in the voice is modulated by unconsciously recognized facial expressions but not by unconsciously recognized affective pictures. *Proceedings of the National Academy of Sciences USA, 99,* 4121–4126.

de Gelder, B., & Vroomen, J. (2000). The perception of emotions by ear and by eye. *Cognition and Emotion, 14,* 289–311.

de Haan, M., Nelson, C. A., Gunnar, M. R., & Tout, K. A. (1998). Hemispheric differences in brain activity related to the recognition of emotional expressions by 5-year-old children. *Developmental Neuropsychology, 14,* 495–518.

Dimberg, U., Thunberg, M., & Elmehed, K. (2000). Unconscious facial reactions to emotional facial expressions. *Psychological Science, 11,* 86–89.

Durand, K., Gallay, M., Seigneuric, A., Robichon, F., & Baudouin, J. Y. (2007). The development of facial emotion recognition: The role of configural information. *Journal of Experimental Child Psychology, 97,* 14–27.

Edwards, K. (1998). The face of time: Temporal cues in facial expressions of emotion. *Psychological Science, 9,* 270–276.

Ethofer, T., Pourtois, G., & Wildgruber, D. (2006). Investigating audiovisual integration of emotional signals in the human brain. *Progress in Brain Research, 156,* 345–361.

Farroni, T., Csibra, G., Simion, F., & Johnson, M. H. (2002). Eye contact detection in humans from birth. *Proceedings of the National Academy of Sciences USA, 99,* 9602–9605.

Field, T., Woodson, R., Greenberg, R., & Cohen, D. (1982). Discrimination and imitation of facial expressions by neonates. *Science, 218,* 179–181.

Flom, R., & Bahrick, L. E. (2007). The development of infant discrimination of affect in multimodal and unimodal stimulation: The role of intersensory redundancy. *Developmental Psychology, 43,* 238–252.

Gobbini, M. I., & Haxby, J. V. (2007). Neural systems for recognition of familiar faces. *Neuropsychologia, 45,* 32–41.

Goren, C. C., Sarty, M., & Wu, P. Y. (1975). Visual following and pattern discrimination of face-like stimuli by newborn infants. *Pediatrics, 56,* 544–549.

Grezes, J., Fonlupt, P., Bertenthal, B., Delon-Martin, C., Segebarth, C.. & Decety, J. (2001). Does perception of biological motion rely on specific brain regions? *NeuroImage, 13,* 775–785.

Grossmannn, T., Striano, T., & Friederici, A. D. (2005). Infants' electric brain responses to emotional prosody. *NeuroReport, 16,* 1825–1828.

Grossmannn, T., Striano, T., & Friederici, A. D. (2006). Crossmodal integration of emotional information from face and voice in the infant brain. *Developmental Science, 9,* 309–315.

Halgren, E., Baudena, P., Heit, G., Clarke, J. M., Marinkovic, K., & Clarke, M. (1994). Spatio-temporal stages in face and word processing: I. Depth-recorded potentials in the human occipital, temporal and parietal lobes. *Journal of Physiology (Paris), 88,* 1–50.

Harmer, C. J., Thilo, K. V., Rothwell, J. C., & Goodwin, G. M. (2001). Transcranial magnetic stimulation of medial-frontal cortex impairs the processing of angry facial expressions. *Nature Neuroscience, 4*, 17–18.

Hasselmo, M. E., Rolls, E. T., & Baylis, G. C. (1989). The role of expression and identity in the face-selective responses of neurons in the temporal visual cortex of the monkey. *Behavioural Brain Research, 32*, 203–218.

Haxby, J. V., Hoffman, E. A., & Gobbini, M. K. (2000). The distributed human neural system for face perception. *Trends in Cognitive Sciences, 4*, 223–233.

Herba, C. M., Benson, P., Landau, S., Russell, T., Goodwin, C., Lemche, E., et al. (2008). Impact of familiarity upon children's developing facial expression recognition. *Journal of Child Psychology and Psychiatry, 49*, 201–210

Hess, U., & Blairy, S. (2001). Facial mimicry and emotional contagion to dynamic emotional facial expressions and their influence on decoding accuracy. *International Journal of Psychophysiology, 40*, 129–141.

Hess, U., Kappas, A., & Scherer, K. (1988). Multichannel communication of emotion: Synthetic signal production. In K. Scherer (Ed.), *Facets of emotion* (pp. 161–182). Hillsdale, NJ: Erlbaum.

Hoehl, S., Palumbo, L., Heinisch, C., & Striano, T. (2008). Infants' attention is biased by emotional expressions and eye gaze direction. *NeuroReport, 26*, 579–582.

Hoffman, E. A., & Haxby, J. V. (2000). Distinct representations of eye gaze and identity in the distributed human neural system for face perception. *Nature Neuroscience, 3*, 80–84.

Johansson, G. (1973). Visual perception of biological motion and a model for its analysis. *Perception and Psychophysics, 14*, 201–211.

Johnson, M. H. (2005). Subcortical face processing. *Nature Reviews Neuroscience, 6*, 766–774.

Johnson, M. H., Griffin, R., Csibra, G., Halit, H., Farroni, T., de Haan, M., et al. (2005). The emergence of the social brain network. *Development and Psychopathology, 17*, 599–619.

Kahana-Kalman, R., & Walker-Andrews, A. S. (2001). The role of person familiarity in young infants' perception of emotional expressions. *Child Development, 72*, 352–369.

Kanwisher, N., McDermott, J., & Chun, M. M. (1997). The fusiform face area: A module in human extrastriate cortex specialized for face perception. *Journal of Neuroscience, 17*, 4302–4311.

Kaplan, P. S., Zarlengo-Strouse, P., Kirk, L. C., & Angel, C. L. (1997). Selective and nonselective associations between speech segments and faces in human infants. *Developmental Psychology, 33*, 990–999.

Kotsoni, E., de Haan, M., & Johnson, M. H. (2001). Categorical perception of facial expressions by 7-month-olds. *Perception, 30*, 1115–1125.

Kreifelts, B., Ethofer, T., Grodd, W., Erb, M., & Wildgruber, D. (2007). Audiovisual integration of emotional signals in voice and face: An event-related fMRI study. *NeuroImage, 37*, 1444–1456.

LaBar, K. S., Crupain, M. J., Voyvodic, J. T., & McCarthy, G. (2003). Dynamic perception of facial affect and facial identity in the human brain. *Cerebral Cortex, 13*, 1023–1033.

LaBarbera, J. D., Izard, C. E., Vietze, P., & Parisi, S. A. (1976). Four- and six-month-old infants' visual responses to joy, anger and neutral expressions. *Child Development, 47*, 535–538.

Lang, P. J., Bradley, M. M., & Cuthbert, B. N. (1990). Emotion, attention and the startle reflex. *Psychological Review, 97*, 377–395.

LeDoux, J. E., Sakaguchi, A., Iwata, J., & Reis, D. J. (1986). Interruption of projections from the medial geniculate body to an archi-neostriatal field disrupts the classical conditioning of emotional responses to acoustic stimuli. *Neuroscience, 17*, 615–627.

Leppanen, J. M., Moulson, C. M., Vogel-Farley, V. K., & Nelson, C. A. (2007). An ERP study of emotional face processing in the adult and infant brain. *Child Development, 78*, 232–245.

Liu, L., Ioannides, A. A., & Streit, M. (1999). Single trial analysis of neurophysiological correlates of the recognition of complex objects and facial expressions of emotion. *Brain Topography, 11*, 291–303.

Lobaugh, N. J., Gibson, E., & Taylor, M. J. (2006). Children recruit distinct neural systems for implicit emotional face processing. *NeuroReport, 17*, 215–219.

Massaro, D. W., & Egan, P. B. (1996). Perceiving affect from the voice and the face. *Psychonomic Bulletin and Review, 3*, 215–221.

Meletti, S., Benuzzi, F., Nichelli, P., & Tassinaria, C. A. (2003). Damage to the hippocampal–amygdala formation during early infancy and recognition of fearful faces: Neuropsychological and fMRI evidence in subjects with temporal lobe epilepsy. *Annals of the New York Academy of Sciences, 1000*, 385–388.

Mitchell, T. V., & Neville, H. J. (2004). Asynchronies in the development of electrophysiological responses to motion and color. *Journal of Cognitive Neuroscience, 16*, 1363–1374.

Monk, C. S., McClure, E. B., Nelson, E. E., Zarahn, E., Bilder, R. M., Leibenluft, E., et al. (2003). Adolescent immaturity in attention-related brain engagement to emotional facial expressions. *NeuroImage, 20*, 420–428.

Morris, J. S., de Gelder, B., Weiskrantz, L., & Dolan, R. J. (2001). Differential extrageniculostriate and amygdala responses to presentation of emotional faces in a cortically blind field. *Brain, 124*, 1241–1252.

Morris, J. S., Friston, K. J., Buchel, C., Frith, C. D., Young, A. W., Calder, A. J., et al. (1998). A neuromodulatory role for the human amygdala in processing emotional facial expressions. *Brain, 121*, 47–57.

Morris, J. S., Öhman, A., & Dolan, R. J. (1999). A subcortical pathway to the right amygdala mediating 'unseen' fear. *Proceedings of the National Academy of Sciences USA, 96*, 1680–1685.

Morton, J., & Johnson, M. H. (1991). CONSPEC and CONLERN: A two-process theory of infant face recognition. *Psychological Review, 98*, 164–181.

Nelson, C. A., & de Haan, M. (1996). Neural correlates of infants' visual responsiveness to facial expressions of emotion. *Developmental Psychobiology, 29*, 577–595.

Nelson, C. A., & Dolgin, K. (1985). The generalized discrimination of facial expressions by seven-month-old infants. *Child Development, 56*, 58–61.

Nelson, C. A., & Horowitz, F. D. (1983). The perception of facial expressions and stimulus motion by two- and five-month-old infants using holographic stimuli. *Child Development, 54*, 868–877.

Nelson, C. A., Morse, P. A., & Leavitt, L. A. (1979). Recognition of facial expressions by seven-month-old infants. *Child Development, 50*, 1239–1242.

Oram, M. W., & Perrett, D. I. (1994). Responses of anterior superior temporal polysensory neurons to 'biological motion' stimuli. *Journal of Cognitive Neuroscience, 6*, 99–116.

Oster, J., & Ewy, R. (1980). *Discrimination of sad vs. happy fades by 4-month-olds: When is a smile seen as a smile?* Unpublished manuscript, University of Pennsylvania.

Perrett, D. I., Smith, P. A., Potter, D. D., Mistlin, A. J., Head, A. S., Milner, A. D., et al. (1985). Visual cells in the temporal cortex sensitive to face view and gaze direction. *Proceedings of the Royal Society of London, Series B, Biological Sciences, 223*, 293–317.

Pike, G. E., Kemp, R. I., Towell, N. A., & Phillips, K. C. (1997). Recognizing moving faces: The relative contribution of motion and perspective view information. *Visual Cognition, 4*, 409–437.

Pourtois, G., de Gelder, B., Bol, A., & Crommelinck, M. (2005). Perception of faces and voices and of their combination in the human brain. *Cortex, 41*, 49–59.

Puce, A., Epling, J. A., Thompson, J. C., & Carrick, O. K. (2007). Neural responses elicited to face motion and vocalization pairing. *Neuropsychologia, 45*, 93–106.

Sato, W., Kochiyama, T., Yoshikawa, S., & Matsumara, M. (2001). Emotional expression boosts early visual processing of the face: ERP recording and its decomposition by independent components analysis. *NeuroReport, 12*, 709–714.

Sato, W., Kochiyama, T., Yoshikawa, S., Naito, E., & Matsumara, M. (2004). Enhanced neural activity in response to dynamic facial expressions of emotion: An fMRI study. *Brain Research: Cognitive Brain Research, 45*, 174–194.

Shackman, J. E., & Pollak, S. D. (2005). Experiential influences on multimodal perception of emotion. *Child Development, 76*, 1116–1126.

Simion, F., Valenza, E., Umiltà, C., & Dalla Barba, B. (1998). Preferential orienting to faces in newborns: A temporal–nasal asymmetry. *Journal of Experimental Psychology: Human Perception and Performance, 24,* 1399–1405.

Slater, A. (2001). Visual perception. In G. Bremner & A. Fogel (Eds.), *Blackwell handbook of infant development* (pp. 5–34). Malden, MA: Blackwell.

Soken, N. H., & Pick, A. D. (1992). Intermodal perception of happy and angry expressive behaviors by seven-month-old infants. *Child Development, 63,* 787–795.

Soken, N. H., & Pick A. D. (1999). Infants' perception of dynamic affective expressions: Do infants distinguish specific expressions? *Child Development, 70,* 1275–1282.

Spencer, J., O'Brien, J., Johnston, A., & Hill, H. (2006). Infants' discrimination of faces by using biological motion cues. *Perception, 35,* 79–89.

Steede, L. L., Tree, J. J., & Hole, G. J. (2007). I can't recognize your face but I can recognize its movement. *Cognitive Neuropsychology, 24,* 451–466.

Tamietto, M., & de Gelder, B. (2008). Affective blindsight in the intact brain: Neural interhemispheric summation for unseen fearful expressions. *Neuropsychologia, 46,* 820–828.

Thomas, L. A., DeBellis, M. D., Graham, R., & LaBar, K. S. (2007). Development of emotional facial recognition in late childhood and adolescence. *Developmental Science, 10,* 547–558.

Todd, R. M., Lewis, M. D., Meusel, L. A., & Zelazo, P. D. (2008). The time course of social-emotional processing in early childhood. *Neuropsychologia, 46,* 595–613.

Vaish, A., & Striano, T. (2004). Is visual reference necessary?: Contributions of facial versus vocal cues in 12-month-olds' social referencing behavior. *Developmental Science, 7,* 261–269.

Vuilleumier, P., & Pourtois, G. (2007). Distributed and interactive brain mechanisms during emotion face perception: Evidence from functional neuroimaging. *Neuropsychologia, 45,* 174–194.

Wade, A. M., Lawrence, K., Mandy, W., & Skuse, D. (2006). Charting the development of emotion recognition from 6 years of age. *Journal of Applied Statistics, 33,* 297–315.

Walker, A. (1982). Intermodal perception of expressive behaviors by human infants. *Journal of Experimental Child Psychology, 33,* 514–535.

Walker-Andrews, A. (1997). Infants' perception of expressive behaviors: Differentiation of multimodal information. *Psychological Bulletin, 121,* 437–456.

Wang, A. T., Lee, S. S., Sigman, M., & Dapretto, M. (2007). Reading affect in the face and voice: Neural correlates of interpreting communicative intent in children and adolescents with autism spectrum disorders. *Archives of General Psychiatry, 64,* 698–708.

Wattam-Bell, J. (1996). Visual motion processing in one-month-old infants: Habituation experiments. *Vision Research, 36,* 1679–1685.

Wildgruber, D., Ackermann, H., Kreifelts, B., & Ethofer, T. (2006). Cerebral processing of linguistic and emotional prosody: fMRI studies. *Progress in Brain Research, 156,* 249–268.

Yang, T. T., Simons, A. N., Matthews, S. C., Tapert, S. F., Bischoff-Grethe, A., Frank, G. K. W., et al. (2007). Increased amygdala activation is related to heart rate during emotion processing in adolescent subjects. *Neuroscience Letters, 428,* 109–114.

Young-Browne, G., Rosenfield, H. M., & Horowitz, F. D. (1977). Infant discrimination of facial expressions. *Child Development, 48,* 555–562.

Yurgelun-Todd, D. A., & Killgore, W. D. (2006). Fear-related activity in the prefrontal cortex increases with age during adolescence: A preliminary fMRI study. *Neuroscience Letters, 406,* 194–199.

CHAPTER 7

The Development of Social Information Gathering in Infancy

A Model of Neural Substrates and Developmental Mechanisms

LESLIE J. CARVER
LAUREN CORNEW

Over the course of early development, infants and toddlers make use of familiar adults to guide their behavior in unusual, unanticipated, or novel situations. Previous research in this area has focused primarily on behavioral developments in diverse individual aspects of social information gathering, such as sharing attention, theory of mind, and emotion regulation. In this chapter, we propose a model for the development of social information gathering from a developmental social-cognitive neuroscience perspective. This model focuses on the relations among the component parts that are involved in social information gathering, and on how these relations change across development. We describe research on the underlying neural basis of different aspects of social information gathering, and propose that experience and associative abilities guide the formation of a complex system in the brain that in turn shapes later social cognition. Because infants do not use language to communicate, much of their social information gathering comes through sharing attention with, and receiving verbal and nonverbal signals from, adults. Infants look to adults when they encounter novel events, and use adults' facial and vocal expressions to regulate their own behavior. Studies have established that the behavioral patterns associated with this skill, known as "social referencing," undergo substantial development over the course of infancy. However, little is understood about the brain basis of these patterns and about how the ontogenetic development of brain systems affects their emergence. Baldwin and Moses (1996) have proposed a developmental model to account for the increase in sophistication in information gathering across the second year of life. In this chapter, we expand on this developmental view by presenting a model of the development of a network of brain areas dedicated to social information gathering, with the aim of providing a framework around which future developmental studies might be designed (see Figure 7.1). We begin by introducing the various competencies that constitute joint attention and social referencing. We then describe what is known or suspected about the development of the brain systems underlying the various skills involved in these two crucial means of social information gathering. We include discussions of brain areas involved in sharing attention, which is important for seeking social information; emotion recognition and associative learning, which are important for relating emotional information provided by others to novel events; and emotion regulation, which is important for using emotional information provided by others for regulating one's own behavior. A primary goal of this chapter is to integrate findings from the aforementioned research areas, in order to furnish a more comprehensive picture of the neural underpinnings of early social-cognitive development. At the end of the chapter, we provide preliminary evidence from studies conducted in our laboratory that assesses the multiple components of social referencing and the relations among them.

Social Information Gathering

Research investigating the development of social information gathering suggests that several distinct behaviors play important roles. For instance, Feinman, Roberts, Hsieh, Sawyer, and Swanson (1992) claimed that the following three elements are involved in gathering nonverbal information about novel situations and stimuli in a sophisticated way. First, the infant seeks information by triadic sharing of attention with an adult on a third object. Second, if the adult expresses emotion regarding the novel stimulus, the infant must recognize the emotion, understand what it means, and relate it to the novel stimulus. Finally, the infant must use the information provided by the adult to regulate his or her own behavior and emotions.

Additional support for a multicomponent view of social information gathering comes from experiments that have differentiated between the processes of simply responding to, and actively seeking, social information. Baldwin and Moses (1996) have proposed that whereas 12-month-olds may be sophisticated "consumers" of social information, they may not be "seekers" of such information. Consistent with this idea, an important distinction in the model we propose is between *early* social information gathering, in which infants respond differentially to various emotions (e.g., by matching an adult's emotion or avoiding an object that the adult has tagged with negative emotion), and *later* social information gathering, in which infants actively seek information from an adult. In the early stage, infants' responses can be seen as driven by emotional contagion, or in some cases even by the relative novelty of seeing an adult display negative emotion. However, in the later-emerging stage, social information seeking is more volitional, and behavior regulation occurs because of a true understanding of the other's knowledge of the situation, as well as of the meaning of emotional displays.[1] Despite some research regarding the development of social information

---

[1] Baldwin and Moses (1996) emphasize this distinction by employing the term "social referencing" to refer to any effect of another person's emotion on an infant's behavior, regardless of the infant's level of understanding concerning the mind of the source. According to this idea, 12-month-olds engage in social referencing. However, they use the term "social information gathering" to refer to the more sophisticated way in which 18-month-olds seek and respond to emotional information. In this chapter, we use the terms "social referencing" and "social information gathering" interchangeably, and instead distinguish between primitive and more advanced behaviors of this kind by referring to "early" versus "later" forms of these social-cognitive abilities.

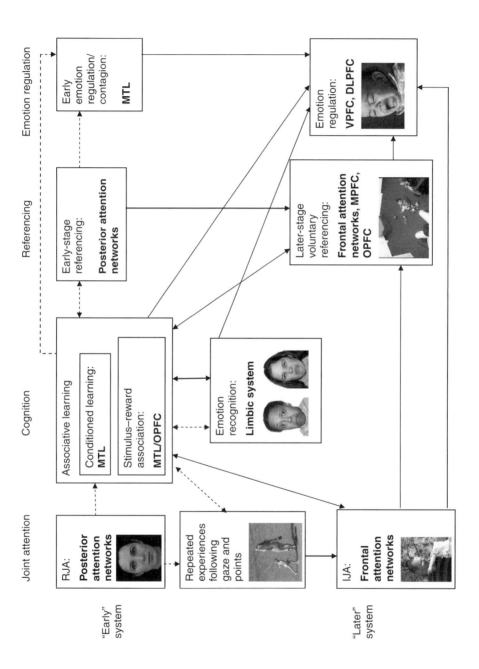

Joint attention   Cognition   Referencing   Emotion regulation

FIGURE 7.1. Proposed model for development of social referencing. In this model, early gaze and point following, repeated numerous times, leads to the development of initiating joint attention. This proposed mechanism works through associative and stimulus–reward systems to increase infants' looks to caregivers in novel situations. As infants increasingly look to caregivers and see their emotional expressions, they learn the meaning of those expressions, and begin to use them to regulate their behavior. Associative structures are, in this view, the "glue" that links early, automatic referencing and emotional regulation through contagion to later-developing, voluntary referencing systems. RJA, responding to joint attention; IJA, initiating joint attention; MTL, medial temporal lobe; OPFC, orbital prefrontal cortex; MPFC, medial prefrontal cortex; VPFC, ventral prefrontal cortex; DLPFC, dorsolateral prefrontal cortex.

124

gathering, little is understood about how the individual elements emerge and how they relate to one another. In the following sections, we describe the components and what is known or hypothesized about the development of their neural bases.

Joint Attention

One of the more important elements of social referencing is the ability to share attention with an adult to a third object, event, or person, which manifests itself as the infant's alternating gaze between the adult and the third object. Rudimentary aspects of this ability, termed "joint attention," can be seen relatively early in development. In certain situations, infants as young as 3 months of age appear to follow the gaze of adults (D'Entremont, Hains, & Muir, 1997; see also Grossmann & Farroni, Chapter 5, this volume, for a discussion of perception of eye gaze in infancy). In one study, 3-month-old infants responded differently to adults who shared attention with them than to adults who looked only to an object without alternating gaze between the infant and the object (Striano & Stahl, 2005). Nevertheless, despite its early emergence, considerable development is seen in attention sharing. Infants 10 and 11 months of age, but not 9-month-olds, followed the head turn of an adult who had her eyes open (and therefore might be looking at something interesting), but did not follow the head turn of an adult whose eyes were closed (Brooks & Meltzoff, 2005). In other research, infants 12–13 months of age followed gaze more when adults were wearing headbands than when their eyes were covered with a blindfold (Brooks & Meltzoff, 2002; D'Entremont & Morgan, 2006). Thus substantial development occurs from early precursors of joint attention, such as gaze following, through "true" joint attention, in which infants follow another's gaze with the understanding that the person is an observer of something potentially interesting (see Moore & Corkum, 1994, for a review).

There are a few data regarding the neural correlates of the development of initiating and responding to joint attention. Striano, Reid, and Hoehl (2006) measured infants' brain activity in response to pictures that were displayed on a video screen. An adult either attended to the stimulus without looking at the infant, or alternated gaze between the infant and the pictured object. The results indicated that the amplitude of the negative component (Nc), a developmental event-related potential (ERP) component thought to index attention to salient events, was greater when adults shared attention with infants than when adults only attended to the screen. These results suggest that infants' attentional resources are facilitated and focused by shared attention with an adult. Mundy, Card, and Fox (2000) have suggested that initiating joint attention (IJA; i.e., initiating alternating looks between an adult and an external stimulus) and responding to joint attention (RJA; i.e., following an adult's gaze or point) involve separate, distinguishable neural networks. Frontal EEG data were associated with infants' IJA bids on the Early Social Communication Scales, a behavioral instrument developed to measure joint attention. In contrast, RJA was associated with parietal electroencephalographic (EEG) activation. Mundy and colleagues suggested that initiating joint attention may uniquely involve working memory and representational demands, as well as the emergence of frontally mediated attention networks (Rueda, Posner, & Rothbart, 2005), whereas RJA may reflect the functioning of an earlier maturing posterior attention network (Rueda et al., 2005). Mundy et al. (2000) have proposed a developmental sequence in which these posterior attention systems lead into and contribute to the differentiation of frontally mediated patterns of joint attention seen by the middle of the second year of life (see Van Hecke & Mundy, 2007, for a review).

Theory of Mind and Joint Attention

A sophisticated use of joint attention in order to seek information from another person implies some understanding of others' capacity for having such information (see Baldwin & Moses, 1996). This understanding, termed "theory of mind" (ToM), has been a topic of controversy in the developmental literature, especially with respect to the age at which it emerges (see also Choudhury, Charman, & Blakemore, Chapter 9, this volume). Psychologists traditionally assessed children's ToM through variations of the false-belief task, in which children are asked to make judgments about what a character in a story believes to be the truth. In the typical false-belief paradigm, children are shown a character who hides an object in one location. Unbeknownst to this character, a second character then moves the object. The child is asked where the object really is located, and where the first character *believes* the object is located.

Findings from studies that used this method (Flavell, 2000; Flavell, Mumme, Green, & Flavell, 1992) tended to indicate that 4-year-olds, but not 3-year-olds, passed the false-belief task; therefore, researchers typically inferred that ToM emerged somewhere between the ages of 3 and 4 years. However, studies that have tracked the emergence of ToM suggest that performance may depend at least partly on task demands, as the age at which children successfully perform some ToM tasks is younger than the formerly established 4-year threshold for the false-belief task. For example, Wellman and Liu (2004) showed that children pass some tests about others' beliefs and desires far sooner than they pass the false-belief test. Other studies have been aimed at reducing or eliminating linguistic and cognitive demands in an effort to obtain a purer measure of children's understanding of intentionality. Results from these experiments demonstrate at least a rudimentary ToM as early as 18 months. For example, infants as young as 18 months of age (Meltzoff, 1995) appear to understand adults' intentions: They successfully imitate adults' intended actions, even when an adult "slips" and fails to carry out an action completely. In another study, infants 18½ months of age and older, but not younger infants, offered an experimenter food that she had previously displayed a desire for and a positive response to, even when the adult's desire was contrary to the infants' own preference (Repacholi & Gopnik, 1997). Thus even infants understand at least something about others' intentions and desires.

Despite the sizeable literature on the behavioral manifestations of ToM in infancy and early childhood, fewer studies have investigated the neural basis of ToM. Of these brain-based inquiries, most have been conducted with adults (see Choudhury et al., Chapter 9, this volume, for a discussion of such studies in adolescents). Adult patients with damage to the dorsolateral prefrontal cortex and/or the amygdala show deficits in tasks designed to test the understanding of others' false beliefs and social faux pas (Stone, Baron-Cohen, Calder, Keane, & Young, 2003; Stone, Baron-Cohen, & Knight, 1998). Studies of healthy adults corroborate the association between ToM and prefrontal function in adults. Imaging studies suggest that areas of the medial frontal and orbitofrontal cortex are likely to be involved in ToM (Fletcher et al., 1995; Gallagher & Frith, 2003; Gallagher et al., 2000). One electrophysiological study also measured brain activity as it related to ToM. Adults were asked to judge where a hidden object really was, or where an observer would think the object was. ERPs revealed a late slow-wave component that differentiated between conditions. This effect was localized, via source estimation, to the left orbitofrontal cortex (Liu, Sabbagh, Gehring, & Wellman, 2004). The waveform elicited was consistent with previous activity that was observed when individuals distinguished between false photographs and false mental beliefs (Sabbagh & Taylor, 2000).

In a developmental version of Liu et al.'s (2004) study, children were shown an animated cartoon character who placed two animals in two separate boxes. Then, with the cartoon protagonist not watching, one of the animals emerged from its box, and either moved into the other box or returned to its own box. Children were asked either where the animal really was, or where the cartoon protagonist thought it was. ERP responses in these children were similar to those observed in the adults in Liu et al.'s (2004) study; in addition, this activity differentiated between 5- and 6-year-olds who passed the false-belief task and those who failed (Liu, Sabbagh, Gehring, & Wellman, in press). Despite the important contribution of this study, additional work is needed to elucidate the development of the brain system involved in ToM. Because of their interrelatedness, such work will be valuable not only for ToM research, but also for the joint attention and social referencing fields.

Emotion Recognition

In addition to joint attention, the ability to recognize emotional expressions and associate them with novel stimuli is important for the development of social referencing. Similar to studies of social information gathering, there is a substantial behavioral literature on the development of emotion recognition, but our understanding of the underlying brain development is nascent (see de Haan & Matheson, Chapter 6, this volume). A great deal more is known about the neural substrates of emotion recognition in adult populations. In this section, we review what is known about the brain basis of emotion recognition, and how this brain function matures in development.

Visual and auditory displays of emotion are salient signals that are crucial for the development of the joint attention and social referencing systems. Most fundamentally, emotion recognition is an adaptive function that keeps humans safe. In dyadic interactions, another person's emotional expressions convey information concerning the other's intentions and how he or she is likely to behave. For instance, an angry display warns of potential aggression, whereas a happy display indicates friendliness or submission. Emotional expressions confer survival value in triadic interactions as well, whether the triad is completed by a third person or by a situation or object. Here affective displays can provide information about stimuli in the environment, as is the case when another's fearful face or vocalization in response to some stimulus signals that the stimulus may be harmful and is worth avoiding. When an infant encounters a novel object, for instance, and seeks information about that object from a trusted adult, the infant must then interpret the adult's emotional reaction to the object in order to garner an understanding of how to react him- or herself.

A substantial amount of research has been aimed at elucidating the brain bases of emotion recognition in adult populations; results from this body of work can be divided on the basis of stimulus presentation modality. In terms of the recognition of emotions in the visual modality, much of which has focused on the processing of emotional facial expressions, the roles of several brain regions have been highlighted. Studies of patients with brain lesions have implicated the basal ganglia in recognizing facial expressions of disgust (Sprengelmeyer et al., 1996), and the amygdala in processing fearful faces (Adolphs et al., 1999). Neuroimaging in healthy individuals has corroborated these findings: The basal ganglia, as well as the insula, are activated during the processing of facial expressions of disgust (Phillips, Fahy, David, & Senior, 1998; Phillips et al., 1997; Sambataro et al., 2006). In addition, the amygdala's role in fear identification is supported by neuroimaging evidence (Morris et al., 1998). ERP studies have shown enhanced processing of fearful faces compared to neutral faces

(Eimer & Holmes, 2002) and faces expressing other basic emotions (Batty & Taylor, 2003). Additional ERP studies have reported enhanced processing of emotional compared to neutral faces (Balconi & Pozzoli, 2003), whereas others have failed to find differences between emotions (Sato, Kochiyama, Yoshikawa, & Matsumura, 2001).

Research on emotion recognition in the auditory modality demonstrates some overlap with visual emotion recognition in regard to the brain areas recruited. For example, bilateral damage to the amygdala has been associated with selective deficits in identification of fearful and angry prosody in spoken sentences (Scott et al., 1997), and nonverbal vocalizations of fear activate neural systems involving the amygdala in healthy people (Morris, Scott, & Dolan, 1999). However, much of the work on emotional prosody has been aimed at testing hemispheric asymmetry hypotheses. Many findings in this area suggest right-hemisphere dominance for recognizing emotion in tone of voice, based on data from lesion studies (Bowers, Coslett, Bauer, Speedie, & Heilman, 1987; Heilman, Scholes, & Watson, 1975; Ross, Thompson, & Yenkosky, 1997), neuroimaging studies (Buchanan et al., 2000; George et al., 1996), and electrophysiological research (Pihan, Altenmüller, & Ackermann, 1997). However, numerous studies have provided conflicting evidence (Pell, 1998; Van Lancker & Sidtis, 1992; Wildgruber et al., 2004) regarding the right hemisphere's role in recognizing affective intonation patterns. Unfortunately, because much of the research in this area has focused on gross hemispheric differences, few attempts at fine-grained localization of function have been made. Furthermore, most studies have addressed the processing of emotional versus nonemotional prosody on a broad level. Therefore, less is known about the degree of overlap in the brain areas recruited for auditory comprehension of specific emotions or of positive versus negative valence. Nevertheless, a few ERP experiments have revealed valence- and emotion-dependent processing effects (Alter et al., 2003; Bostanov & Kotchoubey, 2004).

The literature on the neurodevelopmental origins of emotion perception is noticeably smaller than the adult emotion recognition literature, in large part because of methodological constraints. Functional imaging of older children has revealed greater amygdala activation in response to neutral compared to fearful faces—a pattern opposite to that seen in adults (Thomas et al., 2001), suggesting a protracted developmental progression in the processing of emotional displays. However, infant emotion recognition studies have historically been behavioral in nature. Those that do measure neural activity typically do so via ERPs, as functional neuroimaging is utilized only rarely in infant populations, due to requirements for participants to remain still for extended periods of time. There have also been very few, if any, lesion studies conducted with infants with brain damage; children who suffer early neurological insults are typically not studied until later in childhood (and sometimes not even until adolescence or adulthood).

Behavioral studies of infants' recognition of emotional facial expressions have yielded mixed results concerning both the age at which this capability emerges and the specific emotions that can be discriminated. Under certain experimental conditions, 7-month-old infants have been shown to discriminate happiness from fear, anger (Kestenbaum & Nelson, 1990), and surprise (Ludemann & Nelson, 1988); however, other studies provide contradictory findings for infants of the same age (Phillips, Wagner, Fells, & Lynch, 1990). Similarly, other experiments with 3-month-olds (Barrera & Maurer, 1981) and 5-month-olds (Bornstein & Arterberry, 2003) have yielded conflicting views of the development of emotion identification; some studies suggest that infants this young do in fact demonstrate discrimination, although other experiments indicate that this may be the case only under certain experimental conditions (Schwartz, Izard, & Ansul, 1985).

One potential reason for the inconsistency in results from behavioral research is meth-odological in nature: Most of the studies in this area assess infants' rudimentary ability to recognize emotions by measuring looking time to a novel stimulus, compared to a stimulus to which infants had been habituated immediately before. Numerous studies have reported differences in emotion discrimination depending on which emotion is presented during the habituation phase, which Ludemann and Nelson (1988) interpreted as reflecting infants' degree of familiarity with various emotional expressions in naturalistic settings. This prob-lem complicates the interpretation of results and makes it difficult to discern infants' true emotion recognition capabilities. ERPs can provide an index of neural response to (and potential discrimination of) different emotions in a way that bypasses presentation order issues, because no habituation phase is necessary in ERP experiments.

Noticeably few researchers have used ERPs to tackle questions about infants' emotion recognition skills, compared to those using habituation and other behavioral paradigms. One of the first ERP studies examining facial emotion discrimination during the first year of life (Nelson, 1993) combined habituation and ERP approaches and demonstrated that by 7 months of age, patterns of brain electrical activity differed for happy and fearful faces. However, similar to behavioral evidence, this was apparent only when infants were first habituated to a happy display. In a subsequent set of ERP studies of 7-month-olds, Nelson and de Haan (1996) found differences between responses to happy and fearful faces in the amplitudes of an early positive component, the Nc, and a late positive component. Interest-ingly, when the authors used the same experimental paradigm but compared responses to angry and fearful facial expressions, there were no main effects or interactions involving emotion for the amplitude of the same three ERP components. Despite the apparent lack of discrimination between angry and fearful faces, the same ERP components were elicited when infants viewed all emotional expressions, which Nelson and de Haan (1996) interpreted as reflecting the involvement of the same underlying neural structure(s) for processing all emotional expressions.

In sum, evidence from behavioral and ERP studies suggests that infants become capable of visually discriminating between at least some emotions during the first year of life, and that the processing of several different emotions is accomplished by the same neural structure(s). In one recent study, Leppanen, Moulson, Vogel-Farley, and Nelson (2007) showed that early posterior ERP components that are responsive to emotional information in adults show simi-lar response patterns in 7-month-old infants. These brain responses differentiated fearful faces from both happy and neutral faces, suggesting at least some overlap in brain systems between processing emotional and nonemotional faces. Future research should address the processing of a wider range of facial expressions, in order to determine whether the mecha-nism speculated to underlie visual processing of happy, angry, and fearful expressions also underlies the processing of other emotions.

In terms of the development of emotion recognition in the auditory modality, few studies to date have investigated the processing of emotional prosody during infancy. Nevertheless, this is an especially important area of research, because the auditory system develops earlier in ontogeny than the visual system (Joseph, 2000); therefore, emotional prosody (along with nonlinguistic emotional vocalizations) typically constitutes humans' first exposure to others' emotionality. Also, before infants are capable of understanding the meaning of specific words in an utterance, they probably garner the general meaning via prosody. Indeed, behavioral studies of social referencing indicate that recognition of emotion in the auditory domain may be more important for the development of social referencing than the ability to recognize emotions visually. Mumme, Fernald, and Herrera (1996) tested 12-month-old infants in a social referencing paradigm in which infants were presented with a novel toy, and their caregiver provided one of three emotional signals regarding the toy (happy, fearful, or neutral). Infants who received negative information only through vocal expressions showed changes in affect, looking behavior to the caregiver, and behavior regulation (e.g., avoiding the novel toy). Infants who received the emotional information only through facial expressions did not show this pattern. This finding suggests that emotion recognition in the auditory modality is at least as important as, if not more important than, recognition in the visual modality for social referencing.

Nevertheless, we are aware of only one published study to date that has employed ERPs to examine the development of emotional prosody recognition. In this study (Grossmann, Striano, & Friederici, 2005), 7-month-old infants' brain activity was recorded as they heard semantically neutral words spoken with happy, angry, and neutral intonation. Results indicated that between 300 and 600 msec following stimulus onset, there was a negative shift in the ERP waveform for angry prosody, which was significantly different in amplitude from the waveforms for happy and neutral prosody. In addition, the amplitude of the positive slow-wave (PSW) component was greater for both happy and angry prosody than for neutral prosody, but the happy and angry conditions did not differ from one another, suggesting greater processing of emotional than of to nonemotional speech.

Investigations of infants' ability to recognize emotional displays in visual and auditory modalities continue to elucidate the developmental trajectory of social cognition. Nevertheless, the emotional displays typically encountered in daily life are multimodal in nature; that is, visual and auditory information is presented simultaneously. There is a behavioral literature on multimodal emotion perception in infancy, wherein variations of preferential looking techniques have been utilized. Findings from these studies indicate that by 7 months of age, infants discriminate happy, sad, angry, and interested emotions (Soken & Pick, 1992, 1999). A more recent study (Montague & Walker-Andrews, 2001) used the game of peekaboo as a means to test 4-month-old infants' multimodal emotion recognition capabilities, with results suggesting discrimination of happy from angry and fearful displays even at this young age. A recent ERP study provides the only known evidence to date concerning the neural processing associated with processing emotion multimodally. Grossmann, Striano, and Friederici (2006) presented 7-month-old infants with words spoken with either happy or angry intonation while they viewed an image of an emotionally congruent or incongruent facial expression. Data analysis revealed that the amplitude of two ERP components, the Nc and a positive component, were sensitive to the emotional congruency between the visual and auditory displays. These results demonstrate that by seven months of age, infants recognize happiness and anger across modalities and have some understanding that an emotional expression in one modality should match that in another modality; that is, a happy voice should co-occur with a happy face, and an angry voice should co-occur with an angry face.

Taken together, the results from studies of emotion perception indicate that infants are capable of discriminating some emotions by the second half of the first year of life—a time point that precedes later-stage voluntary social referencing. This is not surprising, as differentially responding to objects and situations based on facial and vocal expressions of various emotions presupposes the ability to discriminate between them. However, discriminating between displays of different emotions on a perceptual level (e.g., on the basis of physical cues, such as the movements and configurations associated with anger vs. happiness) does not necessarily imply an understanding of the meanings of those emotions. This distinction has been explained as the difference between the discrimination and recognition of emotions (e.g., Walker-Andrews, 1997): Discrimination is rooted purely in the physical features of the emotional signals, whereas recognition entails a cognitive appreciation of their behavioral and situational implications. Whereas early-stage social referencing could occur in the absence of true recognition of emotions, it is likely that later-stage referencing reflects more advanced recognition processes. It is unclear exactly when in ontogeny perceptual discrimination evolves into true cognitive recognition of emotions, and the neurodevelopmental mechanisms at work are not well understood. However, given the functional link between emotion recognition and social referencing, it appears plausible that some of the same brain regions are involved in both. In fact, there may even be a reciprocal relationship between them that promotes the advancement from perceptual discrimination of emotions to an understanding of what they mean.

Association between Emotions and Objects

Another important aspect of social referencing entails associating adults' emotions with the objects that elicit them. Even very young infants have been shown to learn associations between pairs of stimuli (Herbert, Eckerman, & Stanton, 2003). These learned associations could potentially occur via two distinct mechanisms. The first mechanism is implicit, involving brain systems used for associative learning, such as the cerebellum, hippocampus, and other structures involved in conditioned learning. In one study, infants were trained using a standard conditioning paradigm, a long-interstimulus-interval paradigm, and a trace conditioning paradigm. In the standard conditioning paradigm, a tone was paired with the presentation of an air puff designed to elicit a conditioned eyeblink response. In the long-delay conditioning paradigm, the conditioned and unconditioned stimuli were still paired, but the time between the presentation of each stimulus was increased (i.e., the interstimulus interval was longer) compared to the standard delayed conditioning paradigm. Trace conditioning entailed the presentation of conditioned and unconditioned stimuli at different times, without overlap, and with an intervening period during which no stimulus was presented (Herbert et al., 2003). Trace conditioning has been shown to differ from standard classical conditioning in that it involves later-maturing structures such as the prefrontal cortex, in addition to the hippocampus and other subcortical structures that are important for conditioned learning (Herbert et al., 2003). In this study, infants learned associations in each of the conditions, although their learning was more effective (i.e., they learned the association more quickly) during a standard conditioning paradigm than in either a long-interstimulus-interval condition or a trace conditioning paradigm. This finding suggests that associative learning is available via implicit means to even very young infants, and supports the idea that associative learning may serve as a mechanism by which infants learn about relations between sharing attention and attention-eliciting stimuli, as well as between emotional information and emotion-eliciting objects.

In addition to implicit association, young infants may come to pair stimuli with one another through explicit association. This process may rely on a memory system that, similar to the system is thought to underlie implicit learned associations, includes the hippocampus. However, this second system also depends on cortical areas surrounding the hippocampus, as well as the prefrontal cortex (Henke, Buck, Weber, & Wieser, 1997). It is not clear at this point whether the implicit or explicit types of associative learning, or both, are at work in social referencing. One possibility is that early in development, associations between emotions and the objects that elicit them occur at an implicit level through conditioning, whereas later in development infants form more explicit representations of these relations.

Research has established that infants as young as 12 months of age are able to determine the referent to which adults' emotions are directed in social referencing contexts. For example, Moses, Baldwin, Rosicky, and Tidball (2001) tested whether 12- and 18-month-olds were able to determine the referential intent of an adult's emotional outburst by manipulating whether an "in-view" (visible to an infant) or "out-of-view" adult displayed the outburst. The 18-month-olds, but not the 12-month-olds, regulated their behavior in accordance with the visible adult's display more than with the out-of-view adult's display, suggesting that the older, but not younger, infants understood the referential link. In a second study of 12- and 18-month-olds, an adult displayed emotion while an infant and the adult looked at one of two toys. The adult directed an emotional display either toward the same toy to which the infant was attending, or toward a different toy. Regardless of whether the adult's attentional focus matched the infant's, all infants (12- and 18-month-olds) displayed more positive affect toward the toy, examined the toy more, and looked more frequently at the toy when the adult displayed positive emotion toward it than when the adult displayed negative emotion toward it. This finding suggests that even 12-month-olds were able to determine the focus of the adult's emotional display.

Moses et al.'s (2001) study focused on infants' understanding of the referential intent behind an adult's outburst—that is, of what the emotion was directed toward. Another important aspect of association is the ability to maintain this referential information after it has been received. If the goal of social information gathering is that infants learn something about the positive or negative attributes of objects, it is important for them to maintain that learning beyond the original learning context, in order to apply it in future situations. In another study investigating association in social referencing, the authors examined infants' ability to maintain information about a novel stimulus over a delay. Hertenstein and Campos (2004) exposed 11- and 14-month-old infants to emotional messages about novel objects, and examined their behavior after a delay. The 14-month-olds, but not the 11-month-olds, regulated their behavior in a way that was consistent with the emotional message (e.g., they were slower to touch stimuli related to negative emotion, but approached stimuli associated with positive emotion) after the delay interval. This finding suggests that, at least for 14-month-old infants, some ability to associate emotional messages with stimuli and to maintain that association is apparent and lasts at least 60 minutes. However, this study did not directly measure infants' associative abilities; rather, it measured infants' ability to use those representations to regulate behavior. Further research should focus on directly measuring infants' associative abilities and their role in social referencing. One way of doing this would be to incorporate measures of attention and memory into assessments of social referencing (see below).

Regulation of Behavior and Emotion

In addition to seeking information, recognizing emotions, and associating those emotions with novel stimuli, the final component involved in social referencing is behavior regulation. Studies in which adults have been asked to regulate their emotions provide recent evidence that frontal areas are involved in this function. For example, anticipation of a reward is associated with activation of a circuit including the nucleus accumbens and medial prefrontal cortex in both adults (Knutson, Taylor, Kaufman, Peterson, & Glover, 2005) and adolescents (Bjork et al., 2004; see Ernst & Spear, Chapter 17, this volume, for a discussion of brain reward systems). In these studies, participants received a cue that was predictive of a later reward, and pressed a button in response to an unrelated target. The critical measure was brain activation in response to cues that indicated varying magnitude and probability of reward. The nucleus accumbens in particular was associated with anticipating of the size of a monetary reward, whereas the medial prefrontal cortex was related to anticipating of the probability of reward. Although similar areas of activation were seen in adults and adolescents in these studies, areas associated with reward anticipation were less activated in adolescents than in adults. No differences were seen in responses once the reward had been granted. This study makes clear that there are long-term changes in brain circuits related to anticipating an emotional "payoff," and that these circuits do not completely reach adult-like levels of functioning even by adolescence.

In another study (Ochsner et al., 2004), participants were asked to regulate emotion by attributing it to themselves ("How does this picture make you feel?") or to a different person ("How does the person depicted feel?"). Areas in the medial prefrontal cortex and left temporal cortex were selectively activated when emotion was attributed to oneself, but attributing emotion to another (which should serve to regulate an observer's emotion) was related to different, more lateral areas in prefrontal cortex, as well as medial occipital cortex. Thus the available brain data suggest that prefrontal cortex areas are integrally involved in regulating one's behavior and emotion, and that development in these areas continues well into adolescence. However, studies of behavior make clear that some ability to regulate behavior is apparent before frontal systems are fully developed.

Individual differences in emotion regulation abilities are also reflected in differences in brain activation. In one study, individuals who differed in their tendency to focus on the negative when considering the self or life interpretations (i.e., trait rumination) were asked to increase or decrease their negative affect in response to negative and neutral images (Ray et al., 2005). Participants regulated behavior by either relating to the pictures personally (e.g., imagining themselves or someone close to them as the figure depicted in the photo) or relating to the pictures as external witnesses. When asked to increase their negative affect, individuals who tended to show more negative trait rumination exhibited more activation in the left amygdala and left prefrontal cortex to neutral pictures. In response to negative pictures, greater rumination was related to right amygdala activation. These results suggest that individual differences in self-directed emotional behavior can interact with emotion regulation systems in the brain.

One problem that is readily apparent in attempting to understand the neurodevelopmental trajectory of emotion regulation is the lack of developmental data in the area. There is a substantial literature on the development of emotion regulation behaviors in older children, including effects of parents, schools, and other influences on regulatory skills, and even some speculation about brain mechanisms that may underlie such development (see Zeman, Cassano, Perry-Parrish, & Stegall, 2006, for a review). However, few developmental studies have attempted to measure the neural correlates of emotion regulation directly. In one of the few studies to do so, Lewis, Lamm, Segalowitz, Stieben, and Zelazo (2006) tested 5- to 16-year-olds on a task requiring inhibition of responses in a go/no-go paradigm. Children were motivated to perform well on the task in order to win a desirable prize. Their emotion was then manipulated by the inclusion of a phase in the game where it appeared that they were going to "lose" and end up with a less desirable prize (at the end of the study, feedback was changed so that all children received positive feedback and were given the desirable prize). Younger children especially exhibited impulsive response patterns during the emotion induction block. Using ERPs, the authors also measured brain responses during the different phases of the task. The N2 component of the ERP, associated with response inhibition, varied with emotion induction for the older children in the sample, suggesting that they increased the effort with which they regulated emotion. Source localization implicated the anterior prefrontal cortex in emotion regulation, especially in the older children, who showed brain activity patterns consistent with effortful emotion regulation.

Although Lewis et al.'s (2006) study provides evidence for age-related changes in emotion regulation, additional work is necessary to elucidate the full trajectory of these changes. For instance, research on the neural correlates of emotion regulation in infancy could provide additional insight into the mechanisms underlying the development of this skill. Such infant work has the potential to be especially informative, given the vast extent of prefrontal cortical development during the first years of life.

Interaction among Elements Involved in Social Referencing

We and our colleagues have begun to investigate the development of the proposed social referencing system by measuring referencing, association, and brain activity in infancy. In contrast to previous research, the goal of these studies has been to investigate social referencing beyond looking at infants' ability to regulate behavior in response to adults' emotion. That is, we aim to clarify the maturational patterns associated with all aspects of social referencing, as well as the interactions among them. In combination with behavioral measures, we utilize ERPs to tap into the underlying neural processing. Carver and Vaccaro (2007) developed a paradigm in which they observed infants' referencing behavior and emotion regulation behaviorally, and measured infants' association between objects and emotion through ERPs. The first part of this study was very similar to previous research on social referencing in infancy: Infants were shown novel, ambiguous stimuli that elicited social referencing on pilot testing. In response to joint attention bids (alternating looks between the novel objects and an adult—either an infant's caregiver or the experimenter), infants were provided with negative (disgust), positive (happy), or neutral information about each object. Variables of interest included the amount of time it took infants to look at an adult (a caregiver or an experimenter) after the presentation of the novel stimulus; behavioral responses to the novel object (attention, reaching, touching, looking at, etc.); and behaviors directed toward the caregiver (approaching, looking, etc).

After infants were exposed to the novel objects and associated emotional messages, Carver and Vaccaro (2007) measured their brain activity in response to pictures of the objects. Twelve-month-old infants showed larger amplitudes of the Nc to stimuli that had been associated with adults' negative emotional expressions. This result suggests that for 12-month-olds, negative emotion increases the attentional salience of the objects. In addition, infants' brain activity in response to the negatively tagged object was correlated with several aspects of both their referencing behavior and emotional and behavioral regulation, whereas those two behavioral measures were weakly if at all related to each other. Even though infants' referencing behavior and emotion regulation were measured in the same setting and at the same time, their brain activity, which was measured separately and in a different room, was more strongly associated with both information seeking and regulation than they were with each other. This study raised several important questions regarding what mechanisms are at play in infants' referencing and emotion regulation behavior at this age, as infants' brain activity was inconsistent with behaviors that have often been reported in social referencing situations. They attended *more* to the object associated with negative emotion, whereas most behavioral studies have found an avoidance of negatively tagged stimuli. In addition, if infants were seeking emotional information *so that* they could regulate their behavior, we would predict a strong relationship between information seeking and regulation, which was not seen.

To begin to examine these issues, we have since conducted a study using the identical paradigm with 18-month-old infants (Cornew, Dobkins, Akshoomoff, McCleery, & Carver, 2008). Preliminary results from this study suggest several notable differences in brain activity, behavior, information seeking, and the relations among them. In contrast to the pattern seen in 12-month-olds in the Carver and Vaccaro (2007) study, 18-month-olds' ERP responses showed the greatest amplitudes to pictures of stimuli associated with neutral emotion. The next largest amplitudes were to pictures of stimuli in the negative condition, and the smallest amplitudes were observed in the positive condition. One possible interpretation of these data follows from studies of emotion processing in older children. As reviewed above, until at least age 10 or so, children appear to process emotionally neutral information and negative information similarly. Children's brain activation, especially in the amygdala, in response to neutral faces is similar to (and may even exceed) activation to negative faces (Thomas et al., 2001). In addition, Thomas and colleagues recorded children's identification of fearful versus neutral faces, and found that children were less accurate at identifying neutral than fearful faces. Thus 18-month-olds may be viewed as showing a more mature (or at least a more "older-child-like") pattern of responding to emotion than younger infants. This suggests that something other than true emotional understanding drives 12-month-olds' emotion in a social referencing context. Perhaps 12-month-olds' behavior is driven primarily by emotional contagion, whereas older infants use a more sophisticated understanding of what the emotions mean to regulate their behavior.

In addition, infants' overall patterns of emotion regulation were more robust then in the Carver and Vaccaro (2007) study: Whereas 12-month-olds' affect primarily changed by increasing in negativity following negative adult emotion, and failed to differentiate between positive emotion and either negative or neutral emotion, 18-month-olds showed an increase in positivity after positive emotion displayed by adults, compared to both negative and neutral emotion (see Figure 7.2). One possible explanation for this difference in response patterns is that whereas 12-month-olds primarily respond to highly salient negative emotion, they may not fully understand the meaning of that emotion. Data from studies of locomotor development suggest that with increasing independent locomotion, infants are exposed to increasing levels of negative emotion. Infants who crawl, for example, have caregivers who are more likely to display negative emotion than the caregivers of noncrawlers are (Campos et al., 2000). Since infants have been mobile for a shorter time at 12 months than at 18 months, it is reasonable to expect that 12-month-olds are likely to see fewer displays of negative emotion than 18-month-olds. Thus, through the course of development, infants may increasingly understand the meanings of emotions as they are communicated, and may gradually come to use them to guide behavior because of an understanding of the emotional message. In addition, we propose that the ability to associate emotions with novel stimuli, which was strongly linked to behavior in 12-month-olds (Carver & Vaccaro, 2007), acts as a bridge to drive more sophisticated patterns of social referencing later in development.

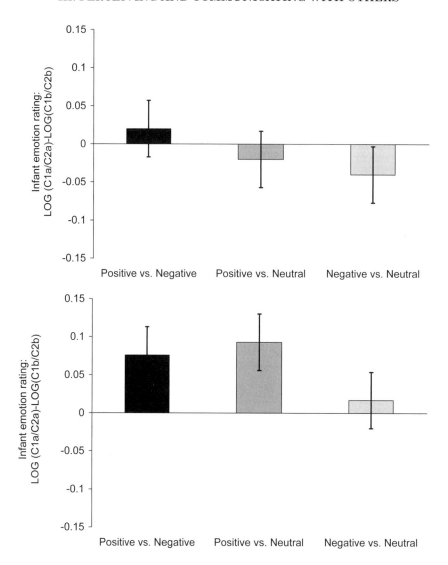

FIGURE 7.2. Log difference scores in infants' emotional displays in response to adult positive and negative (black bars), positive and neutral (striped bars), and negative and neutral (gray bars) emotion. Scores greater than zero indicate a change in infants' emotion from before the emotional signal in one condition (C1b: condition 1, before signal) to after the adult provides the emotional signal (C1a: condition 1, after signal), which is significantly different from a second emotional condition (C2b: condition 2, before signal; C2a: condition 2, after signal). Twelve-month-olds (top) showed a greater increase in negative affect following negative compared to neutral signals, but did not change their affect differentially between positive and either negative or neutral conditions. Eighteen-month-olds (bottom) showed a greater increase in positive emotion following positive compared to both negative and neutral signals, but there was no significant difference between negative and neutral conditions. Data in the top graph from Carver and Vaccaro (2007).

Summary of Our Model and Future Directions

Our proposed model for the development of social referencing incorporates the development of each component that is important for sophisticated information seeking in infants. We propose that a complex neural system underlies social information gathering, and that the component parts of this system emerge separately and gradually through the course of development. Early-maturing systems for responding to joint attention and associating stimuli with one another first support a primitive referencing system maintained primarily by posterior attention networks (Mundy et al., 2000). In this early system, emotion regulation is driven primarily by emotional contagion, rather than by a full-fledged understanding of the meaning of emotion. With repeated experiences of sharing attention, forming associations between others' emotions and novel stimuli, and automatic regulation, infants come to initiate joint attention; this leads to voluntary social referencing, or information-seeking behavior. Also with repeated experience in associating stimuli with emotions, infants come to understand the meaning behind emotional signals. Through this course of development, the system comes to drive intentional, voluntary social information seeking.

Taken together, our results suggest that there is significant development in social referencing, the elements that constitute it, and the relations among them between 12 and 18 months of age. Most of the links described in the model presented here are hypothetical. Further understanding of the mechanisms behind development in social information seeking will require longitudinal studies that track the emergence of these behaviors, their neural bases, and their relations to one another over time as development progresses.

References

Adolphs, R., Tranel, D., Hamann, S., Young, A. W., Calder, A. J., Phelps, E. A., et al. (1999). Recognition of facial emotion in nine individuals with bilateral amygdala damage. *Neuropsychologia, 37,* 1111–1117.

Alter, K., Rank, E., Kotz, S. A., Toepel, U., Besson, M., Schirmer, A., et al. (2003). Affective encoding in the speech signal and in event-related brain potentials. *Speech Communication, 40,* 61–70.

Balconi, M., & Pozzoli, U. (2003). Face-selective processing and the effect of pleasant and unpleasant emotional expressions on ERP correlates. *International Journal of Psychophysiology, 49,* 67–74.

Baldwin, D. A., & Moses, L. J. (1996). The ontogeny of social information gathering. *Child Development, 67,* 1915–1939.

Barrera, M. E., & Maurer, D. (1981). The perception of facial expressions by the three-month-old. *Child Development, 52,* 203–206.

Batty, M., & Taylor, M. J. (2003). Early processing of the six basic facial emotional expressions. *Cognitive Brain Research, 17,* 613–620.

Bjork, J. M., Knutson, B., Fong, G. W., Caggiano, D. M., Bennett, S. M., & Hommer, D. W. (2004). Incentive-elicited brain activation in adolescents: Similiarities and differences from young adults. *Journal of Neuroscience, 24,* 1793–1802.

Bornstein, M. H., & Arterberry, M. E. (2003). Recognition, discrimination and categorization of smiling by 5-month-old infants. *Developmental Science, 6,* 585–599.

Bostanov, V., & Kotchoubey, B. (2004). Recognition of affective prosody: Continuous wavelet measures of event-related brain potentials to emotional exclamations. *Psychophysiology, 41,* 259–268.

Bowers, D., Coslett, H. B., Bauer, R., Speedie, L., & Heilman, K. (1987). Comprehension of emotional prosody following unilateral hemispheric lesions: Processing defect versus distraction defect. *Neuropsychologia, 25,* 317–328.

Brooks, R., & Meltzoff, A. N. (2002). The importance of eyes: How infants interpret adult looking behavior. *Developmental Psychology, 38*, 958–966.

Brooks, R., & Meltzoff, A. N. (2005). The development of gaze following and its relation to language. *Developmental Science, 8*, 535–543.

Buchanan, T. W., Lutz, K., Mirzazade, S., Specht, K., Shah, N. J., Zilles, K., et al. (2000). Recognition of emotional prosody and verbal components of spoken language: An fMRI study. *Cognitive Brain Research, 9*, 227–238.

Campos, J. J., Anderson, D. I., Barbu-Roth, M. A., Hubbard, E. M., Hertenstein, M. J., & Witherington, D. (2000). Travel broadens the mind. *Infancy, 1*(2), 149–219.

Carver, L. J., & Vaccaro, B. G. (2007). Twelve-month-old infants allocate increased neural resources to stimuli associated with negative parental expressions. *Developmental Psychology, 43*, 54–69.

Cornew, L., Dobkins, K., Akshoomoff, N. A., McCleery, J., & Carver, L. J. (2008). [Social referencing in infants at risk for autism: Evidence for a broader autism phenotype at 18 months of age.] Unpublished raw data.

D'Entremont, B., Hains, S. M. J., & Muir, D. W. (1997). A demonstration of gaze following in 3- to 6-month-olds. *Infant Behavior and Development, 20*, 569–572.

D'Entremont, B. A., & Morgan, R. (2006). Experience with visual barriers and its effects on subsequent gaze-following in 12- to 13-month-olds. *British Journal of Developmental Psychology, 24*, 465–475.

Eimer, M., & Holmes, A. (2002). An ERP study on the time course of emotional face processing. *NeuroReport, 13*, 427–431.

Feinman, S., Roberts, D., Hsieh, K.-F., Sawyer, D., & Swanson, D. (1992). A critical review of social referencing in infancy. In S. Feinman (Ed.), *Social referencing and the social construction of reality in infancy* (pp. 15–54). New York: Plenum Press.

Flavell, J. H. (2000). Development of children's knowledge about the mental world. *International Journal of Behavioral Development, 24*, 15–23.

Flavell, J. H., Mumme, D. L., Green, F. L., & Flavell, E. R. (1992). Young children's understanding of different types of beliefs. *Child Development, 63*, 960–977.

Fletcher, P. C., Happé, F., Frith, U., Baker, S. C., Dolan, R. J., Frackowiak, R. S. J., et al. (1995). Other minds in the brain: A functional imaging study of "theory of mind" in story comprehension. *Cognition, 57*, 109–128.

Gallagher, H. L., & Frith, C. D. (2003). Functional imaging of 'theory of mind.' *Trends in Cognitive Sciences, 7*, 77–83.

Gallagher, H. L., Happé, F., Brunswick, N., Fletcher, P. C., Frith, U., & Frith, C. D. (2000). Reading the mind in cartoons and stories: An fMRI study of 'theory of mind' in verbal and nonverbal tasks. *Neuropsychologia, 38*, 11–21.

George, M. S., Parekh, P., Rosinsky, N., Ketter, T. A., Kimbrell, T. A., Heilman, K. M., et al. (1996). Understanding emotional prosody activates right hemisphere regions. *Archives of Neurology, 53*, 665–670.

Grossmann, T., Striano, T., & Friederici, A. D. (2005). Infants' electric brain responses to emotional prosody. *NeuroReport, 16*, 1825–1828.

Grossmann, T., Striano, T., & Friederici, A. (2006). Crossmodal integration of emotional information from face and voice in the infant brain. *Developmental Science, 9*, 309–315.

Heilman, K. M., Scholes, R., & Watson, R. T. (1975). Auditory affective agnosia: Disturbed comprehension of affective speech. *Journal of Neurology, Neurosurgery and Psychiatry, 38*, 69–72.

Henke, K., Buck, A., Weber, B., & Wieser, H. G. (1997). Human hippocampus establishes associations in memory. *Hippocampus, 7*, 249–256.

Herbert, J., Eckerman, C., & Stanton, M. (2003). The ontogeny of human learning in delay, long-delay, and trace eyeblink conditioning. *Behavioral Neuroscience, 117*, 1196–1210.

Hertenstein, M. J., & Campos, J. J. (2004). The retention effects of an adult's emotional displays on infant behavior. *Child Development, 75,* 595–613.

Joseph, R. (2000). Fetal brain behavior and cognitive development. *Developmental Review, 20,* 81–98.

Kestenbaum, R., & Nelson, C. A. (1990). The recognition and categorization of upright and inverted emotional expressions by 7-month-old infants. *Infant Behavior and Development, 13,* 497–511.

Knutson, B., Taylor, J., Kaufman, M., Peterson, R., & Glover, G. (2005). Distributed neural representation of expected value. *Journal of Neuroscience, 25,* 4806–4812.

Leppanen, J. M., Moulson, M. C., Vogel-Farley, V. K., & Nelson, C. A. (2007). An ERP study of emotional face processing in the adult and infant brain. *Child Development, 89,* 232–245.

Lewis, M. D., Lamm, C., Segalowitz, S. J., Stieben, J., & Zelazo, P. D. (2006). Neurophysiological correlates of emotion regulation in children and adolescents. *Journal of Cognitive Neuroscience, 18,* 430–443.

Liu, D., Sabbagh, M. A., Gehring, W. J., & Wellman, H. M. (2004). Decoupling beliefs from reality in the brain: An ERP study of theory of mind. *NeuroReport, 15,* 991–995.

Liu, D., Sabbagh, M. A., Gehring, W. J., & Wellman, H. M. (in press). Neural correlates of children's theory of mind development. *Child Development.*

Ludemann, P. M., & Nelson, C. A. (1988). Categorical representation of facial expressions by 7-month-old infants. *Developmental Psychology, 24,* 492–501.

Meltzoff, A. N. (1995). Understanding the intentions of others: Re-enactment of intended acts by 18-month-old children. *Developmental Psychology, 31,* 838–850.

Montague, D. P. F., & Walker-Andrews, A. S. (2001). Peekaboo: A new look at infants' perception of emotion expressions. *Developmental Psychology, 37,* 826–838.

Moore, C., & Corkum, V. (1994). Social understanding at the end of the first year of life. *Developmental Review, 14,* 349–372.

Morris, J. S., Friston, K. J., Büchel, C., Frith, C. D., Young, A. W., Calder, A. J., et al. (1998). A neuromodulatory role for the human amygdala in processing emotional facial expressions. *Brain, 121,* 47–57.

Morris, J. S., Scott, S. K., & Dolan, R. J. (1999). Saying it with feeling: Neural responses to emotional vocalizations. *Neuropsychologia, 37,* 1155–1163.

Moses, L. J., Baldwin, D. A., Rosicky, J. G., & Tidball, G. (2001). Evidence for referential understanding in the emotions domain at twelve and eighteen months. *Child Development, 72,* 718–735.

Mumme, D. L., Fernald, A., & Herrera, C. (1996). Infants' responses to facial and vocal emotional signals in a social referencing paradigm. *Child Development, 67,* 3219–3237.

Mundy, P., Card, J., & Fox, N. A. (2000). EEG correlates of the development of infant joint attention skills. *Developmental Psychobiology, 69,* 325–338.

Nelson, C. A. (1993). The recognition of facial expressions in infancy: Behavioral and electrophysiological evidence. In B. de Boysson-Bardies, S. de Schonen, P. Jusczyk, & P. MacNeilage (Eds.), *Developmental neurocognition: Speech and face processing in the first year of life* (pp. 187–198). Dordrecht, The Netherlands: Kluwer Academic.

Nelson, C. A., & de Haan, M. (1996). Neural correlates of infants' visual responsiveness to facial expressions of emotion. *Developmental Psychobiology, 29,* 577–595.

Ochsner, K. N., Knierim, K., Ludlow, D. H., Hanelin, J., Ramachandran, T., Glover, G., et al. (2004). Reflecting upon feelings: An fMRI study of neural systems supporting the attribution of emotion to self and other. *Journal of Cognitive Neuroscience, 16,* 1746–1772.

Pell, M. D. (1998). Recognition of prosody following unilateral brain lesion: Influence of functional and structural attributes of prosodic contours. *Neuropsychologia, 36,* 701–715.

Phillips, M. L., Fahy, T., David, A. S., & Senior, C. (1998). Disgust: The forgotten emotion of psychiatry. *British Journal of Psychiatry, 172,* 373–375.

Phillips, M. L., Young, A. W., Senior, C., Brammer, M., Andrews, C., Calder, A. J., et al. (1997). A specific neural substrate for perceiving facial expressions of disgust. *Nature, 389*(6650), 495–498.

Phillips, R. D., Wagner, S. H., Fells, C. A., & Lynch, M. (1990). Do infants recognize emotion in facial expressions?: Categorical and "metaphorical" evidence. *Infant Behavior and Development, 13,* 71–84.

Pihan, H., Altenmüller, E., & Ackermann, H. (1997). The cortical processing of perceived emotion: A DC-potential study on affective speech prosody. *NeuroReport, 8,* 623–627.

Ray, R. D., Ochsner, K. N., Cooper, J. C., Robertson, E. R., Gabrieli, J. D. E., & Gross, J. J. (2005). Individual differences in trait rumination and the neural systems supporting cognitive reappraisal. *Cognitive, Affective, and Behavioral Neuroscience, 5,* 156–168.

Repacholi, B. M., & Gopnik, A. (1997). Early reasoning about desires: Evidence from 14- and 18-month-olds. *Developmental Psychology, 33,* 12–21.

Ross, E. D., Thompson, R. D., & Yenkosky, J. (1997). Lateralization of affective prosody in brain and the callosal integration of hemispheric language functions. *Brain and Language, 56,* 27–54.

Rueda, M. R., Posner, M. I., & Rothbart, M. K. (2005). The development of executive attention: Contributions to the emergence of self-regulation. *Developmental Neuropsychology, 28,* 573–594.

Sabbagh, M. A., & Taylor, M. J. (2000). Neural correlates of the theory-of-mind reasoning: An event-related potential study. *Psychological Science, 11,* 46–50.

Sambataro, F., Dimalta, S., Di Giorgio, A., Taurisano, P., Blasi, G., Scarabino, T., et al. (2006). Preferential responses in amygdala and insula during presentation of facial contempt and disgust. *European Journal of Neuroscience, 24,* 2355–2362.

Sato, W., Kochiyama, T., Yoshikawa, S., & Matsumura, M. (2001). Emotional expression boosts early visual processing of the face: ERP recording and its decomposition by independent component analysis. *NeuroReport, 12,* 709–714.

Schwartz, G. M., Izard, C. E., & Ansul, S. E. (1985). The 5-month-old's ability to discriminate facial expressions of emotion. *Infant Behavior and Development, 8,* 65–77.

Scott, S. K., Young, A. W., Calder, A. J., Hellawell, D. J., Aggleton, J. P., & Johnson, M. (1997). Impaired auditory recognition of fear and anger following bilateral amygdala lesions. *Nature, 385,* 254–257.

Soken, N. H., & Pick, A. D. (1992). Intermodal perception of happy and angry expressive behaviors by seven-month-old infants. *Child Development, 63,* 787–795.

Soken, N. H., & Pick, A. D. (1999). Infants' perception of dynamic affective expressions: Do infants distinguish specific expressions? *Child Development, 70,* 1275–1282.

Sprengelmeyer, R., Young, A. W., Calder, A. J., Karnat, A., Lange, H., Homberg, V., et al. (1996). Loss of disgust: Perception of faces and emotions in Huntington's disease. *Brain, 119,* 1647–1665.

Stone, V. E., Baron-Cohen, S., Calder, A., Keane, J., & Young, A. (2003). Acquired theory of mind impairments in individuals with bilateral amygdala lesions. *Neuropsychologia, 41,* 209–220.

Stone, V. E., Baron-Cohen, S., & Knight, R. T. (1998). Frontal lobe contributions to theory of mind. *Journal of Cognitive Neuroscience, 10,* 640–656.

Striano, T., Reid, V. M., & Hoehl, S. (2006). Neural mechanisms of joint attention in infancy. *European Journal of Neuroscience, 23,* 2819–2823.

Striano, T., & Stahl, D. (2005). Sensitivity to triadic attention in early infancy. *Developmental Science, 8,* 333–343.

Thomas, K. M., Drevets, K. M., Whalen, P. J., Eccard, C. H., Dahl, R. E., Ryan, N. D., et al. (2001). Amygdala response to facial expression in children and adults. *Biological Psychiatry, 49,* 309–316.

Van Hecke, A. V., & Mundy, P. (2007). Neural systems and the development of gaze following and related joint attention skills. In R. A. Flom, K. Lee, & D. W. Muir (Eds.), *Gaze-following: Its development and significance* (pp. 17–51). Mahwah, NJ: Erlbaum.

Van Lancker, D., & Sidtis, J. J. (1992). The identification of affective-prosodic stimuli by left- and right-hemisphere-damaged subjects: All errors are not created equal. *Journal of Speech and Hearing Research, 35*, 963–970.

Walker-Andrews, A. S. (1997). Infants' perception of expressive behaviors: Differentiation of multimodal information. *Psychological Bulletin, 121*, 437–456.

Wellman, H. M., & Liu, D. (2004). Scaling of theory-of-mind tasks. *Child Development, 75*, 523–541.

Wildgruber, D., Hertrich, I., Riecker, A., Erb, M., Anders, S., Grodd, W., et al. (2004). Distinct frontal regions subserve evaluation of linguistic and emotional aspects of speech intonation. *Cerebral Cortex, 14*, 1384–1389.

Zeman, J., Cassano, M., Perry-Parrish, C., & Stegall, S. (2006). Emotion regulation in children and adolescents. *Journal of Developmental and Behavioral Pediatrics, 27*, 155–168.

CHAPTER 8

Imitation as a Stepping Stone to Empathy

JEAN DECETY
MEGHAN MEYER

The study of both imitation and empathy was traditionally reserved for developmental and social psychologists. However, in the past decade, progress in cognitive neuroscience has begun to identify the neural circuits underlying these phenomena. Psychologists, clinicians, and neuroscientists all emphasize the social importance of imitation and empathy in day-to-day social interactions, highlighting their role in the development of theory of mind and intersubjectivity. However, these literatures largely overlook the tie, both phenomenological and neurological, between imitation and empathy. The goal of this chapter is to bring together findings from developmental science and cognitive neuroscience on the topic of imitation and empathy. We draw from findings on typically developing humans, as well as on individuals with autism spectrum disorders (ASD), as these disorders are characterized by deficits in both imitation and empathy (see Dawson, Sterling, & Faja, Chapter 22, this volume, for additional discussion of ASD). We also draw to a lesser extent from findings on individuals with conduct problems and persons with schizophrenia.

In this chapter, we examine the behavior of imitation and empathy within the neuropsychological framework of shared representations between the self and other. Decety (2006, 2007) has proposed a model of empathy that draws on bottom-up processing of shared motor representations, which are navigated by parietal areas known to be crucial in differentiating self- and other-perspectives; all of these can be regulated by top-down executive processing, in which the perceiver's motivation, intentions, and self-regulation influence the extent of an empathic experience (see Plate 8.1).

We place imitation within this framework, because it is also proposed to be grounded in shared motor representations between self and other (Meltzoff & Decety, 2003), as well as regulated by executive functions (Decety, 2006). Moreover, imitation has been theorized to provide a basis for the child's developing sense of agency, self, and self–other differentiation, which are also characteristics involved in empathy. Thus imitation and empathy are closely linked phenomenological experiences, and we believe it is time to address the extent to which they are related. Imitation and empathy are not redundant psychological constructs characterizing the same neurological process. They are instead partially distinct, though interrelated. Studying the development of these two abilities contributes to neuroscience research by providing theoretical and ecologically valid perspectives.

However, although we demonstrate in the following sections that imitation and empathy rely in part on commensurate neural bases, they are not entirely analogous. At the phenomenological level, we propose that there is a distinction between imitation and empathy. However, we demonstrate that motor mimicry, which is unconscious and automatic, plays a key role in the development of emotion sharing, which represents the primary aspect of empathy. Of course, imitation can be more than unconscious mimicry. The second, higher-level layer of imitation is termed here "intentional imitation," which is conscious and not automatic. Interestingly, this level of imitation also relates to components of empathy—that is, the sense of agency and perspective taking. Theorists purport that intentional imitation provides the basis for both the imitator's and imitatee's sense of agency (i.e., ownership over actions). In other words, when one child intentionally imitates another, doing so fosters both children's sense of who is directing whose action. This sense of agency is necessary for developing and disentangling a sense of self- and other-awareness during an empathic experience. In turn, an individual can better gauge the perspective of the observed target.

A Definition of Imitation

Theorists of human imitation acknowledge that it is an adaptive way to learn skills and develop a sense of self; an important form of communication; and a milestone in the development of intentional communication in particular, and of social interaction more generally (e.g., Rogers, 1999). Despite the fact that there are many definitions of "imitation" and many disagreements over these definitions, most scholars agree that it is a natural mechanism involving perception and action coupling for mapping one's own behavior onto the behavior observed in others, and that it has great adaptive value. However, such a minimal definition of imitation does not fully account for its role in human social development. In addition, in humans, imitation is likely to depend on the observer's motivation to attend to the model's behavior.

In this chapter, rather than considering imitation as a simple matching mechanism, we view it as a complex construct—one that includes such subcomponents as perception–action coupling, visual attention, short-term memory, body schema, mental state attribution, and agency. These components rely on distributed network connectivity, as demonstrated by several brain imaging studies that are discussed here. Each component of the network computes different aspects of imitative behavior, and together the network orchestrates the task. The role of neuroscience is to help identify the mechanisms responsible for imitation and decipher the contribution of each of its components. However, the various forms of imitation—ranging from copying a movement after seeing it performed to reproducing an action intentionally offline—may well constitute a continuum from simple acts to complex ones, from unconscious mimicry to intentional reproduction, and from familiar actions to novel actions. This view of imitation is compatible with the idea of a hierarchy of mechanisms. In fact, we believe that major disagreements between theorists seem to arise from the level (or definition) of imitation employed (e.g., simple vs. complex actions, immediate vs. differed imitation). But again, one may view these various behaviors on a continuum that ranges from response facilitation (i.e., the automatic tendency to reproduce an observed movement) to learning a new behavior that is not present in the motor repertoire. We argue that the basic neural mechanism unifying this variety of phenomena relies on the direct perception–action coupling mechanism, also called the "mirror system." Such a system provides a key foundation for the building of imitative and mind-reading competencies. However, additional computational mechanisms are necessary to account for their full maturation.

A Definition of Empathy

Like "imitation," "empathy" is a loaded term with various definitions in the literature. Broadly construed, "empathy" has been defined as an affective response stemming from the understanding of another's emotional state or condition, and similar to what the other person is feeling or would be expected to feel in the given situation (Eisenberg, Shea, Carlo, & Knight, 1991). In line with this conception, empathy can represent an interaction between two individuals, with one experiencing and sharing the feeling of the other. Other theorists more narrowly define empathy as one specific set of congruent emotions—those feelings that are more other-focused than self-focused (Batson, Fultz, & Schoenrade, 1987).

In the following sections, we discuss the affective and cognitive components of empathy. We first review the automatic proclivity to share emotions with others, and the cognitive processes of perspective taking and executive control, which allow individuals to be aware of their intentions and feelings and to maintain separate perspectives on self and other. Here we consider the construct of empathy within an overarching conceptual framework, in which empathy involves parallel and distributed processing in a number of dissociable computational mechanisms (see Plate 8.1). Shared neural circuits, self-awareness, mental flexibility, and emotion regulation constitute the basic macrocomponents of empathy; these are mediated by specific neural systems, including aspects of the prefrontal cortex, the anterior insula, and frontoparietal networks. Consequently, this model assumes that dysfunction in any of these macrocomponents may lead to an alteration of empathic behavior and produce a particular social disorder, depending on which aspect is disrupted.

It should be noted that the experience of empathy does not necessarily entail imitation or mimicry. However, if imitation takes place, it can be a stepping stone in the development of empathy. Full empathic experiences do not mature until 2–4 years of age—a time in which executive functions rapidly develop. However, the initial basis of empathy is emotion sharing, which can be observed in infancy. Emotion sharing stems from motor mimicry (i.e., unconscious imitation). Complex imitation, which involves intentional and conscious behavior matching, is theoretically tied to developing a sense of agency. Agency plays a role in empathy, as it facilitates the differentiation of self-executive from observed behaviors: Knowing whose actions (including emotions) belong to whom preserves individuals from overidentifying with the observed target, which would otherwise lead to empathic distress.

Thus imitation and empathy are intimately intertwined. If emotion sharing fails to take place, developmental disorders follow suit, as can be observed in individuals with antisocial personality disorder or in those with ASD (Decety, Jackson, & Brunet, 2007).

Emotion Sharing

The Automaticity of Emotion Sharing

The basic building block of empathy is emotion sharing, and this process is facilitated by motor mimicry (i.e., a form of unconscious and automatic mirroring of the other person's behavior). Bodily expressions help humans and other animals communicate various types of information to members of their species.

Emotional expression not only informs an individual of another's subjective (and physiological) experience, but also serves as a means of maintaining emotional reciprocity among dyads and groups. "Emotional contagion"—defined as the tendency to automatically mimic and synchronize facial expressions, vocalizations, postures, and movements with those of another person, and consequently to converge emotionally with the other (Hatfield, Cacioppo, & Rapson, 1993)—is a social phenomenon of shared emotional expression that occurs at a basic level outside of conscious awareness.

Development of Motor Mimicry

In a classic developmental study, Meltzoff and Moore (1977) showed that infants less than 1 hour old can mimic human actions (e.g., tongue protrusion). The findings suggest that infants enter the world with an innate sociability, and that this predisposition is grounded in action–perception coupling of sensorimotor information of one's own behaviors with others' behavior. Such imitative behaviors led Trevarthen (1979) to propose that infants are endowed with an innate "primary subjectivity." That is, infants have access to others' emotional states via perceiving the other persons' physical actions and facial expressions. From infancy, complex facial motor patterns permit infants to match facial emotion expressions with those of others (e.g., Field, Woodson, Greenberg, & Cohen, 1982). Very young infants are able to send emotional signals and to receive and detect the emotional signals sent by others. Shortly after birth, healthy infants can convey facial expressions of interest, sadness, and disgust (Field, 1989). Likewise, discrete facial expressions of emotion have been identified in newborns, including joy, interest, disgust, and distress (Izard, 1982). These findings suggest that subcomponents of full emotional expressions are present at birth, supporting the possibility that these processes are hard-wired in the brain.

Developmental data suggest that the mechanism subserving emotion sharing between infant and caretaker is immediately present from birth (e.g., Rogeness, Cepeda, Macedo, Fischer, & Harris, 1990). Newborns are innately and highly attuned to other people and motivated to interact socially with others. From the earliest months of their lives, infants engage with other people and with the actions and feelings expressed through other people's bodies (e.g., Hobson, 2002). Such a mechanism is grounded in the automatic perception–action coupling of sensorimotor information, which seems hard-wired in the brain. This mimicry between self and other is critical for many facets of social functioning. For instance, it facilitates attachment by regulating one's own emotions and providing information about the other's emotional state.

Perception–Action Coupling Mechanism

The automatic mapping between self and other is supported by considerable empirical literature in the domain of perception and action, which has been marshaled under the prominent common-coding theory. This theory claims that somewhere in the chain of operation leading from perception to action, the system generates certain derivatives of stimulation and certain antecedents of action, which are commensurate in the sense that they share the same system of representational dimensions (Prinz, 1997). The core assumption of the common-coding theory is that actions are coded in terms of the perceivable effects (i.e., the distal perceptual events) they should generate. Performing a movement leaves behind a bidirectional association between the motor pattern it was generated by and the sensory effects that it produces. Such an association can then be used "backward" to retrieve a movement by anticipating its effects (Hommel, Musseler, Aschersleben, & Prinz, 2001).

Perception–action codes are also accessible during action observation, and perception activates action representations to the degree that the perceived and the represented actions are similar. Such a mechanism has also been proposed to account for emotion sharing and its contribution to the experience of empathy (Preston & de Waal, 2002). In the context of emotion processing, it is posited that perception of emotions activates in the observer the neural mechanisms that are responsible for the generation of similar emotions. It should be noted that a similar mechanism was previously proposed to account for emotional contagion. Indeed, Hatfield, Cacioppo, and Rapson (1994) argue that people catch the emotions of others as a result of afferent feedback generated by elementary motor mimicry of others' expressive behavior, which produces a simultaneous matching emotional experience.

The motor mimicry involved in emotional contagion is supported by research using facial electromyography (EMG). In one study, participants were exposed very briefly (56 msec) to pictures of happy or angry facial expressions while EMG was recorded from their facial muscles (Sonnby-Borgström, Jönsson, & Svensson, 2003). The results demonstrated facial mimicry, despite the fact that the participants were unaware of the stimuli. Furthermore, this effect was stronger for the participants who scored higher on self-reports of empathy.

Another study (Niedenthal, Brauer, Halberstadt, & Innes-Ker, 2001) indicates that facial mimicry plays an imperative role in the processing of emotional expression. Participants watched one facial expression morph into another, and were asked to detect when the expression changed. Some participants were free to mimic the expressions, whereas others were prevented from imitating them by being required to hold a pencil laterally between their lips and teeth. Participants who were free to mimic detected the changes in emotional expression earlier and more efficiently for any facial expression than did participants who were prevented from imitating the expressions.

In neuroscience, direct evidence for this perception–action coupling comes from electrophysiological recordings in monkeys. Neurons with sensorimotor properties, known as "mirror neurons," have been identified in the ventral premotor and posterior parietal cortices. These mirror neurons fire during both goal-directed actions and observations of the same actions performed by another individual. Most mirror neurons show a clear congruence between the visual actions they respond to and the motor responses they code (Rizzolatti, Fogassi, & Gallese, 2001). These neurons are part of a circuit that reciprocally connects the posterior superior temporal gyrus (in which neurons respond to the sight of actions made by others), the posterior parietal cortex, and the ventral premotor cortex.

A mirror neuron system also seems to exist in humans. Numerous functional neuroimaging experiments indicate that the neural circuits involved in action representation (in the posterior parietal and premotor cortices) overlap with those activated when actions are observed (see Jackson & Decety, 2004, for a review). In addition, several neuroimaging studies have shown that a similar neural network is reliably activated during imagining one's own action, imagining another's action, and imitating actions performed by a model (Decety & Grèzes, 2006). Notably, a functional magnetic resonance imaging (fMRI) study found that similar cortical areas were engaged when individuals observed or imitated emotional facial expressions (Carr, Iacoboni, Dubeau, Mazziotta, & Lenzi, 2003). Within this network, there was greater activity during imitation than during observation of emotions in premotor areas including the inferior frontal cortex, as well as in the superior temporal cortex, insula, and amygdala. Such shared circuits reflect an automatic transformation of other people's behavior (actions or emotions) into the neural representation of one's own behavior, and provides a functional bridge between first- and third-person perspectives, culminating in empathic experience (Decety & Sommerville, 2003).

Evidence That Social Deficits Can Be Tied to Poor Perception–Action Coupling

As suggested above, an effective means to measure the role of perception–action coupling in emotion sharing is EMG recording of the activation of specific facial muscles in response to viewing other people's facial expressions. Facial mimicry has been defined narrowly as congruent facial reactions to the emotional facial displays of others (Hess & Blairy, 2001). More broadly construed, emotional contagion is an affective state that matches another person's emotional display. Thus facial mimicry can be conceived of as a physical manifestation of emotional contagion, and it occurs at an automatic level in response to viewing others' emotions (Bush, Barr, McHugo, & Lanzetta, 1989). As also mentioned above, motor mimicry and emotional contagion are analogous with basic imitation, which is also defined as the automatic and nonconscious replication of behavior.

Individuals with ASD are often reported to lack automatic mimicry of facial expressions. A recent study of adolescent and adult individuals with ASD and controls measured their automatic and voluntary mimicry of emotional facial expressions via EMG recordings of the cheek and brow muscle regions, while participants viewed still photographs of happy, angry, and neutral facial expressions (McIntosh, Reichmann-Decker, Winkielman, & Wilbarger, 2006). The cheek and brow muscles of individuals with ASD failed to activate in response to the videos, indicating that they did not automatically mimic the facial expressions, whereas the muscles of the normally developing controls showed activation. In attempt to examine a potential link between mirror neuron dysfunction and developmental delay of social-cognitive skills, one fMRI study found a lack of activation in the inferior frontal gyrus (a key mirror neuron area) in children with ASD as compared to controls during the observation and imitation of basic facial emotion expressions (Dapretto et al., 2006). Difficulties in mimicking other people's emotional expressions may prevent individuals with ASD from experiencing the afferent feedback that provides information of what others are feeling (Rogers, 1999).

The shared neural representations account suggests that problems with one's own motor or body schema may undermine capacities for understanding others. Consequently, it is possible that developmental problems involving sensorimotor processes may have an effect on the capabilities that make up "primary intersubjectivity," or the ability to react contingently to others' emotional expressions (Trevarthen & Aitken, 2001) and therefore to resonate emotionally with others. It thus seems plausible that the social and sensorimotor problems of individuals with ASD may in part reflect a disturbed motor representation matching system at the neuronal level. This speculation not only helps explain problems in primary intersubjectivity, but also other sensorimotor symptoms of autism (repetitious and odd movements, and possibly echolalia).

Self–Other Awareness and Agency: Mediating Whose Emotions Belong to Whom

The fact that the observation of an emotion elicits the activation of analogous motor representation in the observer raises this question: Why is there not complete overlap between internally generated and externally engendered motor representations? Indeed, developmental research highlights that infants map their own and others' actions commensurately. Although some evidence suggests that infants can discriminate their own from others' bodily states (Rochat & Striano, 2000) and emotional states (Martin & Clark, 1987), an inclination to merge concepts of the self and other persists in infancy. Indeed, affective sharing among infants highlights the overlap between self and other in infancy. These findings imply that the process of "translating" one's own actions into those of another is immediate, and that the perceptions of one's own and the other's perspectives are essentially co-occurring, not moving from the self to the other as previously believed. Indeed, inhibiting the self-perspective from overlapping with the other's perspective persists through adulthood, as adults tend to fail in regulating extension of first-person subjective states to third-person parties (Royzman, Cassidy, & Baron, 2003), and a large research effort is dedicated to discerning why egocentric bias persists.

Implications of intentional imitation may help account for how individuals determine who performs a given action and who is observing—and, accordingly, who is the possessor of a given subjective state and who is an observer. Childhood imitation is known to bring to conscious awareness what has been termed a sense of "agency." Researchers in neuroscience and developmental science use the term "agency" to describe the ability to recognize oneself as the agent of an action, thought, or desire, which is crucial for attributing a behavior to its proper agent. Through complex, intentional imitation, children learn that they are the initiators of their replicating behavior. Likewise, children who are intentionally imitated notice that their actions are being replicated by the imitators. Through this process, a child grasps a sense that the "other" is an intentional, goal-directed agent like the self, but that the other's actions are dissociable from the child's own, though easily replicated.

Although intentional imitation brings a sense of agency to the forefront of conscious awareness, an implicit agentive stance may be manifested early in infancy. From birth, infants learn to be effective in relation to objects and events. Within hours following birth, neonates can learn to suck in certain ways and to apply specific pressures on a dummy pacifier to hear their mothers' voices or see their mothers' faces (e.g., DeCasper & Fifer, 1980), suggesting that infants manifest a sense of themselves as agentive in the environment. Furthermore, by 2 months of age infants also show positive affect, such as smiling and other expressions of pleasure, when they accomplish causing an auditory and visual event (activating a music box by pulling a cord attached to a limb). When the cord is then furtively disconnected from the box, hindering infants' effectiveness, they switch expressions from pleasure to anger (Rochat & Striano, 2000).

This agentive capacity is critical for empathy: In a complete empathic experience, affective sharing must be modulated and monitored by the sense of whose feelings belong to whom (Decety & Jackson, 2004). Furthermore, self-awareness generally and agency in particular are crucial aspects in promoting a selfless regard for the other, rather than a selfish desire to escape aversive arousal.

In sum, the studies reviewed indicate not only that perception–action coupling, which leads to emotional expression, is hard-wired in the brain, but that a sense of self, agency, and self–other distinction emerge early in infancy as well. The propensity to perceive others as "like the self" could potentially be detrimental for the altruistic function of empathy; self–other merging causes personal distress, not prosocial helping behavior (e.g., Lamm, Batson, & Decety, 2007). A sense of agency, however, helps children discriminate self-produced actions from other-produced actions. Importantly, the development of agency in children is greatly bolstered by the emergence of intentional imitation.

Cognitive Neuroscience of Self–Other Awareness and Agency

One way in which cognitive neuroscience can contribute to the study of the self and others is to help conceptually define the distinct dimensions, aspects, and characteristics of processing of the self and others, to help address the potential separability or relatedness of each component part of self-processing. It has been proposed that nonoverlapping parts of the neural circuit mediating shared representations (i.e., the areas that are activated for self-processing and not for other-processing) generate a specific signal for each form of representation (Jeannerod, 1999). This set of signals involved in the comparison between self-generated actions and actions observed from others ultimately allows the attribution of agency. It has also been suggested that the dynamics of neural activation within the shared cortical network constitute an important aspect of distinguishing one's own actions from the actions of others (Decety & Jackson, 2004; Decety & Grèzes, 2006). Furthermore, the fact that the onset of the hemodynamic signal is earlier for self-processing than for other-processing (e.g., Jackson, Brunet, Meltzoff, & Decety, 2006a) can be considered as a neural signature of the privileged and readily accessible self-perspective.

Accumulating evidence from neuroimaging studies of both healthy individuals and psychiatric populations, as well as lesion studies of neurological patients, indicates that the right inferior parietal cortex, at the junction with the posterior temporal cortex (the temporoparietal junction, or TPJ), plays a critical role in the distinction between self-produced actions and actions generated by others (e.g., Jackson & Decety, 2004). The TPJ is a heteromodal association cortex; it integrates input from the lateral and posterior thalamus, as well as visual, auditory, somesthetic, and limbic areas. It has reciprocal connections to the prefrontal cortex and to the temporal lobes. Because of these anatomical characteristics, this region is a key neural locus for self-processing: It is involved in multisensory body-related information processing, as well as in the processing of phenomenological and cognitive aspects of the self (Blanke & Arzy, 2005). Its lesion can produce a variety of disorders associated with body knowledge and self-awareness, such as anosognosia, asomatognosia, and somatoparaphrenia. For instance, Blanke, Ortigue, Landis, and Seeck (2002) demonstrated that out-of-body experiences (i.e., the experience of dissociation of self from body) can be induced by electrical stimulation of the right TPJ.

In addition, several functional imaging studies point out the involvement of the right inferior parietal lobule in the process of agency. Attribution of action to another agent has been associated with specific increased activity in the right inferior parietal lobe. In one fMRI study, Farrer and Frith (2002) instructed participants to use a joystick to drive a circle along a T-shaped path. They were told that the circle would be driven either by themselves or by the experimenter. In the former case, subjects were requested to drive the circle, to be aware that they drove the circle, and thus to mentally attribute the action seen on the screen to themselves. In the latter case, they were also requested to perform the task, but they were aware that the experimenter drove the action seen on the screen. The results showed that being aware of causing an action was associated with activation in the anterior insula, whereas being aware of not causing the action and attributing it to another person was associated with activation in the right inferior parietal cortex.

Interestingly, individuals experiencing incorrect agency judgments feel that some outside force is creating their own actions. One neuroimaging study found hyperactivity in the right inferior parietal lobule when patients with schizophrenia experienced alien control during a movement selection task (Spence et al., 1997). Delusions of control may arise because of a disconnection between frontal brain regions (where actions are initiated) and parietal regions (where the current and predicted states of limbs are represented).

Another study used a device that allowed modifying a participant's degree of control over the movements of a virtual hand presented on a screen (Farrer et al., 2003). Experimental conditions varied the degree of distortion of the visual feedback provided to the participants about their own movements. Results demonstrated a gradation in hemodynamic activity of the right inferior parietal lobule that paralleled the degree of mismatch between the executed movements and the visual reafference. Strikingly, such a pattern of neural response was not detected in individuals with schizophrenia who were scanned during the same procedure (Farrer et al., 2004). Instead, an aberrant relationship between these individuals' degree of control over the movements and the hemodynamic activity was found in the right inferior parietal cortex, and no modulation in the insular cortex.

The right inferior parietal cortex was also found to be activated in a study where participants mentally simulated actions from someone else's perspective but not from their own (Ruby & Decety, 2001). Similarly, this region was specifically involved when participants imagined how another person would feel in everyday life situations that elicit social emotions (Ruby & Decety, 2004) or in painful experiences (e.g., Jackson et al., 2006a; Lamm et al., 2007), but not when they imagined these situations for themselves. Such findings point to the similarity of the neural mechanisms that account for the correct attribution of actions, emotions, pain, and thoughts to their respective agents when one mentally simulates actions for oneself or for another individual. Furthermore, they support a role for the right TPJ not only in mental state processing, but also in lower-level processing of socially relevant stimuli (see Decety & Lamm, 2007, for a meta-analysis on the role of the TPJ in social interaction).

Empathy Demands Top-Down Regulation of the First-Person Perspective

Given the shared representations of one's own emotional states and those of others, as well as the similarities in brain circuits during first- and third-person perspective taking, it would seem difficult not to experience emotional distress while viewing another's distressed state—

and personal distress does not contribute to the empathic process (Batson et al., 2003). Indeed, distress in the self may hinder one's inclination to soothe. Interestingly, empathic distress can be traced to infancy, as 1-day-old infants selectively cry in response to the vocal character-istics of another infant's cry; this finding led to the speculation that from birth infants are endowed with an innate precursor of empathic distress (Hoffman, 1975).

Adopting another's perspective, which is a higher-order cognitive task that relies on executive functions, is integral to human empathy and is linked to the development of moral reasoning (Kohlberg, 1976) and altruism (Batson, 1991), as well as to a decreased likelihood of interpersonal aggression (Eisenberg, Spinrad, & Sadovsky, 2006). Perspective taking is truly an amazing human capacity, and is more difficult than one may assume. Regulating personal self-knowledge poses particular difficulty for both human children and adults. A cognitive neuroscience study demonstrated the primacy of the self-perspective. Jackson, Meltzoff, and Decety (2006b) asked participants to adopt either a first-person perspective or a third-person perspective while imitating observed actions. Structures related to motor representations, including those in the somatosensory cortex, recruited greater activation during the first-person perspective than during the third-person perspective, thus implying the immediacy of the first-person experience over that of the third-person perspective.

Skills acquired via intentional imitation, however, help children (by about age 4) and adults better gauge the perspectives of third parties (though errors often persist). As men-tioned above, both basic imitation and intentional imitation are theorized to facilitate the development of intersubjectivity in children. Similar to theory of mind, "intersubjectivity" is the capacity to estimate another individual's mental and emotional state. Thus both basic and intentional imitation—via the former's relation with intersubjectivity, and via the latter's relation with both intersubjectivity and agency—play pivotal roles in empathy. Clearly, then, empathy cannot be conceptualized without the inclusion of imitation.

Perspective Taking Induces Empathic Concern

Of special interest are findings from social psychology that document the distinction between imagining the other and imagining oneself in a negative situation (Batson, Early, & Salva-rini, 1997). These studies show that the former may evoke empathic concern (i.e., an other-oriented response congruent with the perceived distress of the person in need), while the latter induces both empathic concern and personal distress (i.e., a self-oriented aversive emotional response, such as anxiety or discomfort). This observation may help explain why empathy, or sharing someone else's emotion, need not yield prosocial behavior. If perceiving another person in an emotionally or physically painful circumstance elicits personal distress, then the observer may not fully attend to the other's experience and as a result may lack sympathetic behaviors.

The effect of perspective taking to generate empathic concern was documented in a study conducted by Stotland (1969). In his experiment, participants viewed an individual whose hand was strapped in a machine that participants were told generated painful heat. One group of subjects was instructed to watch the target person carefully; another group of participants was instructed to imagine the way the target felt; and the third group was instructed to imagine themselves in the target's situation. Physiological measures (palm sweating and vasoconstriction) and verbal assessments of empathy demonstrated that the deliberate acts of imagination yielded a greater response than did passive viewing. Empa-thy specifically seems to be sensitive to perspective taking, as demonstrated by a series of studies demonstrating that perspective-taking instructions are effective in inducing empathy (Batson et al., 1997) and that empathy-inducing conditions do not compromise the distinction between the self and other (Batson et al., 1997; however, see Cialdini, Brown, Lewis, Luce, & Neuberg, 1997, for a different account of empathy and self–other merging).

A recent study (Lamm et al., 2007) investigated the distinction between empathic concern and personal distress by combining a number of behavioral measures and event-related fMRI. Participants were asked to watch a series of video clips featuring patients undergoing painful medical treatment, with the instruction either to put themselves explicitly in the shoes of the patient or, in another condition, to focus their attention on the feelings and reactions of the patient. The behavioral measures confirmed previous social-psychological findings that projecting oneself into an aversive situation leads to higher personal distress and lower empathic concern, whereas focusing on the emotional and behavioral reactions of another's plight is accompanied by higher empathic concern and lower personal distress. The neuroimaging data were also consistent with such findings. The self-perspective evoked stronger hemodynamic responses in brain regions involved in coding the motivational/affective dimensions of pain, including bilateral insular cortices and anterior medial cingulate cortex. In addition, the self-perspective led to stronger activation in the amygdala, a limbic structure that plays a critical role in fear-related behaviors, such as the evaluation of actual or potential threats. Interestingly, the amygdala receives nociceptive information from the spinoparabrachial pain system and the insula, and its activity appears closely tied to the context and level of aversiveness of the perceived stimuli. Imagining oneself to be in a painful and potentially dangerous situation thus triggers a stronger fearful and/or aversive response than imagining someone else to be in the same situation. It is particularly worth noting that the insular activation was located in the middorsal section; this part of the insula plays a role in coding the sensorimotor aspects of painful stimulation, and it has strong connections with the basal ganglia, in which activity was also higher when the self-perspective was adopted. Therefore, activity in this aspect of the insula possibly reflects the simulation of the sensory aspects of the painful experience. Such a simulation might lead both to the mobilization of motor areas (including the supplementary motor area) in order to prepare defensive or withdrawal behaviors, and to interoceptive monitoring associated with autonomic changes evoked by this simulation process.

Disorders of Empathy and Deficits in Perspective Taking

Children with empathy deficits also show deficits in executive function. A series of studies found that when an experimenter feigned distress in a room where children were playing, children with ASD looked to the experimenter much less than either healthy children or children with mental retardation did (e.g., Corona, Dissanayake, Arbelle, Wellington, & Sigman, 1998). However, when Blair (1999) replicated such studies, but controlled for executive function demands of attention, children with ASD performed like healthy children: When the experimenter's feigned distress was unambiguous and took place under conditions of low distractibility, children with ASD showed autonomic responses similar to those of controls. In studies measuring facial mimicry, when individuals with ASD are provided with ample time, they do show affective compensatory tactics to accomplish emotion reading. Moreover, in emotion recognition tasks, individuals with ASD show activation in brain areas related to intentional attentional provision and categorization instead of automatic processing (Hall,

Szechtman, & Nahmias, 2003). These data indicate that alongside bottom-up information-processing deficits (e.g., mimicry), top-down executive control is also impaired in individuals with autism. The deficit in cognitive flexibility may contribute to the apathetic behaviors characteristic of autism.

Violent offenders, as well as children with aggressive behavior problems, experience deficits in empathy, though the lack of empathy is manifested in behavior differing from that observed in individuals with ASD. The former groups respond aggressively to others' distress (Arsenio & Lemerise, 2001), while the latter group simply lacks prosocial behavior. The distinction can be understood as the difference between hostility and apathy, both of which are categorized as unempathetic in the traditional sense, though one is "active" and the other is "passive." Individuals with ASD seem to lack either the interest or the capacity to resonate emotionally with others or to engage in intersubjective transactions. In contrast, children with developmental aggression disorders react aggressively to the observation of others' distress.

Of particular interest are the aggressive and nonempathetic reactions of children who have been colloquially been termed "bullies." This subpopulation falls within the antisocial category, but their aggressive reactions seem to be specific to peers in distress. Dautenhahn and Woods (2003) have proposed a model to account for the empathy deficits observed in bullies. In this model, bullies (as well as adults with antisocial personality disorder) are not aggressive because of poor emotion processing, theory of mind, or perspective taking; instead, they have heightened goal-directed behavior, often with an aggressive, antisocial goal to inflict personal distress on others. This model is particularly intriguing, as it questions previously held beliefs that young males with conduct problems experience weak regulation (Gill & Calkins, 2003). It also has particular implications for the role of perspective taking in empathy, because it emphasizes that perspective-taking ability is not sufficient for regulating an egocentric bias; personal motivation (i.e., the desire to induce distress in another, vs. the desire to offer prosocial help) plays a crucial role. In other words, a bully or an individual with antisocial personality disorder has the capability for an empathic experience, but the goal to elicit distress. This hypothesis is supported by a recent fMRI study, in which youth with aggressive conduct disorder were compared with age-, sex-, and raced-matched participants during the perception of pain in others (Decety, Michalska, Akitsuki, & Lahey, 2009). In both groups the perception of others in pain was associated with activation of the pain matrix, including the ACC, insula, somatosensory cortex, supplementary motor area, and periaqueductal gray. The pain matrix was activated to a significantly greater extent in participants with CD, who also showed significantly greater amygdala and ventral striatum activation, which may indicate that they enjoyed watching other people in pain.

Executive functions not only facilitate perspective taking, but also control attention and metacognitive capacities, both of which facilitate prosocial responding in reaction to another's distress. Children first demonstrate responses to the distress of others with other-focused behaviors (such as concern, attention to the distress of the other, cognitive exploration of the event, and prosocial interventions) by about the second year of life. At this age children manifest a self-concept and self-conscious emotions, and children's reparative behaviors after they cause distress in another also emerge (Zahn-Waxler, Radke-Yarrow, & Wagner, 1992). A longitudinal study of young children's development of concern for others' distress showed that prosocial behaviors, such as hugs and pats, emerged about the beginning of the second year of life, increasing in intensity throughout this year and sometimes providing self-comfort.

However, by the end of the second year, prosocial behaviors appeared to be more appropriate to the victims' needs and were not necessarily self-serving; children's emotions also appeared to be better regulated (Radke-Yarrow & Zahn-Waxler, 1984).

The ability to regulate emotions may be subject to individual differences, and may interact with the degree to which individuals experience emotions. Eisenberg et al. (1994) proposed a model suggesting an interaction between the intensity with which emotions are experienced and the extent to which individuals can regulate their emotions. In line with their model, a multimethod analysis of empathy-related responses (including self-report, facial, and heart rate responses) suggests that increased emotional intensity and decreased regulation predict personal distress. Emotion contagion corresponds with moderate emotionality when regulation is controlled, and perspective taking corresponds with high regulation and high emotionality only when perspective taking is controlled. These interactions are first seen in infancy, as findings from infant development demonstrate that 4-month-olds low in self-regulation are prone to personal distress at 12 months of age (Ungerer et al., 1990). In childhood, individuals with increased levels of emotional intensity (based on self-report, teacher and parent reports, and autonomic measurements) and weak regulation are prone to personal distress in response to another's predicament, as they become overwhelmed by their vicariously induced negative emotions (Eisenberg et al., 1994).

In summary, the maturation of executive functions by 2 years of age allows for the shared representations of self and other to be mediated, and the development of prosocial behaviors follows at this stage. On the other hand, if executive functioning is not intact, self- and other-perspectives may not be regulated, and individuals may overidentify or underidentify with an observed target. In the case of childhood aggression and conduct problems, it is likely that either over- or underregulation contributes to empathy deficits, though other factors (such as the environmental context and past experiences) also contribute to aggressive behaviors.

Conclusions

In this chapter, we have argued that imitation and empathy are inextricably linked. The findings that basic motor mimicry and complex intentional imitation provide a basis for empathy development emphasize the accuracy of viewing empathy, and social resonance in general, within a perception–action framework—as, at its core, imitation is the reenacting of a perception. This chapter has also addressed the corresponding deficits in imitation and empathy in certain disorders, particularly ASD, schizophrenia, and conduct problems. At an applications level, imitation may serve as an excellent remediation technique for enhancing empathy in these populations.

It is worth noting that empathy and imitation cannot be entirely conceived within an experimental or neuroscientific context. Personality traits, temperaments, and cultural norms of emotional display also contribute to the degree to which empathy may be not only experienced by the observer (e.g., Posner & Rothbart, 2000), but also modulated or even inhibited. In development, girls are more prone to comprehend emotional display rules than boys, particularly in cultures in which feminine roles demand more management and control of emotion display. Likewise, children from cultures that promote reciprocal relations and cooperation tend to be better at perspective-taking tasks than children living in individualistic cultures (Eisenberg, Bridget, & Shepard, 1997). Social psychologists emphasize the role of situational context as opposed to personality in the experience of empathy (or the absence of it), although many recognize the combination of situation and personality as the best predictor of social behavior (Fiske, 2004). Because situational context is important, creating ecologically valid situations in a laboratory setting poses a challenge. Thus designing ecologically valid experiments remains a challenging process, especially with children.

A current trend in cognitive neuroscience is studying the interaction between affect and cognition. Imitation and empathy exemplify this complex process, as motor mimicry is a stepping stone to emotion sharing, which in turn affects the projected mental state content during perspective taking. In addition, affective and social-cognitive developmental neuroscience offer promising insights for our understanding of both typical and psychopathological social behavior.

Acknowledgment

The writing of this chapter was supported by an NSF (BCS-0718480) award to Jean Decety.

References

Arsenio, W. F., & Lemerise, E. A. (2001). Varieties of childhood bullying: Values, emotion, processes, and social competence. *Review of Social Development, 10,* 59–73.

Batson, C. D. (1991). *The altruism question: Toward a social-psychological answer.* Hillsdale, NJ: Erlbaum.

Batson, C. D., Early, S., & Salvarini, G. (1997). Perspective taking: Imagining how another feels versus imagining how you would feel. *Personality and Social Personality Bulletin, 23,* 751–775.

Batson, C. D., Fultz, J., & Schoenrade, P. A. (1987). Distress and empathy: Two qualitatively distinct vicarious emotions with different motivational consequences. *Journal of Personality, 55,* 19–39.

Batson, C. D., Lishner, D. A., Carpenter, A., Dulin, L., Harjusola-Webb, S., Stocks, E. L., et al. (2003). As you would have them do unto you: Does imagining yourself in the other's place stimulate moral action? *Personality and Social Psychology Bulletin, 29,* 1190–1201.

Blair, R. J. R. (1999). Psycho-physiological responsiveness to the distress of others in children with autism. *Personality and Individual Differences, 26,* 477–485.

Blanke, O., & Arzy, S. (2005). The out-of-body experience: Distributed self-processing at the temporo-parietal junction. *The Neuroscientist, 11,* 16–24.

Blanke, O., Ortigue, S., Landis, T., & Seeck, M. (2002). Stimulating illusory own-body perceptions: The part of the brain that can induce out-of-body experiences has been located. *Nature, 419,* 269–270.

Bush, L. K., Barr, C. L., McHugo, G. J., & Lanzetta, J. T. (1989). The effects of facial control and facial mimicry on subjective reactions to comedy routines. *Motivation and Emotion, 13,* 31–52.

Carr, L., Iacoboni, M., Dubeau, M. C., Mazziotta, J. C., & Lenzi, G. L. (2003). Neural mechanisms of empathy in humans: A relay from neural systems for imitation to limbic areas. *Proceedings of the National Academy of Sciences USA, 100,* 5497–5502.

Cialdini, R., Brown, S., Lewis, B., Luce, C., & Neuberg, S. (1997). Reinterpreting the empathy–altruism relationship: When one into one equals oneness. *Journal of Personality and Social Psychology, 73,* 481–494.

Corona, C., Dissanayake, C., Arbelle, A., Wellington, P., & Sigman, M. (1998). Is affect aversive to young children with autism?: Behavioral and cardiac response to experimenter distress. *Child Development, 69,* 1494–1502.

Dapretto, M., Davies, M. S., Pfeifer, J. H., Scott, A. A., Sigman, M., Bookheimer, S. Y., et al. (2006).

Understanding emotions in others: Mirror neuron dysfunction in children with autism spectrum disorders. *Nature Neuroscience, 9,* 28–31.

Dautenhahn, K., & Woods, S. (2003). Possible connections between bullying behaviour, empathy, and imitation. *Proceedings of the Second International Symposium on Imitation in Animals and Artifacts, The Society for the Study of Artificial Intelligence and Simulation of Behaviour, 7,* 68–77.

DeCasper, A. J., & Fifer, W. P. (1980). Of human bonding: Newborns prefer their mothers' voices. *Science, 208*(4448), 1174–1176.

Decety, J. (2006). A cognitive neuroscience view of imitation. In S. Rogers & J. Williams (Eds.), *Imitation and the social mind: Autism and typical development* (pp. 251–274). New York: Guilford Press.

Decety, J. (2007). A social cognitive neuroscience model of human empathy. In E. Harmon-Jones & P. Winkielman (Eds.), *Social neuroscience: Integrating biological and psychological explanations of social behavior* (pp. 246–270). New York: Guilford Press.

Decety, J., & Grèzes, J. (2006). The power of simulation: Imagining one's own and other's behavior. *Brain Research, 1079,* 4–14.

Decety, J., & Jackson, P. L. (2004). The functional architecture of human empathy. *Behavioral and Cognitive Neuroscience Reviews, 3,* 71–100.

Decety, J., Jackson, P. L., & Brunet, E. (2007). The cognitive neuropsychology of empathy. In T. F. Farrow & P. W. Woodruff (Eds.), *Empathy in mental illness* (pp. 239–260). Cambridge, UK: Cambridge University Press.

Decety, J., & Lamm, C. (2007). The role of the right temporoparietal junction in social interaction: How low-level computational processes contribute to meta-cognition. *The Neuroscientist, 13*(6), 580–593.

Decety, J., Michalska, K. J., Akitsuki, Y., & Lahey, B. B. (2009). Atypical empathic responses in adolescents with aggressive conduct disorder: A functional MRI investigation. *Biological Psychology, 80*(2), 203–211.

Decety, J., & Sommerville, J. A. (2003). Shared representations between self and others: A social cognitive neuroscience view. *Trends in Cognitive Sciences, 7,* 527–533.

Eisenberg, N., Bridget, C. M., & Shepard, S. (1997). The development of empathic accuracy. In W. Ickes (Ed.), *Empathic accuracy* (pp. 194–215). New York: Guilford Press.

Eisenberg, N., Fabes, R. A., Murphy, B., Karbon, M., Maszk, P., Smith, M., et al. (1994). The relations of emotionality and regulation to dispositional and situational empathy-related responding. *Journal of Personality and Social Psychology, 66,* 776–797.

Eisenberg, N., Shea, C. L., Carlo, G., & Knight, G. P. (1991). Empathy-related responding and cognition: A "chicken and the egg" dilemma. In W. M. Kurtines & J. L. Gewirtz (Eds.), *Handbook of moral behavior and development: Vol. 2. Research* (pp. 63–88). Hillsdale, NJ: Erlbaum.

Eisenberg, N., Spinrad, T. L., & Sadovsky, A. (2006). Empathy-related responding in children. In M. Killen & J. Smetana (Eds.), *Handbook of moral development* (pp. 517–549). Mahwah, NJ: Erlbaum.

Farrer, C., Franck, N., Frith, C. D., Decety, J., Damato, T., & Jeannerod, M. (2004). Neural correlates of action attribution in schizophrenia. *Psychiatry Research: Neuroimaging, 131,* 31–44.

Farrer, C., Franck, N., Georgieff, N., Frith, C. D., Decety, J., & Jeannerod, M. (2003). Modulating the experience of agency: A positron emission tomography study. *NeuroImage, 18,* 324–333.

Farrer, C., & Frith, C. D. (2002). Experiencing oneself vs. another person as being the cause of an action: The neural correlates of the experience of agency. *NeuroImage, 15,* 596–603.

Field, T. M. (1989). Individual and maturational differences in infant expressivity. In N. Eisenberg (Ed.), *Empathy and related emotional responses* (pp. 9–23). San Francisco: Jossey-Bass.

Field, T. M., Woodson, R., Greenberg, R., & Cohen, D. (1982). Discrimination and imitation of facial expression by neonates. *Science, 219,* 179–181.

Fiske, S. (2004). *Social beings: A core motives approach to social psychology.* New York: Wiley.

Gill, K. L., & Calkins, S. (2003). Do aggressive/destructive toddlers lack concern for others?: Behavioral and physiological indicators or empathic responding in 2-year-old children. *Development and Psychopathology, 15,* 55–71.

Hall, G. B. C., Szechtman, H., & Nahmias, C. (2003). Enhanced salience and emotion recognition in autism: A PET study. *American Journal of Psychiatry, 160,* 1439–1441.

Hatfield, E., Cacioppo, J. T., & Rapson, R. L. (1993). Emotional contagion. *Current Directions in Psychological Science, 2,* 96–99.

Hatfield, E., Cacioppo, J. T., & Rapson, R. L. (1994). *Emotional contagion.* New York: Cambridge University Press.

Hess, U., & Blairy, S. (2001). Facial mimicry and emotional contagion to dynamic emotional facial expressions and their influence on decoding accuracy. *International Journal of Psychophysiology, 40,* 129–141.

Hobson, R. P. (2002). *The cradle of thought.* London: Macmillan.

Hoffman, M. L. (1975). Developmental synthesis of affect and cognition and its implications of altruistic motivation. *Developmental Psychology, 23,* 97–104.

Hommel, B., Musseler, J., Aschersleben, G., & Prinz, W. (2001). The theory of event coding: A framework for perception and action. *Behavioral and Brain Sciences, 24,* 849–878.

Izard, C. E. (1982). *Measuring emotions in infants and young children.* New York: Cambridge University Press.

Jackson, P. L., Brunet, E., Meltzoff, A. N., & Decety, J. (2006a). Empathy examined through the neural mechanisms involved in imagining how I feel versus how you feel pain: An event-related fMRI study. *Neuropsychologia, 44,* 752–761.

Jackson, P. L., & Decety, J. (2004). Motor cognition: A new paradigm to investigate social interactions. *Current Opinion in Neurobiology, 14,* 1–5.

Jackson, P. L., Meltzoff, A. N., & Decety, J. (2006b). Neural circuits involved in imitation and perspective-taking. *NeuroImage, 31,* 429–439.

Jeannerod, M. (1999). To act or not to act: Perspective on the representation of actions. *Quarterly Journal of Experimental Psychology, 42,* 1–29.

Kohlberg, L. (1976). Moral stages and moralization: The cognitive-developmental approach. In T. Lickona (Ed.), *Moral development and behavior* (pp. 31–53). New York: Holt, Rinehart & Winston.

Lamm, C., Batson, C. D., & Decety, J. (2007). The neural substrate of human empathy: Effects of perspective-taking and cognitive appraisal. *Journal of Cognitive Neuroscience, 19,* 42–58.

Martin, G. B., & Clark, R. D. (1987). Distress crying in neonates: Species and peer specificity. *Developmental Psychology, 18,* 3–9.

McIntosh, D. N., Reichmann-Decker, A., Winkielman, P., & Wilbarger, J. L. (2006). When the social mirror breaks: Deficits in automatic, but not voluntary mimicry of emotional facial expressions in autism. *Developmental Science, 9,* 295–302.

Meltzoff, A. N., & Decety, J. (2003). What imitation tells us about social cognition: A rapprochement between developmental psychology and cognitive neuroscience. *Philosophical Transactions of the Royal Society of London, Series B, Biological Sciences, 358,* 491–500.

Meltzoff, A. N., & Moore, M. K. (1977). Imitation of facial and manual gestures by human neonates. *Science, 198,* 75–78.

Niedenthal, P. M., Brauer, M., Halberstadt, J., & Innes-Ker, A. (2001). When did her smile drop?: Facial mimicry and the influence of emotional state on the detection of change in emotional expression. *Cognition and Emotion, 15,* 853–864.

Posner, M. I., & Rothbart, M. K. (2000). Developing mechanisms of self-regulation. *Development and Psychopathology, 12,* 427–441.

Preston, S. D., & de Waal, F. B. M. (2002). Empathy: Its ultimate and proximate bases. *Behavioral and Brain Sciences, 25,* 1–72.

Prinz, W. (1997). Perception and action planning. *European Journal of Cognitive Psychology, 9,* 129–154.

Radke-Yarrow, M., & Zahn-Waxler, C. (1984). Roots, motives, and patterning in children's prosocial behavior. In E. Staub, K. D. Bartal, J. Karylowski, & J. Raykowski (Eds.), *The development and maintenance of prosocial behavior: International perspectives on positive morality* (pp. 81–99). New York: Plenum Press.

Rizzolatti, G., Fogassi, L., & Gallese, V. (2001). Neurophysiological mechanisms underlying the understanding and the imitation of action. *Nature Reviews Neuroscience, 2,* 661–670.

Rochat, P., & Striano, T. (2000). Perceived self in infancy. *Infant Behavior and Development, 23,* 513–530.

Rogeness, G. A., Cepeda, C., Macedo, C., Fischer, C., & Harris, W. (1990). Differences in heart rate and blood pressure in children with conduct disorder, major depression, and separation anxiety. *Psychiatry Research, 33,* 199–206.

Rogers, S. J. (1999). An examination of the imitation deficit in autism: The roles of imitation and executive function. In J. Nadel & G. Butterworth (Eds.), *Imitation in infancy* (pp. 254–283). New York: Cambridge University Press.

Royzman, E. B., Cassidy, K. M., & Baron, J. (2003). I know, you know: Epistemic egocentrism in children and adults. *Review of General Psychology, 7,* 38–65.

Ruby, P., & Decety, J. (2001). Effect of subjective perspective taking during simulation of action: A PET investigation of agency. *Nature Neuroscience, 4,* 546–550.

Ruby, P., & Decety, J. (2004). How would you feel versus how do you think she would feel?: A neuroimaging study of perspective-taking with social emotions. *Journal of Cognitive Neuroscience, 16,* 988–999.

Sonnby-Borgström, M., Jönsson, P., & Svensson, O. (2003). Emotional empathy as related to mimicry reactions at different levels of information processing. *Journal of Nonverbal Behavior, 27*(1), 3–23.

Spence, S. A., Brooks, D. J., Hirsch, S. R., Liddle, P. F., Meehan, J., & Grasby, P. M. (1997). A PET study of voluntary movement in schizophrenic patients experiencing passivity phenomena (delusions of alien control). *Brain, 120,* 1997–2011.

Stotland, E. (1969). Exploratory investigations of empathy. In L. Berkowitz (Ed.), *Advances in experimental social psychology* (Vol. 4, pp. 271–314). New York: Academic Press.

Trevarthen, C. (1979). Communication and cooperation in early infancy. In M. Bullowa (Ed.), *Before speech: The beginning of human communication* (pp. 321–347). Cambridge, UK: Cambridge University Press.

Trevarthen, C., & Aitken, K. J. (2001). Infant intersubjectivity: Research, theory, and clinical applications. *Journal of Child Psychiatry, 42,* 3–48.

Ungerer, J. A., Dolby, R., Waters, B., Barnett, B., Kelk, N., & Lwein, V. (1990). The early development of empathy: Self-regulation and individual differences in the first year. *Motivational Emotion, 14,* 93–106.

Zahn-Waxler, C. L., Radke-Yarrow, M., Wagner, E., & Chapman, M. (1992). Development of concern for others. *Developmental Psychology, 28,* 126–136.

CHAPTER 9

Mentalizing and Development during Adolescence

SUPARNA CHOUDHURY
TONY CHARMAN
SARAH-JAYNE BLAKEMORE

Research suggests that the human brain may be adaptive to the behaviorally demanding social environment during adolescence. In the past few years, several pioneering experiments have investigated the development of the brain and cognitive processes during this period of life. Even though brain adaptation can occur throughout the lifespan, the maturational phases during early life—that is, during fetal development, childhood, and adolescence—are thought to be the most dramatic (Toga, Thompson, & Sowell, 2006). The first section of this chapter summarizes key findings in social neuroscience that have shed light on how the human brain processes sensations, actions, and emotions so that the meanings of experiences can be shared with others. We then discuss how social cognitive skills develop, focusing on findings from behavioral and neuroimaging studies of social-emotion processing during adolescence.

The Social Brain

Social cognition encompasses any cognitive process that involves conspecifics, either at a group level or on a one-to-one basis. Social neuroscience encompasses the empirical study of the neural mechanisms underlying social-cognitive processes. One key question is whether the general cognitive processes involved in perception, language, memory, and attention are sufficient to explain social competence, or whether, over and above these general processes, there are processes specific to social interaction. The field of social neuroscience is still relatively new, but builds on a variety of well-established disciplines (including social, developmental, and cognitive psychology; evolutionary biology; neuropsychology; and computer science), each providing a solid basis of relevant research. In this section, we describe two areas of social neuroscience that have received attention recently: understanding others' actions and understanding others' minds.

Understanding Others' Actions

In the past decade, neurophysiological research has provided evidence of a brain system that decodes conspecifics' actions and may contribute to the understanding of other people's intentions, goals, and desires. Mirror neurons, found in the ventral premotor cortex of macaque monkeys, are activated both when a monkey executes grasping actions and when it observes someone else (or another monkey) making grasping actions (Rizzolatti, Fogassi, & Gallese, 2001). Mirror neurons appear to distinguish between biological and nonbiological actions, responding only to the observation of hand–object interactions and not to the same action if performed by a mechanical tool, such as a pair of pliers (unless a monkey has been trained to understand the function of a particular tool; Ferrari, Rozzi, & Fogassi, 2005).

Following the discovery of mirror neurons in monkeys, researchers have provided increasing evidence that a large proportion of the human motor system is activated by the mere observation of action (Rizzolatti et al., 1996). Brain imaging studies have revealed that the motor activation to observed action is functionally specific: The premotor cortex and parietal cortex are activated in a somatotopic manner, according to the modality of the action being observed (Buccino et al., 2001). In addition, observing an action affects the peripheral motor system in the specific muscles that are used in the action being observed (Fadiga, Fogassi, Pavesi, & Rizzolatti, 1995). Observing another person's actions also influences one's own ongoing movements. Recent evidence suggests that observing an action interferes with one's own actions when these are different from those being observed (Kilner, Paulignan, & Blakemore, 2003). This interference effect seems to be specific for observed human actions; observing a robot making a movement does not interfere with ongoing actions.

The study of the mirror system provides an example of an attempt to identify the neurophysiological activity that underlies the ability to understand the meaning of one's own and another person's actions. This class of mechanism may be fundamental to a number of higher-level social processes, where the actions of other agents are interpreted in such a way that they directly influence one's own actions. This is the case in the attribution of intentions to others and oneself, the ability to imitate as well as to teach others, and "theory of mind" (see Gallese, 2006).

Mentalizing: Understanding Others' Minds

We humans are naturally endowed with the capacity to infer the intentions, desires, emotions, and beliefs of other people from minimal information. Not only are we typically able to decouple our beliefs from reality, but we can also distinguish other people's thoughts from our own and perceive their mental states from the mere observation of a gesture or a gaze. This intuitive social insight has been labeled "theory of mind" (Premack & Woodruff, 1978), and the capacity we use to understand other minds is known as "mentalizing" (Fletcher et al., 1995; Frith & Frith, 2003, 2006). So automatic and pervasive is this mentalizing mechanism that ordinary adults feel compelled to attribute intentions and other psychological motives to animated abstract shapes, simply on the basis of their movement patterns (Heider & Simmel, 1944).

How the Brain Understands Other Minds

The ability to surmise mental states from simply watching an agent's action or gesture seems to rely on a network of areas in the brain. Several different types of paradigms have been used to test the neural basis of mentalizing, involving cartoon strips, stories, moving shapes, and interactive games. Consistently, a network including the temporal poles, the posterior superior temporal sulcus (STS)/temporoparietal junction (TPJ), and the medial prefrontal cortex (MPFC) have been associated with tasks in which participants are required to mentalize (Frith & Frith, 2003).

In one of the first studies, participants were scanned while they performed story comprehension tasks that required the attribution of mental states. In one such story, participants had to work out that the protagonist's action (a robber giving himself up to the police) was based on his false assumption about a policeman's beliefs (i.e., that the policeman knew he had robbed a shop). This task required mental state attribution, because the beliefs of the robber and policeman were not made explicit in the story. When compared with comprehension tasks involving paragraphs made up of unlinked sentences, the theory-of-mind task produced activation in the STS, MPFC, and temporal poles (Fletcher et al., 1995).

Using the same stories and functional magnetic resonance imaging (fMRI), Gallagher et al. (2000) found that the MPFC, STS, and temporal poles were bilaterally activated by the theory-of-mind stories. In addition, these regions were activated by nonverbal cartoons that involved mental state attribution in order to be understood. Similarly, the STS and MPFC were activated in a positron emission tomography (PET) study in which participants were instructed to choose the right ending for a series of cartoons only when this required mental state attribution, and not when it required physical reasoning (Brunet, Sarfati, Hardy-Bayle, & Decety, 2000).

An implicit mentalizing task involves showing participants animations of moving shapes. This task has established that subjects feel compelled to attribute intentions and other psychological motives to animated abstract shapes, simply on the basis of their movement patterns (Heider & Simmel, 1944). Castelli, Happé, Frith, and Frith (2000), using PET, contrasted brain activity when subjects observed animations in which the movements of two triangles were designed to evoke mental state attributions (e.g., one triangle surprising or mocking the other), and when they observed animations where the triangles moved randomly and did not evoke such attributions. Like other imaging studies of mentalizing, this comparison revealed activation in the same system, including the MPFC, STS, and temporal poles.

Interactive games that involve spontaneous mentalizing have also been used in brain imaging experiments. For example, in one study, volunteers were scanned while they played a game similar to the "prisoner's dilemma" game with another person (McCabe, Houser, Ryan, Smith, & Trouard, 2001). In this game, mutual cooperation between players increased the amount of money that could be won. In the comparison task, the volunteers were told that they were playing with a computer that used fixed rules. In fact, in both conditions, subjects were playing with a computer. A comparison of brain activation when subjects believed they were playing with another person and when they believed they were playing with a computer revealed activity within the MPFC.

MPFC was also activated when participants played "stone–paper–scissors," a competitive game in which success depends upon predicting what the other player will do next (Gallagher, Jack, Roepstorff, & Frith, 2002). Again, volunteers were told that they were either playing against another person or playing against a computer (in fact, they were playing with a computer, and the sequence of the opponent's moves was the same in both conditions). Participants described guessing and second-guessing their opponent's responses, and felt that they could understand and "go along with" what their opponent, but not the computer, was doing. The MPFC was activated only when the volunteers believed that they were playing with another person.

Components of the mentalizing network are also engaged when participants make semantic decisions about other people's characteristics. For example, making semantic judgments about people was associated with activity in the MPFC, right TPJ, STS, and fusiform gyrus (Mitchell, Heatherton, & Macrae, 2002). A subsequent fMRI study investigated the specificity of these neural representations for knowledge about other people, by comparing brain activity during semantic judgments about other people and about dogs (Mitchell, Banaji, & Macrae, 2005). For each of these targets, participants were required to make judgments about the applicability of words to psychological states or to physical parts of the body of the target (person or dog). Two important findings emerged from the imaging data: first, that higher MPFC activation was associated with judgments of psychological states relative to body parts, and, second, that this MPFC activity effect extended to dogs as well as people. This suggests that the MPFC has a specific role for understanding mental states, regardless of the target (Mitchell et al., 2005).

There is emerging neuroimaging evidence for the differential roles of each of these areas. First, the STS/TPJ have been implicated in the ability to see the world from another person's point of view (Aichhorn, Perner, Kronbichler, Staffen, & Ladurner, 2006) (e.g., "What does Mary think from her perspective?"). Most likely, this is because of its role in eye movement observation (Pelphrey, Morris, & McCarthy, 2004) and in the representation of the body in space (Blanke et al., 2005). The TPJ is associated with false-belief tasks (Apperly, Samson, Chiavarino, & Humphreys, 2004), and the temporal poles are involved in the storage and use of knowledge of the world (Funnell, 2001) (e.g., "What would Mary *tend* to think, given the circumstances?"). Finally, a large number of studies have linked social-cognitive processing with the MPFC, including the anterior cingulate cortex. Based on experimental data and an analysis of the anatomical connectivity of the MPFC, Amodio and Frith (2006) have recently suggested that self-reflection, mentalizing, and person perception are all associated with the MPFC. Different subdivisions are dedicated to different processes: The caudal regions are associated with monitoring actions; the orbital regions are linked with outcomes; and the more anterior regions are related to metacognitive processes, or "thinking about thinking" (e.g., "What would Mary think, given how she feels, in the context of how I feel?").

Perspective Taking and the Brain

The ability to take another's perspective is crucial for successful social communication. In order to reason about others, and to understand what they think, feel, or believe, it is necessary to step into their "mental shoes" and take their perspective (Gallese & Goldman, 1998). Perspective taking includes awareness of one's own subjective space or mental states ("first-person perspective," or 1PP) and requires the ability to ascribe viewpoints, mental states, or emotions to another person ("third-person perspective," or 3PP). It is thus related to first-order theory of mind, in that it involves surmising what another person is thinking or feeling (Harris, 1989).

Functional neuroimaging studies have revealed that the MPFC, inferior parietal lobe (IPL), and STS are associated with making the distinction between 3PP and 1PP at the motor (Ruby & Decety, 2001), visuospatial (Vogeley et al., 2004), conceptual (Ruby & Decety, 2003; Vogeley et al., 2001), and emotional (Ruby & Decety, 2004) levels. Although common activations have been shown in 1PP and 3PP conditions in prefrontal and parietal areas (Ruby & Decety, 2001; Vogeley et al., 2004), and common deactivations in such areas as lateral superior temporal cortex (Vogeley et al., 2004), differences in activity between the two perspective conditions have also been reported. Taking someone else's perspective—whether it involves thinking about how another person would think or feel, or imagining the person making an action, relative to one's own perspective—was associated with increased activity in medial superior frontal gyrus, left STS, left temporal pole, and right IPL (Ruby & Decety, 2001, 2003, 2004). The IPL has also been implicated in the distinction between self and other at the sensorimotor level (Farrer & Frith, 2002) as well as at a higher social-cognitive level (Uddin, Molnar-Szakacs, Zaidel, & Iacoboni, 2006).

The Development of Mentalizing during Young Childhood

Since the seminal paper by Premack and Woodruff (1978), much theoretical and empirical work in the fields of psychology and philosophy has investigated the human ability to mentalize, and this work includes the development of experimental paradigms to test this capacity (Baron-Cohen, Leslie, & Frith, 1985; Dennett, 1987; Wimmer & Perner, 1983). Autism, a developmental social communication disorder characterized by "mindblindness" or the inability to mentalize, has provided key insights about the processes we typically use to understand other minds (Baron-Cohen, 1995).

The false-belief test has been a crucial paradigm for investigating the development of mentalizing (Dennett, 1987; Wimmer & Perner, 1983). Wimmer and Perner designed the first false-belief test, in which a character called Maxi places a piece of chocolate in a kitchen cupboard and then leaves the room. While he is out, his mother enters the room and moves his chocolate into a drawer. Maxi then returns to the room, and the key question is where Maxi will look for the chocolate—in the cupboard or in the drawer (Wimmer & Perner, 1983). The task therefore taps the ability to decouple one's own knowledge from another's knowledge, as well as the other person's belief from reality. Several variants of this task have since been designed, and such experiments suggest that, children show signs of understanding the scenario at age 4, performing at levels higher than chance, and, are able to understand it without any problems by age 6. Also by age 6, children are competent in attributing beliefs about another person's belief—so-called second-order false-belief tasks (Perner & Wimmer, 1985).

Passing a test about another person's false belief seems to be closely correlated with passing tests about one's own false beliefs, suggesting shared representations for the self and others (Gopnik & Astington, 1988). It has been shown that children who are able to report their own mental states can also report the mental states of others. Specifically, children were presented with a box of sweets and then shown that there were actually pencils inside it. They were asked, "What [will] Nicky think is in the box?" and then "When you first saw the box, before we opened it, what did you think was inside it?" The ability to answer correctly questions relating to the self was significantly correlated with the ability to answer questions about another person (Gopnik & Astington, 1988). Theory-of-mind tasks may therefore be measures not only of the awareness of others' mental states, but also of awareness of one's own mind.

Although several studies have reiterated findings about the development of mentalizing, demonstrating similar ages for these milestones, less research has been directed to the next major phase of neural plasticity and social development—adolescence. As discussed in the next section, regions of the brain linked to the capacity to mentalize are subject to structural change during adolescence. If studies have shown that the ability to mentalize develops in childhood at about the age of 4, what are the cognitive consequences of the continued development of its underlying neural circuitry?

Brain Development during Adolescence

Histological Studies of the Adolescent Brain

Until recently, very little was known about brain development during adolescence. The notion that the brain continues to develop after childhood is relatively new. Experiments on animals, starting in the 1950s, showed that sensory regions of the brain go through "sensitive periods" soon after birth, during which time environmental stimulation appears to be crucial for normal brain development and normal perceptual development to occur (e.g., Hubel & Wiesel, 1962). It was not until the late 1970s that research on postmortem human brains revealed that some brain areas continue to develop well beyond early childhood, which in turn suggested that sensitive periods for the human brain may be more protracted than previously thought. It was shown that cellular events take different trajectories in different areas of the human brain (Huttenlocher, 1979). Synaptic density in the visual cortex reaches a peak during the fourth postnatal month, and is followed by the elimination of synapses and the stabilization of synaptic density to adult levels before the age of 4. In contrast, the structure of the PFC undergoes significant changes during puberty and adolescence. Two main changes from before to after puberty have been revealed in the brain. As a neuron develops, a layer of myelin is formed around its extension, or axon, from supporting glial cells. Myelin acts as an insulator and massively increases the speed of transmission (up to 100-fold) of electrical impulses from neuron to neuron. Whereas sensory and motor brain regions become fully myelinated in the first few years of life, axons in the frontal cortex continue to be myelinated well into adolescence (Yakovlev & Lecours, 1967). The implication of this research is that the transmission speed of neural information in the frontal cortex should increase throughout childhood and adolescence.

The second difference between the brains of prepubescent children and those of adolescents pertains to changes in synaptic density in the PFC. Early in postnatal development, the brain begins to form new synapses, so that the synaptic density (the number of synapses per unit volume of brain tissue) greatly exceeds adult levels. This process of synaptic proliferation, called "synaptogenesis," can last up to several months or years, depending on the species of animal and the brain region. These early peaks in synaptic density are followed by a period of synaptic elimination (or "pruning"), in which frequently used connections are strengthened and infrequently used connections are eliminated. This experience-dependent process, which occurs over a period of years, reduces the overall synaptic density to adult levels.

Histological studies of monkey and human PFC have shown that synaptogenesis and synaptic pruning in this area have a particularly protracted time course. These studies show that there is a proliferation of synapses in the subgranular layers of the PFC during childhood and again at puberty, followed by a plateau phase and a subsequent elimination and reorganization of prefrontal synaptic connections after puberty (Huttenlocher, 1979; Woo, Pucak, Kye, Matus, & Lewis, 1997). According to these data, synaptic pruning occurs throughout adolescence in the human brain, and results in a net decrease in synaptic density in the MPFC during this time. The focus of the rest of this chapter is the cognitive implications of this protracted period of synaptogenesis in the PFC at the onset of puberty, and the process of synaptic pruning that follows it throughout adolescence.

MRI Studies of Adolescent Brain Development

The scarcity of postmortem child and adolescent brains meant that knowledge of the adolescent brain was until recently extremely scanty. However, since the advent of MRI, a number of brain imaging studies using large samples of participants have provided further evidence of the ongoing maturation of the cortex into adolescence and even into adulthood. Since the first *in vivo* studies reflecting that, in spite of somewhat larger total brain volumes, adults have less gray matter (GM) than children (Jernigan & Tallal, 1990), several MRI studies have been performed in the past few years to investigate the development of the structure of the brain during childhood and adolescence in humans (see Paus, 2005). Increases in white matter (WM), and decreases in GM, at different time points have been consistently reported (see Figure 9.1).

Linear Increases in WM during Adolescence

One of the most consistent findings from these MRI studies is that there is a steady linear increase in WM in certain brain regions during childhood and adolescence (see Figure 9.1A). This change has been highlighted in both frontal and parietal regions (Sowell et al., 1999). Myelin appears white in MRI scans, and therefore the increase in WM and decrease in GM with age were interpreted as reflecting increased axonal myelination in the frontal cortex.

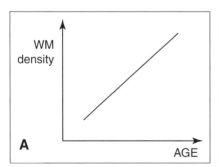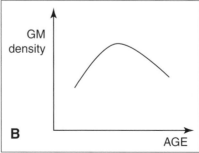

FIGURE 9.1. WM and GM development with age in PFC. (A) Linear development of WM with increasing age. This is thought to reflect a linear increase in myelination around axons. (B) Nonlinear development of GM. This is thought to reflect synaptogenesis followed by synaptic pruning.

The increased WM density and decreased GM density in frontal as well as parietal cortex throughout adolescence have now been demonstrated by several studies carried out by different research groups with large groups of participants (e.g., Giedd et al., 1996, 1999).

Nonlinear Decreases in GM during Adolescence

Whereas the increase in WM seems to be linear across all brain areas, the changes in GM density appear to follow a region-specific, nonlinear pattern. As the following studies have shown, its pattern of development in certain brain regions follows an inverted-U shape (see Figure 9.1B).

Giedd et al. (1999) performed a MRI study on 145 healthy boys and girls ranging in age from about 4 to 22 years. Individual growth patterns revealed heterochronous GM development during adolescence. Changes in the frontal and parietal regions were similarly pronounced. The volume of GM in the frontal lobe increased during preadolescence, with a peak occurring at about 12 years for males and 11 years for females. This was followed by a decline during postadolescence. Similarly, parietal lobe GM volume increased during the preadolescent stage to a peak at about 12 years for males and 10 years for females, and this was followed by a decline during postadolescence. GM development in the temporal lobes was also nonlinear, but the peak was reached later at about 17 years. Whereas frontal and parietal cortex development is relatively rapid during adolescence (Sowell, Thompson, Tessner, & Toga, 2001), GM volume in the superior temporal cortex, including the STS, steadily declines during adolescence and across the lifetime, following an inverted-U curve and reaching maturity relatively late (Gogtay et al., 2004; Toga et al., 2006).

In a study of participants between 4 and 21 years of age, frontal lobe maturation was shown to occur in a back-to-front direction, starting in the primary motor cortex (the precentral gyrus), then extending anteriorly over the superior and inferior frontal gyri (Gogtay et al., 2004). The PFC was shown to develop relatively late. In the posterior half of the brain, the maturation began in the primary sensory area, spreading laterally over the rest of the parietal lobe. Lateral temporal lobes were the last to mature. The authors noted that the temporal lobes showed a distinct maturation pattern: Whereas the temporal poles matured early, other areas, including the superior temporal gyrus and STS, matured latest in the brain. It was suggested that the sequence of structural maturation corresponds to the regionally relevant milestones in cognitive development. In other words, those brain areas associated with more basic cognitive skills, such as motor and sensory functions, mature first, followed by brain areas (including parietal cortex) linked to spatial orientation and attention, and finally by the regions related to executive function (PFC).

The MRI results demonstrating a nonlinear decrease in GM in various brain regions throughout adolescence have been interpreted in two ways. First, it is likely that axonal myelination results in an increase in WM and a simultaneous decrease in GM as viewed by MRI. A second, additional explanation is that the GM changes reflect the synaptic reorganization that occurs at the onset of, and after, puberty (Huttenlocher, 1979). Thus the increase in GM apparent at the onset of puberty (Giedd et al., 1999) may reflect a wave of synapse proliferation at this time. In other words, the increase in GM at puberty has been interpreted to reflect a sudden increase in the number of synapses. At some point after puberty, there is a process of refinement, so that these excess synapses are eliminated (Huttenlocher, 1979). It is speculated that this synaptic pruning is reflected by a steady decline in GM density seen in MRI (see Blakemore & Choudhury, 2006). If these brain areas undergoing structural change are associated with various cognitive skills, what are the implications for the development of executive function and social cognition?

Social-Cognitive Development during Adolescence

To date, few studies have investigated the social-cognitive consequences of adolescent brain development. Given that areas including the MPFC and STS are subject to protracted development, it might be predicted that social-cognitive abilities that impinge on the functioning of these brain areas would also develop. The period of adolescence involves new social encounters and heightened awareness and interest in other people.

The Development of Perspective Taking during Adolescence

Anecdotal evidence and self-report data from studies in Western cultures indicate that children seem to become progressively self-conscious and concerned with other people's opinions as they go through puberty and the period of adolescence (Berzonsky & Adams, 2003). The psychosocial context for adolescents in these cultural settings is markedly different from that for children or adults. For example, relationships with peers, family, and society go through distinct developments during this time. At the same time, the school context involves an intense socialization process during which adolescents become increasingly aware of the perspectives of classmates, teachers, and other societal influences. This transitional period seems to involve both the establishment of a sense of self and a process of orienting toward others.

The emergence of the social self seems to be marked by a period of heightened self-consciousness, during which adolescents are thought to become preoccupied with other people's concerns about their own actions, thoughts, and appearance. Several social-psychological studies have investigated changes in social thinking during adolescence, and emphasize that it is marked by a focus on "what other people think" (e.g., Lapsley & Murphey, 1985).

Given that the social environment often dramatically changes during adolescence, and that the brain undergoes a restructuring process, it might be expected that social-cognitive abilities such as perspective taking develop during adolescence as well. We recently investigated the development of perspective taking during adolescence (Choudhury, Blakemore, & Charman, 2006). We tested preadolescent children (age 9 years), adolescents (age 13 years), and adults (age 24 years) on a perspective-taking task that required each participant to imagine either how he or she would feel (1PP) or how a protagonist would feel (3PP) in various scenarios (see Figure 9.2). The participant was asked to choose one of two possible emotional faces in answer to each question, as quickly as possible. The results demonstrated that the difference in reaction time between 1PP and 3PP (ΔRT) decreased significantly with age. As shown in Figure 9.3, ΔRT among both groups of younger participants was larger and spread almost equally in both directions (i.e., 3PP > 1PP and 1PP > 3PP), whereas among adults there was little difference, with ΔRT values clustering around the zero mark of the difference scale (i.e., 3PP = 1PP). This finding suggests that the efficiency of perspective taking develops during adolescence, perhaps in parallel with the underlying neural circuitry.

If we assume that among the age groups tested, the adults were the most experienced in social interaction and had mature frontal and parietal neural circuitry, then a low ΔRT (3PP = 1PP) is likely to indicate the highest proficiency in perspective taking. In contrast, the

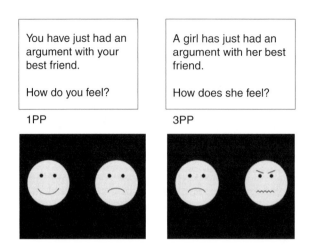

FIGURE 9.2. Design of task: Two perspectives (1PP and 3PP). See text for details. From Choudhury, Blakemore, and Charman (2006). Copyright 2006 by Oxford University Press. Reprinted by permission.

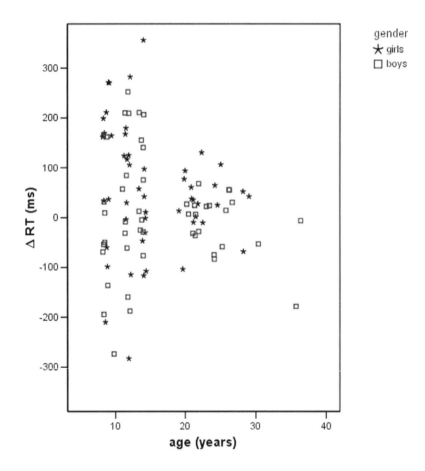

FIGURE 9.3. Difference between 3PP and 1PP (ΔRT) against age, showing direction of ΔRT. Differences are larger and more scattered in both directions for the younger participants, particularly the preadolescents. With increasing age, differences get closer to zero. From Choudhury, Blakemore, and Charman (2006). Copyright 2006 by Oxford University Press. Reprinted by permission.

most pronounced difference in RT between 1PP and 3PP, seen in the preadolescent group, would therefore indicate relatively inefficient processing. Therefore, it might be speculated that prior to adolescence, the unsystematic direction of ΔRT reflects an immature cognitive mechanism for perspective taking.

Whether this response pattern among preadolescents was a result of a relative difficulty in differentiating between 1PP and 3PP, or whether children of this age group are less inclined (or find it more difficult) to step into another person's "mental shoes," requires further investigation. The differences between age groups may also have been influenced by differences in social experience. Compared with children and adolescents, adults are generally more skilled at instinctively inferring the perspectives of other people. Perhaps adults show no difference between RTs for 1PP and 3PP as a result of their mature neural circuitry supporting social cognition, as well as their greater social experience.

Social Emotion Processing

The importance of evaluating other people may be associated with increased attention to socially salient stimuli, particularly faces, and the processing of emotional information. Recognition of facial expressions of emotion is one area of social cognition that has been investigated during adolescence (see Herba & Phillips, 2004, and de Haan & Matheson, Chapter 6, this volume). fMRI studies support the notion that development of certain brain regions is linked to the development of emotion processing. In an fMRI study of adolescents ages 13–17, the perception of happy faces compared with neutral faces was associated with significant bilateral amygdala activation (Yang, Menon, Reid, Gotlib, & Reiss, 2003). The effect of age was addressed by Thomas et al. (2001) in their investigation of amygdalar response to fearful facial expressions in two groups: children (mean age 11 years) and adults (mean age 24). Adults (relative to children) demonstrated greater amygdala activation to fearful facial expressions, whereas children (relative to adults) showed greater amygdala activation to neutral faces. It was argued that the children perceived the neutral faces as more ambiguous than the fearful facial expressions, with resulting increases in amygdala activation to the neutral faces.

Sex differences in amygdala-mediated cognitive development have also been reported to occur during adolescence. For example, females, but not males, showed increased activation in dorsolateral PFC in response to fearful faces between childhood and adolescence (Killgore, Oki, & Yurgelun-Todd, 2001). This result was replicated in a recent study, showing increased activity in PFC (bilaterally for girls, right-sided for boys) with age in response to fear perception (Yurgelun-Todd & Killgore, 2006). These fMRI results suggest that both the brain's emotion processing and cognitive appraisal systems develop during adolescence. This development has previously been interpreted in the context of the social information-processing network model (Nelson, Leibenluft, McClure, & Pine, 2005).

The Social Information-Processing Network Model

This model posits that social information processing occurs by way of three interacting neural "nodes," which afford the detection of social stimuli that are then integrated to a larger emotional and cognitive framework (see Crone & Westenberg, Chapter 19, this volume, for additional discussion of this model). Nelson et al. (2005) propose that the "detection node" (comprising the intraparietal sulcus, the STS, and the fusiform face area, as well as temporal and occipital regions) deciphers social properties of the stimulus, such as biological motion. The "affective node" (comprising limbic areas of the brain, including the amygdala, ventral striatum, hypothalamus, and orbital frontal cortex) is then thought to process the emotional significance of the social stimulus. Finally, the "cognitive–regulatory node" (consisting of much of the PFC) is responsible for theory of mind, impulse inhibition, and goal-directed behavior. The development of the nodes—that is, the connections between them, the innervation by gonadal steroid receptors, and the maturation of the neural substrates themselves—during adolescence is proposed to explain the development of social-cognitive behaviors.

A recent fMRI study used an irony comprehension task to investigate the development of communicative intent, and found that children (ages between 9 and 14) engaged frontal regions (MPFC and left inferior frontal gyrus) more than did adults in this task (Wang, Lee, Sigman, & Dapretto, 2006). A similar result was obtained in a study in which we investigated the development during adolescence of the neural network underlying thinking about intentions (Blakemore, den Ouden, Choudhury, & Frith, 2007). In this study, 19 adolescent participants (ages 12.1–18.1 years), and 11 adults (ages 22.4–37.8 years), were scanned using fMRI. A factorial design was employed, with age group as a between-subjects factor and causality (intentional or physical) as a within-subjects factor. In both adults and adolescents, answering questions about intentional causality versus physical causality activated the mentalizing network, including the MPFC, STS, and temporal poles. In addition, there was a significant interaction between group and task in the MPFC. During intentional relative to physical causality, adolescents activated part of the MPFC more than did adults, and adults activated part of the right STS more than did adolescents. These results suggest that the neural strategy for thinking about intentions changes between adolescence and adulthood. Although the same neural network is active, the relative roles of the different areas change, with activity moving from anterior (MPFC) regions to posterior (temporal) regions with age.

Thus functional imaging studies to date have reported mixed findings with respect to changes in frontal activity during social cognition tasks with age. However, there is a hint that activity during certain social cognition tasks may increase during childhood and decrease between adolescence and adulthood.

Conclusions

Although face processing is an example of an area of social-cognitive development during adolescence that has received attention in recent years, very little is known about how other aspects of social cognition change during the teenage years. It appears paradoxical that some of the brain regions involved in social cognition undergo such dramatic development into adolescence, when the functions mediated by these regions (e.g., mentalizing) appear to mature much earlier. If an ability (such as passing a theory-of-mind task) is accomplished by early to middle childhood, it is unlikely that it will undergo dramatic changes beyond that time. One possibility is that neural development during adolescence influences more subtle abilities, such as the capacity to modulate social-cognitive processes in the context of everyday life. Another possibility is that tasks tapping into implicit social-cognitive processes may be more likely to undergo change during adolescence. We have recently found some evidence for this in the domain of action imagery (Choudhury, Charman, Bird, & Blakemore, 2007a, 2007b). However, in the realm of development of social cognition during adolescence, there is a large gap in knowledge waiting to be filled.

References

Aichhorn, M., Perner, J., Kronbichler, M., Staffen, W., & Ladurner, G. (2006). Do visual perspective tasks need theory of mind? *NeuroImage, 30,* 1059–1068.

Amodio, D. M., & Frith, C. D. (2006). Meeting of minds: The medial frontal cortex and social cognition. *Nature Reviews Neuroscience, 7,* 268–277.

Apperly, I. A., Samson, D., Chiavarino, C., & Humphreys, G. W. (2004). Frontal and temporo-parietal lobe contributions to theory of mind: Neuropsychological evidence from a false-belief task with reduced language and executive demands. *Journal of Cognitive Neuroscience, 16,* 1773–1784.

Baron-Cohen, S. (1995). *Mind blindness: An essay on autism and theory of mind.* Cambridge, MA: MIT Press.

Baron-Cohen, S., Leslie, A. M., & Frith, U. (1985). Does the autistic child have a "theory of mind"? *Cognition, 21,* 37–46.

Berzonsky, M. D., & Adams, G. R. (2003). *The Blackwell handbook of adolescence.* Oxford: Blackwell.

Blakemore, S. J., & Choudhury, S. (2006). Development of the adolescent brain: Implications for executive function and social cognition. *Journal of Child Psychology and Psychiatry, 47,* 296–312.

Blakemore, S. J., den Ouden, H. E. M., Choudhury, S., & Frith, C. (2007). Adolescent development of the neural circuitry for thinking about intentions. *Social Cognitive and Affective Neuroscience, 2,* 130–139.

Blanke, O., Mohr, C., Michel, C. M., Pascual-Leone, A., Brugger, P., Seeck, M., et al. (2005). Linking out-of-body experience and self processing to mental own-body imagery at the temporoparietal junction. *Journal of Neuroscience, 25,* 550–557.

Brunet, E., Sarfati, Y., Hardy-Bayle, M. C., & Decety, J. (2000). A PET investigation of the attribution of intentions with a nonverbal task. *NeuroImage, 11,* 157–166.

Buccino, G., Binkofski, F., Fink, G. R., Fadiga, L., Fogassi, L., Gallese, V., et al. (2001). Action observation activates premotor and parietal areas in somatotopic manner: An fMRI study. *European Journal of Neuroscience, 13,* 400–404.

Castelli, F., Happé, F., Frith, U., & Frith, C. (2000). Movement and mind: A functional imaging study of perception and interpretation of complex intentional movement patterns. *NeuroImage, 12,* 314–325.

Choudhury, S., Blakemore, S.-J., & Charman, T. (2006). Social cognitive development during adolescence. *Social Cognitive and Affective Neuroscience, 1,* 165–174.

Choudhury, S., Charman, T., Bird, V., & Blakemore, S.-J. (2007a). Adolescent development of motor imagery in a visually guided pointing task. *Consciousness and Cognition, 16*(4), 886–896.

Choudhury, S., Charman, T., Bird, V., & Blakemore, S.-J. (2007b). Development of action representation during adolescence. *Neuropsychologia, 45,* 255–262.

Dennett, D. (1987). *The intentional stance.* Cambridge, MA: MIT Press/Bradford Books.

Fadiga, L., Fogassi, L., Pavesi, G., & Rizzolatti, G. (1995). Motor facilitation during action observation: A magnetic stimulation study. *Journal of Neurophysiology, 73,* 2608–2611.

Farrer, C., & Frith, C. D. (2002). Experiencing oneself vs. another person as being the cause of an action: The neural correlates of the experience of agency. *NeuroImage, 15,* 596–603.

Ferrari, P. F., Rozzi, S., & Fogassi, L. (2005). Mirror neurons responding to observation of actions made with tools in monkey ventral premotor cortex. *Journal of Cognitive Neuroscience, 17*(2), 212–226.

Fletcher, P. C., Happé, F., Frith, U., Baker, S. C., Dolan, R. J., Frackowiak, R. S. J., et al. (1995). Other minds in the brain: A functional imaging study of 'theory of mind' in story comprehension. *Cognition, 57,* 109–128.

Frith, C. D., & Frith, U. (2006). The neural basis of mentalizing. *Neuron, 50,* 531–4.

Frith, U., & Frith, C. D. (2003). Development and neurophysiology of mentalising. *Philosophical Transactions of the Royal Society of London, Series B, Biological Sciences, 358,* 459–473.

Funnell, E. (2001). Evidence for scripts in semantic dementia: Implications for theories of semantic memory. *Cognitive Neuropsychology, 18*, 323–341.

Gallagher, H. L., Happé, F., Brunswick, N., Fletcher, P. C., Frith, U., & Frith, C. D. (2000). Reading the mind in cartoons and stories: An fMRI study of theory of mind in verbal and nonverbal tasks. *Neuropsychologia, 38*, 11–21.

Gallagher, H. L., Jack, A. I., Roepstorff, A., & Frith, C. D. (2002). Imaging the intentional stance in a competitive game. *NeuroImage, 16*, 814–821.

Gallese, V. (2006). Intentional attunement: A neurophysiological perspective on social cognition and its disruption in autism. *Brain Research, 1079*, 15–24.

Gallese, V., & Goldman, A. (1998). Mirror neurons and the simulation theory of mindreading. *Trends in Cognitive Sciences, 2*, 493–501.

Giedd, J. N., Blumenthal, J., Jeffries, N. O., Castellanos, F. X., Liu, H., Zijdenbos, A., et al. (1999). Brain development during childhood and adolescence: A longitudinal MRI study. *Nature Neuroscience, 2*(10), 861–863.

Giedd, J. N., Snell, J. W., Lange, N., Rajapakse, J. C., Kaysen, D., Vaituzis, A. C., et al. (1996). Quantitative magnetic resonance imaging of human brain development: Ages 4–18. *Cerebral Cortex, 6*, 551–560.

Gogtay, N., Giedd, J. N., Lusk, L., Hayashi, K. M., Greenstein, D., Vaituzis, A. C., et al. (2004). Dynamic mapping of human cortical development during childhood through early adulthood. *Proceedings of the National Academy of Sciences USA, 101*, 8174–8179.

Gopnik, A., & Astington, J. W. (1988). Children's understanding of representational change and its relation to the understanding of false belief and the appearance–reality distinction. *Child Development, 59*, 26–37.

Harris, P. (1989). *Children and emotion.* Oxford: Blackwell.

Heider, F., & Simmel, M. (1944). An experimental study of apparent behavior. *American Journal of Psychology, 57*, 243–249.

Herba, C., & Phillips, M. (2004). Annotation: Development of facial expression recognition from childhood to adolescence: Behavioral and neurological perspectives. *Journal of Child Psychology and Psychiatry, 45*, 1185–1198.

Hubel, D. N., & Wiesel, T. N. (1962). Receptive fields, binocular interactions and functional architecture in the cat's visual cortex. *Journal of Physiology, 160*, 106–154.

Huttenlocher, P. R. (1979). Synaptic density in human frontal cortex: Developmental changes and effects of aging. *Brain Research, 163*, 195–205.

Jernigan, T. L., & Tallal, P. (1990). Late childhood changes in brain morphology observable with MRI. *Developmental Medicine and Child Neurology, 32*, 379–385.

Killgore, W. D. S., Oki, M., & Yurgelun-Todd, D. A. (2001). Sex-specific developmental changes in amygdale responses to affective faces. *NeuroReport, 12*, 427–433.

Kilner, J. M., Paulignan, Y., & Blakemore, S.-J. (2003). An interference effect of observed biological movement on action. *Current Biology, 13*(6), 2–5.

Lapsley, D. K., & Murphy, M. N. (1985). Another look at the theoretical assumptions of adolescent egocentrism. *Developmental Review, 5*, 201–217.

McCabe, K., Houser, D., Ryan, L., Smith, V., & Trouard, T. (2001). A functional imaging study of cooperation in two-person reciprocal exchange. *Proceedings of the National Academy of Sciences USA, 98*, 11832–11835.

Mitchell, J. P., Banaji, M. R., & Macrae, C. N. (2005). General and specific contributions of the medial prefrontal cortex to knowledge about mental states. *NeuroImage, 28*, 757–762.

Mitchell, J. P., Heatherton, T. F., & Macrae, C. N. (2002). Distinct neural systems subserve person and object knowledge. *Proceedings of the National Academy of Sciences USA, 99*, 15238–15243.

Nelson, E., Leibenluft, E., McClure, E. B., & Pine, D. S. (2005). The social re-orientation of adolescence: A neuroscience perspective on the process and its relation to psychopathology. *Psychological Medicine, 35,* 63–74.

Paus, T. (2005). Mapping brain maturation and cognitive development during adolescence. *Trends in Cognitive Sciences, 9,* 60–68.

Pelphrey, K. A., Morris, J. P., & McCarthy, G. (2004). Grasping the intentions of others: The perceived intentionality of an action influences activity in the superior temporal sulcus during social perception. *Journal of Cognitive Neuroscience, 16,* 1706–1716.

Perner, J., & Wimmer, H. (1985). 'John thinks that Mary thinks that . . . ': Attribution of second-order beliefs by 5- to 10-year-old children. *Journal of Experimental Child Psychology, 39,* 437–471.

Premack, D., & Woodruff, G. (1978). Does the chimpanzee have a theory of mind? *Behavioral and Brain Sciences, 1,* 515–526.

Rizzolatti, G., Fadiga, L., Matelli, M., Bettinardi, V., Perani, D., & Fazio, F. (1996). Localization of grasp representations in humans by PET: 1. Observation versus execution. *Experimental Brain Research, 111,* 246–252.

Rizzolatti, G., Fogassi, L., & Gallese, V. (2001). Neurophysiological mechanisms underlying the understanding and imitation of action. *Nature Reviews Neuroscience, 2*(9), 661–670.

Ruby, P., & Decety, J. (2001). Effect of subjective perspective taking during simulation of action: A PET investigation of agency. *Nature Neuroscience, 4,* 546–550.

Ruby, P., & Decety, J. (2003). What you believe versus what you think they believe: A neuroimaging study of conceptual perspective-taking. *European Journal of Neuroscience, 17,* 2475–2480.

Ruby, P., & Decety, J. (2004). How would you feel versus how do you think she would feel?: A neuroimaging study of perspective-taking with social emotions. *Journal of Cognitive Neuroscience, 16,* 988–999.

Sowell, E. R., Thompson, P. M., Holmes, C. J., Batth, R., Jernigan, T. L., & Toga, A. W. (1999). Localizing age-related changes in brain structure between childhood and adolescence using statistical parametric mapping. *NeuroImage, 6,* 587–597.

Sowell, E. R., Thompson, P. M., Tessner, K. D., & Toga, A. W. (2001). Mapping continued brain growth and gray matter density reduction in dorsal frontal cortex: Inverse relationships during postadolescent brain maturation. *Journal of Neuroscience, 21,* 8819–8829.

Thomas, K. M., Drevets, W. C., Whalen, P. J., Eccard, C. H., Dahl, R. E., Ryan, N. D., et al. (2001). Amygdala response to facial expressions in children and adults. *Biological Psychiatry, 49,* 309–316.

Toga, A. W., Thompson, P. M., & Sowell, E. R. (2006). Mapping brain maturation. *Trends in Neurosciences, 29,* 148–159.

Uddin, L., Molnar-Szakacs, I., Zaidel, E., & Iacoboni, M. (2006). rTMS to the right inferior parietal lobule disrupts self–other discrimination. *Social Cognitive and Affective Neuroscience, 1,* 65–71.

Vogeley, K., Bussfeld, P., Newen, A., Herrmann, S., Happé, F., Falkai, P., et al. (2001). Mind reading: Neural mechanisms of theory of mind and self-perspective. *NeuroImage, 14,* 170–181.

Vogeley, K., May, M., Ritzl, A., Falkai, P., Zilles, K., & Fink, G. R. (2004). Neural correlates of first-person perspective as one constituent of human self-consciousness. *Journal of Cognitive Neuroscience, 16,* 817–827.

Wang, A. T., Lee, S. S., Sigman, M., & Dapretto, M. (2006). Developmental changes in the neural basis of interpreting communicative intent. *Social Cognitive Affective Neuroscience, 1,* 107–121.

Wimmer, H., & Perner, J. (1983). Beliefs about beliefs: Representation and constraining function of wrong beliefs in young children's understanding of deception. *Cognition, 13,* 103–128.

Woo, T. U., Pucak, M. L., Kye, C. H., Matus, C. V., & Lewis, D. A. (1997). Peripubertal refinement of the intrinsic and associational circuitry in monkey prefrontal cortex. *Neuroscience, 80,* 1149–1158.

Yakovlev, P. A., & Lecours, I. R. (1967). The myelogenetic cycles of regional maturation of the brain. In A. Minkowski (Ed.), *Regional development of the brain in early life* (pp. 3–70). Oxford: Blackwell.

Yang, T. T., Menon, V., Reid, A. J., Gotlib, I. H., & Reiss, A. L. (2003). Amygdalar activation associated with happy facial expressions in adolescents: A 3-T functional MRI study. *Journal of American Academy of Child and Adolescent Psychiatry, 42,* 979–985.

Yurgelun-Todd, D. A., & Killgore, W. D. (2006). Fear-related activity in the prefrontal cortex increases with age during adolescence: A preliminary fMRI study. *Neuroscience Letters, 406,* 194–199.

CHAPTER 10

Early Communicative Development and the Social Brain

DEBRA MILLS
BARBARA T. CONBOY

The development of communication skills is one of the most important achievements of early childhood—one that allows the child to participate fully in a social world. For several decades, developmental psychologists and psycholinguists have conducted studies to better understand the processes by which typically developing children succeed in this remarkable achievement. Much of that research has focused on the attainment of component skills and shifts in abilities, often referred to as "developmental milestones." Examples of language milestones occurring in the first 3 years of life include changes in the perception of phonetic contrasts in infants' native versus non-native language, associating words with meanings, producing the first words, combining words into two- or three-word utterances, and speaking in full sentences. Yet the development of the underlying brain systems that precede, accompany, or follow these achievements is not well understood.

Recently, developmental cognitive neuroscientists have begun to apply neural imaging methods to focus on prospective changes in the organization of the neural processes that accompany developmental changes in language abilities. Methods such as event-related potentials (ERPs), electroencephalography (EEG), magnetoencephalography (MEG), near-infrared spectroscopy (NIRS), and functional magnetic resonance imaging (fMRI) have been used to examine initial brain biases and experiential effects on cerebral specializations for speech perception and early word learning during the course of primary language acquisition. The vast majority of the studies reviewed in this chapter used the ERP technique. ERPs are averages of epochs of electrical activity time-locked to specific events. The ERP technique has excellent temporal resolution and provides a direct measure of changes in neural activity over time on a millisecond-by-millisecond basis. ERPs are safe, are noninvasive, do not require an overt response, and are currently the most practical method for studying brain activity linked to cognitive processes in awake infants.

The chapter is organized around two main areas pertaining to the neurobiology of language development. A central issue throughout the chapter is the effect of experience on brain and language development. Of particular interest here are behavioral and neurobiological studies suggesting that social interactions may play a critical role in setting up specializations for the perception of phonological categories and vocabulary development. In the first part of the chapter, we review the literature on speech processing in very young infants, and consider how neurobiological studies of phoneme processing during the first year of life have elucidated the effects of language experience on speech perception. The second part reviews developmental studies of brain activity linked to word recognition and semantic processing of words and gestures. The brief final section addresses the role of social and cultural influences on the neurobiology of early language development.

Development of Speech Perception

In order to communicate in the spoken modality, infants need to become attuned to the features of speech that are important for their native language. Typically developing infants with intact hearing begin this process with auditory-perceptual abilities that are subsequently shaped by experience with their native language (see Kuhl et al., 2008; Werker & Tees, 1999). How this process unfolds is not completely understood, but several findings have been established and replicated over the past three decades of research on the topic. Infants as young as 1 to 4 months of age discriminate consonants in a categorical manner (e.g., Eimas, Siqueland, Jusczyk, & Vigorito, 1971); the boundaries used to draw these categories are not initially influenced by the ambient language, but seem to be "natural" category boundaries based on innate auditory abilities that are found across several mammalian species (e.g., Kuhl & Miller, 1975; Kuhl & Padden, 1983). At these young ages, and even for several more months, infants can discriminate many different types of speech contrasts—including ones that do not distinguish meaning in their native language, referred to as "non-native contrasts" (Aslin, Pisoni, Hennessy, & Perey, 1981; Best & McRoberts, 2003; Best, McRoberts, & Sithole, 1988; Conboy, Rivera-Gaxiola, Klarman, Aksoylu, & Kuhl, 2005; Kuhl et al., 2006; Lasky, Syrdal-Lasky, & Klein, 1975; Streeter, 1976; Trehub, 1976; Werker, Gilbert, Humphrey, & Tees, 1981; Werker & Lalonde, 1988; Werker & Tees, 1984). However, over the next several months, the effects of native-language learning are increasingly noted. A "tuning out" of non-native phonetic cues is evident by 6 months for vowels (Kuhl, Williams, Lacerda, Stevens, & Lindblom, 1992; Polka & Werker, 1994) and between 8 and 12 months for consonants (Best et al., 1988; Best, McRoberts, LaFleur, & Silver-Isenstadt, 1995; Conboy et al., 2005; Eilers, Gavin, & Wilson, 1979; Kuhl et al., 2006; Lalonde & Werker, 1995; Werker & Tees, 1984). During this same time period, infants' ability to perceive native speech contrasts may also improve (Kuhl et al., 2006). Thus English-learning infants improve in discriminating the initial consonants of the English words *rock* and *lock*, which are native contrasts for them, whereas Japanese-learning infants decline in their discrimination of the same speech sounds, which are non-native contrasts for them (Kuhl et al., 2006; see also Tsao, Liu, & Kuhl, 2006).

Individual Variability in the Development of Speech Perception

The development of native-language perceptual abilities, which involves both a "tuning out" of irrelevant non-native phonetic information and a "tuning in" of relevant native phonetic information, may be thought of as a developmental milestone in language acquisition. Attainment of this milestone is not a discrete event that parents can record in their infant's diary, like the first spoken word; rather, it is a more gradual, continuous process that can only be probed with sensitive experimental measures. As with other developmental milestones, the attainment of native-language speech perception shows variability across infants. All typically developing infants eventually master this important skill, but they do so at varying rates, for reasons not yet completely understood. Individual variability across infants has been noted in both non-native and native discrimination ability. Infants who are further along in the process of tuning out phonetic information that is irrelevant for their native language (i.e., worse performance on non-native contrasts) show more advanced skills on several nonlinguistic cognitive control tasks (Conboy, Sommerville, & Kuhl, 2008b; Lalonde & Werker, 1995). Infants who perform better on speech discrimination tests of native contrasts show better vocabulary skills, both concomitantly (Conboy et al., 2005, 2008b) and at later ages (Kuhl, Conboy, Padden, Nelson, & Pruitt, 2005b; Tsao, Liu, & Kuhl, 2004). Although these correlational studies do not clearly show a causal relationship between speech perception and vocabulary learning or nonlinguistic cognitive factors, they imply that there may be functional relationships across these domains of learning. They may represent continuity across domains of learning, or the fact that learning from similar types of experience is essential for each (see also Jusczyk, 1997; Werker & Curtin, 2005). For example, the process of forming word representations relies on speech perception skills, but it may also affect the types of acoustic information infants attend to in their native language. At the same time, an increasing domain-general ability to control and inhibit attention to irrelevant cues may be necessary for infants to tune out the acoustic information that is not used to distinguish meaning in their native language (Conboy et al., 2008b; Diamond, Werker, & Lalonde, 1994).

Social Factors in the Development of Speech Perception

It is likely that social factors also play a role in the development of speech perception, although this is not yet clearly established. Language is naturally acquired in social contexts, and variations in these contexts—for example, in rates of contingent responses between caregivers and infants—have been linked to later cognitive and language outcomes (e.g., Baumwell, Tamis-LeMonda, & Bornstein, 1997; Beckwith, Cohen, Kopp, Parmelee, & Marcy, 1976; Beckwith & Cohen, 1989; Belsky, Goode, & Most, 1980; Bloom, 1993; Bornstein, 1989; Bornstein, Miyake, Azuma, Tamis-LeMonda, & Toda, 1990; Bornstein & Tamis-LeMonda, 1989, 1997; Bornstein, Tamis-LeMonda, & Haynes, 1999; Carew & Clarke-Stewart, 1980; Clarke-Stewart, 1973; Landry, Smith, Miller-Loncar, & Swank, 1997; Landry, Smith, Swank, Assel, & Vellet, 2001; Nicely, Tamis-LeMonda, & Bornstein, 1999; Tamis-LeMonda, Bornstein, Baumwell, & Damast, 1996; Tamis-LeMonda, Bornstein, Marc, & Baumwell, 2001). Given this, it would not be surprising if the language experience required for the types of phonetic learning that support future language acquisition also had a highly social nature.

One recent study directly tested the importance of a social context for phonetic learning, using quasi-naturalistic input (Kuhl, Tsao, & Liu, 2003). In this work, American infants from monolingual English-speaking homes received 5 hours of conversational exposure to a second language between 9 and 10 months of age. The language was delivered via twelve 25-minute play sessions spread out over approximately 1 month (e.g., three sessions per week). During the play sessions, native Mandarin speakers addressed these infants in Mandarin while showing them picture books and toys. Infants were tested at the end of the study, using a conditioned head-turn speech discrimination test of a Mandarin fricative–affricate contrast. Infants who had participated in these live exposure settings showed discrimination of the Mandarin contrast at the same levels as infants from monolingual Mandarin-speaking homes in Taiwan. Infants in a control group, however, did not show the same level of discrimination of the Mandarin contrast. Thus this study established that infants can learn a non-native contrast, given even a short amount of complex naturalistic exposure to the non-native language.

Most important to the above-mentioned results were two comparison groups of infants who received the same amounts of Mandarin exposure, from the same speakers, but in a passive context rather than a live one. One group of infants watched DVDs of the Mandarin speakers, and the other group listened while the same DVDs were played without the visual display. Both groups were tested on the Mandarin speech contrast at the end of the study, with methods identical to those used for the infants who participated in the live exposure group. However, these infants performed at the same levels as infants who had not received any Mandarin input. In sum, the study results suggest an important role for social factors in phonetic learning that takes place in complex naturalistic situations.

Kuhl and colleagues (Conboy & Kuhl, 2007; Kuhl, 1998, 2007; Kuhl et al., 2003) have suggested that learning is enhanced in social situations because of arousal factors, social engagement with providers of the input, and/or a process involving the referential nature of language learning. In either case, social interaction may serve to heighten infants' attention to relevant linguistic cues in the input. Such cues may occur in both the auditory and visual domains. In the Kuhl et al. (2003) study, for example, attention to the speaker's face may have been enhanced in the live situation, but not in the more passive video-viewing situation. Such audiovisual speaker cues have been shown to be important for phonetic learning in infancy (e.g., Kuhl & Meltzoff, 1982, 1996). In a recent study in which infants were exposed to Spanish in live sessions similar to those used in Kuhl et al. (2003), phonetic learning was linked to levels of joint engagement, as rated from infants' eye gaze alternations between the adult speakers' faces and objects of reference (Conboy, Brooks, Taylor, Meltzoff, & Kuhl, 2008a).

Although the findings reviewed above underscore the role of social interaction in phonetic learning, it is also true that controlled exposure to only a few minutes of non-native speech sound contrasts, in the absence of a social context, has been shown to induce learning in young infants (e.g., Maye, Weiss, & Aslin, 2008; Maye, Werker, & Gerken, 2002; McMurray & Aslin, 2005). For example, Maye et al. (2008) exposed 8-month-old infants to distributions of prevoiced vs. short-lag stop consonant contrasts, which are non-native for English (either /da/–/ta/ or /ga/–/ka/), and then tested them on the contrast that involved the other place of articulation, using a habituation–dishabituation paradigm. Infants not only learned the trained information, but were also able to generalize the voicing feature to a different place of articulation, indicating more abstract learning. In this study and in previous work, Maye and colleagues showed that input providing clear distributional cues is most effective for rapid infant learning. However, it is not clear whether such learning resembles what happens in the real world, or whether it is robust or long-lived. Certainly in the real world, and even in quasi-naturalistic play sessions such as those used in the Kuhl et al. (2003) and Conboy et al. (2008a) studies, infants are faced with a much more complex learning situation than in experiments in which they are familiarized with controlled stimuli. The real-world situation infants face is one in which regularities across units of speech must be extracted from noisy input, characterized by between-speaker variability in pitch and within-speaker variability caused by phonetic context, coarticulation, affect, and the like. This complexity may result in more robust learning than simpler input may. Infants are remarkably good at ignoring extralinguistic speaker variability (e.g., Dehaene-Lambertz & Peña, 2001; Jusczyk, Pisoni, & Mullennix, 1992; Kuhl, 1983), and such variability may actually facilitate language learning (Singh, 2008).

During naturalistic interactions between infants and caregivers, phonetic and other linguistic features are imbued with meaning. Even before infants have access to referential meaning, they may come to appreciate the cultural meaning of speech sounds that are embedded in mutually attuned, face-to-face protoconversations with their caregivers. By engaging in such interactions with infants, adults provide a community of meaning (Rommetveit, 1998). Furthermore, when infants learn to understand others as intentional agents and to share perception of communicative intentions—skills that are arguably crucial for various aspects of language acquisition (Akhtar & Tomasello, 1998; Tomasello, 1999)—their attention to the phonetic information contained in particular communicative acts may be enhanced. This is not to say that infants can integrate all of the relevant information present in the signal at once; recent studies have suggested that young infants ignore fine phonetic detail in words in certain structured learning situations (Mills et al., 2004; Stager & Werker, 1997; Werker, Fennell, Corcoran, & Stager, 2002; see below). Caregiver communicative acts that are contingent upon infants' behaviors may be most effective in influencing infants' attention to phonetic detail. Additional cues provided by caregivers in such situations—for example, the exaggeration of particular acoustic features in infant-directed speech—may further enhance learning (Fernald, 1984; Fernald & Kuhl, 1989; Kuhl et al., 1997; Liu, Kuhl, & Tsao, 2003; Liu, Tsao, & Kuhl, 2007). Infants' behaviors may in turn enhance the quality of the input they receive from caregivers (Conboy & Kuhl, 2007). It is not yet clear whether these are universal processes. Although there are cultures in which face-to-face communication between infants and adults is rare, infants in those communities are surrounded by the speech of others, and their actions or interests may be commented on by others in a contingent fashion (Rogoff, 2003; Schieffelin, 1991).

Neural Correlates of the Development of Speech Perception

There has been some debate as to whether the human perceptual system is predisposed for language acquisition at birth, or shaped by experience with language. The gradual attunement of the infant's perceptual system with the native-language environment suggests a prominent role for experience. Neural imaging techniques have shed some light on developmental changes in the way the infant brain handles phonetic information. In particular, ERPs have been used to explore the development of speech processing systems in infants, because they can be recorded while an infant listens passively. A commonly used method of studying speech perception is the "auditory oddball" experiment, in which ERPs are recorded to a frequently occurring "standard" stimulus and a less frequent "deviant" stimulus. Such paradigms elicit a large negative effect to the deviant stimulus, known as "mismatch negativity" (MMN), when subjects are not required to respond overtly to the change (Näätänen et al., 1997). The MMN is believed to reflect the brain's automatic change detection response (Näätänen et al., 1997; Näätänen, Gaillard, & Mäntysalo, 1978). Source localization studies have indicated generators in both auditory and frontal cortex, reflecting the formation of traces in auditory sensory memory and subsequent involuntary preattentional switches to the deviant stimulus, respectively (Näätänen, 1999). Speech perception can be tested by using minimal-pair syllables that differ by a single phonetic feature in the consonant or vowel (e.g., /pa/ vs. /ta/ differ in the place-of-articulation feature). In adults, the sources of the MMN elicited by minimal phoneme pairs are neural generators in left auditory cortex (Näätänen et al., 1997; Rinne et al., 1999). Given these properties, and the added benefit that testing can be completed without an overt response from subjects, the MMN is a useful tool for studying the development of language-specific phonetic representations (see Cheour, Leppanen, & Kraus, 2000, and Näätänen, 1999).

A number of studies have demonstrated mismatch responses in infants and young children, though not reliably the MMN. Cheour-Luhtanen et al. (1995) reported a component resembling the MMN in sleeping newborns while they listened to vowel contrasts, peaking at approximately 200–250 msec after stimulus onset. Subsequent studies have shown increased negativity in similar time windows to the deviant versus standard stimulus throughout the first year, for vowel contrasts (Cheour et al., 1997, 1998a, 1998b; Cheour-Luhtanen et al., 1996; Friederici, Friedrich, & Weber, 2002) and for consonant contrasts (Dehaene-Lambertz & Baillet,1998; Kuhl et al., 2008; Pang et al., 1998; Rivera-Gaxiola, Klarman, García-Sierra, & Kuhl, 2005a; Rivera-Gaxiola, Silva-Pereyra, & Kuhl, 2005b; Rivera-Gaxiola et al., 2007). The effects reported for infants in those studies had longer latencies and different scalp distributions than those reported for adults (for a review, see Cheour et al., 2000).

ERPs have also been used to study changes in the brain's response to phonetic units over the first year. Cheour et al. (1998b) recorded ERPs to Finnish and Estonian vowel contrasts in Finnish infants at 6 and 12 months. At 6 months, infants showed a similar mismatch response to both vowel contrasts (i.e., regardless of language experience), but by 12 months the mismatch response was greatly attenuated for the non-native (Estonian) contrast, compared to that observed for 12-month-old Estonian infants. Other studies using consonant contrasts, have shown similar shifts between 7 and 11 months (Kuhl et al., 2008; Rivera-Gaxiola et al., 2005a).

In infants, ERP effects to a detected phonetic change do not always take the form of the mismatch negativity. In some cases, a positive-going wave has been observed to the deviant stimulus (Dehaene-Lambertz, 2000; Dehaene-Lambertz & Baillet, 1998; Dehaene-Lambertz & Dehaene, 1994; Rivera-Gaxiola et al., 2005a, 2005b, 2007). Rivera-Gaxiola et al. (2005a, 2005b, 2007) noted that at 11 months, some infants showed an enhanced positive deflection to a non-native-language contrast in an earlier time window (P150–250 effect), rather than the later negativity (N250–550 effect). Many of these "P-responders" also showed the positive effect to the native-language contrast at 7 months, although by 11 months all infants showed the negative effect to that contrast. For the infants who showed the negative effect to the non-native contrast at 11 months ("N-responders"), the effect was smaller than for the native contrast. In addition to differing in polarity and latency, the P150–250 and N250–550 effects differed in scalp distribution, suggesting that they index distinct discriminatory processes. The P150–250 amplitudes were largest over frontocentral sites, whereas the N250–550 amplitudes were largest over parietal sites. Further evidence that these are distinct discriminatory components is provided by data linking these components to later language outcomes (see next section, below).

Shifts in the polarity and scalp distributions of ERPs to native versus non-native speech contrasts over the first year suggest that linguistic information is processed by a dynamic system that changes with language experience. This view is supported by a recent study that used MEG to record infants' discrimination of speech and nonspeech contrasts soon after birth, at 6 months of age, and at 12 months (Imada et al., 2006). MEG provides more precise spatial information than ERP, because volume conduction does not affect the scalp-recorded magnetic fields measured by MEG, as it does with the scalp-recorded electrical fields measured by ERP. In this study, infants showed discrimination of the speech and nonspeech contrasts at all three ages, and activation was noted in the left superior temporal (ST) cortex, regardless of age. (Note that in this study, only left hemisphere activations were measured; thus the relative contributions of the left and right hemispheres to infant speech and nonspeech processing were not examined.) However, there were age-related differences in activation in the left inferior frontal (IF) cortex, commonly known as Broca's area, which is involved in the motoric aspects of speech. This activation was noted by 6 months, though at a later latency for speech than for nonspeech stimuli. The researchers suggested that a coupling of activity between the IF and ST cortices for speech stimuli emerges by about 6 months of age, in tandem with other advances in early language development (such as canonical babbling, verbal imitation, and language-specific phonetic discrimination). They further suggested that these advances in brain function and behavior may be mediated by the maturation of a "mirror neuron" system, which binds action and perception. It has been suggested elsewhere that the mirror neuron system provides the social foundation for early language acquisition (see Bråten, 2007). The results challenge accounts of continuity between the initial and mature states of phoneme processing. For example, a recent fMRI study of sleeping infants while they listened to sentences indicated activation in Broca's area as early as 3 months of age (Dehaene-Lambertz et al., 2006a). The authors of this study argued that Broca's area is innately biased to drive the learning of complex motor speech sequences through interactions with the perceptual system.

Studies that have focused on functional asymmetries for speech processing have indicated interactions between "networks" present at birth and experience with language. Dehaene-Lambertz and Dehaene (1994) found a left-hemisphere asymmetry in the ERP mismatch response of 3-month-old infants who were presented with a place-of-articulation contrast (/ba/–/ga/). In that study, infants showed a bilateral mismatch response (larger amplitude to the deviant versus standard by about 400 msec after the onset of the deviant syllable), but the effect was larger over the left hemisphere than the right. The authors interpreted the mismatch effect as a response that was sensitive to phonetic features of the stimuli, and further suggested generators in the temporal cortex. However, they also noted that there was individual variability across infants in the direction of the asymmetry, with some infants showing larger effects over the right than the left hemisphere. Thus their results did not support a sharp division between the left and right hemispheres for processing phonetic information.

In a follow-up study, Dehaene-Lambertz and Baillet (1998) reported that, like adults, 3-month-old infants showed a larger mismatch response to contrasts that cross a phoneme category for their native language than for within-category contrasts, but this effect did not interact with hemisphere. However, the spatial distributions of the two types of contrast were nonidentical, suggesting separate functional networks. Similar to previous results, these results suggested bilateral generators in temporal cortex for both types of speech contrasts. Thus, in spite of overall greater activity to all syllables at left- than at right-hemisphere electrode sites, there was no evidence of a left-hemisphere bias in the phonemic discrimination response in infants. Moreover, Dehaene-Lambertz (2000) reported that 4-month-old infants did not show greater left-hemisphere activity when discriminating speech contrasts versus nonspeech contrasts.

The results presented above as well as other evidence suggest that the functional brain asymmetries for speech processing noted in adults are not innately specified, but rather emerge with development. Although both structural asymmetries (such as a larger left planum temporale) and functional asymmetries have been noted during the processing of speech early in life, it is not clear whether such asymmetries reflect an early specialization for speech processing, or the ability of the left temporal cortical areas to process many types of stimuli characterized by rapid transitions in acoustic information, of which speech is one (see Dehaene-Lambertz, Hertz-Pannier, & Dubois, 2006b). For example, fMRI studies of 3-month-old infants have indicated that both forward speech and backward speech activate left-hemisphere regions to a greater extent than they do right-hemisphere regions (Dehaene-Lambertz, Dehaene, & Hertz-Pannier, 2002).

Speech Perception and Vocabulary Development

The development of language-specific speech perception plays an important role in communication by allowing infants to form word representations (Werker & Yeung, 2005). Several studies have linked early phonetic perception to later language outcomes. Rivera-Gaxiola et al. (2005b) found that infants who showed a P150–250 effect for their non-native contrast at 11 months had higher subsequent vocabulary scores from 18 to 30 months than did infants who showed the N250–550 effect for their non-native contrast. Kuhl et al. (2008) recorded ERPs at 7.5 months and collected parent report measures of language skills at 14, 18, 24, and 30 months. Results indicated that infants' mismatch responses for the native and non-native contrasts were differentially associated with language skills between 14 and 30 months. Infants showing a larger negative mismatch response (negativity to the deviant vs. the standard stimulus) for their native language contrast at 7.5 months produced a larger number of words at 18 and 24 months, faster vocabulary growth from 14 to 30 months, more complex sentences at 24 months, and longer utterances at 24 and 30 months than did those who either did not show the mismatch effect or showed greater positivity to the deviant versus the standard stimulus. In contrast, infants who showed a larger negative mismatch response for the non-native contrast produced fewer words and less complex sentences at 24 months, slower growth in vocabulary size over the whole time period, and shorter utterances at 30 months. Relatedly, Molfese and colleagues (Molfese, 2000; Molfese & Molfese, 1985, 1997; Molfese, Molfese, & Espy, 1999) showed that ERPs recorded to syllables shortly after birth were associated with language scores at 3, 5, and 8 years and with reading disabilities at 8 years. Moreover, maturation of the ERP response to speech and nonspeech stimuli from 1 to 8 years was related to reading scores at 8 years (Espy, Molfese, Molfese, & Modglin, 2004).

Behavioral measures of speech discrimination in infancy have also been linked to later vocabulary and/or utterance length and complexity. Tsao et al. (2004) tested 6-month-old infants from monolingual English-speaking homes on a vowel contrast, using a conditioned head-turn paradigm, and subsequently recorded the infants' language skills between 14 and 30 months. The results indicated that the 6-month head-turn scores were positively correlated with later vocabulary size, utterance length, and utterance complexity. Kuhl et al. (2005b) tested 7.5-month-old monolingual English infants on native and non-native contrasts, using the head-turn paradigm, and recorded language skills from 14 and 30 months. In striking similarity to the ERP study described above (in which the same phonetic contrasts were used with many of the same infants), the head-turn scores for the native contrast were positively correlated with later language scores, and the head-turn scores for the non-native contrast were negatively correlated with later language scores. Conboy et al. (2005) conducted a double-target conditioned head-turn test with 7.5- and 11-month-old infants. The infants were tested on both contrasts simultaneously, so that performance factors such as fatigue and inattentiveness would be expected to affect both contrasts equally. At both ages there were individual differences in performance across contrasts, and these were linked to 11-month receptive vocabulary sizes. The infants who showed better performance for the native than for the non-native contrast were reported to comprehend more words than those who processed both contrasts at similar levels.

In sum, both ERP and behavioral measures of speech sound contrast discrimination during the first year of life predict subsequent achievements in language development over the next several years. Further research is needed to determine whether early attunement to the relevant features of speech sounds for an infant's native language serves as a "bootstrapping" mechanism for learning at the word level (see Werker & Yeung, 2005), or whether the relationships between rates of learning in each of these domains derive solely from other factors, such as amounts and types of input and more general social-cognitive abilities. Most likely, multiple factors interact to produce successful language development: Attention to relevant linguistic details (and resistance to distracting information), the ability to process multiple sources of information at once, and levels of social engagement with caregivers probably all influence the uptake of information used by infants to form representations of words. At the same time, caregivers respond to cues from their infants and provide appropriate input that facilitates further learning.

As mentioned previously, even though very young infants can discriminate fine differences in phonetic detail, they may not use that information when mapping sounds onto meaning (Stager & Werker, 1997; Werker, Cohen, Lloyd, Casasola, & Stager, 1998; Werker et al., 2002). Using a habituation–switch looking-preference task, Werker et al. (2002) showed that 14-month-old infants would confuse phonetically similar nonwords (e.g., *bih* vs. *dih*) when these were paired with an object. In contrast, more experienced word learners at 20 months of age distinguished between similar sounding nonwords and mapped them correctly. The performance of 17-month-olds fell between that of 14- and 20-month-olds. Of particular interest was that measures of both vocabulary comprehension and production were positively correlated with performance on the task. However, in that study, interpretation of the results depended on the younger infants' failure to make a discrimination. ERPs do not require an overt response and may provide a more sensitive measure of discrimination. Mills et al. (2004) used ERPs to examine processing of phonetic details for comprehended words and minimal-pair mispronunciations from 14 to 20 months. The results were consistent with the behavioral findings, in that at 14 months ERPs to known words (e.g., *bear*) and their minimal-pair mispronunciations (e.g., *gare*) both differed in amplitude from ERPs to the dissimilar nonwords (e.g., *neem*), but were identical to each other (Figure 10.1). This suggested that for inexperienced word learners, minimal-pair mispronunciations were processed as known words. In contrast, at 20 months, ERPs to known words differed from those to both the mispronunciations and the dissimilar nonwords. Mills et al. interpreted the findings as showing that when cognitive load is increased, as is the case in mapping words onto meaning, it is difficult for an inexperienced word learner to keep all of the phonological details in mind. As the child becomes a more accomplished word learner, access to phonological details becomes more readily available. Thus, in addition to the findings (discussed above)

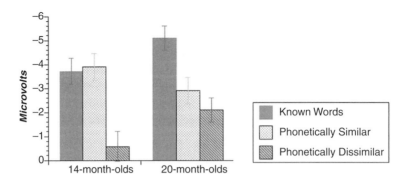

FIGURE 10.1. N200–400 mean area in 14- and 20-month-old infants listening to known words, minimal-pair mispronunciations (phonetically similar), and dissimilar nonwords (phonetically dissimilar). Data from Mills et al. (2004).

that speech perception abilities predict later language development, language experience is an important factor when infants map sounds onto meaning.

Brain and Vocabulary Development

One of the ways the social world influences language development is through the modification of speech used in talking to very young children. Across most cultures, men, women, and older children use a more melodic rhythm and simplify grammatical structures directed toward infants, in a form of speech called "motherese," "baby talk," or "infant-directed speech" (IDS). In this section, we examine how the special acoustic and affective properties of IDS influence the development of brain activity in response to spoken language.

IDS and Emotional Prosody

IDS is characteristically higher and more variable in pitch, has a slower tempo with elongated vowels, and includes more repetitions of key words than adult-directed speech (ADS) (Fernald et al., 1989; Garnica, 1977; Grieser & Kuhl, 1988). A number of studies have shown that from a very early age, infants prefer IDS to ADS (e.g., Cooper & Aslin, 1990; Fernald, 1984; Fernald & Kuhl, 1987; Werker & McLeod, 1989). Functionally, IDS is thought to focus infants' attention on potentially meaningful words, to enhance word segmentation, to convey affective information, and to facilitate word learning (Fernald, 1992). Children's preference for highly exaggerated pitch changes with age, as does the proposed role IDS plays in language processing.

Zangl and Mills (2007) examined brain activity to IDS versus ADS to familiar and unfamiliar words in typically developing infants at ages 6 and 13 months. They hypothesized that age-related changes in brain activity to IDS relative to ADS would reflect developmental differences in the underlying function of IDS in attentional and semantic processing at 6 and 13 months, respectively. The study tested the hypothesis that if IDS primarily serves to increase attention to potentially meaningful words, then words presented in IDS at 6 months would elicit increased amplitudes in ERP components linked to attention, such as the negative component Nc. The Nc displays a frontal distribution and is thought to index attention to and integration of the stimulus (de Haan & Nelson, 1997; Reynolds & Richards, 2005). The amplitude of the Nc increases with increased allocation of attention (Nelson & Monk, 2001). In addition, Zangl and Mills hypothesized that if IDS facilitates word comprehension, words presented in IDS at 13 months would elicit greater-amplitude ERP components associated with word recognition and meaning. An alternative hypothesis was that the physical characteristics of IDS would elicit larger-amplitude ERPs, regardless of children's age or familiarity with the words.

The results showed that for familiar words at both 6 and 13 months, IDS relative to ADS elicited a more negative-going wave from 600 to 800 msec, called the N600–800, over frontal, anterior temporal, temporal, and parietal regions (Figure 10.2). The timing and distribution of the N600–800 were consistent with the timing and distribution of the Nc, linked to attentional processing in other studies. This suggested that at both ages IDS serves to boost neural activity linked to attention to familiar words for which a child may already have some phonological representation. At 13 months, unfamiliar words presented in IDS also elicited a larger N600–800 than did those presented in ADS. That is, older, more experienced word learners may take advantage of the special properties of IDS for both familiar and novel acoustic–phonetic representations. In addition, at 13 but not at 6 months, the N200–400, which has been linked to word recognition, was larger to familiar words in IDS than to unfamiliar words in IDS over temporal and parietal regions of the left hemisphere. Although this study did not directly examine the effect of IDS on word learning, this result has some implications for the role of prosody in processing familiar phonological representations. Zangl and Mills have proposed that the acoustic features of IDS may facilitate early word learning by increasing neural activity to highly familiar and potentially meaningful words. If the increased activity occurs in temporal congruity with adult social attention to an object, IDS may serve to increase the strength of the neural association between the word and its referent.

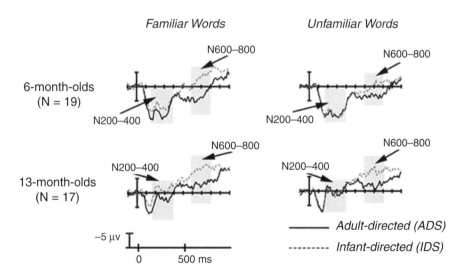

FIGURE 10.2. ERPs to infant-directed speech (IDS) versus adult-directed speech (ADS). Data from Zangl and Mills (2007).

Developmental changes in brain activity linked to processing prosodic information have also been observed using NIRS with Japanese infants 3 to 10 months of age (Homae, Watanabe, Nakano, & Taga, 2007). NIRS measures the hemodynamic response linked to neural activity by measuring the intensity of reflected light in the near-infrared range that has been projected through the skull and into the brain. In agreement with Zangl and Mills (2007), these authors found age-related changes in the distribution and responsiveness of cortical activation linked to processing prosodic information over frontal and temporoparietal regions. In the first study (Homae, Watanabe, Nakano, Asakawa, & Taga, 2006), 3-month-olds showed increased activity to normal relative to flattened speech. At 10 months the opposite pattern was observed, with greater activation to flattened speech (Homae et al., 2007). Developmental changes also included increased activity over prefrontal regions at 10 but not at 3 months of age. The NIRS results were interpreted as reflecting developmental changes in attention and integration of prosodic information in comparing flattened speech to the more familiar normal speech. Although this study did not directly examine activity to IDS versus ADS, the results have implications for the spatial localization of the ERP results linked to processing prosodic information.

Another proposed function of IDS is to convey meaning about affect and emotion. Behavioral studies suggest that Infants' preference for IDS over ADS may be associated more with the happy emotional affect characteristic of IDS than with other cues, such as mean pitch and fundamental frequency (Singh, Morgan, & Best, 2002; Trainor, Austin, & Dejardins, 2000). In a series of five looking-preference experiments in which affect and. IDS versus ADS speech register were modulated, Singh et al. (2002) showed that 6-month-old infants' preference for IDS over ADS disappeared when emotional prosody (affect) was held constant. Two electrophysiological studies have examined infants' brain activity to speech varying in affective quality. Santesso, Schmidt, and Trainor (2007) examined EEG responses to IDS that varied in emotional content. For 6-month-old infants, EEG power in the 4- to 8-Hz range over frontal regions was larger in response to fear than to surprise or to love/comfort. The difference between EEG to surprise and to love/comfort only approached significance, but EEG to love/comfort did not differ from baseline. Because EEG power is inversely correlated with activity, lower power denotes increased activity. In contrast to previous studies showing right frontal activity related to processing negative emotional stimuli (Davidson, 2000), Santesso et al. did not find patterns of EEG asymmetry associated with IDS for positive versus negative affect. The authors interpreted the findings as suggesting that increased EEG power is related to cognitive and attentional properties, as well as to intensity of the IDS.

The only ERP study examining processing of emotional prosody in infants found that angry voices elicited a greater negativity from 300 to 500 msec and a larger positive slow wave from 500 to 1000 msec than did happy or neutral prosody (Grossmann, Striano, & Friederici, 2005). Happy stimuli also elicited a positive slow wave relative to neutral stimulus, suggesting increased activity to the emotional intensity of the stimuli, but not to positive versus negative valence. Both the EEG and ERP studies suggested greater allocation of attention to negative emotional tone than to positive tone. In contrast, the Zangl and Mills (2007) study found increased negativity to IDS than to ADS in the 600- to 800-msec time window. This suggests greater allocation of attention to IDS. The Zangl and Mills study did not evaluate the stimuli for emotional content, but IDS stimuli are characteristically rated as happier sounding than ADS at the younger age. The Zangl and Mills study also only found modulation of ERP amplitudes for familiar words. It is possible that the other two studies did not find increased negativity to positive emotional stimuli because the words were not specifically familiar to the children. Additional research is needed to examine separable brain responses to speech register versus emotional content, in an ERP paradigm similar to that used in the behavioral study by Singh et al. (2002). The role that speech register and emotional prosody play in brain activity linked to word recognition also needs further study.

Word Recognition

One of the earliest challenges infants face in understanding spoken language is parsing individual words from continuous speech. Unlike written language, in which blank spaces mark word boundaries, continuous spoken language has no equivalent auditory cues to mark the onsets and offsets of words. Yet at about 7 months of age, even before speech perception has been tuned to their native language, infants are able to segment familiar words from continuous speech. Using the head-turn preference procedure (HPP), Jusczyk and Aslin (1995) showed that 7-month-olds exposed to English could recognize familiar words from continuous speech. Infants familiarized with two words preferred to listen to subsequent passages with sentences containing those words than to similar passages that did not contain the familiarized words. The HPP has been used extensively to examine developmental changes in word segmentation abilities in infants exposed to a variety of different languages (for a review, see Jusczyk, 1997). Mechanisms proposed to underlie infants' abilities to segment words from continuous speech include attunement to the statistical regularities of syllabic transitions within versus between words (Saffran & Theissen, 2003), the ability to use acoustic stress cues correlated with the beginnings of words (Theissen & Saffran, 2007), and the use of highly familiar words to segment unfamiliar speech (Bortfeld, Morgan, Golinkoff, & Rathbun, 2005).

Although there has been considerable research on word segmentation using behavioral methods, very little is known about the development of brain activity linked to word recognition. ERPs have been used to study word segmentation in adults in both English (Sanders & Neville, 2003) and French (Nazzi, Iakimova, Bertoncini, Serres, & de Schonen, 2008). Kooijman, Hagoort, and Cutler (2005) provided the first electrophysiological evidence showing signatures of brain activity linked to segmenting newly familiarized word forms from continuous speech in 10-month-old infants exposed to Dutch. Infants were exposed to 10 repetitions of two-syllable Dutch words with a strong–weak stress pattern, presented in isolation in a familiarization phase. ERPs from 200 to 500 msec became less positive (increased negative voltage) with increasing exposure to the words presented in isolation. During the subsequent test phase, infants listened to auditory sentences presented in continuous speech. ERPs time-locked to the previously familiarized words also showed a larger negativity than those to unfamiliar words, but only over the left hemisphere. Although the duration of the two-syllable words was approximately 720 msec, ERP differences to familiar versus unfamiliar words were apparent by 340–370 msec after word onset. That is, brain activity linked to recognition was evident by the end of the first syllable. A recent preliminary report with 7-month-olds suggests that similar ERP patterns were observed, but with a different distribution over the scalp (Kooijman, Johnson, & Cutler, 2008). This is in contrast to behavioral findings for 7.5-month-old Dutch infants with the HPP (Kooijman et al., 2008). The behavioral studies did not show evidence of word segmentation until 9 months of age, even when the ERP and behavioral tasks used the same stimuli (Kooijman et al., 2008). The authors interpreted the ERP results as reflecting a brain response that is the precursor to the overt head-turn behavior. Taken together, the ERP results go beyond the behavioral findings by providing information about the timing and developmental changes in the organization of brain activity linked to the segmentation process.

Recognition of familiar words from a child's environment, rather than those trained in a laboratory setting, are of particular interest for later vocabulary development. Preference looking procedures have shown that infants as young as 6 months of age can use highly familiar names such as "Mommy" and their own names to segment adjacent unfamiliar words from continuous speech (Bortfeld et al., 2005), and show evidence of mapping these highly familiar names onto meaning (Tincoff & Jusczyk, 1999). Using the HPP, Vihman, Nakai, DePaolis, and Halle (2004) found evidence of recognition of multiple untrained words at 11 but not at 9 months. Vihman et al. argue that at this young age, the reaction to familiar words reflects recognition of word forms, but not necessarily comprehension of word meanings.

Electrophysiological studies show discrimination of words whose meanings were understood from unfamiliar words by 200–500 msec after word onset in children between 12 and 20 months of age (Conboy & Mills, 2006; Mills, Coffey-Corina, & Neville, 1993, 1997; Molfese, 1989; Molfese, Morse, & Peters, 1990; Molfese, Wetzel, & Gill, 1993). These findings suggested that recognition effects are apparent even before the end of the word. In these studies each of the words was presented several times, and the early onset may in part have been due to repetition during the testing session. Thierry, Vihman, and Roberts (2003) modified the Mills et al. (1997) paradigm and presented 11-month-olds with 56 unfamiliar words and 56 words familiar to children in that age range (Hamilton, Plunkett, & Schafer, 2001). Consistent with the findings by Mills and colleagues, they found an increased negativity to familiar versus unfamiliar words, peaking at about 200 msec. However, in contrast to Mills et al. and Molfese et al., they interpreted the results as reflecting the MMN. As noted earlier, the MMN is an obligatory response to a change in acoustic input and is thought to reflect auditory sensory memory. Thierry et al.'s explanation was that most of the children only understood a small subset of the words; therefore, the low-probability familiar words elicited an MMN relative to the high-probability unknown words. In an extension of that study, using a combined HPP and ERP technique in Welsh- and English-speaking infants, the N2 effect to familiar word forms was observed by 10 months of age in monolingual English infants and by 11 months in bilingual English–Welsh infants (Vihman, Thierry, Lum, Keren-Portnoy, & Martin, 2007). The ERP effects were correlated with the HPP findings, showing an effect of familiarity on N2 amplitudes.

In spite of the similarities of the N2 familiarization effect in peak latency and polarity between the Vihman et al. and Mills et al. studies, it is unlikely that they reflect the same underlying functional significance. First, in the Vihman et al. (2007) study, the N2 familiarity effect was absent by 12 months of age. The authors concluded that by 12 months of age, the infants understood a large enough proportion of the words to eliminate the oddball familiarity effect. Second, the number of known versus unknown words was balanced in the Mills et al. studies, as the word lists were tailored to each child's comprehension vocabulary. Third, in the Mills et al. studies, one-third of the stimuli were backward words—that is, known word files played backward to control for some of the same physical characteristics of speech. The backward words did not elicit an N2. If the N2 was a response to differences in the probability of phonological familiarity, one would expect the lower-probability backward words to elicit an MMN. Finally, the distribution of the N200–400 (Mills et al., 1997) was largest over temporal and parietal regions, in contrast to the N2 (Vihman et al., 2003, 2007), which was largest over right anterior regions and more consistent with the distribution of an MMN effect.

More recently, the development of brain activity linked to familiarity was examined in monolingual English infants from 3 to 11 months of age (preliminary results reported in Sheehan & Mills, 2008). In this study, familiarity was determined by parental rating based on the number of times a child heard a word; it did not reflect word comprehension. A word was scored as familiar if a parent rated the word as heard by an infant several times a day or several times per week. In contrast, unfamiliar words were rated as rarely or never heard by the child. The ERPs from infants ages 3–4, 6–8, and 9–11 months are shown in Figure 10.3.

At 9–11 months, an N200–400 was observed to both familiar and unfamiliar words, but not to backward words. There was also a correlation with vocabulary development. Consistent with the earlier findings with 13-month-olds, in the 9- to 11-month-old infants whose parents rated the familiar words as comprehended by the children, the N200–500 was larger for the familiar (known) words than for the unfamiliar words. No such differences were observed in the children whose parents said that the familiar words were heard but not understood by the children. From 6 to 8 months, unfamiliar and backward words elicited large positivities, and only familiar words elicited negative-going peaks from 200 to 500 msec. In contrast, at 3–4 months, all three word types elicited a large positivity between 175 and 550 msec. This positive-going wave was significantly larger for familiar than for unfamiliar or backward words. The lack of the N200–400 responses in 3-month-olds suggests an immature brain response to familiar speech. However, the significant difference in the amplitude of the positive-going wave suggests a familiarity effect even at 3 months. This is the youngest age at which any

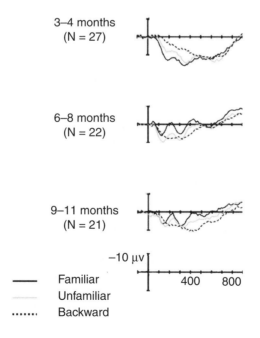

FIGURE 10.3. Word recognition ERPs in infants ages 3–4, 6–8, and 9–11 months. Data from Sheehan and Mills (2008).

form of word recognition has been reported in studies using ERP or behavioral methods. These findings are inconsistent with the lack of familiarity effects observed in the Vihman et al. studies prior to 10 months of age. In the Sheehan and Mills (2008) study, the words were each repeated six times and were rated as highly familiar to the infants, thus increasing the likelihood of finding familiarity effects.

Recognition of phonologically specified word forms through familiarization in laboratory training studies or recognition of familiar words from a child's environment, as described in the previous sections, does not necessitate comprehension of the meanings of the words. In the remainder of the chapter, we focus on developmental changes associated with semantic processing of meaningful words and gestures.

Meaning and Single-Word Processing

By their first birthdays, most children understand several words, intentionally point at objects to request labels, and have even started to say a few words. The earliest mapping of word–object associations has been demonstrated in infants as young as 6–8 months of age (Gogate, Bolzani, & Bentancourt, 2006; Tincoff & Jusczyk, 1999). By 13–15 months of age, children show evidence of "fast-mapping" words to meanings with a limited number of exposures (Schafer & Plunkett, 1998; Woodward, Markman, & Fitzsimmons, 1994), along with a sharp increase in the number of words understood. At 18–20 months infants typically show a rapid increase in the production of words, often called the "vocabulary spurt" or "naming explosion." The extent to which these putative language milestones are associated with qualitative changes in underlying cognitive abilities has been a topic of considerable interest and debate (for a review, see Plunkett, Sinha, Møller, & Strandsby, 1992). In their early ERP studies of single-word processing, Mills and colleagues addressed the question of whether the attainment of different language milestones was associated with a concomitant change in the organization of brain activity to known and unknown words (Mills, Coffey-Corina, & Neville, 1994; Mills et al., 1993, 1997). They reasoned that if dramatic changes in vocabulary were linked to a qualitative shift in cognitive processes, children would show concomitant changes in the organization of language-relevant brain activity.

Of particular interest was the phenomenon of the marked acceleration in vocabulary size that often occurs when a child can say 50–100 words; again, this "vocabulary burst" typically occurs at about 18–20 months. Mills et al. contrasted patterns of brain activity to words whose meanings were understood by the children, according to parental report and laboratory measures, with the patterns to words whose meanings the children did not know; they used two age groups designed to measure changes before and after the vocabulary spurt (i.e., 13–17 and 20 months). The approach taken to isolate the differential effects of language experience versus maturation was to study children who were the same age but differed in language abilities, and to contrast these results with those for children who were matched on language skills but differed in chronological age. The results showed that at 13–17 months, when children had small vocabularies, the amplitudes of ERPs from 200 to 400 msec were larger to known than to unknown words. These ERP differences were broadly distributed over anterior and posterior regions of both the left and right hemispheres. In contrast, for 20-month-olds, who had vocabularies over 150 words, ERP differences from 200 to 400 msec were more focally distributed over temporal and parietal regions of the left hemisphere. To determine the effects of vocabulary size independently of chronological age, ERPs were examined based on a median split of the groups for the number of words comprehended at 13–17 months, and number of words produced at 20 months. The results from high- versus low-vocabulary groups at each age showed that this effect was linked to vocabulary size rather than to chronological age per se. Mills et al. raised the working hypothesis that observed differences in the distribution of brain activity to known versus unknown words were linked to the marked changes in language abilities between 13 and 20 months. Moreover, they suggested that the results were consistent with a qualitative shift in the way children process known words before versus after the vocabulary spurt.

The hypothesis that increasing vocabulary size is associated with concomitant changes in the lateral distribution of ERP differences to known versus unknown words was further supported by subsequent studies of late talkers at 20 and 28–30 months (Mills, Conboy, & Paton, 2005a; Mills & Neville, 1997); for 20-month-old toddlers learning two languages at the same time (Conboy & Mills, 2006); and for content versus relational words in children ages 20, 28–30, and 36 months (St. George & Mills, 2001). However, an alternative explanation to the hypothesis of qualitative reorganization is that the strength of the word–object associations as reflected in amount of experience with individual words, rather than absolute vocabulary size, plays a role in the lateral distribution of specific ERP components. This question was addressed in an ERP study of novel word learning in 20-month-olds with large versus small vocabularies (Mills, Plunkett, Prat, & Shafer, 2005b). ERPs to novel words that were paired with an object during a training phase were compared to ERPs to novel words that were simply repeated the same number of times, to control for phonological familiarity of the word form. Two competing hypotheses were tested. If the left-lateralized distribution of brain activity observed at 20 months in the previous studies was linked to the amount of experience with individual words, then newly learned words should show a bilateral difference in ERPs to "newly learned" versus "simply repeated" novel words, similar to the known versus unknown difference for 13- to 17-month-olds in the Mills et al. (1997) study. However, if postspurt vocabulary size was associated with a shift in the way 20-month-olds process words, then the ERP differences to newly learned versus simply repeated novel words should be lateralized to temporal and parietal regions of the left hemisphere. After training, the N200–500 to the novel words that were paired with an object was larger than the N200–500 to the untrained novel words. This result was important in supporting the position that meaning rather than phonological familiarity modulates the amplitude of this ERP component. Consistent with the individual-word-learning hypothesis, the N200–500 difference effect to newly learned versus simply repeated words was bilaterally distributed. To test the hypothesis further, the ERPs were compared for the groups of 20-month-old infants with high and low vocabulary sizes. Both high and low producers showed significant ERP differences over the left and right hemispheres; however, for the high-production group, a left-greater-than-right asymmetry emerged. This suggested that the left-hemisphere asymmetry in the high production group was linked to faster learning rates. Indeed, across all of these studies, the rate of learning (as indexed by percentile rankings relative to same-age peers) was important in the lateral distribution of ERPs. That is, faster learners with higher percentile rankings showed more lateral asymmetries than slower learners, regardless of age or absolute vocabulary size.

A naturally occurring phenomenon in which vocabulary size varies within the same child, while maturation is held constant, is the case of simultaneous bilingualism. Bilingual infants typically learn each of their languages within different social contexts, and therefore tend to develop each language at different rates. We (Conboy & Mills, 2006) found that bilingual 20-month-old infants showed bilateral N200–500 ERP effects to known versus unknown words in each of their languages. In contrast to the findings for monolingual infants, these effects were larger over right frontal and anterior temporal electrode sites for infants with larger overall vocabularies. However, this asymmetry was only observed for ERPs to words in the dominant language (i.e., the language in which the infants produced more words). These results cannot be due to different rates of brain maturation for slow versus fast language learners, because different patterns of lateralization occurred within the same brains. Instead, they must be due to differing rates of development across the two languages of the same infants, as well as across infants. We further compared two groups of infants who heard words either in a randomly mixed language paradigm or in blocks of 50 words of each language at a time, and who were matched on vocabulary sizes in each language. The right asymmetry of the known–unknown word effects were present only in those infants tested in the mixed-language condition. From these results, we hypothesized that the right frontal effects reflected the recruitment of additional cognitive processes for the more demanding mixed-language condition. The absence of those effects in infants tested in the single-language condition, as well as in the nondominant language of children tested in the mixed-language condition and in children with smaller vocabularies, further suggests that language experience interacts with task demands. Thus, although the original hypothesis that brain organization changes with a particular language milestone may have been too simple, our recent work is consistent with the position that cerebral specializations for language emerge as a function of the experience of learning language, and to some extent depend on the rate of learning. We do not mean to suggest that these findings are consistent with a pattern of progressive lateralization that occurs as a function of brain maturation. Instead, the studies suggest that emerging functional specializations are dynamic, change as a function of the immediate task, and interact with a variety of processes across domains.

In sum, this set of studies suggests that lateral asymmetries emerge as a function of rapid learning and more automatic processes. More experienced word learners reach a higher level of proficiency at a faster rate than do inexperienced word learners. Therefore, the observed changes from bilateral to left lateralized ERPs in monolingual children before versus after the vocabulary spurt do not necessarily reflect qualitative changes in the way children process words, but may be due to an interaction between familiarity with specific word meanings and faster rates of learning.

On a methodological note, the words in this set of studies were presented as single words without objects, pictures, or accompanying phrases. Even very young children show evidence of understanding the meanings of words in the absence of perceptual support. We believe that the negative-going waveform in the 200- to 500-msec range indexes overlapping neural activity linked to both word recognition and word meaning (see Sheehan & Mills, 2008, for a complete discussion of this argument). However, another approach to studying semantic processing is to examine cross-modal integration of auditory words paired with objects or visual scenes. In the next section, we review developmental studies of brain activity linked to semantic processing of early words and gestures.

Cross-Modal Processing of Word Meaning

ERP studies of semantic processing in adults have demonstrated that contextual information modulates the neural response to words. Semantic processing is reflected in a negative-going component starting at 200 msec and peaking at about 400 msec, called the N400. The amplitude of the N400 is increased when a meaningful stimulus (such as a word or picture) violates the preceding context, and is decreased when the stimulus is congruent with the context. For example, the word *dog* preceded by the picture of a dog (congruent condition) would elicit a small N400, whereas the word *dog* preceded by a picture of a chair (incongruent condition) would elicit a larger N400.

The picture–word semantic congruency paradigm has been used to examine brain activity linked to semantic processing in young infants (Friedrich & Friederici, 2004, 2005a, 2005b, 2006). The studies included children at 12, 14, and 19 months of age. The stimuli consisted of pictures of common objects; words that named or did not name the object in the picture (congruent or incongruent words); pseudowords; and nonwords. This review focuses on the results for congruent versus incongruent words. Pictures were presented for 4 sec, followed by one of the types of words 900 msec after the onset of each picture. All age groups showed a negative response to congruent relative to incongruent words, starting at 150–500 msec after onset. This negativity was interpreted as a phonological–lexical priming effect similar to that observed in adults. The later N400 effect, which was larger to incongruent words, was observed at 19 and 14 months but not at 12 months. This suggested that the neural mechanisms involved in producing the N400 are not yet mature at 12 months. In addition, at 19 months children with age-adequate language skills showed both the lexical priming and N400 effects to congruent versus incongruent words. By contrast, 19-month-olds who scored more than one standard deviation below the mean on a standardized measure of comprehension and production showed the lexical priming effect to incongruent words, but not an N400. Also, children who showed age-typical ERP patterns at 19 months had better language production scores at 30 months than did children with lower language scores who did not show these ERP patterns. The results were interpreted as showing a direct relationship between development of early comprehension mechanisms and later production abilities.

Using a similar cross-modal paradigm, Mills and colleagues (as reported in Mills & Sheehan, 2007) investigated the role of working memory in semantic processing in 13-, 20-, and 36-month-olds and adults. The first experiment used a picture–word paradigm similar to that in the studies discussed above. However, there were some differences between the methods used in this study and those used by Friedrich and Friederici. First, the word lists were tailored to each child, so that only words whose meanings the child understood were included in the study. Second, trials were eliminated from the analysis if the child did not see the pictures. Third, to increase the number of trials that could be presented in a short period of time, each picture was presented for 1500 msec. At 500 msec after the picture appeared, a word was presented that was either congruent or incongruent with the picture. Each picture and word were presented twice, once as a match and once as a mismatch. ERPs were examined for the presence of the N400 component. All age groups showed an N400-like response that was larger to the incongruent than to the congruent condition from 200 msec after word onset to the end of the 1000-msec epoch, and this response was broadly distributed across the scalp. For the two youngest groups, a median split for vocabulary size was used to divide the groups into high and low comprehenders at 13 months and high and low producers at 20 months. Unlike the findings by Friedrich and Friederici, the onset, duration, and distribution of the N400 effect did not differ for the high- and low-vocabulary groups.

It is possible that the methodological differences between the studies could account for some of the differences in the results. In the studies by Friedrich and Friederici (2004, 2005a, 2005b, 2006) the infants, especially the low comprehenders, did not understand the meanings of all of the words. According to parental report, the average proportion of word comprehension was 76% of 44 words (Friedrich & Friederici, 2004). In addition, participants were only required to attend to 70% of the trials to be included, and missed an average of 15–17% of the trials, which were not excluded from the final analysis unless they contained other movement or eye artifact. In our studies, only known words were used, and all of the trials in the analysis were ones in which a child saw the pictures. Also, trials were not included if the infant did not see the pictures. Our results suggest that if infants understand the meanings of all the words and see all the pictures, the onset of the N400 response is at the same latency as the adult N400, even for 13-month-olds. These findings suggest that the neural systems that mediate the N400 in adults and older children are active even for early word–object associations. Additional research is needed to examine the same effects in 12-month-olds, using only words whose meanings are understood by the children. Another puzzling difference in the results is that we did not find an early negative congruency effect at any age. It is possible that the shorter time period between the onset of the picture and the word (i.e., 500 msec in our study and 900 msec in the Friedrich and Friederici studies) did not allow sufficient processing time in working memory to elicit this effect. That is, 900 msec may be sufficient time for young infants to see a picture and generate a word, but 500 msec may not be sufficient time. Another possibility is that our studies did not include pseudowords or illegal nonwords, and the addition of these word types in the Friedrich and Friederici studies may have drawn more attention to the beginning phonemes or elicited a familiarity-based MMN response to the lower-probability known words, as in the Vihman et al. studies discussed earlier.

The second experiment we manipulated working memory demands by presenting the word prior to the picture. In the first study, each word was presented while the picture was in view, decreasing any working memory load; in this second experiment, each word was presented 500 msec prior to the onset of the picture. One interpretation of the vocabulary spurt is that the marked changes in vocabulary size are related to increasing memory abilities (Gershkoff-Stowe, 2002). In this task, increases in the flexibility of working memory abilities would allow an infant to hold the word in mind longer and compare the word with the picture in working memory. It was predicted that vocabulary size would have an effect on the N400 response with increased working memory load. Experienced word learners with a larger vocabulary size would have a lower working memory load due to highly familiar word–object associations, allowing for more in-depth processing of related associations. Thus it was predicted that the 20-month-olds would show an N400 effect in this study. In contrast, it was predicted that children with weaker word–object associations and a less mature working memory system at 13 months might not show the N400 semantic priming effect.

The results showed that adults, 3-year-olds, and 20-month-olds showed an N400 response to the anomalous word–picture pairs. Unlike the findings for the simultaneous picture–word paradigm, the latency and distribution of the N400 effect showed clear developmental differences even for these age groups. At the younger ages, the N400 response increased in the onset and peak latency of the effect. Adults and 3-year-olds showed a larger N400 response to incongruent words from 200 to 600 msec after a picture appeared. At 20 months, the onset was later and was more limited in distribution. The N400 effect was not observed for the 13-month-olds as a group. However, the 13-month-olds with larger comprehension vocabularies (>100 words) showed an N400 response similar to that seen in the 20-month-olds. In contrast, 13-month-olds with smaller vocabularies (<100 words) did not show a significant N400 effect. We know from a variety of different memory tasks that 13-month-olds can keep something in mind for much longer than 500 msec. We believe that for this age group, a picture may interfere with and replace the representation of a word. That is, the children could not keep the word in mind and simultaneously compare it to the subsequent picture. There are two possible explanations for why the 13-month-old high comprehenders did show the N400 effect. The 13-month-old children with larger vocabularies may have had better working memory systems and/or stronger word–object associations that reduced working memory load. These findings are consistent with Gershkoff-Stowe's (2002) position that memory abilities are important in vocabulary development between 13 and 20 months. More generally, the findings lend support to the position that domain-general processes such as working memory influence patterns of language-relevant brain activity.

The extent to which language is domain-specific or emerges from more domain-general cognitive processes is a central topic in cognitive neuroscience and is particularly relevant to theories of development. One area of social-communicative development that is thought to reflect shared domain-general processes with language is the use of gestures. Bates and Dick (2002) have argued that similar domain-general neural systems mediate language and gesture processing in infants and adults. It has also been suggested that gestures played an important function in the evolution of language (Arbib, 2005; Kelly et al., 2002; Rizzolatti & Arbib, 1998). In the next section, we examine evidence for shared neural systems in processing meaning in words and gestures.

Neural Activity to Meaningful Words and Gestures

Shortly before infants begin to talk until about the time of the vocabulary spurt, they may use gestures in much the same way as words to refer to an object or action (Acredolo & Goodwyn, 1988; Bretherton et al., 1981). For example, infants may use the word *car* to label a car, but may use a representational gesture (such as moving the hand back and forth as if rolling a car) to refer to the object instead. The period of using gestures to name objects is limited to early development between 12 and 18 months of age. At about the time of the vocabulary burst, words and gestures take on divergent communicative functions reflecting the conventions of a child's language. Infants show a gradual change in the use of gestural communication as they gain experience with language, with words replacing symbolic gestures over time (Acredolo & Goodwyn, 1988; Bates & Dick, 2002; Bretherton et al., 1981; Iverson, Capirci, & Caselli, 1994; Namy, Campbell, & Tomasello, 2004; Namy & Waxman, 1998, 2002).

Sheehan, Namy, and Mills (2007) used ERPs to study developmental changes from 18 to 26 months in patterns of brain activity linked to processing of words and gestures, using a cross-modal priming task similar to the word–picture paradigm described in the section above. The stimuli consisted of a movie clip of an actor either saying a word or producing a gesture, followed by a picture that matched or did not match the preceding word or gesture. Of particular interest were developmental changes in the patterns of brain activity linked to semantic processing of words compared to gestures. As in the study described above, an N400 priming effect was predicted for incongruent pictures following videos of words. Similarities in the timing and distribution of the N400 response to semantic violations for gestures and words would be taken as evidence for shared neural systems. In contrast, differ-

ences in the occurrence, timing, and/or distribution of N400 effects between the trials beginning with a word and the trials beginning with a gesture would indicate that processing of
these differing stimuli was being mediated by nonidentical neural systems. In addition, due
to developmental changes in the way 18- versus 26-month-olds use gestures to communicate,
Sheehan et al. predicted developmental differences in the N400 response to incongruent
pictures following gestures but not words.

The results for the 18-month-olds indicated an N400 semantic congruency effect from
400 to 600 msec to pictures preceded by words or gestures. These results were consistent
with the hypothesis that early in development, similar neural systems are used to process
gestures and words. In contrast, at 26 months the congruency effect was limited to pictures
preceded by words (Figure 10.4). The different patterns of activation for words and gestures
at 26 months are consistent with the hypothesis that children's use of gestures takes on less
of a symbolic and more of a deictic function over the course of early development (Iverson,
Capirci, Longobardi, & Caselli, 1999; Masur, 1982). Namy and colleagues (Namy et al., 2004;
Namy & Waxman, 1998) suggest that children may accept a gesture to name a novel object
at 18 but not at 26 months, because they have adapted to the conventional roles of words and
gestures in communication. One explanation is that at 26 months the semantic association
between the gesture and the object is weakened, relative to the association between the word
and the picture. Thus gestures may be less predictive of a specific object than are words,
resulting in greater neural activity on the congruent trials and diminishing the incongruency
N400 effect. These findings illustrate how dynamic changes in the use of communicative
symbols across development shape the organization of neural activity linked to semantic
processing.

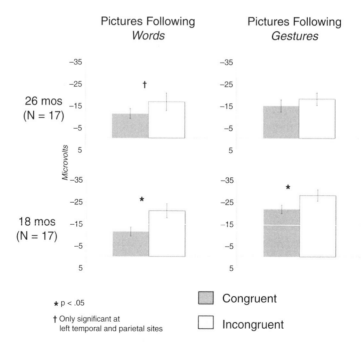

FIGURE 10.4. N200–400 mean amplitudes to congruent versus incongruent pictures following words
versus gestures, presented to 18- and 26-month-olds. Data from Sheehan, Namy, and Mills (2007).

Social Factors in Brain and Language Development

Even before a child is born, the social world can influence early preferences for the infant's mother's voice and the rhythm of what will be the infant's native language (DeCasper, & Fifer, 1980; Mehler, Bertoncini, Barriere, & Jassik-Gerschenfeld, 1978). Baldwin and Meyer (2007) discuss three ways in which social factors influence language development: social input, social responsiveness, and social understanding. Social input provides a variety of meaningful cues for language development. One example is IDS, in which adults speak more slowly, use prosodic cues, and repeat important words. In this chapter, we have reviewed a set of studies showing that language experience shapes phonological categories specific to a child's native language, and that early speech perception abilities predict later language development. Kuhl and colleagues (Conboy & Kuhl, 2007; Kuhl, 2007; Kuhl et al., 2003; Kuhl et al., 2008) suggest that social interaction is essential for the transition from an initial universal state of language processing to one that is language-specific. In her native-language magnet theory—expanded, Kuhl (2007; Kuhl et al., 2008) suggests that although distributional properties of the language input affect learning in the first phase of speech perception, attention to social cues is necessary for infants to become native listeners.

One of the critical ways social factors influence language learning is by attention to and preference for IDS. The exaggeration of phonetic features in IDS facilitates learning by helping infants attend to relevant information and ignore irrelevant information (Liu et al., 2003, 2007). Xiao, Bitsko, Sheehan, Larson, and Mills (2008) have reviewed studies showing developmental changes in brain activity to IDS and emotional prosody. In an extension of the Zangl and Mills (2007) paper, we have examined brain activity to IDS in 6-month-old infants who have altered early social input because their mothers are depressed. Depressed mothers use less of IDS and other social cues (such as smiling and approval) than do healthy mothers. We propose that mothers with depression do not use IDS in a reliable manner, and therefore that their children will have difficulty taking advantage of the important role IDS plays in phonological perception, word segmentation, and later vocabulary development. Our preliminary results suggest that 6-month-old infants of depressed mothers do not show increased activity to IDS even for familiar words, compared to 6-month-olds with healthy mothers. Interestingly, increased brain activity to IDS relative to ADS was strongly and only correlated with mothers' use of IDS in a 3-minute observation for both depressed and healthy mothers. Of particular concern, follow-up measures at 13 and 20 months showed that infants of depressed mothers lagged behind their same-age peers in vocabulary development, with the majority of children scoring at about the 25th percentile on both comprehension and production measures of vocabulary size. Infants of depressed mothers have altered social input, which puts them at risk for later language delay.

A second form of social influence on language development involves a child's responsiveness to social input. One example of this influence is that children with autism who lack a preference for IDS over nonspeech analogues also show abnormal ERP responses to speech syllables and increased delay in verbal scores, compared to those who prefer IDS to nonspeech stimuli (Kuhl, Coffey-Corina, Padden, & Dawson, 2005a).

A third way in which social factors influence language development is social understanding. Behavioral studies of word learning suggest strong correlations among such factors as joint attention, imitation, and detecting the goals and intentions of the speaker. Recent work by Conboy et al. (2008a) used ERP measures of phonetic discrimination and word processing to examine how joint engagement/attention with adults providing input in a second language at 9–10 months would influence the processing of second-language stimuli. In this research, infants from monolingual English-speaking homes were provided with laboratory experience in Spanish through a series of twelve 25-minute play sessions spread across 1 month's time. Thus the input provided to the infants was designed to simulate the naturalistic social contexts in which infants typically experience language in the real world. Infants were tested on an ERP speech paradigm that assessed discrimination of a Spanish and an English phonetic contrast, and on an ERP familiar–unfamiliar word paradigm (such as the ones described above) using Spanish and English words, both before and after the Spanish exposure sessions. Infants showed an MMN-like response to both the English and Spanish phonetic contrasts after exposure to Spanish (vs. a response only to English prior to the Spanish exposure). Of particular interest is that the amplitude of this effect was linked to the rates of infants' eye gaze shifts between toys and the faces of the Spanish-speaking adults who showed the infants the toys while speaking about the toys (Conboy et al., 2008a). Similarly, infants who showed ERP N200–500 effects to words used during the Spanish sessions versus unexposed words had higher rates of eye gaze shifts than those who did not show the ERP effects (Conboy et al., 2008a).

We interpret these gaze shift behaviors as reflections of infants' emerging understanding of the Spanish-language tutors as partners in communication regarding the toys. Infants who showed higher levels of joint engagement with their language tutors appeared to make better use of the language input provided by those tutors, as reflected in their brain activity to phonetic contrasts and to words. These are the first data that provide a link between infant–adult joint engagement and phonetic learning, and the first data to link such behaviors to the development of the brain systems that mediate language. Further work is needed to better understand the relationship between the neural bases of social abilities and those that mediate language abilities. This is a critical direction for the field, and there is much work to be done.

References

Acredolo, L. P., & Goodwyn, S. W. (1988). Symbolic gesturing in normal infants. *Child Development, 59,* 450–466.

Akhtar, N., & Tomasello, M. (1998). Intersubjectivity in early language learning and use. In S. Bråten (Ed.), *Intersubjective communication and emotion in early ontogeny* (pp. 316–335). Cambridge, UK: Cambridge University Press.

Arbib, M. A. (2005). From monkey-like action recognition to human language: An evolutionary framework for neurolinguistics. *Behavioral and Brain Sciences, 28,* 105–167.

Aslin, R. N., Pisoni, D. B., Hennessy, B. L., & Perey, A. J. (1981). Discrimination of voice onset time by human infants: New findings and implications for the effects of early experience. *Child Development, 52,* 1135–1145.

Baldwin, D., & Meyer, M., (2007). How inherently social is language? In E. Hoff & M. Shatz (Eds.), *Blackwell handbook of language development* (pp. 87–106). Malden, MA: Blackwell.

Bates, E., & Dick, F. (2002). Language, gesture, and the developing brain. *Developmental Psychobiology, 40,* 293–310.

Baumwell, L., Tamis-LeMonda, C. S., & Bornstein, M. H. (1997). Maternal verbal sensitivity and child language comprehension. *Infant Behavior and Development, 20*(2), 247–258.

Beckwith, L., & Cohen, S. E. (1989). Maternal responsiveness with preterm infants and later competency. In M. H. Bornstein (Ed.), *New directions for child development: No. 43. Maternal responsiveness: Characteristics and consequences* (pp. 75–87). San Francisco: Jossey-Bass.

Beckwith, L., Cohen, S. E., Kopp, C. B., Parmelee, A. H., & Marcy, T. G. (1976). Caregiver–infant interaction and early cognitive development in preterm infants. *Child Development, 47,* 579–587.

Belsky, J., Goode, M. K., & Most, R. K. (1980). Maternal stimulation and infant exploratory competence: Cross-sectional, correlational, and experimental analysis. *Child Development, 51*(4), 1168–1178.

Best, C. T., & McRoberts, G. (2003). Infant perception of non-native consonant contrasts that adults assimilate in different ways. *Language and Speech, 46,* 183–216.

Best, C. T., McRoberts, G. W., LaFleur, R., & Silver-Isenstadt, J. (1995). Divergent developmental patterns for infants' perception of two nonnative consonant contrasts. *Infant Behavior and Development, 18*(3), 339–350.

Best, C. T., McRoberts, G. W., & Sithole, N. M. (1988). Examination of perceptual reorganization for non-native speech contrasts: Zulu click discrimination by English-speaking adults and infants. *Journal of Experimental Psychology: Human Perception and Performance, 14*(3), 345–360.

Bloom, L. (1993). *The transition from infancy to language.* New York: Cambridge University Press.

Bornstein, M. H. (Ed.). (1989). *New directions for child development: No. 43. Maternal responsiveness: Characteristics and consequences.* San Francisco: Jossey-Bass.

Bornstein, M. H., Miyake, K., Azuma, H., Tamis-LeMonda, C. S., & Toda, S. (1990). Responsiveness in Japanese mothers: Consequences and characteristics. In *Annual Report of the Research and Clinical Center for Child Development* (Vol. 12, pp. 15–26). Sapporo, Japan: Hokkaido University.

Bornstein, M. H., & Tamis-LeMonda, C. S. (1989). Maternal responsiveness and cognitive development in children. In M. H. Bornstein (Ed.), *New directions for child development: No. 43. Maternal responsiveness: Characteristics and consequences* (pp. 49–61). San Francisco: Jossey-Bass.

Bornstein, M. H., & Tamis-LeMonda, C. S. (1997). Mothers' responsiveness in infancy and their toddlers' attention span, symbolic play, and language comprehension: Specific predictive relations. *Infant Behavior and Development, 20,* 283–296.

Bornstein, M. H., Tamis-LeMonda, C. S., & Haynes, O. M. (1999). First words in the second year: Continuity, stability, and models of concurrent and predictive correspondence in vocabulary and verbal responsiveness across age and context. *Infant Behavior and Development, 22*(1), 65–85.

Bortfeld, H., Morgan, J. L., Golinkoff, R. M., & Rathbun, K. (2005). Mommy and me: Familiar names help launch babies into speech-stream segmentation. *Psychological Science, 16,* 298–304.

Bråten, S. L. (Ed.). (2007). *On being moved: From mirror neurons to empathy.* Amsterdam: John Benjamins.

Bretherton, I., Bates, E., McNew, S., Shore, C., Williamson, C., & Beeghly-Smith, M. (1981). Comprehension and production of symbols in infancy: An experimental study. *Developmental Psychology, 17*(6), 728–736.

Carew, J. V., & Clarke-Stewart, K. A. (1980). Experience and the development of intelligence in young children at home and in day care. *Monographs of the Society for Research in Child Development, 45*(6–7, Serial No. 187).

Cheour, M., Alho, K., Ceponiene, R., Reinikainen, K., Sainio, K., Pohjavuori, M., et al. (1998a). Maturation of mismatch negativity in infants. *International Journal of Psychophysiology, 29*(2), 217–226.

Cheour, M., Alho, K., Sainio, K., Reinikainen, K., Renlund, M., Aaltonen, O., et al. (1997). The mismatch negativity to changes in speech sounds at the age of three months. *Developmental Neuropsychology, 13*(2), 167–174.

Cheour, M., Ceponiene, R., Lehtokoski, A., Luuk, A., Allik, J., Alho, K., et al. (1998b). Development of language-specific phoneme representations in the infant brain. *Nature Neuroscience, 1,* 351–353.

Cheour, M., Leppanen, P., & Kraus, N. (2000). Mismatch negativity (MMN) as a tool for investigating auditory discrimination and sensory memory in infants and children. *Clinical Neurophysiology, 111,* 4–16.

Cheour-Luhtanen, M., Alho, K., Kujala, T., Sainio, K., Reinikainen, K., Renlund, M., et al. (1995). Mismatch negativity indicates vowel discrimination in newborns. *Hearing Research, 82*, 53–58.

Cheour-Luhtanen, M., Alho, K., Sainio, K., Rinne, T., Reinikainen, K., Pohjavuori, M., et al. (1996). The ontogenetically earliest discriminative response of the human brain. *Psychophysiology, 33*, 478–481.

Clarke-Stewart, K. A. (1973). Interactions between mothers and their young children: Characteristics and consequences. *Monographs of the Society for Research in Child Development, 38*(6–7, Serial No. 153).

Conboy, B. T., Brooks, R., Taylor, M., Meltzoff, A., & Kuhl, P. K. (2008a, March). *Joint engagement with language tutors predicts brain and behavioral responses to second-language phonetic stimuli.* Poster presented at the Biennial Meeting of the International Society for Infant Studies, Vancouver, British Columbia, Canada.

Conboy, B. T., & Kuhl, P. K. (2007). Developing a culturally specific way of listening through social interaction. In S. L. Bråten (Ed.), *On being moved: From mirror neurons to empathy* (pp. 175–199). Amsterdam: John Benjamins.

Conboy, B. T., & Mills, D. L. (2006). Two languages, one developing brain: Event-related potentials to words in bilingual toddlers. *Developmental Science, 9*, F1–F12.

Conboy, B. T., Rivera-Gaxiola, M., Klarman, L., Aksoylu, E., & Kuhl, P. (2005). Associations between native and nonnative speech sound discrimination and language development at the end of the first year. In A. Brugos, M. R. Clark-Cotton, & S. Ha (Eds.), *Supplement to the Proceedings of the 29th Boston University Conference on Language Development.* Retrieved from *www.bu.edu/linguistics/APPLIED/BUCLD/supp29.html*

Conboy, B. T., Sommerville, J., & Kuhl, P. K. (2008b). Cognitive control factors in speech perception at 11 months. *Developmental Psychology, 44*(5), 1505–1512.

Cooper, R. P., & Aslin, R. N. (1990). Preference of infant-directed speech in the first month after birth. *Child Development, 61*, 1584–1595.

Davidson, R. J. (2000). Affective style, psychopathology, and resilience: Brain mechanisms and plasticity. *American Psychologist, 55*, 1196–1214.

DeCasper, A. J., & Fifer, W. P. (1980). Of human bonding: Newborns prefer their mother's voices. *Science, 208*, 1174–1176.

de Haan, M., & Nelson, C. A. (1997). Recognition of the mother's face by six-month-old infants: A neurobehavioral study. *Child Development, 68*, 187–210.

Dehaene-Lambertz, G. (2000). Cerebral specialization for speech and non-speech stimuli in infants. *Journal of Cognitive Neuroscience, 12*(3), 449–460.

Dehaene-Lambertz, G., & Baillet, S. (1998). A phonological representation in the infant brain. *NeuroReport, 9*, 1885–1888.

Dehaene-Lambertz, G., & Dehaene, S. (1994). Speed and cerebral correlates of syllable discrimination in infants. *Nature, 370*, 292–295.

Dehaene-Lambertz, G., Dehaene, S., & Hertz-Pannier, L. (2002). Functional neuroimaging of speech perception in infants. *Science, 298*, 2013–2015.

Dehaene-Lambertz, G., Hertz-Pannier, L., & Dubois, J. (2006a). Nature and nurture in language acquisition: Anatomical and functional brain-imaging studies in infants. *Trends in Neurosciences, 29*(7), 367–373.

Dehaene-Lambertz, G., Hertz-Pannier, L., Dubois, J., Mériaux, S., Roche, A., Sigman, M., et al. (2006a). Functional organization of perisylvian activation during presentation of sentences in preverbal infants. *Proceedings of the National Academy of Sciences USA, 103*, 14240–14245.

Dehaene-Lambertz, G., & Peña, M. (2001). Electrophysiological evidence for automatic phonetic processing in neonates. *NeuroReport, 12*, 3155–3158.

Diamond, A., Werker, J. F., & Lalonde, C. (1994). Toward understanding commonalities in the development of object search, detour navigation, categorization, and speech perception. In G. Dawson & K. W. Fischer (Eds.), *Human behavior and the developing brain* (pp. 380–426). New York: Guilford Press.

Eilers, R. E., Gavin, W., & Wilson, W. R. (1979). Linguistic experience and phonemic perception in infancy: A crosslinguistic study. *Child Development, 50*(1), 14–18.

Eimas, P. D., Siqueland, E. R., Jusczyk, P., & Vigorito, J. (1971). Speech perception in infants. *Science, 171*, 303–306.

Espy, K. A., Molfese, D. L., Molfese, V. J., & Modglin, A. (2004). Development of auditory event-related potentials in young children and relations to word-level reading abilities at age 8 years. *Annals of Dyslexia, 54*(1), 9–38.

Fernald, A. (1984). The perceptual and affective salience of mothers' speech to infants. In L. Feagans, C. Garvey, R. Golinkoff, M. T. Greenberg, C. Harding, & J. Bohannon (Eds.), *The origins and growth of communication* (pp. 5–29). Norwood, NJ: Ablex.

Fernald, A. (1992). Human maternal vocalizations to infants as biologically relevant signals: An evolutionary perspective. In J. H. Barkow, L. Cosmides, & J. Tooby (Eds.), *The adapted mind: Evolutionary psychology and the generation of culture* (pp. 391–428). New York: Oxford University Press.

Fernald, A., & Kuhl, P. (1987). Acoustic determinants of infant preference for motherese speech. *Infant Behavior and Development, 10*(3), 279–293.

Fernald, A., Taeschner, T., Dunn, J., Papousek, M., Boysson-Bardies, B., & Fukui, I. (1989). A cross-language study of prosodic modifications in mothers' and fathers' speech to preverbal infants. *Journal of Child Language, 16*, 477–501.

Friederici, A. D., Friedrich, M., & Weber, C. (2002). Neural manifestation of cognitive and precognitive mismatch detection in early infancy. *NeuroReport, 13*(10), 1251–1254.

Friedrich, M., & Friederici, A. D. (2004). N400-like semantic incongruity effect in 19-month-olds: Processing known words in picture contexts. *Journal of Cognitive Neuroscience, 16*(8), 1465–1477.

Friedrich, M., & Friederici, A. D. (2005a). Lexical priming and semantic integration reflected in the event-related potential of 14-month-olds. *NeuroReport, 16*, 653–656.

Friedrich, M., & Friederici, A. D. (2005b). Phonotactic knowledge and lexical–semantic processing in one-year-olds: Brain responses to words and nonsense words in picture contexts. *Journal of Cognitive Neuroscience, 17*, 1785–1802.

Friedrich, M., & Friederici, A. D. (2006). Early N400 development and later language acquisition. *Psychophysiology, 43*, 1–12.

Garnica, O. (1977). Some prosodic and paralinguistic features of speech to young children. In C. E. Snow & C. A. Ferguson (Eds.), *Talking to children: Language input and acquisition* (pp. 63–68). Cambridge, UK: Cambridge University Press.

Gershkoff-Stowe, L. (2002). Object naming, vocabulary growth, and the development of word retrieval abilities. *Journal of Memory and Language, 46*, 665–687.

Gogate, L. J., Bolzani, L. H., & Betancourt, E. A. (2006). Attention to maternal multimodal naming by 6- to 8-month-old infants and learning of word–object relations. *Infancy, 9*, 259–288.

Grieser, D. L., & Kuhl, P. (1988). Maternal speech to infants in a tonal language: Support for universal features in motherese. *Developmental Psychology, 24*, 14–20.

Grossmann, T., Striano, T., & Friederici, A. D. (2005). Infants' electric brain responses to emotional prosody. *NeuroReport, 16*, 1825–1828.

Hamilton, A., Plunkett, K., & Schafer, G. (2000). Infant vocabulary development assessed with a British communicative development inventory. *Journal of Child Language, 27*, 689–705.

Homae, F., Watanabe, H., Nakano, T., Asakawa, K., & Taga, G. (2006). The right hemisphere of sleeping infant perceives sentential prosody. *Neuroscience Research, 54*, 276–280.

Homae, F., Watanabe, H., Nakano, T., & Taga, G. (2007). Prosodic processing in the developing brain. *Neuroscience Research, 59*, 29–39.

Imada, T., Zhang, Y., Cheour, M., Taulu, S., Ahonen, A., & Kuhl, P. K. (2006). Infant speech perception activates Broca's area: A developmental magnetoencephalography study. *NeuroReport, 17*, 957–962.

Iverson, J. M., Capirci, O., & Caselli, M. (1994). From communication to language in two modalities. *Cognitive Development, 9*, 23–43.

Iverson, J. M., Capirci, O., Longobardi, E., & Caselli, M. C. (1999). Gesturing in mother–child interactions. *Cognitive Development, 14,* 57–75.

Jusczyk, P. W. (1997). *The discovery of spoken language.* Cambridge, MA: MIT Press.

Jusczyk, P. W., & Aslin, R. N. (1995). Infants' detection of the sound patterns of words in fluent speech. *Cognitive Psychology, 29,* 1–23.

Jusczyk, P. W., Pisoni, D. B., & Mullennix, J. (1992). Some consequences of stimulus variability on speech processing by 2-month-old infants. *Cognition, 43*(3), 253–291.

Kelly, S. D., Iverson, J. M., Terranova, J., Niego, J., Hopkins, M., & Goldsmith, C. (2002). Putting language back in the body: Speech and gesture on three time frames. *Developmental Neuropsychology, 22*(1), 323–349.

Kooijman, V., Hagoort, P., & Cutler, A. (2005). Electrophysiological evidence for prelinguistic infants' word recognition in continuous speech. *Cognitive Brain Research, 24,* 109–116.

Kooijman, V., Johnson, E. K., & Cutler, A. (2008). Reflections on reflections of infant word learning. In A. Friederici & G. Thierry (Eds.), *Trends in language acquisition research: Vol. 5. Early language development: Bridging brain and behaviour* (pp. 91–114). Amsterdam: John Benjamins.

Kuhl, P. K. (1983). Perception of auditory equivalence classes for speech in early infancy. *Infant Behavior and Development, 6,* 263–285.

Kuhl, P. K. (1998). Language, culture and intersubjectivity: The creation of shared perception. In S. Bråten (Ed.), *Intersubjective communication and emotion in early ontogeny* (pp. 297–315). Cambridge, UK: Cambridge University Press.

Kuhl, P. K. (2007). Is speech learning 'gated' by the social brain? *Developmental Science, 10,* 110–120.

Kuhl, P. K., Andruski, J., Christovich, I., Christovich, L., Kozhevnikova, E., Ryskina, V., et al. (1997). Cross-language analysis of phonetic units in language addressed to infants. *Science, 277,* 684–686.

Kuhl, P. K., Coffey-Corina, S., Padden, D., & Dawson, G. (2005a). Links between social and linguistic processing of speech in preschool children with autism: Behavioral and electrophysiological measures. *Developmental Science, 8,* F1–F12.

Kuhl, P. K., Conboy, B. T., Coffey-Corina, S., Padden, D., Rivera-Gaxiola, M., & Nelson, T. (2008). Phonetic learning as a pathway to language: New data and native language magnet theory—expanded (NLM-e). *Philosophical Transactions of the Royal Society of London, Series B, Biological Sciences, 363*(1493), 979–1000.

Kuhl, P. K., Conboy, B. T., Padden, D., Nelson, T., & Pruitt, J. (2005b). Early speech perception and later language development: Implications for the "critical period." *Language Learning and Development, 1,* 237–264.

Kuhl, P. K., & Meltzoff, A. N. (1982). The bimodal perception of speech in infancy. *Science, 218,* 1138–1141.

Kuhl, P. K., & Meltzoff, A. N. (1996). Infant vocalizations in response to speech: Vocal imitation and developmental change. *Journal of the Acoustical Society of America, 100*(4, Pt. 1), 2425–2438.

Kuhl, P. K., & Miller, J. D. (1975). Speech perception by the chinchilla: Voiced–voiceless distinction in alveolar plosive consonants. *Science, 190,* 69–72.

Kuhl, P. K., & Padden, D. M. (1983). Enhanced discriminability at the phonetic boundaries for the place feature in macaques. *Journal of the Acoustical Society of America, 73,* 1003–1010.

Kuhl, P. K., Stevens, E., Hayashi, A., Deguchi, T., Kiritani, S., & Iverson, P. (2006). Infants show a facilitation effect for native language phonetic perception between 6 and 12 months. *Developmental Science, 9,* F13–F21.

Kuhl, P. K., Tsao, F. M., & Liu, H. M. (2003). Foreign-language experience in infancy: Effects of short-term exposure and social interaction on phonetic learning. *Proceedings of the National Academy of Sciences USA, 100*(15), 9096–9101.

Kuhl, P. K., Williams, K. A., Lacerda, F., Stevens, K. N., & Lindblom, B. (1992). Linguistic experience alters phonetic perception in infants by 6 months of age. *Science, 255,* 606–608.

Lalonde, C., & Werker, J. (1995). Cognitive influence on cross-language speech perception in infancy. *Infant Behavior and Development, 18,* 459–475.

Landry, S. H., Smith, K. E., Miller-Loncar, C. L., & Swank, P. R. (1997). Predicting cognitive-language and social growth curves from early maternal behaviors in children at varying degrees of biological risk. *Developmental Psychology, 33*(6), 1040–1053.

Landry, S. H., Smith, K. E., Swank, P. R., Assel, M. A., & Vellet, S. (2001). Does early responsive parenting have a special importance for children's development or is consistency across early childhood necessary? *Developmental Psychology, 37*(3), 387–403.

Lasky, R. E., Syrdal-Lasky, A., & Klein, R. E. (1975). VOT discrimination by four to six and a half month old infants from Spanish environments. *Journal of Experimental Child Psychology, 20,* 215–225.

Liu, H.-M., Kuhl, P. K., & Tsao, F.-M. (2003). An association between mothers' speech clarity and infants' speech discrimination skills. *Developmental Science, 6*(3), F1–F10.

Liu, H.-M., Tsao, F.-M., & Kuhl, P. K. (2007). Acoustic analysis of lexical tone in Mandarin infant-directed speech. *Developmental Psychology, 43*(4), 912–917.

Masur, E. F. (1982). Mothers' responses to infants' object-related gestures: Influences on lexical development. *Journal of Child Language, 9*(1), 23–30.

Maye, J., Weiss, D. J., & Aslin, R. N. (2008). Statistical phonetic learning in infants: facilitation and feature generalization. *Developmental Science, 11*(1), 122–134.

Maye, J., Werker, J. F., & Gerken, L. (2002). Infant sensitivity to distributional information can affect phonetic discrimination. *Cognition, 82,* B101–B111.

McMurray, B., & Aslin, R. N. (2005). Infants are sensitive to within-category variation in speech perception. *Cognition, 95,* B15–B26.

Mehler, J., Bertoncini, J., Barriere, M., & Jassik-Gerschenfeld, D. (1978). Infant recognition of mother's voice. *Perception, 7,* 491–497.

Mills, D. L., Coffey-Corina, S. A., & Neville, H. J. (1993). Language acquisition and cerebral specialization in 20-month-old infants. *Journal of Cognitive Neuroscience, 5,* 326–342.

Mills, D. L., Coffey-Corina, S. A., & Neville, H. J. (1994). Variability in cerebral organization during primary language acquisition. In G. Dawson & K. Fischer (Eds.), *Human behavior and the developing brain* (pp. 427–455). New York: Guilford Press.

Mills, D. L., Coffey-Corina, S. A., & Neville, H. J. (1997). Language comprehension and cerebral specialization from 13–20 months. *Developmental Neuropsychology, 13,* 397–446.

Mills, D. L., Conboy, B., & Paton, C. (2005a). Do changes in brain organization reflect shifts in symbolic functioning? In L. Namy (Ed.), *Symbol use and symbolic representation* (pp. 123–153). Mahwah, NJ: Erlbaum.

Mills, D. L., & Neville, H. (1997). Electrophysiological studies of language and language impairment. *Seminars in Pediatric Neurology, 4,* 125–134.

Mills, D. L., Plunkett, K., Prat, C., & Schafer, G. (2005b). Watching the infant brain learn words: Effects of language and experience. *Cognitive Development, 20,* 19–31.

Mills, D. L., Prat, C., Zangl, R., Stager, C. L., Neville, H. J., & Werker, J. F. (2004). Language experience and the organization of brain activity to phonetically similar words: ERP evidence from 14- and 20-month-olds. *Journal of Cognitive Neuroscience, 16*(8), 1452–1464.

Mills, D. L., & Sheehan, L. (2007). Experience and developmental changes in the organization of language-relevant brain activity. In D. Coch, K. W. Fischer, & G. Dawson (Eds.), *Human behavior, learning, and the developing brain: Typical development* (pp. 183–218). New York: Guilford Press.

Molfese, D. L. (1989). Electrophysiological correlates of word meanings in 14-month-old human infants. *Developmental Neuropsychology, 5,* 79–103.

Molfese, D. L. (2000). Predicting dyslexia at 8 years of age using neonatal brain responses. *Brain and Language, 72,* 238–245.

Molfese, D. L., & Molfese, V. J. (1985). Electrophysiological indices of auditory discrimination in newborn infants: The bases for predicting later language development? *Infant Behavior and Development, 8*(2), 197–211.

Molfese, D. L., & Molfese, V. J. (1997). Discrimination of language skills at five years of age using event-related potentials recorded at birth. *Developmental Neuropsychology, 13*, 135–156.

Molfese, D. L., Molfese, V. J., & Espy, K. A. (1999). The predictive use of event-related potentials in language development and the treatment of language disorders. *Developmental Neuropsychology, 16*(3), 373–377.

Molfese, D. L., Morse, P. A., & Peters, C. J. (1990). Auditory evoked responses to names for different objects: Cross-modal processing as a basis for infant language acquisition. *Developmental Psychology, 26*, 780–795.

Molfese, D. L., Wetzel, W. F., & Gill, L. A. (1993). Known versus unknown word discriminations in 12-month-old human infants: Electrophysiological correlates. *Developmental Neuropsychology, 9*, 241–258.

Näätänen, R. (1999). The perception of speech sounds by the human brain as reflected by the mismatch negativity (MMN) and its magnetic equivalent (MMNm). *Psychophysiology, 38*, 1–21.

Näätänen, R., Gaillard, A. W. K., & Mäntysalo, S. (1978). Early selective-attention effect on evoked potential reinterpreted. *Acta Psychologica, 42*, 313–329.

Näätänen, R., Lehtokoski, A., Lennes, M., Cheour, M., Huotilainen, M., Iivonen, A., et al. (1997). Language-specific phoneme representations revealed by electric and magnetic brain responses. *Nature, 385*, 432–434.

Namy, L., Campbell, A., & Tomasello, T. (2004). The changing role of iconicity in non-verbal symbol learning: A U-shaped trajectory in the acquisition of arbitrary gestures. *Journal of Cognition and Development, 5*(1), 37–57.

Namy, L., & Waxman, S. (1998). Words and gestures: Infants' interpretations of different forms of symbolic reference. *Child Development, 69*, 295–308.

Namy, L., & Waxman, S. (2002). Patterns of spontaneous production of novel words and gestures within an experimental setting in children ages 1:6 and 2:2. *Journal of Child Language, 29*, 911–921.

Nazzi, T., Iakimova, G., Bertoncini, S. M., Serres, J., & de Schonen, S. (2008). Behavioral and electrophysiological exploration of early word segmentation in French. In A. Friederici & G. Thierry (Eds.), *Trends in language acquisition research: Vol. 5. Early language development: Bridging brain and behaviour* (pp. 65–89) Amsterdam: John Benjamins.

Nelson, C. A., & Monk, C. S. (2001). The use of event-related potentials in the study of cognitive development. In C. Nelson & M. Luciana (Eds.), *Handbook of developmental cognitive neuroscience* (pp. 125–136). Cambridge, MA: MIT Press.

Nicely, P., Tamis-LeMonda, C. S., & Bornstein, M. H. (1999). Mothers' attuned responses to infant affect expressivity promote earlier achievement of language milestones. *Infant Behavior and Development, 22*(4), 557–568.

Pang, E., Edmonds, G., Desjardins, R., Khan, S., Trainor, L., & Taylor, M. (1998). Mismatch negativity to speech stimuli in 8-month-old infants and adults. *International Journal of Psychophysiology, 29*, 227–236.

Plunkett, K., Sinha, C., Møller, M. F., & Strandsby, O. (1992). Symbol grounding or the emergence of symbols?: Vocabulary growth in children and a connectionist net. *Connection Science, 4*, 293–312.

Polka, L., & Werker, J. F. (1994). Developmental changes in perception of non-native vowel contrasts. *Journal of Experimental Psychology: Human Perception and Performance, 20*, 421–435.

Reynolds, G. D., & Richards, J. E. (2005). Familiarization, attention, and recognition memory in infancy: An event-related potential and cortical source localization study. *Developmental Psychology, 41*, 598–615.

Rinne, T., Alho, K., Alku, P., Holi, M., Sinkkonen, J., Virtanen, J., et al. (1999). Analysis of speech sounds is left-hemisphere predominant at 100–150 ms after sound onset. *NeuroReport, 10*(5), 1113–1117.

Rivera-Gaxiola, M., Klarman, L., Garcia-Sierra, A., & Kuhl, P. K. (2005). Neural patterns to speech and vocabulary growth in American infants. *NeuroReport, 16*, 495–498.

Rivera-Gaxiola, M., Silva-Pereyra, J., Klarman, L., Garcia-Sierra, A., Lara-Ayala, L., Cadena-Salazar, C., et al. (2007). Principal component analyses and scalp distribution of the auditory P150–250 and the N250–550 to speech contrasts in Mexican and American infants. *Developmental Neuropsychology, 31*, 363–378.

Rivera-Gaxiola, M., Silva-Pereyra, J., & Kuhl, P. K. (2005b). Brain potentials to native and non-native speech contrasts in 7- and 11-month-old American infants. *Developmental Science, 8*, 162–172.

Rizzolatti, G., & Arbib, M. (1998). Language within our grasp. *Trends in Neurosciences, 21*(5), 188–194.

Rogoff, B. (2003). *The cultural nature of human development*. London: Oxford University Press.

Rommetveit, R. (1998). Intersubjective attunement and linguistically mediated meaning in discourse. In S. Bråten (Ed.), *Intersubjective communication and emotion in early ontogeny* (pp. 354–371). Cambridge, UK: Cambridge University Press.

Saffran, J. R., & Thiessen, E. D. (2003). Pattern induction by infant language learners. *Developmental Psychology, 39*, 484–494.

Sanders, L. D., & Neville, H. J. (2003). An ERP study of continuous speech processing: II. Segmentation, semantics, and syntax in non-native speakers. *Cognitive Brain Research, 15*, 214–227.

Santesso, D. L., Schmidt, L. A., & Trainor, L. J. (2007). Frontal brain electrical activity (EEG) and heart rate in response to affective infant-directed (ID) speech in 9-month-old infants. *Brain and Cognition, 65*, 14–21.

Schafer, G., & Plunkett, K. (1998). Rapid word learning by fifteen-month-olds under tightly controlled conditions. *Child Development, 69*, 309–320.

Schieffelin, B. B. (1991). *The give and take of everyday life: Language socialization of Kaluli children*. Cambridge, UK: Cambridge University Press.

Sheehan, L., & Mills, D. L. (2008). The effects of early word learning on brain development. In A. Friederici & G. Thierry (Eds.), *Trends in language acquisition research: Vol. 5. Early language development: Bridging brain and behaviour* (pp. 161–190). Amsterdam: John Benjamins.

Sheehan, L., Namy, L., & Mills, D. L. (2007). Developmental changes in neural activity to familiar words and gestures. *Brain and Language, 101*, 246–259.

Singh, L. (2008). Influences of high and low variability on infant word recognition. *Cognition, 106*(2), 833–870.

Singh, L., Morgan, J. L., & Best, C. T. (2002). Infants' listening preferences: Baby talk or happy talk? *Infancy, 3*, 365–394.

Stager, C. L., & Werker, J. F. (1997). Infants listen for more phonetic detail in speech perception than in word-learning tasks. *Nature, 388*, 381–382.

St. George, M., & Mills, D. L. (2001). Electrophysiological studies of language development. In J. Weissenborn & B. Hoehle (Eds.), *Language acquisition and language disorders* (pp. 247–259). Amsterdam: John Benjamins.

Streeter, L. A. (1976). Language perception of 2-month-old infants shows effects of both innate mechanisms and experience. *Nature, 259*, 39–41.

Tamis-LeMonda, C. S., Bornstein, M. H., Baumwell, L., & Damast, A. M. (1996). Responsive parenting in the second year: Specific influences on children's language and play. *Early Development and Parenting, 5*(4), 173–183.

Tamis-LeMonda, C. S., Bornstein, M. H., Marc, H., & Baumwell, L. (2001). Maternal responsiveness and children's achievement of language milestones. *Child Development, 72*(3), 748–767.

Thierry, G., Vihman, M., & Roberts, M. (2003). Familiar words capture the attention of 11-month-olds in less than 250 ms. *NeuroReport, 14*, 2307–2310.

Thiessen, E. D., & Saffran, J. R. (2007). Learning to learn: Infants' acquisition of stress-based strategies for word segmentation. *Language Learning and Development, 3*, 73–100.

Tincoff, R., & Jusczyk, P. W. (1999). Some beginnings of word comprehension in 6-month-olds. *Psychological Science, 10*, 172–175.

Tomasello, M. (1999). Having intentions, understanding intentions, and understanding communicative intentions. In P. D. Zelazo, J. W. Astington, & D. R. Olson (Eds.), *Developing theories of intention: Social understanding and self-control* (pp. 63–75). Mahwah, NJ: Erlbaum.

Trainor, L. J., Austin, C. M., & Desjardins, R. N. (2000). Is infant-directed speech prosody a result of the vocal expression of emotion? *Psychological Science, 11,* 188–195.

Trehub, S. E. (1976). The discrimination of foreign speech contrasts by infants and adults. *Child Development, 47,* 466–472.

Tsao, F. M., Liu, H. M., & Kuhl, P. K. (2004). Speech perception in infancy predicts language development in the second year of life: A longitudinal study. *Child Development, 75,* 1067–1084.

Tsao, F. M., Liu, H. M., & Kuhl, P. K. (2006). Perception of native and non-native affricate–fricative contrasts: Cross-language tests on adults and infants. *Journal of the Acoustical Society of America, 120*(4), 2285–2294.

Vihman, M. M., Nakai, S., DePaolis, R. A., & Halle, P. (2004). The role of accentual pattern in early lexical representation. *Journal of Memory and Language, 50,* 336–353.

Vihman, M. M., Thierry, G., Lum, J., Keren-Portnoy, T., & Martin, P. (2007). Onset of word form recognition in English, Welsh, and English–Welsh bilingual infants. *Applied Psycholinguistics, 28,* 475–493.

Werker, J. F., Cohen, L. B., Lloyd, V. L., Casasola, M., & Stager, C. L. (1998). Acquisition of word-object associations by 14-month-old infants. *Developmental Psychology, 34,* 1289–1309.

Werker, J. F., & Curtin, S. (2005). PRIMIR: A developmental model of speech processing. *Language Learning and Development, 1,* 197–234.

Werker, J. F., Fennell, C. T., Corcoran, K. M., & Stager, C. L. (2002). Infants ability to learn phonetically similar words: Effects of age and vocabulary size. *Infancy, 3*(1), 1–30.

Werker, J. F., Gilbert, J. H., Humphrey, K., & Tees, R. C. (1981). Developmental aspects of cross-language speech perception. *Child Development, 52*(1), 349–355.

Werker, J. F., & Lalonde, C. (1988). Cross-language speech perception: Initial capabilities and developmental change. *Developmental Psychology, 24,* 672–683.

Werker, J. F., & McLeod, P. J. (1989). Infant preference for both male and female infant-directed talk: A developmental study of attentional and affective responsiveness. *Canadian Journal of Psychology, 43,* 230–246.

Werker, J. F., & Tees, R. (1984). Cross-language speech perception: Evidence for perceptual reorganization during the first year of life. *Infant Behavior and Development, 7,* 49–63.

Werker, J. F., & Tees, R. C. (1999). Influences on infant speech processing: Toward a new synthesis. *Annual Review of Psychology, 50,* 509–535.

Werker, J. F., & Yeung, H. H. (2005). Infant speech perception bootstraps word learning. *Trends in Cognitive Sciences, 9*(11), 519–527.

Woodward, A., Markman, E. M., & Fitzsimmons, C. M. (1994). Rapid word learning in 13- and 18-month-olds. *Developmental Psychology, 30,* 533–566.

Xiao, Y., Bitsko, R., Sheehan, E., Larson, M., & Mills, D. (2008, April). *Maternal emotion and use of infant-directed speech affects infant behavior and brain activity.* Poster presented at the International Conference on Infancy Studies, Vancouver, Canada.

Zangl, R., & Mills, D. L. (2007). Brain activity to infant versus adult directed speech in 6- and 13-month olds. *Infancy, 11,* 31–62.

CHAPTER 11

Evolutionary Origins of Social Communication

MASAKO MYOWA-YAMAKOSHI
MASAKI TOMONAGA

The Development of Imitation: A Comparative Cognitive Science Approach

Imitation: The Key to Human Intelligence

As reflected in the aphorism "Monkey see, monkey do," imitation is often regarded negatively in everyday life, as connoting a lack of independence or creativity. The fact that we can actually imitate the actions of others easily and without being aware of it supports this view.

However, the phenomenon of imitation is not that simplistic. It is actually a complex process that is particularly important in supporting the high-level intelligence of humans. In order to imitate, we humans must (1) appropriately extract the key elements from the information of others' body movements, which constantly change over time; (2) mentally assemble the extracted elements; and (3) control our own bodies to replicate the actions accurately. Why did humans evolve this complicated ability to imitate? In order to answer this question, we must consider the adaptive significance of imitation for human survival.

Previous studies of imitation's functions have typically focused on two aspects. The first of these involves the methods and techniques of making and using tools, which are not genetically transmitted. By imitating the actions of others, an observer can efficiently acquire the adaptive skills specific to his or her group, without depending solely on trial and error or on individual learning. Furthermore, imitation is an extremely important mechanism that permits such information to be transmitted across generations as aspects of culture. It is believed that imitation has enabled humans to acquire accumulated information from earlier generations and transmit them to future generations in an accurate and stable form (Bandura, 1986; Galef, 1991; Tomasello, Kruger, & Ratner, 1993).

The second aspect of imitation function is to ensure successful communication with others in order to ensure the survival of group members. Effective communication requires the ability to predict and understand the behavior of others with the same mental state as oneself. This ability to understand the intentions, goals, beliefs, and thoughts of others is known as "theory of mind" (Premack & Woodruff, 1978), and it is considered to be a product of evolution that has enabled humans to survive in complex social environments (Byrne & Whiten, 1988, Humphrey, 1984; see also Choudhury, Charman, & Blakemore, Chapter 9, this volume, for further discussion of theory of mind in humans).

Thus imitation is believed to serve as the foundation for the development of these kinds of mental structures. By imitating the body movements of others, human infants learn that they share certain physical traits with others, and compare their own experience of performing the same actions with the observed actions of others. By doing this, infants develop a mental state of awareness that links them to the mental states of the other individuals (Meltzoff & Gopnik, 1993). In addition, imitation is believed to play an especially important role in the imaginative ability to conceive of an object not presently visible (Werner & Kaplan, 1969), and to serve as a prerequisite for representing an object by abstract thought (as in language; Piaget, 1962).

Imitating Animals: Humans versus Chimpanzees

Of all the animals currently living on Earth, the species most systematically similar to the human (*Homo sapiens*) is the chimpanzee (*Pan troglodytes*). From the mid-1990s until the present day, comparative cognitive science, which aims to solve the mystery behind the unique origins of the human intellect, has clearly demonstrated that the saying "Monkey see, monkey do" is untrue. Surprisingly, research has shown that body imitation is difficult not just for monkeys, but for chimpanzees as well (Custance, Whiten, & Bard, 1995; Myowa-Yamakoshi & Matsuzawa, 1999, 2000; Nagell, Olguin, & Tomasello, 1993; Whiten, Custance, Gomez, Teixidor, & Bard, 1996).

Let us consider a specific example of the extent to which chimpanzees are poor at imitation. Myowa-Yamakoshi and Matsuzawa (1999) performed a study investigating the degree to which imitation in adult chimpanzees is limited, compared to that in humans. In face-to-face situations between a chimpanzee and a human model, the latter used an object to perform meaningless actions that were not related to the intended function of the object, as shown in Figure 11.1. In order to determine the actions that are particularly difficult for chimpanzees to imitate, 48 action types were presented, with the following variables: (1) the number of objects to manipulate and the direction of manipulation (e.g., manipulating one object,

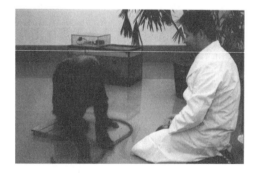

FIGURE 11.1. A chimpanzee (Chloé) performing a demonstrated action (rolling the hose) in the one-object condition.

manipulating one object toward self, manipulating one object toward another object); and (2) the motor pattern—that is, patterns that were familiar (e.g., hitting, pushing, poking, etc.) or not familiar (e.g., stroking, rolling, etc.) to the chimpanzees.

The results revealed minimal reproduction (imitation) of the actions that were presented to the chimpanzees without practice (5.4% of the total actions). With regard to factors related to the difficulty of the reproduced actions, the chimpanzees found it more difficult to perform actions that involved manipulating one object than those that involved manipulating one object toward another object or manipulating one object toward oneself (Figure 11.2). Moreover, they reproduced actions involving familiar motor patterns more easily than they did the unfamiliar motor patterns. However, it is important to note that the frequency of imitation even for the actions involving the familiar motor patterns was very low.

It appears that the style of recognizing the actions of others is different in chimpanzees than in humans. For the chimpanzees in this study, information about an object that was being manipulated by a model (e.g., shape, direction, function) served as a useful clue for reproduction (Horner, Whiten, Flynn, & de Waal, 2006; Whiten, Horner, & de Waal, 2005). However, information about the body movements of another individual (motor patterns) did not easily translate into clues for reproducing the actions. As proof of this, when these chimpanzees only observed the actions, there was very minimal imitation even of the motor patterns that are part of a chimpanzee's normal repertoire. For example, even with regard to two familiar actions involving the same hitting motor pattern, chimpanzees found it more difficult to reproduce hitting a box with one hand (manipulating one object) than hitting a box with a stick (manipulating one object with another object). The results indicated that when chimpanzees are imitating the actions of others, they pay less attention to body movements and more to the objects being manipulated. Apparently it is very difficult for chimpanzees to appropriately map the visual images of the body movements of another individual in correspondence to the images of their own body movements. Thus we can surmise that imitation is an extremely important capability that must have developed at some point after the human and chimpanzee lineages branched apart, approximately 6 million years ago.

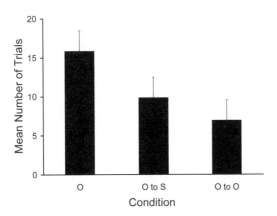

FIGURE 11.2. Mean number of trials (plus the standard error) required to perform the demonstrated actions in each of the three conditions. O, the one-object condition; O to S, the one-object-to-self condition; O to O, the one-object-to-another condition. From Myowa-Yamakoshi and Matsuzawa (1999). Copyright 1999 by the American Psychological Association. Reprinted by permission.

Can Humans Imitate from the Time of Birth?

When and how did the ability to imitate—an ability that seems to be unique to humans—develop? Such questions surrounding the development of individual imitation have drawn the interest of many developmental psychologists. The findings of Meltzoff and Moore (1977) attracted exceptional attention: Their study showed that even newborn humans possess the ability to perform imitation with parts of their bodies that they cannot even see themselves.

Meltzoff and Moore demonstrated that newborn human infants were able to imitate some facial expressions, such as protruding their tongues or opening their mouths, and they referred to this behavior as "neonatal imitation." Based on this fact, they have hypothesized that humans possess an inherent body-mapping capability, and have named this "active inter-modal mapping" (AIM). It is believed that this inherent body-mapping capability develops with age and eventually allows humans to develop the ability to imitate more complex behaviors.

Many follow-up studies have been conducted in other cultures, and there is no dearth of researchers challenging Meltzoff and Moore's (1977) premise that neonatal imitation is the origin of imitation. One reason for this is the assertion that the neonatal imitative response disappears or is lessened at approximately 2 months after birth and then reappears at about 1 year (Abravanel & Sigafoos, 1984; Fontaine, 1984). In other words, there is no reason to explain why neonatal imitation must temporarily disappear. Moreover, since only one type of imitative response of facial expression (i.e., tongue protrusion) was observed, some assert that neonatal imitation is not imitation at all, but merely a reflection, an innate releasing mechanism, or a form of exploratory behavior in response to interesting/arousing stimuli (Jacobson, 1979; Jones, 1996, 2006).

Is neonatal imitation really the origin of imitation? Or is it a sensory system with origins different from the form of imitation that becomes active by about the first birthday? There has been no clear answer to this puzzle.

Nevertheless, recent comparative cognitive science research offers a few clues to solving this puzzle. We have asserted above that adult chimpanzees are not very adept at body mapping. If neonatal imitation is the origin of imitation behavior, and if AIM—which enables imitation—is an ability unique to humans, then imitation immediately after birth should be present only in humans. In order to verify this theory, using roughly the same method as that employed by Meltzoff and Moore (1977), we performed comparative experiments on the newborn infants of several primate species: chimpanzees, gibbons, Japanese macaques, and squirrel monkeys (Myowa-Yamakoshi, 2006; Myowa-Yamakoshi, Tomonaga, Tanaka, & Matsuzawa, 2004).

In these experiments, a human model displayed various expressions to primate infants (such as tongue protrusion, mouth opening, and lip protrusion) as soon as possible after the infants were born. As shown in Figure 11.3, clear proof of the existence of neonatal imitation was found only in newborn chimpanzees. The chimpanzees were able to differentiate and reproduce the expressions of other individuals from a week after birth. No evidence of neonatal imitation was found in the other three primate species that were tested. Neonatal imitation has been demonstrated to be shared by humans and chimpanzees (see also Bard, 2007; recent research has revealed evidence of neonatal imitation in rhesus macaques as well, and the evolutionary origins of neonatal imitation continues to be a matter of debate [Ferrari et al., 2006]).

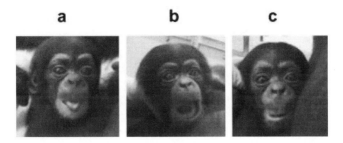

FIGURE 11.3. The imitative responses of the three demonstrated facial gestures: (a) tongue protrusion, (b) mouth opening, and (c) lip protrusion (Pal, 2 weeks of age). From Myowa-Yamakoshi, Tomonaga, Tanaka, and Matsuzawa (2004). Copyright 2004 by Blackwell Publishing Ltd. Reprinted by permission.

Mechanisms Supporting Neonatal Imitation

When comparing the imitation abilities of humans and chimpanzees, we note first and foremost that the ability of adult chimpanzees to map the bodies of others onto their own body images is limited; as a result, they have difficulty with imitation, whereas neonatal imitation-level mapping limited to a certain body region is possible even among newborn chimpanzees. However, we found that this chimpanzee neonatal imitation disappeared 9 weeks after birth (Myowa-Yamakoshi et al., 2004). Even after several years, no more complex imitative ability approaching that of human children developed among these chimpanzees (Myowa-Yamakoshi, 2006). Our series of studies supports the premise that this gap in imitation development between the two species occurs because the mechanism of neonatal imitation differs from that of later imitation, which appears at approximately 8–12 months of age in humans.

We believe that neonatal imitation cannot be explained by primordial reflex theory. This is because human and chimpanzee infants are able to differentiate and imitate at least two different expressions—namely, tongue protrusion and mouth opening. However, if neonatal imitation and later imitation do not develop in a connected way, as we assert here, then what are the mechanisms through which these forms of imitation are realized? Recently, a new and interesting theory has begun to gain popularity; this theory suggests that it is possible to perform imitative actions limited to the neonatal imitation level without performing body mapping between one's own body and another's.

Baron-Cohen (1996) has asserted that human neonates do not perceive stimuli via specific sensory modalities with specialized cortical nerve pathways (e.g., vision, tactile, auditory). Instead, he has argued that the stimuli are perceived in a synesthetic, amodal state on the subcortical level. Let us consider an example to illustrate this. A pacifier with bumps is given to a blindfolded human infant to suck. Then, when the blindfold is removed and the infant is presented with two pacifiers—one with bumps, and the other without—the infant stares longer at the pacifier with bumps (Meltzoff & Borton, 1979). Thus it appears that human infants perceive stimulus traits without separating sight and touch. According to Baron-Cohen (1996), this phenomenon is what lies behind the unique amodal sensory style of newborn infants.

Let us examine this phenomenon in further detail. Amodal sensory perception continues until nerve pathways develop on the cortical levels for each modality, at approximately 6–8 weeks of age (Baron-Cohen, 1996). Neonates may perceive physical stimuli in a unified state on the basis of various dimensions. For example, they may perceive the strength or speed of a stimulus from its energy; time or spatial frequency from its texture; long-period structure from its rhythm; or position or direction of movement from its orientation (flow) from the perspective of their own bodies. Within this amodal sensory system, newborns are thought to perceive moving stimuli (i.e., movement orientation or the position of stimuli, such as tongues protruding or lips parting widely). Furthermore, it is suggested that neonatal imitation is the result of infants' synesthetically outputting this information onto their own bodies (Baron-Cohen, 1996; Kojima, 2005). Research that supports this view will be important in the future.

Imitation as a Form of Communication, and Its Evolutionary Origins

In the first section of this chapter, we have suggested that imitation ability beyond the level evidenced in neonatal imitation is a trait specific to humans. From approximately the time of their first birthday, humans begin regularly imitating complex behaviors, such as the use of novel tools or pantomimes involving the entire body.

The different ways in which humans and chimpanzees recognize (input) the behavior of others may be one reason for the difference in the imitation abilities of humans and chimpanzees. How do humans and chimpanzees perceive the actions of others? Furthermore, how do their cognitive styles change developmentally? Below, we compare humans and chimpanzees from these perspectives, and we probe into the mechanisms responsible for the uniqueness of imitation in humans.

How Are the Actions of Others Regarded?: Two Views

Currently, there are two main views regarding the cognitive processes of humans observing the actions of others. As asserted by Meltzoff and his colleagues, the first of these is the idea that an understanding of the mental state of another person is developed through automatic and direct inherent mapping of the other's body movements onto those of one's own body. Its origins are in the neonatal imitation evidenced immediately after birth, in which basically all actions of another individual are imitated and actually experienced with one's own body. This leads to a simulation of another individual's perspective through oneself. The idea is that through imitation immediately after birth, infants attribute their own mental states to the mind of the actor.

The second view differs from that of Meltzoff and colleagues. According to this view, human infants interpret the actions of others as being goal-directed (Csibra, Gergely, Biro, Koos, & Brockbank, 1999; Gergely, Nádasdy, Csibra, & Bíró, 1995). In other words, human infants interpret and imitate the actions of others in attempts to achieve goals efficiently. Researchers supporting this view argue as follows: If body movement mapping between oneself and others is automatically possible immediately after birth, then we humans should be able to imitate the complex behaviors of others with a level of precision rivaling a mirror image. Moreover, it should not be possible for imitation responses to appear in different forms under different contexts. Researchers supporting this view present data discrediting this theory and do not support the AIM theory.

Gergely, Bekkerling, and Király (2002) presented one of two actions to 14-month-old infants. In each action, a model was covered with a blanket; however, the manner in which the blanket covered the model was different for each action. In the first action, the model was completely covered with the blanket, including the model's hands, and was unable to use either hand. In the second action, the model was covered with the blanket, albeit only up to the wrists, and was able to use both hands freely. While covered with a blanket in one of these two manners, models performed the unusual action of pressing with their foreheads a light box placed on a table that lit up when pressed (Figure 11.4). After 1 week, this light box was given to the infants. The results showed that 69% of the infants who watched the action of the model whose hands were free used their foreheads in the same manner to turn on the light box. However, among the infants who saw the actions of the model whose hands were covered by the blanket, only 21% pressed the box with their foreheads, and most used their hands. According to Gergely et al., infants reached the rational conclusion that the models were unable to light the box by using their hands and used their foreheads instead. They surmise that infants therefore chose the most rational action when reproducing the observed action themselves, which was to use their hands in order to achieve the goal of lighting the box.

There is one more study demonstrating that when humans imitate the actions of others, they do this not solely on the basis of mapping, but select their own body movements based on a rational criterion. Bekkering and Wohlschläger (2000) presented human infants with a model performing the action of touching an ear. The action was very simple: Either the left or the right ear was touched with either the left or the right hand. The results, however, were intriguing. The infants could touch the correct ear in each case; however, when imitating the action in which the ear was to be touched with the opposing hand, they used the hand on the same side to touch their own ears. In other words, imitation faithfully based on mapping was not observed.

Gergely and colleagues have termed this rational cognitive style observed by about 12 months of age "teleological stance." Since the rational criteria to achieve the target action differ according to the context, the types of action recognition are numerous, depending on the underlying situations (Csibra et al., 1999; Gergely et al., 1995). However, the type of cog-

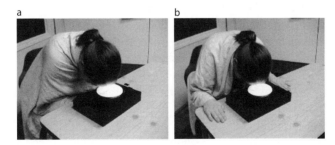

FIGURE 11.4. Experimental situation in Gergely, Bekkerling, and Király (2002). The model who could freely use hands (the hands-free condition; b) and the model whose hands were covered by the blanket (the hands-occupied condition; a) were required to switch on the light by touching the lamp with their foreheads. From Gergely et al. (2002). Copyright 2002 by Nature Publishing Group. Reprinted by permission from Macmillan Publishers Ltd.

nitive style at this age is not associated with the mental states of the other individual, such as their intentions, desires, or beliefs; rather, it is thought to be limited to a nonmentalistic level of explaining and predicting the goal of the action (Gergely & Csibra, 2003).

The Development of Social Communication and Imitation

We have proposed that the mechanism of neonatal imitation differs from that of later imitation, which appears at approximately 8–12 months of age in humans. If this is the case, it should be valid to interpret this subsequent imitation as the expression of the formation of the cognitive style described by Gergely and colleagues.

In reality, however, humans are able to imitate behaviors even before 1 year of age if the action is sufficiently simple (Piaget, 1962). For example, if a model claps his or her hands before a 2-month old infant, although the infant is unable to bring the hands together in the same manner, he or she will aggressively wave both hands back and forth. The infant will exhibit a similar motion of the fingers if presented with a hand opening and closing.

Social communication based on the dyadic relationship (with an object or with a person) occurs increasingly from the fourth month after birth. Infants also begin to engage in triadic exchanges with a person who is manipulating objects. Infants learn during this period to manipulate objects in simple ways, such as picking up toys by themselves and raising them up to their mouths, or searching for objects with their fingers. In some cases they may manipulate objects on their own, but in the vast majority of cases objects are manipulated during communication with others. We are all familiar with such interaction, such as reaching out to accept a toy offered by another individual, or being handed a toy and then invited to engage in a mini-game of tug-of-war. The two individuals experience the action together through the same object and through both their bodies. Imitation based on a triadic relationship that is guided by an adult enables a more faithful reexperiencing in the infant of the other's action than when imitation is based on dyadic relationships.

From approximately 9 months of age, imitation through turn-taking interaction begins to be frequently observed in the context of play. Moreover, infants begin not only to imitate the actions of others endlessly, but also to recognize that others are imitating them. According to Meltzoff (1990), 9-month-old infants prefer those who are imitating them to those who are not. Just after 1 year, infants become more capable of judging whether their own behavior is the same as that of others. When others imitate their actions, the infants not only pay attention, but may also suddenly stop what they are doing and begin doing something different, as if to "trick" the imitators intentionally.

According to the view of Gergely and colleagues, this developmental sequence is valuable in providing clues to the reason for the complex imitation beyond the level of neonatal imitation. However, in our opinion, it is too hasty to conclude that the development of the unique human imitation ability that depends on the cognitive ability to identify the goals of others' actions appears dramatically by about the first year after birth.

In the year after birth, until they become able to predict the goals of others' actions, human infants learn each day with guidance from others how to communicate through imitation. This is discussed later in more detail (see "Development and Evolution of Imitation Ability," below); however, it has been recently shown that this type of communication is only observed among humans. In this chapter, we develop the view that the type of communication that humans experience in approximately the year following birth is precisely the key to the development of the unique human ability to recognize and imitate others' actions. We discuss the idea that the development of imitation ability does not merely have the results and outcomes that appear in the actions of an individual; we also discuss this ability in relation to the developmental changes in the contents of communication experienced by infants.

Sharing Body Experiences

Try pinching your arm. You experience more than just the pain of your arm being pinched. When you touch your own body while pinching, the sensation of your finger pinching your arm is also present. There is a bidirectional sensation (i.e., from finger to arm, and from arm to finger). However, the pain in the arm is unidirectional when you are being pinched by another individual or when you bump into an item, for instance. On the surface, the two types of actions appear to be similar, but you experience each of these types in completely different ways. The different sensations experienced in each of these cases is defined by the profile or outline of your body, and serves as an important clue to drawing the border between yourself and another. Through repeated body experiences such as these, human infants become aware of how their own action can affect the external world (i.e., the results of their actions). Neisser (1991, 1995) terms this level the "ecological self."

As discussed earlier, from approximately 4 months of age, humans increasingly participate in triadic "self–object–other" communication. Through this type of communication, infants experience in conjunction with others the functions of objects and the results of manipulations. After a mother shakes a rattle before her baby, she gives the rattle to the baby to hold and shake. After placing food in her child's mouth, she says, "Mmm, good," and eats her food together with the child. Infants do not stop at visually observing the actions of others in front of them, but also experience and share those actions through proprioception.

The process that follows is what is important. Actions that at first were merely presented or done to an infant, and perceived by the infant, are remembered and organized as awareness based on the body sensory experiences of the infant him- or herself (i.e., the infant's ecological self). Through repetitions of this self-awareness, the infant becomes able to differentiate between his or her own actions and those of others, while at the same time coming to realize that many aspects are in fact shared between the two. By substituting his or her own body experiences for the actions of another individual, the infant projects his or her own mental state onto that individual. In other words, by merely watching the actions of others, the infant can gain access to the intentional mental states behind those actions.

From birth, humans communicate through sharing body sensory experiences with others. These experiences enable the formation of complex body-motion-mapping structures on a level far higher than neonatal imitation. This promotes the formation of a cognitive style that can deal with the actions of others and the actions of the self, and can make it easier to predict the goal of an action. Sharing the goal of a predicted action with others also helps to promote efficient communication. As the infant performs increased shared imitation with others as appropriate for various situations, mapping occurs in greater detail. Thus the teleological stance and AIM must develop in a complex interaction. Moreover, these abilities are not acquired through sudden and dramatic changes; they are thought to develop gradually through the accumulation of shared body experiences with others based on triadic communication, beginning shortly after birth (Figure 11.5).

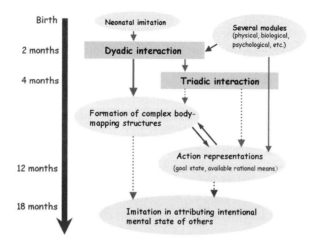

FIGURE 11.5. The developmental relationship between social communication and the ability to understand others' actions. The dotted lines indicate differences between humans and chimpanzees.

Development and Evolution of Imitation Ability

It is likely that the differences in imitation ability between humans and chimpanzees are closely related to the differences in their development of communication based on dyadic and triadic relationships. As in humans, dyadic interaction is observed in chimpanzees until approximately 2 months of age. Furthermore, as humans do, chimpanzees prefer directed-gaze faces to averted-gaze faces even moments after birth (Myowa-Yamakoshi, Tomonaga, Tanaka, & Matsuzawa, 2003). From about 2 months after birth, chimpanzees begin to reach out actively to interact with others; they gaze mutually and smile frequently or emit vocal sounds toward their caretakers (Bard et al., 2005; Mizuno, Takeshita, & Matsuzawa, 2006). Humans proceed to develop communication based on "self–object–other" triadic relationships (Trevarthen & Hubley, 1978; Rochat, 2001), and differences in communication between the two species become more pronounced from this point. It is very rare for chimpanzees to interact triadically (Tomonaga, 2006; Tomonaga et al., 2004).

For instance, when human infants encounter an unfamiliar object, they repeatedly alternate their gaze between the object and their mothers. They attempt to obtain additional information about the object from their mothers' expressions or mood. This behavior is called "social referencing" and becomes common at about 9 months after birth (see Carver & Cornew, Chapter 7, this volume, for further discussion of social referencing in humans). Furthermore, human mothers actively introduce objects into interactions with their children and act like mirrors before them. Human infants learn about objects from the relationships of those objects with others. In other words, they understand objects through the perspective of others (Tomasello, 1999).

On the other hand, this kind of communication is rarely observed between mother and infant chimpanzees even after 3 years of age. There are frequent instances wherein an infant chimpanzee comes near the mother who is manipulating an unfamiliar object and food and attempts to touch them (Hirata & Celli, 2003; Ueno & Matsuzawa, 2005). However, unlike their human counterparts, chimpanzee infants have never been observed to intentionally bring an object to show their mothers, or to call for their mothers in order to attract their attention toward the object. Moreover, chimpanzee mothers do not exhibit the behavior of intentionally presenting a novel object before their infants or attempting to get their infants to hold the object. Chimpanzee mothers rarely ever interact with their offspring through objects or use objects in shared actions (Matsuzawa, 2007). As compared to human infants, chimpanzee infants have virtually no opportunity to learn about an object's function or usage by experiencing the object together with others. One might even state that chimpanzee infants do not learn about objects from the perspective of others, but instead acquire this information through a unique process of isolated observational learning and trial and error (Figure 11.6; Matsuzawa, 1996, 2006).

Even chimpanzees can predict to a certain extent the goal of another's action by observing that action (e.g., Call, Carpenter, & Tomasello, 2005, Myowa-Yamakoshi & Matsuzawa, 2000; Tomasello, Carpenter, Call, Behne, & Moll, 2005). However, their cognitive style differs from that of humans. As mentioned earlier (see "Imitating Animals: Humans versus Chimpanzees"), we have found that chimpanzees are less likely to focus on the details of another's body movements involved in a manipulation; instead, they pay more attention to the direction in which the object is being manipulated. Chimpanzees appear to predict the purpose of an action from their knowledge of the object's traits and function, the causal relationship between these, and the change of state that they experience.

For example, the results of the Myowa-Yamakoshi and Matsuzawa (1999) study of chimpanzees' imitation ability suggest that this is true. The incorrect imitation responses shown by these chimpanzees had clear traits. Chimpanzees have a strong grasp of one-to-one relationships between specific, habitual (functional) motor patterns and familiar objects. When the chimpanzees in this study were shown an action in which a model manipulated an object in a manner inconsistent with such a motor pattern, they did not reproduce the pattern according to the context. For instance, chimpanzees familiar with a brush were unable to manipulate it by using motor patterns other than brushing. Even after observing the (meaningless) action in which the brush was used in a hitting motion pattern, they could only relentlessly repeat their familiar response of brushing.

FIGURE 11.6. Infant chimpanzees exhibit strong interest in their mothers' actions, both in captivity (left) and in the wild (right). However, chimpanzee mothers rarely use objects in shared actions with their children or teach them the actions, as human mothers do.

Humans, on the other hand, pay attention not only to the information about an object, but also to the body movements of an individual manipulating the object. Since humans, as they grow and develop, encounter especially frequent opportunities to share sensory body experiences with others, they are able to understand the mental states of others by incorporating the body movements of others into their own body experiences. Humans can predict the goal of an observed individual's action without being restricted by a specific context, and can share this in order to communicate. If human infants were shown an unfamiliar action, such as a hitting movement with a brush, they would perhaps faithfully imitate that action. Humans can avoid being bound to the functions or characteristics of a familiar object, and can direct attention to the body movements or facial expressions of another individual. From body movement information, they can flexibly read the intentional mental states of others according to the context, such as "Let's have fun playing around with a brush." What do humans and nonhuman primates properly choose to imitate from the actions of others, depending on the various contexts of communication? The answer to this question is the key to understanding the uniqueness of human imitation and its adaptive significance.

Another unsolved issue concerns the visuomotor mechanisms each species has available for imitation. As mentioned earlier, no strong behavioral evidence of imitation has been found in monkeys. On the other hand, clear evidence of kinesthetic–visual matching has been found at a neuronal level in the macaque brain. Mirror neurons, found in the ventral premotor area F5 (Rizzolatti, Fadiga, Gallese, & Fogassi, 1996) and the inferior parietal cortex (Gallese, Fadiga, Fogassi, & Rizzolatti, 2002) of the macaque, discharge both when a monkey performs an action, and when it sees a similar action being performed by an experimenter or another monkey. Paukner, Anderson, Borelli, Visalberghi, and Ferrari (2005) recently revealed that pigtailed macaques (*Macaca nemestrina*) are capable of recognizing when they are being imitated. Similar results for imitation recognition have been found in chimpanzees, although they have been limited to "object-related" actions (Nielsen, Collier-Baker, Davis, & Suddendorf, 2004). These studies suggest that the species differences in imitation ability may be especially related to the cognitive-producing (output) processes of imitation (e.g., planning skills, memory capacities, and inhibitory control; Paukner et al., 2005). How should such species differences in producing imitation be interpreted at the neuronal level? Are there any species differences in the mirror systems among monkeys, chimpanzees, and humans? Further comparative cognitive studies on the development of both recognizing and producing imitation can help to answer these questions.

Acknowledgments

We would like to thank T. Matsuzawa, H. Takeshita, M. Tanaka, S. Hirata, N. Nakashima, A. Ueno, Y. Mizuno, S. Okamoto, T. Imura, M. Hayashi, K. Matsubayashi, K. Kumazaki, N. Maeda, A. Kato, G. Hatano, O. Takenaka, K. A. Bard, M. Sakai, and K. Yuri. The research reported here and the preparation of this chapter were financially supported by Grants-in-Aid for Scientific Research from the Japan Society for the Promotion of Science (JSPS) and the Ministry of Education Culture, Sports, Science and Technology (MEXT) (Nos. 12002009 and 16002001 to T. Matsuzawa, 13610086 to M. Tomonaga, 09207105 to G. Hatano, 10CE2005 to O. Takenaka, 16203034 to H. Takeshita, 16683003 and 19680013 to M. Myowa-Yamakoshi); the MEXT Grant-in-Aid for the 21st Century COE Programs (Nos. A14 and D10 to Kyoto University); the research fellowship to M. Myowa-Yamakoshi from JSPS for Young Scientists; the JSPS core-to-core program HOPE; and the JSPS Grant-in-Aid for Creative Scientific Research, "Synthetic Study of Imitation in Humans and Robots," to T. Sato.

References

Abravanel, E., & Sigafoos, A. D. (1984). Exploring the presence of imitation during early infancy. *Child Development, 55*, 381–392.

Bandura, A. (1986). *Social foundations of thought and action: A social cognitive theory.* Englewood Cliffs, NJ: Prentice-Hall.

Bard, K. A. (2007). Neonatal imitation in chimpanzees (*Pan troglodytes*) tested with two paradigms. *Animal Cognition, 10*(2), 233–242.

Bard, K. A., Myowa-Yamakoshi, M., Tomonaga, M., Tanaka, M., Quinn, J., Costall, A., et al. (2005). Group differences in the mutual gaze of chimpanzees (*Pan troglodytes*). *Developmental Psychology, 41*, 616–624.

Baron-Cohen, S. (1996). Is there a normal phase of synaesthesia in development? *Psyche, 2*(27), *http://psyche.cs.monash.edu.au/v2/psyche-2-27-baron_cohenhtml.*

Bekkering, H., & Wohlschläger, A. (2000). Imitation of gestures in children is goal-directed. *Quarterly Journal of Experimental Psychology, 53A*, 153–164.

Byrne, R. W., & Whiten, A. (Eds.). (1988). *Machiavellian intelligence: Social expertise and the evolution of intellect in monkeys, apes, and humans.* New York: Oxford University Press.

Call, J., Carpenter, M., & Tomasello, M. (2005). Copying results and copying actions in the process of social learning: Chimpanzees (*Pan troglodytes*) and human children (*Homo sapiens*). *Animal Cognition, 8*, 151–163.

Csibra, G., Gergely, G., Biro, S., Koos, O., & Brockbank, M. (1999). Goal attribution without agency cues: The perception of 'pure reason' in infancy. *Cognition, 72*, 237–267.

Custance, D. M., Whiten, A., & Bard, K. A. (1995). Can young chimpanzees (*Pan troglodytes*) imitate arbitrary actions?: Hayes and Hayes (1952) revisited. *Behaviour, 132*, 839–858.

Ferrari, P. F., Visalberghi, E., Paukner, A., Fogassi, L., Ruggiero, A., & Suomi, S. J. (2006). Neonatal imitation in rhesus macaques. *PLoS Biology, 4*, 1501–1508.

Fontaine, R. (1984). Imitative skills between birth and six months. *Infant Behavior and Development, 7*, 323–333.

Galef, B. G., Jr. (1991). The question of animal culture. *Human Nature, 3*, 157–178.

Gallese, V., Fadiga, L., Fogassi, L., & Rizzolatti, G. (2002). Action representation in the inferior parietal lobe. In W. Prinz & B. Hommel (Eds.), *Common mechanisms in perception and action* (pp. 334–355). Oxford: Oxford University Press.

Gergely, G., Bekkering, H., & Király, I. (2002). Rational imitation in preverbal infants. *Nature, 415*, 755.

Gergely, G., & Csibra, G. (2003). Teleological reasoning in infancy: The naive theory of rational action. *Trends in Cognitive Sciences, 7*, 287–292.

Gergely, G., Nádasdy, Z., Csibra, G., & Bíró, S. (1995). Taking the intentional stance at 12 months of age. *Cognition, 56*, 165–193.

Hirata, S., & Celli, M. (2003). Role of mothers in the acquisition of tool-use behaviors by captive infant chimpanzees. *Animal Cognition, 6*, 235–244.

Horner, V., Whiten, A., Flynn, E., & de Waal, F. B. M. (2006). Faithful replication of foraging techniques along cultural transmission chains by chimpanzees and children. *Proceedings of the National Academy of Sciences USA, 103*(37), 13878–13883.

Humphrey, N. (1984). *Consciousness regained: Chapters in the development of mind.* Oxford: Oxford University Press.

Jacobson, S. W. (1979). Matching behavior in the young infant. *Child Development, 50*, 425–430.

Jones, S. S. (1996). Imitation or exploration?: Young infants' matching of adults' oral gestures. *Child Development, 67*, 1952–1969.

Jones, S. S. (2006). Exploration or imitation?: The effect of music on 4-week-old infants' tongue protrusions. *Infant Behavior and Development, 29*, 126–130.

Kojima, H. (2005). [Imitation and formation of social communication from the perspective of developmental robotics]. *Journal of the Society of Biomechanics, 29*, 26–30 (in Japanese).

Matsuzawa, T. (1996). Chimpanzee intelligence in nature and in captivity: Isomorphism of symbol use and tool use. In W. McGrew, L. F. Marchant, & T. Nishida (Eds.), *Great ape societies* (pp. 196–209). Cambridge, UK: Cambridge University Press.

Matsuzawa, T. (2006). Evolutionary origins of the human mother–infant relationship. In T. Matsuzawa, M. Tomonaga, & M. Tanaka (Eds.), *Cognitive development in chimpanzees* (pp. 127–141). Tokyo: Springer-Verlag.

Matsuzawa, T. (2007). Comparative cognitive development. *Developmental Science, 10,* 97–103.

Meltzoff, A. N. (1990). Foundations for developing a concept of self: The role of imitation in relating self to other and the value of social mirroring, social modeling, and self practice in infancy. In D. Cicchetti & M. Beeghly (Eds.), *The self in transition: Infancy to childhood* (pp. 139–164). Chicago: University of Chicago Press.

Meltzoff, A. N., & Borton, R. W. (1979). Intermodal matching by human neonates. *Nature, 282,* 403–404.

Meltzoff, A. N., & Gopnik, A. (1993). The role of imitation in understanding persons and developing a theory of mind. In S. Baron-Cohen, H. Tager-Flusberg, & D. Cohen (Eds.), *Understanding other minds: Perspectives from autism* (pp. 335–366). New York: Oxford University Press.

Meltzoff, A. N., & Moore, M. K. (1977). Imitation of facial and manual gestures by newborn infants. *Science, 198,* 75–78.

Mizuno, Y., Takeshita, H., & Matsuzawa, T. (2006). Behavior of infant chimpanzees during the night in the first four months of life: Smiling and suckling in relation to behavioral state. *Infancy, 9,* 215–234.

Myowa-Yamakoshi, M. (2006). How and when do chimpanzees acquire the ability to imitate? In T. Matsuzawa, M. Tomonaga, & M. Tanaka (Eds.), *Cognitive development in chimpanzees* (pp. 214–232). Tokyo: Springer-Verlag.

Myowa-Yamakoshi, M., & Matsuzawa, T. (1999). Factors influencing imitation of manipulatory actions in chimpanzees (*Pan troglodytes*). *Journal of Comparative Psychology, 113,* 128–136.

Myowa-Yamakoshi, M., & Matsuzawa, T. (2000). Imitation of intentional manipulatory actions in chimpanzees (*Pan troglodytes*). *Journal of Comparative Psychology, 114,* 381–391.

Myowa-Yamakoshi, M., Tomonaga, M., Tanaka, M., & Matsuzawa, T. (2003). Preference for human direct gaze in infant chimpanzees (*Pan troglodytes*). *Cognition, 89,* B53–B64.

Myowa-Yamakoshi, M., Tomonaga, M., Tanaka, M., & Matsuzawa, T. (2004). Neonatal imitation in chimpanzees (*Pan troglodytes*). *Developmental Science, 7,* 437–442.

Nagell, K., Olguin, R., & Tomasello, M. (1993). Processes of social learning in the imitative learning of chimpanzees and human children. *Journal of Comparative Psychology, 107,* 174–186.

Neisser, U. (1991). Two perceptually given aspects of the self and their development. *Developmental Review, 11,* 197–209.

Neisser, U. (1995). Criteria for an ecological self. In P. Rochat (Ed.), *Advances in psychology: Vol. 112. The self in infancy: Theory and research* (pp. 17–34). Amsterdam: Elsevier.

Nielsen, M., Collier-Baker, E., Davis, J. M., & Suddendorf, T. (2004). Imitation recognition in a captive chimpanzee (*Pan troglodytes*). *Animal Cognition, 8,* 31–36.

Paukner, A., Anderson, J. R., Borelli, E., Visalberghi, E., & Ferrari, P. F. (2005). Macaques (*Macaca nemestrina*) recognize when they are being imitated. *Biology Letters, 1,* 219–222.

Piaget, J. (1962). *Play, dreams and imitation in childhood.* New York: Norton.

Premack, D., & Woodruff, G. (1978). Does the chimpanzee have a theory of mind? *Behavioral and Brain Sciences, 1,* 515–526.

Rizzolatti, G., Fadiga, L., Gallese, V., & Fogassi, L. (1996). Premotor cortex and the recognition of motor actions. *Cognitive Brain Research, 3,* 131–141.

Rochat, P. (2001). *The infant's world.* Cambridge, MA: Harvard University Press.

Tomasello, M. (1999). *The cultural origins of human cognition.* Cambridge, MA: Harvard University Press.

Tomasello, M., Carpenter, M., Call, J., Behne, T., & Moll, H. (2005). Understanding and sharing intentions: The origins of cultural cognition. *Behavioral and Brain Sciences, 28,* 1–17.

Tomasello, M., Kruger, A. C., & Ratner, H. H. (1993). Cultural learning. *Behavioral and Brain Sciences, 16,* 495–552.

Tomonaga, M. (2006). Development of chimpanzee social cognition in the first 2 years of life. In T. Matsuzawa, M. Tomonaga, & M. Tanaka (Eds.), *Cognitive development in chimpanzees* (pp. 182–197). Tokyo: Springer-Verlag.

Tomonaga, M., Tanaka, M., Matsuzawa, T., Myowa-Yamakoshi, M., Kosugi, D., Mizuno, Y., et al. (2004). Development of social cognition in chimpanzees (*Pan troglodytes*): Face recognition, smiling, mutual gaze, gaze following and the lack of triadic interactions. *Japanese Psychological Research, 46,* 227–235.

Trevarthen, C., & Hubley, P. (1978). Secondary intersubjectivity: Confidence, confiding and acts of meaning in the first year. In A. Lock (Ed.), *Action, gesture and symbol* (pp. 183–229). London: Academic Press.

Ueno, A., & Matsuzawa, T. (2005). Response to novel food in infant chimpanzees: Do infants refer to mothers before ingesting food on their own? *Behavioral Processes, 68,* 85–90.

Werner, H., & Kaplan, B. (1963). *Symbol formation: An organismic-developmental approach to language and the expression of thought.* New York: Wiley.

Whiten, A., Custance, D. M., Gomez, J.-C., Teixidor, P., & Bard, K. A. (1996). Imitative learning of artificial fruit processing in children (*Homo sapiens*) and chimpanzees (*Pan troglodytes*). *Journal of Comparative Psychology, 110,* 3–14.

Whiten, A., Horner, V., & de Waal, F. B. M. (2005). Conformity to cultural norms of tool use in chimpanzees. *Nature, 437*(7059), 737–740.

PART IV

RELATIONSHIPS

CHAPTER 12

Attachment and the Comparative Psychobiology of Mothering

ANDREA GONZALEZ
LESLIE ATKINSON
ALISON S. FLEMING

Wₑ survey the literature on the neuropsychology and neuroanatomy of mothering within an attachment framework. Our comparative approach focuses on rats and humans. We review aspects of attachment theory pertinent to the psychobiology of caregiving (viz., control systems, goal-corrected behavior, and internal working models), as these relate to motivation and executive functions. We discuss motivational aspects (approach–avoidance) and attentional aspects (attentional set shifting, selective attention) of maternal responsiveness, emphasizing the medial preoptic area (MPOA) and the nucleus accumbens (NAC) (involved in motivation), and the prefrontal cortex (PFC) and anterior cingulate cortex (ACC) (involved in attention). We argue that these areas function integratively to fashion caregiving behavior. Variations in motivation and executive functions presumably underlie different patterns of caregiving.

Attachment Theory and Mothering: Implications for Psychobiology

In humans, early mother–infant interactions provide a baby with the social experiences that initiate a trajectory influencing later relationships. Central to attachment theory in humans is the reciprocity of the mother–infant relationship (Bowlby, 1969). For the infant, the goal of attachment behaviors is to reduce stress and reinstate a sense of security through some form of contact with the caregiver. For the caregiver, attachment behaviors are directed at regulating the infant's arousal and optimizing his or her sense of security by cooperating with the infant's attachment bids. These reciprocal behaviors are rooted in phylogeny and ontogeny.

225

Despite this reciprocity, however, attachment theory focuses on the infant's behavioral systems; the study of caregiving behavior remains derivative, a means of explaining infant attachment patterns, but without intrinsic interest (George & Solomon, 1999). Of particular relevance in this regard, George and Solomon (1999) called for further investigation of the neurological mechanisms underlying caregiving.

If we are to understand the neurobiology of caregiving within an attachment framework, then three constructs are crucial: "control systems," "goal-corrected behavior," and "internal working models" (Bowlby, 1969). Control systems serve to maintain equilibrium through feedback. For example, should an infant wander too far from his or her caregiver, this triggers discomfort in the latter, who works toward retrieving the baby; that is, the caregiver's control system serves to motivate maintenance of the infant within optimal distance. The actual retrieval process involves goal-corrected behavior. The caregiver, through a series of feedback-based strategies, recovers the infant. The strategies are variable and flexible; only the set goal of retrieval is inviolate.

However, goal-corrected behavior is only possible if organisms build internal, small-scale models of their environments (Bowlby, 1969). These models enable prediction, planning, and change. Of particular relevance here are internal working models of attachment. These models are "tolerably accurate" representations of the self *vis-à-vis* attachment figures, continuously constructed from infancy. George and Solomon (1999) speculated that these mental representations are transduced into caregiving models, such that children whose caregivers were consistently insensitive, consistently sensitive, or inconsistently sensitive typically become parents with distant (preferring emotional and physical distance from their children), flexible (enjoying closeness to their children, but also valuing the children's independence), or close (preferring highly dependent children) caregiving patterns, respectively. Children who experience unresolved trauma typically become "disabled" as caregivers (they abdicate their protective role, at least occasionally). In terms of control systems, it is interesting to note, for example, that a caregiver with a distant preference may tolerate greater distance from the infant than other caregivers do, and may experience discomfort when the infant is too close.

Two points should be made about the preceding paragraphs. First, Bowlby (1969) developed a theory of cybernetic (control) systems organized as substrategies (goal-corrected behaviors) within larger hierarchies (internal working models). Control systems are environmentally driven, reflexive, stereotyped, and predictable. From a neurobiological perspective, they are "bottom-up," automatic, and subcortical. By contrast, goal-corrected behaviors and internal working models are purposive, flexible, and feedback-responsive; they are goal-corrected because the "effects of performance are continuously reported back to a central regulating apparatus" (Bowlby, 1969, p. 66). In neurobiological terms, these processes are "top-down," executive, and prefrontal. The PFC comes into play "when we need to use the 'rules of the game,' internal representations of goals, and the means to achieve them" (Miller & Cohen, 2001, p. 168). Consistent with Bowlby's (1969) theorizing, we argue that the integrated functioning of subcortical and prefrontal areas is what underpins caregiving behavior.

The second point is that there are between-species differences in the range and flexibility of caregiving behaviors, largely due to the balance of subcortical and prefrontal involvement. There are also differences within species, although these differences are far less pronounced. Furthermore, the range of between- and within-species variability is likely to depend on the degree of prefrontal involvement. For example, the expression of maternal care in rats consists of nest building, crouching, licking, and pup retrieval. Although there is variability in the frequency and quality of these behaviors, they are nevertheless relatively invariant or stereotyped. It is therefore not surprising that they are largely driven by subcortical circuitry (Numan, Fleming, & Levy, 2006).

By contrast, reciprocity between a human caregiver and infant incorporates multiple components, including shared focus of attention, temporal coordination, and contingency—all goal-corrected behaviors coordinated primarily by the caregiver (Isabella & Belsky, 1991). In particular, parental responsiveness involves behaviors that are contingent, timely, and appropriate (Ainsworth, Blehar, Waters, & Wall, 1978). Although, as in the rat, the motivation to engage in such complex behavior must be subcortically mediated, the PFC plays a disproportionate role. The PFC "is critical in situations when the mappings between sensory inputs, thoughts, and actions ... are rapidly changing" (Miller & Cohen, 2001, p. 168). The requisite coordination of sensory inputs, thoughts, and actions demands the use of internal representations of goals and the means to achieve them. This is the "cardinal function" (Miller & Cohen, 2001, p. 168) of the PFC.

Moreover, human caregivers evince variability not only in the range of caregiving behaviors potentially at their disposal, but in the quality and quantity of care they provide. Human caregivers may abdicate caregiving responsibility completely, may be capable of only close or distal caregiving, or may maintain flexibility (George & Solomon, 1999). If prefrontal and subcortical areas are involved in human caregiving, then this variability suggests that differences in executive functioning and subcortical motivation, and in their interaction, underlie differential within-species parenting.

Maternal Motivation

Maternal Motivation Involves Approach

The onset of maternal behavior at parturition involves shifts in hedonic reactions to infants, which permit mothers to approach their young. New mother rats differ from nonmothers in that they are attracted to infant olfactory and auditory cues. For instance, when presented with a choice between a pup odor and an odor of a diestrous female, virgins prefer the latter, whereas new mothers prefer the pup odor (Fleming & Rosenblatt, 1974; Kinsley & Bridges, 1990). These changes at birth are manifested in many mammalian species, including humans, and are affected by the hormones of parturition (Numan et al., 2006). With little experience of their infants, new human mothers find the infant body odor more attractive than do nonmothers and come to recognize their own infants by their odor (Corter & Fleming, 2002). Visual cues also enhance mothering. In parents, the sight of their own babies' crying or smiling causes heart rate (HR) deceleration and then acceleration when they are viewing silent videotapes of their infants (Wiesenfeld & Klorman, 1978). Mothers responded with HR acceleration when the gaze of an unfamiliar infant was directed toward them, but did not display this arousal when the infant was looking away (Leavitt & Donovan, 1979). At the behavioral level, the infant's gaze evokes the mother's gaze and leads to *en face* behavior, which Klaus, Trause, and Kennell (1975) have described as species-typical maternal behavior. As such, from an attachment perspective, *en face* behavior would be considered cybernetically controlled (and subcortical). Such studies indicate that infant visual stimuli powerfully elicit maternal behavior and are important as precursors and components of infant–mother interaction. These hedonic effects are related to the amount of experience mothers have with their infants and to their hormonal profile during the early postpartum period (Corter & Fleming, 2002). In both rats and humans, several infant stimuli orient mothers toward their infants.

Maternal Responsiveness Involves Absence of Avoidance

Enhanced approach also depends on the reduction of avoidance and active inhibition. Prior to pregnancy and parturition, female rats are neophobic and withdraw from novel stimuli, including pups (Fleming & Rosenblatt, 1974). Hormonal changes in late pregnancy and parturition produce a shift, such that stimuli that formerly produced withdrawal now produce approach responding (Orpen, Furman, Wong, & Fleming, 1987; Pedersen, Ascher, Monroe, & Prange, 1982). With the litter's birth, neophobia in general and withdrawal in particular are reduced, and mothers tolerate close proximity to the pups.

Similarly, human mothers develop appropriate physiological and behavioral responses to their infants' cries, which are often experienced by nonparents as aversive. Infants' cries signal need; mothers often respond by approaching (Stallings, Fleming, Corter, Worthman, & Steiner, 2001). The acoustic and emotion-inducing properties of these signals are well characterized (LaGasse, Neal, & Lester, 2005), and are more easily identified and discriminated as mothers gain experience with them.

Stallings et al. (2001) explored the subjective, autonomic and endocrine responses to recorded "pain" and "hunger" cries of unfamiliar infants in nonmothers versus new mothers. Nonmothers tended to experience emotions that would discourage approach responses, whereas new mothers showed the opposite. For instance, in comparison to first-time mothers, nonmothers were less sympathetic and less alerted to the intense "pain" cries of infants (but not to "hunger" cries). Furthermore, in comparison to more sympathetic mothers, less sympathetic mothers showed reduced baseline HR and salivary stress hormones both before and while listening to recorded cries (Stallings et al., 2001).

So again, among humans, the inhibition of withdrawal responding in nonmothers and its replacement by approach responding in mothers occur, just as they do in rats. However, the shift is not as clear-cut in humans, where juvenile caregiving and aunt "alloparenting" are common (Hrdy, 2000). Nevertheless, women vary in the degree of positive feelings and behavior they exhibit toward infants or infant cues, and these differences are evident in patterns of brain activation.

Maternal Responsiveness

As discussed, maternal sensitivity involves behaviors that are goal-corrected so as to be contingent, timely, and appropriate. These behaviors are complex and involve cross-situation adaptations. During play, components of sensitivity include structuring interactions, following the infant's lead, and adjusting the interaction so as to remain attuned to the infant. Under conditions of infant distress, four aspects of response include noticing the distress signal, intervening, and the timing and effectiveness of the interventions. If the infant is not soothed by the intervention, the parent must switch interventions. At the same time, the parent must balance interactions with the infant against competing stimuli and demands. These behaviors involve coordination across varied situations in response to multiple signals. Furthermore, the requisite coordination demands the use of internal representations of goals and the means to achieve them—primary functions of the PFC. As Miller and Cohen (2001) have pointed out, "Two classic tasks illustrate this point: the Stroop task and the Wisconsin [Card Sorting Test] (WCST)" (p. 168).

Here we review the evidence implicating executive function in caregiving at the neuropsychological, neuroanatomical, and neurochemical levels, focusing on attention tasks (particularly the Stroop and WCST). We concentrate first on how mothers' attentional capabilities are translated into their ability to respond sensitively to their infants. Attention as a behavioral characteristic is central to both maternal sensitivity (Atkinson et al., 1995, 2000; Gonzalez & Fleming, 2008) and executive functioning.

Executive Functioning

We have assessed relations between attention and maternal behavior with attentional-set-shifting tasks and measures of sustained attention. To assess attentional set shifting in rats and humans, we have utilized a modified version of the WCST, an intradimensional/extradimensional (ID/ED) shift task (Birrell & Brown, 2000; Downes et al., 1989). Among humans, we have also explored performance on an emotional Stroop task with neutral and negative attachment/emotion conditions.

The attentional-set-shifting task involves two measures of shifting. ID shifting occurs when new stimuli are presented, but the subject must continue to choose the same perceptual dimension. ED shifting occurs when contingencies change, and responding entails switching attention between two perceptual dimensions (Robbins, 2007). ED shifting requires cognitive flexibility (shifting from one cognitive set to another); ID shifting requires perceptual flexibility and establishment of an "attentional set," shifting from one exemplar to another within the same stimulus dimension (e.g., shape) (Robbins, 2007). Rats or humans who obtain higher numbers of errors in switching strategies once the "rule" has changed have problems with attentional flexibility. Both functions are likely to be central to sensitive mothering.

Attention and Mothering

In rats, attention and executive functioning are linked to qualitative aspects of maternal behavior. Dams that were more easily distracted and less attentive during interactions with their litters took longer to reach criterion on an attentional-set-shifting task and exhibited lower prepulse inhibition (a measure of sensorimotor gating) than did mothers that were more attentive and engaged in more licking behaviors (Lovic & Fleming, 2004). Lovic, Palombo, Kraemer, and Fleming (2008) found that dams exhibiting impulsive or hyperactive behaviors also mothered less adequately.

Our human research verifies relations between cognitive flexibility and maternal responsiveness. Using the Cambridge Neuropsychological Test Automated Battery, we found that mothers who responded with fewer errors on the ED shifting task at 2–6 months postpartum (indicating greater cognitive flexibility) interacted more sensitively with their infants, as assessed by the Ainsworth (1979) attachment scales, and showed more contingent responding to infant cues (Gonzalez & Fleming, 2008). In this same sample, we found no relation between maternal sensitivity and a mother's ability to sustain her attention, the number of errors on the ID shifting component, or reversals. Taken together, these results suggest that maternal sensitivity is specifically related to a mother's ability to switch attention and change strategies, and not her ability to maintain attention to a stimulus or to form an attentional set.

In a study examining selective attention and mothers' attachment to their infants, Atkinson et al. (2009) administered the Adult Attachment Interview (AAI; Main & Goldwyn, 1984) and the Working Model of the Child Interview (WMCI; Zeanah et al., 1993) to mothers prenatally. At 12 months, their infants were observed in the Strange Situation (Ainsworth et al., 1978). At 6 and 12 months, mothers were administered the emotional Stroop task (Williams, Mathews, & MacLeod, 1996), with negative attachment/emotion words (e.g., *lonely*) as the target condition. Mothers classified as disorganized on the AAI (unresolved) and WMCI (irrational fear of loss), and mothers with disorganized babies, were compared to mothers classified as organized and to mothers with organized babies. Organized mothers showed a cohesive, although not necessarily optimal, attachment strategy. Disorganized mothers showed lapses in the monitoring of reasoning, discourse, and behavior. Organized mothers, or the mothers of organized infants, showed a coherent strategy of distant, close, or flexible caregiving approaches. Disorganized mothers were disabled with respect to caregiving (George & Solomon, 1999). Atkinson et al. (2009) also found that disorganized attachment was related to relative Stroop reaction time. That is, unlike organized mothers, disorganized mothers responded to negative attachment/emotion stimuli more slowly than to neutral stimuli; relative speed of response was negatively related to the number of times the dyad was classified as disorganized; and change in relative Stroop response time from 6 to 12 months was related to the match–mismatch status of mother and infant attachment classifications (e.g., if a mother classified as disorganized prenatally had an organized infant at 12 months, her reaction time to target stimuli from 6 to 12 months had probably improved).

Furthermore, Atkinson, Goldberg, Leung, and Benoit (2005) showed that emotional Stroop reaction time was related to maternal sensitivity: Mothers who showed greater delay in responding to target stimuli, as compared to neutral stimuli, showed less sensitivity toward their infants. These data are consistent with work (Atkinson et al., 1995) showing that maternal attentional strategies are related to maternal sensitivity. Specifically, mothers who reported avoidant cognitive coping strategies tended to be insensitive toward their babies, while mothers who were highly vigilant reported high distress and, as a result, low sensitivity. These results suggest that attentional mechanisms are central to maternal sensitivity.

Neuroendocrinology Associated with Mothering and Attention

Maternal behaviors are expressed initially in response to hormonal changes associated with gestation and parturition. In most species, the timing of the expression of maternal behavior depends on the coordinated change in hormones that occurs at parturition (Numan et al., 2006; Rosenblatt, 2002). Addressing the suite of hormones underlying maternal behavior is beyond the scope of this chapter (for reviews, see Numan et al., 2006; Rosenblatt, 2002). Here we review the hypothalamic–pituitary–adrenal (HPA) axis and glucocorticoids, related to "motivational" aspects of caregiving and positive appraisal of young, and to maternal sensitivity and related executive processes.

The HPA axis is activated during stress and is involved in behavioral, emotional, and cognitive processes (Erickson, Drevets, & Schulkin, 2003). Cortisol acts in an "inverted-U shaped" manner on many of these systems, such that moderate levels are optimal, while extremely high or low concentrations have adverse outcomes (Charmandari, Kino, & Chrousos, 2004). In caregiving, cortisol is involved in influencing how mothers experience their infants (Corter & Fleming, 2002), as well as in variations in licking behavior and the inter-generational transmission of maternal licking and stress reactivity in rats (Champagne & Meaney, 2001). In humans, the HPA system is associated with stress reactivity within infant–caregiver attachment relations. Securely attached babies secrete less cortisol than insecurely attached babies in response to the Strange Situation, and disorganized infants secrete more cortisol than all other groups (Hertsgaard, Gunnar, Erickson, & Nachmias, 1995; Spangler & Grossmann, 1993). The HPA axis is most easily engaged in temperamentally fearful infants, such that cortisol differences are revealed when these infants need to use their attachment figures to feel safe (Gunnar, Broderson, Krueger, & Rigatuso, 1996; Nachmias, Gunnar, Man-gelsdorf, Parritz, & Buss, 1996).

Our research has linked glucocorticoids to quality of maternal behaviors in both rats and humans. In rats, adrenalectomies in late pregnancy decreased (but did not abolish) mater-nal behavior and the retention of the maternal experience, and corticosterone replacement reversed these effects (Graham, Rees, Steiner, & Fleming, 2006; Rees, Panesar, Steiner, & Fleming, 2004, 2006). The primary effects were on mothers' licking behaviors. Consistent with a positive role of the glucocorticoids in the modulation of maternal behavior during the early postpartum period in mother rats, elevations of baseline cortisol on postpartum days 2 and 3 in human mothers were associated with more affectionate behaviors, greater attrac-tion to infant odors, and more sympathy in response to infant cries (Fleming, Ruble, Krieger, & Wong, 1997; Stallings et al., 2001). A shift occurs at 2–6 months postpartum, such that elevated levels of cortisol are associated with negative rather than positive maternal feelings and behaviors (Krpan, Coombs, Zinga, Steiner, & Fleming, 2005; Gonzalez, Steiner, & Flem-ing, 2008). The mechanism by which the influence of cortisol affects maternal responsive-ness across the postpartum period is unknown. However, one possible mechanism involves the impact of glucocorticoids on neural plasticity early in the postpartum period (Leuner, Mirescu, Noiman, & Gould, 2007). This influence is less likely to play a role later in the postpartum period.

In addition to its role in caregiving systems, the HPA axis is associated with mood, cog-nition, and attention (Erickson et al., 2003). Depending on various factors, cortisol may be positively related to arousal, attention, perceptual functioning, and vigilance (as occurs early in the postpartum period), or negatively related to these endpoints (Henry, 1992). Later in the postpartum period, higher levels of cortisol are associated with increased response latencies to emotional Stroop stimuli (Atkinson et al., 2005) and more errors in ED shifting (Gonzalez & Fleming, 2008). Furthermore, increased levels of cortisol and number of errors on the ED shifting task and increased response latencies on the Stroop were negatively associated with maternal sensitivity (Atkinson et al., 2005; Gonzalez & Fleming, 2008). Mothers with high levels of cortisol had impaired cognitive flexibility, longer latencies to target Stroop stimuli, and less sensitivity with their infants. These are the first studies linking neuropsychological and endocrinological correlates to human maternal behavior.

Neuroanatomy of Maternal Motivation: Onset and Expression

Here we discuss the functional neuroanatomy of mothering; we concentrate on animal work, subcortical systems, and maternal motivation, the foci of most research to date. We provide more limited discussion of the neuroanatomy of human maternal behavior, emphasizing the

PFC as it regulates maternal sensitivity and the behavioral systems that influence mother–young interactions, primarily through the attentional system.

The neuroanatomy of the systems and circuits that regulate maternal behavior is best described in terms of the functionality of its component parts (Li & Fleming, 2002; Numan et al., 2006). Consistent with this formulation of how the behavioral systems are organized, we find that the functional neuroanatomy of mothering comprises excitatory and inhibitory neural systems whose activity affects the "final common path" for the expression of the behavior. Below we discuss the functional relation between behavioral and neural systems as though they exist at the same level of analysis.

Neuroanatomy and Excitatory Systems

Medial Preoptic Area

As shown in Figure 12.1, in several mammalian species the MPOA is the nuclear group central to the expression of maternal behavior (Numan et al., 2006). When a new mother is

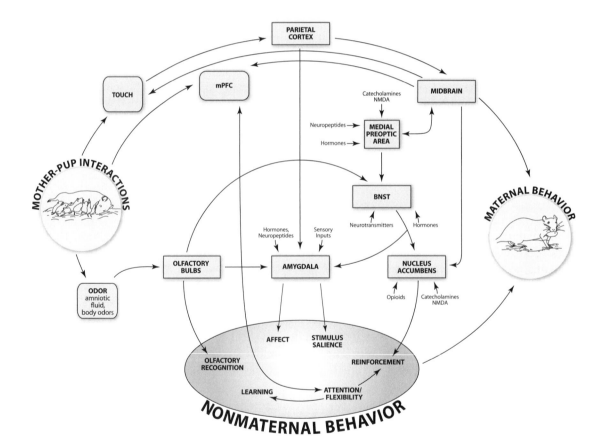

FIGURE 12.1. Functional neuroanatomy mediating maternal and related behaviors in mammals. Neuroanatomical structures include the olfactory bulbs, amygdala, nucleus accumbens, bed nucleus of the stria terminalis (BNST), medial preoptic area, midbrain, parietal cortex, and medial prefrontal cortex (mPFC). NMDA, N-methyl-D-aspartate. Adapted from Fleming, O'Day, and Kraemer (1999). Copyright 1999 by Elsevier. Adapted by permission.

hormonally primed at parturition, hormones bind to receptors for estradiol, progesterone, prolactin, and oxytocin in cells in the MPOA, increasing the probability of their activation. When mothers engage in their repertoire of maternal behaviors, cells within the MPOA and the ventral bed nucleus of the stria terminal is (BNST) activate c-fos (Numan & Numan, 1994; Fleming, Suh, Korsmit, & Rusak, 1994; Fleming & Korsmit, 1996) and undergo structural change with maternal experience (Featherstone, Fleming, & Ivy, 2000; Kinsley et al., 2006). Lesions of the MPOA/ventral BNST eliminate maternal behavior, whereas electrical stimulation of this site activates the maternal behavior to foster pups in animals that would normally not be responsive (Li & Fleming, 2002). The MPOA and its downstream brainstem projections is an excitatory neural system that functions as the "final common path" for the expression of maternal behavior. Feeding into this system are other excitatory systems whose activation is important for mothers' attraction to their pups at parturition, and for the species-typical coordination and contingent responding between mother and young (Fleming et al., 1994; Li & Fleming, 2002; Numan, 2007).

Nucleus Accumbens and Dopamine System

The dopamine system is involved in functions including cognition, movement, attachment, and motivation/reinforcement associated with food, sex, drugs, and maternal behavior (Nieoullon & Coquerel, 2003; Schultz, 2006; see Figures 12.2a and 12.2b).

Evidence for involvement of these excitatory systems in maternal behavior comes from animal work using lesions, pharmacological agents, and techniques to measure extracellular neurotransmitters in the brain. These studies demonstrate that dopamine is involved in the normal expression of maternal behavior, reinforcement processes, and attention (Champagne et al., 2004; Chudasama & Robbins, 2006; Hansen, Harthon, Wallen, Lofberg, & Svensson, 1991a, 1991b; Li & Fleming, 2002; Numan, 2007). For example, lesions to the NAC, which receives dopamine input from the midbrain, retard mothers' initial pup retrieval responses and block formation of maternal memory, possibly by altering the emotional valence of the conditioned pup stimulus (Keer & Stern, 1999; Li & Fleming, 2002). The administration of dopamine blockers to the NAC or the MPOA has the same effect (Keer & Stern, 1999; Miller & Lonstein, 2005), while simple exposure of maternally experienced animals to pups as opposed to food results in a rapid release of dopamine from the NAC (Afonso, Grella, Chatterjee, & Fleming, 2008; Champagne et al., 2004). Studies associating the role of dopamine with human maternal behavior are not available, given the invasive techniques necessary to measure dopamine levels in humans. As discussed below, evidence instead is provided by human imaging studies with mothers, implicating a similar pattern of activation in reward circuitry.

Neuroanatomy and Inhibitory Systems

The change from avoidance to approach that occurs with experience with pups and/or with hormones associated with pregnancy and birth is accomplished by a disinhibition of an "inhibitory" brain system acting on the MPOA. Figure 12.1 shows that the inhibitory system includes the olfactory bulbs, the amygdala, the stria terminalis, and the ventromedial hypothalamus (not shown) (Li & Fleming, 2002; Numan, 2007). Lesions and stimulation to nuclear groups or pathways within this system result in disinhibition and inhibition, respectively, of the expression of maternal behavior (Li & Fleming, 2002; Sheehan, Paul, Amaral, Numan, & Numan, 2001). Hence lesions of this system both reduce the animals' natural withdrawal behavior and general neophobia (so that animals are willing to remain in close

(a)

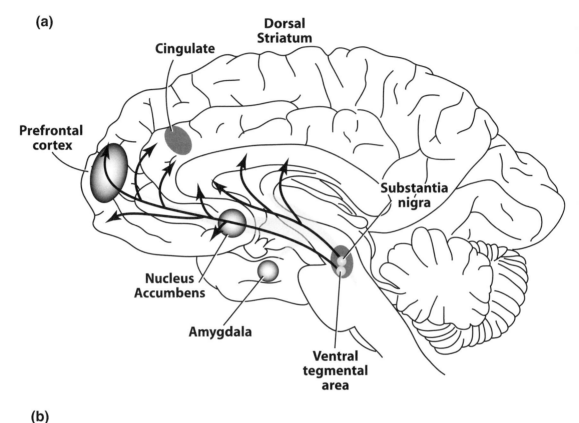

(b)

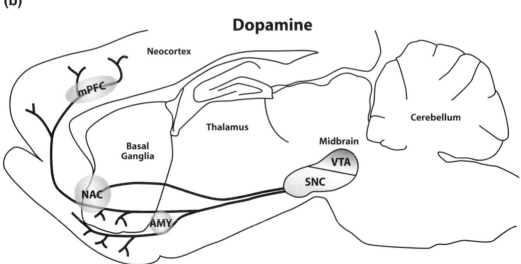

FIGURE 12.2. Dopamine (DA) pathways in the (a) human brain and (b) rat brain. DA cell bodies originating in the ventral tegmental area (VTA) of the brain project to the ventral striatum, with major efferents to the nucleus accumbens (NAC), and the medial prefrontal cortex (mPFC) in response to various stimuli. AMY, amygdala; SNC, substantia nigra complex.

proximity to pups) and remove an inhibitory input to the MPOA system (Li & Fleming, 2002; Numan, 2007), whereas electrical stimulation of components of this same system activates withdrawal and "fear" responding and activates inhibitory action of the MPOA (Morgan, Watchus, & Fleming, 1997).

Neuroanatomy Associated with Human Maternal Behavior

In humans, understanding of the neural bases of maternal behavior derives primarily from functional magnetic resonance imaging (fMRI) studies where mothers are presented with pictures of either their own infants or familiar or unfamiliar infants (Bartels & Zeki, 2004; Leibenluft, Gobbini, Harrison, & Haxy, 2004; Nitschke et al., 2004; Strathearn, Li, & Montague, 2005); recordings of infant cries (Lorberbaum et al., 1999; Seifritz et al., 2003; Swain, Lorberbaum, Kose, & Strathearn, 2007) or infant laughter (Seifritz et al., 2003); or videotapes of infants (Ranote et al., 2004). These studies focus on the effects of infant cues on brain activity in regions that mediate positive and negative emotions (ACC, orbital frontal cortex [OFC—positive], amygdala); stimulus salience or reinforcement (striatum/NAC); cognitive and executive functions (dorsolateral PFC, medial PFC, and ACC); and memory (NAC, medial PFC). In response to infant laughing and crying, mothers show a different pattern of activity than do nonmothers in regions involved in emotion (Seifritz et al., 2003). Mothers show a greater amygdaloid response to infant cries than infant laughs, whereas nonmothers show the opposite (Seifritz et al., 2003). Other studies demonstrate that regions underlying maternal motivation include areas specific to maternal behavior that are activated by rewarding social stimuli. For instance, Bartels and Zeki (2004) presented either infant-related pictures or pictures of romantic partners to women, and found increases in neural activity in some brain regions for both stimuli sets (striatum and middle insula, ACC), whereas in other brain regions activation was more specialized to infant stimuli (OFC, periaqueductal gray). Similarly, unique patterns of activation were seen in the ventral striatum, thalamus, and NAC, and in the amygdala, BNST, and hippocampus, in mothers viewing their own infants' pictures versus pictures of familiar but other infants (Popeski et al., 2008; Strathearn et al., 2005). In humans (as in other animals), brain regions involved in processing emotion, reinforcement, and stimulus salience not specific to maternal behavior, but necessary for its occurrence, are also activated by infant cues.

Neuroanatomy of Executive Function and Its Relation to Maternal Sensitivity

Few studies have assessed the role of the cortex in mothering. Here we first focus on cortical systems involved in general processes of attention shifting and selective attention. This is followed by a brief discussion of fMRI studies that explore comparable brain systems during presentation of infant-related stimuli in mothers.

Neuroanatomy Associated with Attention: Set Shifting and Selective Attention

Imaging and lesion research employing the attentional-set-shifting (ID/ED) task demonstrates double-dissociable patterns of impairment between ED shifting (across modalities) and reversals of the conditioned stimulus (within modalities) across species. In marmoset monkeys and rats, lesions to the OFC impair reversal learning, whereas ED shifting is impaired by lesions to the lateral PFC in monkeys and the medial PFC in rats (Birrell & Brown, 2000; McAlonan & Brown, 2003; Robbins, 2007).

Human imaging studies confirm the involvement of dissociable areas of the PFC in attentional-set-shifting tasks. Using positron emission tomography to study ID/ED shifting, Rogers, Andrews, Grasby, Brooks, and Robbins (2000) found that ED shifts were correlated with activation of the dorsolateral PFC and putamen, supporting the hypothesis that ED shifts are dependent on the frontostriatal system. Activation of other areas include the ventromedial PFC, OFC, and a number of nonfrontal areas (including parietal cortex, basal ganglia, and occipital cortices) (Alvarez & Emory, 2006; Barcelo, 2001).

The Stroop is associated with increased activation in the ACC and the dorsolateral PFC (Alvarez & Emory, 2006; Cabeza & Nyberg, 2000; Compton et al., 2003). Specifically, the ACC acts as "central executor" in selective attention, coordinating and integrating multiple pathways used to read and name colors (Peterson et al., 1999); the dorsolateral PFC is associated with the need to maintain an attention set that selects task-relevant information in the presence of salient distractors (Banich et al., 2000; Compton et al., 2003). In addition to these two areas, the emotional Stroop task differentially activates the OFC and posterior brain regions (Compton et al., 2003).

Neuroanatomy Associated with Maternal Sequences and Maternal Sensitivity

Although the PFC has not been intensively studied in rats, there is evidence that it is not involved in initiating maternal response or in the motivation to respond, but is involved in the temporal organization of maternal behaviors (Dietrich & Allen, 1998; Schneider & Koch, 2005). Medial PFC lesions in female rats results in disorganized and persistent retrieving; for example, animals pick pups up but fail to release them (Stamm, 1955; Slotnick, 1967). Afonso, Sisson, Lovic, and Fleming (2007) found that lesions of the medial PFC prior to mating had no effect on gestation, parturition, or initiation of maternal responding, but disrupted the frequency, duration, and execution of maternal behavior sequences associated with licking, hovering, and retrieving.

As discussed earlier, the importance of planning, cognitive flexibility, and selective attention for mothering is greater in humans than in rodents. Therefore, the role of the PFC as a regulatory mechanism in organizing mothering is likely to be more central in humans. With respect to the circuitry involved in cognitive flexibility and selective attention (performance on the ID/ED shift task and the Stroop), imaging studies have demonstrated differential activation and deactivation patterns in areas associated with these tasks (including the ACC, dorsolateral PFC, and NAC) when mothers are viewing pictures of their own versus other children and hearing infant cries (Bartels & Zeki, 2004; Leibenluft et al., 2004; Lorberbaum et al., 2002; Popeski et al., 2008; Ranote et al., 2004). Other overlapping areas of interest that are activated in response to infant cues include the OFC, the medial PFC, and the ventral PFC.

It is difficult to extrapolate the meaning of differential activation and deactivation of these areas, given that mothers were instructed to passively apprehend the presented stimuli and did not actively respond to, judge, or manipulate them. In addition, the methodologies across studies differed in terms of the nature of the stimuli (cries vs. pictures), the types of contrasts (own vs. other infants), and the ages of the infants. However, it is clear that these areas are involved in the emotional and cognitive processing of infant-relevant stimuli. It is possible that viewing pictures of one's own infant versus an unfamiliar or familiar infant, or listening to infant cries versus white noise, triggers a set of thoughts and intentions that were not explored in the studies reviewed here because of their methodological designs. These areas are also associated with other systems involved in emotion regulation, empathy, and theory of mind (Adolphs, 2006; Beer & Ochsner, 2006; Decety & Sommerville, 2003; Frith & Frith, 2006; Lieberman, 2007; Völlm et al., 2006) and may be activated in coordination with other areas, contributing to these higher-level processes.

We know little about the functional neuroanatomy of maternal behavior through fMRI studies. Most of the studies have small samples and do not have clear hypotheses regarding specific brain function. However, what has been done demonstrates a point we have made above: In humans (as in other animals), brain regions that are involved in processing emotion, reinforcement, stimulus salience, executive tasks, and cognitive flexibility, not specific to maternal behavior but necessary for its occurrence, are also activated by infant cues. Future studies should focus on gaining a greater understanding of the phenomenological experience associated with viewing pictures of one's own infant and connecting these neural activation patterns with behaviors. One longitudinal study, currently in progress, aims to link maternal attachment patterns (assessed via the AAI) to various fMRI responses to infant cues, and to further associate these patterns with infant behavior (Strathearn, 2007). Strathearn hypothesizes that different attachment patterns will be distinguishable by differing patterns of brain activation. Future studies need to determine whether differential maternal patterns of neural activation contribute to differences in parenting behaviors, sensitivity, and infants' attachment classification.

Conclusions and Implications

We have discussed the neurobiology of parenting within an attachment framework. We have argued that Bowlby's constructs of control systems, goal-corrected partnerships, and internal working models incorporate both motivational systems, underpinned by subcortical areas, and executive functions, mediated by the PFC. We have argued that subcortical and prefrontal areas mediate maternal behavior in concert, with nonhuman species depending relatively less, and humans depending relatively more, on prefrontal function. In particular, the subcortical areas include the excitatory (MPOA) and inhibitory (ventromedial hypothalamus, amygdala) systems. The prefrontal areas include the medial PFC, dorsolateral PFC, and the ACC.

We mention several provisos and questions, however. First, we do not understand the relation between motivational (hedonic) and executive (attentional) functions in the regulation of mothering. These issues need further analysis. Second, we have not fully covered all brain areas involved in parenting. Third, we do not assume that the aforementioned systems are dedicated solely to mothering; we recognize that they are part of more general systems that also serve caregiving. Fourth, we acknowledge that the dorsolateral PFC and dorsal ACC are not exclusive to the attentional tasks described above, but have been implicated in a broad range of cognitive demands, including working memory, episodic memory, response selection, and problem solving (Duncan & Owen, 2000). Finally, we are not arguing that the dorsolateral PFC and ACC are the most important areas involved in the executive functions associated with mothering; future caregiving research should tap into various regions using neuropsychological tests associated with known brain regions and known functions, such as empathy and theory of mind (see Figure 12.3).

Although neuroscientists have explored many neurobiological factors that explain varied parenting in animals, there have been few studies on the role of the cortex and mothering. What has been done strongly suggests that the cortex intersects with the limbic system in mediating hypothalamic control of mothering. In contrast to the work in other animals, less work has been done in humans on either cortical or subcortical brain mechanisms of parenting. From an attachment perspective, this remains the central challenge: What are the neurobiological underpinnings of differential parenting? The advantage of looking at this within an attachment framework is that attachment provides a taxonomic system within which one may expect to find psychobiological differences, and which maps directly onto infant classification patterns. This latter is a particularly important consideration, given that parenting styles are transgenerational. All mothers have similar neural circuitry that contributes to their caregiving behavior and attachment processes. It is probable that the patterns of activation are different, thereby producing varied caregiving models. Furthermore, it is likely that these differing activation systems develop early in life from different experiences with caregivers. The animal literature suggests that early caregiving shapes the developing neurocircuitry, producing the intergenerational transmission of caregiving patterns (Champagne & Meaney, 2001; Gonzalez, Lovic, Ward, Wainwright, & Fleming, 2001).

Dopamine and glucocorticoids serve to integrate function across the neurocircuitry of the aforementioned areas. We would speculate that these areas and processes will be implicated in differential caregiving with future research. For example, inferential evidence exists that the cortisol stress response may differentiate between parents with disabled caregiving and parents with flexible, close, and distant caregiving patterns. As outlined earlier, the relationships of both glucocorticoids and dopamine with cognitive function follow an inverted-U-shaped function (Arnsten & Li, 2005). Perhaps flexible caregiving occurs at the peak of the inverted U at "optimal" levels of cortisol and dopamine, and other caregiving patterns fall at either one of the two extremes. This may explain differences in stress responses, caregiving behaviors (sensitivity), and executive functions, as well as in such higher-order processes as empathy and theory of mind.

Finally, the present analysis provides an interesting framework within which to study a host of high-risk populations where mothering is less than optimal and where we suspect neural or neurochemical mediators, such as mothering among young teenagers or among depressed, anxious, or stressed women. The clinical and practical implications of this formulation are considerable.

Acknowledgments

Much of the research summarized here was funded by grants from the Canadian Institutes of Health Research (CIHR), the National Science Engineering Research Council of Canada (NSERC), and the Social Sciences and Humanities Research Council (SSHRC) to Alison S. Fleming, and by the CIHR and SSHRC to Leslie Atkinson. Funding was provided to Andrea Gonzalez by a student training fellowship from the Maternal Adversity, Vulnerability, and Neurodevelopment Project. We thank Alison Dias for her fantastic graphics, and Hope Eaton for assistance with references.

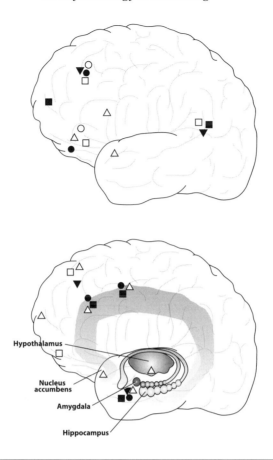

Activation—empathy (Völlm et al., 2006)

Activation on theory of mind tasks (Beer & Ochsner, 2006; Decety & Sommerville, 2003; Frith & Frith, 2006; Lieberman, 2007; Völlm et al., 2006)

Activation on Stroop test (Alvarez & Emory, 2006; Bush, Luu, & Posner, 2000; Cabeza & Nyberg; 2000; Compton et al., 2003)

Activation on test of cognitive flexibility (ID/ED and WCST) (Alvarez & Emory, 2006; Barcelo, 2001; Rogers et al., 2000)

Deactivation in mothers in response to infant cues (cries and pictures)

Activation in mothers in response to infant cues (cries and pcitures) (Bartels & Zeki, 2004; Lorberbaum et al., 2002; Nitzchke et al., 2004; Popeski et al., 2008; Seifritz et al., 2003; Strathearn, Li, Fonagy, & Montague, 2008; Swain et al., 2007)

FIGURE 12.3. Overlapping clusters of activation and deactivation in various neuroimaging studies with mothers in response to infant cues (cries and pictures), tests of selective attention and cognitive flexibility (Stroop, ID/ED shifting, and WCST), and tests of empathy and theory of mind. Top: Lateral surface view. Bottom: Midsagittal view. Shapes are placed schematically within a region and are not meant to indicate precise locations within a region.

References

Adolphs, R. (2006). How do we know the minds of others?: Domain-specificity, simulation, and enactive social cognition. *Brain Research, 1079*(1), 25–35.

Afonso, V. M., Grella, S. L., Chatterjee, D., & Fleming, A. S. (2008). *Maternal experiences affect mesolimbic dopaminergic neuronal activity in the non-lactating rat: A microdialysis study.* Manuscript submitted for publication.

Afonso, V., Sisson, M., Lovic, V., & Fleming, A. S. (2007). The effects of medial prefrontal cortex lesions on maternal behaviour in Sprague Dawley rats. *Behavioral Neuroscience, 121,* 515–526.

Ainsworth, M. D. S. (1979). Attachment as related to mother–infant interaction. *Advances in Study of Behavior, 9,* 1–49.

Ainsworth, M. D. S., Blehar, M. C., Waters, E., & Wall, S. (1978). *Patterns of attachment: A psychological study of the Strange Situation.* Hillsdale, NJ: Erlbaum.

Alvarez, J. A., & Emory, E. (2006). Executive function and the frontal lobes: A meta-analytic review. *Neuropsychology Review, 16,* 17–42.

Arnsten, A. F., & Li, B. M. (2005). Neurobiology of executive functions: Catecholamine influences on prefrontal cortical functions. *Biological Psychiatry, 57*(11), 1377–1384.

Atkinson, L., Goldberg, S., Leung, E., & Benoit, D. (2005). *Cortisol stress response in infancy: Maternal physiological, cognitive, and behavioural correlates.* Paper presented at the meeting of the International Society of Psychoneuroendocrinology, Montréal.

Atkinson, L., Leung, E., Goldberg, S., Benoit, D., Poulton, L., Myhal, N., et al. (2009). Attachment and selective attention: Disorganization and emotional Stroop reaction time. *Development and Psychopathology, 21,* 99–126.

Atkinson, L., Paglia, A., Coolbear, J., Niccols, A., Poulton, L., Leung, E., et al. (2000). L'evaluation de la sensibilité maternelle dans le contexte de la sécurité d'attachment: Une meta-analyse [Assessing maternal sensitivity in the context of attachment security: A meta-analysis]. In G. M. Tarabulsy, S. Larose, D. R. Pederson, & G. Moran (Eds.), *Attachement et développement: Le role des premières relations dans le développement humain [Attachment and development: The role of first relationships in human development].* Québec: Presses de l'Université du Québec.

Atkinson, L., Scott, B., Chisholm, V. C., Blackwell, J., Dickens, S. E., Tamm, F., et al. (1995). Cognitive coping, affective distress, and maternal sensitivity: Mothers of children with Down syndrome. *Developmental Psychology, 31,* 668–679.

Banich, M. T., Milham, M. P., Atchley, R. A., Cohen, N. J., Webb, A., Wszalek, T., et al. (2000). Prefrontal regions play a predominant role in imposing an attentional 'set': Evidence from fMRI. *Brain Research: Cognitive Brain Research, 10*(1–2), 1–9.

Barcelo, F. (2001). Does the Wisconsin Card Sorting Test measure prefrontal function? *Spanish Journal of Psychology, 4*(1), 79–100.

Bartels, A., & Zeki, S. (2004). The neural correlates of maternal and romantic love. *NeuroImage, 21*(3), 1155–1166.

Beer, J. S., & Ochsner, K. N. (2006). Social cognition: A multi-level analysis. *Brain Research, 1079*(1), 98–105.

Birrell, J. M., & Brown, V. J. (2000). Medial frontal cortex mediates perceptual attentional set shifting in the rat. *Journal of Neuroscience, 20*(11), 4320–4324.

Bowlby, J. (1969). *Attachment and loss: Vol. 1. Attachment.* Sydney: Pimlico.

Bush, G., Luu, P., & Posner, M. I. (2000). Cognitive and emotional influences in anterior cingulate cortex. *Trends in Cognitive Sciences, 4*(6), 215–222.

Cabeza, R., & Nyberg, L. (2000). Neural bases of learning and memory: Functional neuroimaging evidence. *Current Opinion in Neurology, 13*(4), 415–421.

Champagne, F. A., Chretien, P., Stevenson, C. W., Zhang, T. Y., Gratton, A., & Meaney, M. J. (2004). Variations in nucleus accumbens dopamine associated with individual differences in maternal behavior in the rat. *Journal of Neuroscience, 24*(17), 4113–4123.

Champagne, F. A., & Meaney, M. J. (2001). Like mother, like daughter: Evidence for non-genomic transmission of parental behavior and stress responsivity. *Progress in Brain Research, 133,* 287–302.

Charmandari, E., Kino, T., & Chrousos, G. P. (2004). Glucocorticoids and their actions: An introduction. *Annals of the New York Academy of Sciences, 1024,* 1–8.

Chudasama, Y., & Robbins, T. W. (2006). Functions of frontostriatal systems in cognition: Comparative neuropsychopharmacological studies in rats, monkeys and humans. *Biological Psychiatry, 73*(1), 19–38.

Compton, R. J., Banich, M. T., Mohanty, A., Milham, M. P., Herrington, J., Miller, G. A., et al. (2003). Paying attention to emotion: An fMRI investigation of cognitive and emotional Stroop tasks. *Cognitive, Affective, and Behavioral Neuroscience, 3*(2), 81–96.

Corter, C., & Fleming, A. S. (2002). Psychobiology of maternal behavior in human beings. In M. H. Bornstein (Ed.), *Handbook of parenting: Vol. 2. Biology and ecology of parenting* (2nd ed., pp. 141–182). Mahwah, NJ: Erlbaum.

Decety, J., & Sommerville, J. A. (2003). Shared representations between self and other: A social cognitive neuroscience view. *Trends in Cognitive Sciences, 7*(12), 527–533.

Dietrich, A., & Allen, J. D. (1998). Functional dissociation of the prefrontal cortex and the hippocampus in timing behavior. *Behavioral Neuroscience, 112*(5), 1043–1047.

Downes, J. J., Roberts, A. C., Sahakian, B. J., Evenden, J. L., Morris, R. G., & Robbins, T. W. (1989). Impaired extra-dimensional shift performance in medicated and unmedicated Parkinson's disease: Evidence for a specific attentional dysfunction. *Neuropsychologia, 27*(11–12), 1329–1343.

Duncan, J., & Owen, A. M. (2000). Common regions of the human frontal lobe recruited by diverse cognitive demands. *Trends in Neurosciences, 23*(10), 475–483.

Erickson, K., Drevets, W., & Schulkin, J. (2003). Glucocorticoid regulation of diverse cognitive functions in normal and pathological emotional states. *Neuroscience and Biobehavioral Reviews, 27*(3), 233–246.

Featherstone, R. E., Fleming, A. S., & Ivy, G. O. (2000). Plasticity in the maternal circuit: Effects of experience and partum condition on brain astrocyte number in female rats. *Behavioral Neuroscience, 114*(1), 158–172.

Fleming, A. S., & Korsmit, M. (1996). Plasticity in the maternal circuit: Effects of maternal experience on Fos-Lir in hypothalamic, limbic, and cortical structures in the postpartum rat. *Behavioral Neuroscience, 110*(3), 567–582.

Fleming, A. S., O'Day, D. H., & Kraemer, G. W. (1999). Neurobiology of mother–infant interactions: Experience and central nervous system plasticity across development and generations. *Neuroscience and Biobehavioral Reviews, 23*(5), 673–685.

Fleming, A. S., & Rosenblatt, J. S. (1974). Maternal behavior in the virgin and lactating rat. *Journal of Comparative Physiology and Psychology, 86*(5), 957–972.

Fleming, A. S., Ruble, D., Krieger, H., & Wong, P. Y. (1997). Hormonal and experiential correlates of maternal responsiveness during pregnancy and the puerperium in human mothers. *Hormones and Behavior, 31*(2), 145–158.

Fleming, A. S., Suh, E. J., Korsmit, M., & Rusak, B. (1994). Activation of Fos-like immunoreactivity in the medial preoptic area and limbic structures by maternal and social interactions in rats. *Behavioral Neuroscience, 108*(4), 724–734.

Frith, C. D., & Frith, U. (2006). The neural basis of mentalizing. *Neuron, 50*(4), 531–534.

George, C., & Solomon, J. (1999). Attachment and caregiving: The caregiving behavioral system. In J. Cassidy & P. R. Shaver (Eds.), *Handbook of attachment: Theory, research, and clinical applications* (pp. 649–670). New York: Guilford Press.

Gonzalez, A., & Fleming, A. S. (2008). *Neurobiology and neuropsychology as mediators between early maltreatment and parenting behaviours.* Manuscript submitted for publication.

Gonzalez, A., Lovic, V., Ward, G. R., Wainwright, P. E., & Fleming, A. S. (2001). Intergenerational effects of complete maternal deprivation and replacement stimulation on maternal behavior and emotionality in female rats. *Developmental Psychobiology, 38*(1), 11–32.

Gonzalez, A., Steiner, M., & Fleming, A. S. (2008). *Depressed mothers show altered physiological responsiveness to infant cries.* Manuscript in preparation.

Graham, M. D., Rees, S. L., Steiner, M., & Fleming, A. S. (2006). The effects of adrenalectomy and corticosterone replacement on maternal memory in postpartum rats. *Hormones and Behavior, 49*(3), 353–361.

Gunnar, M. R., Brodersen, L., Krueger, K., & Rigatuso, J. (1996). Dampening of adrenocortical responses during infancy: Normative changes and individual differences. *Child Development, 67,* 877–889.

Hansen, S., Harthon, C., Wallin, E., Lofberg, L., & Svensson, K. (1991a). The effects of 6-OHDA-induced dopamine depletions in the ventral or dorsal striatum on maternal and sexual behavior in the female rat. *Pharmacology, Biochemistry and Behavior, 39*(1), 71–77.

Hansen, S., Harthon, C., Wallin, E., Lofberg, L., & Svensson, K. (1991b). Mesotelencephalic dopamine system and reproductive behavior in the female rat: Effects of ventral tegmental 6-hydroxydopamine lesions on maternal and sexual responsiveness. *Behavioral Neuroscience, 105*(4), 588–598.

Henry, J. P. (1992). Biological basis of the stress response. *Integrative Physiology and Behavioral Science, 27,* 66–83.

Hertsgaard, L., Gunnar, M., Erickson, M. F., & Nachmias, M. (1995). Adrenocortical responses to the Strange Situation in infants with disorganized/disoriented attachment relationships. *Child Development, 66*(4), 1100–1106.

Hrdy, S. B. (2000). *Mother nature: Maternal instincts and how they shape the human species.* New York: Ballantine Books.

Isabella, R. A., & Belsky, J. (1991). Interactional synchrony and the origins of infant–mother attachment: A replication study. *Child Development, 62*(2), 373–384.

Keer, S. E., & Stern, J. M. (1999). Dopamine receptor blockade in the nucleus accumbens inhibits maternal retrieval and licking, but enhances nursing behavior in lactating rats. *Physiology and Behavior, 67*(5), 659–669.

Kinsley, C. H., & Bridges, R. S. (1990). Morphine treatment and reproductive condition alter olfactory preferences for pup and adult male odors in female rats. *Developmental Psychobiology, 23*(4), 331–347.

Kinsley, C. H., Trainer, R., Stafisso-Sandoz, G., Quadros, P., Marcus, L. K., Hearon, C., et al. (2006). Motherhood and the hormones of pregnancy modify concentrations of hippocampal neuronal dendritic spines. *Hormones and Behavior, 49*(2), 131–142.

Klaus, M. H., Trause, M., & Kennell, J. H. (1975). *Does human maternal behavior after birth show a characteristic pattern?: Parent–infant interaction.* Paper presented at the CIBA Foundation, Amsterdam.

Krpan, K. M., Coombs, R., Zinga, D., Steiner, M., & Fleming, A. S. (2005). Experiential and hormonal correlates of maternal behavior in teen and adult mothers. *Hormones and Behavior, 47*(1), 112–122.

LaGasse, L. L., Neal, A. R., & Lester, B. M. (2005). Assessment of infant cry: Acoustic cry analysis and parental perception. *Mental Retardation and Developmental Disability Research Reviews, 11*(1), 83–93.

Leavitt, L., & Donovan, W. (1979). Perceived infant temperament, focus of control, and maternal physiological response to infant gaze. *Journal of Research in Personality, 13,* 267–278.

Leibenluft, E., Gobbini, M. I., Harrison, T., & Haxby, J. V. (2004). Mothers' neural activation in response to pictures of their children and other children. *Biological Psychiatry, 56*(4), 225–232.

Leuner, B., Mirescu, C., Noiman, L., & Gould, E. (2007). Maternal experience inhibits the production of immature neurons in the hippocampus during the postpartum period through elevations in adrenal steroids. *Hippocampus, 17,* 434–442.

Li, M., & Fleming, A. S. (2002). Psychobiology of maternal behavior and its early determinants in non-human mammals. In M. H. Bornstein (Ed.), *Handbook of parenting: Vol. 2. Biology and ecology of parenting* (2nd ed., pp. 61–97). Mahwah, NJ: Erlbaum.

Lieberman, M. D. (2007). Social cognitive neuroscience: A review of core processes. *Annual Review of Psychology, 58*, 259–289.

Lorberbaum, J. P., Newman, J. D., Dubno, J. R., Horwitz, A. R., Nahas, Z., Teneback, C. C., et al. (1999). Feasibility of using fMRI to study mothers responding to infant cries. *Depression and Anxiety, 10*(3), 99–104.

Lorberbaum, J. P., Newman, J. D., Horwitz, A. R., Dubno, J. R., Lydiard, R. B., Hamner, M. B., et al. (2002). A potential role for thalamocingulate circuitry in human maternal behavior. *Biological Psychiatry, 51*(6), 431–445.

Lovic, V., & Fleming, A. S. (2004). Artificially-reared female rats show reduced prepulse inhibition and deficits in the attentional set shifting task: Reversal of effects with maternal-like licking stimulation. *Behavioural Brain Research, 148*(1–2), 209–219.

Lovic, V., Palombo, D. J., Kraemer, G. W., & Fleming, A. S. (2008). *Relationship between motor impulsiveness and maternal behaviour.* Manuscript in preparation.

Main, M., & Goldwyn, R. (1984). Predicting rejection of her infant from mother's representation of her own experience: Implications for the abused–abusing intergenerational cycle. *Child Abuse and Neglect, 8*(2), 203–217.

McAlonan, K., & Brown, V. J. (2003). Orbital prefrontal cortex mediates reversal learning and not attentional set shifting in the rat. *Behavioural Brain Research, 146*(1–2), 97–103.

Miller, E. K., & Cohen, J. D. (2001). An integrative theory of prefrontal cortex function. *Annual Review of Neuroscience, 24*, 167–202.

Miller, S. M., & Lonstein, J. S. (2005). Dopamine D1 and D2 receptor antagonism in the preoptic area produces different effects on maternal behavior in lactating rats. *Behavioral Neuroscience, 119*(4), 1072–1083.

Morgan, H. D., Watchus, J. A., & Fleming, A. S. (1997). The effects of electrical stimulation of the medial preoptic area and the medial amygdala on maternal responsiveness in female rats. *Annals of the New York Academy of Sciences, 807*, 602–605.

Nachmias, M., Gunnar, M., Mangelsdorf, S., Parritz, R. H., & Buss, K. (1996). Behavioral inhibition and stress reactivity: The moderating role of attachment security. *Child Development, 67*(2), 508–522.

Nieoullon, A., & Coquerel, A. (2003). Dopamine: A key regulator to adapt action, emotion, motivation and cognition. *Current Opinion in Neurology, 16*(Suppl. 2), S3–S9.

Nitschke, J. B., Nelson, E. E., Rusch, B. D., Fox, A. S., Oakes, T. R., & Davidson, R. J. (2004). Orbitofrontal cortex tracks positive mood in mothers viewing pictures of their newborn infants. *NeuroImage, 21*(2), 583–592.

Numan, M. (2007). Motivational systems and the neural circuitry of maternal behavior in the rat. *Developmental Psychobiology, 49*(1), 12–21.

Numan, M., Fleming, A. S., & Levy, F. (2006). Maternal behavior. In J. D. Neill (Ed.), *Knobil and Neill's physiology of reproduction* (3rd ed., pp. 1921–1993). San Diego, CA: Academic Press.

Numan, M., & Numan, M. J. (1994). Expression of *fos*-like immunoreactivity in the preoptic area of maternally behaving virgin and postpartum rats. *Behavioral Neuroscience, 108*(2), 379–394.

Orpen, B. G., Furman, N., Wong, P. Y., & Fleming, A. S. (1987). Hormonal influences on the duration of postpartum maternal responsiveness in the rat. *Physiology and Behavior, 40*(3), 307–315.

Pedersen, C. A., Ascher, J. A., Monroe, Y. L., & Prange, A. J. J. (1982). Oxytocin induces maternal behavior in virgin female rats. *Science, 216*(4546), 648–650.

Peterson, B. S., Skudlarski, P., Gatenby, J. C., Zhang, H., Anderson, A. W., & Gore, J. C. (1999). An fMRI study of Stroop word–color interference: Evidence for cingulate subregions subserving multiple distributed attentional systems. *Biological Psychiatry, 45*(10), 1237–1258.

Popeski, N., Scherling, C., Fleming, A. S., Lydon, J., Pruessner, J. C., & Meaney, M. J. (2008). *Maternal adversity alters patterns of neural activation in response to infant cues*. Manuscript in preparation.

Ranote, S., Elliott, R., Abel, K. M., Mitchell, R., Deakin, J. F., & Appleby, L. (2004). The neural basis of maternal responsiveness to infants: An fMRI study. *NeuroReport, 15*(11), 1825–1829.

Rees, S., Panesar, S., Steiner, M., & Fleming, A. (2004). The effects of adrenalectomy and corticosterone replacement on maternal behavior in the postpartum rat. *Hormones and Behavior, 46*(4), 411–419.

Rees, S., Panesar, S., Steiner, M., & Fleming, A. (2006). The effects of adrenalectomy and corticosterone replacement on induction of maternal behavior in the virgin female rat. *Hormones and Behavior, 49*(3), 337–345.

Robbins, T. W. (2007). Shifting and stopping: Fronto-striatal substrates, neurochemical modulation and clinical implications. *Philosophical Transactions of the Royal Society of London, Series B, Biological Sciences, 362*, 917–932.

Rogers, R. D., Andrews, T. C., Grasby, P. M., Brooks, D. J., & Robbins, T. W. (2000). Contrasting cortical and subcortical activations produced by attentional set-shifting and reversal learning in humans. *Journal of Cognitive Neuroscience, 12*(1), 142–162.

Rosenblatt, J. S. (2002). Hormonal bases of parenting in mammals. In M. H. Bornstein (Ed.), *Handbook of parenting: Vol. 2. Biology and ecology of parenting* (2nd ed., pp. 141–182). Mahwah, NJ: Erlbaum.

Schneider, M., & Koch, M. (2005). Behavioral and morphological alterations following neonatal excitotoxic lesions of the medial prefrontal cortex in rats. *Experimental Neurology, 195*(1), 185–198.

Schultz, W. (2006). Behavioral theories and the neurophysiology of reward. *Annual Review of Psychology, 57*, 87–115.

Seifritz, E., Esposito, F., Neuhoff, J. G., Luthi, A., Mustovic, H., Dammann, G., et al. (2003). Differential sex-independent amygdala response to infant crying and laughing in parents versus nonparents. *Biological Psychiatry, 54*(12), 1367–1375.

Sheehan, T., Paul, M., Amaral, E., Numan, M. J., & Numan, M. (2001). Evidence that the medial amygdala projects to the anterior/ventromedial hypothalamic nuclei to inhibit maternal behavior in rats. *Neuroscience, 106*, 341–356.

Slotnick, B. M. (1967). Disturbances of maternal behavior in the rat following lesions of the cingulate cortex. *Behaviour, 29*(2), 204–236.

Spangler, G., & Grossmann, K. E. (1993). Biobehavioral organization in securely and insecurely attached infants. *Child Development, 64*(5), 1439–1450.

Stallings, J., Fleming, A. S., Corter, C., Worthman, C., & Steiner, M. (2001). The effects of infant cries and odors on sympathy, cortisol, and autonomic responses in new mothers and nonpostpartum women. *Parenting: Science and Practice, 1*, 71–100.

Stamm, J. S. (1955). The function of the median cerebral cortex in maternal behavior of rats. *Journal of Comparative Physiology and Psychology, 48*(4), 347–356.

Strathearn, L. (2007). Exploring the neurobiology of attachment. In L. C. Mayes, P. Fonagy, & M. Target (Eds.), *Developmental science and psychoanalysis* (pp. 117–142). London: Karnac Press.

Strathearn, L., Li, J., Fonagy, P., & Montague, P.R. (2008). What's in a smile?: Maternal brain responses to infant facial cues. *Pediatrics, 122*(1), 40–51.

Strathearn, L., Li, J., & Montague, P. R. (2005). An fMRI study of maternal mentalization: Having the baby's mind in mind. *NeuroImage, 26*(Suppl. 1), S25.

Swain, J. E., Lorberbaum, J. P., Kose, S., & Strathearn, L. (2007). Brain basis of early parent–infant interactions: Psychology, physiology, and *in vivo* functional neuroimaging studies. *Journal of Child Psychology and Psychiatry, 48*(3–4), 262–287.

Völlm, B. A., Taylor, A. N., Richardson, P., Corcoran, R., Stirling, J., McKie, S., et al. (2006). Neuronal correlates of theory of mind and empathy: A functional magnetic resonance imaging study in a nonverbal task. *NeuroImage, 29*(1), 90–98.

Wiesenfeld, A., & Klorman, R. (1978). The mother's psychophysiological reactions to contrasting affective expressions by her own and an unfamiliar infant. *Developmental Psychology, 14*, 294–304.

Williams, J. M., Mathews, A., & MacLeod, C. (1996). The emotional Stroop task and psychopathology. *Psychological Bulletin, 120*(1), 3–24.

Zeanah, C. H., Benoit, D., Barton, M., Regan, C., Hirshberg, L. M., & Lipsitt, L. P. (1993). Representations of attachment in mothers and their one-year-old infants. *Journal of the American Academy of Child and Adolescent Psychiatry, 32*(2), 278–286.

CHAPTER 13

Neuroendocrine Mechanisms of Social Bonds and Child–Parent Attachment, from the Child's Perspective

KAREN L. BALES
C. SUE CARTER

Attachment and bonding are hypothetical constructs. No one has ever seen a social bond or directly measured its strength. However, there is no doubt that such processes do exist and that they are found not only in humans, but also in other species. Social relationships, including those structured around social bonds and emotional attachments, can be a source of life's greatest pleasures and deepest pain. From birth onward, social interactions and social bonds are essential for normal development in humans and many animals; yet we know remarkably little about the mechanisms underlying these processes.

The purpose of this chapter is to examine the development and expression of social bonds from a neuroendocrine perspective. At present, a limited number of neurochemicals and their neural targets have been implicated in social bonding (see also Marazziti, Chapter 14, and Wommack, Liu, & Wang, Chapter 15, this volume). We focus here on the role of these molecules in social behavior and social bonding generally, and, where possible, in the formation of infant–caregiver bonds specifically.

Interest in attachment and bonding originated in several different scientific disciplines, with different research traditions and different operational definitions for these constructs (Carter et al., 2005). Developmental psychologists have defined "attachment" as a propensity of one individual (usually a child) to bond to another who is viewed as stronger and wiser. This approach, which came to be known as "attachment theory," focused on the child's response to its mother, and attachment was commonly studied in children capable of walking and behaviorally expressing their response to a caretaker. Another point of view, coming primarily from pediatrics, hypothesized that human mothers are programmed to develop "bonds" with their own infants (Klaus & Kennell, 1982; Kennell & Klaus, 1998). The neural mechanisms through which infants become attached to their caretakers are less well identified than those involved in maternal bonding (see Gonzalez, Atkinson, & Fleming, Chapter 12, this volume). Animal researchers, who have done most of the work on neural mechanisms in this field, have often used the terms "attachment" and "bonding" interchangeably (Carter & Keverne, 2002). However, common to all of these definitions is the occurrence of selective social behaviors or emotional responses toward another individual.

Attachment Theory from the Child's Perspective

For many years, attachment theory—as articulated originally by Bowlby (1973, 1982) and operationalized by Ainsworth, Blehar, Waters, and Wall (1978)—has been the dominant theoretical framework used to study and explain the formation and persistence of selective social behaviors in humans. In this model, attachment security and the use of the mother as a "secure base" are conceptualized as the central organizing features of an infant's early social environment. "Internal working models" are constructs proposed to explain the later formation and maintenance of social relationships (Bowlby, 1982; Bretherton, 1997), including healthy social functioning (Sroufe, Carlson, Levy, & Egeland, 1999). Under certain circumstances, the capacity or opportunity to socially engage and form social bonds may remain unexpressed or be defectively expressed. The absence of secure or sensitive infant–caregiver relationships also may have consequences for subsequent social behavior and for the capacity to regulate physical and emotional reactions to stressors. There is evidence that inconsistent early social experiences lead to disturbed relationships in later life (O'Conner, 2005). The absence of normal social relationships and social support increases the risk for mental and physical illness (Uchino, Cacioppo, & Kiecolt-Glaser, 1996; Uchino, 2006). Various mental disorders, such as autism and schizophrenia, are characterized by deficits in social behaviors and in some cases by atypical attachment and bonding (Baird, Cass, & Slonims, 2003; Kirkpatrick, 1997; Carter, 2007). In contrast, perceived social bonds (or the physiological changes associated with such bonds) may be protective in the face of both emotional and physical challenges (Sachser, Durschlag, & Hirzel, 1998; Altemus, Deuster, Galliven, Carter, & Gold, 1995; Altemus et al., 2001).

The studies of Harry Harlow were also influential in thinking about mother–infant attachment (e.g., Harlow, 1961). By using models of restricted social experience during development for rhesus monkeys, he demonstrated long-lasting detrimental effects of social deprivation (Harlow, 1964). In addition, Harlow and colleagues found that it was possible to socialize previously isolated animals through the use of infant "therapists," reducing some of the effects of early deprivation (Harlow & Suomi, 1971).

Comparative Perspectives
on Neuroendocrine Processes of Attachment

Both Bowlby (1982) and Harlow (1961) rejected the idea that infants form attachments to caregivers solely on the basis of nourishment or oral gratification. In most mammalian models, it is difficult to disentangle these concepts, because for newborns mothers are the source of both food and all other infant care. However, a series of studies in the South American titi monkey (*Callicebus cupreus*) has bolstered Bowlby's and Harlow's positions by distinguishing between the effects of feeding and other types of caregiving on attachment. Titi monkeys are a monogamous species; males and females form pair bonds, reflected in large amounts of time spent in proximity, tail-twining behavior, and aggression toward same-sex conspecifics (Mason, 1966, 1968, 1974; Mason & Mendoza, 1998). Fathers are the primary carriers of infants, although this can vary on an individual basis from family to family (Mendoza & Mason, 1986; Welker & Schafer-Witt, 1987). Titi monkeys thus provide an excellent opportunity to compare attachment to a primary caregiver (i.e., the father) with attachment to the sole source of food (i.e., the mother). As it turned out, Mason, Mendoza, and colleagues found that when an infant's mother was removed from the home cage, but the father was still present, the infant exhibited neither a rise in the stress hormone cortisol nor an increase in vocalizations; however, if the father was removed while the mother was still present, the infant demonstrated both a rise in cortisol and an increase in vocalizations (Hoffman, Mendoza, Hennessy, & Mason, 1995). These findings thus demonstrated the infant's "attachment" to its father, or at least its capacity to selectively recognize the father's absence.

An additional consideration in the comparative study of attachment is that in many animal models, attachment bonds are not symmetric; that is, infants may be attached to their parents, but the parents do not display a reciprocal attachment to the infants (Mason & Mendoza, 1998). Titi monkeys also serve as an example of this phenomenon (Mendoza et al., 1986). In many rodent species, offspring are freely cross-fostered, and mothers do not appear to give differential care to their own versus unrelated pups (Hayes, O'Bryan, Christiansen, & Solomon, 2004). Other species, such as sheep and Old World monkeys (e.g., macaques) and apes, can undoubtedly distinguish their own from other offspring, and in many cases have strong or even lifelong emotional bonds (Kendrick et al., 1997; Mason & Mendoza, 1998). To some extent, attachment may be predicted by social organization even within closely related species (Hennessy, Bullinger, Neisen, Kaiser, & Sachser, 2006; Shapiro & Insel, 1990). In a pair-living guinea pig species, mothers, but not unfamiliar females, buffered infants' cortisol response to a novel environment; the same was not true for a harem-living species (Hennessy et al., 2006). Infant prairie voles (a socially monogamous rodent species), when separated from their mothers, display higher levels of ultrasonic vocalizations and a greater increase in plasma corticosterone than do montane vole pups (a species in which adults are solitary) (Shapiro & Insel, 1990). As described below, closely related rodents exhibiting species differences in sociality have proven particularly useful in the analysis of the neurobiological substrates of attachment.

Defining Attachment and Bonding

Attempts to operationally define "attachment" and "bonding" have focused on selective social behaviors, or emotional responses toward another individual. Responses to separation and reunion, such as distress vocalizations or attempts to reunite with a partner after separation, were among the earliest methods for describing the presence or absence of a social bond or an attachment (Panksepp, 1998). However, such responses may not be specific to attachment or bonding.

The concept of "attachment" also might be defined as a construct that unifies all of the processes underlying and maintaining social relationships; these processes can include the needs for warmth, food, and buffering against stress (Hofer, 2006). Each of these needs may be met in an infant mammal through association with its mother, and each of the physiological changes occurring during separation may be regulated through different aspects of that association. For instance, changes in physical activity due to separation can be reversed by warmth, while changes in cardiac activity can be reversed by infusion of milk (for a review, see Hofer, 2006). This also allows us to deconstruct the concept of attachment into more measurable units. In addition, it introduces the concept of "hidden regulators" (Hofer, 1978)—multiple processes, including the physiological status of the infant or caretaker, that may be guiding changes in the infant–caretaker relationship over development.

Mechanistic studies examining infants' attachments to their parents or caretakers only rarely describe selective responses—possibly either because in many species infants do not have the capacity to discriminate their own parents, or because the measures that have been used (such as "distress" cries) are insufficiently sensitive to permit such identification. Therefore, most neuroendocrine research to date has focused on the response of infants to separation or to an environment made novel by the absence of a familiar presence, usually via measuring infant cries. This is only one potential measure of attachment. However, the altricial state and small size of many young rodents often makes other measures (e.g., preference behaviors, following behavior, or indices of emotional responses) difficult to collect. In the absence of sensitive measures, it is difficult to differentiate general state changes in infants, such as increases in anxiety, from specific responses to the absence of a parent. Studies of more precocial infants such as guinea pigs or primates (Hennessy, 2003), in which a wider range of measures may be possible, have been particularly useful; we reference such studies whenever possible. However, it is important to note that altricial animals, born in a less developed state, also may differ from more precocial species in the degree to which they form selective relationships with their caretakers or the mechanisms through which bonds are formed.

Proximate Mechanisms for Attachment and Bonding

The mechanisms underlying social bonding and attachment are most easily understood in the context of their evolved functions. The mechanisms associated with social bonds are ancient, based on neural circuitry and endocrine processes rooted deep in mammalian evolution. The nature and timing of these processes, along with their ultimate (evolutionary) and proximate (ontogenetic, epigenetic and physiological) causes, have increasingly become subjects of neurobiological research.

Proximate processes necessary for social bonding, as indexed by selective social behaviors, are species-typical and shaped by evolution. These same processes can be quite plastic, and are subject to ontogenetic and epigenetic processes. Thus cognitive and emotional experiences can, over time, produce individual differences in the development of the capacity to show social bonds and in the expression of these bonds. Human research on the endocrinology of social attachment has been largely focused on the analysis of the endocrine consequences of the presence or absence of such bonds (Gunnar, 2005), rather than on the mechanisms responsible for allowing social bonds to form (Carter, 1998; Carter & Keverne, 2002).

Animal research suggests that specific hormones and neurotransmitters, acting on definable (if at present incompletely understood) pathways, play a role in the formation of social bonds (see Gonzalez et al., Chapter 12, and Wommack et al., Chapter 15, this volume). In general, recent human research in this field relies on imaging, with limited resolution. This work also does not provide information on specific neurochemicals, relying instead on inferences from animal studies (Bartels & Zeki, 2000, 2004; Swain, Leckman, Mayes, Feldman, & Schultz, 2004).

Research in animals suggests that at least some of the mechanisms underlying selective social behaviors are based on neuroendocrine systems shared in part with systems responsible for homeostasis and coping with challenge. For this reason, the presence or absence of social bonds also has widespread effects on health and well-being in the face of life's challenges (Carter, 1998). The physical and emotional status of the body, involving both the central and autonomic nervous systems, also can alter an individual's readiness to engage in social behaviors and eventually form attachments (Porges, 2005). Moreover, the physiological substrates for social bonds are shared with other processes, including those responsible for reproduction and the management of "stress" responses.

Especially important to the discovery of the neural substrates of social bonding has been the analysis of maternal behavior in general, particularly in species (e.g., sheep) in which studies have analyzed the neurochemical basis of selective responses of mothers toward newborns (Carter & Keverne, 2002). Studies of maternal bonding in turn provided clues regarding the neuroendocrine mechanisms necessary for the formation or expression of other kinds of social bonds.

Among the neuropeptide hormones that have been implicated in social bond formation are oxytocin (OT), arginine vasopressin (AVP), and corticotropin-releasing factor (CRF). Under optimal conditions, these peptides may facilitate social engagement and other positive social interactions. However, these same neuropeptides also influence social fear and anxiety. In addition, the salience of a relationship may be reinforced by neural mechanisms that are shared with reward and pleasure and also with reactivity to stressful conditions, including dopamine (DA) and the endogenous opioids. Endogenous opioids also have been implicated in the separation responses of young animals (Panksepp, 1998).

Animal Models for Social Bonding

Particularly useful in understanding the neurobiology of social bonds has been evidence that certain mammalian species have the capacity to form long-lasting social bonds (Kleiman, 1977). The study of species capable of pair bonding and sometimes described as "socially monogamous" has allowed the experimental analysis of proximate mechanisms underlying social bonds. Prairie voles (*Microtus ochrogaster*) are small microtine rodents native to the American Midwest; like titi monkeys, they display the characteristics of social monogamy, including a social preference for a single animal of the opposite sex (a pair bond) and biparental care, including male parenting behavior (Carter, DeVries, & Getz, 1995). In addition, monogamous species (including prairie voles) are often cooperative breeders, with older offspring remaining in the natal group and aiding in caring for their younger siblings (Carter & Roberts, 1997). Here we review research on the neuroendocrine basis of adult attachments, providing when possible comparisons to factors that may regulate infant–caregiver attachment (see also Marazziti, Chapter 14, and Wommack et al., Chapter 15, this volume).

Neurochemicals and the Formation of Adult Social Bonds

OT and AVP are both nine-amino-acid peptides that, though unique to mammals, bear many functional similarities to related molecules with at least some similar functions in other taxa (isotocin and mesotocin for OT, vasotocin for AVP) (Chang & Hsu, 2004). In mammals, OT is involved in uterine contractions during labor and is crucial to mammary contractions during lactation (Zingg, 2002; Nishimori, Young, Guo, Wang, & Insel, 1996); AVP regulates water balance and vasodilation (Berecek, 1991), as well as territoriality and defensive aggression (Winslow, Hastings, Carter, Harbaugh, & Insel, 1993). Both neuropeptides are also known to be critical to the process of social attachment formation (see below), perhaps in part through their role in the formation of "social memory" (Winslow & Insel, 2004). Social memory, or the ability to remember the identity of another individual, is crucial to the ability to form a long-term, selective bond. Several animal models with defective OT and AVP regulation show deficits in social memory—for example, Brattleboro rats (which lack AVP) (Engelmann, Wotjak, Neumann, Ludwig, & Landgraf, 1996); male (but not female) mice made genetically deficient for the AVP V1a receptor type (i.e., AVP V1a "knockouts") (Bielsky, Hu, Ren, Terwilliger, & Young, 2005a; Bielsky, Hu, Szegda, Westphal, & Young, 2004; Bielsky, Hu, & Young, 2005b); and OT knockout mice (Ferguson et al., 2000). OT and AVP are also involved in the response to stress in species-specific ways; in general, however, AVP is released during stress and acts as a secretagogue of adrenocorticotropic hormone (ACTH), while OT, which may be released by either positive or negative experiences, is often shown to have calming effects, promoting a return to homeostasis (Carter & Keverne, 2002; Carter, 1998; Ebner, Bosch, Kromer, Singewald, & Neumann, 2005).

Evidence implicating OT and AVP in adult pair bonding has come from two sources. The distributions of OT and AVP V1a receptors differ in monogamous versus polygynous species (Insel, Wang, & Ferris, 1994; Insel & Shapiro, 1992; Witt, Carter, & Insel, 1991). Studies in adults have experimentally demonstrated the involvement of OT in the formation of pair bonds in female prairie voles (Williams, Insel, Harbaugh, & Carter, 1994; Cho, DeVries, Williams, & Carter, 1999) and the involvement of AVP in the formation of pair bonds in males (Winslow et al., 1993; Cho et al., 1999; Lim, Hammock, & Young, 2004; Young, Murphy Young, & Hammock, 2005). Although there is evidence for sex differences in endogenous OT and AVP (Carter, 2007), it is important to note that there is considerable "cross-talk" between these compounds, and it is likely that both OT and AVP have behavioral effects in both males and females (Cho et al., 1999; Bales, Kim, Lewis-Reese, & Carter, 2004b).

DA has also been implicated in the process of pair-bond formation (Aragona, Liu, Curtis, Stephan, & Wang, 2003; Aragona et al., 2006), particularly via the colocalization of DA D2 receptors with OT receptors in the nucleus accumbens in females (Liu & Wang, 2003) and with AVP V1a receptors in the ventral pallidum in males (Aragona et al., 2003).

Several other neuropeptides have also been implicated in the formation of pair bonds in adults. These include CRF, a central factor in the regulation of the hypothalamic–pituitary–adrenal (HPA) axis. CRF also exhibits species differences in its receptor distribution between monogamous and polygynous species (Lim, Nair, & Young, 2005b), and CRF has a dose-dependent capacity to facilitate partner preference formation in male voles (DeVries, Guptaa, Cardillo, Cho, & Carter, 2002). There are at present no published data on CRF's effects on pair bonding in females. Cocaine- and amphetamine-related transcript (CART) peptide, which is heavily involved in the regulation of feeding behavior and reward/drug addiction (Dominguez et al., 2004), also differs in distribution between prairie (monogamous) and meadow (polygamous) voles (Hunter, Lim, Philpot, Young, & Kuhar, 2005). CART and CRF, as well as other ligands that activate CRF receptors (Bale & Vale, 2004), represent promising additional candidates for roles in the neuroendocrine basis of attachment and its development.

The Neurochemical Basis of Infant–Caregiver Bonds

In animal studies, an infant's attachment to its caregiver has often been indexed indirectly—either through changes in stress hormones (glucocorticoids—corticosterone in rodents, cortisol in humans and nonhuman primates), or through an increase in distress vocalizations (ultrasonic or audible). A third, less frequently used type of measure, preference testing, assesses the infant's preference for an attachment figure or other alternatives (Mason & Mendoza, 1998) and is used more often in primate studies. Many of the same hormones implicated in adult attachment bonds (OT, AVP, CRF, DA) have also been studied in relation to separation-induced distress vocalizations.

OT is a prime candidate in the formation of infant–mother attachment, and several studies support the hypothesis that OT may play a role in the response of an infant to separation from its caretaker. OT is present in breast milk (Leake, Weitzman, & Fisher, 1981), and can be released by contact and by warmth (Uvnas-Moberg, 1998). The act of suckling induces the release of OT. For example, in dairy calves, suckling resulted in higher levels of OT than drinking the mother's milk from a bucket did (Lupoli, Johansson, Uvnas-Moberg, & Svennersten-Sjaunja, 2001). Manipulations of OT have been shown to affect separation vocalizations in chicks (Panksepp, 1996) and in rodent pups. Another line of research has used animals made genetically deficient for OT (OT knockout mice) to test the hypothesis that OT plays a role in the separation response. As predicted, OT knockout mice demonstrated lower rates of vocalization on separation from their mothers (Winslow et al., 2000). When an oxytocin antagonist (OTA, used to block OT receptors) was administered to prairie vole pups in a single postnatal dosage, female (but not male pups) showed reduced numbers of ultrasonic vocalizations (Kramer, Cushing, & Carter, 2003). Administration of neonatal OT, while not affecting the corticosterone response to separation, did tend to dampen basal corticosterone in females when compared to a saline control, at least in animals tested at 8 days of age (Kramer et al., 2003).

OT is not the only neurochemical capable of influencing separation responses. The administration of AVP also reduces separation-induced vocalizations in infant Sprague–Dawley rats (Winslow & Insel, 1993). Blocking the AVP receptor (V1a subtype) did not reduce normal rates of stress-induced vocalizations, although it did reduce the effect of AVP on calling (Winslow & Insel, 1993). These and other studies continue to implicate AVP in sociality.

DA, the third major hormone/neurotransmitter implicated in adult pair bonding, has also been shown to modulate attachment behavior in infant animals. Research on the role of DA has focused recently on possible effects through binding to specific DA receptor subtypes (D1, D2, or D3). Maternal separation leads to an increase in DA binding to the D1 receptor in *Octodon degus* (Ziabreva, Schnabel, Poeggel, & Braun, 2003); the presence of the mother's call can prevent this up-regulation (Ziabreva et al., 2003). A D1 receptor agonist, a D2/D3 receptor agonist, and a D3 receptor agonist all reduced separation vocalization in rat pups (Dastur, McGregor, & Brown, 1999). These studies are difficult to interpret, because D1 and D2/D3 antagonists did not block the D1 and D2/D3 agonist effects. The D2/D3 receptor antagonist inhibited the D3 agonist effects (Dastur et al., 1999). In another study in rats, a selective D2 antagonist blocked maternal modulation of infant vocalizations, without affecting calling rates in infants not exposed to their mother (Muller, Brunelli, Moore, Myers, & Shair, 2005). Cerebrospinal fluid measurements of homovanillic acid, a DA metabolite, were lower in rhesus macaque infants during periods of separation from their mothers (Erickson et al., 2005). These results generally implicate DA in the response to separation. However, the roles of the various DA receptor subtypes need to be clarified.

CRF has also been shown to modulate distress vocalizations in many species. However, like those of DA, the effects of CRF are complex—in part because CRF acts on several types of receptors, and these receptors may have functionally opposing, potentially homeostatic actions. For example, acute, centrally administered CRF can increase vocalizations and cause a rise in corticosterone (Tachibana et al., 2004), while CRF type 1 receptor antagonists reduce vocalizations (Iijima & Chaki, 2006). The effects of CRF vary by context, and the effects are stronger under more acute stress (Dirks et al., 2002). In other studies, CRF has been shown to cause a suppression of vocalizations in isolated rat and guinea pig pups (Hennessy, Becker, & O'Neil, 1991; Insel & Harbaugh, 1989). Apparent discrepancies among studies also may be due in part to the length of isolation or the dosages used. The generalization of the role of CRF to primates remains to be determined as well. For example, in one primate study in which infants were separated from their mothers, no effect of CRF on vocalizations was found (Kalin, Shelton, & Barksdale, 1989).

Opioids have likewise been extensively studied in relation to the regulation of separation vocalizations (for reviews, see Nelson & Panksepp, 1998; Panksepp, 1998; Panksepp, Herman, Vilberg, Bishop, & DeEskinazi, 1980). These studies tested the hypothesis that opioid antagonism should result in higher vocalizations and increased need for proximity to caregivers, whereas proper caregiving should result in opioid release (Panksepp, 1998; Weller & Feldman, 2003; Herman & Panksepp, 1978, 1981). Direct evidence regarding the role of opioids in adult pair bonding is rare, although connections have been hypothesized (Depue & Morrone-Strupinsky, 2005). One study in prairie voles found that a high dose (10 mg/kg) of morphine reduced huddling behavior in pairs of prairie voles, but also affected their activity patterns (less time in locomotion and more time stationary). The same study found no effects of naloxone, an opioid antagonist, on time spent huddling by pair mates (Shapiro, Meyer, & Dewsbury, 1989). Opioid effects on behavior are not well understood, but are likely to be at least as complex as CRF and DA effects, since there are several types of opioids and several types of receptors, with many potential interactions. In addition, opioids have an important role in the regulation of neuropeptides, including OT and AVP (Brunton, Sabatier, Leng, & Russell, 2006; Russell et al., 1989), through which they might be expected to influence social bonding. The hypothesis that pleasure and pain, both of which are modulated by opioids, are involved in attachment and bonding continues to be attractive.

A summary of the direction of hormonal effects on adult pair bonding versus infant vocalizations is displayed in Figure 13.1. A positive effect on infant attachment may be reflected in a decrease in separation-induced vocalizations.

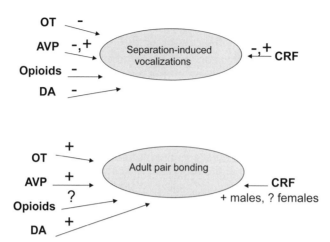

FIGURE 13.1. Summary of hormone interactions with an index of infant attachment (separation vocalizations) and adult attachment (pair bonding).

Long-Term Effects of Developmental Manipulations of OT and AVP

One of the basic assumptions of developmental social psychology is that individual differences in the attachment relationship to caregivers during early development are stable into adulthood and are reflected in different styles of adult attachment (Thompson & Raikes, 2003). Animal studies may also offer insight into the role of development in the capacity to form social bonds.

Because prairie voles are among the few mammalian laboratory species that display adult attachment bonds, they have proven especially useful in developmental studies of affiliative behaviors. These include behaviors that are associated with attachment behavior (pair bonding) and other behaviors, such as "alloparenting" (care of infants not one's own), that may be indicative of an animal's general tendency to be social. This species has been the subject of a series of recent studies of the long-term effects of various neurochemical and behavioral manipulations on subsequent social behaviors (Carter, 2003). For example, in prairie voles, developmental exposure to OT or an OTA has long-lasting consequences (Bales & Carter, 2003a, 2003b; Bales, Abdelnabi, Cushing, Ottinger, & Carter, 2004a; Bales, Pfeifer, & Carter, 2004c; Bales et al., 2007b, 2007c). These studies used neonatal dosages (3 μg OT and 0.3 μg OTA) and treatment paradigms identical to those in the Kramer et al. (2003) study described earlier. However, rather than being studied as infants, instead pups were returned to their parents after injection and then tested 60–90 days later, as adults. A single neonatal treatment with OT (3.0 μg) was associated in later life with a facilitation of pair bonding in male prairie voles (Bales & Carter, 2003b). In contrast, a single neonatal treatment with OTA (0.3 μg) drastically reduced male alloparenting behavior and increased pup attacks (Bales et al., 2004c). The only behavioral effect seen on females at this comparatively low dosage was an increase in mate-guarding behavior (same-sex aggression following exposure to a male) (Bales & Carter, 2003a). At a higher dosage (6 μg), neonatal OT did facilitate pair bonding in female voles as well (Bales et al., 2007c).

These long-term behavioral consequences of a short-term developmental manipulation appear to be mediated by long-lasting changes in the OT and AVP system and their receptors, especially the AVP receptor. In females, neonatal treatment with either OT or OTA resulted in a significant increase in production of hypothalamic OT, measured at 21 days of age; in contrast, in males but not females, treatment with OTA resulted in a reduction in hypothalamic AVP (Yamamoto et al., 2004).

Early manipulations of OT or OTA also have long-lasting and sexually dimorphic effects on the AVP (V1a) receptor. In males, neonatal treatments that blocked OT (i.e., OTA) led to reductions in AVP (V1a) receptor binding in many neural regions associated with pair bonding and parenting behavior, while increasing OT led to increases in many of the same regions. In females, on the other hand, a single neonatal exposure to OT produced reductions in AVP (V1a) receptors (Bales et al., 2007b). The AVP (V1a) receptor is sensitive to even brief changes in OT in neonatal life, and these effects differ in males and females. These changes are also consistent with behavioral data suggesting that males of many species are especially sensitive to changes in AVP, particularly in brain regions that are necessary for pair bonding and male parental behavior. The particular dependence of the AVP receptor and male social behavior on neonatal hormones may help to explain males' vulnerability to various disorders of sociality, such as autism (Carter, 2007).

Developmental manipulations of AVP have also been studied in relation to the long-term development of pair bonding and intrasexual aggression (Stribley, 1998). Prairie vole pups were exposed to a range of doses of AVP for the first 7 days after birth. Early AVP exposure led to long-term increases in same-sex aggressive behavior, especially in males, although no effects of early AVP were found on the development of pair-bonding behavior (Stribley & Carter, 1999). The mechanisms underlying these effects, and especially their possible effects on peptide receptors, remain to be explored.

Of particular theoretical importance is the possibility that early social experiences, such as those between an infant and its caregiver, may also have long-lasting effects on the neural systems responsible for later sociality. For example, parental caregiving style, crucial to the formation of secure or insecure attachments, could directly exert long-term effects on social bonding via changes in peptides, such as OT and AVP, or their receptors.

Studies in rats have been especially productive in describing the neurobiological effects of early social experiences (Meaney & Szyf, 2005). For example, mother rats that lick and groom their infants at a high rate produce female pups with high levels of OT receptors in certain neural regions, and male pups with high levels of AVP receptors (Francis, Young, Meaney, & Insel, 2002). These naturally occurring differences in maternal behavior can be transmitted intergenerationally, possibly through epigenetic changes in specific genes associated with sociality and/or fear (Francis, Diorio, Liu, & Meaney, 1999; Pedersen & Boccia, 2002). However, because rats do not show the selective behaviors indicative of social bonds, the role of differential parenting on behaviors directly relevant to bonding has not been examined in this model.

In prairie voles—rodents with the remarkable capacity to form pair bonds, as noted throughout this chapter—changes in husbandry can produce long-term changes in the capacity for attachment (Bales, Lewis-Reese, Pfeifer, Kramer, & Carter, 2007a), probably mediated by changes in parental behavior (Tyler, Michel, Bales, & Carter, 2005). In a series of studies, we have found that surprisingly small differences in early handling produce long-lasting changes in behavior, similar to those seen after early peptide manipulations. For example, if during routine cage changes, parents and their young are transferred by being picked up (manipulated, or MAN1) or by being transferred in a cup (unmanipulated, or MAN0), the MAN1 parents display higher levels of pup-directed behavior following the manipulation (Tyler et al., 2005). The female offspring of MAN0 parents grow up with deficits in the ability to form pair bonds, while male MAN0 offspring grow up displaying lower levels of alloparental behavior (Bales et al., 2007a). The neural mechanisms for these effects are under study. However, these studies offer strong support for the importance of early experience in later sociality, and specifically in the capacity to form social bonds.

Long-Term Developmental Effects of Other Hormones Implicated in Attachment

Of the other hormones implicated in adult attachment and affiliative behaviors, those best studied developmentally are stress-related hormones, including glucocorticoids and CRF. Developmental exposure to glucocorticoids—whether administered exogenously (Pechnick et al., 2006) or caused by maternal separation, prenatal stress, or variations in care during infancy (Smith, Kim, Vanoers, & Levine, 1997; Vazquez et al., 2006; Levine, 2002)—can cause long-term changes in functioning of the HPA axis. It is possible that long-term changes in anxiety and reactivity to stress contribute to long-term effects of early attachment on adult parenting or bonding behavior. However, these effects are likely to be both complex and sexually dimorphic. For example, in adults, anxiety and stress mediate sociality and the ability to form social bonds (Fleming & Leubke, 1981; DeVries, 2002; DeVries, DeVries, Taymans, & Carter, 1995; DeVries, DeVries, Taymans, & Carter, 1996; Fleming & Corter, 1995). Corticosterone administered developmentally to prairie voles on postnatal days 1–6 resulted in a reduction in alloparenting behavior in females, although it did not affect alloparenting in males (Roberts, Zullo, & Carter, 1996). The specific effects of developmental exposure to corticosterone or stressors on pair-bonding behavior in prairie voles remain to be studied, although early or even prenatal stressors have been shown in many studies to affect social behavior in other species (Kaiser & Sachser, 1998, 2005).

Developmental administration of low dosages of a mixed DA D1/D2 antagonist to monogamous zebra finches caused reductions in female-directed singing and courtship behavior (Harding, 2004). Developmental studies of DA manipulations in mammals thus far have not focused on long-term effects on attachment-related behaviors.

Neuroendocrine Processes of Attachment and Developmental Psychopathology

It has been proposed that attachment theory may contribute to our understanding of psychopathology. For example, early secure or insecure attachment may have consequences for subsequent adolescent or adult psychopathologies of both internalizing and externalizing types (Sroufe et al., 1999). Autism spectrum disorders are developmental disorders characterized by abnormal social behavior and the inability to form social bonds, either through lack of interaction with caregivers or through indiscriminate friendliness (Baird et al., 2003; see also Dawson, Sterling, & Faja, Chapter 22, this volume). There is one report that plasma OT is low in children with autism (Modahl et al., 1998), possibly due to abnormal metabolism and an associated increase in the OT precursor (Green et al., 2001). It has also been proposed, though not yet supported, that the use of pitocin (artificial OT) during labor and delivery could contribute to the development of autism (Rojas Wahl, 2004). The potential relationship between OT and autism has been recently reviewed (Carter, 2007; Insel, O'Brien, & Leckman, 1999; Lim, Bielsky, & Young, 2005a). OT is also being tested as a potential therapy for the social deficits displayed in autism (Hollander et al., 2007; Bartz & Hollander, 2006). However, many questions remain regarding the relative importance of specific pharmacological manipulations in human behavior.

The associations between OT and AVP on the one hand, and the development of psychopathologies in adolescence (Bales & Carter, 2007) and adulthood (Carter & Altemus, 2005) on the other hand, have also been recently reviewed. However, human studies of peptides entail multiple methodological problems. Most studies can examine OT and AVP levels only in plasma or in cerebrospinal fluid, which are at best indirect indices of central peptide function and make it difficult to distinguish between cause and effect. Studies of the associations between early attachment style and the development of psychopathology present many of the same problems (Sroufe et al., 1999). It is of course difficult to disentangle attachment from other correlated variables. In addition, developmental studies are essential, and noninvasive methods for studying developmental mechanisms are not currently available.

Conclusions

Although many basic issues remain unresolved, the data to date suggest that neurochemicals crucial to formation of the adult bond may also be involved in infant–parent bonds, and more generally in the social behaviors that characterize the interactions between young animals and their caretakers. In animal models, several critical neurochemical systems, including OT, AVP, CRF, and opioids, have been implicated in both adult pair bonding and infant–caregiver bonding. However, theories regarding how these hormones interact to form these attachments are still emerging. Very few studies test more than one hormone at a time. Dosages used in different studies are inconsistent, and dose–response studies are rare. Some results for individual hormones are contradictory, although it is important to note that neuropeptides tend to show nonlinear, U-shaped dose–response curves, which may complicate results and comparisons among studies using different dependent variables. Further complexity arises from the fact that many hormones, such as OT and AVP, can affect each other's receptors; these interactions are also dose-dependent. In addition, sex differences are routine findings in these studies. At present, more refined methodologies are necessary to identify specific neurochemical effects on attachment behaviors, as well as the specific receptors underlying these responses. Finally, more refinement in behavioral studies is essential to determine whether the various agents act directly on neural circuits involved in a particular behavior, and/or may be acting on other, less specific systems (such as the response to novelty, hunger, hypothermia, or other forms of discomfort).

Acknowledgments

We are grateful to our many colleagues who conducted the studies described here, and for research support from the National Institute of Child Health and Human Development (Grant No. PO1 38490 to C. Sue Carter, National Research Service Award No. 08702 to Karen L. Bales); the National Institute of Mental Health (Grant No. RO1 073022 to both of us); the National Alliance for Autism Research (to C. Sue Carter); the Good Nature Institute (to Karen L. Bales); and the National Science Foundation (Grant No. 0437523 to Karen L. Bales).

References

Ainsworth, M. D. S., Blehar, M. C., Waters, E., & Wall, S. (1978). *Patterns of attachment.* Hillsdale, NJ: Erlbaum.

Altemus, M., Deuster, P. A., Galliven, E., Carter, C. S., & Gold, P. W. (1995). Suppression of hypothalamic–pituitary–adrenal axis responses to stress in lactating women. *Journal of Clinical Endocrinology and Metabolism, 80,* 2954–2959.

Altemus, M., Redwine, L. S., Leong, Y. M., Frye, C. A., Porges, S. W., & Carter, C. S. (2001). Responses to laboratory psychosocial stress in postpartum women. *Psychosomatic Medicine, 63,* 814–821.

Aragona, B. J., Liu, Y., Curtis, T., Stephan, F. K., & Wang, Z. X. (2003). A critical role for nucleus accumbens dopamine in partner-preference formation in male prairie voles. *Journal of Neuroscience, 23,* 3483–3490.

Aragona, B. J., Liu, Y., Yu, Y. J., Curtis, J. T., Detwiler, J. M., Insel, T. R., et al. (2006). Nucleus accumbens dopamine differentially mediates the formation and maintenance of monogamous pair bonds. *Nature Neuroscience, 9,* 133–139.

Baird, G., Cass, H., & Slonims, V. (2003). Diagnosis of autism. *British Medical Journal, 327,* 488–493.

Bale, T. L., & Vale, W. W. (2004). CRF and CRF receptors: Role in stress responsivity and other behaviors. *Annual Review of Pharmacology and Toxicology, 44,* 525–557.

Bales, K. L., Abdelnabi, M., Cushing, B. S., Ottinger, M. A., & Carter, C. S. (2004a). Effects of neonatal oxytocin manipulations on male reproductive potential in prairie voles. *Physiology and Behavior, 81,* 519–526.

Bales, K. L., & Carter, C. S. (2003a). Sex differences and developmental effects of oxytocin on aggression and social behavior in prairie voles (*Microtus ochrogaster*). *Hormones and Behavior, 44,* 178–184.

Bales, K. L., & Carter, C. S. (2003b). Developmental exposure to oxytocin facilitates partner preferences in male prairie voles (*Microtus ochrogaster*). *Behavioral Neuroscience, 117,* 854–859.

Bales, K. L., & Carter, C. S. (2007). Neuropeptides and the development of social behaviors: Implications for adolescent psychopathology. In D. Romer & E. Walker (Eds.), *Adolescent psychopathology and the developing brain: Integrating brain and prevention science* (pp. 173–196). Oxford: Oxford University Press.

Bales, K. L., Kim, A. J., Lewis-Reese, A. D., & Carter, C. S. (2004b). Both oxytocin and vasopressin may influence alloparental behavior in male prairie voles. *Hormones and Behavior, 45,* 354–361.

Bales, K. L., Lewis-Reese, A. D., Pfeifer, L. A., Kramer, K. M., & Carter, C. S. (2007a). Early experience affects the traits of monogamy in a sexually dimorphic manner. *Developmental Psychobiology, 49,* 335–342.

Bales, K. L., Pfeifer, L. A., & Carter, C. S. (2004c). Sex differences and effects of manipulations of oxytocin on alloparenting and anxiety in prairie voles. *Developmental Psychobiology, 44,* 123–131.

Bales, K. L., Plotsky, P. M., Young, L. J., Lim, M. M., Grotte, N. D., Ferrer, E., et al. (2007b). Neonatal oxytocin manipulations have long-lasting, sexually dimorphic effects on vasopressin receptors. *Neuroscience, 144,* 38–45.

Bales, K. L., van Westerhuyzen, J. A., Lewis-Reese, A. D., Grotte, N. D., Lanter, J. A., & Carter, C. S. (2007c). Oxytocin has dose-dependent developmental effects on pair-bonding and alloparental care in female prairie voles. *Hormones and Behavior, 52*(2), 274–279.

Bartels, A., & Zeki, S. (2000). The neural basis of romantic love. *NeuroReport, 11,* 3829–3834.

Bartels, A., & Zeki, S. (2004). The neural correlates of maternal and romantic love. *NeuroImage, 21,* 1155–1166.

Bartz, J., & Hollander, E. (2006). The neuroscience of affiliation: Forging links between basic and clinical research on neuropeptides and social behavior. *Hormones and Behavior, 50,* 518–528.

Berecek, K. H. (1991). Role of vasopressin in central cardiovascular regulation. In G. Kunos & J. Ciriello (Eds.), *Central neural mechanisms in cardiovascular regulation* (Vol. 2, pp. 1–34). Boston: Birkhauser.

Bielsky, I. F., Hu, S.-B., Ren, X., Terwilliger, E. F., & Young, L. J. (2005a). The V1a vasopressin receptor is necessary and sufficient for normal social recognition: A gene replacement study. *Neuron, 47,* 503–513.

Bielsky, I. F., Hu, S.-B., Szegda, K. L., Westphal, H., & Young, L. J. (2004). Profound impairment in social recognition and reduction in anxiety-like behavior in vasopressin V1a receptor knockout mice. *Neuropsychopharmacology, 29,* 483–493.

Bielsky, I. F., Hu, S.-B., & Young, L. J. (2005b). Sexual dimorphism in the vasopressin system: Lack of an altered behavioral phenotype in female V1a receptor knockout mice. *Behavioural Brain Research, 164,* 132–136.

Bowlby, J. (1973). *Attachment and loss: Vol. 2. Separation: Anxiety and anger.* New York: Basic Books.

Bowlby, J. (1982). *Attachment and loss: Vol. 1. Attachment* (2nd ed.). New York: Basic Books.

Bretherton, I. (1997). Bowlby's legacy to developmental psychology. *Child Psychiatry and Human Development, 28,* 33–43.

Brunton, P. J., Sabatier, N., Leng, G., & Russell, J. A. (2006). Suppressed oxytocin neuron responses to immune challenge in late pregnant rats: A role for endogenous opioids. *European Journal of Neuroscience, 23,* 1241–1247.

Carter, C. S. (1998). Neuroendocrine perspectives on social attachment and love. *Psychoneuroendocrinology, 23,* 779–818.

Carter, C. S. (2003). Developmental consequences of oxytocin. *Physiology and Behavior, 79,* 383–397.

Carter, C. S. (2007). Sex differences in oxytocin and vasopressin: Implications for autism spectrum disorders? *Behavioural Brain Research, 176,* 170–186.

Carter, C. S., Ahnert, L., Grossmann, K. E., Hrdy, S. B., Lamb, M. E., Porges, S. W., et al. (Eds.). (2005). *Attachment and bonding: A new synthesis.* Cambridge, MA: MIT Press.

Carter, C. S., & Altemus, M. (2005). Oxytocin, vasopressin, and depression. In J. A. den Boer, M. S. George, & G. J. ter Horst (Eds.), *Current and future developments in psychopharmacology* (pp. 201–216). Amsterdam: Benecke.

Carter, C. S., DeVries, A. C., & Getz, L. L. (1995). Physiological substrates of mammalian monogamy: The prairie vole model. *Neuroscience and Biobehavioral Reviews, 19,* 303–314.

Carter, C. S., & Keverne, E. B. (2002). The neurobiology of social affiliation and pair bonding. In D. W. Pfaff, A. P. Arnold, A. M. Etgen, S. E. Fahrbach, & R. T. Rubin (Eds.), *Hormones, brain, and behavior* (Vol. 1, pp. 299–337). San Diego, CA: Academic Press.

Carter, C. S., & Roberts, R. L. (1997). The psychobiological basis of cooperative breeding in rodents. In N. G. Solomon & J. A. French (Eds.), *Cooperative breeding in mammals* (pp. 231–266). New York: Cambridge University Press.

Chang, C. L., & Hsu, S. Y. (2004). Ancient evolution of stress-regulating peptides in vertebrates. *Peptides, 25,* 1681–1688.

Cho, M. M., DeVries, A. C., Williams, J. R., & Carter, C. S. (1999). The effects of oxytocin and vasopressin on partner preferences in male and female prairie voles (*Microtus ochrogaster*). *Behavioral Neuroscience, 113,* 1071–1079.

Dastur, F. N., McGregor, I. S., & Brown, R. E. (1999). Dopaminergic modulation of rat pup ultrasonic vocalizations. *European Journal of Pharmacology, 382,* 53–67.

Depue, R. A., & Morrone-Strupinsky, J. V. (2005). A neurobehavioral model of affiliative bonding: Implications for conceptualizing a human trait of affiliation. *Behavioral and Brain Sciences, 28,* 313–350.

DeVries, A. C. (2002). Interaction among social environment, the hypothalamic–pituitary–adrenal axis, and behavior. *Hormones and Behavior, 41,* 405–413.

DeVries, A. C., DeVries, M. B., Taymans, S., & Carter, C. S. (1995). The modulation of pair bonding by corticosterone in female prairie voles (*Microtus ochrogaster*). *Proceedings of the National Academy of Sciences USA, 92,* 7744–7748.

DeVries, A. C., DeVries, M. B., Taymans, S. E., & Carter, C. S. (1996). The effects of stress on social preferences are sexually dimorphic in prairie voles. *Proceedings of the National Academy of Sciences USA, 93,* 11980–11984.

DeVries, A. C., Guptaa, T., Cardillo, S., Cho, M., & Carter, C. S. (2002). Corticotropin-releasing factor induces social preferences in male prairie voles. *Psychoneuroendocrinology, 27,* 705–714.

Dirks, A., Fish, E. W., Kikusui, T., van der Gugten, J., Groenink, L., Olivier, B., et al. (2002). Effects of corticotropin-releasing hormone on distress vocalizations and locomotion in maternally separated mouse pups. *Pharmacology, Biochemistry and Behavior, 72,* 993–999.

Dominguez, G., Vicentic, A., del Giudice, E. M., Jaworski, J., Hunter, R. G., & Kuhar, M. J. (2004). CART peptides: Modulators of mesolimbic dopamine, feeding, and stress. *Annals of the New York Academy of Sciences, 1025,* 363–369.

Ebner, K., Bosch, O. J., Kromer, S. A., Singewald, N., & Neumann, I. D. (2005). Release of oxytocin in the rat central amygdala modulates stress-coping behavior and the release of excitatory amino acids. *Neuropsychopharmacology, 30,* 223–230.

Engelmann, M., Wotjak, C. T., Neumann, I., Ludwig, M., & Landgraf, R. (1996). Behavioral consequences of intracerebral vasopressin and oxytocin: Focus on learning and memory. *Neuroscience and Biobehavioral Reviews, 20,* 341–358.

Erickson, K., Gabry, K. E., Lindell, S., Champoux, M., Schulkin, J., Gold, P., et al. (2005). Social withdrawal behaviors in nonhuman primates and changes in neuroendocrine and monoamine concentrations during a separation paradigm. *Developmental Psychobiology, 46,* 331–339.

Ferguson, J. N., Young, L. J., Hearn, E. F., Matzuk, M. M., Insel, T. R., & Winslow, J. T. (2000). Social amnesia in mice lacking the oxytocin gene. *Nature Genetics, 25,* 284–288.

Fleming, A. S., & Corter, C. M. (1995). Psychobiology of maternal behavior in nonhuman mammals. In M. H. Bornstein (Ed.), *Handbook of parenting: Vol. 2. Biology and ecology of parenting* (pp. 59–85). Mahwah, NJ: Erlbaum.

Fleming, A. S., & Leubke, C. (1981). Timidity prevents the virgin female rat from being a good mother: Emotionality differences between nulliparous and parturient females. *Physiology and Behavior, 27,* 863–868.

Francis, D. D., Diorio, J., Liu, D., & Meaney, M. J. (1999). Nongenomic transmission across generations of maternal behavior and stress responses in the rat. *Science, 286,* 1155–1158.

Francis, D. D., Young, L. J., Meaney, M. J., & Insel, T. R. (2002). Naturally occurring differences in maternal care are associated with the expression of oxytocin and vasopressin (V1a) receptors: Gender differences. *Journal of Neuroendocrinology, 14,* 349–353.

Green, L. A., Fein, D., Modahl, C., Feinstein, C., Waterhouse, L., & Morris, M. (2001). Oxytocin and autistic disorder: Alterations in peptide forms. *Biological Psychiatry, 50,* 609–613.

Gunnar, M. R. (2005). Attachment and stress in early development: Does attachment add to the potency of social regulators of early stress? In C. S. Carter, L. Ahnert, K. E. Grossmann, S. B. Hrdy, M. E. Lamb, S. W. Porges, et al. (Eds.), *Attachment and bonding: A new synthesis* (pp. 245–255). Cambridge, MA: MIT Press.

Harding, C. F. (2004). Brief alteration in dopaminergic function during development causes deficits in adult reproductive behavior. *Journal of Neurobiology, 61,* 301–308.

Harlow, H. F. (1961). The development of affectional patterns in infant monkeys. In B. M. Foss (Ed.), *Determinants of infant behavior* (pp. 75–97). New York: Wiley.

Harlow, H. F. (1964). Early social deprivation and later behavior in the monkey. In A. Abrams, H. H. Garner, & J. E. P. Tomal (Eds.), *Unfinished tasks in the behavioral sciences* (pp. 154–173). Baltimore: Williams & Wilkins.

Harlow, H. F., & Suomi, S. J. (1971). Social recovery of isolation-reared monkeys. *Proceedings of the National Academy of Sciences USA, 68,* 1534–1538.

Hayes, L. D., O'Bryan, E., Christiansen, A. M., & Solomon, N. G. (2004). Temporal changes in mother–offspring discrimination in the prairie vole (*Microtus ochrogaster*). *Ethology, Ecology and Evolution, 16,* 145–156.

Hennessy, M. B. (2003). Enduring maternal influences in a precocial rodent. *Developmental Psychobiology, 42,* 225–236.

Hennessy, M. B., Becker, L. A., & O'Neil, D. R. (1991). Peripherally administered CRH suppresses the vocalizations of isolated guinea pig pups. *Physiology and Behavior, 50,* 17–22.

Hennessy, M. B., Bullinger, K. L., Neisen, G., Kaiser, S., & Sachser, N. (2006). Social organization predicts nature of infant–adult interactions in two species of wild guinea pigs (*Cavis aperea* and *Galea monasteriensis*). *Journal of Comparative Psychology, 120,* 12–18.

Herman, B. H., & Panksepp, J. (1978). Effects of morphine and naloxone on separation distress and approach attachment: Evidence of opiate mediation of social effect. *Pharmacology, Biochemistry and Behavior, 9,* 213–220.

Herman, B. H., & Panksepp, J. (1981). Ascending endorphin inhibition of distress vocalization. *Science, 211,* 1060–1062.

Hofer, M. A. (1978). Hidden regulatory processes in early social relationships. In P. P. G. Bateson & P. H. Klopfer (Eds.), *Perspectives in ethology* (Vol. 3, pp. 135–166). New York: Plenum Press.

Hofer, M. A. (2006). Psychobiological roots of early attachment. *Current Directions in Psychological Science, 15,* 84–88.

Hoffman, K. A., Mendoza, S. P., Hennessy, M. B., & Mason, W. A. (1995). Responses of infant titi monkeys, *Callicebus moloch,* to removal of one or both parents: Evidence for paternal attachment. *Developmental Psychobiology, 28,* 399–407.

Hollander, E., Bartz, J., Chaplin, W., Phillips, A., Sumner, J., Soorya, L., et al. (2007). Oxytocin increases retention of social cognition in autism. *Biological Psychiatry, 61*(4), 498–503.

Hunter, R. G., Lim, M. M., Philpot, K. B., Young, L. J., & Kuhar, M. J. (2005). Species differences in brain distribution of CART mRNA and CART peptide between prairie and meadow voles. *Brain Research, 1048*(1–2), 12–23.

Iijima, M., & Chaki, S. (2006). Separation-induced ultrasonic vocalization in rat pups: Further pharmacological characterization. *Pharmacology, Biochemistry and Behavior, 82,* 652–657.

Insel, T. R., & Harbaugh, C. R. (1989). Central administration of corticotropin releasing factor alters rat pup isolation calls. *Pharmacology, Biochemistry and Behavior, 32,* 197–201.

Insel, T. R., O'Brien, D. J., & Leckman, J. F. (1999). Oxytocin, vasopressin, and autism: Is there a connection? *Biological Psychiatry, 45,* 145–157.

Insel, T. R., & Shapiro, L. E. (1992). Oxytocin receptor distribution reflects social organization in monogamous and polygamous voles. *Proceedings of the National Academy of Sciences USA, 89,* 5981–5985.

Insel, T. R., Wang, Z. X., & Ferris, C. F. (1994). Patterns of brain vasopressin receptor distribution associated with social organization in microtine rodents. *Journal of Neuroscience, 14,* 5381–5392.

Kaiser, S., & Sachser, N. (1998). The social environment during pregnancy and lactation affects the female offsprings' endocrine status and behaviour in guinea pigs. *Physiology and Behavior, 63,* 361–366.

Kaiser, S., & Sachser, N. (2005). The effects of prenatal social stress on behavior: Mechanisms and function. *Neuroscience and Biobehavioral Reviews, 29,* 283–294.

Kalin, N. H., Shelton, S. E., & Barksdale, C. M. (1989). Behavioral and physiologic effects of CRH administered to infant primates undergoing maternal separation. *Neuropsychopharmacology, 2,* 97–104.

Kendrick, K. M., da Costa, A. P. C., Broad, K. D., Ohkura, S., Guevera, R., Levy, F., et al. (1997). Neural control of maternal behaviour and olfactory recognition of offspring. *Brain Research Bulletin, 44*, 383–395.

Kennell, J. H., & Klaus, M. H. (1998). Bonding: Recent observations that alter perinatal care. *Pediatrics in Review, 19*, 4–12.

Kirkpatrick, B. (1997). Affiliation and neuropsychiatric disorders: The deficit syndrome of schizophrenia. *Integrative Neurobiology of Affiliation, 807*, 455–468.

Klaus, M. H., & Kennell, J. H. (1982). *Parent–infant bonding.* St. Louis, MO: Mosby.

Kleiman, D. G. (1977). Monogamy in mammals. *Quarterly Review of Biology, 52*, 39–69.

Kramer, K. M., Cushing, B. S., & Carter, C. S. (2003). Developmental effects of oxytocin on stress response: Single versus repeated exposure. *Physiology and Behavior, 79*, 775–782.

Leake, R. D., Weitzman, R. E., & Fisher, D. A. (1981). Oxytocin concentrations during the neonatal period. *Biology of the Neonate, 39*, 127–131.

Levine, S. (2002). Enduring effects of early experience on adult behavior. In D. W. Pfaff, A. P. Arnold, A. M. Etgen, S. E. Fahrbach, & R. T. Rubin (Eds.), *Hormones, brain and behavior* (Vol. 4, pp. 535–542). San Diego, CA: Academic Press.

Lim, M. M., Bielsky, I. F., & Young, L. J. (2005a). Neuropeptides and the social brain: Potential rodent models of autism. *International Journal of Developmental Neuroscience, 23*, 235–243.

Lim, M. M., Hammock, E. A. D., & Young, L. J. (2004). The role of vasopressin in the genetic and neural regulation of monogamy. *Journal of Neuroendocrinology, 16*, 325–332.

Lim, M. M., Nair, H. P., & Young, L. J. (2005b). Species and sex differences in brain distribution of corticotropin-releasing factor receptor subtypes 1 and 2 in monogamous and promiscuous vole species. *Journal of Comparative Neurology, 487*, 75–92.

Liu, Y., & Wang, Z. X. (2003). Nucleus accumbens oxytocin and dopamine interact to regulate pair bond formation in female prairie voles. *Neuroscience, 121*, 537–544.

Lupoli, B., Johansson, B., Uvnas-Moberg, K., & Svennersten-Sjaunja, K. (2001). Effect of suckling on the release of oxytocin, prolactin, cortisol, gastrin, cholecystokinin, somatostatin and insulin in dairy cows and their calves. *Journal of Dairy Research, 68*, 175–187.

Mason, W. A. (1966). Social organization of the South American monkey, *Callicebus moloch*: A preliminary report. *Tulane Studies in Zoology, 13*, 23–28.

Mason, W. A. (1968). Use of space by *Callicebus* groups. In P. C. Jay (Ed.), *Primates: Studies in adaptation and variability* (pp. 200–216). New York: Holt, Rinehart & Winston.

Mason, W. A. (1974). Comparative studies of *Callicebus* and *Saimiri*: Behaviour of male–female pairs. *Folia Primatologica, 22*, 1–8.

Mason, W. A., & Mendoza, S. P. (1998). Generic aspects of primate attachments: Parents, offspring and mates. *Psychoneuroendocrinology, 23*, 765–778.

Meaney, M. J., & Szyf, M. (2005). Environmental programming of stress responses through DNA methylation: Life at the interface between a dynamic environment and a fixed genome. *Dialogues in Clinical Neuroscience, 7*, 103–123.

Mendoza, S. P., & Mason, W. A. (1986). Parental division of labour and differentiation of attachments in a monogamous primate (*Callicebus cupreus*). *Animal Behaviour, 34*, 1336–1347.

Modahl, C., Green, L. A., Fein, D., Morris, M., Waterhouse, L., Feinstein, C., et al. (1998). Plasma oxytocin levels in autistic children. *Biological Psychiatry, 43*, 270–277.

Muller, J. M., Brunelli, S. A., Moore, H., Myers, M. M., & Shair, H. N. (2005). Maternally modulated infant separation responses are regulated by D2-family dopamine receptors. *Behavioral Neuroscience, 119*, 1384–1388.

Nelson, E. E., & Panksepp, J. (1998). Brain substrates of infant–mother attachment: Contributions of opioids, oxytocin, and norepinephrine. *Neuroscience and Biobehavioral Reviews, 22*, 437–452.

Nishimori, K., Young, L. J., Guo, Q. X., Wang, Z. X., & Insel, T. R. (1996). Oxytocin is required for nursing, but is not essential for parturition or reproductive behavior. *Proceedings of the National Academy of Sciences USA, 93*, 11699–11704.

O'Conner, T. G. (2005). Attachment disturbances associated with early severe deprivation. In C. S. Carter, L. Ahnert, K. E. Grossmann, S. B. Hrdy, M. E. Lamb, S. W. Porges, et al. (Eds.), *Attachment and bonding: A new synthesis* (pp. 257–267). Cambridge, MA: MIT Press.

Panksepp, J. (1996). Affective neuroscience: A paradigm to study the animate circuits for human emotions. In R. D. Kavanaugh, B. Zimmerberg, & S. Fein (Eds.), *Emotions: Interdisciplinary perspectives* (pp. 29–60). Mahwah, NJ: Erlbaum.

Panksepp, J. (1998). *Affective neuroscience: The foundations of human and animal emotions.* New York: Oxford University Press.

Panksepp, J., Herman, B. H., Vilberg, T., Bishop, P., & DeEskinazi, F. G. (1980). Endogenous opioids and social behavior. *Neuroscience and Biobehavioral Reviews, 4,* 473–487.

Pechnick, R. N., Kariagina, A., Hartvig, E., Bresee, C. J., Poland, R. E., & Chesnokova, V. M. (2006). Developmental exposure to corticosterone: Behavioral changes and differential effects on leukemia inhibitory factor (LIF) and corticotropin-releasing hormone (CRH) gene expression in the mouse. *Psychopharmacology, 185,* 76–83.

Pedersen, C. A., & Boccia, M. L. (2002). Oxytocin links mothering received, mothering bestowed, and adult stress responses. *Stress, 5,* 259–267.

Porges, S. W. (2005). The role of social engagement in attachment and bonding: A phylogenetic perspective. In C. S. Carter, L. Ahnert, K. E. Grossmann, S. B. Hrdy, M. E. Lamb, S. W. Porges, et al. (Eds.), *Attachment and bonding: A new synthesis* (pp. 33–54). Cambridge, MA: MIT Press.

Roberts, R. L., Zullo, A. S., & Carter, C. S. (1996). Perinatal steroid treatments alter alloparental and affiliative behavior in prairie voles. *Hormones and Behavior, 30,* 576–582.

Rojas Wahl, R. U. (2004). Could oxytocin administration during labor contribute to autism and related behavioral disorders?: A look at the literature. *Medical Hypotheses, 63,* 456–460.

Russell, J. A., Gosden, R. G., Humphreys, E. M., Cutting, R., Fitzsimons, N., Johnston, V., et al. (1989). Interruption of parturition in rats by morphine: A result of inhibition of oxytocin secretion. *Journal of Endocrinology, 121,* 521–536.

Sachser, N., Durschlag, M., & Hirzel, D. (1998). Social relationships and the management of stress. *Psychoneuroendocrinology, 23,* 891–904.

Shapiro, L. E., & Insel, T. R. (1990). Infant's response to social separation reflects adult differences in affiliative behavior: A comparative developmental study in prairie and montane voles. *Developmental Psychobiology, 23,* 375–393.

Shapiro, L. E., Meyer, M. E., & Dewsbury, D. A. (1989). Affiliative behavior in voles: Effects of morphine, naloxone, and cross-fostering. *Physiology and Behavior, 46,* 719–723.

Smith, M. A., Kim, S. Y., Vanoers, H. J. J., & Levine, S. (1997). Maternal deprivation and stress induce immediate early genes in the infant rat brain. *Endocrinology, 138,* 4622–4628.

Sroufe, L. A., Carlson, E. A., Levy, A. K., & Egeland, B. (1999). Implications of attachment theory for developmental psychopathology. *Development and Psychopathology, 11,* 1–13.

Stribley, J. M. (1998). *The role of arginine vasopressin in the development of prairie vole social behaviors.* Unpublished doctoral dissertation, University of Maryland, College Park.

Stribley, J. M., & Carter, C. S. (1999). Developmental exposure to vasopressin increases aggression in adult prairie voles. *Proceedings of the National Academy of Sciences USA, 96,* 12601–12604.

Swain, J. E., Leckman, J. F., Mayes, L. C., Feldman, R., & Schultz, R. T. (2004). Functional brain imaging: The development of human parent–infant attachment. *Neuropsychopharmacology, 29,* S195.

Tachibana, T., Saito, E.-S., Saito, S., Tomonaga, S., Denbow, D. M., & Furuse, M. (2004). Comparison of brain arginine–vasotocin and corticotrophin-releasing factor for physiological responses in chicks. *Neuroscience Letters, 360,* 165–169.

Thompson, R. A., & Raikes, H. A. (2003). Toward the next quarter-century: Conceptual and methodological challenges for attachment theory. *Development and Psychopathology, 15,* 691–718.

Tyler, A. N., Michel, G. F., Bales, K. L., & Carter, C. S. (2005). Do brief early disturbances of parents affect parental care in the bi-parental prairie vole (*Microtus ochrogaster*)? *Developmental Psychobiology, 47*, 451.

Uchino, B. N. (2006). Social support and health: A review of physiological processes potentially underlying links to disease outcomes. *Journal of Behavioral Medicine, 29*, 377–387.

Uchino, B. N., Cacioppo, J. T., & Kiecolt-Glaser, J. K. (1996). The relationship between social support and physiological processes: A review with emphasis on underlying mechanisms and implications for health. *Psychological Bulletin, 119*, 488–531.

Uvnas-Moberg, K. (1998). Oxytocin may mediate the benefits of positive social interactions and emotions. *Psychoneuroendocrinology, 23*, 819–835.

Vazquez, D. M., Bailey, C., Dent, G. W., Okimoto, D. K., Steffek, A., Lopez, J. F., et al. (2006). Brain corticotropin-releasing hormone (CRH) circuits in the developing rat: Effect of maternal deprivation. *Brain Research, 1121*, 83–94.

Welker, C., & Schafer-Witt, C. (1987). On the carrying behaviour of basic South American primates. *Human Evolution, 2*, 459–473.

Weller, A., & Feldman, R. (2003). Emotion regulation and touch in infants: The role of cholecystokinin and opioids. *Peptides, 24*, 779–788.

Williams, J. R., Insel, T. R., Harbaugh, C. R., & Carter, C. S. (1994). Oxytocin centrally administered facilitates formation of a partner preference in female prairie voles (*Microtus ochrogaster*). *Journal of Neuroendocrinology, 6*, 247–250.

Winslow, J. T., Hastings, N., Carter, C. S., Harbaugh, C. R., & Insel, T. R. (1993). A role for central vasopressin in pair bonding in monogamous prairie voles. *Nature, 365*, 545–548.

Winslow, J. T., Hearn, E. F., Ferguson, J., Young, L. J., Matzuk, M. M., & Insel, T. R. (2000). Infant vocalization, adult aggression, and fear behavior of an oxytocin null mutant mouse. *Hormones and Behavior, 37*, 145–155.

Winslow, J. T., & Insel, T. R. (1993). Effects of central vasopressin administration to infant rats. *European Journal of Pharmacology, 233*, 101–107.

Winslow, J. T., & Insel, T. R. (2004). Neuroendocrine basis of social recognition. *Current Opinion in Neurobiology, 14*, 248–253.

Witt, D. M., Carter, C. S., & Insel, T. R. (1991). Oxytocin receptor binding in female prairie voles: Endogenous and exogenous estradiol stimulation. *Journal of Neuroendocrinology, 3*, 155–161.

Yamamoto, Y., Cushing, B. S., Kramer, K. M., Epperson, P. D., Hoffman, G. E., & Carter, C. S. (2004). Neonatal manipulations of oxytocin alter expression of oxytocin and vasopressin immunoreactive cells in the paraventricular nucleus of the hypothalamus in a gender specific manner. *Neuroscience, 125*, 947–955.

Young, L. J., Murphy Young, A. Z., & Hammock, E. A. (2005). Anatomy and neurochemistry of the pair bond. *Journal of Comparative Neurology, 493*, 51–57.

Ziabreva, I., Schnabel, R., Poeggel, G., & Braun, K. (2003). Mother's voice "buffers" separation-induced receptor changes in the prefrontal cortex of *Octodon degus*. *Neuroscience, 119*, 433–441.

Zingg, H. H. (2002). Oxytocin. In D. W. Pfaff, A. P. Arnold, A. M. Etgen, S. E. Fahrbach, & R. T. Rubin (Eds.), *Hormones, brain, and behavior* (Vol. 3, pp. 779–802). San Diego, CA: Academic Press.

CHAPTER 14

Neurobiology and Hormonal Aspects of Romantic Relationships

DONATELLA MARAZZITI

A romantic relationship is a form of bonding that is crucial for the survival of the human species. Such a relationship can be defined as one established with a partner (Hazan & Shaver, 1987) through specific selection of him or her from among individuals outside the family (Marazziti, 2002). Although breeding with genetically unrelated individuals produces healthier offspring, being attracted to nonfamilial conspecifics also creates a significant paradox. In the landscape of our evolution, we humans lived in small, genetically related social groups. People to whom we were not genetically related were, generally speaking, strangers. A romantic relationship thus must involve psychological strategies that enable each of us to overcome neophobia and to mate with and create a strong, often lifelong bond with a stranger. These strategies involve emotions, behaviors, and subjective awareness—the sum of which constitutes, perhaps, the essence of what we call "love."

Some components of these complex processes can also be identified in other mammals; however, what renders human romantic relationships unusual is that pair bonding in our species is related not only to reproduction, but also to the creation of group structures, social organizations, and interactions, the ultimate goal of which is to provide safe environments for the rearing of children from infancy to autonomy. Creating stable structures lasting over many years for the protection and rearing of children is essential, as our offspring, compared to those of other mammals, are the weakest and require the longest care. Nature must provide mechanisms of sufficient complexity for assuring that our romantic focus is limited to one partner and to permit a sense of safety to be formed and maintained over those many years, while rewarding us with that feeling of pleasure and completeness that we call "love." Therefore, from an evolutionary perspective it is not surprising that specific brain mechanisms have evolved to accomplish this goal, so that romantic relationships are not regulated by chance, but rather by well-established biological processes.

Over the past two decades, evolutionary theories have played a major role in promoting an increasing interest among scientists in studying the possible biological bases of love. This increase in the scientific study of love is striking, because for a long time, love (like other emotions) was not considered a worthy topic of experimental science. Hence research on love was neglected, or hindered by the skeptical belief that those trying to study love were wasting their time and resources (Carter, 1998). Although scientific disdain was a major obstacle, the subjectivity of love also created major methodological problems for research in this field (Porges, 1997). Simply put, scientific methods appeared inappropriate to the exploration of emotions, which always exhibit peculiarities linked to particular individuals. In addition, the subtlety and subjectivity of different components of this complex feeling we call "love," expressed so well in the poetic language of writers, novelists, and psychologists, proved elusive even to eminent scientists such as Harlow (1952). Fortunately, the growth of neuroscience in the past 20 years has revolutionized this area of study and profoundly mitigated the original skepticism. As LeDoux (2000) has noted, "It may ... be able to help the resurrection of emotion research by providing a strategy that allows its study independent of subjective emotional experience." Therefore neuroscience, instead of promoting the death of the study of emotions (as once predicted), has provided reliable tools for their exploration. The aim of this chapter is to present a comprehensive review of the neurobiological substrates of those processes that, all together, constitute the feeling we recognize as love. Particular emphasis is given to my colleagues' and my contributions in this area, and to some speculative models that may constitute a starting point for deeper investigations of the biological mechanisms underlying love.

Neurobiological Correlates of Attraction

Attraction represents the starting phase of a romantic relationship. Attraction is present in all cultures and societies where it has been investigated, and therefore it is believed to have strong biological roots that are genetically transmitted (Jankoviak & Fischer, 1992). Most of the times, attraction is a sudden and unpredictable experience that has the specific aim of facilitating the bonding between two unrelated individuals, since the more distant genetically different the parents are, the healthier the next generation will be. The importance of genetic distance in stimulating attraction is supported by data showing that young girls found boys who were more different from them genetically more attractive (McCoy & Pitino, 2002).

According to evolutionary theories, attraction is believed to last between 6 months and 3 years, which is considered sufficient time for a man to stay close to a woman in order for her to become pregnant and provide primary care to a newborn (Fisher, 1992). No biological correlate of the onset of human attraction is available. Pheromones play a role in mate selection and subsequent sexual behavior in a variety of mammals, but their role in regulating human mating behavior is still debated (Weller, 1998; Keverne, 2002). Pheromones produced in the armpits do appear to play a role in orchestrating menstrual synchrony among women sharing the same environment. Although these substances are not perceived as having any particular odor, they appear to affect the length of the menstrual cycle, probably through their influence on gonadal hormones. In addition, convergent data suggest the existence of signaling pheromones in women that influence their choice of a sexual partner. However, it is premature to draw conclusions about the effects of pheromones on sexual attractiveness, because these studies were conducted under laboratory conditions and may not be generalized to attraction under real-world conditions (Chen & Haviland-Jones, 1999; Jacob, Garcia, Hayreh, & McClintock, 2002).

I would like to propose a model for attraction that is mainly speculative, although based on models that have already been proposed for basic emotions, such as fear or anxiety. "If we want to understand feelings, it is likely going to be necessary to figure out how the more basic system works" (LeDoux, 2000). Along this line, my colleagues and I have argued that attraction is a basic emotion, because it is fundamental for the survival of the species, and that as such it may share some neurobiological substrates with fear and anxiety. We hypothesize that different triggers (hormonal changes, life events, etc.) may alter human brain chemistry or functioning, and may predispose the brain to become susceptible to stimuli coming from another individual. That is, heightened emotionality may increase the propensity to become attracted to and to fall in love with a potential mate (Marazziti & Cassano, 2003). The main stimuli would be visual, given the importance of vision for human beings, but auditory, tactile, gustatory, and olfactory stimuli may also contribute to such attraction. Such stimuli, after first being processed in the thalamus, are generally split into two main groups: One is directed immediately to the amygdala, through a short pathway, while the other takes a longer route from the thalamus to the cortex and then to the amygdala (Figure 14.1). The amygdala, when activated, modulates a series of responses in different cerebral areas and peripheral organs; the changes in neurovegetative functions constitute the bases for the subjective feeling of the

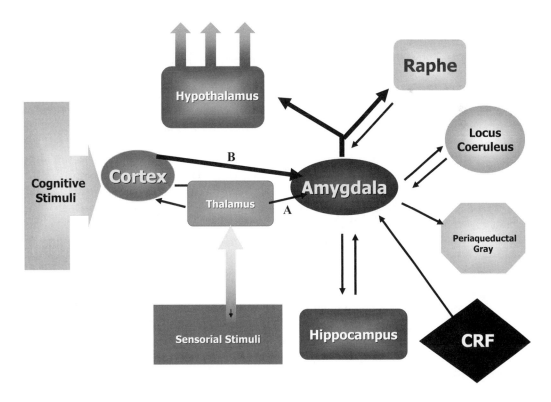

FIGURE 14.1. The neural pathways hypothetically involved in the process of attraction. CRF, corticotropin-releasing factor.

emotion (Damasio, 2003). The cortex, in turn, is informed by the amygdala itself of its functioning and discriminates the "quality" (fear, joy, falling in love) of its activation, so that an individual becomes aware of the feeling. At this point, the person labels the feeling as love, rather than as fear or anxiety. Recognition of the feeling state permits more sophisticated planning of appropriate strategies, such as those that will enable him or her to meet the potential partner again (Marazziti & Cassano, 2003). Although these strategies are voluntary, the first process is not; for this reason, it is difficult (although not impossible) to counteract it, and for the same reason, it may not be easily described.

Attraction to one individual over another is not casual, because a person will choose the one who is able to evoke positive states related to early experiences recorded in the hippocampus, a structure connected to the amygdala. This is in agreement with psychological constructs, which have shown the crucial role of early experiences in the formation of the so-called "love maps" that seem to have a profound effect on attraction to a partner.

Attraction has distinctive features. It is generally characterized by an altered mental state, with mood elation characterized by the sensation of being full of energy and strength, as well as loss of appetite and sleeplessness. These features resemble the hypomanic phase of bipolar disorder (Liebowitz, 1983; Marazziti & Cassano, 2003). Attraction also can be characterized by mood swings from depression to joy, depending on the partner's response; again, these are quite similar to the opposing phases of bipolar disorder. It has therefore been suggested that the neurochemical abnormalities reported in bipolar disorder, such as increased functioning of the norepinephrine and dopamine systems, may also be involved in the attraction phase of romantic relationships. Over 25 years ago, Liebowitz (1983) proposed a direct role of phenylethylamine, a trace neurotransmitter similar to the amphetamines, in attraction. Although intriguing, this hypothesis was never tested.

Attraction is also characterized by intense craving for the partner and specific behavioral patterns designed to evoke a reciprocal response. These patterns are similar to grooming behavior and even to compulsions, so that they may perhaps be due to increased levels of dopamine and decreased concentrations of serotonin (Marazziti & Cassano, 2003). Moreover, a significant shift of consciousness also occurs; specifically, a person becomes certain that the romantic partner is the most extraordinary individual in the world. This certainty is coupled with an exclusive focusing of attention on the partner, which leads to a decreased interest in the person's normal routine and in mundane activities. However, the characteristic that is considered the "core" feature of attraction (Fisher, 1992) consists of intrusive thoughts regarding the partner, which resemble the obsessions typical of obsessive–compulsive disorder (OCD) (Marazziti, Akiskal, Rossi, & Cassano, 1999).

We have argued that these cognitions about the object of attraction are the pure "human" component of this love feeling, with no animal parallel. Notably, these cognitive features of love may exist and persist even when no actual bonding occurs and in the absence of sexual consummation. Such instances include sublimation through art or religious/mystical experiences. We propose that this component of love typical of the early phase of the relationship, can be conceptualized dimensionally. In models such as this one, the dimensions are basic products of the functioning of the mind or brain (e.g., dimensions of aggression or mood); they are sometimes attributed simplistically (see Changeux, 1985) to the functioning of single neurotransmitters. Although dimensional models are speculative, they can be helpful by providing research tools for exploring the relations between normal and pathological states, as such theoretical frameworks permit the exploration of borderline or not fully expressed conditions (Jenike, 1990; Hollander, 1993; Cassano et al., 1997). Overlapping features between the overvalued ideas in OCD and those typical of a romantic relationship or parental love have been widely reported (Leckman et al., 1999). Furthermore, both romantic attraction and parental love have been related to a physiological obsessive state, although, according to the specific personality structure, manic, depressive, or paranoid features can also be present (Netter, 1989; Forward & Craig, 1991; Griffin-Shelley, 1991).

We have suggested that the overlap between obsessive, overvalued ideations and romantic attraction can be conceptualized along the dimensions of "certainty–uncertainty" and "insight–no insight." The "certainty–uncertainty" dimension may be mainly related to the serotonin system. We formulated this hypothesis on the basis of the results of a study examining subjects who had recently "fallen in love" and were still at the romantic phase of the relationship, prior to the onset of sexual intercourse. Among these subjects, the density of the platelet serotonin transporter—a reliable peripheral model of the same structure present in presynaptic serotonergic neurons (Lesch & Mossner, 1998)—was found to be lower than in "normal" control subjects and similar to that observed in a group of patients with OCD (Marazziti et al., 1999; Table 14.1). In our study, "normal" was defined as healthy and either involved in a long-term love relationship or single. We would therefore hypothesize that a continuum of obsessive–compulsive ideation, from uncertainty/insight through the prevalent ideation typical of romantic lovers (certainty/insight) to delusional states (certainty/ no insight), may be involved in romantic attraction. This argument is based on observations of shifts from normal to pathological states in expressions of love, which in some instances can reach the severity of paranoid states, as documented in the literature on romantic love (Forward & Craig, 1991; Griffin-Shelley, 1991). Therefore, we agree with others (Leckman et al., 1999; Kane, 1987; O'Dwyer & Marks, 2000) that love and pair bonding on the one hand, and pathological conditions such as OCD and paranoia on the other, can be related to the same neurobiological systems. The risk of becoming fully "obsessive" or "paranoid" about the partner may be the cost paid, in evolutionary terms, in order to increase the likelihood of bonding and faithfulness in romantic relationships. Probably this sort of transitory madness has evolved in order to render human beings more prone to forming romantic bonds with unrelated individuals, permitting them to overcome neophobia and anxiety about separation from their natal families, and willing to leave the safe "nest" represented by their usual settings. Really, humans must be "crazy" to forget all their fears when they meet "that special person," and to abandon all their reluctance to reveal their most intimate aspects. Interestingly, our study showed that the serotonin abnormalities did not last: Serotonin platelet numbers in our subjects returned to normal values 12–18 months from the beginning of their

TABLE 14.1. [^{3}H] Paroxetine binding parameters in Subjects with a Romantic Love, Patients with OCD, and Healthy Subjects

Group	B_{max}	K_d
Subjects in love	625 ± 219*	0.24 ± 0.27
Patients with OCD	736 ± 457*	0.16 ± 0.18
Healthy subjects	1324 ± 486	0.17 ± 0.28

Note. Binding parameters are B_{max} (fmol/mg protein, mean ± *SD*) and K_d (nM, mean ± *SD*). Analysis of variance, followed by "post hoc" procedure: subjects in love = patients with OCD < healthy subjects. Data from Marazziti, Akiskal, Rossi, and Cassano (1999).
*Significant at $p < .001$.

relationships (Marazziti et al., 1999). Notably, this finding is consistent with anthropological studies reporting that attraction lasts no longer than 3 years (Jankoviak & Fischer, 1992).

Consistent with our model that falling in love exhibits similarities to the emotions evoked in stressful conditions are the observations that stressful and threatening situations may facilitate the onset of new social bonds and intimate ties (Panksepp, 1992). The literature relevant to humans in this regard is meager, albeit in agreement with animal findings, which suggest that the activation of the hypothalamic–pituitary–adrenal (HPA) axis may facilitate bonding in other mammals. In humans, these conditions may also accompany the initial phases of falling in love (Chiodera et al., 1991). Given the paucity of data in this field, we evaluated the levels of some stress and gonadal hormones in a homogeneous group of 24 subjects of both sexes who were in the early, romantic phase of a relationship, and compared them with those of subjects who were single or were already in a long-lasting relationship (Marazziti & Canale, 2004; Table 14.2). Several findings from this study are noteworthy. First, we did find that the cortisol levels of the subjects in love were higher than those of the control subjects. This condition of "hypercortisolemia" was interpreted as a nonspecific indicator of some changes occurring during the early phase of a relationship, reflecting the stress associated with the initiation of a social contact. We observed no difference in cortisol levels between women and men, but this is perhaps not surprising, given the indications these levels represent rather unspecific reactions to different triggers. Changes in the hypothalamic–pituitary–gonadal axis, however, exhibited sex-differentiated impacts. Although luteinizing hormone (LH), estradiol, progesterone, dehydroepiandrosterone (DHEA), and androstenedione levels did not differ between men and women, testosterone levels were lower in men and higher in women at the early stages of relationships than in the comparison subjects. Although these sex-linked changes in testosterone levels did not reach pathological levels, all subjects showed this pattern of change, as if falling in love tended to eliminate some differences between the sexes temporarily. It is possible that decreases in testosterone levels serve to soften some male features in men and, in parallel, to increase them in women. It is tempting to link the changes in testosterone levels to changes in aggressive traits that move in different directions in the two sexes (Zitzmann & Nieschlag, 2001); however, apart from

TABLE 14.2. Hormonal Levels in Subjects in the Early Stage of Falling in Love and in Control Subjects

Hormone	Subjects in love M	Subjects in love F	Control subjects M	Control subjects F
FSH	3.2 ± 1.1^	8.1 ± 4.2	9.3 ± 3.8	9.1 ± 3.1
LH	6.9 ± 2.3	12.3 ± 3.4	7.1 ± 2.8	10 ± 4.3
Estradiol	<50	170 ± 23	<50	145 ± 32
Progesterone	<0.2	0.57 ± 0.3	<0.2	0.55 ± 0.3
Testosterone	4.1 ± 1.0*	1.2 ± 0.4**	6.8 ± 2.1	0.6 + 0.2
DHEA	2736 ± 1122	2232 ± 986	2450 ± 1000	2315 ± 980
Cortisol	224 ± 21°	243 ± 41°°	165 ± 21	172 ± 44
Androstenedione	2.0 ± 1.0	2.1 ± 0.7	2.1 ± 0.7	1.9 ± 0.7

Note. M, male; F, female; FSH, follicle-stimulating hormone; LH, luteinizing hormone; DHEA, dihydroxyepiandrosterone. Data from Marazziti and Canale (2004).
^Significant at $p < .0001$; *significant at $p < .003$; **significant at $p < .001$; °significant at $p < .001$; °°significant at $p < .0001$.

a few observations, we have no data substantiating this hypothesis. Thus it requires future exploration. We also observed a decrease in levels of follicle-stimulating hormone (FSH) in the men who recently had fallen in love—a finding for which we have no explanation, apart from the possibility that it represents another marker of hypothalamic involvement in the process of falling in love.

We measured cortisol, testosterone, and FSH levels again 12–18 months later in the 16 (of 24) individuals who were still in the same relationship but were no longer in the same mental state they had reported at the first assessment. They now described themselves as feeling calmer and no longer "obsessed" with their partners. Notably, their hormone levels were no different from those of the control group. This finding would suggest that the hormonal changes we observed are reversible, state-dependent, and probably related to some physical and/or psychological features typically associated with falling in love. In conclusion, our study would suggest that falling in love is stressful and is characterized by specific hormonal patterns, one of which, involving testosterone, seems to show a sex-related specificity.

The stress related to falling in love may also explain the report of increased concentrations of nerve growth factor, which is a potent antianxiety agent, in romantic lovers (Enzo et al., 2006). Some studies carried out *in vivo*, by means of functional magnetic resonance imaging (fMRI), support our model of attraction as well, and in addition suggest that this process is linked to reward/motivational systems (Bartels & Zeki, 2004; Aron et al., 2005).

Neurobiological Correlates of Attachment

If the process of falling in love is successful—that is to say, the partner reciprocates the person's feelings, and the two individuals start a romantic relationship that continues—there is a marked change in subjective feelings over time. Mood becomes more stable, anxiety is reduced, and the mind is free from obsessive thoughts regarding the partner. Probably the chemical storm that has flooded the brain has faded away. Indeed, if it continued it would be quite uneconomical and exhausting, because it would involve an extreme release of neurotransmitters and hyperstimulation of receptors that could not be tolerated too long. However, it is more correct to say that attraction, rather than diminishing if the relationship goes on, is replaced by another process—specifically, one of attachment (Jankoviak & Fischer, 1992; Marazziti & Cassano, 2003). Romantic attachment is fundamental for keeping two individuals together once the flame of passionate love dims. It can thus be defined as the "glue" necessary condition for tolerating a partner for a long time and for continuing a successful relationship.

Generally speaking, "attachment" can be defined as a social process involving the emotional relationship between an individual and the attachment object. Developing over a certain period of time, it is characterized by the need for propinquity; by behavioral and physiological signs of distress and agitation when separation from the attachment object occurs; and by reduction of these signs after reunion. Other distinctive features include its selectivity, which distinguishes it from nonspecific social relationships, and its duration, which is generally long and not transient.

For a long time, distress at separation was the only aspect of attachment examined biologically, which led to a focus on opioids (Panksepp, 1982). More recently, attention has broadened to include a focus on the positive valence associated with attachment (i.e., positive feelings and rewards). Separation anxiety and positively valenced aspects of attachment may involve different neural systems. As Insel (1997) stated, "there is no obvious reason for which attachment and separation should be subserved by the same neural system." Accordingly, the last decade has seen an accumulation of data highlighting a key role for neuropeptides such as oxytocin (OT) and arginine vasopressin (AVP) in the initiation and maintenance of infant attachment, maternal behavior, and pair bonding (Marazziti et al., 2006a).

OT and AVP are both peptides that consist of nine amino acids (nonapeptides) and differ from each other in only two amino acids. Therefore, they may be considered to have evolved from a common, ancestral peptide (Acher, 1996). Interestingly, although related to other substances present in lower animals, they are found only in mammals. A growing body of evidence implicates OT in the mediation of complex social behaviors. It may be no coincidence that this peptide has been implicated in prototypically mammalian functions, such as milk ejection during nursing (Wakerley & Lincoln, 1973), uterine contraction during labor, and sexual behavior (Carmichael et al., 1987; Carter, 1992).

OT and AVP synthesis occurs in the paraventricular (PVN) and in the supraoptic nuclei (SPN) of the hypothalamus, which project to the posterior pituitary and also to the limbic system and to different autonomic centers. These projections probably provide the anatomical basis for the somatic changes related to the behaviors regulated by these peptides (Buijs, 1978; Brownstein, Russell, & Gainer, 1980). OT interacts with one specific receptor, and AVP with at least three receptor subtypes widely distributed in the brain (Tribollet, Charpak, Schmidt, Dubois-Dauphin, & Dreifuss, 1989; Ostrowski, 1998).

The development of adult–adult pair bonds is certainly the least studied form of attachment from a neurobiological perspective. The relative paucity of studies can be attributed to the absence of pair bonds in commonly used laboratory animals, such as rats and mice. By definition, pair bonds occur in monogamous animals, and only about 3% of mammals are currently considered monogamous. The percentage of primates that are monogamous, however, is considerably higher, perhaps as high as 15% (Van Schaik & De Visser, 1990).

Prairie voles (*Microtus ochrogaster*) and montane voles (*Microtus montanus*) provide an intriguing natural experiment for studying the neural substrates of pair bonding (Insel, 1997; see also Bales & Carter, Chapter 13, and Wommack, Liu, & Wang, Chapter 15, this volume). Montane voles are remarkably similar to the prairie voles in physical appearance and share many features of their behavior; however, they differ consistently on measures of social behavior. The prairie vole is a mouse-sized rodent that is usually found in multigenerational family groups with a single breeding pair (Carter, DeVries, & Getz, 1995). They manifest the classic features of monogamy: A breeding pair shares the same nest and territory, where the partners are in frequent contact; males participate in parental care; and intruders of either sex are rejected. Following the death of one partner, a new mate is accepted by the survivor only about 20% of the time. In contrast, montane voles are generally found in isolated burrows, show little interest in social contact, and are clearly not monogamous.

Many studies have investigated whether prairie and montane voles differ in central pathways for OT. There is strong evidence that they do, exhibiting differences in the neural distribution of receptors for both peptides and different expression of receptors for these peptides within these pathways (Insel, Wang, & Ferris, 1994). In the prairie vole, OT receptors are found in brain regions associated with reward (the nucleus accumbens and prelimbic cortex), suggesting that OT may have reinforcing properties selectively in this species. Conversely, receptors in the lateral septum, found only in the montane vole, may be responsible for the effects of OT on self-grooming—effects that are observed in the montane vole but not the prairie vole. Furthermore, other vole species (pine voles and meadow voles) selected for analogous differences in social organization (i.e., monogamous vs. nonmonogamous) manifest similar differences in receptor distribution for both OT and AVP (Insel et al., 1994). Finally, after parturition, when the female montane vole becomes briefly parental, the pattern of OT receptor binding changes to resemble the pattern observed in the highly parental prairie vole.

A critical requirement for social behavior is the ability of animals to identify conspecifics (Insel & Fernald, 2004). Neural pathways employing the nonapeptides AVP and OT appear to play a particularly prominent role in both social learning and recognition ("social memory"). For example, OT supports, on the basis of olfactory stimuli, the onset of a partner preference in rats (Popik & Van Ree, 1993). OT seems involved in acquisition rather than in consolidation of social bonds. One study found that mice specially bred for a deficiency of OT (OT "knockout" mice) failed to recognize previously encountered conspecifics; in addition, central administration of OT before the first contact, but not afterwards, restored normal social learning and recognition (Dantzer, Bluthe, Koob, & Le Moal, 1987).

The regulation of social behavior requires not only the recognition of familiar conspecifics, but also modification of behaviors that may affect the likelihood and consequences of a social encounter. For example, anxiety and novelty avoidance may be expected to reduce the likelihood of approaching a conspecific. Learning and memory mechanisms may also affect social behavior by modifying the impact of prior social encounters on an individual's behavioral responses. It is interesting that although AVP and OT are both required for social recognition, they differentially regulate anxiety-like behavior and avoidance learning. Specifically, administration of AVP and activation of AVP V1a receptors increase anxiety-like behaviors in male mice, whereas OT decreases them (Bielsky & Young, 2004). Recently, the amygdala has been proposed as a candidate site at which AVP and OT may exert their opposing effects on anxiety-like behavior and avoidance learning (Huber, Veinante, & Stoop, 2005). AVP V1a receptors and OT receptors are expressed within distinct subregions of the central nucleus of the amygdala, and the two neuropeptides interact in a manner that could produce opposing effects on neuronal activity.

The fact that AVP and OT facilitate social recognition, but produce distinct and sometimes opposite effects on other behaviors, may be explained by the need to regulate gender differences in social behaviors that require differential modulation of anxiety-like behavior, avoidance learning, or aggression. For example, OT regulates sexual behavior and social interactions in both males and females, and maternal behavior in females (Gimpl & Fahrenholz, 2001)—all of which require an inhibition of novelty avoidance, suppression of prior social avoidance learning, and decreased aggression. In contrast, AVP promotes behavior modifications that would influence the formation of territories and dominance hierarchies, which are characteristic components of male social behavior. In addition, interactions among OT, AVP, and glucocorticoids could provide substrates for dynamic changes in social behaviors. Non-noxious sensory stimulation associated with friendly social interaction induces a response pattern involving sedation, relaxation, and decreased sympathoadrenal activity. It is suggested that OT released from parvocellular neurons in the PVN in response to non-noxious stimulation integrates this response pattern at the hypothalamic level. The health-promoting aspect of friendly and supportive relationships may be a consequence of repetitive exposure to non-noxious sensory stimulation (Uvnäs-Moberg, 1998).

Adding to information on the neural systems involved in pair bonding, Curtis and Wang (2005) examined c-fos (a transcription factor) expression in brain areas implicated in social behavior in voles. They hypothesized that the presence of c-fos protein after a period of time sufficient for pair bonding to occur may indicate brain areas that are especially important in pair-bond formation; elevated levels of c-fos immunoreactivity have been found in the medial and cortical amygdala, medial preoptic area, and bed nucleus of the stria terminalis in females that mated several times over a 6-hour period, as compared to a variety of unmated controls. These results have been compared with data obtained in OT knockout mice. Wild-type (WT) and OT knockout mice showed similar neuronal activation in the olfactory bulbs, piriform cortex, cortical amygdala, and lateral septum. However, WT mice, but not OT knockout mice, exhibited an induction of c-fos in the medial amygdala. Projection sites of the medial amygdala also failed to show c-fos induction in the OT knockout mice. OT knockout mice, but not WT mice, showed dramatic increases in c-fos in the somatosensory cortex and the hippocampus, suggesting alternative processing of social cues in these animals. Still other findings of interest concern the different roles that OT plays in male and female voles. Central OT administration in female, but not in male, prairie voles facilitates the development of a partner preference in the absence of mating (Winslow, Hastings, Carter, Harbaugh, & Insel, 1993). A selective OT antagonist given centrally before mating blocks formation of the partner preference without interfering with mating (Insel & Hulihan, 1995).

The relevance of these intriguing findings for humans, and especially for romantic attachment and love, has not been clarified as yet (Insel & Young, 2001; Young & Wang, 2004). OT receptors in the human brain are mainly distributed in the substantia nigra and globus pallidus. Together with the anterior cingulate and medial insula, these are areas that have been shown to be activated in adults looking at pictures of their partners and in mothers looking at pictures of their children (Bartels & Zeki, 2004). This pattern of activation overlaps with the pattern found during cocaine-induced euphoria (Young, Lim, Gingrich, & Insel, 2001; Young, 2002), and further supports the notion of a link between romantic attachment and reward pathways (Insel, 2003). In addition, OT administration in humans has been shown to increase trust—again supporting the involvement of the amygdala, a central component of the neurocircuitry of fear and social cognition, which has been linked to trust and highly expresses OT receptors (Kosfeld, Heinrichs, Zack, Fischbacher, & Fehr, 2005). Consistent with this argument, a recent double-blind study that used fMRI to visualize amygdala activation by fear-inducing visual stimuli (Kirsch et al., 2005) showed that human amygdala function is strongly modulated by OT. Compared with placebo, OT potently reduced activation of the amygdala and reduced coupling of the amygdala to brainstem regions implicated in autonomic and behavioral manifestations of fear. This effect was located on the level of the midbrain and encompassed both the periaqueductal gray and the reticular formation, which are prominent among the brainstem areas to which the central nucleus of the amygdala projects (LeDoux, 2000), and which mediate fear behavior and arousal (LeDoux, Iwata, Cicchetti, & Reis, 1988).

In agreement with these findings, autonomic response to aversive pictures has been reported previously to be reduced under OT administration (Pitman, Orr, & Lasko, 1993). Nonetheless, it is of interest to note that OT administration did not affect self-reports of psychological state. This finding agrees with the observations of Kosfeld et al. (2005), who also did not find an effect of OT on measured calmness and mood, and showed that actual social interaction was necessary to bring out the OT effect at the level of behavior. That is, the neural effect of the neuropeptide on behavior was evident in the social context, but not when subjects rated themselves in isolation. Moreover, the reduction in amygdala activation was more significant for socially relevant stimuli (faces) than for the less socially relevant scenes. Such differential impairment of amygdala signaling related to the social relevance of the stimuli is in agreement with emerging data from primate lesion studies (Prather et al., 2001), and with human data indicating that social and nonsocial fear may depend on dissociable neural systems (Meyer-Lindenberg et al., 2005).

There are also data indicating that OT plays an important role in human sexual behavior. It is well documented that levels of circulating OT increase during sexual stimulation and arousal, and peak during orgasm, in both men and women (Carmichael et al., 1987; Carter, 1992). Plasma OT and AVP concentrations were measured in men during sexual arousal and ejaculation, and plasma AVP was found to be significantly increased during arousal (Murphy & Dwarte, 1987). However, mean plasma OT rose about fivefold at ejaculation, falling back to basal concentrations within 30 minutes. AVP concentrations had already returned to basal levels at the time of ejaculation and remained stable thereafter. Men who took the opioid antagonist naloxone before self-stimulation had reductions in both OT secretion and the degree of arousal and orgasm. In a study of women, peak levels of serum OT were measured at or shortly after orgasm (Blaicher et al., 1999). The intensity of muscular contractions during orgasm in both men and women was highly correlated with OT plasma levels (Carmichael, Warburton, Dixen, & Davidson, 1994); this suggests that some of OT's effects may be related to its ability to stimulate the contraction of smooth muscles in the genital/pelvic area. Enhanced sexual arousal and orgasm intensity have also been reported in a case study of a woman during intranasal administration of OT (Anderson-Hunt & Dennerstein, 1994, 1995), which described the case of a woman who, about 2 hours after the use of a synthetic OT spray, noticed copious vaginal transudate and a subsequent intense sexual desire. This response could be elicited only while she was taking daily doses of an oral contraceptive with estrogenic and progestogenic actions, and might have been caused through direct effects on sexual organs or sensory nerve sensitivity.

OT is not only released in humans during sexual activity; it is also released in response to different relaxation techniques, and therefore it is believed to be one of the mediators of the decrease in stress responses (Carter, 1992). Consequently, it has been proposed that it may underlie the benefits of positive relationships in promoting health, such as the lower prevalence of cardiovascular diseases or depression in individuals with stable partners (Uvnäs-Moberg, 1998).

Following this line of research, we recently carried out a study designed to explore the association between plasma OT levels and romantic attachment in a group of 45 healthy subjects, with the hypothesis that some links might exist between OT and certain characteristics of attachment and/or features of the relationship (Marazziti et al., 2006b). The subjects were 12 men and 33 women, all of whom volunteered for the study. In terms of marital status, 32 (71.1%) individuals were single, 12 (26.7%) were married, and 1 (2.2%) was divorced. Thirty-three subjects had a current romantic relationship, with a mean duration of 80.5 months (ranging from a minimum of 1 month to a maximum of 25 years); the remaining 12 had no current relationship. Although it is still debated whether the measurement of OT in plasma is a reliable mirror of OT in the brain (Insel & Shapiro, 1992; Keverne & Kendrick, 1992), we used it because it is a relatively easy measure to obtain in normal populations. It was difficult and took a long time to set up an assay sensitive enough to measure the low basal levels of OT in humans, but in the end we were successful. The OT plasma levels, representing the mean $+ SD$ of three evaluations performed within 1 hour, ranged between 0.13 ± 0.02 and 4.59 ± 0.01 pg/ml (overall ± SD = 1.53 ± 1.18), with no difference between women and men.

The major finding of our study was the presence of a statistically significant and positive correlation between OT plasma levels and scores on the Anxiety scale of Experiences in

Close Relationships (ECR), a self-report questionnaire measuring adult romantic attachment (Brennan, Clark, & Shaver, 1998). Without further study, it is not possible to conclude from our data whether the OT levels were a consequence or a cause of the anxiety measured by the ECR. However, in line with the majority of available findings (Carter, 1992; Uvnäs-Moberg, 1997; Insel & Young, 2001; Heinrichs, Baumgartner, Kirschbaum, & Ehlert, 2003), we would suggest tentatively that the former might have been the case and that OT may serve to help to counteract anxiety—or at least that form of anxious stress associated with romantic attachment and concomitant deep concerns over whether the relationship will continue.

Although our findings are in agreement with previous animal studies of possible role of OT in social bonding, they undoubtedly represent the first report of a direct link between OT and the state of anxiety that is associated with romantic attachment in humans. Previously, the relationship between OT and anxiety was sustained only by the indirect evidence that basal OT levels were correlated with measures of anxiety, aggression, guilt, and suspicion, along with evidence that noise stress provoked the release of the neuropeptide in highly emotional women (Uvnäs-Moberg, Arn, Theorell, & Jonsson, 1991). A more recent study has reported that low plasma OT levels seem to be typical of individuals with low anxiety traits (Turner et al., 2002). Pursuing this line of thought, romantic relationships, and perhaps social relationships in general, could be interpreted as amounting to stress conditions, both acute and chronic, depending on the phase (Gillath, Bunge, Shaver, Wendelken, & Mikulincer, 2005). The role of OT would seem generally to be that of keeping anxiety levels under control to a point where they are no longer harmful, but may nevertheless lead to strategies and behaviors that are best suited to ensuring a partner's continued proximity, both during the first stages of the romance and subsequently. Interestingly, there is evidence that low OT concentrations are linked with pain syndromes, such as fibromyalgia (Anderberg & Uvnäs-Moberg, 2000) and abdominal pain (Alfven, 2004). OT may thus be considered an essential element in securing the rewarding effects of a romantic relationship, as a result of its increasing a prospective sexual partner's willingness to accept the risk deriving from social contacts, because it moderates the impact of anxiety associated with those risks (Kosfeld et al., 2005). Of course, if particularly vulnerable individuals are excessively affected by a relationship itself or by other events, these delicate mechanisms may be maladaptive, in the sense that such subjects may become too anxious and thus cross the line between normal and pathological states—even to the point of developing a full-blown psychiatric disorder.

Conclusions

In this chapter, some of the major findings on the neurobiology of romantic relationships and love have been reviewed. This is an emerging and intriguing field of research, which only recently has become the topic of intensive scientific investigation through the application of advanced neuroscientific methods. The accumulating evidence, albeit preliminary and fragmentary, has permitted us to develop an initial framework and hypotheses that await testing. The available data converge strongly on the notion that romantic relationships and love may emerge from the functioning of highly evolved and structured neural systems and from the bidirectional "conversation" of these systems with peripheral organs. Naturally, this is only the beginning of the story; my colleagues and I believe that, instead of decreasing the poetry of love, it will enhance our wonder at the integrated efforts of nature to ensure that our species might survive and receive benefits during this process.

References

Acher, R. (1996). Molecular evolution of fish neurohypophyseal hormones: Neutral and selective evolutionary mechanisms. *Genetics and Comprehensive Endocrinology, 102*, 167–172.

Alfven, G. (2004). Plasma oxytocin in children with recurrent abdominal pain. *Journal of Pediatric Gastroenterology and Nutrition, 38*, 513–517.

Anderberg, U. M., & Uvnäs-Moberg, K. (2000). Plasma oxytocin level s in female fibromyalgia syndrome patients. *Journal of Rheumatology, 59*, 373–379.

Anderson-Hunt, M., & Dennerstein, L. (1994). Increased female sexual response after oxytocin. *British Medical Journal, 309*, 929.

Anderson-Hunt, M., & Dennerstein, L. (1995). Oxytocin and female sexuality. *Gynecologic and Obstetric Investigation, 40*(4), 217–211.

Aron, A., Fisher, H., Mashek, D. J., Strong, G., Li, H., & Brown, L. L. (2005). Reward, motivation and emotion systems related to early stage romantic love. *Journal of Neurophysiology, 94*, 327–337.

Bartels, A., & Zeki, S. (2004). The neural correlates of romantic and maternal love. *NeuroImage, 21*, 1155–1166.

Bielsky, I. F., & Young, L. J. (2004). Oxytocin, vasopressin, and social recognition in mammals. *Peptides, 25*(9), 1565–1574.

Blaicher, W., Gruber, D., Bieglmayer, C., Blaicher, A. M., Knogler, W., & Huber, J. C. (1999). The role of oxytocin in relation to female sexual arousal. *Gynecologic and Obstetric Investigation, 47*(2), 125–126.

Brennan, K. A., Clark, C. L., & Shaver, P. R. (1998). Self-report measurement of adult attachment: An integrative overview. In J. A. Simpson & W. S. Rholes (Eds.), *Attachment theory and close relationships* (pp. 46–76). New York: Guilford Press.

Brownstein, M. J., Russell, J. T., & Gainer, H. (1980). Synthesis, transport and release of posterior pituitary hormones. *Science, 207*, 373–378.

Buijs, R. (1978). Intra- and extra-hypothalamic vasopressin and oxytocin pathways in the rat: Pathways to the limbic system, medulla oblongata and spinal cord. *Cell and Tissue Research, 252*, 355–365.

Carmichael, M. S., Humbert, R., Dixen, J., Palmisano, G., Greenleaf, W., & Davidson, J. M. (1987). Plasma oxytocin increases in the human sexual response. *Journal of Clinical Endocrinology and Metabolism, 64*, 27–31.

Carmichael, M. S., Warburton, V. L., Dixen, J., & Davidson, J. M. (1994). Relationships among cardiovascular, muscular, and oxytocin responses during human sexual activity. *Archives of Sexual Behavior, 23*, 59–79.

Carter, C. S. (1992). Oxytocin and sexual behavior. *Neuroscience and Biobehavioral Reviews, 16*, 131–144.

Carter, C. S. (1998). Neuroendocrine perspectives on social attachment and love. *Psychoneuroendocrinolology, 23*, 779–818.

Carter, C. S., DeVries, A. C., & Getz, L. L. (1995). Physiological substrates of mammalian monogamy: The prairie vole model. *Neuroscience and Biobehavioral Reviews, 19*(2), 303–314.

Cassano, G. B., Michelini, S., Shear, M. K., Coli, E., Maser, J. D., & Frank, E. (1997). The panic–agoraphobic spectrum: A descriptive approach to the assessment and treatment of subtle symptoms. *American Journal of Psychiatry, 154*, 27–38.

Changeux, J. P. (1985). *L'homme neuronal*. Paris: Fayard.

Chen, D., & Haviland-Jones, J. (1999). Rapid mood change and human odors. *Physiology and Behavior, 68*(1–2), 241–250.

Chiodera, P., Volpi, R., Capretti, L., Marchesi, C., D'Amato, L., De Ferri, A., et al. (1991). Effect of estrogen or insulin-induced hypoglycemia on plasma oxytocin levels in bulimia and anorexia nervosa. *Metabolism, 40*, 1226–1230.

Curtis, J. T., & Wang, Z. (2005). Ventral tegmental area involvement in pair bonding in male prairie voles. *Physiology and Behavior, 86*(3), 338–346.

Damasio, A. (2003). Feelings of emotion and the self. *Annals of the New York Academy of Sciences, 1001,* 253–261.

Dantzer, R., Bluthe, R. M., Koob, G. F., & Le Moal, M. (1987). Modulation of social memory in male rats by neurohypophysial peptides. *Psychopharmacology, 91,* 363–368.

Enzo, E., Politi, P., Bianchi, M., Minoretti, P., Bertona, M., & Geroldi, D. (2006). Raised plasma nerve growth factor levels associated with early-stage romantic love. *Psychoneuroendocrinology, 3,* 288–294.

Fisher, H. E. (1992). *Anatomy of love.* New York: Norton.

Forward, S., & Craig, B. (1991). *Obsessive love: When passion holds you prisoner.* New York: Bantam Books.

Gillath, O., Bunge, S. A., Shaver, P. R., Wendelken, C., & Mikulincer, M. (2005). Attachment-style differences in the ability to suppress negative thoughts: Exploring the neural correlates. *NeuroImage, 28*(4), 835–847.

Gimpl, G., & Fahrenholz, F. (2001). The oxytocin receptor system: Structure, function, and regulation. *Physiological Reviews, 81,* 629–683.

Griffin-Shelley, E. (1991). *Sex and love: Addiction, treatment and recovery.* Westport, CT: Praeger.

Harlow, H. F. (1952). Learning. *Annual Review of Psychology, 3,* 29–54.

Hazan, C., & Shaver, P. (1987). Romantic love conceptualized as an attachment process. *Journal of Personality and Social Psychology, 52,* 511–524.

Heinrichs, M., Baumgartner, T., Kirschbaum, C., & Ehlert, U. (2003). Social support and oxytocin interact to suppress cortisol and subjective responses to psychosocial stress. *Biological Psychiatry, 54*(12), 1389–1398.

Hollander, E. (Ed.). (1993). *Obsessive–compulsive-related disorders.* Washington, DC: American Psychiatric Press.

Huber, D., Veinante, P., & Stoop, R. (2005). Vasopressin and oxytocin excite distinct neuronal populations in the central amygdala. *Science, 308,* 245–248.

Insel, T. R. (1997). A neurobiological basis of social attachment. *American Journal of Psychiatry, 154,* 726–735.

Insel, T. R. (2003). Is social attachment an addictive disorder? *Physiology and Behavior, 79,* 351–357.

Insel, T. R., & Fernald, R. D. (2004). How the brain processes social information: Searching for the social brain. *Annual Review of Neuroscience, 27,* 697–722.

Insel, T. R., & Hulihan, T. J. (1995). A gender-specific mechanism for pair bonding: Oxytocin and partner preference formation in monogamous voles. *Behavioral Neuroscience, 109*(4), 782–789.

Insel, T. R., & Shapiro, L. E. (1992). Oxytocin receptor distribution reflects social organization in monogamous and polygamous voles. *Proceedings of the National Academy of Sciences USA, 89,* 5981–5985.

Insel, T. R., Wang, Z. X., & Ferris, C. F. (1994). Patterns of brain vasopressin receptor distribution associated with social organization in microtine rodents. *Journal of Neuroscience, 14*(9), 381–392.

Insel, T. R., & Young, L. J. (2001). The neurobiology of attachment. *Nature Reviews Neuroscience, 2,* 129–136.

Jacob, S., Garcia, S., Hayreh, D., & McClintock, M. K. (2002). Psychological effects of musky compounds: Comparison of androstadienone with androstenol and muscone. *Hormones and Behavior, 42*(3), 274–283.

Jankoviak, W. R., & Fischer, E. F. (1992). A cross-cultural perspective on romantic love. *Ethology, 31,* 149–155.

Jenike, M. A. (1990). Illnesses related to obsessive–compulsive disorder. In M. A. Jenike, L. B. Baer, &

W. E. Minichiello (Eds.), *Obsessive–compulsive disorders: Theory and management* (pp. 39–60). Chicago: Year Book Medical.

Kane, J. M. (1987). Treatment of schizophrenia. *Schizophrenia Bulletin, 13,* 133–156.

Keverne, E. B. (2002). Pheromones, vomeronasal function, and gender-specific behavior. *Cell, 108*(6), 735–738.

Keverne, E. B., & Kendrick, K. M. (1992). Oxytocin facilitation of maternal behavior in sheep. *Annals of the New York Academy of Sciences, 652,* 83–101.

Kirsch, P., Esslinger, C., Chen, Q., Mier, D., Lis, S., Siddhanti, S., et al. (2005). Oxytocin modulates neural circuitry for social cognition and fear in humans. *Journal of Neuroscience, 25,* 11489–11493.

Kosfeld, M., Heinrichs, M., Zack, P. J., Fischbacher, U., & Fehr, E. (2005). Oxytocin increases trust in humans. *Nature, 435,* 673–676.

Leckman, J. F., Mayes, L. C., Feldman, R., Evans, D. W., King, R. A., & Cohen, D. J. (1999). Early parental preoccupations and behaviors and their possible relationship to the symptoms of obsessive–compulsive disorder. *Acta Psychiatrica Scandinavica, 100*(Suppl. 396), 1–26.

LeDoux, J. E. (2000). Emotion circuits in the brain. *Annual Review of Neuroscience, 23,* 155–184.

LeDoux, J. E., Iwata, J., Cicchetti, P., & Reis, D. J. (1988). Different projections of the central amygdaloid nucleus mediate autonomic and behavioral correlates of conditioned fear. *Journal of Neuroscience, 8,* 2517–2529.

Lesch, K. P., & Mossner, R. (1998). Genetically driven variation in serotonin uptake: Is there a link to affective spectrum, neurodevelopmental, and neurodegenerative disorders? *Biological Psychiatry, 1,* 179–192.

Liebowitz, M. R. (1983). *The chemistry of love.* Boston: Little, Brown.

Marazziti, D. (2002). *La natura dell'amore.* Milan, Italy: Rizzoli.

Marazziti, D., Akiskal, H. S., Rossi, A., & Cassano, G. B. (1999). Alteration of the platelet serotonin transporter in romantic love. *Psychological Medicine, 29,* 741–745.

Marazziti, D., Bani, A., Casamassima, F., Catena, M., Consoli, G., Gesi, C., et al. (2006a) Oxytocin: An old hormone for new avenues. *Clinical Neuropsychiatry, 3,* 302–321.

Marazziti, D., & Canale, D. (2004). Hormonal changes when falling in love. *Psychoneuroendocrinology, 29,* 931–936.

Marazziti, D., & Cassano, G. B. (2003). The neurobiology of attraction. *Journal of Endocrinological Investigation, 26,* 58–61.

Marazziti, D., Dell'Osso, B., Baroni, S., Mungai, F., Catena, M., Pucci, P., et al. (2006b). A relationship between oxytocin and anxiety of romantic attachment. *Clinical and Practical Epidemiology of Mental Health, 2,* 28–32.

McCoy, N. L., & Pitino, L. (2002). Pheromonal influences on socio-sexual behavior in young women. *Physiology and Behavior, 75,* 367–375.

Meyer-Lindenberg, A., Hariri, A. R., Munoz, K. E., Mervis, C. B., Mattay, V. S., Morris, C. A., et al. (2005). Neural correlates of genetically abnormal social cognition in Williams syndrome. *Nature Neuroscience, 8,* 991–993.

Murphy, C. R., & Dwarte, D. M. (1987). Increase in cholesterol in the apical plasma membrane of uterine epithelial cells during early pregnancy in the rat. *Acta Anatomica, 128,* 76–79.

Netter, A. (1989). The agony of passion. *Revue Française de Psychanalyse, 2,* 33–39.

O'Dwyer, A. M., & Marks, I. (2000). Obsessive–compulsive disorder and delusions revisited. *British Journal of Psychiatry, 174,* 281–284.

Ostrowski, N. L. (1998). Oxytocin receptor mRNA expression in rat brain: Implications for behavioral integration and reproductive success. *Psychoneuroendocrinology, 23,* 989–1004.

Panksepp, J. (1982). Toward a psychobiological theory of emotions. *Behavioural Brain Research, 5,* 407–467.

Panksepp, J. (1992). Oxytocin effects on emotional processes: Separation distress, social bonding and relationships to psychiatric disorders. *Annals of the New York Academy of Sciences, 652,* 243–252.

Pitman, R. K., Orr, S. P., & Lasko, N. B. (1993). Effects of intranasal vasopressin and oxytocin on physiologic responding during personal combat imagery in Vietnam veterans with posttraumatic stress disorder. *Psychiatry Research, 48,* 107–117.

Popik, P., & Van Ree, J. M. (1993). Social transmission of flavored tea preferences: Facilitation by a vasopressin analog and oxytocin. *Behavioral and Neural Biology, 59,* 63–68.

Porges, S. W. (1997). Emotions: An evolutionary by-product of the neural regulation of the autonomous nervous system. *Annals of the New York Academy of Sciences, 807,* 62–77.

Prather, M. D., Lavenex, P., Mauldin-Jourdain, M. L., Mason, W. A., Capitanio, J. P., Mendoza, S. P., et al. (2001). Increased social fear and decreased fear of objects in monkeys with neonatal amygdala lesions. *Neuroscience, 106,* 653–658.

Tribollet, E., Charpak, S., Schmidt, A., Dubois-Dauphin, M., & Dreifuss, J. J. (1989). Appearance and transient expression of oxytocin receptors in fetal, infant and peripubertal rat brain studies by autoradiography and electrophysiology. *Journal of Neuroscience, 9,* 1764–1773.

Turner, R. A., Altemus, M., Yip, D. N., Kupferman, E., Fletcher, D., Bostrom, A., et al. (2002). Effects of emotion on oxytocin, prolactin, and ACTH in women. *Stress, 5*(4), 269–276.

Uvnäs-Moberg, K. (1997). Oxytocin linked antistress effects: The relaxation and growth response. *Acta Physiologica Scandinavica, 161*(Suppl. 640), 38–42.

Uvnäs-Moberg, K. (1998). Oxytocin may mediate the benefit of positive social interaction and emotions. *Psychoneuroendocrinology, 23,* 819–835.

Uvnäs-Moberg, K., Arn, I., Theorell, T., & Jonsson, C. O. (1991). Personality traits in a group of individuals with functional disorders of the gastrointestinal tract and their correlation with gastrin, somatostatin and oxytocin levels. *Journal of Psychosomatic Research, 35,* 515–523.

Van Schaik, C. P., & De Visser, J. A. (1990). Fragile sons or harassed daughters?: Sex differences in mortality among juvenile primates. *Folia Primatolologica, 55,* 10–23.

Wakerley, J. B., & Lincoln, D. W. (1973). The milk-ejection reflex of the rat: A 20- to 40-fold acceleration in the firing of paraventricular neurones during oxytocin release. *Journal of Endocrinology, 57,* 477–493.

Weller, A. (1998). Communication through body odour. *Nature, 392,* 126–127.

Winslow, J. T., Hastings, N., Carter, C. S., Harbaugh, C. R., & Insel, T. R. (1993). A role for central vasopressin in pair bonding in monogamous prairie voles. *Nature, 365,* 545–548.

Young, L. J. (2002). The neurobiology of social recognition, approach and avoidance. *Biological Psychiatry, 51,* 18–26.

Young, L. J., Lim, M. M., Gingrich, B., & Insel, T. R. (2001). Cellular mechanisms of social attachment. *Hormones and Behavior, 40,* 133–138.

Young, L. J., & Wang, Z. (2004). The neurobiology of pair bonding. *Nature Neuroscience, 7,* 1048–1056.

Zitzmann, M., & Nieschlag, E. (2001). Testosterone levels in healthy men and the relation to behavioural and physical characteristics: Facts and constructs. *European Journal of Endocrinology, 144*(3), 183–197.

CHAPTER 15

Animal Models
of Romantic Relationships

JOEL C. WOMMACK
YAN LIU
ZUOXIN WANG

In human societies, romantic relationships serve as strong social attachments that help to form the basis of family units. As a result, the establishment of such a relationship affects numerous aspects of an individual's life, including parenting style and the well-being of children. For example, children raised by both parents often have better outcomes than children raised by a single parent do. Therefore, studying the biological mechanisms that regulate romantic relationships provides information not simply limited to pair bonding, but with potential implications for understanding and improving child health and development.

Monogamy is a life strategy in which males and females form long-term mating partnerships, often referred to as "pair bond." This life strategy is observed across various taxa, from birds to fishes (Whiteman & Cote, 2004) to amphibians (Gillette, Jaeger, & Peterson, 2000) to mammals (Kleiman, 1977), although the occurrence of monogamous systems varies. For example, it is estimated that as many as 90% of bird species employ a monogamous life strategy, whereas only 5% of mammals are monogamous (Kleiman, 1977).

The characteristics of monogamous relationships often differ from species to species. As such, the term "monogamy" is not synonymous for all pair-living species. Therefore, a more detailed terminology system has been used to classify distinct categories of monogamy. First, "social monogamy" refers to a situation where a male and a female share a common territory and often spend most of their time in close proximity (Reichard, 2003). Although extra-pair copulations may occur in this system, individuals typically prefer to mate with their partners. "Sexual monogamy" refers to a system in which a male and a female form an exclusive mating pair (Reichard, 2003).

Regardless of whether a species is socially or sexually monogamous, pair-bonded animals engage in a number of common behaviors. Obviously, formation of a preferred mate choice is an important factor in pair bonding. This key aspect of the pair bond is often observed as a strong affiliative relationship between members of the breeding pair, and it requires an ability to form individual recognition and social memories. At the same time, rejection of unfamiliar individuals via mate guarding or territorial defense serves to maintain the established pair bond. These behaviors involve rejecting other potential mating partners, as well as denying access to mating competition. In addition, biparental care is often a component of a pair bond. Unlike promiscuous species, monogamous animals share parental responsibilities with their mates. It has been argued that paternal care enhances the survival of the offspring and leads to increased fitness (Ribble, 1992). As such, each of these behaviors probably contributes to the establishment of monogamy as an evolutionarily adaptive life strategy.

Despite species variations, monogamy is likely to be regulated by similar physiological mechanisms. Particularly among mammalian species, the neurochemical processes regulating pair-bond formation and maintenance show many similarities (Insel & Young, 2000). However, difficulties have arisen in establishing animal models for laboratory research on monogamy. Therefore, careful investigations of a few model species are likely to provide insight into the physiology underlying monogamy in other mammalian species (Curtis & Wang, 2003). As such, the scope of this chapter is limited to a few laboratory rodent species in an attempt to outline the neurobiological mechanisms that regulate pair bonding in mammals. We focus mainly on the prairie vole, as it has been used extensively to investigate the mechanisms involved in pair-bond formation (see also Bales & Carter, Chapter 13, and Marazziti, Chapter 14, this volume). In addition, some available data in other species are discussed.

Prairie Voles as a Model Species

Prairie voles (*Microtus ochrogaster*) are socially monogamous rodents indigenous to the central grasslands of the United States. In their natural environment, male and female prairie voles form pair bonds and share a nest with their offspring (Getz & Hofmann, 1986; Getz, Carter, & Gavish, 1981; Getz, McGuire, Pizzuto, Hofmann, & Frase, 1993). For example, pair-bonded male and female prairie voles share a variety of responsibilities that include nest building, nest guarding, and certain aspects of parental care (e.g., pup retrieval and huddling) (Getz & Carter, 1996; Gruder-Adams & Getz, 1985; Thomas & Birney, 1979). Unlike some species that form monogamous relationship for the duration of a single breeding season, prairie voles form lifelong attachments to their partners (Getz & Carter, 1996). Interestingly, the pair bonds formed by prairie voles are so remarkably enduring in nature that, after the death of a partner, the remaining member of the pair rarely finds a replacement mate (Getz et al., 1981, 1993; Carter, DeVries, & Getz, 1995).

Prairie voles have proven not only to be useful for field observations of monogamous relationships, but also to provide excellent opportunities for laboratory research on social attachment. Although voles may be somewhat of a nontraditional laboratory species (even though they are becoming more and more mainstream), they reliably breed in captivity and require amounts of laboratory maintenance similar to those for other rodent species (Ranson, 2003). Moreover, the various behavioral aspects encompassing a pair bond, such as partner preference, selective aggression, and biparental care, are reliably expressed in captive vole populations (Carter & Getz, 1993; Dewsbury, 1987; Getz & Carter 1996; McGuire & Novak, 1984; Oliveras & Novak, 1986; Thomas & Birney, 1979). Carefully designed tests for these behavioral aspects of monogamy have allowed researchers to uncover the factors necessary for pair-bond formation. For example, a pair-bonded vole spends significantly more time with its mate than with an unfamiliar conspecific. This phenomenon is known as "partner preference" and can easily be tested in the laboratory by giving an animal a "choice test" following experiences, such as mating, that lead to pair-bond formation (Williams, Catania, & Carter, 1992). This behavioral test for partner preference was first developed in C. Sue Carter's laboratory (Williams et al., 1992). Although several other laboratories have now adopted this procedure in unique ways, key aspects of the procedure remain the same. Specifically, partner preference tests utilize a common apparatus consisting of a central chamber connected by tubing to identical side chambers (Figure 15.1A). During the test, a familiar partner and a stranger are tethered in separate side chambers, while the experimental animal is free to move between chambers and to interact with either conspecific. Partner preference is

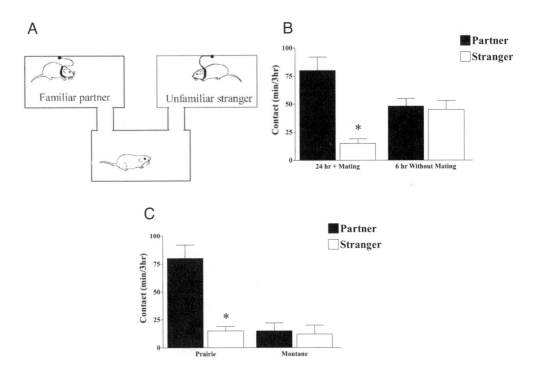

FIGURE 15.1. (A) The standard apparatus for testing partner preferences consists of three cages. A central cage for the experimental animal is connected to two other cages; the familiar partner is tethered in one of these, and a stranger is tethered in the other. This setup gives the experimental animal equal access to each other animal. Typically, partner preference tests last for 3 hours. Later, partner preference formation is assessed by video analysis of side-by-side contact. (B) When animals are allowed 24 hours of cohabitation with mating, they spend more time in side-by-side contact with their partners, indicating partner preference formation. In contrast, 6 hours of cohabitation without mating is insufficient for partner preference formation. (C) Following 24 hours of cohabitation with mating, monogamous prairie voles reliably spend more time with their familiar partners, whereas promiscuous vole species, such as montane voles, do not show partner preference.

observed when the experimental animal spends significantly more time in side-by-side contact with the familiar partner than with the stranger (Williams et al., 1992).

Although prairie voles reliably form partner preferences in the laboratory, certain conditions must be met for this to occur. Following 24 hours of cohabitation and mating, prairie voles display a preference for side-by-side contact with their familiar partners (Williams et al., 1992; Winslow, Hastings, Carter, Harbaugh, & Insel, 1993; Figure 15.1B). In contrast, 6 hours of cohabitation with a heterosexual conspecific in the absence of mating is insufficient to induce partner preference. As such, this behavioral paradigm has been extensively used to investigate physiological mechanisms underlying pair-bond formation (Winslow et al., 1993; Insel & Hulihan, 1995). Importantly, the carefully combined use of the 24- and 6-hour behavioral exposures has allowed researchers to identify the factors that are necessary and sufficient for partner preference formation. For example, one can test the necessity of a factor by manipulating the action of that factor prior to and/or during a 24-hour period of cohabitation with mating. In contrast, by manipulating the action of a factor prior to and/or during a 6-hour period of cohabitation without mating, researchers can identify mechanisms that are sufficient to induce partner preference formation.

Another advantage of the prairie vole as a model species for the study of monogamy is the readiness of comparative studies. Whereas prairie voles are socially monogamous, other closely related vole species employ promiscuous life strategies. For example, meadow voles (*Microtus pennsylvanicus*) and montane voles (*Microtus montanus*) are promiscuous species that engage in different species-specific social behaviors. Compared to prairie voles, these species do not form mating-induced partner preferences (Figure 15.1C), show low levels of affiliative behavior, and do not engage in biparental care (Dewsbury, 1987; Insel & Hulihan, 1995; Jannett, 1982; McGuire & Novak, 1984; Wang & Novak, 1992). Yet, despite these differences in sociality, montane, meadow, and prairie voles show strong similarities in nonsocial behaviors (Tamarin, 1985). As such, careful comparisons between these species have provided numerous insights into the neurobiology of pair bonding.

The Role of Neuropeptides in Pair Bonding

Early investigations into the neural control of monogamous behavior focused on the neuropeptides arginine vasopressin (AVP) and oxytocin (OT; see Marazziti, Chapter 14, this volume, for a discussion of these peptides in human pair bonding). These neuropeptides were speculated to play a role in pair bonding for several reasons. First, AVP and OT are involved in a number of social behaviors, such as sexual behavior (Argiolas et al., 1989; Carter, 1992), aggression (Delville, Mansour, & Ferris, 1996; Ferris & Potegal, 1988), and paternal behaviors (Insel, 1990; Pedersen, Ascher, Prange, & Monroe, 1982)—all of which are commonly displayed by monogamous species. Additional studies on learning and memory provided evidence indicating that these peptides modulate social memories, such as individual recognition (Bluthe, Koob, & Dantzer, 1991; Engelmann, Wotjak, Neuman, Ludwig, & Landgraf, 1996; Ferguson, Young, & Insel, 2002; Le Moal, Dantzer, Michaud, & Koob, 1987; Popik & van Ree, 1998), which plays an important role in pair bonding (Carter, 1992). Because a pair bond is a complex behavioral phenomenon incorporating each of the above-mentioned aspects of behavioral and/or cognitive functions, these peptides were hypothesized to play a strong role in the establishment of a monogamous pair bond.

The initial studies on AVP and OT took a comparative approach to examining species and sex differences in AVP- and OT-producing cells. Interestingly, virtually no species differences have been found regarding AVP and OT cell populations or projections between monogamous and nonmonogamous voles. In addition, males and females have similar numbers of OT cells in the hypothalamus, the bed nucleus of the stria terminalis (BNST), and the medial preoptic area (MPOA) (Wang, Zhou, Hulihan, & Insel, 1996). AVP cells are found in the hypothalamus, BNST, and medial amygdala in vole species (Wang et al., 1996). It is known that AVP-producing cells in the BNST and medial amygdala project to various forebrain regions, such as the lateral septum (LS) and lateral habenular nucleus (LH). Interestingly, key sex differences have been found: Males have significantly more AVP cells within the BNST and medial amygdala, and higher AVP fiber densities in the LS and LH (Bamshad, Novak, & De Vries, 1993, 1994; Wang, Ferris, & De Vries, 1994; Wang et al., 1996), than do females. These sex differences are testosterone-dependent, as cell numbers and fiber densities decrease following gonadectomy (Wang & De Vries, 1993). Moreover, the effect of social experience on the central AVP system has also been found to be sexually dimorphic. Specifically, male, but not female, prairie voles show increased AVP messenger RNA (mRNA) expression within the BNST and decreased AVP-immunoreactive density in the LS following 3 days of mating and cohabitation (Bamshad et al., 1994; Wang et al., 1994). Importantly, the decreased AVP-immunoreactive fiber density within the LS could be indicative of sustained AVP release. Under this interpretation of the results, it was hypothesized that local AVP release could be involved in pair-bond formation.

Significant species differences have been observed for AVP and OT receptor distributions. For example, compared to nonmonogamous montane voles, prairie voles have higher levels of AVP V1a receptor binding and mRNA expression within the BNST, ventral pallidum (VP), and thalamus (Insel, Wang, & Ferris, 1994; Wang, Young, Liu, & Insel, 1997b; Young, Winslow, Nilsen, & Insel, 1997; Figure 15.2). In contrast, montane voles have higher levels of AVP V1a receptor binding and mRNA labeling in the LS and ventromedial hypothalamus than do prairie voles (Insel et al., 1994; Wang et al., 1997b; Young et al., 1997), indicating species differences in regional brain responsiveness to AVP. Similar species differences have also been observed for OT receptors. Specifically, prairie voles have more OT receptors in the nucleus accumbens (NAC), prelimbic cortex, BNST, and lateral portions of the amygdala, while montane voles express higher OT receptor densities within the ventromedial hypothalamus, LS, and cortical amygdala (Insel & Shapiro, 1992; Young, Huot, Nilsen, Wang, & Insel, 1996). These species differences are specific to AVP and OT receptors, as prairie and montane voles have comparable distributions of mu-opioid and benzodiazepine receptors (Insel & Shapiro, 1992). Interestingly, no major sex differences have been observed for AVP and OT receptors, indicating that separate mechanisms regulate receptor distribution and cell and fiber density. Nevertheless, the differences observed between monogamous and nonmonogamous vole species indicate potential involvement of these neuropeptides in species-specific social behaviors.

Indeed, pharmacological studies have provided evidence that both AVP and OT are directly involved in pair-bond formation. In one study of male prairie voles, intracerebroventricular (ICV) administration of an AVP V1a receptor antagonist blocked partner preference formation induced by 24 hours of mating and cohabitation with a female (Winslow et al., 1993). In contrast, ICV administration of an AVP V1a receptor agonist induced partner preference formation and enhanced selective aggression toward unfamiliar females in male voles that cohabitated with a female for 6 hours without mating (Winslow et al., 1993). In this study, neither OT receptor agonists nor antagonists were found to affect partner preference formation or selective aggression in male prairie voles. However, later studies in females showed

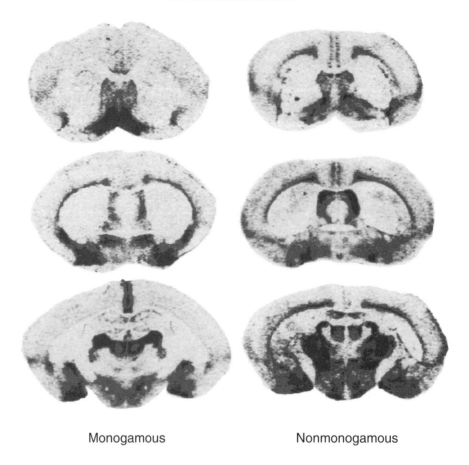

Monogamous   Nonmonogamous

FIGURE 15.2. AVP V1a receptor distributions differ between monogamous and nonmonogamous vole species. Photomicrographs of V1a receptor binding show species differences between prairie and montane voles in numerous forebrain areas.

that OT, but not AVP, controlled their partner preference formation (Insel & Hulihan, 1995; Williams, Insel, Harbaugh, & Carter, 1994). When coupled with the sex differences in AVP neuroanatomy, these results led to the hypothesis that the neuropeptidergic control of partner preference formation is sexually dimorphic, with AVP controlling partner preference in males and OT regulating the same behavior in females. Additional studies, however, disproved this initial hypothesis by expanding the range of drug doses, altering methods of drug delivery, and modifying the behavioral paradigms (Cho, DeVries, Williams, & Carter, 1999; Liu, Curtis, & Wang, 2001). For example, site-specific administration of an OT antagonist into the LS blocks partner preference in male prairie voles (Liu et al., 2001). It is currently understood that AVP and OT affect pair bonding in both male and female prairie voles, with males being more responsive to AVP and females more responsive to OT.

A number of studies have investigated the roles of AVP and OT in specific brain areas. Previous studies showed that mating decreases the density of AVP fibers in the LS in male prairie voles and increases AVP mRNA expression in the BNST, an area projecting to the LS (Bamshad et al., 1994; Wang et al., 1994). As this is indicative of enhanced AVP release in the

LS, it was hypothesized that released AVP in the LS plays an important role in pair-bond formation (Wang, Young, De Vries, & Insel, 1998). This hypothesis was confirmed, as infusions of AVP into the LS induce partner preference formation following 6 hours of cohabitation (Figure 15.3A), whereas microinjections of an AVP V1a antagonist into the LS block partner preference formation induced by 24 hours of mating in male prairie voles (Figure 15.3B; Liu et al., 2001). Interestingly, microinjections of AVP into the LS of other rodent species enhance social memory formation (Bluthe et al., 1991; Engelmann et al., 1996; Ferguson et al., 2002; Le Moal et al., 1987; Popik & van Ree, 1998). In addition, c-fos, a marker of neuronal activation, is increased within the LS during mating-induced selective aggression toward conspecific strangers (Wang, Hulihan, & Insel, 1997a; Gobrogge, Liu, Jia, & Wang, 2007; Figure 15.3C). Together, these data suggest that AVP in the LS plays a role in (1) pair-bond formation by establishing partner recognition, and (2) pair-bond maintenance via its involvement in mating-induced selective aggression. Moreover, microinjection of an OT receptor antagonist into the LS blocks partner preference formation in male prairie voles (Liu et al., 2001). This

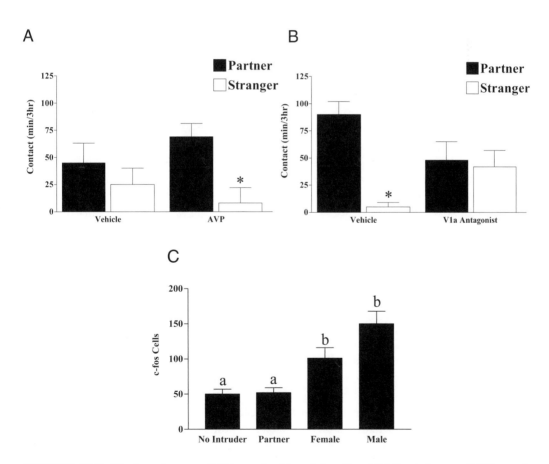

FIGURE 15.3. The lateral septum (LS) is involved in pair bonding. (A) Microinfusions of AVP into the LS can induce pair-bond formation following 6 hours of cohabitation without mating (B). The effects of AVP are mediated through V1a receptors, as microinfusion of a V1a antagonist into the LS blocks partner preference formation following 24 hours of cohabitation with mating. (C) In addition, social stimuli alter neuronal activation, as shown through c-fos labeling, within the LS.

latter finding not only underscores the importance of the LS in the control of pair bonding, but also presents the possibility of a site-specific interaction between AVP and OT.

Additional studies have shown that AVP and OT also regulate pair-bond formation in numerous other brain areas. For example, partner preference formation in female prairie voles is blocked by administration of an OT antagonist into the NAC and medial prefrontal cortex (mPFC) (Young, Lim, Gingrich, & Insel, 2001; Liu & Wang, 2003). Administration of an AVP V1a receptor antagonist into the VP blocks partner preference formation in male prairie voles (Lim & Young, 2004). Because pair bonding is a complex behavioral phenomenon, it is to be expected that multiple brain regions are involved in pair bonding. Future studies will be required to build a more global, circuit-based understanding of how AVP and OT regulate monogamy-associated behaviors.

Other approaches have been used to investigate the genetic mechanisms underlying pair bonding. As previously mentioned, monogamous prairie voles show more V1a receptor binding and mRNA expression in the VP than do promiscuous meadow or montane voles (Insel et al., 1994; Wang et al., 1997b; Young et al., 1997). Genetic analysis indicates that this species difference in V1a receptor expression may have a genetic basis. Specifically, monogamous vole species have a number of repetitive microsatellite DNA in the 5′ promotor region of the V1a receptor gene (Young, Nilsen, Waymire, MacGregor, & Insel, 1999). Interestingly, transgenic mice that express the prairie vole V1a receptor gene show prairie-vole-like receptor distribution patterns (Young et al., 1999). To test the hypothesis that the differences in the 5′ regulatory region of the V1a receptor were directly related to species differences in reproductive life strategies, V1a receptors were overexpressed via viral vector-mediated transfer in the VP of nonmonogamous meadow voles. The results showed that meadow voles overexpressing the V1a receptor in the VP developed a partner preference in response to AVP administration (Lim et al., 2004). Although it is debatable whether differences in the V1a gene are responsible for monogamy across all mammalian species (Fink, Excoffier, & Heckel, 2006), these data were the first to show that pair bonding can be regulated by a single gene.

The actions of AVP and OT are not limited to regulating prosocial behaviors in adults. These neuropeptides may also exert organizational affects during postnatal neural development (Boer, Quak, de Vries, & Heinsbroek, 1994; Cushing & Kramer, 2005; Shapiro & Insel, 1989; Wang & Young, 1997; Yoshimura, Kimure, Watanabe, & Kiyama, 1996). For example, peripheral administration of either OT or OT antagonists during the early postnatal period produces long-lasting effects on a variety of behaviors, including female sexual behavior (Cushing, Levine, & Cushing, 2005), stress responsiveness (Kramer, Cushing, & Carter, 2003), and partner preference formation (Bales & Carter, 2003). AVP is also likely to influence the development of social behaviors, as male prairie voles neonatally treated with AVP show increased aggression during adulthood (Stribley & Carter, 1999). In addition, early OT treatment results in increased numbers of AVP and OT cells in the paraventricular nucleus of the hypothalamus in juvenile prairie voles, indicating that the behavioral effects and early peptide exposure are related to neural plasticity within these systems (Yamamoto et al., 2004). Early OT treatment also alters estrogen receptor alpha expression in areas involved in social behavior (Yamamoto, Carter, & Cushing 2006). This latter finding suggests that the organizational effects of AVP and OT are more than simply effects of self-regulation. Moreover, the presence of estrogen receptor alpha in select brain regions has been hypothesized to increase prosocial behaviors in voles (Cushing & Kramer, 2005). As such, early activity of the AVP and OT systems can organize the central nervous system and facilitate the expression of the monogamous phenotype.

The Role of Dopamine in Pair Bonding

Dopamine (DA) is a monoaminergic neurotransmitter that regulates a broad range of cognitive and emotional processes, including those involved in social behaviors. For example, DA is involved in aspects of motivation, learning, and memory, as well as in the rewarding effects of natural substances, such as food and drugs of abuse/addiction (Azzara, Bodnar, Delamater, & Sclafani, 2001; Kivastik, Vuorikallas, Piepponen, Zharkovsky, & Ahtee, 1996; Saal, Dong, Bonci, & Malenka, 2003). In addition, this neurotransmitter has been implicated in a number of social contexts, varying from sexual behaviors (Dominguez & Hull, 2005; Pfaus et al., 1990; Pfaus, Damsma, Wenkstern, & Fibiger, 1995) to social stress (Lucas, et al., 2004; Wommack, Salinas, Melloni, & Delville, 2004; Vegas, Beitia, Sanchez-Martin, Arregi, & Azpiroz, 2004) and parental behaviors (Keer & Stern, 1999; Lonstein, 2002). Recently, the DA system has been shown to play a prominent role in the formation and maintenance of a monogamous pair bond in prairie voles (Aragona, Liu, Curtis, Stephan, & Wang, 2003; Aragona et al., 2006; Gingrich, Liu, Cascio, Wang, & Insel, 2000; Wang et al., 1999).

Much of the research on DA's control of pair bonding has stemmed from evidence of this neurotransmitter's involvement in sexual behaviors. Copulation and other reproductive behaviors increase DA levels in various forebrain regions, including areas involved in social interactions and/or reward (Dominguez & Hull, 2005; Pfaus et al., 1990, 1995). Specifically, studies in rats have shown that copulation results in increased DA levels within the MPOA, a region largely associated with social behaviors (Dominguez & Hull, 2005). Copulation also causes increased DA levels within the NAC (Mermelstein & Becker, 1995; Pfaus et al., 1990, 1995), a brain region implicated in learning, memory, and reward (Ikemoto & Panksepp, 1999; Salamone & Correa, 2002; Self, Choi, Simmons, Walker, & Smagula, 2004). When investigators considered the potential contributions of these factors to the formation of a social attachment, it became reasonable to hypothesize that DA could play a strong role in pair-bond formation.

Initial studies showed that mating, an essential factor in pair-bond formation, caused increases in DA activity within the NAC of both male and female prairie voles (Aragona et al., 2003; Gingrich et al., 2000). Moreover, peripheral injections of a nonselective DA receptor antagonist blocked pair-bond formation, while administration of a general DA agonist induced partner preference formation even in the absence of mating (Aragona et al., 2003; Gingrich et al., 2000). Additional studies have shown a more precise and intricate understanding of DA's involvement in pair bonding. First, the effects of DA on pair-bond formation are site-specific to the NAC. Intra-NAC microinjection of a DA antagonist blocks partner preference formation (Figure 15.4A). In contrast, DA agonists induce partner preference formation when microinjected into the NAC but not the dorsal striatum (Figure 15.4B; Aragona et al., 2003). Moreover, the NAC is not a homogeneous area. Rather, this nucleus consists of anatomically and functionally distinct subregions known as the NAC shell and core (Di Chiara, 2002; Zahm, 1999). These subregions are differentially involved in pair bonding, as administration of a DA agonist induces partner preference formation when injected into the shell but not the core (Figure 15.4C). Moreover, careful examination of injection sites has shown that the effects of DA are restricted to the rostral portion of the NAC shell (Aragona et al., 2006).

The effects of DA on partner preference are not limited to the NAC. Rather, these effects are likely to reflect coordinated activity of a larger DA circuit that includes the NAC, the ventral tegmental area (VTA), and the mPFC. For example, the VTA is a hindbrain region

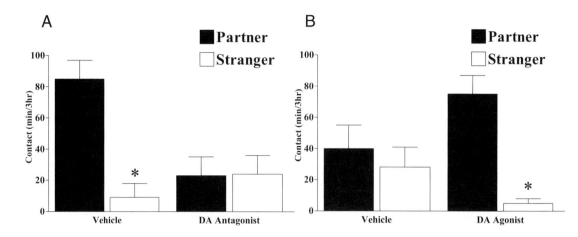

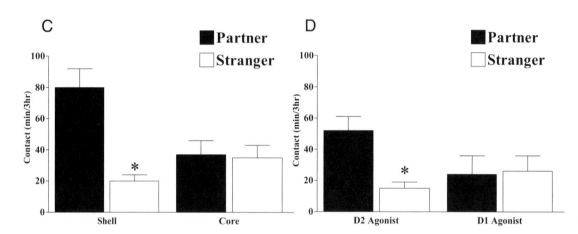

FIGURE 15.4. The effects of dopamine (DA) in the nucleus accumbens (NAC) on partner preference formation are region-specific and differentially regulated by DA receptor subtypes. (A) Microinjection of DA antagonist into the NAC blocks partner preference following 24 hours with mating, whereas (B) microinjection of a DA agonist induces partner preference formation following 6 hours of cohabitation without mating. (C) The effects of DA in the NAC are specific to subregions, as microinjection of a DA agonist induces partner preference formation when administered into the NAC shell but not the core. (D) Finally, the effects of DA are differentially mediated by D1 and D2 receptor subtypes. Microinjection of a D2 agonist into the NAC induces partner preference formation following 6 hours without mating, whereas treatment with a D1 agonist blocks partner preference formation following 24 hours with mating.

that sends DAergic projections to the NAC and the mPFC (Swanson, 1982). Pharmacological manipulations that reduce the inhibitory control of the VTA by blocking glutamate or gamma-aminobutyric acid (GABA) receptors induce pair-bond formation even in the absence of mating (Curtis & Wang, 2005a). The mPFC is a cortical region that also receives DAergic projections for the VTA (Swanson, 1982). Importantly, the mPFC shows considerable anatomical and functional connectivity with the mesolimbic DA system (e.g., the VTA and NAC) (Yang, Seamans, & Gorelova, 1999; Tzschentke, 2000; Sesack & Carr, 2002). As the mPFC is reciprocally connected to the VTA and sends projections to the NAC, this area is well positioned to influence mesolimbic DA release and may play a critical role in partner preference formation. Indeed, DA receptor blockade within the mPFC does not inhibit partner preference formation (Gingrich et al., 2000). Instead, DA receptor antagonism in the mPFC can induce pair-bond formation in the absence of mating (Smeltzer, Curtis, Aragona, & Wang, 2005). These suggest that the mesocortical and mesolimbic DA systems function in a coordinated manner during partner preference formation.

The effects of DA on partner preference formation are differentially regulated by distinct DA receptor subtypes. Briefly, DA receptors are distinguishable by their effects on intracellular signaling pathways. Activation of the D1 receptor family increases cyclic adenosine monophosphate (cAMP) activity, while D2 receptors inhibit cAMP activity (Missale, Nash, Robinson, Jaber, & Caron, 1998). These opposing actions create the potential for differential involvement in neural processing (West & Grace, 2002). Studies focusing on the NAC have shown that the effects of DA on pair-bond formation are primarily regulated by D2 receptors. Site-specific administration of a D2 receptor antagonist into the NAC blocks partner preference formation (Gingrich et al., 2000). Conversely, microinjection of a D2 agonist into the NAC induces partner preference formation (Gingrich et al., 2000; Aragona et al., 2006; Figure 15.4D). Together, these data show a positive relationship between D2 receptor activation and partner preference formation. D1 receptor activation, on the other hand, is negatively related to partner preference formation (Aragona et al., 2006). When concurrently administered into the NAC, D1 agonists block the effects of D2 agonists on partner preference formation, and D1 receptor activation alone blocks partner preference formation induced by mating (Aragona et al., 2006). Following pair-bond formation, male prairie voles become highly aggressive toward unfamiliar females (Carter et al., 1995). This phenomenon, known as "mate guarding," is thought to be important for preserving the pair bond, as it leads to the rejection of potential new mates. Interestingly, increased D1 receptor binding in the NAC was correlated with an increased level of selective aggression in male prairie voles that were pair-bonded for 2 weeks, and blockade of these D1 receptors inhibited selective aggression, suggesting that DA is also important for maintaining the pair bond (Aragona et al., 2006).

Interactions between DA and Neuropeptides

Species-specific interactions between DA and other neural modulators are found to be essential for pair bonding (Wang & Aragona, 2004; Curtis, Liu, Aragona, & Wang, 2006). For example, sexual behavior increases NAC DA not only in monogamous prairie voles, but in promiscuous rodent species, such as rats (Mermelstein & Becker, 1995; Pfaus et al., 1990, 1995). However, prairie voles show considerably higher OT receptor binding in the NAC than promiscuous meadow voles do (Insel & Shapiro, 1992). Blockade of either DA or OT receptors in the NAC inhibits partner preference formation (Young et al., 2001; Aragona et al., 2006). Moreover, activation of both DA and OT receptors within the NAC is required for partner preference formation (Liu & Wang, 2003). This interaction is probably not limited to the NAC. The mPFC also shows species differences in DA and OT receptor binding (Smeltzer, Curtis, Aragona, & Wang, 2006). As both DA and OT have been shown to regulate partner preference formation at the level of the mPFC, it is possible that these two factors also interact in this brain region (Young et al., 2001; Smeltzer et al., 2005).

DA also interacts with AVP to produce pair bonds. Prairie voles show more AVP V1a receptor binding in the VP than promiscuous voles do (Young et al., 1997). AVP V1a overexpression within the VP of meadow voles facilitates partner preference formation (Lim et al., 2004). Interestingly, this effect is blocked when animals are pretreated with a D2-specific antagonist prior to cohabitation and mating, indicating interactions between the DA and AVP systems in the regulation of pair bonding (Lim et al., 2004). The precise nature of this interaction is not yet fully understood. The VP is influenced by DAergic transmission in the NAC and also receives direct DA projections from the VTA (McBride, Murphy, & Ikemoto, 1999). Therefore, future investigations will be required to determine whether the interaction between DA and AVP is direct or indirect in this area.

Stress and Pair Bonding

Studies have also shown that pair bonding is influenced by stress. Corticosterone, the primary glucocorticoid in prairie voles (Hastings, Orchinik, Aubourg, & McEwen, 1999), has been shown to exert sexually dimorphic effects on partner preference formation (DeVries, DeVries, Taymans, & Carter, 1995, 1996). In males, adrenalectomy inhibits partner preference formation, whereas corticosterone treatment induces it (DeVries et al., 1996). The same study reported that corticosterone produces the opposite effect on partner preference formation in females. In addition, corticosteroid receptor antagonists have been found to induce partner preference in female prairie voles (Curtis & Wang, 2005b). Interestingly, this effect can be reversed by treatment with DA receptor antagonists, indicating potential interactions between the DA system and stress in the regulation of pair bonding (Curtis & Wang, 2005b).

The interaction between stress and pair bonding is not limited to corticosteroids. Corticotropin-releasing factor (CRF) is a stress-related neuropeptide that has been shown to regulate reward- and anxiety-related behaviors, as well as social interactions (Bale & Vale, 2004; Farrokhi et al., 2004; Jasnow, Banks, Owens, & Huhman, 1999; Sillaber et al., 2002). In male prairie voles, ICV administration of CRF induces partner preference formation without affecting locomotor or anxiety-related behaviors (DeVries, Guptaa, Cardillo, Cho, & Carter, 2002). A more recent study showed that selective CRF type I and type II receptor agonists can induce partner preference in male prairie voles site-specifically within the NAC (Lim et al., 2007). Species differences between monogamous and nonmonogamous voles have been reported for CRF receptor densities in the NAC (Lim, Nair, & Young, 2005), as they have for AVP and OT.

California Mice as a Model Species

In addition to voles, the genus *Peromyscus*, commonly referred to as "deer mice," are also useful model species for the study of monogamy. This genus includes more than 50 different species that exhibit a variety of reproductive strategies, including promiscuity and social and sexual monogamy (McCabe & Blanchard, 1950; Ribble, 2003). The California mouse (*Peromyscus californicus*) is a well-studied member of this genus that exhibits social and sexual monogamy, biparental care, and selective aggression (Gubernick & Nordby, 1993; Ribble, 2003). DNA analysis from field animals has shown that extra-pair fertilizations do not exist in this species (Ribble, 1991). Pairs in the wild typically remain together until the death of a partner, after which the surviving member will find a new mate (Ribble, 1992).

As voles do, species from this genus reliably reproduce and establish pair bonds in a laboratory setting. Importantly, the high degree of fidelity observed in the field has also been observed in carefully controlled laboratory experiments (Gubernick & Nordby, 1993). When an experimental paradigm similar to the partner preference test for voles was used, it was found that male and female pair-bonded California mice rarely engaged in extra-pair copulations (Gubernick & Nordby, 1993). When tested in the presence or absence of their mates, males did not copulate with a stranger. Instead, the pair-bonded animals spent a great deal of time next to their partners.

The majority of the studies on social behaviors of this highly monogamous species have focused on paternal behavior and aggression. Male California mice invest a great deal of effort in paternal behavior. Reports indicate that males and females engage in equal degrees of parental behaviors, with the exception of nursing (Dudley, 1974a; Gubernick & Alberts, 1987). These behaviors include carrying, licking, and huddling with the pups, and males begin expressing these behaviors on the first day of parturition (Dudley, 1974a; Gubernick & Alberts, 1987). Compared to females, males engage in pup licking with a higher frequency (Gubernick & Alberts, 1987). Males also spend a considerable amount of time carrying pups and building and attending to the nest (Gubernick & Alberts, 1987). Laboratory studies have shown that males continue to engage in these behaviors even in the absence of their females, and that the paternal care they provide significantly enhances pup survival rates when the ambient temperature is low (Dudley, 1974b). In contrast, absence of paternal care due to male removal decreases pup survival rates (Gubernick & Teferi, 2000).

Previous studies have investigated the endocrine correlates of paternal behavior in male California mice. For instance, one study found that paternal behavior in this species was inhibited following castration and restored with testosterone replacement, implicating the involvement of androgens in this behavior (Trainor & Marler, 2001). Interestingly, females paired with castrated males displayed more maternal behaviors than females paired with sham-operated males, suggesting that these animals possess a degree of plasticity in maternal behaviors that may allow them to care adequately for their pups when they are paired with suboptimal mates. A follow-up study showed that the effects of testosterone were mediated by its aromatization to estradiol, as estradiol replacement restored paternal behavior in castrated fathers (Trainor & Marler, 2002). Furthermore, a postpartum increase in male prolactin levels has been hypothesized to be involved in the onset of paternal behavior in California mice (Gubernick & Nelson, 1989).

Studies on the neural correlates of paternal behavior have shown that the MPOA is a key brain area. First, the MPOA has been shown to regulate parental behavior in male rats (Rosenblatt & Ceus, 1998). In California mice, the MPOA is larger in females than in males, but this sexual dimorphism is abolished shortly after parturition (Gubernick, Sengelaub, & Kurz, 1993). Additional studies have shown that lesions of this area inhibit paternal behaviors in California mice (Lee & Brown, 2002). Interestingly, the involvement of this area in paternal behaviors may also be mediated by the aromatization of testosterone to estrogen. Male

California mice that were allowed to become paternal showed increased aromatase activity within the MPOA (Trainor, Bird, Alday, Schlinger, & Marler, 2003).

AVP, which has been implicated in the monogamous life strategy and behaviors in prairie voles, has also been shown to be important in the monogamous life strategy in California mice. When compared to a promiscuous *Peromyscus* species (the white-footed mouse, *Peromyscus leucopus*), California mice have higher levels of AVP immunostaining in the BNST and AVP receptor binding in the LS (Bester-Meredith, Young, & Marler, 1999). Moreover, California mice that were cross-fostered and reared by white-footed mice showed decreased AVP immunoreactivity in the BNST in comparison to in-fostered controls (Bester-Meredith & Marler, 2001). These cross-fostered California mice also showed deficiencies in paternal behavior, as evidenced by a decline in pup retrieval (Bester-Meredith & Marler, 2003a). Furthermore, the degree to which pup retrieval was impaired was correlated with AVP immunoreactivity in the BNST, as males with lower levels of AVP staining in this area showed lower levels of paternal care (Bester-Meredith & Marler, 2003b). It has been suggested that paternal care causes changes in AVP within the BNST, which in turn shapes the development of paternal and aggressive behaviors in this species (Frazier, Trainor, Cravens, Whitney, & Marler, 2006). Therefore, species-specific behaviors in this monogamous species may be programmed by experiences during early development.

Although studies in this species have provided useful information about other aspects of pair bonding, such as paternal care, there is considerable potential for this model to provide insights into the mechanisms controlling pair bonding. When combined with data from voles, investigations into pair bonding in California mice could greatly expand our understanding of this topic.

Implications and Future Directions

Voles and deer mice have both proven to be excellent model organisms for investigation of the neurobiology of pair bonding. Reliable laboratory tests and carefully planned experiments have demonstrated roles for several neurotransmitters in the control of this complex social phenomenon. (Table 15.1 summarizes these findings for prairie voles.) Undoubtedly, more studies will be needed to further integrate the various neural pathways involved in pair bonding. These studies will probably use a combination of neuroanatomical, neurochemical, and molecular techniques to understand how certain factors affect pair bonding at the level of the circuit as well as intracellular mechanisms. In addition, further investigation into sex and individual differences will provide insight into the experience- and genetics-based factors that lead to the establishment and maintenance of pair-bond formation (Hammock & Young, 2005; Lim et al., 2004; Phelps & Young, 2003). Another concept that emerges from the DAergic control of pair bonding is its striking parallels to Daergic control of drug abuse and addiction. Both mating and drug use enhance DAergic transmission in the NAC, and repeated or prolonged exposures to these respective stimuli produce lasting changes in the DA system that maintain access to the stimulus. Not surprisingly, it has been hypothesized that love is an addictive disorder (Insel, 2003). Although there are limited data on pair bonding and substance abuse, detailed studies on this topic are currently being conducted.

In all, understanding the basic mechanisms controlling monogamous relationships and investigating their overlap with systems involved in nonsocial behaviors should serve to pro-

TABLE 15.1. Physiological Factors Influencing Partner Preference Formation in Prairie Voles

Physiological factor	Males	Females	Brain area	References
Cort	↑	↓	N/A	Curtis & Wang (2005b); DeVries et al. (1995, 1996)
AVP	↑	↑	↑ ICV; ↑ LS; ↑ VP	Cho et al. (1999); Liu et al. (2001); Lim et al. (2004); Winslow et al. (1993)
OT	↑	↑	↑ ICV; ↑ NAC; ↑ mPFC	Cho et al. (1999); Liu et al. (2003); Williams et al. (1994); Young et al. (2001)
CRF	↑	?	↑ NAC	DeVries et al. (2002); Lim et al. (2007)
DA	↑	↑	↑ NAC; ↓ mPFC	Aragona et al. (2003, 2006); Gingrich et al. (2000); Smeltzer et al. (2005); Wang et al. (1999)
GABA	↓	?	↓ VTA	Curtis & Wang (2005a)
Glu	↓	?	↓ VTA	Curtis & Wang (2005a)

Note. Upward arrows indicate a positive relationship between the factor and partner preference formation; downward arrows indicate a negative relationship. ICV, intracerebroventricular injection; AVP, arginine vasopressin; OT, oxytocin; CRF, corticotropin-releasing factor; DA, dopamine; GABA, gamma-aminobutyric acid; Glu, glutamate; mPFC, medial prefrontal cortex; LS, lateral septum; VP, ventral pallidum; NAC, nucleus accumbens; VTA, ventral tegmental area.

vide valuable insights with direct relevance to issues concerning human behavior and health. Moreover, potential discoveries from this line of research will enable us to understand various aspects of human life, from marriages to parenting strategies. Because these factors greatly influence child health and development, this area of research can be expected to expand our understanding of human social development and improve child outcomes.

References

Aragona, B. J., Liu, Y., Curtis, J. T., Stephan, F. K., & Wang, Z. (2003). A critical role for nucleus accumbens dopamine in partner-preference formation in male prairie voles. *Journal of Neuroscience, 23,* 3483–3490.

Aragona, B. J., Liu, Y., Yu, Y. J., Curtis, J. T., Detwiler, J. M., Insel, T. R., et al. (2006). Nucleus accumbens dopamine differentially mediates the formation and maintenance of monogamous pair bonds. *Nature Neuroscience, 9,* 133–139.

Argiolas, A., Collu, M., D'Aquila, P., Gessa, G. L., Melis, M. R., & Serra, G. (1989). Apomorphine stimulation of male copulatory behavior is prevented by the oxytocin antagonist d(CH2)5 Tyr(Me)-Orn8-vasotocin in rats. *Pharmacology, Biochemistry and Behavior, 33,* 81–83.

Azzara, A. V., Bodnar, R. J., Delamater, A. R., & Sclafani, A. (2001). D1 but not D2 receptor antagonism blocks the acquisition of flavor preference conditioned by intragastric carbohydrate infusions. *Pharmacology, Biochemistry and Behavior, 68,* 709–720.

Bale, T. L., & Vale, W. W. (2004). CRF and CRF receptors: Role in stress responsivity and other behaviors. *Annual Review of Pharmacology and Toxicology, 44,* 525–557.

Bales, K. L., & Carter, C. S. (2003). Developmental exposure to oxytocin facilitates partner preferences in male prairie voles. *Behavioral Neuroscience, 117,* 854–859.

Bamshad, M., Novak, M. A., & De Vries, G. J. (1993). Sex and species differences in the vasopressin innervation of sexually naïve and parental prairie voles, *Microtus ochrogaster* and meadow voles, *Microtus pennsylvanicus*. *Journal of Neuroendocrinology, 5,* 247–255.

Bamshad, M., Novak, M., & De Vries, G. J. (1994). Cohabitation alters vasopressin innervation and paternal behavior in prairie voles (*Microtus ochrogaster*). *Physiology and Behavior, 56,* 751–758.

Bester-Meredith, J. K., & Marler, C. A. (2001). Vasopressin and aggression in cross-fostered California mice (*Peromyscus californicus*) and white-footed mice (*Peromyscus leucopus*). *Hormones and Behavior, 40,* 51–64.

Bester-Meredith, J. K., & Marler, C. A. (2003a). The association between male offspring aggression and paternal and maternal behavior of *Peromyscus* mice. *Ethology, 109,* 797–808.

Bester-Meredith, J. K., & Marler, C. A. (2003b). Vasopressin and the transmission of paternal behavior across generations in mated, cross-fostered *Peromyscus* mice. *Behavioral Neuroscience, 117,* 455–463.

Bester-Meredith, J. K., Young, L. J., & Marler, C. A. (1999). Species differences in paternal behavior and aggression and their associations with vasopressin immunoreactivity and receptors. *Hormones and Behavior, 36,* 25–38.

Bluthe, R. M., Koob, G. F., & Dantzer, R. (1991). Hypertonic saline mimics the effects of vasopressin and social recognition in rats. *Behavioral Pharmacology, 2,* 513–516.

Boer, G. J., Quak, J., de Vries, M. C., & Heinsbroek, R. P. W. (1994). Mild sustained effects of neonatal vasopressin and oxytocin treatment on brain growth and behavior of the rat. *Peptides, 15,* 229–236.

Carter, C. S. (1992). Oxytocin and sexual behavior. *Neuroscience and Biobehavioral Reviews, 16,* 131–144.

Carter, C. S., DeVries, A. C., & Getz, L. L. (1995). Physiological substrates of mammalian monogamy: The prairie vole model. *Neuroscience and Biobehavioral Reviews, 19,* 303–314.

Carter, C. S., & Getz, L. L. (1993). Monogamy and the prairie vole. *Scientific American, 268,* 100–106.

Cho, M. M., DeVries, A. C., Williams, J. R., & Carter, C. S. (1999). The effects of oxytocin and vasopressin on partner preference formation in male and female prairie voles (*Microtus ochrogaster*). *Behavioral Neuroscience, 113,* 1071–1079.

Curtis, J. T., Liu, Y., Aragona, B. J., & Wang, Z. (2006). Dopamine and monogamy. *Brain Research, 1126,* 76–90.

Curtis, J. T., & Wang, Z. (2003). The neurochemistry of pair bonding. *Current Directions in Psychological Science, 12,* 49–53.

Curtis, J. T., & Wang, Z. (2005a). Ventral tegmental area involvement in pair bonding in male prairie voles. *Physiology and Behavior, 86,* 338–346.

Curtis, J. T., & Wang, Z. (2005b). Glucocorticoid receptor involvement in pair bonding in female prairie voles: The effects of acute blockade and interactions with central dopamine reward systems. *Neuroscience, 134,* 369–376.

Cushing, B. S., & Kramer, K. M. (2005). Mechanisms underlying epigenetic effects of early social experience: The role of neuropeptides and steroids. *Neuroscience and Biobehavioral Reviews, 29,* 1089–1105.

Cushing, B. S., Levine, K., & Cushing, N. L. (2005). Neonatal manipulation of oxytocin affects female reproductive behavior and success. *Hormones and Behavior, 47,* 22–28.

Delville, Y., Mansour, K. M., & Ferris, C. F. (1996). Testosterone facilitates aggression by modulating vasopressin receptors in the hypothalamus. *Physiology and Behavior, 60,* 25–29.

DeVries, A. C., DeVries, M. B., Taymans, S., & Carter, C. S. (1995). Modulation of pair bonding in female prairie voles (*Microtus ochrogaster*) by corticosterone. *Proceedings of the National Academy of Sciences USA, 92,* 7744–7748.

DeVries, A. C., DeVries, M. B., Taymans, S. E., & Carter, C. S. (1996). The effects of stress on social preferences are sexually dimorphic in prairie voles. *Proceedings of the National Academy of Sciences USA, 93,* 11980–11984.

DeVries, A. C., Guptaa, T., Cardillo, S., Cho, M., & Carter, C. S. (2002). Corticotropin-releasing factor induces social preference in male prairie voles. *Psychoneuroendocrinology, 27,* 705–714.

Dewsbury, D. A. (1987). The comparative psychology on monogamy. *Nebraska Symposium on Motivation, 35,* 1–50.

Di Chiara, G. (2002). Nucleus accumbens shell and core dopamine: Differential role in behavior and addiction. *Behavioural Brain Research, 137,* 75–114.

Dominguez, J. M., & Hull, E. M. (2005). Dopamine, the medial preoptic area, and male sexual behavior. *Physiology and Behavior, 86,* 356–368.

Dudley, D. (1974a). Contributions of paternal care to the growth and development of the young in *Peromyscus californicus. Behavioral Biology, 11,* 155–166.

Dudley, D. (1974b). Paternal behavior in the California mouse, *Peromyscus californicus. Behavioral Biology, 11,* 247–252.

Engelmann, M., Wotjak, C. T., Neuman, I., Ludwig, M., & Landgraf, R. (1996). Behavioral consequences of intracerebral vasopressin and oxytocin: Focus on learning and memory. *Neuroscience and Biobehavioral Reviews, 20,* 341–358.

Farrokhi, C., Blanchard, D. C., Griebel, G., Yang, M., Gonzales, C., Markham, C., et al. (2004). Effects of CRF1 antagonist SSR12543a on aggressive behavior in hamsters. *Pharmacology, Biochemistry and Behavior, 77,* 465–469.

Ferguson, J. N., Young, L. J., & Insel, T. R. (2002). The neuroendorine basis of social recognition. *Frontiers in Neuroendocrinology, 23,* 200–224.

Ferris, C. F., & Potegal, M. (1988). Vasopressin receptor blockade in the anterior hypothalamus suppresses aggression in hamsters. *Physiology and Behavior, 44,* 235–239.

Fink, S., Excoffier, L., & Heckel, G. (2006). Mammalian monogamy is not controlled by a single gene. *Proceeding of the National Academy of Sciences USA, 103,* 10956–10960.

Frazier, C. R., Trainor, B. C., Cravens, C. J., Whitney, T. K., & Marler, C. A. (2006). Paternal behavior influences development of aggression and vasopressin expression in male California mice offspring. *Hormones and Behavior, 50,* 699–707.

Getz, L. L., & Carter, C. S. (1996). Prairie-vole partnerships. *American Scientist, 84,* 56–62.

Getz, L. L., Carter, C. S., & Gavish, L. (1981). The mating system of the prairie vole *Microtus ochrogaster*: Field and laboratory evidence for pair bonding. *Behavioral Ecology and Sociobiology, 8,* 189–194.

Getz, L. L., & Hofmann, J. E. (1986). Social organization in free living prairie voles, *Microtus ochrogaster. Behavioral Ecology and Sociobiology, 18,* 275–282.

Getz, L. L., McGuire, B., Pizzuto, T., Hofmann, J. E., & Frase, B. (1993). Social organization of the prairie vole (*Microtus ochrogaster*). *Journal of Mammalogy, 74,* 44–58.

Gillette, J. R., Jaeger, R. G., & Peterson, M. G. (2000). Social monogamy in a territorial salamander. *Animal Behavior, 59,* 1241–1250.

Gingrich, B., Liu, Y., Cascio, C., Wang, Z., & Insel, T. R. (2000). Dopamine D2 receptors in the nucleus accumbens are important for social attachment in female prairie voles (*Microtus ochrogaster*). *Behavioral Neuroscience, 114,* 173–183.

Gobrogge, K. L., Liu, Y., Jia, X., & Wang, Z. (2007). Anterior hypothalamic neural activation and neurochemical associations with aggression in pair bonded male prairie voles. *Journal of Comparative Neurology, 502,* 1109–1122.

Gruder-Adams, S., & Getz, L. L. (1985). Comparisons of mating systems and parental behavior in *Microtus ochrogaster* and *Microtus pennsylvanicus. Journal of Mammalogy, 66,* 165–167.

Gubernick, D. J., & Alberts, J. R. (1987). The biparental care system of the California mouse, *Peromyscus californicus. Journal of Comparative Psychology, 101,* 169–177.

Gubernick, D. J., & Nelson, R. J. (1989). Prolactin and paternal behavior in the biparental California mouse, *Peromyscus californicus. Hormones and Behavior, 23,* 203–210.

Gubernick, D. J., & Nordby, J. C. (1993). Mechanisms of sexual fidelity in the monogamous California mouse, *Peromyscus californicus*. *Behavioral Ecology and Sociobiology, 32,* 211–219.

Gubernick, D. J., Sengelaub, D. R., & Kurz, E. M. (1993). A neuroanatomical correlate of paternal and maternal behavior in the biparental California mouse (*Peromyscus californicus*). *Behavioral Neuroscience, 107,* 194–201.

Gubernick, D. J., & Teferi, T. (2000). Adaptive significance of male parental care in a monogamous mammal. *Proceedings of the Royal Society of London, Series B, Biological Sciences, 267,* 147–150.

Hammock, E. A., & Young, L. J. (2005). Microsatellite instability generates diversity in brain and sociobehavioral traits. *Science, 308,* 1630–1634.

Hastings, N. B., Orchinik, M., Aubourg, M. V., & McEwen, B. S. (1999). Pharmacological characterization of central and peripheral type I and type II adrenal steroid receptors in the prairie vole, a glucocorticoid-resistant rodent. *Endocrinology, 140,* 4459–4469.

Ikemoto, S., & Panksepp, J. (1999). The role of nucleus accumbens dopamine in motivated behavior: A unifying interpretation with special reference to reward-seeking. *Brain Research: Brain Research Reviews, 31,* 6–41.

Insel, T. R. (1990). Oxytocin and maternal behavior. In K. A. Krasnegor & B. S. Bridges (Eds.), *Mammalian parenting* (pp. 260–279). New York: Oxford University Press.

Insel, T. R. (2003). Is social attachment an addictive disorder? *Physiology and Behavior, 79,* 351–357.

Insel, T. R., & Hulihan, T. J. (1995). A gender-specific mechanism for pair bonding: Oxytocin and partner preference formation in monogamous voles. *Behavioral Neuroscience, 109,* 782–789.

Insel, T. R., & Shapiro, L. E. (1992). Oxytocin receptor distribution reflects social organization in monogamous and promiscuous voles. *Proceedings of the National Academy of Sciences USA, 89,* 5981–5985.

Insel, T. R., Wang, Z., & Ferris, C. F. (1994). Patterns of brain vasopressin receptor distribution associated with social organization in microtine rodents. *Journal of Neuroscience, 14,* 5381–5392.

Insel, T. R., & Young, L. J. (2000). Neuropeptides and the evolution of social behavior. *Current Opinion in Neurobiology, 10,* 784–789.

Jannett, F. J. (1982). Nesting patterns of adult voles, *Microtus montanus,* in field populations. *Journal of Mammalogy, 63,* 495–498.

Jasnow, A. M., Banks, M. C., Owens, E. C., & Huhman, K. L. (1999). Differential effects of two corticotropin-releasing factor antagonists on conditioned defeat in male Syrian hamsters (*Mesocricetus auratus*). *Brain Research, 846,* 122–128.

Keer, S. E., & Stern, J. M. (1999). Dopamine receptor blockade in the nucleus accumbens inhibits maternal retrieval and licking, but enhances nursing behavior in lactating rats. *Physiology and Behavior, 67,* 659–669.

Kivastik, T., Vuorikallas, K., Piepponen, T. P., Zharkovsky, A., & Ahtee, L. (1996). Morphine- and cocaine-induced conditioned place preference: Effects of quinpirole and preclamol. *Pharmacology, Biochemistry and Behavior, 54,* 787–792.

Klciman, D. G. (1977). Monogamy in mammals. *Quarterly Review of Biology, 52,* 39–69.

Kramer, K. M., Cushing, B. S., & Carter, C. S. (2003). Developmental effects of oxytocin on stress response: Acute versus repeated exposure. *Physiology and Behavior, 79,* 775–782.

Le Moal, F., Dantzer, R., Michaud, B., & Koob, G. F. (1987). Centrally injected arginine vasopressin (AVP) facilitates social memory in rats. *Neuroscience Letters, 77,* 353–359.

Lee, A. W., & Brown, R. E. (2002). Medial preoptic lesions disrupt parental behavior in both male and female California mice (*Peromyscus californicus*). *Behavioral Neuroscience, 116,* 968–975.

Lim, M. M., Liu, Y., Ryabinin, A. E., Bai, Y., Wang, Z., & Young, L. J. (2007). CRF receptors in the nucleus accumbens modulate partner preference in prairie voles. *Hormones and Behavior, 51,* 508–515..

Lim, M. M., Nair, H. P., & Young, L. J. (2005). Species and sex differences in brain distribution of corticotropin-releasing factor subtypes 1 and 2 in monogamous and promiscuous vole species. *Journal of Comparative Neurology, 487*, 75–92.

Lim, M. M., Wang, Z., Olazabal, D. E., Ren, X., Terwilliger, E. F., & Young, L. J. (2004). Enhanced partner preference formation in a promiscuous species by manipulating the expression of a single gene. *Nature, 429*, 754–757.

Lim, M. M., & Young, L. J. (2004). Vasopressin-dependent neural circuits underlying pair bond formation in the monogamous prairie vole. *Neuroscience, 125*, 35–45.

Liu, Y., Curtis, J. T., & Wang, Z. (2001). Vasopressin in the lateral septum regulates pair bond formation in male prairie voles (*Microtus ochrogaster*). *Behavioral Neuroscience, 115*, 910–919.

Liu, Y., & Wang, Z. X. (2003). Nucleus accumbens oxytocin and dopamine interact to regulate pair bond formation in female prairie voles. *Neuroscience, 121*, 537–544.

Lonstein, J. S. (2002). Effects of dopamine receptor antagonism with haloperidol on nurturing behavior in the biparental prairie vole. *Pharmacology, Biochemistry and Behavior, 74*, 11–19.

Lucas, L. R., Celen, Z., Tamashiro, K. L., Blanchard, R. J., Blanchard, D. C., Markham, C., et al. (2004). Repeated exposure to social stress has long-term effects on indirect markers of dopaminergic activity in brain regions associated with motivated behavior. *Neuroscience, 124*, 449–457.

McBride, W. J., Murphy, J. M., & Ikemoto, S. (1999). Localization of brain reinforcement mechanisms: Intracranial self-administration and intracranial place-conditioning studies. *Behavioural Brain Research, 101*, 129–152.

McCabe, T. T., & Blanchard, B. D. (1950). *Three species of Peromyscus*. Santa Barbara, CA: Rood.

McGuire, B., & Novak, M. (1984). A comparison of maternal behavior in the meadow vole (*Microtus pennsylvanicus*), prairie vole (*M. ochrogaster*) and pine vole (*M. pinetorum*). *Animal Behaviour, 32*, 1132–1141.

Mermelstein, P. G., & Becker, J. B. (1995). Increased extracellular dopamine in the nucleus accumbens and striatum of the female during paced copulatory behavior. *Behavioral Neuroscience, 109*, 354–365.

Missale, C., Nash, S. R., Robinson, S. W., Jaber, M., & Caron, M. G. (1998). Dopamine receptors: From structure to function. *Physiological Reviews, 78*, 189–225.

Oliveras, D., & Novak, M. (1986). A comparison of paternal behavior in the meadow vole *Microtus pennsylvanicus*, the pine vole *M. pinetorum* and the prairie vole *M. ochrogaster*. *Animal Behaviour, 34*, 519–526.

Pedersen, C. A., Ascher, J. A., Prange, A. J. J., & Monroe, Y. L. (1982). Oxytocin induces maternal behavior in virgin female rats. *Science, 216*, 648–650.

Pfaus, J. G., Damsma, G., Nomikos, G. G., Wenkstern, D. G., Blaha, C. D., Phillips, A. G., et al. (1990). Sexual behavior enhances central dopamine transmission in the male rat. *Brain Research, 530*, 345–348.

Pfaus, J. G., Damsma, G., Wenkstern, D., & Fibiger, H. C. (1995). Sexual activity increases dopamine transmission in the nucleus accumbens and striatum of female rats. *Brain Research, 693*, 21–30.

Phelps, S. M., & Young, L. J. (2003). Extraordinary diversity in vasopressin (V1a) receptor distributions among wild prairie voles (*Microtus ochrogaster*): Patterns of variation and covariation. *Journal of Comparative Neurology, 466*, 564–576.

Popik, P., & van Ree, J. M. (1998). Neurohypophyseal peptides and social recognition in rats. *Progress in Brain Research, 119*, 415–436.

Ranson, R. M. (2003). The field vole (*Microtus*) as a laboratory animal. *Journal of Animal Ecology, 3*, 70–76.

Reichard, U. H. (2003). Monogamy: Past and present. In U. H. Reichard & C. Boesch (Eds.), *Monogamy: Mating strategies and partnerships in birds, humans and other mammals* (pp. 3–25). New York: Cambridge University Press.

Ribble, D. O. (1991). The monogamous mating system of *Peromyscus californicus* as revealed by DNA fingerprinting. *Behavioral Ecology and Sociobiology, 29*, 161–166.

Ribble, D. O. (1992). Lifetime reproductive success and its correlates in the monogamous rodent, *Peromyscus californicus*. *Journal of Animal Ecology, 61*, 457–468.

Ribble, D. O. (2003). The evolution of social and reproductive monogamy in *Peromyscus*: Evidence from *Peromyscus californicus* (the California mouse). In U. H. Reichard & C. Boesch (Eds.), *Monogamy: Mating strategies and partnerships in birds, humans and other mammals* (pp. 81–92). New York: Cambridge University Press.

Rosenblatt, J. S., & Ceus, K. (1998). Estrogen implants in the medial preoptic area stimulate maternal behavior in male rats. *Hormones and Behavior, 33*, 23–30.

Saal, D., Dong, Y., Bonci, A., & Malenka, R. C. (2003). Drugs of abuse and stress trigger a common synaptic adaptation in dopamine neurons. *Neuron, 37*, 577–582.

Salamone, J. D., & Correa, M. (2002). Motivational views of reinforcement: Implications for understanding the behavioral functions of nucleus accumbens dopamine. *Behavioural Brain Research, 137*, 3–25.

Self, D. W., Choi, K. H., Simmons, D., Walker, J. R., & Smagula, C. S. (2004). Extinction training regulates neuroadaptive responses to withdrawal from chronic cocaine administration. *Learning and Memory, 11*, 648–657.

Sesack, S. R., & Carr, D. B. (2002). Selective prefrontal cortex inputs to dopamine cells: Implications for schizophrenia. *Physiology and Behavior, 77*, 513–517.

Shapiro, L. E., & Insel, T. R. (1989). Ontogeny of oxytocin receptors in the rat forebrain: A quantitative study. *Synapse, 4*, 259–266.

Sillaber, I., Rammes, G., Zimmermann, S., Mahal, B., Zieglgansberger, W., Wurst, W., et al. (2002). Enhanced and delayed stress-induced alcohol drinking in mice lacking functional CRH1 receptors. *Science, 296*, 931–933.

Smeltzer, M. D., Curtis, J. T., Aragona, B. J., & Wang, Z. (2005). Oxytocin and dopamine in the medial prefrontal cortex of prairie voles affects social attachment. *Society for Neuroscience Abstracts*, Abstract Viewer No. 420.9.

Smeltzer, M. D., Curtis, J. T., Aragona, B. J., & Wang, Z. (2006). Dopamine, oxytocin, and vasopressin receptor binding in the medial prefrontal cortex of monogamous and promiscuous voles. *Neuroscience Letters, 394*, 146–151.

Stribley, J. M., & Carter, C. S. (1999). Developmental exposure to vasopressin increases aggression in adult prairie voles. *Proceedings of the National Academy of Sciences USA, 96*, 12601–12604.

Swanson, L. W. (1982). The projections of the ventral tegmental area and adjacent regions: A combined fluorescent retrograde tracer and immunofluoresence study in the rat. *Brain Research Bulletin, 9*, 321–353.

Tamarin, R. (Ed.). (1985). *Biology of the New World Microtus* (American Society of Mammalogists Special Publication No. 8). American Society of Mammalogists.

Thomas, J. A., & Birney, E. C. (1979). Parental care and mating system of the prairie vole, *Microtus ochrogaster*. *Behavioral Ecology and Sociobiology, 5*, 171–186.

Trainor, B. C., Bird, I. M., Alday, N. A., Schlinger, B. A., & Marler, C. A. (2003). Variation in aromatase activity in the medial preoptic area and plasma progesterone is associated with the onset of paternal behavior. *Neuroendocrinology, 78*, 36–44.

Trainor, B. C., & Marler, C. A. (2001). Testosterone, paternal behavior, and aggression in monogamous California mouse (*Peromyscus californicus*). *Hormones and Behavior, 40*, 32–42.

Trainor, B. C., & Marler, C. A. (2002). Testosterone promotes paternal behavior in a monogamous mammal via conversion to oestrogen. *Proceedings of the Royal Society of London, Series B, Biological Sciences, 269*, 823–829.

Tzschentke, T. M. (2000). The medial prefrontal cortex as a part of the brain reward system. *Amino Acids, 19*, 211–219.

Vegas, O., Beitia, G., Sanchez-Martin, J. R., Arregi, A., & Azpiroz, A. (2004). Behavioural and neurochemical responses in mice bearing tumors submitted to social stress. *Behavioural Brain Research, 155*, 125–134.

Wang, Z., & Aragona, B. J. (2004). Neurochemical regulation of pair bonding in male prairie voles. *Physiology and Behavior, 83*, 319–328.

Wang, Z., & De Vries, G. J. (1993). Testosterone effects on paternal behavior and vasopressin immunoreactive projections in prairie voles (*Microtus ochrogaster*). *Brain Research, 631*, 156–160.

Wang, Z., Ferris, C. F., & De Vries, G. J. (1994). Role of septal vasopressin innervation in paternal behavior in prairie voles (*Microtus ochrogaster*). *Proceedings of the National Academy of Sciences USA, 91*, 400–404.

Wang, Z., Hulihan, T. J., & Insel, T. R. (1997a). Sexual and social experience is associated with different patterns of behavior and neural activation in male prairie voles. *Brain Research, 767*, 321–332.

Wang, Z., & Novak, M. A. (1992). The influence of social environment on parental behavior and pup development in meadow voles (*Microtus pennsylvanicus*) and prairie voles (*Microtus ochrogaster*). *Journal of Comparative Psychology, 106*, 163–171.

Wang, Z., & Young, L. J. (1997). Ontogeny of oxytocin and vasopressin receptor binding in the lateral septum in the prairie and montane voles. *Developmental Brain Research, 104*, 191–195.

Wang, Z., Young, L. J., De Vries, G. J., & Insel, T. R. (1998). Voles and vasopressin: A review of molecular, cellular and behavioral studies of pair bonding and paternal behavior. *Progress in Brain Research, 119*, 479–495.

Wang, Z., Young, L. J., Liu, Y., & Insel, T. R. (1997b). Species differences in vasopressin receptor binding are evident early in development: Comparative anatomic studies in prairie and montane voles. *Journal of Comparative Neurology, 378*, 535–546.

Wang, Z., Yu, G., Cascio, C., Liu, Y., Gingrich, B., & Insel, T. R. (1999). Dopamine D2 receptor-mediated regulation of partner preferences in female prairie voles (*Microtus ochrogaster*). *Behavioral Neuroscience, 113*, 602–611.

Wang, Z., Zhou, L., Hulihan, T. J., & Insel, T. R. (1996). Immunoreactivity of central vasopressin and oxytocin pathways in microtine rodents: A quantitative comparative study. *Journal of Comparative Neurology, 366*, 726–737.

West, A. R., & Grace, A. A. (2002). Opposite influences of endogenous dopamine D-1 and D-2 receptor activation on activity states and electrophysiological properties of striatal neurons: Studies combining *in vivo* intracellular recordings and reverse microdialysis. *Journal of Neuroscience, 22*, 294–304.

Whiteman, E. A., & Cote, I. M. (2004). Monogamy in marine fishes. *Biological Reviews of the Cambridge Philosophical Society, 79*, 351–375.

Williams, J. R., Catania, K. C., & Carter, C. S. (1992). Development of partner preferences in female prairie voles (*Microtus ochrogaster*): The role of social and sexual experience. *Hormones and Behavior, 26*, 339–349.

Williams, J. R., Insel, T. R., Harbaugh, C. R., & Carter, C. S. (1994). Oxytocin administered centrally facilitates formation of a partner preference in female prairie voles (*Microtus ochrogaster*). *Journal of Neuroendocrinology, 6*, 247–250.

Winslow, J. T., Hastings, N., Carter, C. S., Harbaugh, C. R., & Insel, T. R. (1993). A role for central vasopressin in pair bonding in monogamous prairie voles. *Nature, 365*, 545–548.

Wommack, J. C., Salinas, A., Melloni, R. H., Jr., & Delville, Y. (2004). Behavioural and neuroendocrine adaptations to repeated stress during puberty in male golden hamsters. *Journal of Neuroendocrinology, 17*, 781–787.

Yamamoto, Y., Carter, C. S., & Cushing, B. S. (2006). Neonatal manipulation of oxytocin affects expression of estrogen receptor alpha. *Neuroscience, 137*, 157–164.

Yamamoto, Y., Cushing, B. S., Kramer, K. M., Epperson, P., Hoffman, G. E., & Carter, C. S. (2004). Neonatal manipulations of oxytocin alter expression of oxytocin and vasopressin immunoreactive cells in the paraventricular nucleus of the hypothalamus in a gender specific manner. *Neuroscience, 125*, 947–955.

Yang, C. R., Seamans, J. K., & Gorelova, N. (1999). Developing a neuronal model for the pathophysiology of schizophrenia based on the nature of electrophysiological actions of dopamine in the medial prefrontal cortex. *Neuropsychopharmacology, 21,* 161–194.

Yoshimura, R., Kimure, T., Watanabe, D., & Kiyama, H. (1996). Differential expression of oxytocin receptor mRNA in the developing rat brain. *Neuroscience Research, 24,* 291–304.

Young, L. J., Huot, B., Nilsen, R., Wang, Z., & Insel, T. R. (1996). Species differences in central oxytocin receptor gene expression: Comparative analysis of promoter sequences. *Journal of Neuroendocrinology, 8,* 777–783.

Young, L. J., Lim, M. M., Gingrich, B., & Insel, T. R. (2001). Cellular mechanisms of social attachment. *Hormones and Behavior, 40,* 133–138.

Young, L. J., Nilsen, R., Waymire, K. G., MacGregor, G. R., & Insel, T. R. (1999). Increased affiliative response to vasopressin in mice expressing the V1a receptor from a monogamous vole. *Nature, 400,* 766–768.

Young, L. J., Winslow, J. T., Nilsen, R., & Insel, T. R. (1997). Species differences in V1a receptor gene expression in monogamous and nonmonogamous voles: Behavioral consequences. *Behavioral Neuroscience, 111,* 599–605.

Zahm, D. S. (1999). Functional–anatomical implications of the nucleus accumbens core and shell subterritories. *Annals of the New York Academy of Sciences, 877,* 113–128.

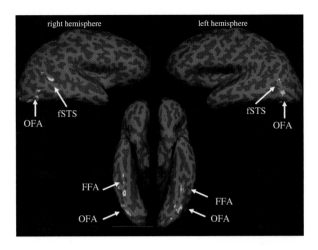

PLATE 4.1. Face-selective activation (faces < objects, *p* < .0001) on an inflated brain of one subject, shown from lateral and ventral views of the right and left hemispheres. Three face-selective regions are typically found: the FFA in the fusiform gyrus along the ventral part of the brain, the OFA in the lateral occipital area, and the fSTS in the posterior region of the superior temporal sulcus. From Kanwisher and Yovel (2006). Copyright 2006 by the Royal Society of London. Reprinted by permission.

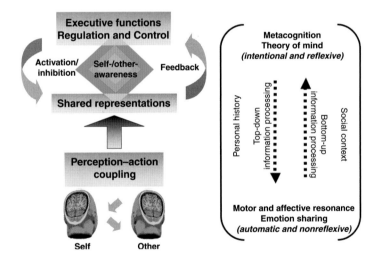

PLATE 8.1. Schematic representation of the mechanisms underpinning the experience of empathy. Two dimensions interact: (1) bottom-up information processing (i.e., direct matching between perception and action), and (2) top-down information processing (i.e., regulation and control). The bottom-up processing level, which is automatically activated (unless inhibited) by perceptual input, accounts for emotion sharing, which leads to the implicit recognition that others are like us. This aspect is hard-wired and function right after birth. The top-down level, which overlaps with the notion of executive control, is implemented in the prefrontal cortex and develops gradually during childhood. Executive control regulates both cognition and emotion, notably through selective attention and self-regulation. This metacognitive level is continuously updated by bottom-up information, and in return controls the lower level by providing top-down input. Thus top-down regulation, through executive functions, modulates bottom-up information and adds flexibility, allowing the individual to be less dependent on external cues. The metacognitive feedback plays a crucial role in taking into account one's own mental competence in order to react (or not) to the affective states of others. Self-awareness and other-awareness are important aspects of this model. The computational mechanism of self–other distinction is crucial for the higher-level cognitive processing involved in social cognition, such as empathy and theory of mind. Both empathy and theory of mind involve the ability to distinguish simultaneously between different possible perspectives on the same situation.

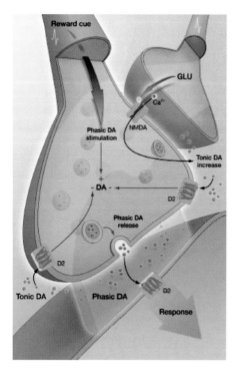

PLATE 17.1. Cartoon of a dopaminergic synapse, representing mechanisms of the tonic–phasic DA theory (Grace, 1995). "Phasic" DA stimulation refers to the rapid and transient increase in *intrasynaptic* DA levels. This rapid increase stems from the acute release of DA in the synapse triggered by action potentials (green arrows). These action potentials, which correspond to burst firing of dopaminergic neurons, are elicited by natural rewards and reward predictive cues. "Tonic" DA stimulation, in contrast, refers to substantially lower levels of extracellular *extrasynaptic* DA, which binds to the extrasynpatic DA autoreceptors. These autoreceptors antagonize phasic DA release (red arrows). Most importantly, tonic DA levels are increased by glutaminergic afferents from the prefrontal cortex and possibly other mesocorticolimbic regions such as the amygdala. GLU, glutamate; NMDA, *N*-methyl-D-aspartate.

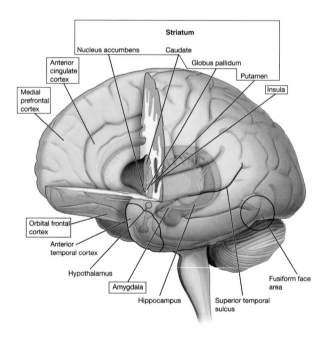

PLATE 17.2. Major nodes of the reward circuitry. They include the amygdala as the translator of affective value; the insula and orbital frontal cortices as the integrators of the current value of reward; the striatum as the promoter of action; the anterior cingulate cortex as the detector of errors of estimation or conflict/planning for behavioral responses; and the medial prefrontal cortex as the evaluator of the self-related significance of stimuli. These structures are represented within a schematic brain. Other regions of the temporal cortex are also detailed, to provide a more comprehensive map of the brain.

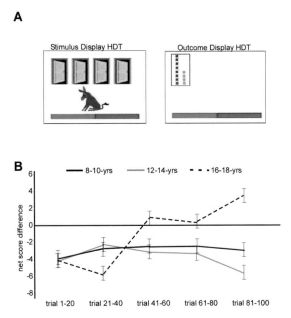

PLATE 19.1. (A) Stimulus and outcome displays of the Hungry Donkey Task (HDT), a child analogue of the Iowa Gambling Task. (B) Behavioral data for three age groups, plotted across five task blocks. The net score difference represents the difference score of the number of advantageous minus the number of disadvantageous choices. Older adolescents, but not young adolescents and children, learned to make long-term advantageous decisions. From Crone and van der Molen (2007). Copyright 2007 by Taylor and Francis. Reprinted by permission.

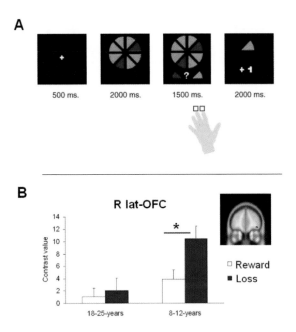

PLATE 19.2. (A) Stimulus display of the Cake Task. (B) Brain activation in a region of interest in lateral OFC in response to reward and loss. Children showed larger neural activity in response to loss than adults did. From van Leijenhorst, Crone, and Bunge (2006). Copyright 2006 by Elsevier. Reprinted by permission.

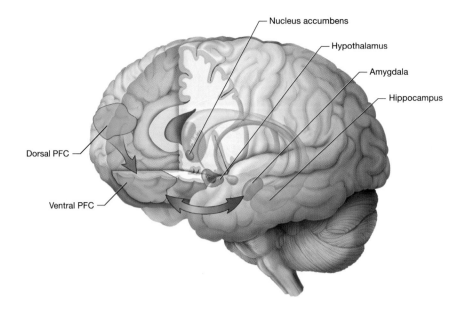

PLATE 20.1. Brain regions implicated in MDD. Key structures implicated in MDD are indicated. Coloring and arrows illustrate how components of the PFC interact in regulating amygdala activity.

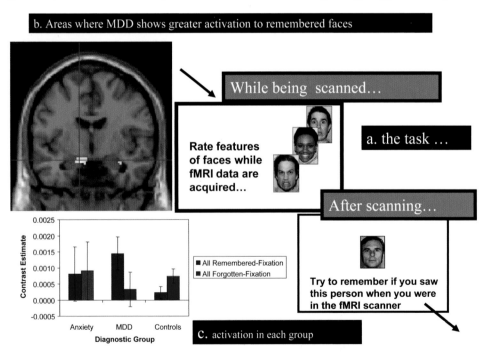

PLATE 20.2. Summary of findings from one recent fMRI study in adolescents with MDD ($n = 10$), anxiety disorders but not MDD ($n = 11$), or no psychopathology ($n = 23$). Plate 20.2a summarizes details from the task, in which subjects underwent fMRI scanning while they viewed and rated aspects of emotionally evocative faces, not knowing that they would receive a subsequent memory test. Subjects were then shown neutral expressions of people, some of whom were viewed while making evocative expressions during scanning, and subjects were asked to indicate which subjects were seen previously. Plate 20.2b maps a coronal section at the amygdala brain regions where subjects with MDD differed from healthy subjects, and Plate 20.2c shows levels of activation for the region indicated by the cross-hairs in Plate 20.2b. As shown, only the patients with MDD showed greater amygdala activation to remembered, relative to forgotten, faces. Data from Roberson-Nay et al. (2006).

PART V

REGULATORY SYSTEMS
Motivation and Emotion

CHAPTER 16

Temperament and Affect Vulnerability

Behavioral, Electrocortical, and Neuroimaging Perspectives

LOUIS A. SCHMIDT
MICHELLE K. JETHA

The search for biological origins, correlates, and outcomes of individual differences in temperament has a long and rich history; it spans centuries and diverse fields, including medicine, science, philosophy, theology, and literature (Zuckerman, 2005). Galen (c. 130–200 A.D.), a Greek physician, was the first to propose a scientific theory of temperament from a biological perspective (see Kagan, 1994). Galen related individual differences in temperament to variability in concentrations of four fundamental bodily fluids, or "humors" (i.e., *flegma*, *chole*, *melanchole*, and *sanguis*). To this day, Galen's constitutional typology of temperament remains influential in current personality theory and structure, as we continue to expand and validate his fundamental treatise put forth long ago in his opus *On the Natural Faculties*.

Although current theories and conceptualizations of temperament do not include variable concentrations of bodily humors, recent work in this domain is rooted in Galen's early notion and has benefited largely from theoretical and methodological advances in the fields of genetics, biology, and neuroscience. Its vitality is evidenced in part here with the range of chapters published in this volume, as well as recent special journal issues devoted to this very topic (Adolphs & Skuse, 2006; Schmidt, 2003, 2007). Today, theoretically derived hypotheses regarding brain–behavior relations can be tested in pediatric populations by using many different noninvasive psychophysiological measures (see Schmidt & Segalowitz, 2008, for a review). This multimeasure, multimethod approach has provided the developmental commu-

305

nity with rich sources of converging evidence as to the possible neural mechanisms underlying typical and atypical socioemotional development. In so doing, it has permitted theories about the biology of temperament and socioemotional development to be refined and
expanded, allowing for the derivation of new hypotheses and consolidation of old ones.

"Temperament" has been defined as constitutional differences in reactivity and self-
regulation, influenced over time by heredity, maturation, and experience (Rothbart & Derryberry, 1981; Rothbart & Bates, 1998, 2006). It is viewed as biological in origin (Buss &
Plomin, 1984) and is identified more closely with reactive style than with goal-oriented purposeful behavior (Thomas & Chess, 1977). Individual differences in temperament are considered stable and enduring tendencies to react cognitively, behaviorally, and emotionally
with a characteristic style and intensity to life experiences, and are believed to be predictive
of developmental psychopathology (Rothbart & Bates, 1998, 2006). Inherent in this definition is the notion that some temperament styles may be more "vulnerable" than others to
different affect states, possibly conferring risk for developmental psychopathology. In this
chapter, we focus our discussion on individual differences in temperament, and consider
how temperament may bias emotional processing and influence typical and atypical socioemotional development, as evidenced by behavioral and multiple brain-based measures. We
provide a review of literature on the topic, focusing on studies of adults and children that
have employed behavioral measures, electroencephalogram (EEG) asymmetry, event-related
potentials (ERPs), and functional magnetic resonance imaging (fMRI).

Behavioral Perspectives

Infant Temperament

The study of temperament in North America has a long and rich history. Allport (1937) provided an early description of temperament over 70 years ago. Allport viewed temperament as
the characteristic phenomena of an individual's emotional nature, comprising susceptibility
to emotional stimulation, strength and speed of response, the quality of the person's pre-
vailing mood, and fluctuation and intensity of mood. The study of childhood temperament
in North America then waned for the next several decades; given the dominance first of
learning theory and then of the cognitive revolution in psychology, the idea of innate disposi-
tions in accounting for behavior was relatively ignored for some time. Although the study of
temperament in Europe remained alive and well during this period, it eventually began to
reemerge in North America during the 1960s with the work of Alexander Thomas and Stella
Chess.

Thomas and Chess, in their seminal work related to the New York Longitudinal Study
(Thomas & Chess, 1981; Thomas, Chess, & Birch, 1968), provided descriptions of newborns' behavior patterns along nine dimensions of temperament: activity level, rhythmicity,
approach–withdrawal, adaptability, intensity, sensory threshold, mood, distractibility, and
attention span/persistence. These infant temperament dimensions remain relatively stable
throughout infant and child development, and at the same time interact with environmental
influences to shape later personality. Furthermore, some temperamental and environmental
interactions place individuals at differential risk for developing subsequent behavioral problems and affective disorders.

During the last two decades, there has been considerable interest in the issue of temperamental contributions to socioemotional behavior and developmental psychopathology,

and in systematic investigation of the issues described by Thomas and Chess (1981). Two researchers have played important roles here for different reasons. Mary Rothbart and her colleagues (for reviews, see Rothbart, 2007; Rothbart & Bates, 2006; Rothbart & Posner, 2006) at the University of Oregon extended the work of Thomas and Chess by providing the developmental community with a refined theoretical framework for studying temperament along two, rather than nine, dimensions: reactivity and regulation. Rothbart's group has been instrumental in developing highly reliable and valid measures to index maternal perceptions of infant and child temperament (e.g., Rothbart, Ahadi, Hershey, & Fisher, 2001). She and her colleagues have done seminal work on the influence of temperament on emotion regulation, effort control, and attention (for reviews, see Rothbart, 2007; Rothbart & Bates, 2006; Rothbart & Posner, 2006; Rothbart & Sheese, 2007). This corpus of work has revealed that individual differences in temperament, present at birth, are associated with distinct neural correlates that influence executive functioning and emotion regulation skills. Some children lack the effortful control and attentional flexibility needed for reacting in stressful situations. These children may be at risk for developmental psychopathology.

The work of Jerome Kagan and his colleagues at Harvard University has also played an important role. Kagan's work is important for at least three reasons. First, he anchored temperament and behavior to a fundamental approach–avoidance heuristic, allowing researchers to derive testable hypotheses. Second, Kagan moved the field from a reliance on subjective measures to direct observation of behavior in the laboratory, providing more scientific objectivity. Third, Kagan and colleagues noted that the child's biology plays a particularly important role in directing socioemotional development; the influences of the environment are not negligible, but for Kagan these influences are not as powerful as the child's biological constitution present at birth.

Kagan and the Harvard group followed longitudinally a cohort of infants who were all healthy, normally developing, primarily European American, and mainly from middle-class backgrounds. Kagan and his colleagues (Kagan, 1994, 1997, 1999; Kagan & Snidman, 1991a, 1991b) noted that individual differences in these infants' early reactions to novelty were predictive of developmental outcome and later personality. Very early in postnatal life, the infants exhibited individual differences in approach or withdrawal tendencies, inhibition to the unfamiliar, and the degree to which they were aroused by environmental stimuli. For example, infants who exhibited a high degree of motor activity and distress in response to the presentation of novel auditory and visual stimuli during the first 2 years of postnatal life were shy and timid preschoolers. Infants who displayed a high degree of motor activity and positive affect in response to these same stimuli were sociable preschoolers.

Kagan (1994, 1999) speculated that individual differences in infant reactivity to novelty may be linked to sensitivity in forebrain circuits involved in the processing and regulation of emotion. Children who become easily distressed during the presentation of novel stimuli may have a lower threshold for arousal in forebrain areas, particularly the central nucleus of the amygdala. This hypothesis was based largely on findings from studies of animals that the amygdala plays an important role in the regulation and maintenance of conditioned fear (see LeDoux, 1996). This idea was recently empirically tested and supported with fMRI measures in human adults who 20 years earlier were classified as behaviorally inhibited (see Schwartz, Wright, Shin, Kagan, & Rauch, 2003; also discussed later).

Others have also examined the contribution of temperament to socioemotional development. Kochanska (1993) has noted that temperament—specifically, fear and effortful control—contributes to compliance and appropriate behavioral conduct. That is, infants must be able to effectively regulate their emotional responses and frustration in order to internalize societal standards for conduct. Infants who are easily frustrated and fail to modulate their emotions may respond with noncompliance and employ this strategy in subsequent situations of emotional arousal. These children are likely to be at risk for externalizing problems. More recently, Stifter, Spinrad, and Braungart-Rieker (1999) examined compliance in groups of infants classified by reactivity and emotion regulation skills. Stifter and her colleagues found that infants who exhibited high levels of reactivity and low levels of regulation at 5 months displayed more defiance/noncompliance than all other infants. Infants characterized as low-reactive/high in regulation or low-reactive/low in regulation were likely to use passive noncompliance (i.e., ignoring requests), while infants who were high-reactive/high in regulation were the least defiant. These tendencies remained relatively stable over the first 2 years after birth.

Adult Personality

The idea that temperament and affect are interrelated has a long and rich history, particularly in the adult personality literature—dating to Allport (1937) and Murray (1938), and then later to Eysenck (1947, 1953, 1967). More recently, several behavioral studies have examined temperament dimensions in relation to susceptibility to positive and negative affect (Costa & McCrae, 1980) and to intensity of emotional experience (Gross, Sutton, & Ketelaar, 1998; Larsen & Ketelaar, 1991) in adults. These authors found that the specific temperament dimensions of extraversion and neuroticism described by Eysenck (1967) may predispose an individual to experience more positive or more negative affect, respectively (Costa & McCrae, 1980; Emmons & Diener, 1985; Meyer & Shack, 1989; Warr, Barter, & Brownbridge, 1983; Watson & Clark, 1984). For example, in a series of studies conducted to assess relations between temperament and affect, Costa and McCrae (1980) found that extraversion and neuroticism were related to self-reports of positive and negative affect, respectively, in daily life. In addition, these measures of temperament predicted self-report of positive and negative affect up to 10 years later. Additional studies have examined extraversion and neuroticism in response to mood induction procedures and have reported similar associations (Gross et al., 1998; Larsen & Ketelaar, 1991).

In the study by Larsen and Ketelaar (1991), the relations between extraversion and neuroticism on the one hand, and increased susceptibility to positive and negative mood induction on the other, were assessed. The mood induction procedure comprised having participants read a written scenario designed to elicit a positive, negative, or neutral mood, and then vividly imagine being in that scenario. This procedure was then repeated with a different scenario of the same hedonic tone, after which participants were administered a mood rating scale. The results of this investigation revealed relations between neuroticism and heightened emotional reactivity to negative mood induction, and between extraversion and heightened emotional reactivity to positive mood induction. On the basis of these results, the authors proposed that extraversion and neuroticism are related to an increased vulnerability to positive and negative affective states, respectively.

Gross et al. (1998) expanded on earlier designs in order to clarify the association of extraversion and neuroticism with dispositional mood and with magnitude of affective reactivity. In this study, state levels of affect (i.e., affective reactivity) were measured before, during, and after mood-inducing film clips and related to dimensions of extraversion and neuroticism, as well as to dispositional positive and negative affect as measured by the Positive and Negative

Affect Schedule (Watson, Clark, & Tellegen, 1988). Positive and negative affective states were related to measures of extraversion and neuroticism and to dispositional positive and negative affect in predicted directions; however, the strongest associations were noted between positive affective states and measures of extraversion, and between negative affective states and measures of neuroticism. Collectively, the findings from these behavioral studies suggest that individual differences in temperament influence vulnerability to experiencing more positive or more negative emotion. These behavioral studies have served as a platform for researchers to investigate the biological basis of differential sensitivity to emotion processing by examining the functional neurocircuitry underlying affective reactivity in association with temperament/personality dimensions in infants, children, and adults. Work in this area of inquiry has come from a variety of sources, to which we next turn.

Electrocortical Perspectives:
I. Frontal Electroencephalogram (EEG) Asymmetry Studies

A Model

Over two decades ago, Fox and Davidson (Fox & Davidson, 1984; for more recent reviews, see Davidson, 1993, 2000; Fox, 1991, 1994) articulated a frontal activation model of emotion. They argued that emotions are organized around approach–withdrawal motivational tendencies, which are differentially lateralized at the level of the cerebral cortex. The left frontal cortical area is involved in positive, approach-related emotions (e.g., happiness, joy, interest), whereas the right frontal cortical area is involved in negative, withdrawal-related emotions (e.g., fear, sadness, disgust). Fox and Davidson then used regional measures of brain electrical activity (EEG) to index brain responses during the concurrent presentation of affective stimuli. As predicted, greater relative right frontal EEG activity was noted during the presentation of visual stimuli used to elicit negative emotions, while greater relative left frontal EEG activity was noted during the presentation of visual stimuli used to elicit positive emotions. These same basic patterns of frontal EEG responses to affective stimuli have been noted across different sensory modalities and different ages (for reviews, see Fox & Davidson, 1987, 1988; Santesso, Schmidt, & Trainor, 2007; Schmidt & Trainor, 2001). The functional neurocircuitry underlying affective reactivity with respect to approach- and withdrawal-related motivational systems has been intensively investigated over the past two decades (for reviews, see Davidson, Ekman, Saron, Senulis, & Friesen, 1990; Davidson, Jackson, & Kalin, 2000a).

Davidson and Fox (see Davidson, 1993, 2000; Fox, 1991, 1994) later theorized that patterns of resting or "tonic" lateralized alpha activity (i.e., 8–13 Hz) in the frontal hemispheres are predictive of individual differences in "affective style"—a term that Davidson (2000) has used to refer to the broad range of variability in dispositional mood and affective reactivity across individuals. Individual differences in patterns of frontal activation are presumed to reflect differences in excitability of forebrain limbic circuitry involved in the maintenance of emotion. The frontal regions, particularly regions of the prefrontal cortex, are believed to be involved in the regulation of subcortical sites, such as the amygdala, a region known to be intimately involved with withdrawal-related negative affect (LeDoux, Iwata, Cicchetti, & Reis, 1988). More specifically, left-sided activation of the prefrontal cortex may exert an inhibitory effect on the activation of the amygdala, thereby dampening the duration of negative affect in response to negatively valenced stimuli (Davidson, 1998). Positron emission tomography (PET) studies are consistent with this idea, indicating that in healthy human participants, glucose metabolism in regions of the left prefrontal cortex is inversely related to metabolic activity in the amygdala (Abercrombie et al., 1998). More recent studies have suggested that activity in the ventromedial prefrontal cortex may suppress amygdala activity by mediating connections between other areas of the frontal cortex (e.g., the frontal polar cortex) and the amygdala (Urry et al., 2006). These findings and other studies suggest that individual differences in patterns of prefrontal activity at rest may mediate the regulation of the time course of emotional responding, such that individuals showing greater relative left prefrontal activity recover more rapidly from negative affect than those individuals with greater relative right prefrontal activity do (see Davidson, 2000).

The Evidence

Over the past three decades, almost 100 empirical studies have investigated the relation between patterns of resting frontal EEG asymmetry and the experience, expression, regulation of emotion and its role in affective style (see Coan & Allen, 2004, for a review). Overall, there is strong empirical support for the Davidson and Fox model across studies of infants, children, and healthy adults, as well as some clinical populations. Individuals who exhibit greater relative right frontal EEG activity at rest are shy, socially withdrawn, and anxious/ depressed. Individuals who exhibit greater relative left frontal EEG activity at rest are social, extraverted, and outgoing.

Patterns of resting or tonic asymmetric EEG in frontal regions are presumed to reflect a predisposition to experience a certain valence (i.e., positive vs. negative) and intensity of affective stimuli. For example, when EEG was measured prior to the presentation of emotionally evoking film clips, adult participants who presented with greater left-sided prefrontal activity at baseline reported experiencing more positive affect in response to positive emotional film clips, whereas those who presented with greater right-sided prefrontal activity at baseline reported experiencing more negative affect in response to negative emotional film clips (Tomarken, Davidson, & Henriques, 1990; see also Jones & Fox, 1992). Additional investigations have focused on individual differences in affective reactivity in response to social stress in infants and children.

Infants who exhibited greater relative right resting frontal EEG activity were more likely to cry at a stranger's approach and at separation from their mothers than infants who exhibited greater relative left frontal EEG activity (Davidson & Fox, 1989; Fox, Bell, & Jones, 1992). Interestingly, another study also found evidence that prefrontal asymmetry predicted the magnitude of recovery following a negative emotional stimulus (Larson, Sutton, & Davidson, 1998). Collectively, these findings may be interpreted to indicate that greater relative left frontal resting EEG activity reflects a tendency to experience positive affect and to effectively regulate the duration of negative affect, whereas greater relative right frontal resting EEG activity may reflect a tendency to experience negative affect and to ineffectively regulate the duration of this affect.

The pattern of resting and reactive frontal EEG activity has also been examined in children and adults with temperamental styles (i.e., shy and social) hypothesized to be characterized and maintained by the experience of negative or positive emotion, respectively, in response to social stimuli. For example, greater activity in the right than in the left prefrontal cortex during resting or baseline conditions predicted temperamental inhibition (a behavioral precursor of shyness) in infants (Calkins, Fox, & Marshall, 1996) and children (Fox et al., 1995; Fox, Henderson, Rubin, Calkins, & Schmidt, 2001), as well as internalizing (vs. externalizing) related problems in preschoolers (Fox, Schmidt, Calkins, Rubin, & Coplan, 1996; Santesso, Reker, Schmidt, & Segalowitz, 2006a). Schmidt and his colleagues have conducted a series of studies with shy children and adults. Theall-Honey and Schmidt (2006) found that temperamentally shy preschoolers exhibited greater relative right central activity during the presentation of fear-eliciting video clips at age 4 than their nonshy counterparts did. In addition, temperamentally shy school-age children displayed a greater increase in right (but not left) frontal EEG activity, a greater increase in heart rate, and a greater decrease in heart rate variability than medium- or low-shyness children in response to a task designed to elicit self-presentation anxiety at age 7 (Schmidt, Fox, Schulkin, & Gold, 1999). Temperamentally shy adults exhibited greater relative resting right frontal EEG activity, whereas their social counterparts exhibited greater relative left frontal EEG activity at rest (Schmidt, 1999), and this pattern of frontal activity was exaggerated in these temperamental styles when participants were presented with social challenge (Schmidt & Fox, 1994). Greater relative resting left frontal EEG activity has also been found in healthy adults who scored high on measures of behavioral approach (Sutton & Davidson, 1997) and sensation seeking (Santesso et al., 2008).

Stability Studies

If the frontal EEG alpha asymmetry is used as a trait-like index of dispositional affective style in individuals, then the metric should remain moderately stable over time. In contrast to the relatively large literature investigating frontal activity patterns in relation to temperament and emotion, only a few studies have assessed the stability of these measures over time and context. Several studies have noted very good short-term stability over time in frontal EEG asymmetry and power measures in second-by-second stability in healthy 9-month-old infants (Schmidt, 2008) and 7-year-old children (Schmidt, 1996), and across 3 weeks in healthy adults (Tomarken, Davidson, Wheeler, & Kinney, 1992). Another study noted modest stability across a longer period of time than the studies above (i.e., 8- and 16-week intervals) in frontal alpha asymmetry in nonmedicated clinically depressed individuals, and also found that changes in asymmetry were not related to changes in clinical state (Allen, Urry, Hitt, & Coan, 2004). Two recent studies have examined long-term stability of the frontal asymmetry metric over a 1- to 3-year interval in individuals with a history of depression (Vuga et al., 2006) and with schizophrenia (Jetha, Schmidt, & Goldberg, in press-a) and reported moderate stability of these measures, which were not related to concurrent or history of symptoms.

Two other studies have examined frontal EEG alpha asymmetries over different contexts (i.e., wakefulness and various sleep stages). Both of these studies reported modest relations between wakefulness and sleep stages, particularly during the rapid-eye-movement stage (the stage most analogous to waking alpha activity), in healthy adults (Schmidt, Cote, Santesso, & Milner, 2003) and in a group of patients undergoing polysomnographic evaluation for possible sleep disorders (Benca et al., 1999).

Clinical Populations

In addition to studies investigating individual differences in temperament and affective processing in healthy adults and children, the Davidson and Fox model has been applied to studies of different clinical populations. Although these studies did not focus on individual differences in the processing of emotion per se, their results are relevant to the present review. Specifically, inherent to the diagnosis of certain clinical disorders, such as mood and anxiety disorders, is the experience of either increased or decreased dispositional affect. For example, a defining characteristic of anxiety disorders is heightened emotional reactivity in general or in response to specific emotional stimuli, whereas a defining characteristic of depression is an inability to experience positive affect. Therefore, it has been useful to examine this theoretical model in patients diagnosed with either anxiety disorders or depression, compared to healthy controls.

Studies examining resting EEG activity in individuals diagnosed with depression versus never-depressed controls reported greater right prefrontal activity in the former than in the latter (Henriques & Davidson, 1990, 1991). Importantly, greater relative right activity was reported to be a function of low activity in the left frontal region (Henriques & Davidson, 1991). It has been suggested, based on these patterns of results and on Davidson's model, that depression may result from lack of approach and positive affect due to hypoactivation of the left frontal region. This finding is consistent with the idea that depression may be the absence of an ability to experience positive affect rather than the result of increased negative affect, which would be reflected in right frontal hyperactivation, according to the frontal asymmetry model (Henriques & Davidson, 1991).

Researchers have also examined electrocortical and autonomic activity in individuals with social phobia during anticipation of public speaking (Davidson, Marshall, Tomarken, & Henriques, 2000b). Individuals with social phobia showed greater relative right prefrontal activity and increased heart rate than controls during anticipation of public speaking, although these differences were not evident during baseline conditions. As noted earlier, this same pattern of increased right frontal brain activity and heart rate during anticipation of social presentation has been noted in typically developing temperamentally shy children (Schmidt et al., 1999) and adults (Schmidt & Fox, 1994). Individuals diagnosed with panic disorder are also reported to exhibit greater relative right frontal activity than controls during resting conditions and in response to anxiety-provoking stimuli (Wiedemann et al., 1999).

We recently found the predicted relations between greater relative right frontal EEG activity and temperamental shyness, and between greater relative left frontal EEG activity and temperamental sociability, in a stable group of community outpatients with schizophrenia (Jetha, Schmidt, & Goldberg, in press-b). Interestingly, these predicted relations between resting frontal EEG activity and temperament only emerged after levels of patient symptoms were statistically controlled for. Concurrent mood is also known to obscure the relations between temperament and resting measures of frontal EEG asymmetry. In another recent study of adults, Beaton et al. (2008a) noted the predicted relation between temperamental shyness and greater relative right frontal EEG activity at rest only when concurrent depressive mood was statistically controlled for. Accordingly, certain clinical populations appear predisposed to experience more positive or more negative emotion than controls, and this predisposition is reflected in temperament and patterns of resting frontal brain activity, but the predicted relations between resting brain electrical activity and temperament may be obscured by concurrent mood and/or patient symptoms.

Other studies have examined the pattern of resting brain activity in infants and children of depressed mothers. In a series of studies, Field and her colleagues found that depressed mothers and their infants exhibit atypical behavioral and frontal EEG patterns. Depressed mothers are known to display less positive affect and reduced levels of stimulation when interacting with their infants (Cohn, Matias, Tronick, Connell, & Lyons-Ruth, 1986; Cohn & Tronick, 1989; Field, 1986; Field et al., 1988), and these behavioral symptoms are apparently transmitted to the infants. Infants of depressed mothers are known to display less positive affect and increased irritability (Cohn et al., 1986; Cohn & Tronick, 1989; Field, 1986; Field et al., 1985), and greater relative right frontal EEG activity (a marker of stress) (Dawson, Klinger, Panagiotides, Hill, & Spieker, 1992; Field, Fox, Pickens, & Nawrocki, 1995), than infants of nondepressed mothers. This frontal EEG asymmetry can develop as early as 1 month of age in infants of depressed mothers (Jones, Field, Fox, Lundy, & Davalos, 1997b), and it appears to remain stable for the first several years of the children's lives (Jones, Field, Davalos, & Pickens, 1997a).

Other studies have investigated regional brain activity in disorders marked by emotional hyporesponsiveness and lack of anticipatory fear (i.e., antisocial personality disorder and psychopathy). In one study, increased left frontal EEG activity was associated with a decreased likelihood of the diagnosis of antisocial personality disorder (Deckel, Hesselbrock, & Bauer, 1996). Persons with psychopathy have also been found to have an increased proportion of mixed-handedness relative to individuals without psychopathy, suggesting anomalous cerebral asymmetry in the former (Mayer & Kosson, 2000). Future studies should include the investigation of individual differences within different clinical populations (i.e., temperament dimensions) to assess the contribution of temperament to affective processes in atypical development.

Some Caveats

There are at least four caveats to the frontal EEG asymmetry–emotion model that warrant discussion. First, the model is primarily concerned with differentiating between positive and negative *valence* of emotion experience, and among individual differences in the styles or characteristic tendencies of responding with more positive or more negative emotion. The model is, however, limited in its description and prediction of another major characteristic of emotion—namely, *intensity*. Dawson (1994; Dawson et al., 1992), and Schmidt (1999; Schmidt & Fox, 1994), and their colleagues have theorized that overall frontal alpha power, which is inversely related to cortical activity (Davidson & Tomarken, 1989), is related to the intensity of emotional experience. These authors have speculated that individual differences in affective style may be distinguished by overall alpha power in the frontal regions during resting conditions, and that lower overall alpha power (i.e., increased activity) in the frontal regions may reflect a predisposition towards greater intensity of emotional experience. Dawson (1994; Dawson et al., 1992) found that infants who exhibited lower overall frontal EEG alpha power (i.e., higher activity) were more easily distressed by maternal separation and a stranger's approach than infants with low overall frontal activity were. Schmidt (1999) also found that low overall frontal EEG alpha power (i.e., higher activity) predicted healthy adults who were "conflicted" in their social affiliation styles, in that they reported being both highly shy and highly social (i.e., they expressed a high desire for social affiliation, yet experienced high anxiety and withdrawal-related behavior during social engagement). Schmidt has suggested that the greater overall frontal activity specific to this "conflicted" group may reflect a tendency to experience more intense emotion as a result of a heightened approach–withdrawal conflict.

Second, although the use of frontal EEG asymmetry metric as a measure of individual differences in affective responding has received much empirical support over the past 25 years, a few studies have failed to find the relation between resting left and right frontal EEG asymmetry and experimentally evoked "states" of positive and negative affect, respectively (see Coan & Allen, 2004, for a review). In reconciling these inconsistencies, some have suggested that frontal asymmetry may not reflect a predisposition to positive and negative emotion per se, but more general approach–withdrawal motivational tendencies (Harmon-Jones, 2004).

Third, as described above, if the frontal EEG asymmetry metric is to be considered trait-like, it needs to be stable across time and context. Some investigators have failed to find stability over time in resting frontal EEG asymmetry (Hagemann, Naumann, Thayer, & Bartussek, 2002).

Fourth, EEG measures, while providing excellent temporal resolution, provide little information regarding spatial resolution. Determining the exact source of the EEG signal and what the signal actually reflects is difficult. Future work in this area, using current source density mapping and fMRI measures (discussed later), may provide converging pieces of corroborating evidence.

Electrocortical Perspectives: II. ERP Studies

What Are Event-Related Potentials (ERPs)?

ERPs are small voltage fluctuations (i.e., a few microvolts) in brain electrical activity that are associated with a given event (see Gunnar & de Haan, Chapter 2, this volume, for a further description of ERP methods) . ERPs are thought to arise from the summation of synchronous postsynaptic neural activity of cortical pyramidal cells. Because the voltage fluctuations of ERPs are so small, many samples of ERPs that are time-locked to stimulus presentation are averaged together to increase signal and to reduce background noise. Averaging multiple samples produces a voltage × time function, which has a number of positive and negative peaks (see Luck, 2005, for a review of ERP technique). These peaks are described in terms of their spatial location on the scalp, their polarity (N for negative and P for positive potential), and their latency in relation to an event. For example, the N170 component of an ERP associated with the passive viewing of facial stimuli is a negative potential measured at occipitotemporal electrode sites, and peaking at approximately 170 msec after stimulus onset.

The use of noninvasive ERP methodology provides several alternatives over the use of behavioral measures and asymmetry measures derived from continuous EEG data (described in the previous section) in the study of cognitive and affective processes (de Haan, 2008; Luck, 2005). For instance, behavioral responses result from a large number of cognitive processes and are therefore difficult to attribute to specific processes, although the functional significance of a behavioral response is usually quite clear. Continuous EEG measures are not time-locked, so little can be inferred about the relation between the processing of a particular stimulus and the brain physiology tied to that single event. ERPs, however, provide a continuous measure of processing from the beginning of stimulus onset and proceeding either until a response is given, or until the next stimulus is presented, depending on task requirements. This online processing can be monitored with high temporal resolution on the order of 1 msec, which has advantages over high-spatial-resolution/low-temporal-resolution methodologies such as PET and fMRI. One disadvantage of the ERP technique, however, is that researchers cannot determine the functional significance of ERP components without relying on a host of assumptions and inferences.

ERP and Temperamental Differences

An emerging literature involves the use of ERP measures to understand the relations between temperament and socioemotional processes in typical development. Fox and a colleague have examined temperamental contributions to children's performance in an emotional word-processing task in two separate studies of typically developing children. In an initial study, Perez-Edgar and Fox (2003) found that processing of positive and negative words showed differences in processing across ERP components. Negative words appeared to tax attentional and processing resources, as reflected in P300 amplitude compared with negative words. In a second study, Perez-Edgar and Fox (2007) noted a relation between individual differences in temperament and affect processing, as evidenced on several ERP components. Children rated high in soothability and attentional control showed slower behavioral responses to socially negative words. Children low in attentional control displayed processing differences from other children in the later, more cognitive ERP components to both positive and negative words.

Recently, Schmidt and colleagues have examined the relation between temperament and information processing in typically developing children, using ERP measures. In two studies, they examined such children's temperament in relation to error-related negativity (ERN)—an ERP component that is a negative wave occurring at about 100 msec following an error, and is thought to be generated in the anterior cingulate cortex with changes developmentally (Santesso, Segalowitz, & Schmidt, 2006b). In one study, they found that 10-year-old children who scored high on the Child Behavior Checklist (CBCL) measure of obsessive–compulsive behaviors exhibited greater ERN responses to errors made on a visual flanker task than children scoring low on the CBCL measure did (Santesso, Segalowitz, & Schmidt, 2006c). The pattern of ERN response was interpreted to mean that the children with obsessive–compulsive tendencies may have been more concerned about making errors than the children scoring low on the CBCL measure were. In another study, they found that children who were characterized by low socialization (i.e., who were antisocial) showed few to no ERN responses to errors made on a visual flanker task, in contrast to children who were high in socialization (i.e., who were prosocial) (Santesso, Segalowitz, & Schmidt, 2005). This pattern of ERN response was interpreted to mean that the children who were antisocial may have had little concern about making errors.

Still other studies have examined ERP responses in infants to the presentation of novel stimuli. Infant reactions to novelty are presumed to be early precursors of individual differences in temperament and developmental outcome. As noted earlier, Kagan and Snidman (1991a, 1991b) found that a subset (10–15%) of infants who exhibited a high degree of motor activity and distress in response to novel auditory and visual stimuli were likely to become behaviorally inhibited toddlers and shy and socially withdrawn preschoolers. Subsequent studies found that such infants exhibited exaggerated startle EMG responses to a stranger approach condition (Schmidt & Fox, 1998) and distinct ERP responses to novel auditory stimuli (Marshall, Reeb, & Fox, in press) at 9 months of age. In other research, Gunnar and Nelson (1994) recorded ERPs from 1-year-old infants during the presentation of sets of familiar faces presented frequently and infrequently, and a set of novel faces presented infrequently. A normative ERP pattern from this sample in response to the faces was computed from data collected at the frontal and central sites. This metric was related to infants' separation distress, emotionality, and cortisol levels. Infants scoring higher on the normative ERP metric were more distressed during separation and were reported by their parents to smile and laugh more, and the infants had lower cortisol concentrations during the ERP testing. Interestingly, some infants exhibited a positive slow wave to the infrequent novel faces—the normative pattern exhibited by the infants to the infrequent familiar faces.

Neuroimaging Perspectives

A number of studies conducted within the last decade have used neuroimaging to examine the relation between temperament and affect. Unlike EEG asymmetry and ERP measures, fMRI measures provide more accurate spatial resolution of complex brain dynamics. However, the fMRI measures are typically more invasive and, accordingly, more difficult to use with nonclinical pediatric populations, although this is changing.

In two studies of healthy adults, Canli and his colleagues used fMRI to measure brain activation to emotion experience (Canli et al., 2001) and to the perception of emotional stimuli (Canli, Sivers, Whitfield, Gotlib, & Gabrieli, 2002), in relation to self-reports of personality characteristics (i.e., extraversion, neuroticism, openness, agreeableness, and conscientiousness) as measured by the NEO Five-Factor Inventory (Costa & McCrae, 1991). In the first study, Canli et al. (2001) examined the relation between personality traits and brain reactivity in regions known to be involved in the processing of emotion (i.e., prefrontal cortex, anterior cingulate, insula, and amygdala) during the mood induction of positive and negative affect. Stimuli comprised blocks of positively or negatively valenced pictures. Results indicated that extraversion was associated with increased activation in numerous cortical and subcortical locations, including the right amygdala, during the processing of positively valenced, but not negatively valenced, pictures; by contrast, neuroticism was associated with increased activation in the frontal and temporal lobes during the processing of negatively valenced, but not positively valenced, pictures. This study was the first to explore the relations between personality trait factors and regional brain activity as measured with fMRI. However, findings from this study must be interpreted with caution. Limitations and concerns include the omission of a neutral condition. The interpretations of results are hence ambiguous, as the findings of increased activation for positively valenced stimuli may also be interpreted as a decrease in activity for negatively valenced stimuli. Additional limitations included the use of a small and gender-biased sample, which included only 14 female participants.

In a follow-up study, Canli et al. (2002) examined the relations between personality trait factors and amygdala activation in response to faces expressing fear, happiness, sadness, and anger. Results from this study indicated that across participants, increased amygdala activation was associated with the perception of faces expressing fear, but not with that of faces expressing other emotions. These results are consistent with well-established findings suggesting that the amygdala is intimately involved with the processing of fear-related stimuli, so that direct projections from sensory perception pathways to this subcortical region have evolved to ensure rapid detection of and response to danger (LeDoux, 1996). Although decades of research in this area clearly suggest an adaptive response to fear that is universal in nature, it should be noted that certain factors, such as biological predispositions and/or conditioning to fear-related stimuli, are thought to contribute to individual differences in the magnitude and recovery of this response (LeDoux, 1996). When individual differences were considered in the study by Canli et al. (2002), extraversion was found to predict left-lateralized amygdala activation, specifically in response to the perception of happy faces.

Beaton et al. (2008b) recently examined the relation between temperament and processing of neutral faces in a group of healthy adults, using fMRI measures. Unlike Canli and his colleagues, Beaton et al. selected individuals who reported high levels of shyness and high levels of boldness, and used neutral face stimuli. The shy–bold continuum is a fundamental behavioral trait observed across human and nonhuman species (Wilson, Clark, Coleman, & Dearstyne, 1994). Individual differences along this continuum are known to be related to human psychopathology (Davidson, 1998, 2000). These individual differences are presumed to arise from, and to be maintained by, different neural systems and differences in excitability of forebrain limbic areas involved in the evaluation of stimulus salience (Davidson, 2000; Fox, 1991; Kinsbourne, 1978). To test this hypothesis, Beaton et al. conducted an event-related fMRI study in which the brains of shy and bold adults were scanned during the presentation of two groups of neutral faces: faces of strangers and of personally familiar individuals. We found that shy adults exhibited greater bilateral amygdala activation during the presentation of strangers' faces, and greater left amygdala activation during personally familiar faces, than their bold counterparts did. Bold adults exhibited greater bilateral nucleus accumbens activation in response to personally familiar faces than did shy adults. These findings suggest that there are distinct neural substrates underlying and maintaining individual differences along a shy–bold continuum in humans.

Two other recent fMRI studies have examined temperamental shyness in relation to affect in young adults (Schwartz et al., 2003) and adolescents (Guyer et al., 2006) originally selected for temperamental inhibition as children. Kagan and his colleagues reported greater bilateral amygdalar blood-oxygen-level-dependent signal in response to novel versus newly familiar faces in healthy young adults who were characterized as temperamentally inhibited toddlers (Schwartz et al., 2003). Most recently, Fox and his colleagues noted greater striatal activation to positive incentives in adolescents who were categorized as behaviorally inhibited and temperamentally shy in early childhood (Guyer et al., 2006). This latter study appears to be the first study to examine the relation between temperament and affect in typically developing children by using fMRI measures.

Summary and Conclusions

Converging evidence from the studies of behavior, resting frontal EEG asymmetry, ERPs, and fMRI reviewed above suggests that individual differences in temperament may bias affect processing in nonclinical populations and some clinical populations, and that these relations may be moderated by differences in underlying functional neurocircuitry. Behavioral and biological studies of the temperament dimensions characterized by approach and withdrawal indicate that these temperamental styles are differentially associated with differences in dispositional affect and affective reactivity, as well as with variations in brain activity in response to emotional experience and perception.

Although the exact mechanisms underlying typical and atypical socioemotional development in humans are unknown, a wealth of research suggests that a child comes into the world with a particular dispositional style and pattern of affiliative behavior, which are moderated by individual differences in temperament and underlying neurocircuitry, and influenced by a multitude of environmental and social factors (see Kagan, 1999, for a review). The interaction of biological predisposition and environmental influences over time will shape and determine whether an individual will react with timidity and nervous apprehension or with enthusiasm and exuberance in social situations and to life events.

Acknowledgments

Preparation of this chapter was supported in part by grants from the Natural Science and Engineering Research Council of Canada, and the Social Sciences and Humanities Research Council of Canada, to Louis A. Schmidt.

References

Abercombie, H. C., Schaefer, S. M., Larson, C. L., Oakes, T. R., Lindgren, K. A., Holden, J. E., et al. (1998). Metabolic rate in the right amygdala predicts negative affect in depressed patients. *NeuroReport, 9,* 3301–3307.

Adolphs, R., & Skuse, D. (Eds.). (2006). Genetic, comparative and cognitive studies of social behavior [Special issue]. *Social Cognitive and Affective Neuroscience, 1*(3), 163–164.

Allen, J. J., Urry, H. L., Hitt, S. K., & Coan, J. A. (2004). The stability of resting frontal electroencephalographic asymmetry in depression. *Psychophysiology, 41,* 269–280.

Allport, G. (1937). *Personality: A psychological interpretation.* New York: Holt.

Beaton, E. A., Schmidt, L. A., Ashbaugh, A. R., Santesso, D. L., Antony, M. M., & McCabe, R. E. (2008a). Resting and reactive frontal brain electrical activity (EEG) among a non-clinical sample of socially anxious adults: Does concurrent depressive mood matter? *Neuropsychiatric Disease and Treatment, 4,* 187–192.

Beaton, E. A., Schmidt, L. A., Schulkin, J., Antony, M. M., Swinson, R. P., & Hall, G. B. (2008b). Different neural responses to stranger and personally familiar faces in shy and bold adults. *Behavioral Neuroscience, 122,* 704–709.

Benca, R. M., Obermeyer, W. H., Larson, C. L., Yun, B., Dolski, I., Kleist, K. D., et al. (1999). EEG alpha power and alpha power asymmetry in sleep and wakefulness. *Psychophysiology, 36,* 430–436.

Buss, A. H., & Plomin, R. (1984). *Temperament: Early developing personality traits.* Hillsdale, NJ: Erlbaum.

Calkins, S. D., Fox, N. A., & Marshall, T. R. (1996). Behavioral and physiological antecedents of inhibited and uninhibited behavior. *Child Development, 67,* 523–540.

Canli, T., Sivers, H., Whitfield, S. L., Gotlib, I. H., & Gabrieli, J. D. E. (2002). Amygdala response to happy faces as a function of extraversion. *Science, 296,* 2191.

Canli, T., Zhao, Z., Desmond, J. E., Kang, E., Gross, J., & Gabrieli, J. D. E. (2001). An fMRI study of personality influences on brain reactivity to emotional stimuli. *Behavioral Neuroscience, 115,* 33–42.

Coan, J. A., & Allen, J. J. B. (2004). Frontal EEG asymmetry as a moderator and mediator of emotion. *Biological Psychology, 67,* 7–49.

Cohn, J. F., Matias, R., Tronick, E. Z., Connell, D., & Lyons-Ruth, D. (1986). Face-to-face interactions of depressed mothers and their infants. In E. Z. Tronick & T. Field (Eds.), *Maternal depression and infant disturbance* (pp. 31–45). San Francisco: Jossey-Bass.

Cohn, J. F., & Tronick, E. Z. (1989). Specificity of infant's response to mothers' affective behavior. *Journal of the American Academy of Child and Adolescent Psychiatry, 28,* 242–248.

Costa, P. T., & McCrae, R. R. (1980). Influence of extraversion and neuroticism on subjective well-being: Happy and unhappy people. *Journal of Personality and Social Psychology, 38,* 668–678.

Costa, P. T., & McCrae, R. R. (1991). *Professional manual for the revised NEO Personality Inventory and NEO Five-Factor Inventory.* Odessa, FL: Psychological Assessment Resources.

Davidson, R. J. (1993). The neuropsychology of emotion and affective style. In M. Lewis & J. M. Haviland (Eds.), *Handbook of emotions* (pp. 143–154). New York: Guilford Press.

Davidson, R. J. (1998). Affective style and affective disorders: Perspectives from affective neuroscience. *Cognition and Emotion, 12,* 307–330.

Davidson, R. J. (2000). Affective style, psychopathology, and resilience: Brain mechanisms and plasticity. *American Psychologist, 55,* 1196–1214.

Davidson, R. J., Ekman, P., Saron, C. D., Senulis, J. A., & Friesen, W. V. (1990). Approach/withdrawal and cerebral asymmetry: Emotional expression and brain physiology. I. *Journal of Personality and Social Psychology, 58,* 330–341.

Davidson, R. J., & Fox, N. A. (1989). The relation between tonic EEG asymmetry and ten-month-olds emotional response to separation. *Journal of Abnormal Psychology, 98,* 127–131.

Davidson, R. J., Jackson, D. C., & Kalin, N. H. (2000a). Emotion, plasticity, context, and regulation: Perspectives from affective neuroscience. *Psychological Bulletin, 126,* 890–909.

Davidson, R. J., Marshall, J. R., Tomarken, A. J., & Henriques, J. B. (2000b). While a phobic waits: Regional brain electrical and autonomic activity in social phobics during anticipation of public speaking. *Biological Psychiatry, 47,* 85–95.

Davidson, R. J., & Tomarken, A. J. (1989). Laterality and emotion: An electrophysiological approach. In F. Boller & J. Grafman (Eds.), *Handbook of neuropsychology* (pp. 419–441). Amsterdam: Elsevier.

Dawson, G. (1994). Frontal electroencephalographic correlates of individual differences in emotional expression in infants. In N. A. Fox (Ed.), The development of emotion regulation: Biological and behavioral considerations. *Monographs of the Society for Research in Child Development, 59*(2–3, Serial No. 240), 135–151.

Dawson, G., Klinger, L. G., Panagiotides, H., Hill, D., & Spieker, S. (1992). Frontal lobe activity and affective behavior in infants of mothers with depressive symptoms. *Child Development, 63,* 725–737.

Deckel, A. W., Hesselbrock, V., & Bauer, L. (1996). Antisocial personality disorder, childhood delinquency, and frontal brain functioning: EEG and neuropsychological findings. *Journal of Clinical Psychology, 52,* 639–650.

de Haan, M. (2008). Event-related potential (ERP) measures in visual development research. In L. A. Schmidt & S. J. Segalowitz (Eds.), *Developmental psychophysiology: Theory, systems, and methods* (pp. 103–126). New York: Cambridge University Press.

Emmons, R. A., & Diener, E. (1985). Personality correlates of subjective well-being. *Personality and Social Psychology Bulletin, 11,* 89–97.

Eysenck, H. J. (1947). *Dimensions of personality.* London: Routledge & Kegan Paul.

Eysenck, H. J. (1953). *The structure of human personality.* London: Methuen.

Eysenck, H. J. (1967). *The biological basis of personality.* Springfield, IL: Thomas.

Field, T. (1986). Models of reactive and chronic depression in infancy. In E. Z. Tronick & T. Field (Eds.), *Maternal depression and infant disturbance* (pp. 47–60). San Francisco: Jossey-Bass.

Field, T., Fox, N. A., Pickens, J., & Nawrocki, T. (1995). Relative right frontal EEG activation in 3- to 6-month-old infants of depressed mothers. *Developmental Psychology, 31,* 358–363.

Field, T., Healy, B., Goldstein, S., Perry, S., Bendall, D., Schanberg, S., et al. (1988). Infants of depressed mothers show "depressed" behavior even with nondepressed adults. *Child Development, 59,* 1569–1579.

Field, T., Sandberg, D., Garcia, R., Vega-Lahr, N., Goldstein, S., & Guy, L. (1985). Prenatal problems, post-partum depression, and early mother–infant interaction. *Developmental Psychology, 12,* 1152–1156.

Fox, N. A. (1991). If it's not left, it's right: Electroencephalogram asymmetry and the development of emotion. *American Psychologist, 46,* 863–872.

Fox, N. A. (1994). Dynamic cerebral processes underlying emotion regulation. In N. A. Fox (Ed.), The development of emotion regulation: Biological and behavioral considerations. *Monographs of the Society for Research in Child Development, 59*(2–3, Serial No. 240), 152–166.

Fox, N. A., Bell, M. A., & Jones, N. A. (1992). Individual differences in response to stress and cerebral asymmetry. *Developmental Neuropsychology, 8,* 161–184.

Fox, N. A., & Davidson, R. J. (1984). Hemispheric substrates of affect: A developmental model. In

N. A. Fox & R. J. Davidson (Eds.), *The psychobiology of affective development* (pp. 353–382). Hillsdale, NJ: Erlbaum.

Fox, N. A., & Davidson, R. J. (1987). Electroencephalogram asymmetry in response to the approach of a stranger and maternal separation in 10-month-old infants. *Developmental Psychology, 23,* 233–240.

Fox, N. A., & Davidson, R. J. (1988). Patterns of brain electrical activity during facial signs of emotion in 10-month-old infants. *Developmental Psychology, 24,* 230–236.

Fox, N. A., Henderson, H. A., Rubin, K. H., Calkins, S. D., & Schmidt, L. A. (2001). Continuity and discontinuity of behavioral inhibition and exuberance: Psychophysiological and behavioral influences across the first four years of life. *Child Development, 72,* 1–21.

Fox, N. A., Rubin, K. H., Calkins, S. D., Marshall, T. R., Coplan, R. J., Porges, S. W., et al. (1995). Frontal activation asymmetry and social competence at four years of age. *Child Development, 66,* 1770–1784.

Fox, N. A., Schmidt, L. A., Calkins, S. D., Rubin, K. H., & Coplan, R. J. (1996). The role of frontal activation in the regulation and dysregulation of social behavior during the preschool years. *Development and Psychopathology, 8,* 89–102.

Gross, J. J., Sutton, S. K., & Ketelaar, T. V. (1998). Relations between affect and personality: Support for the affect-level and affective-reactivity views. *Personality and Social Psychology Bulletin, 24,* 279–288.

Gunnar, M. R., & Nelson, C. A. (1994). Event-related potentials in year-old infants: Relations with emotionality and cortisol. *Child Development, 65,* 80–94.

Guyer, A. E., Nelson, E. E., Perez-Edgar, K., Hardin, M. G., Roberson-Nay, R., Monk, C. S., et al. (2006). Striatal functional alteration in adolescents characterized by early childhood behavioral inhibition. *Journal of Neuroscience, 26,* 6399–6405.

Hagemann, D., Naumann, E., Thayer, J. F., & Bartussek, D. (2002). Does resting electroencephalograph asymmetry reflect a trait?: An application of latent state–trait theory. *Journal of Personality and Social Psychology, 82,* 619–641.

Harmon-Jones, E. (2004). Contributions from research on anger and cognitive dissonance to understanding the motivational functions of asymmetrical frontal brain activity. *Biological Psychology, 67,* 51–76.

Henriques, J. B., & Davidson, R. J. (1990). Regional brain electrical asymmetries discriminate between previously depressed and healthy controls. *Journal of Abnormal Psychology, 99,* 22–31.

Henriques, J. B., & Davidson, R. J. (1991). Left frontal hypoactivation in depression. *Journal of Abnormal Psychology, 100,* 535–545.

Jetha, M. K., Schmidt, L. A., & Goldberg, J. O. (in press-a). Long-term stability of resting frontal EEG alpha asymmetry and power in a sample of stable community outpatients with schizophrenia. *International Journal of Psychophysiology.*

Jetha, M. K., Schmidt, L. A., & Goldberg, J. O. (in press-b). Resting frontal EEG asymmetry and shyness and sociability in schizophrenia: A pilot study of community-based outpatients. *International Journal of Neuroscience.*

Jones, N. A., Field, T., Davalos, M., & Pickens, J. (1997a). EEG stability in infants/children of depressed mothers. *Child Psychiatry and Human Development, 28,* 59–70.

Jones, N. A., Field, T., Fox, N. A., Lundy, B., & Davalos, M. (1997b). EEG activation in 1-month-old infants of depressed mothers. *Development and Psychopathology, 9,* 491–505.

Jones, N. A., & Fox, N. A. (1992). Electroencephalogram asymmetry during emotionally evocative films and its relation to positive and negative affectivity. *Brain and Cognition, 20,* 280–299.

Kagan, J. (1994). *Galen's prophecy: Temperament in human nature.* New York: Basic Books.

Kagan, J. (1997). Temperament and the reactions to unfamiliarity. *Child Development, 68,* 139–143.

Kagan, J. (1999). The concept of behavioral inhibition. In L. A. Schmidt & J. Schulkin (Eds.), *Extreme fear, shyness, and social phobia: Origins, biological mechanisms, and clinical outcomes* (pp. 3–13). New York: Oxford University Press.

Kagan, J., & Snidman, N. (1991a). Temperamental factors in human development. *American Psychologist, 46,* 856–862.

Kagan, J., & Snidman, N. (1991b). Infant predictors of inhibited and uninhibited profiles. *Psychological Science, 2,* 40–44.

Kinsbourne, M. (1978). *Asymmetrical function of the brain.* New York: Cambridge University Press.

Kochanska, G. (1993). Toward a synthesis of parental socialization and child temperament in early development of conscience. *Child Development, 64,* 325–347.

Larsen, R. J., & Ketelaar, T. (1991). Personality and susceptibility to positive and negative emotional states. *Journal of Personality and Social Psychology, 61,* 132–140.

Larson, C. L., Sutton, S. K., & Davidson, R. J. (1998). Affective style, frontal brain asymmetry and the time course of the emotion-modulated startle response. *Psychophysiology, 37,* 92–101.

LeDoux, J. E. (1996). *The emotional brain.* New York: Simon & Schuster.

LeDoux, J. E., Iwata, J., Cicchetti, P., & Reis, D. J. (1988). Different projections of the central amygdaloid nucleus mediate autonomic and behavioral correlates of conditioned fear. *Journal of Neuroscience, 8,* 2517–2519.

Luck, S. J. (2005). *An introduction to the event-related potential technique.* Cambridge, MA: MIT Press.

Marshall, P. J., Reeb, B. C., & Fox, N. A. (in press). Electrophysiological responses to auditory novelty in temperamentally different 9-month-old infants. *Developmental Science.*

Mayer, A. R., & Kosson, D. S. (2000). Handedness and psychopathy. *Neuropsychiatry, Neuropsychology, and Behavioral Neurology, 4,* 233–238.

Meyer, G. J., & Shack, J. R. (1989). The structural convergence of mood and personality: Evidence for old and new "directions." *Journal of Personality and Social Psychology, 57,* 691–706.

Murray, H. A. (1938). *Explorations in personality.* New York: Oxford University Press.

Perez-Edgar, K., & Fox, N. A. (2003). Individual differences in children's performance during an emotional Stroop task: A behavioral and electrophysiological study. *Brain and Cognition, 52,* 33–51.

Perez-Edgar, K., & Fox, N. A. (2007). Temperamental contributions to children's performance in an emotion-word processing task: A behavioral and electrophysiological study. *Brain and Cognition, 65,* 22–35.

Rothbart, M. K. (2007). Temperament, development, and personality. *Current Directions in Psychological Science, 16,* 207–212.

Rothbart, M. K., Ahadi, S. A., Hershey, K., & Fisher, P. (2001). Investigations of temperament at three to seven years: The Children's Behavior Questionnaire. *Child Development, 72,* 1394–1408.

Rothbart, M. K., & Bates, J. E. (1998). Temperament. In W. Damon (Series Ed.) & N. Eisenberg (Vol. Ed.), *Handbook of child psychology: Vol. 3. Social, emotional, and personality development* (5th ed., pp. 105–176). New York: Wiley.

Rothbart, M. K., & Bates, J. E. (2006). Temperament. In W. Damon & R. Lerner (Series Eds.) & N. Eisenberg (Vol. Ed.), *Handbook of child psychology: Vol. 3. Social, emotional, and personality development* (6th ed., pp. 99–166). Hoboken, NJ: Wiley.

Rothbart, M. K., & Derryberry, D. (1981). Development of individual differences in temperament. In M. E. Lamb & A. L. Brown (Eds.), *Advances in developmental psychology* (Vol. 1, pp. 37–86). Hillsdale, NJ: Erlbaum.

Rothbart, M. K., & Posner, M. I. (2006). Temperament, attention, and developmental psychopathology. In D. Cicchetti & D. Cohen (Eds.), *Developmental psychopathology: Vol. 2. Developmental neuroscience* (2nd ed., pp. 465–501). Hoboken, NJ: Wiley.

Rothbart, M. K., & Sheese, B. E. (2007). Temperament and emotion regulation. In J. J. Gross (Ed.), *Handbook of emotion regulation* (pp. 331–350). New York: Guilford Press.

Santesso, D. L., Reker, D. L., Schmidt, L. A., & Segalowitz, S. J. (2006a). Frontal electroencephalogram activation asymmetry, emotional intelligence, and externalizing behaviors in 10-year-old children. *Child Psychiatry and Human Development, 36*, 311–328.

Santesso, D. L., Schmidt, L. A., & Trainor, L. J. (2007). Frontal brain electrical activity (EEG) and heart rate responses to affective infant-directed (ID) speech in 9-month-old infants. *Brain and Cognition, 65*, 14–21.

Santesso, D. L., Segalowitz, S. J., Ashbaugh, A. R., Antony, M. M., McCabe, R. E., & Schmidt, L. A. (2008). Frontal EEG asymmetry and sensation seeking in young adults. *Biological Psychology, 78*, 164–172.

Santesso, D. L., Segalowitz, S. J., & Schmidt, L. A. (2005). ERP correlates of error monitoring in 10-year olds are related to socialization. *Biological Psychology, 70*, 79–87.

Santesso, D. L., Segalowitz, S. J., & Schmidt, L. A. (2006b). Error-related electrocortical responses in 10-year-old children and young adults. *Developmental Science, 9*, 473–481.

Santesso, D. L., Segalowitz, S. J., & Schmidt, L. A. (2006c). Error-related electrocortical responses are enhanced in children with obsessive–compulsive behaviors. *Developmental Neuropsychology, 29*, 431–445.

Schmidt, L. A. (1996). *The psychophysiology of self-presentation anxiety in seven-year-old children: A multiple measure approach.* Unpublished doctoral dissertation, University of Maryland, College Park.

Schmidt, L. A. (1999). Frontal brain electrical activity in shyness and sociability. *Psychological Science, 10*, 316–320.

Schmidt, L. A. (Ed.). (2003). Affective neuroscience [Special issue]. *Brain and Cognition, 52*(1).

Schmidt, L. A. (Ed.). (2007). Social cognitive and affective neuroscience: Developmental and clinical perspectives [Special issue]. *Brain and Cognition, 65*(1–2).

Schmidt, L. A. (2008). Patterns of second-by-second resting frontal brain (EEG) asymmetry and their relation to heart rate and temperament in 9-month-old human infants. *Personality and Individual Differences, 44*, 216–225.

Schmidt, L. A., Cote, K. A., Santesso, D. L., & Milner, C. E. (2003). Frontal electroencephalogram (EEG) alpha asymmetry during sleep: Stability and its relation to affective style. *Emotion, 3*, 401–407.

Schmidt, L. A., & Fox, N. A. (1994). Patterns of cortical electrophysiology and autonomic activity in adults' shyness and sociability. *Biological Psychology, 38*, 183–198.

Schmidt, L. A., & Fox, N. A. (1998). Fear-potentiated startle responses in temperamentally different human infants. *Developmental Psychobiology, 32*, 113–120.

Schmidt, L. A., Fox, N. A., Schulkin, J., & Gold, P. W. (1999). Behavioral and psychophysiological correlates of self-presentation in temperamentally shy children. *Developmental Psychobiology, 35*, 119–135.

Schmidt, L. A., & Segalowitz, S. J. (Eds.). (2008). *Developmental psychophysiology: Theory, systems, and methods.* New York: Cambridge University Press.

Schmidt, L. A., & Trainor, L. J. (2001). Frontal brain electrical activity (EEG) distinguishes *valence* and *intensity* of musical emotions. *Cognition and Emotion, 15*, 487–500.

Schwartz, C. E., Wright, C. I., Shin, L. M., Kagan, J., & Rauch, S. L. (2003). Inhibited and uninhibited infants "grow up": Adult amygdalar response to novelty. *Science, 300*, 1952–1953.

Stifter, C. A., Spinrad, T. L., & Braungart-Rieker, J. M. (1999). Toward a developmental model of child compliance: The role of emotion regulation in infancy. *Child Development, 70*, 21–32.

Sutton, S. K., & Davidson, R. J. (1997). Prefrontal brain asymmetry: A biological substrate of the behavioral approach and inhibition systems. *Psychological Science, 8*, 204–210.

Theall-Honey, L. A., & Schmidt, L. A. (2006). Do temperamentally shy children process emotion differently than non-shy children?: Behavioral, psychophysiological, and gender differences in reticent preschoolers. *Developmental Psychobiology, 48*, 187–196.

Thomas, A., & Chess, S. (1977). *Temperament and development.* New York: Brunner/Mazel.

Thomas, A., & Chess, S. (1981). The reality of difficult temperament. *Merrill–Palmer Quarterly, 28,* 1–19.

Thomas, A., Chess, S., & Birch, H. G. (1968). *Temperament and behavior: Disorders in children.* New York: New York University Press.

Tomarken, A. J., Davidson, R. J., & Henriques, J. B. (1990). Resting frontal brain asymmetry predicts affective responses to films. *Journal of Personality and Social Psychology, 59,* 791–801.

Tomarken, A. J., Davidson, R. J., Wheeler, R. E., & Kinney, L. (1992). Psychometric properties of resting anterior EEG asymmetry: Temporal stability and internal consistency. *Psychophysiology, 29,* 576–592.

Urry, H. L., van Reekum, C. M., Johnstone, T., Kalin, N. H., Thurow, M. E., Schaefer, H. S., et al. (2006). Amygdala and ventromedial prefrontal cortex are inversely coupled during regulation of negative affect and predict the diurnal pattern of cortisol secretion among older adults. *Journal of Neuroscience, 26,* 4415–4425.

Vuga, M., Fox, N. A., Cohn, J. F., George, C. J., Levenstein, R. M., & Kovacs, M. (2006). Long-term stability of frontal electroencephalographic asymmetry in adults with a history of depression and controls. *International Journal of Psychophysiology, 59,* 107–115.

Warr, P., Barter, J., & Brownbridge, G. (1983). On the independence of positive and negative affect. *Journal of Personality and Social Psychology, 44,* 644–651.

Watson, D., & Clark, L. A. (1984). Negative affectivity: The disposition to experience unpleasant emotional states. *Psychological Bulletin, 95,* 465–490.

Watson, D., Clark, L. A., & Tellegen, A. (1988). Development and validation of brief measures of positive and negative affect: The PANAS scales. *Journal of Personality and Social Psychology, 54,* 1063–1070.

Wiedemann, G., Pauli, P., Dengler, W., Lutzenberger, W., Birbaumer, N., & Buchkremer, G. (1999). Frontal brain asymmetry as a biological substrate of emotions in patients with panic disorders. *Archives of General Psychiatry, 56,* 78–84.

Wilson, D. S., Clark, A. B., Coleman, K., & Dearstyne, T. (1994). Shyness and boldness in humans and other animals. *Trends in Ecology and Evolution, 9,* 442–446.

Zuckerman, M. (2005). *Psychobiology of personality.* Cambridge, UK: Cambridge University Press.

CHAPTER 17

Reward Systems

MONIQUE ERNST
LINDA PATIA SPEAR

This chapter addresses reward processes at the behavioral and neural systems levels. The understanding of the neural basis of reward processes is critically important for the study of developmental social neuroscience. Social stimuli are among the most powerful motivating forces in the deployment of motivated behavior. In addition, early social isolation leads to altered behavior, manifested as increased self-stimulation and avoidance behavior, indicative of reduced novelty-seeking and reward-related behaviors. Such interaction between social and reward processes are supported by a significant overlap of the neural circuitries subserving these processes.

Both animal research and human work are reviewed here. The substantial consistency of current findings across species, and the rich interaction between animal and human research, dominate neuroscience research on reward systems and highlight the interplay of hard-wired, evolution-molded mechanisms.

With respect to neurodevelopment, human research has prominently used functional neuroimaging tools. Whereas a large literature is being accumulated in adults, this methodology presents practical limitations for young children, mostly for those below age 7 years. For this reason, there is a dearth of information on early developmental changes in human reward circuitry. Moreover, despite the critical importance of puberty, very little work with humans (or even with animal models) has tried to tie together hormonal changes during adolescence with aspects of neural maturation of reward systems. These gaps in knowledge emerge clearly in the present work.

It should be noted that this review is by no means exhaustive, given the huge amount of critical work that has been conducted in this field. Thus a number of influential and important studies are omitted, not because of lesser merit, but merely because of parsimony.

First, we define reward processes. Second, we describe what is known of the generic reward circuitry in animals and humans. Third, we address developmental changes that have been observed in this system, and end this section with the proposal of a neurodevelopmental model. Finally, we briefly examine potential interactions between reward-related and socially related neural processes.

Definition

Reward is the crux of goal-directed behavior, and is at the center of research in a large number of disciplines whose mission is to understand human behavior: philosophy, sociology, psychology, economy, finances, education, and of course neuroscience. Within neuroscience, reward occupies a key place in the fields of decision making, addiction, and learning. For this reason, the concept of "reward" may vary as a function of the field of study, which can lead to confusion.

Reward can be viewed as an object or situation that elicits positive feelings, and is thus defined as a function of its hedonic properties. It can also be operationalized as an object or situation that induces approach behavior, and is thus defined as a function of the behavioral responses that it induces. Within the field of learning, a rewarding or "appetitive" stimulus may serve to increase the incidence of immediately preceding behaviors, a property referred to as "positive reinforcement"; this contrasts with punishing or "aversive" stimuli, which suppress the incidence of preceding behaviors, and which are referred to as "negative reinforcers."

Thus reward implies the engagement of attention (attention directed toward rewarding stimuli), emotion (pleasure), motivation (intensity of drive), and motor processes. This multisystem involvement is reflected in the complex findings of functional neuroimaging studies of reward processes. A certain degree of homogeneity is provided by the neurochemical basis of reward function—that is, the dopamine (DA) system. However, here again, the neurochemical, cellular, and molecular implications are not simple. Indeed, components of the reward system are also critically dependent on the function of other neurotransmitters, including their related cascades of intracellular events (Nestler, 2001).

This brief characterization of the neurobiology of reward function has several developmental implications. First, cognitive and affective processes can follow different maturational trajectories, resulting in different behavioral manifestations of reward processes as a function of age (Ernst, Pine, & Hardin, 2006). Similarly, developmental neural changes at the structural (e.g., synaptic density), functional (e.g., efficiency of synaptic function), and neurochemical levels can all affect the development of reward systems. Finally, the environment may differentially affect the various maturational trajectories of each of these components. It may do so with more or less potency as a function of age, based on time windows of greater or lower potential for neural plasticity.

Although even if this cursory review of the factors to be considered in the development of reward processes may appear hopelessly complex, within a given framework some understanding of the neurobiology and behavioral consequences of this development can be achieved. The framework in this chapter is that of cognitive neuroscience and basic animal work. More specifically, we concentrate on the neural circuits that have been associated with the coding of motivation, which is at the interface of affective and motor responses. We attempt to draw some inferences for the biology of social neuroscience.

Models of Basic Reward Neurocircuitry

Basic research to identify the neural substrates of reward has revealed that behaviors motivationally oriented toward natural rewards and drugs of abuse can be partitioned into diverse psychological components, each with sometimes overlapping and often incompletely understood neural substrates (e.g., Baxter & Murray, 2002; Cardinal, Parkinson, Hall, & Everitt, 2002).

Research in addiction—the prototypical behavioral perturbation of reward processes—has been at the forefront of the examination of the basic reward circuitry. One critical component of brain reward processing is the mesocorticolimbic DA system, characterized by projections from DA cell bodies located in the ventral tegmental area (VTA) to forebrain regions that include the nucleus accumbens, amygdala, ventral pallidum, and prefrontal cortex (PFC). These regions in turn project back either directly or indirectly to the VTA, and are highly interconnected with each other in forebrain circuitry that also includes the medial dorsal thalamus and portions of the extended amygdala (Kalivas & Volkow, 2005). A number of hypotheses on the role of DA in reward processes continue to be debated (for reviews, see Nestler, 2001; Schultz, 2006). These models arise from different research perspectives and may not necessarily be mutually exclusive.

An early set of theories postulated that DA is critical for mediating the hedonic impact of rewards (Wise, 1980). According to this view, functional insufficiencies in mesocorticolimbic DA reward pathways result in a "reward deficiency syndrome" (Blum et al., 1995; Gardner, 1999) (Figure 17.1). Such functional hypodopaminergic states have been posited to emerge progressively during repeated abuse of drugs, leading to addiction and an intensification of drug use in an attempt to compensate for this reward (DA) insufficiency (e.g., Volkow, Fowler, Wang, & Goldstein, 2002)

Another prominent theory of the neural bases of reward, the "incentive salience theory" (Berridge, 2000), posits that the mesolimbic DA system and related circuitry provide the motivational drive to direct behavior toward reward-related stimuli (Figure 17.1). That is, enhanced DA function in mesolimbic brain regions such as the nucleus accumbens is thought to play a critical role in the attribution of incentive motivation to stimuli associated with rewards and reinforcers, thereby increasing the extent to which these stimuli are "wanted" (Berridge & Robinson, 1998). There is evidence that this DA-dependent, motivational com-

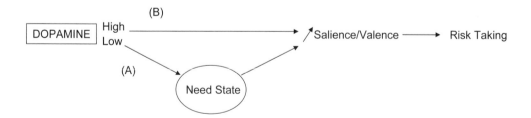

FIGURE 17.1. A schematic representation of two dopaminergic reward models. (A) The reward deficiency model (Blum et al., 1995; Gardner, 1999) is based on a hypodopaminergic state that triggers a "need state" by which stimuli acquire more salience. (B) The incentive salience theory (Berridge, 2000) is based on a hyperdopaminergic state that directly contributes to enhanced motivational drive, and thus increased salience of stimuli.

ponent to reward is distinct from the more affective hedonic response ("liking") generated in response to a rewarding stimulus. Alterations in DA activity do not influence "liking"; rather, the hedonic response to a rewarding stimulus is thought to be related to the actions of opiate, benzodiazepine/gamma-aminobutyric acid (GABA), cannabinoid, and possibly serotonergic systems (Berridge, 2000; Peciña, Smith, & Berridge, 2006). Particular "hedonic hot spots" where opioids recently have been shown to magnify the impact of positive hedonic states include portions of the shell of the nucleus accumbens and the ventral pallidum (Peciña et al., 2006; Tindell, Smith, Peciña, Berridge, & Aldridge, 2006). Here, the incentive salience theory predicts excessive reward-related behavior as a result of elevated DA activity, rather than as a consequence of decreased DA activity as described in the reward deficiency theory.

A third model of the function of dopamine in the reward system, the "reward prediction error," comes from electrophysiological recordings of DA neurons in nonhuman primates (Schultz, Dayan, & Montague, 1997) (Figure 17.2). Firing patterns of DA neurons were found to track the discrepancy between actual and predicted outcomes, such that presentation of

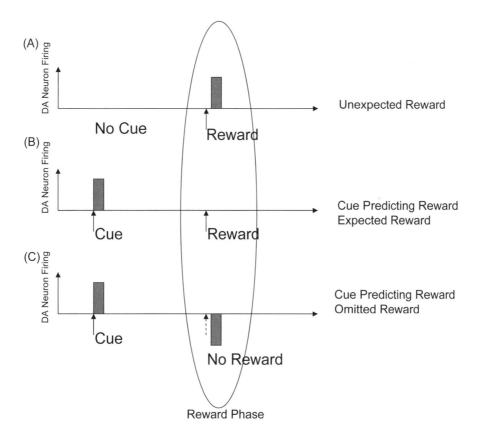

FIGURE 17.2. A schematic representation of the reward prediction error (Schultz, Dayan, & Montague, 1997). Firing patterns of DA neurons were found to track the discrepancy between actual and predicted outcomes, such that presentation of unexpected rewards resulted in increased activity (A) and omissions of expected reward resulted in decreased activity (C). In addition, DA neurons fired at the presentation of a cue that predicted rewards (B).

unexpected rewards resulted in increased activity and omissions of expected reward resulted in decreased activity (Schultz et al., 1997; Tobler, Fiorillo, & Schultz, 2005). This model is critical to learning theories and has led researchers to consider DA changes in response to rewards, or stimuli that predict rewards, as learning signals (Schultz et al., 1997).

Finally, the "tonic–phasic DA theory" (Grace, 1995) has fostered yet another formulation of reward processes (Plate 17.1). Differential functional roles that have been postulated for dopaminergic activity at intrasynaptic versus extrasynaptic levels; these have been termed "phasic" and "tonic" DA activity, respectively (see Thompson, Pogue-Geile, & Grace, 2004). "Phasic" DA stimulation refers to the rapid and transient increase in intrasynaptic DA levels that results from the release of DA in the synapse precipitated by action potentials (see Grace, 1995). Natural rewards and reward-predictive cues are among the stimuli that induce burst firing of dopaminergic neurons and lead to phasic DA release (Schultz, 1998). "Tonic" DA stimulation, in contrast, refers to the much lower levels of extracellular dopamine that are located in extrasynaptic regions. Such extrasynaptic DA activates highly sensitive DA autoreceptors that serve to dampen phasic DA release (see Grace, 1995). Among the stimuli that increase tonic DA levels are glutaminergic afferents from the PFC and possibly other mesocorticolimbic regions, such as the amygdala. Via elevating tonic DA levels, such glutaminergic input is thought to reduce the amount of phasic DA release induced by reward-related stimuli (e.g., Grace, 1991). This is one important route by which prefrontal regions and other portions of the mesocorticolimbic region may influence DA-modulated, reward-directed behaviors.

Emerging from the cognitive neuroscience fields of learning and decision making, efforts are being made to integrate within a coherent model of reward processes the distributed functions of the various nodes of the mesocorticolimbic system (e.g., Ernst & Paulus, 2005). For example, the amygdala has recently been hypothesized to play a role in linking positive emotions to stimuli (i.e., conditioning), which adds to the extant literature implicating this structure in the conditioning processes involving negative emotions (Baxter & Murray, 2002). This function in reward conditioning is also supported by portions of the orbital frontal cortex, providing some redundancy as well as functional complementarity (Baxter & Murray, 2002; Gallagher, McMahan, & Schoenbaum, 1999; Izquierdo & Murray, 2004; Schoenbaum, 2004; Schoenbaum & Roesch, 2005). Functional maps of neural circuits across brain regions have been challenging to trace because of limitations in the approaches available to this type of research until now. These approaches, which include lesion studies, electrophysiological recordings, and microdialysis in animals, can only assess a limited number of structures at a time. The recent advent of functional neuroimaging in humans has provided a radical shift in the apprehension of the studies of distributed neural networks involved in the coding for complex processes. Indeed, such techniques as functional magnetic resonance imaging (fMRI) and positron emission tomography provide indices of whole-brain function associated with specific behavioral, cognitive, or affective processes.

Based on this new line of research, the most current view of the neuroanatomy of the reward circuitry describes a distributed network composed of cortical and subcortical structures, which fulfill specific roles in the use of reward stimuli for adaptive behavior (e.g., Bechara, Damasio, Tranel, & Damasio, 2005; Ernst & Paulus, 2005). This network provides a simplified working model that is based not only on postulated functions of given structures, but also on the efficiency of communications and exchange of information among these structures (Plate 17.2). The major nodes of this network include the amygdala as the translator of affective value; the insula and orbital frontal cortices as the integrators of the current value of reward; the striatum as the promoter of action; the anterior cingulate cortex as the detector of errors of estimation or conflict/planning for behavioral responses; and the medial PFC as the evaluator of the self-related significance of stimuli.

Spatial resolution, however, is a limitation to the drawing of such functional maps. This is a place where "cross-talk" between animal and human research becomes critical. For example, animal work has identified histologically and functionally distinct regions of the nucleus accumbens—that is, the core and the shell areas (Di Chiara, 2002). These subregions have complementary but different functions in the coding of reward stimuli. Not only are we unable to identify these different subregions via functional neuroimaging, but the recognition of the nucleus accumbens as a separate structure with clear boundaries within the ventral striatum of the human brain remains disputed. Similarly, the amygdala is a composite of nuclei, including the basolateral and central nuclei, which play critical but different roles in the coding of affective values of stimuli (Everitt, Cardinal, Hall, Parkinson, & Robbins, 2000). As neuroimaging technology continues to evolve, it is conceivable that the next generation of instruments may provide the necessary spatial resolution to probe the functions of such subnuclei.

Each component of the reward-related network undergoes changes along unique maturational trajectories. The current state of knowledge about these trajectories is quite limited and is in dire need of research. Here again, the animal literature can be of great help, although obvious limitations related to species differences can make the translation of animal findings to human physiology challenging (Levitt, 2003).

Development of Reward-Related Circuitry during Adolescence

Animal Work

The mesocorticolimbic DA system and related forebrain reward-related circuitry undergo considerable development not only during early life, but during adolescence as well. Indeed, within a framework of a developmentally dynamic brain throughout life, evidence has gradually mounted that adolescence is a time of particularly dramatic developmental changes. These alterations include a marked loss of synaptic connections in certain brain regions; a decline in overall brain energy utilization; increases in cortical white matter associated with axon myelination; and a variety of neurochemical, hormonal, and structural modifications (for a review, see Spear, 2007). Interestingly, this adolescence-associated sculpting of the brain is highly conserved, with similar alterations emerging during this developmental transition in humans as in other mammalian species. As discussed below, prominent among the neural regions undergoing ontogenetic changes during adolescence are brain regions previously discussed as being implicated in the processing of rewards and directing reward-related behavior.

Studies in laboratory animals have revealed notable changes in the DA system during adolescence. DA input to the PFC increases in nonhuman primates during adolescence, to peak at levels considerably higher than those seen earlier or later in life (Rosenberg & Lewis, 1995). DA concentrations (Leslie, Robertson, Cutler, & Bennett, 1991) and DA fiber density (Benes, Taylor, & Cunningham, 2000) have also been shown to increase through adolescence in the PFC of rats. In addition, DA modulation within the PFC undergoes important modifications in adolescence. Most impressive is the loss of "buffering capacity" associated with the disappearance of DA autoreceptors in PFC that served a regulatory negative feedback function during the juvenile period (Dumont, Andersen, Thompson, & Teicher, 2004). In PFC slices of adolescent rats, coactivation of D1 DA receptors and N-methyl-D-aspartate (NMDA) receptors fail to yield an adult-typical depolarized "up-state" that is thought to be critical for plasticity and information processing (Tseng & O'Donnell, 2005). These alterations would be consistent with reduced flexibility in adapting to environmental changes in adolescence, at least within this particular component of the PFC. In other ways as well, cortical adaptations may vary during adolescence from those seen in adulthood. For instance, cortical stimulation has been shown to up-regulate $GABA_A$ receptors and to down-regulate glutamate alpha-amino-3-hydroxy-5-methyl-4-isoxasolepropionic acid (AMPA) receptors in adults, with opposite adaptation patterns often emerging in adolescent and younger organisms (Shaw & Lanius, 1992; Shaw & Scarth, 1992).

In contrast to the developmental increases in DA input to the PFC during adolescence, portions of the mesolimbic DA system appear to be streamlined at the same time. Developmental declines of one-third to one-half of the D1 and D2 DA receptors have been reported during adolescence in the striatum (Tarazi & Baldessarini, 2000; Teicher, Andersen, & Hostetter, 1995); similar changes have been reported in the nucleus accumbens by some (Tarazi, Tomasini, & Baldessarini, 1998, 1999), but not all (Teicher et al., 1995), researchers. This subcortical DA receptor pruning contrasts with a more delayed, early adulthood decline in DA receptors in PFC (Andersen, Thompson, Rutstein, Hostetter, & Teicher, 2000).

DA projections to mesocortical (PFC) and mesolimbic/striatal areas show opposite developmental patterns in terms of DA synthesis and turnover, with estimates of basal rates of DA synthesis and turnover higher in the PFC early than late in adolescence and in adulthood, whereas these estimates are lower in striatum early than later in adolescence and adulthood (Anderson, Eisenstat, Shi, & Rubenstein, 1997; Leslie et al., 1991; Teicher et al., 1993). These developmental shifts in DA synthesis and turnover have led to the suggestion that there is an ontogenetic shift in the balance between mesolimbic/striatal and mesocortical DA systems during adolescence (Andersen, 2003; Spear, 2000), with more pronounced DA activity in mesocortical regions in early adolescence shifting toward a greater predominance of mesolimbic/striatal DA activity later in adolescence (e.g., Spear, 2007)

Such region-specific patterns of changes are consistent with the idea of heterogeneity—not only temporal, but also qualitative—in the developmental trajectories of brain structures that act in concert to yield adaptive behavior. As a result, normative behavior is expected to be regulated differently in adolescence than in other ontogenic periods. A possible consequence could be the unique peak of risk-taking behavior during this transitional period of life (Chambers, Taylor, & Potenza, 2003; Ernst et al., 2006; Laviola, Macri, Morley-Fletcher, & Adriani, 2003; Spear, 2000).

Parallels between Human and Animal Work

DA activity during adolescence has not been systematically explored in humans, due to methodological and ethical limitations (Arnold, Zametkin, Caravella, & Korbly, 2000; Munson, Eshel, & Ernst, 2006), with the exception of one early study that used autopsy material. This study revealed an ontogenetic decline in DA receptor binding in striatum over an age range that included adolescence (Seeman et al., 1987). In studies using a number of other indices, however, ontogenetic changes have emerged in reward-related circuitry in human adolescents that are reminiscent of those observed in laboratory animals. For instance, synaptic pruning of presumed glutaminergic excitatory input to the PFC has been reported in adolescent humans (Huttenlocher, 1984) and nonhuman primates (Zecevic, Bourgeois, & Rakic, 1989), whereas developmental declines in cortical binding of the NMDA glutamate receptor are evident in adolescent rats (Insel, Miller, & Gelhard, 1990). Volumetric declines in the PFC have been reported during adolescence in humans (Sowell et al., 1999) as well as in rats (van Eden, Kros, & Uylings, 1990).

Signs of alterations in amygdala function during adolescence have emerged in studies conducted in humans as well as in laboratory animals, although typically very different measures have been used. For instance, studies in laboratory animals have revealed that excitatory input from the basolateral nucleus of the amygdala to the PFC continues to be elaborated during adolescence (Cunningham, Bhattacharyya, & Benes, 2002), and that the amygdala shows a different pattern of stress-induced gene expression during adolescence than during adulthood (Kellogg, Awatramani, & Piekut, 1998). More specifically, in both the medial and cortical nucleus of the amygdala, adolescents show substantially less stress-related increase than adults in Fos, a protein product of the immediate early gene c-fos that is transiently expressed in a variety of neuronal populations in response to activation. These changes are paralleled with human studies suggesting alterations in amygdala function during adolescence revealed by functional neuroimaging studies, as discussed below.

Cognitive Neuroscience Studies

Neuroimaging studies using fMRI techniques follow two main strategies. The first, "structure-driven" strategy assays the function of specific regions or circuits by using cognitive paradigms that are known to recruit the structures under scrutiny. The second, "process-driven" strategy is based on the development of cognitive paradigms that probe specific cognitive or/and affective processes whose neural correlates are being investigated. For example, structure-driven strategies have been used to examine the function of the amygdala, striatum, or anterior cingulate cortex, whereas process-driven strategies have been used to study various aspects of reward-related processes, such as delay discounting, decision making, or learning.

Exposure to emotional stimuli, and particularly to evocative faces, has become the standard approach for probing amygdala function in humans. The few developmental studies of the amygdala have found notable differences in activation patterns between adolescents and adults (or children), although the ontogenetic patterns obtained are not always consistent (Ernst et al., 2005; Guyer et al., 2008; Killgore, Oki, & Yurgelun-Todd, 2001; McClure et al., 2004; Pine et al., 2001; Thomas et al., 2001). The discrepant findings in this literature may be related in part to complications associated with lateralized or sex-specific effects (Killgore & Yurgelun-Todd, 2004; McClure et al., 2004); to differences in emotional content or attentional demands of the paradigms (Monk et al., 2003); and to the lack of spatial resolution necessary to establish foci in specific amygdala nuclei (Zald, 2003). Thus, whereas adolescent-typical alterations in amygdala function have emerged frequently in studies of both humans and laboratory animals, substantially more work is needed to characterize those apparent age-specific effects and their relationship with developmental changes in other portions of the mesocorticolimbic reward circuitry.

Reward-related processes have recently been examined in a huge amount of research using functional neuroimaging. In adults, this research breaks down into two main fields of study: decision making or "neuroeconomics" (e.g., Ernst & Paulus, 2005; Zak, Kurzban, & Matzner, 2004), and learning (for a review, see Martin-Soelch, Linthicum, & Ernst, 2007). With regard to the developmental aspect of reward-related neural correlates, the literature is sparse, although rapidly growing (e.g., see Crone & Westenberg, Chapter 19, this volume, for a discussion of decision making). Based on the available literature in humans and animals, including lesion, electrophysiological, and imaging work, a neurobiological model of the neurodevelopment of reward systems has been proposed (Ernst et al., 2006). This model, the "triadic model," is discussed next.

The Triadic Model

The triadic model, originally designed to provide a framework for the neural basis of the peak onset of risky behaviors in adolescence, is based upon the notion that adaptive motivated behavior requires the functional integration of neural circuits underlying reward-, punishment-, and cognitive-control-related processes. As illustrated in Figure 17.3, the triadic model is constructed within a framework of equilibrium among three systems: the "reward/ approach" system, centered on the ventral striatum; the "punishment/avoidance" system, centered on the amygdala; and the "supervisory" system, which modulates the relative weight of the other two systems, and whose seat is within the medial PFC.

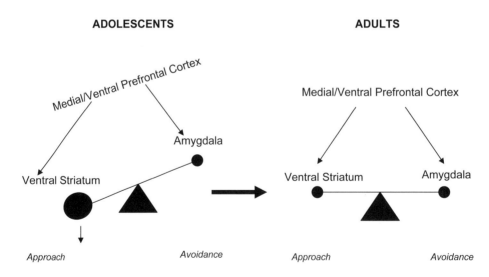

FIGURE 17.3. The triadic developmental model of motivated behavior (Ernst, Pine, & Hardin, 2006) focuses on three structures that are representative of three functional networks dedicated to (1) approach behavior (ventral striatum), (2) avoidance behavior (amygdala), and (3) executive control (medial prefrontal cortex). This model postulates a distinct equilibrium among these three networks in adolescents relative to adults. In adolescents, the balance will be tilted toward approach behavior (e.g., ventral striatal circuits) as a result of a stronger influential role of ventral striatal regions, a relatively lesser impact of amygdala function, and weaker regulatory control from medial prefrontal cortex. Conversely, in adults, the balance will show a better equilibrium between the approach and avoidance functional networks.

It is important to specify that the structures designated to be core components of the model imply the contribution of functionally associated circuits. Rewarding and aversive stimuli are coded and represented within complex distributed and largely overlapping networks. For example, the amygdala and the ventral striatum each contribute to both appetitive and aversive processes in a coordinated fashion. It has been proposed that portions of the amygdala facilitate stimulus–reward learning in close interaction with the nucleus accumbens (Everitt, Cador, & Robbins, 1989). This encoding function, which covers the whole range of positive to negative stimulus values, may be associated with some sensitivity bias; the amygdala may be particularly active in processing negative stimuli (Zald, 2003), and the ventral striatum in processing rewarding stimuli (Ikemoto & Panksepp, 1999). This double characterization of the function of these structures—that is, as involved in the whole "valence space" of stimuli, but also as showing enhanced sensitivity for a given polarity—reflects the heterogeneity of these structures composed of subregions, each of them assuming unique functions associated with distinct webs of connections.

The triadic model proposes that adolescent decision making is characterized by a bias toward approach behavior, reflecting (1) dominance of the reward- over the aversion-related system, and (2) poor cognitive control over the balance between the reward- and aversion-related systems (Ernst et al., 2006). Overall, adolescents do recruit a reward circuit similar to that of adults when responding to rewards or punishments (May et al., 2004). However, in support of the model, the relative engagement of these structures differs between adults and adolescents. Adolescents have been found to activate the ventral striatum more strongly than adults in response to rewarding stimuli (Ernst et al., 2005; Galvan et al., 2006), and to deactivate the amygdala less strongly than adults in response to penalties (Ernst et al., 2005). These findings are congruent with the notion of stronger neural engagement of approach-related systems and lesser engagement of aversion-related systems in adolescents than in adults. During choice making in one study, adolescents engaged the medial PFC, including the anterior cingulate gyrus, and the orbital frontal cortex to a lesser extent than did adults (Eshel, Nelson, Blair, Pine, & Ernst, 2007), suggesting diminished involvement of structures associated with cognitive control in adolescents.

However, discrepant findings have also been reported. Preadolescent children were found to show greater modulation of the medial PFC, including the anterior cingulate gyrus, by reward probability during anticipation relative to adults. Although no age differences were noted during reward anticipation overall, preadolescents showed stronger activation for low than for high probability of reward; in contrast, adults showed no differences (van Leijenhorst, Crone, & Bunge, 2006). In this same study, during the outcome phase, children showed more lateral orbital frontal activation than adults in response to negative outcomes (failure to receive a reward) during the low-probability condition. Only reward probability (and not reward magnitude) was manipulated in this study, and selection did not involve choosing among competing options. This suggests that the developmental findings in this study may relate more closely to processes of uncertainty than to reward per se. In addition, differences in reward processes may also exist between preadolescent children and adolescents (Galvan et al., 2006). Finally, Bjork et al. (2004) used a reaction time task in adolescents and adults, and reported decreased engagement of the ventral striatum in adolescents compared to adults during the period of reward anticipation/motor preparation. This finding could be interpreted in two ways: It could reflect either a weaker reward system or a more efficient reward system, reminiscent of the aforementioned theories of reward deficiency syndrome and incentive salience, respectively. The jury is still out.

Relationships between Reward-Related and Socially Related Neural Processes

As discussed above, the available ontogenetic data, albeit limited, are consistent with the conclusion that adolescence is characterized by notable alterations in mesocorticolimbic circuitry implicated in the processing of reward-related stimuli. These adolescence-associated changes in reward circuitry may also contribute to the shaping of the distinct social landscape typical of the adolescent period.

Among the dramatic changes in social affiliation that occur during adolescence are the increased focus on peer-directed social interactions, increased conflicts with parents, and emerging interest in the opposite sex (e.g., Steinberg, 1989). The increased focus on peer-directed social interactions during adolescence has been documented both in humans (Csikszentmihalyi, Larson, & Prescott, 1977; La Greca, Prinstein, & Fetter, 2001; Larson & Richards, 1991) and in other species (Primus & Kellogg, 1989). Along with the ignition of passion during adolescence, emotions run deeper and become more labile, and the risk for the onset of mood and anxiety disorders reaches a peak (Arnett, 1992; Costello et al., 2002; Dahl, 2004; Ernst et al., 2005; Masten, 2004; Pine, Cohen, Gurley, Brook, & Ma, 1998).

The concomitant changes in both reward-related and social behaviors during adolescence may stem from the overlap of their underlying neural circuitries (Nelson, Leibenluft, McClure, & Pine, 2005). Humans, like many other mammalian species, are social organisms. Social interrelationships take a variety of forms during ontogeny, ranging from the early emergence of attachment behavior in infancy to later peer-directed social interactions, sexual attraction and pair bonding, parental behavior, and aggression. Whereas each type of social stimulus activates a unique pattern of brain circuitry, many prosocial behaviors share common regulatory influences; these include the peptides oxytocin and prolactin, as well as brain opioid systems (Depue & Morrone-Strupinsky, 2005; Panksepp, Nelson, & Bekkedal, 1997). Social stimuli that are affectively pleasing activate mesocorticolimbic DA projections and associated circuitry of the reward system (for reviews, see Depue & Morrone-Strupinsky, 2005; Insel, 2003).

Certain types of social stimuli may be particularly rewarding during adolescence. For instance, in research using a simple animal model of adolescence in the rat, even adolescents that were not socially deprived formed robust conditioned place preferences for an unfamiliar peer, whereas only socially deprived adults found such social interactions rewarding (Douglas, Varlinskaya, & Spear, 2004). The presence or absence of specific social stimuli may also serve as particularly potent contexts and motivational states to influence behaviors directed toward other rewarding stimuli during adolescence. In laboratory animals, for instance, social isolation has been observed to increase the rewarding effects of novelty (Douglas, Varlinskaya, & Spear, 2003), whereas social stress has been shown to lower the rewarding value of sucrose solutions (Rygula et al., 2005).

Social stimuli may also provide a facilitatory context for the expression of risk-taking behaviors during adolescence, with risky behaviors becoming unusually rewarding under these circumstances. For instance, the presence of peers markedly increases such behaviors in human adolescents—a social context effect that has much less impact on risk-related behavior in adults (Gardner & Steinberg, 2005). The interactions between alcohol use and social context provide a well-studied example of this socially based facilitatory effect on reward-related behavior. Adolescents usually drink alcohol socially rather than alone. Such "social drinking" may be promoted in part by the expectation that alcohol will enhance their interpersonal effectiveness (Beck, Thombs, & Summons, 1993; Beck & Treiman, 1996; Brown, Christiansen, & Goldman, 1987). Furthermore, these reward-related and socially related influences on behavior are bidirectional. Adolescents' expectancies for alcohol to induce social facilitation interact reciprocally and positively with drinking experience: The greater the expectancy endorsement, the higher the levels of subsequent drinking, whereas the higher the levels of drinking, the greater the subsequent expectancy endorsement (Smith, Goldman, Greenbaum, & Christiansen, 1995). These apparent bidirectional effects of alcohol and social reward may be partly biologically mediated, in that similar bidirectionality is seen even in simple animal models of adolescence. For instance, alcohol drinking in adolescent rats is enhanced by social contact with intoxicated peers in a dose-dependent manner (Hunt, Holloway, & Scordalakes, 2001), and exposure to low doses of ethanol markedly facilitates social interactions in adolescent (but not adult) rats (e.g., Varlinskaya & Spear, 2002).

Conclusions

In addition to the synergy of social and reward behaviors, the overlap between the neural substrates of the reward and the social coding systems is substantial (Bachevalier & Malkova, 2006; Nelson et al., 2005). Based on these commonalities, we would like to conclude by suggesting that a neurodevelopmental model similar to the triadic model (Figure 17.3) can be entertained to understand some developmental aspects of social information processing.

This proposal reflects recent findings from a study in young adults comparing responses to monetary gains and losses in a social and a nonsocial condition (Nawa, Nelson, Pine, & Ernst, 2008). The merit of this study was that it provided the simplest social condition (i.e., the presence of another player independently involved in the game, and receiving feedback similar to that received by the subject under study). Thus the comparison was between brain responses to outcomes of rewards and penalties when another player was also in the game, and when the subject played alone. Findings suggested a dissociation of the involvement of amygdala-related circuitry and ventral striatum-related circuitry under the two conditions: The amygdala was engaged more significantly during the social condition, and the ventral striatum during the nonsocial condition. This finding was interpreted as reflecting the emergence of an automatic "warning signal" when a social other was present, accompanied by a dampening of activity in the approach behavior system. In adolescents, this balance between these two systems may be altered in favor of a weaker "warning signal" and stronger approach behavior system. This speculation, however, has yet to be examined empirically, and is one of the many promising directions for future research.

This review has provided only a glimpse into the neural basis and ontogeny of reward systems and their interactions with social information processing. Clearly, much more work in this area remains. As illustrated in this chapter, the most promising approaches may stem from the continued and strengthening interactions between basic and clinical neuroscience.

References

Andersen, S. L. (2003). Trajectories of brain development: Point of vulnerability or window of opportunity? *Neuroscience and Biobehavioral Reviews, 27*, 3–18.

Andersen, S. L., Thompson, A. T., Rutstein, M., Hostetter, J. C., & Teicher, M. H. (2000). Dopamine receptor pruning in prefrontal cortex during the periadolescent period in rats. *Synapse, 37,* 167–169.

Anderson, S. A., Eisenstat, D. D., Shi, L., & Rubenstein, J. L. R. (1997). Interneuron migration from basal forebrain to neocortex: Dependence on dix genes. *Science, 278,* 474–476.

Arnett, J. J. (1992). Reckless behavior in adolescence: A developmental perspective. *Developmental Review, 12,* 339–373.

Arnold, L. E., Zametkin, A. J., Caravella, L., & Korbly, N. (2000). Ethical issues in neuroimaging research with children. In M. Ernst & J. Rumsey (Eds.), *Functional neuroimaging in child psychiatry* (pp. 45–58). Cambridge, UK: Cambridge University Press.

Bachevalier, J., & Malkova, L. (2006). The amygdala and development of social cognition: Theoretical comment on Bauman, Toscano, Mason, Lavenex, & Amaral. *Behavioral Neuroscience, 120,* 989–991.

Baxter, M. G., & Murray, E. A. (2002). The amygdala and reward. *Nature Reviews Neuroscience, 3,* 563–573.

Bechara, A., Damasio, H., Tranel, D., & Damasio, A. R. (2005). The Iowa Gambling Task and the somatic marker hypothesis: Some questions and answers. *Trends in Cognitive Sciences, 9,* 159–162.

Beck, K. H., Thombs, D. L., & Summons, T. G. (1993). The social context of drinking scales: Construct validation and relationship to indicants of abuse in an adolescent population. *Addictive Behaviors, 18,* 159–169.

Beck, K. H., & Treiman, K. A. (1996). The relationship of social context of drinking, perceived social norms, and parental influence to various drinking patterns of adolescents. *Addictive Behaviors, 21,* 633–644.

Benes, F. M., Taylor, J. B., & Cunningham, M. C. (2000). Convergence and plasticity of monaminergic systems in the medial prefrontal cortex during the postnatal period: Implications for the development of psychopathology. *Cerebral Cortex, 10,* 1014–1027.

Berridge, K. C. (2000). Measuring hedonic impact in animals and infants: Microstructure of affective taste reactivity patterns. *Neuroscience and Biobehavioral Reviews, 24,* 173–198.

Berridge, K. C., & Robinson, T. E. (1998). What is the role of dopamine in reward: Hedonic impact, reward learning, or incentive salience? *Brain Research: Brain Research Reviews, 28,* 309–369.

Bjork, J. M., Knutson, B., Fong, G. W., Caggiano, D. M., Bennett, S. M., & Hommer, D. W. (2004). Incentive-elicited brain activation in adolescents: Similarities and differences from young adults. *Journal of Neuroscience, 24,* 1793–1802.

Blum, K., Sheridan, P. J., Wood, R. C., Braverman, E. R., Chen, T. J., & Comings, D. E. (1995). Dopamine D2 receptor gene variants: Association and linkage studies in impulsive–addictive–compulsive behaviour. *Pharmacogenetics, 5,* 121–141.

Brown, S. A., Christiansen, B. A., & Goldman, M. S. (1987). The Alcohol Expectancy Questionnaire: An instrument for assessment of adolescent and adult alcohol expectancies. *Journal of Studies on Alcohol, 48,* 483–491.

Cardinal, R. N., Parkinson, J. A., Hall, J., & Everitt, B. J. (2002). Emotion and motivation: The role of the amygdala, ventral striatum, and prefrontal cortex. *Neuroscience and Biobehavioral Reviews, 26,* 321–352.

Chambers, R. A., Taylor, J. R., & Potenza, M .N. (2003). Developmental neurocircuitry of motivation in adolescence: A critical period of addiction vulnerability. *American Journal of Psychiatry, 160,* 1041–1052.

Costello, E. J., Pine, D. S., Hammen, C., March, J. S., Plotsky, P. M., Weissman, M. M., et al. (2002). Development and natural history of mood disorders. *Biological Psychiatry, 52,* 529–542.

Csikszentmihalyi, M., Larson, R., & Prescott, S. (1977). The ecology of adolescent activity and experience. *Journal of Youth and Adolescence, 6,* 281–294.

Cunningham, M. G., Bhattacharyya, S., & Benes, F. M. (2002). Amygdalo-cortical sprouting continues into early adulthood: Implications for the development of normal and abnormal function during adolescence. *Journal of Comparative Neurology, 453*, 116–130.

Dahl, R. E. (2004). Adolescent brain development: A period of vulnerabilities and opportunities [Keynote address]. *Annals of the New York Academy of Sciences, 1021*, 1–22.

Depue, R. A., & Morrone-Strupinsky, J. V. (2005). A neurobehavioral model of affiliative bonding: Implications for conceptualizing a human trait of affiliation. *Behavioral and Brain Sciences, 28*, 313–395.

Di Chiara, G. (2002). Nucleus accumbens shell and core dopamine: Differential role in behavior and addiction. *Behavioural Brain Research, 137*, 75–114.

Douglas, L. A., Varlinskaya, E. I., & Spear, L. P. (2003). Novel object place conditioning in adolescent and adult male and female rats: Effects of social isolation. *Physiology and Behavior, 80*, 317–325.

Douglas, L. A., Varlinskaya, E. I., & Spear, L. P. (2004). Rewarding properties of social interactions in adolescent and adult male and female rats: Impact of social vs. isolate housing of subjects and partners. *Developmental Psychobiology, 45*, 153–162.

Dumont, N. L., Andersen, S. L., Thompson, A. P., & Teicher, M. H. (2004). Transient dopamine synthesis modulation in prefrontal cortex: *In vitro* studies. *Brain Research: Developmental Brain Research, 150*, 163–166.

Ernst, M., Nelson, E. E., Jazbec, S., McClure, E. B., Monk, C. S., Leibenluft, E., et al. (2005). Amygdala and nucleus accumbens in responses to receipt and omission of gains in adults and adolescents. *NeuroImage, 25*, 1279–1291.

Ernst, M., & Paulus, M. P. (2005). Neurobiology of decision making: A selective review from a neurocognitive and clinical perspective. *Biological Psychiatry, 58*(8), 597–604.

Ernst, M., Pine, D. S., & Hardin, M. (2006). Triadic model of the neurobiology of motivated behavior in adolescence. *Psychological Medicine, 36*, 299–312.

Eshel, N., Nelson, E. E., Blair, R. J., Pine, D. S., & Ernst, M. (2007). Neural substrates of choice selection in adults and adolescents: Development of the ventrolateral prefrontal and anterior cingulate cortices. *Neuropsychologia, 45*, 1270–1279.

Everitt, B. J., Cador, M., & Robbins, T. W. (1989). Interactions between the amygdala and ventral striatum in stimulus–reward associations: Studies using a second-order schedule of sexual reinforcement. *Neuroscience, 30*, 63–75.

Everitt, B. J., Cardinal, R. N., Hall, J., Parkinson, J., & Robbins, T. (2000). Differential involvement of amygdala subsystems in appetitive conditioning and drug addiction. In J. Aggleton (Ed.), *The amygdala: A functional analysis* (2nd ed., pp. 353–390). Oxford: Oxford University Press.

Gallagher, M., McMahan, R. W., & Schoenbaum, G. (1999). Orbitofrontal cortex and representation of incentive value in associative learning. *Journal of Neuroscience, 19*, 6610–6614.

Galvan, A., Hare, T. A., Parra, C. E., Penn, J., Voss, H., Glover, G., et al. (2006). Earlier development of the accumbens relative to orbitofrontal cortex might underlie risk-taking behavior in adolescents. *Journal of Neuroscience, 26*, 6885–6892.

Gardner, E. L. (1999). The neurobiology and genetics of addiction: Implications of the reward deficiency syndrome for therapeutic strategies in chemical dependency. In J. Elster (Ed.), *Addiction: Entries and exits* (pp. 57–119). New York: Russell Sage Foundation.

Gardner, M., & Steinberg, L. (2005). Peer influence on risk taking, risk preference, and risky decision making in adolescence and adulthood: An experimental study. *Developmental Psychology, 41*, 625–635.

Grace, A. A. (1991). Phasic versus tonic dopamine release and the modulation of dopamine system responsivity: A hypothesis for the etiology of schizophrenia. *Neuroscience, 41*, 1–24.

Grace, A. A. (1995). The tonic/phasic model of dopamine system regulation: Its relevance for understanding how stimulant abuse can alter basal ganglia function. *Drug and Alcohol Dependence, 37*, 111–129.

Guyer, A. E., Monk, C. S., McClure-Tone, E. B., Nelson, E. E., Roberson-Nay, R., Adler, A. D., et al. (2008). A developmental examination of amygdala response to facial expressions. *Journal of Cognitive Neuroscience, 20*(9), 1565–1582.

Hunt, P. S., Holloway, J. L., & Scordalakes, E. M. (2001). Social interaction with an intoxicated sibling can result in increased intake of ethanol by periadolescent rats. *Developmental Psychobiology, 38,* 101–109.

Huttenlocher, P. R. (1984). Synapse elimination and plasticity in developing human cerebral cortex. *American Journal of Mental Deficiency, 88,* 488–496.

Ikemoto, S., & Panksepp, J. (1999). The role of nucleus accumbens dopamine in motivated behavior: A unifying interpretation with special reference to reward-seeking. *Brain Research: Brain Research Reviews, 31 ,* 6–41.

Insel, T. R. (2003). Is social attachment an addictive disorder? *Physiology and Behavior, 79,* 351–357.

Insel, T. R., Miller, L. P., & Gelhard, R. E. (1990). The ontogeny of excitatory amino acid receptors in rat forebrain: I. N-methyl-D-aspartate and quisqualate receptors. *Neuroscience, 35,* 31–43.

Izquierdo, A., & Murray, E. A. (2004). Combined unilateral lesions of the amygdala and orbital prefrontal cortex impair affective processing in rhesus monkeys. *Journal of Neurophysiology, 91,* 2023–2039.

Kalivas, P. W., & Volkow, N. D. (2005). The neural basis of addiction: A pathology of motivation and choice. *American Journal of Psychiatry, 162,* 1403–1413.

Kellogg, C. K., Awatramani, G. B., & Piekut, D. T. (1998). Adolescent development alters stressor-induced Fos immunoreactivity in rat brain. *Neuroscience, 83,* 681–689.

Killgore, W. D. S., Oki, M., & Yurgelun-Todd, D. A. (2001). Sex-specific developmental changes in amygdala responses to affective faces. *NeuroReport, 12,* 427–433.

Killgore, W. D. S., & Yurgelun-Todd, D. A. (2004). Sex-related developmental differences in the lateralized activation of the prefrontal cortex and amygdala during perception of facial affect. *Perceptual and Motor Skills, 99,* 371–391.

La Greca, A. M., Prinstein, M. J., & Fetter, M. D. (2001). Adolescent peer crowd affiliation: Linkages with health-risk behaviors and close friendships. *Journal of Pediatric Psychology, 26,* 131–143.

Larson, R., & Richards, M. H. (1991). Daily companionship in late childhood and early adolescence: Changing developmental contexts. *Child Development, 62,* 284–300.

Laviola, G., Macri, S., Morley-Fletcher, S., & Adriani, W. (2003). Risk-taking behavior in adolescent mice: Psychobiological determinants and early epigenetic influence. *Neuroscience and Biobehavioral Reviews, 27,* 19–31.

Leslie, C. A., Robertson, M. W., Cutler, A. J., & Bennett, J. P., Jr. (1991). Postnatal development of D_1 dopamine receptors in the medial prefrontal cortex, striatum and nucleus accumbens of normal and neonatal 6-hydroxydopamine treated rats: A quantitative autoradiographic analysis. *Brain Research: Developmental Brain Research, 62,* 109–114.

Levitt, P. (2003). Structural and functional maturation of the developing primate brain. *Journal of Pediatrics, 143,* S35–S45.

Martin-Soelch, C., Linthicum, J., & Ernst, M. (2007). Appetitive conditioning: Neural bases and implications for psychopathology. *Neuroscience and Biobehavioral Reviews, 31*(3), 426–440.

Masten, A. S. (2004). Regulatory processes, risk, and resilience in adolescent development. *Annals of the New York Academy of Sciences, 1021,* 310–319.

May, J. C., Delgado, M. R., Dahl, R. E., Stenger, V. A., Ryan, N. D., Fiez, J. A., et al. (2004). Event-related functional magnetic resonance imaging of reward-related brain circuitry in children and adolescents. *Biological Psychiatry, 55,* 359–366.

McClure, E. B., Monk, C. S., Nelson, E. E., Zarahn, E., Leibenluft, E., Bilder, R. M., et al. (2004). A developmental examination of gender differences in brain engagement during evaluation of threat. *Biological Psychiatry, 55,* 1047–1055.

Monk, C. S., McClure, E. B., Nelson, E. E., Zarahn, E., Bilder, R. M., Leibenluft, E., et al. (2003).

Adolescent immaturity in attention-related brain engagement to emotional facial expressions. *NeuroImage, 20,* 420–428.

Munson, S., Eshel, N., & Ernst, M. (2006). Ethics of PET research in children. In M. Charron (Ed.), *Practical pediatric PET imaging* (pp. 72–91). New York: Springer.

Nawa, N. E., Nelson, E. E., Pine, D. S., & Ernst, M. (2008). Do you make a difference?: Social context in a betting task. *Social Cognitive and Affective Neuroscience, 3*(4), 367–376.

Nelson, E. E., Leibenluft, E., McClure, E. B., & Pine, D. S. (2005). The social re-orientation of adolescence: A neuroscience perspective on the process and its relation to psychopathology. *Psychological Medicine, 35,* 163–174.

Nestler, E. J. (2001). Molecular basis of long-term plasticity underlying addiction. *Nature Reviews Neuroscience, 2,* 119–128.

Panksepp, J., Nelson, E., & Bekkedal, M. (1997). Brain systems for the mediation of social separation-distress and social-reward: Evolutionary antecedents and neuropeptide intermediaries. *Annals of the New York Academy of Sciences, 807,* 78–100.

Peciña, S., Smith, K. S., & Berridge, K. C. (2006). Hedonic hot spots in the brain. *Neuroscientist, 12,* 500–511.

Pine, D. S., Cohen, P., Gurley, D., Brook, J., & Ma, Y. (1998). The risk for early-adulthood anxiety and depressive disorders in adolescents with anxiety and depressive disorders. *Archives of General Psychiatry, 55,* 56–64.

Pine, D. S., Grun, J., Zarahn, E., Fyer, A., Koda, V., Li, W., et al. (2001). Cortical brain regions engaged by masked emotional faces in adolescents and adults: An fMRI study. *Emotion, 1,* 137–147.

Primus, R. J., & Kellogg, C. K. (1989). Pubertal-related changes influence the development of environment-related social interaction in the male rat. *Developmental Psychobiology, 22,* 633–643.

Rosenberg, D. R., & Lewis, D. A. (1995). Postnatal maturation of the dopaminergic innervation of monkey prefrontal and motor cortices: A tyrosine hydroxylase immunohistochemical analysis. *Journal of Comparative Neurology, 358,* 383–400.

Rygula, R., Abumaria, N., Flügge, G., Fuchs, E., Rüther, E., & Havemann-Reinecke, U. (2005). Anhedonia and motivational deficits in rats: Impact of chronic social stress. *Behavioural Brain Research, 162,* 127–134.

Schoenbaum, G. (2004). Affect, action, and ambiguity and the amygdala–orbitofrontal circuit: Focus on "Combined unilateral lesions of the amygdala and orbital prefrontal cortex impair affective processing in rhesus monkeys." *Journal of Neurophysiology, 91,* 1938–1939.

Schoenbaum, G., & Roesch, M. (2005). Orbitofrontal cortex, associative learning, and expectancies. *Neuron, 47,* 633–636.

Schultz, W. (1998). The phasic reward signal of primate dopamine neurons. *Advances in Pharmacology, 42,* 686–690.

Schultz, W. (2006). Behavioral theories and the neurophysiology of reward. *Annual Review of Psychology, 57,* 87–115.

Schultz, W., Dayan, P., & Montague, P. R. (1997). A neural substrate of prediction and reward. *Science, 275,* 1593–1599.

Seeman, P., Bzowej, N. H., Guan, H.-C., Bergeron, C., Becker, L. E., Reynolds, G. P., et al. (1987). Human brain dopamine receptors in children and aging adults. *Synapse, 1,* 399–404.

Shaw, C., & Lanius, R. A. (1992). Cortical AMPA receptors: Age-dependent regulation by cellular depolarization and agonist stimulation. *Brain Research: Developmental Brain Research, 68,* 225–231.

Shaw, C., & Scarth, B. A. (1992). Age-dependent regulation of GABA$_A$ receptors in neocortex. *Brain Research: Molecular Brain Research, 14,* 207–212.

Smith, G. T., Goldman, M. S., Greenbaum, P. E., & Christiansen, B. A. (1995). Expectancy for social facilitation from drinking: The divergent paths of high-expectancy and low-expectancy adolescents. *Journal of Abnormal Psychology, 104,* 32–40.

Sowell, E. R., Thompson, P. M., Holmes, C. J., Batth, R., Jernigan, T. L., & Toga, A. W. (1999). Localizing age-related changes in brain structure between childhood and adolescence using statistical
parametric mapping. *NeuroImage, 9*, 587–597.

Spear, L. P. (2000). The adolescent brain and age-related behavioral manifestations. *Neuroscience and
Behavioral Physiology, 24*, 417–463.

Spear, L. P. (2007). Brain development and adolescent behavior. In D. Coch, G. Dawson, & K. W. Fischer (Eds.), *Human behavior, learning, and the developing brain: Typical development* (pp. 362–
396). New York: Guilford Press.

Steinberg, L. (1989). Pubertal maturation and parent–adolescent distance: An evolutionary perspective. In G. R. Adams, R. Montemayor, & T. P. Gullotta (Eds.), *Advances in adolescent behavior
and development* (pp. 71–97). Newbury Park, CA: Sage.

Tarazi, F. I., & Baldessarini, R. J. (2000). Comparative postnatal development of dopamine D_1, D_2, and
D_4 receptors in rat forebrain. *International Journal of Developmental Neuroscience, 18*, 29–37.

Tarazi, F. I., Tomasini, E. C., & Baldessarini, R. J. (1998). Postnatal development of dopamine D_4-like
receptors in rat forebrain regions: Comparison with D_2-like receptors. *Brain Research: Developmental Brain Research, 110*, 227–233.

Tarazi, F. I., Tomasini, E. C., & Baldessarini, R. J. (1999). Postnatal development of dopamine D_1-like
receptors in rat cortical and striatolimbic brain regions: An autoradiographic study. *Developmental Neuroscience, 21*, 43–49.

Teicher, M. H., Andersen, S. L., & Hostetter, J. C., Jr. (1995). Evidence for dopamine receptor pruning between adolescence and adulthood in striatum but not nucleus accumbens. *Brain Research:
Developmental Brain Research, 89*, 167–172.

Teicher, M. H., Barber, N. I., Gelbard, H. A., Gallitano, A. L., Campbell, A., Marsh, E., et al. (1993).
Developmental differences in acute nigrostriatal and mesocorticolimbic system response to haloperidol. *Neuropsychopharmacology, 9*, 147–156.

Thomas, K. M., Drevets, W. C., Whalen, P. J., Eccard, C. H., Dahl, R. E., Ryan, N. D., et al. (2001).
Amygdala response to facial expressions in children and adults. *Biological Psychiatry, 49*, 309–
316.

Thompson, J. L., Pogue-Geile, M. F., & Grace, A. A. (2004). Developmental pathology, dopamine,
and stress: A model for the age of onset of schizophrenia symptoms. *Schizophrenia Bulletin, 30*,
875–900.

Tindell, A. J., Smith, K. S., Peciña, S., Berridge, K. C., & Aldridge, J. W. (2006). Ventral pallidum
firing codes hedonic reward: When a bad taste turns good. *Journal of Neurophysiology, 96*,
2399–2409.

Tobler, P. N., Fiorillo, C. D., & Schultz, W. (2005). Adaptive coding of reward value by dopamine
neurons. *Science, 307*, 1642–1645.

Tseng, K. Y., & O'Donnell, P. (2005). Post-pubertal emergence of prefrontal cortical up states induced
by D_1–NMDA co-activation. *Cerebral Cortex, 15*, 49–57.

van Eden, C. G., Kros, J. M., & Uylings, H. B. M. (1990). The development of the rat prefrontal cortex:
Its size and development of connections with thalamus, spinal cord and other cortical areas. In H.
B. M. Uylings, C. G. van Eden, J. P. C. De Bruin, M. A. Corner, & M. G. P. Feenstra (Eds.), *The
prefrontal cortex: Its structure, function and pathology* (pp. 169–183). Amsterdam: Elsevier.

van Leijenhorst, L., Crone, E. A., & Bunge, S. A. (2006). Neural correlates of developmental differences in risk estimation and feedback processing. *Neuropsychologia, 44*, 2158–2170.

Varlinskaya, E. I., & Spear, L. P. (2002). Acute effects of ethanol on social behavior of adolescent
and adult rats: Role of familiarity of the test situation. *Alcoholism: Clinical and Experimental
Research, 26*, 1502–1511.

Volkow, N. D., Fowler, J. S., Wang, G.-J., & Goldstein, R. Z. (2002). Role of dopamine, the frontal
cortex and memory circuits in drug addiction: Insight from imaging studies. *Neurobiology of
Learning and Memory, 78*, 610–624.

Wise, R. A. (1980). The dopamine synapse and the notion of "pleasure centers" in the brain. *Trends in Neurosciences, 3,* 91–95.

Zak, P. J., Kurzban, R., & Matzner, W. T. (2004). The neurobiology of trust. *Annals of the New York Academy of Sciences, 1032,* 224–227.

Zald, D. (2003). The human amygdala and the emotional evaluation of sensory stimuli: Review. *Brain Research: Brain Research Reviews, 41,* 88–123.

Zecevic, N., Bourgeois, J.-P., & Rakic, P. (1989). Changes in synaptic density in motor cortex of rhesus monkey during fetal and postnatal life. *Brain Research: Developmental Brain Research, 50,* 11–32.

CHAPTER 18

Social Relationships as Primary Rewards

The Neurobiology of Attachment

LINDA C. MAYES
JESSICA MAGIDSON
C. W. LEJUEZ
SARAH S. NICHOLLS

Anticipating and seeking reward are among the primary motivators of human behavior. Indeed, reward seeking is central to a number of developmental processes, including learning. It is also key to maladaptive behaviors or disorders, such as addictive processes (e.g., drug use, gambling, and overeating). In this chapter, we propose that the need for social attachments is a key, but often underemphasized, primary/basic reward motivating behavior. Humans appetitively seek relationships with others. We argue that this need goes beyond sexual behavior as a basic reward stimulus and extends to the anxiety or fear-reducing quality of the attachment itself—in other words, that social relationships are crucial to adaptive, healthy emotional regulation. We also argue that the threat of loss (with the attendant state of fear and negative arousal) is a powerful signal to activate relationship-seeking behavior, and that, more generally, relationship-seeking behavior comes into play under a range of negative affective states such as anxiety or sadness. Optimally, effective relationships may help an individual manage these negative affective states, but sometimes in such states, an individual can act in destructive (and counterproductive) ways to avoid losing the relationship. In other words, we suggest that—especially when relationships are considered as primary behavioral motivators—there is a constant feedback between seeking the comfort of another for reward, pleasure, and security on the one hand, and using the other to diminish anxiety, fear, and other negative affects (including loneliness and perceived danger from a threatened loss) on the other hand. With optimal development, social relationships may be among the primary motivators for behavior. However, under conditions of early adversity and their attendant impact on later stress response systems or addictive processes, the rewarding properties of attachments either are relatively ineffective in diminishing negative affective states, or are neurobiologically co-opted by more powerful rewards that diminish the rewarding properties of attachments.

In order to make this argument, we integrate operant models of motivational behavior and attachment theory, positing the development of attachment between infant and parent as a primary motivator of early infant behavior that endures into adulthood and influences adult relationship-seeking behavior. We begin with the basic definitions of "reward" in terms of positive and negative reinforcement that we use in this chapter, and follow this with a review of the theory of attachment or social relationships as emotion-regulating systems. We then review the neurobiology of reward systems in general and describe what is currently known about the overlap between those systems and the neural circuitry of attachment, especially regarding parent–infant attachment. We next present a model of the feedback connection between reward and anxiety-regulating systems, in which attachment relationships serve as central mediators of the balance between these systems. We conclude with a discussion of how, for example, early life stress, later abuse, and/or addiction may influence the primary rewarding properties of social relationships.

Rewards, Punishment, and Motivated Behavior

A "reward" is a stimulus that, when presented, results in an increase in intensity and/or persistence in a behavior contingently tied to that reward. This process is referred to as "positive reinforcement," involving basic or primary rewards considered essential for survival (food, water, and sex being among the most basic), with secondary rewards deriving their valence from their learned association with these most basic needs (i.e., conditioned reinforcers). Rewards are associated with the perception of gain, satisfaction, and positive emotions. Rewards induce learning, approach, or appetitive behavior, as in the case of an animal conditioned to press a bar for food or an adult human habitually turning to the coffee shop each morning before work. As a complement to rewards that may be characterized in the context of positive reinforcement, rewards also may involve the process of "negative reinforcement," where the intensity and/or persistence of a behavioral response is increased because it is contingently tied to escape/avoidance from an aversive stimulus. In this case, the individual works to escape/avoid the negative or dangerous state (such as fear or anxiety), as in the case of a conditioned animal seeking to avoid an electric shock or an adult human's sometimes ritualized, sometimes desperate behaviors to avoid reminders of past traumas.

It is also important to understand how these complementary processes may work together in different ways. For example, among theories of drug use, the positive reinforcing or appetitive properties of drugs are important (Koob et al., 2004), as well as the avoidance or dampening of negative affective states. Indeed, in an effort to understand the transition from drug use to dependence, the adult literature has focused on negative reinforcement models (Baker, Piper, McCarthy, Majeskie & Fiore, 2004; Solomon, 1977; Wikler, 1965) that emphasize the reduction or avoidance of aversive internal states as the motivational basis for drug addiction. In the case of either positive or negative reinforcement, motivated behavior reflects learning.

Somewhat counterintuitively, seemingly positive stimuli sometimes may not serve as positive reinforcers, and seemingly aversive stimuli may be met with approach as opposed to escape/avoidance. This could be due to more simplistic factors, such as satiation or availability (e.g., little Johnny may be less likely to complete a chore for candy at his grandparents' home if he has free access to candy at home); however, these counterintuitive instances often result from juxtapositions of rewards and aversive stimuli that produce complex and potentially unexpected behavioral patterns. In one such case, an aversive stimulus can serve as a conditioned reinforcer if it has previously been paired with a reward. As an example, someone experiencing primarily abusive relationships from early childhood may only experience displays of concern or affection from those individuals after the abusive behavior. As a result, abuse comes to signal the likely occurrence of affection. Especially if affection independent of maltreatment is not available from others, abuse becomes a strong (and maybe the only) indicator of impending affection. As such, this pairing of abuse and reward may complicate efforts to avoid future abusive relationships and may impair the individual's participation in future nonabusive relationships. Similarly, consider the drug-seeking behavior of an inner-city resident who uses crack cocaine. The various aversive experiences that often precede obtaining the drug should be aversive and limit drug seeking. However, because these aversive experiences are often directly followed by drug administration, they may become conditioned reinforcers and therefore may come to encourage, as opposed to discourage, future drug seeking. Indeed, such individuals often report that the excitement of the "hustle" to get their drug is something they miss long after becoming abstinent, and sometimes even after the craving for the drug itself has dissipated.

The blend of appetitive and avoidant motivational behaviors in any individual action is captured in a number of personality assessments tapping into preference for or reactivity to novelty and/or stress, reaction to unexpectancies, capacity for pleasure and satisfaction, and reaction to negative emotions (Eysenck, 1985; Larsen & Ketelaar, 1991; O'Malley & Gillette, 1984). These assessments show how individuals vary in their preference for reward or for appetitive or avoidant responses, and a number of investigators have focused on likelihood for psychopathology (such as addictive, anxiety, or conduct disorders) as a function of an individual's organization around avoidant versus hedonic/approach dimensions. That is, understanding these reward processes across individuals also requires an understanding of individual-difference factors that influence the salience of a reward and are linked to early experience and psychopathology. For example, avoiding social interaction in response to social anxiety may be more motivating for an impulsive individual who is unable to think through the longer-term consequences, whereas engaging in risky behavior may be considerably more motivating for a sensation seeker who may value personal safety less because of an abusive history.

Finally, reward-driven behavior is complex in the sense that there may be a number of alternative possibilities. Instead of operating as simple stimulus–response reflex systems with discrete, goal-directed behaviors, motivation or reward processes involve integrating complex information about changing internal states (e.g., emotional states, hunger, pain, sexual arousal) with equally complex cues about environmental possibilities and threats, and deciding upon the most advantageous response (Bechara, 2001; Kalivas, Churchill, & Romanides, 1999). Furthermore, at any one time, multiple goals or needs may be active—hunger, need for social contact, need to reduce anxiety and feel safe—and there must be mechanisms (and an associated neural circuitry) both to prioritize these needs and to select the most advantageous responses that will serve nearly all of the needs at any given time (Chambers, Taylor, & Potenza, 2003). As we discuss in detail below, preclinical and human studies are converging to suggest two interactive neural circuits. The first of these is a primary motivational circuit involving the ventral striatum and prefrontal cortex, with direct input to motor "output" structures (Kalivas et al., 1999). This primary system is supported by a secondary circuitry providing multiple modalities of sensory "input" from a range of structures, with afferent projections into the primary reward or motivational circuit. These structures include the hippocampus and amygdala, providing context-based memory cues and affective information, respectively (Groenewegen et al., 1999; Panksepp, 1998); they also include the hypothalamic and septal nuclei, providing information relevant to states of fear or stress and instinctually driven behaviors, such as hunger, thirst, and self-protection (Swanson, 2000). As we outline in subsequent sections, these same primary and secondary circuits are also involved in initiating new relationships and in responding to salient social cues.

Social Relationships as Rewards and as Regulators of Negative States

Relationships are salient across many species. The evidence is especially compelling for humans and nonhuman primates, and much of the work from the social and developmental sciences approaches the salience of relationships from the standpoint of attachment theory. As originally proposed by John Bowlby (Bowlby, 1969/1982), expanded upon by Mary Ainsworth (Ainsworth, Blehar, Waters, & Wall, 1978), and extensively elaborated in the last two to three decades (e.g., Bretherton, 1985; Bretherton & Munholland, 1999; Cassidy, 1999; Crittenden, 1992; Kobak, 1999; Main, Kaplan, & Cassidy, 1985; Waters, Merrick, Treboux, Crowell, & Albersheim, 2000a), attachment theory has been applied to significant relationships in which one or each partner relies on the other for emotion regulation and safety. The prototypical attachment relationship is the parent–child relationship, but the theory has been extended to a range of adult relationships (Weiss, 1982). Using evolutionary perspectives, Bowlby and those who followed him posited that attachment is an innate biological system promoting proximity seeking between an infant and a specific attachment figure, in order to increase the likelihood of survival to a reproductive age (Sroufe, 2000). In that sense then, proximity-seeking behavior is as much motivated by a need for safety or fear reduction as by a need for closeness to a cared-for individual. Indeed, central to attachment theory is the idea that all human infants become attached to their caregivers—even if the care is harsh or neglectful—because of this basic need for a protective relationship. Depending on the quality of the care, children manifest different patterns of attachment "security–insecurity." Infants of caregivers who are available, responsive, and sensitive to their emotional and physical needs tend to manifest patterns of "secure" attachment; that is, they trust in others' capacity to provide safe, contingent care responsive to their needs. But if the care provided is chaotic, unpredictable, rejecting, or neglectful, or if a caregiver consistently provides non-contingent responses to a child, one of several patterns of "insecure" attachment evolves: The child is either distrustful and/or dismissing of social relationships ("avoidant" attachment), preoccupied with and vigilant toward the caregiver without an attendant reduction of anxiety and fear ("ambivalent" attachment), or chaotic and incoherent in his or her behavior ("disorganized/disoriented" attachment).

These patterns of how relationships are viewed are considered "internal working models" in attachment theory. Although such internal models are shaped and revised in response to ongoing experience, most attachment theorists consider that an individual's pattern of relying on relationships is established in early childhood (Waters, Weinfeld, & Hamilton, 2000b) and remains a significant force shaping the developmental trajectory into adulthood. Longitudinal studies indicate that in the absence of other mitigating factors, securely attached infants demonstrate higher levels of self-esteem and self-reliance, and show more effective self-regulation in the management of their impulses and emotions from preschool through middle childhood and adolescence (Sroufe, 1983; Sroufe, Carlson, & Shulman, 1993). They also form closer relationships and show a better capacity for intimacy. Furthermore, barring intervening severe stress and trauma, secure attachment in infancy is associated with secure attachment in adulthood (van IJzendoorn, 1995). In contrast, insecure patterns of attachment in general and disorganized attachment in particular predict a range of behavioral disturbance, including anxiety disorders, aggression, conduct disorder, and other forms of psychopathology (Carlson, 1998; Sroufe, 1997; Warren, Huston, Egeland, & Sroufe, 1997). Similar classifications of attachment styles capture adult differences in states of mind regarding an individual's overall attachment history (Main, 2000).

Typically, notions of reward and attachment come together in considerations of how attachment relationships serve emotion regulation. Indeed, attachment theory provides a mechanism by which the emotion regulation process becomes internalized as a regulatory system (Fonagy, Gergely, Jurist, & Target, 2002). Bowlby (1969/1982) posited that attachment behaviors arise in the service of increasing proximity or increasing "felt security" (Sroufe & Waters, 1977). Furthermore, Bowlby (1973) presented a substantial body of observational data suggesting that in a child, separation distress serves as an expression of alarm or fear of harm happening to the child's primary caregiver. Schemas of relationships develop during times of stress, when danger cues are heightened—for example, during separations from the primary caregiver, or at times when the infant feels distress at not being relieved sufficiently by the primary caregiver. Stated simply, infants who experience contingent responsiveness to their distress develop internal working models of relationships as key to adaptive emotional regulation. In contrast, individuals with avoidant, ambivalent, or disorganized forms of attachment do not use relationships as readily or as successfully in the service of emotion regulation. Thus it is useful to think of attachment systems as regulating feelings of distress or alarm and as coming into play when individuals feel threatened. The relief of states of tension and distress provide the reward and the motivation for seeking proximity with another person. Simple alarm systems have been well documented in other species and are likely to be part of a common evolutionary heritage (Evans, 1997). Given our history as mammals and primates, we humans are likely to be more acutely attuned to some threats than to others. These expressions of distress on separation or on temporary absence of/failure in caregiving responsiveness are common across the young of many if not most mammalian species.

Brain Systems for Reward

Key to central reward circuitry are limbic structures of the basal forebrain, including the amygdala, hippocampus, prefrontal cortex (PFC), nucleus accumbens (NAC), ventral pallidum (VP), and ventral tegmental area (VTA). Dopaminergic neurons in the VTA modulate information flow through the limbic circuit via projections to the NAC, amygdala, hippocampus, PFC, and VP. Increased dopaminergic transmission in limbic nuclei, particularly the NAC, underlies the reinforcing effect of virtually every abused drug (Self, 1998; Stein et al., 1998). Dopamine is the neurotransmitter that is most closely associated with the reward system in humans. There are five different types of dopamine receptors: D1, D2, D3, D4, and D5. These are classified into two families with similar functions. The D1-like family includes D1 and D5 receptors, which are generally excitatory; the D2-like family includes the other three receptors, which are generally inhibitory. There are also mutant receptors that may be less sensitive to dopamine than normal receptors (Kalat, 2004), and these variations may be key to individual differences in novelty-seeking behavior (Bardo, Donohew, & Harrington, 1996).

Novel situations and salient cues activate the mesolimbic dopamine reward systems through dopamine release into the striatum (Panksepp, 1996). Dopamine release into the primarily motivational circuit, the ventral striatum (NAC) and dorsal striatum (caudate and putamen), operates like a "go" signal and is provoked by excitatory signals from cortex and other dopamine-rich areas, including the VTA and the substantia nigra (Kalivas, 1993; Strafella, Paus, Barrett, & Dagher, 2001). The primary motivational circuit referred to earlier involves parallel loops of axonal projections from the PFC to the ventral striatum (NAC and globus pallidus) to the thalamus and back to the cortex (Masterman & Cummings, 1997; Rolls & Treves, 1998). These cortical–striatal–thalamic–cortical loops are central to the regulation of several different behaviors (Lichter & Cummings, 2001), but especially to motivationally driven behaviors. Specific subregions of the PFC (e.g., anterior cingulate, ventromedial, and dorsolateral regions) project to specific striatal areas that are in turn compartmentalized in their projections to the thalamus (Kolomiets, Deniau, Mailly, Menetrey, & Thierry, 2001). Firing patterns in both the nucleus accumbens and the PFC are regulated by glutamatergic inputs from a secondary circuit, the hippocampus and amygdala (Chambers et al., 2003). Gamma-aminobutyric acid-ergic (GABAergic) inhibitory neurons are also densely concentrated in the striatum and may be in part responsible for encoding a repertoire of motivationally driven behaviors reflecting a combination of facilitated and inhibited responses (Chambers & Self, 2002; Everitt & Wolf, 2002; Kalivas, 1993; O'Donnell, Greene, Pabello, Lewis, & Grace, 1999; Panksepp, 1998; Pennartz, Groenewegen, & Lopez da Silva, 1994; Rolls & Treves, 1998; Strafella et al., 2001; Volkow & Fowler, 2000). Disturbances in reward- or motivation-based behavior, as may be seen in addictions, may reflect an imbalance between primary and secondary circuits and between facilitative and inhibitory neural systems within the primary circuitry (Potenza, 2001).

Dopamine release in the dorsal striatum is primarily associated with the initiation of motor activity and habitual behavior (Sano, Marder, & Dooneief, 1996), whereas release into the ventral striatum is associated with motivational behaviors, rewarding stimuli, and learning new behaviors (Ito, Dalley, Robbins, & Everitt, 2002; Masterman & Cummings, 1997; Panksepp, 1998). A wide variety of potentially rewarding or motivating stimuli increase dopamine release in the NAC, including addictive drugs, primary motivational needs such as food and sex, reward-related situations such as gambling, and stressful or aversive stimuli (Panksepp, 1998; Self & Nestler, 1998). It also appears that seeking and exploring unknown situations and environmental novelty provoke ventral striatal dopamine release (Ljungberg, Apicella, & Schultz, 1992; Panksepp, 1998). Rewards delivered intermittently over repeated trials appear to maintain dopamine cell firing and reward-conditioned behavior (Waelti, Dickinson, & Schultz, 2001), whereas well-learned behaviors or habits performed under expected contingencies are less dependent on ventral striatal dopamine release. Addictive drugs may act similarly to the motivationally stimulating properties of novelty in an individual's environment (Chambers et al., 2003). We might also posit that new relationships may be included in environmental novelty.

Reward-based learning shapes future behavior based on past experiences with novel situations or stimuli. Such learning may involve changes in NAC neurons (Horger et al., 1999), as seen in the changes in cellular proteins for intracellular signaling, gene expression, and cellular structure found with repeated drug-induced dopamine release (Nestler, Barrot, & Self, 2001). Furthermore, changes in morphology of neuronal dendritic trees have been implicated in learning related to dopamine transmission in the NAC and in PFC projections to that same striatal region (Gurden, Tassin, & Jay, 1999; Hyman & Malenka, 2001). These neuroplastic processes may be a part of sensitization, whereby the goal or motivational cue becomes more salient as that reward context is experienced repeatedly (Robinson & Berridge, 1993); thus dopaminergic reward systems may be key to defining the repertoire of what is most salient and rewarding for an individual over time (Chambers et al., 2003).

In addition, there is an overlapping neurobiology between stress and reward. The striatal dopaminergic system is sensitively activated by stressors, and stress increases the risk of using and abusing substances and comfort foods (Epel, Lapidus, McEwen, & Brownell, 2001; Goeders, 2002; Sinha, 2001; Sinha et al., 2005). Stressors can activate dopamine projections to the PFC as well as to the mesolimbic brain regions. Receptors for glucocorticoids (stress hormones) have been identified in the rodent brain on dopaminergic neurons in several areas of the brain, including the NAC. Stress activation of the corticotropin-releasing factor (CRF or CRH), hypothalamic–pituitary adrenal (HPA), and noradrenergic-mediated sympathoadrenal medullary systems increase plasma levels of adrenocorticotropic hormone (ACTH), glucocorticoids (GCs), and catecholamines in animals and humans (Baumann et al., 1995; Heesch et al., 1995; Levy et al., 1991; Mello & Mendelson, 1997; Moldow & Fischman, 1987; Sofuoglu, Nelson, Babb, & Hatsukami, 2001). Acute and repeated stressors that activate the above-mentioned brain stress circuits also increase mesolimbic dopaminergic transmission (Dallman, 2005; DiLeone, Georgescu, & Nestler, 2003; Kalivas & Duffy, 1989; Pecoraro, Reyes, Gomez, Bhargava, & Dallman, 2004; Piazza & Le Moal, 1998; Thierry, Tassin, Blanc, & Glowinski, 1976). Cortisol (corticosterone in animals), an adrenal steroid hormone released during stress, stimulates activity of the mesolimbic dopamine systems (Piazza et al., 1996a, 1996b) and increases self-administration of substances in laboratory animals (Goeders, 1997). Stress-related increases in GC signaling alter energy balance and, combined with insulin, affect brain reward regions (NAC) to amplify incentive motivation and increase the intake of comfort foods (Pecoraro, Gomez, & Dallman, 2005).

Genetic differences in novelty-seeking behavior may be mediated by differences in the mesolimbic dopamine systems. Some suggest that enhanced vulnerability to drug use and abuse is associated with a reward deficiency syndrome. Due to functional deficits in mesolimbic dopamine systems, individuals find reinforcing stimuli less pleasurable than others, leading them to seek out drugs and novelty as a behavioral remediation of reward deficiency. This deficiency may be due to lower-than-normal levels of extracellular dopamine. Another possible cause for this deficiency may be lower D_2 receptor levels caused by the expression of the A_1 allele on the D_2 receptor gene (Spear, 2000). This allele increases the probability that a person will develop alcoholism or will engage in a variety of pleasure-seeking behaviors (Kalat, 2004). Those with an alternate form of the D_4 receptor have been shown to have novelty-seeking personalities (Donohew, Bardo, & Zimmerman, 2004); these lead to higher levels of impulsivity, exploratory behaviors, and quick-temperedness. Studies have shown an extreme excess of D_4 dopamine receptors in persons addicted to heroin, as well as a greater occurrence of this allele in individuals with alcoholism and pathological gambling (Zuckerman & Kuhlman, 2000).

Yet another factor that is linked to novelty seeking is level of monoamine oxidase (MAO). MAO is an enzyme that is involved in the catabolic degradation of the monoamine neurotransmitters dopamine, serotonin, and norepinephrine. There are two forms of MAO: MAOA and MAOB. MAOB is closely tied to the regulations of dopamine, whereas MAOA is more involved in the regulation of serotonin and norepinephrine. MAOB is negatively correlated with sensation seeking (i.e., low levels of MAOB are correlated with high levels of sensation seeking). Platelet levels of MAOB change slowly as a function of age; levels are the lowest in adolescence and increase with age. Levels are also consistently higher in women than men at all ages, corresponding with the fact that men tend to exhibit higher levels of sensation seeking. Low platelet levels of MAOB are linked with higher levels of tobacco, drug, and alcohol use, as well as higher rates of criminal offenses (Zuckerman & Kuhlman, 2000). Low levels of MAOB lead to slower degradation of dopamine, ultimately increasing dopamine levels in the brain. Studies have shown that rats with high levels of novelty seeking have higher levels of dopamine activity in the NAC, both under baseline conditions and during novel stimulation (Dellu, Piazza, Mayo, Le Moal, & Simon, 1996).

The Neural Circuitry of Attachment

Considerable work is now accumulating regarding the overlap between the mesolimbic dopaminergic circuits associated with reward regulation and those associated with parenting behavior (Mayes, Swain, & Leckman, 2005; Swain, Lorberbaum, Kose, & Strathearn, 2007), as well as with other adult–adult relationships, though far fewer studies are currently available for adult attachments (Hazan & Shaver, 1994). There is also an emerging literature on how aspects of stress/fear response systems are specially adapted for early parent–infant attachment.

Reward Circuits and Parenting

Postpartum maternal behavior in the rodent model involves heightened pup-directed behaviors (including licking, grooming, nursing, and retrieval), along with heightened aggression toward intruders at the nest site and a generalized decrease in fearfulness on a range of behavioral tests. The initiation and maintenance of maternal behavior involve a specific neural circuit based in reward and stress response systems (e.g., the amygdala, hippocampus, and striatum). With pregnancy or with repeated exposure to pups, structural and molecular changes occur; in specific limbic, hypothalamic, and midbrain regions most of these changes are not yet completely understood, but they reflect in part an adaptation to the various homeostatic demands associated with maternal care. Many of the same cell groups implicated in the control of maternal behavior have been implicated in the control of ingestive (eating and drinking), thermoregulatory (energy-homeostatic), and social (defensive and sexual) behaviors, as well as in general exploratory or foraging behaviors (with locomotor and orienting components) that are required for obtaining any particular goal object. Many of these same structures are also intimately involved in stress responses (Lopez, Akil, & Watson, 1999). Swanson (2000, p. 123) has conceptualized this set of limbic, hypothalamic, and midbrain nuclei as being the "behavioral control column" that is voluntarily regulated by cerebral projections. Consistent with this formulation, it is readily apparent that motherhood presents a major homeostatic challenge within each of these behavioral domains.

Ascending dopaminergic and noradrengeric systems associated with reward pathways also appear to play a crucial role in facilitating maternal behavior (Koob & Le Moal, 1997). For example, rat dams given microinfusions of the neurotoxin 6-hydroxydopamine in the VTA to destroy catecholaminergic neurons during lactation showed a persistent deficit in pup retrieval, but were not impaired with respect to nursing, nest building, or maternal aggression (Hansen, Harthon, Wallin, Lofberg, & Svensson, 1991a). There also appears to be an important interaction between dopaminergic neurons and oxytocin pathways. Specifically, pup retrieval and assuming a nursing posture over pups were blocked in parturient dams by infusions of an oxytocin antagonist into either the VTA or medial preoptic area (MPOA) (Pedersen, Caldwell, Walker, Ayers, & Mason, 1994).

Classical lesion studies done in rodent model systems (rats, mice, and voles) have implicated the MPOA of the hypothalamus, the ventral part of the bed nucleus of the stria terminalis (BNST), and the lateral septum (LS) as pivotal regions for regulation of pup-directed maternal behavior (Leckman & Herman, 2002; Numan, 1994; Numan & Sheehan, 1997). Estrogen, prolactin, and oxytocin can act on the MPOA to promote maternal behavior (Bridges, Numan, Ronsheim, Mann, & Lupini, 1990; Numan, Rosenblatt, & Kiminsaruk, 1997; Pedersen et al., 1994). Oxytocin is primarily synthesized in the magnocellular secretory neurons of two hypothalamic nuclei, the paraventricular nucleus (PVN) and the supraoptic nucleus (SON). The PVN and SON project to the posterior pituitary gland. Pituitary release of oxytocin into the bloodstream results in milk ejection during nursing and uterine contraction during labor. It has also been shown that oxytocin fibers, which arise from parvocellular neurons in the PVN, project to areas of the limbic system that include the amygdala, BNST, and LS (Sofroniew & Weindl, 1981).

There are several reports that oxytocin facilitates maternal behavior (sensitization) in estrogen-primed nulliparous female rats. Intracerebroventricular administration of oxytocin in virgin female rats induces full maternal behavior within minutes (Pederson & Prange, 1979). Conversely, central injection of an oxytocin antagonist, or a lesion of oxytocin-producing cells in the PVN, suppresses the onset of maternal behavior in postpartum female rats (Insel & Harbaugh, 1989; Van Leengoed, Kerker, & Swanson, 1987). However, these manipulations have no effect on maternal behavior in animals that are permitted several days of postpartum mothering. This result suggests that oxytocin plays an important role in facilitating the onset, rather than the maintenance, of maternal attachment to pups (Pedersen, 1997). Brain areas that may inhibit maternal behavior in rats have also been identified (Sheehan, Cirrito, Numan, & Numan, 2000). For example, the vomeronasal and primary olfactory systems have been identified as brain regions that mediate avoidance behavior in virgin female rats exposed to the odor cues of pups (Fleming, Vaccarino, & Luebke, 1980).

Although the central nervous system events that accompany parental care in humans are largely unknown, it is likely that there is a substantial degree of conservation across mammalian species (Fleming, Steiner, & Corter, 1997), and the available data are consistent with the same circuitry being involved in humans as in rats (Fleming, O'Day, & Kraemer, 1999). For example, Fleming et al. (1997) found that first-time mothers with high levels of circulating cortisol were better able to identify their own infants' odors. In these same primiparous mothers, higher levels of affectionate maternal contact with the infants (affectionate burping, stroking, poking, and hugging) were associated with higher levels of salivary cortisol. Both of these findings support the hypothesis that stress response systems are adaptively activated during the period of heightened maternal sensitivity after the birth of a new infant (see also below).

Studies of human mothers have likewise demonstrated that infant cues, such as facial expressions and cries, activate brain reward regions including the VTA, substantia nigra, and NAC (Bartels & Zeki, 2004; Breiter et al., 1997; Lorberbaum et al., 2002; Squire & Stein, 2003; Strathearn & McClure, 2002). Other activated areas associated with maternal behavior include the midbrain, hypothalamus, dorsal and ventral striatum, LS, and orbital frontal cortex (Nitschke et al., 2004); these findings also parallel preclinical work. Overlapping patterns of activation have been seen in mothers viewing their infants' faces and in adults viewing the faces of romantic partners (Bartels & Zeki, 2000, 2004), suggesting common neural pathways relating to social attachment and affiliation.

Lorberbaum et al. (1999, 2002) provided the first work in this area using baby cries as stimuli and presenting these stimuli to mothers in a functional imaging paradigm. Building on the thalamocingulate theory of maternal behavior in animals developed by MacLean (1990), Lorberbaum et al. predicted that baby cries would selectively activate cingulate, thalamus, medial PFC, and orbital frontal cortex. Mothers who were less than 3.5 months postpartum and were exposed to 30 seconds of a standard baby cry versus white noise stimuli (Lorberbaum et al., 1999), showed increased activity in anterior cingulate and right medial PFC. In a follow-up study of brain activity in breastfeeding first-time mothers 4–8 weeks postpartum listening to a standard baby cry versus intensity- and pattern-matched white noise (Lorberbaum et al., 2002), all of the regions activated were those known to be important for rodent maternal behavior, including midbrain, hypothalamus, striatum, and septal regions (Leckman & Herman, 2002; Numan & Sheehan, 1997). Other groups have demonstrated similar patterns of activation in thalamocortical–basal ganglia circuits in mothers responding to infant cries (Seifritz et al., 2003; Swain et al., 2003, 2004).

In addition to baby cry stimuli, several groups have used baby visual stimuli (Bartels & Zeki, 2004; Leibenluft, Gobbini, Harrison, & Haxby, 2004; Nitschke et al., 2004; Swain et al., 2003, 2004). Hypothesizing that reward circuits are also involved in maternal care, and using photographs of familiar and unfamiliar infants, Bartels and Zeki (2004) reported activations in anterior cingulate, insula, basal ganglia (striatum), and midbrain (periaqueductal gray)—regions potentially mediating the emotionally rewarding aspects of maternal behavior—and a decrease in activity in areas important for negative emotions, avoidance behavior, and social assessment. This may suggest a push–pull mechanism for maternal behavior, in which child stimuli activate reward and shut down avoidance or stress response circuits (Bartels & Zeki, 2004) (see also below). A similar study comparing parents and nonparents responding to infant pictures reported bilateral orbital frontal cortex activations that were correlated with ratings of pleasant mood (Nitschke et al., 2004). Another study, using infant video images, reported activations in amygdala, temporal pole, and occipital regions (Ranote et al., 2004).

Whether neural responses to infant stimuli differ according to past and present parental experience with infants is an important consideration. Several studies have compared parents and nonparents. For example, Seifritz et al. (2003) reported that nonparents activated the right amygdala in response to baby laugh stimuli, but parents did not; with baby cry stimuli, parents activated the right amygdala, but nonparents did not. In research comparing first-time to experienced mothers (Swain et al., 2003, 2004), first-time mothers responding to their own or other babies' cries showed activation in midbrain, basal ganglia, cingulate, amygdala, and insula at 2–4 weeks postpartum, but at 3–4 months postpartum the pattern of activation to similar stimulus sets changed from amygdala and insular activations to medial PFC and hypothalamic activations—findings that may suggest a change or consolidation in neural circuitry as the parent–infant relationship develops. Furthermore, the same research found that the parents with two or more children showed patterns of activation at 2–4 weeks comparable to those of first-time parents at 3–4 months; this finding may suggest sensitization of parental attachment circuits by first-time exposure to the responsibility of caring for an infant (Swain et al., 2003, 2004). Thus it appears that there may be significant differences between parental and nonparental responses to baby signals across sensory modalities, that experience with infants may consolidate neural circuitry, and that first-time parents undergo key changes in reward-related circuitry in the first months postpartum.

Fear Response Systems Supporting Parenting

The circuitry of the social/affiliative systems involved in parenting is similar to or overlaps the circuitry associated with fear systems (Mayes, 2006; Swain et al., 2007). The medial amygdala (MeA) and endogenous opiates, centrally involved in fear conditioning, also play a central role in the regulation of social/affiliative behaviors in general and parenting behaviors in particular among mammalian species. The amygdala complex and associated circuits may be especially key in modulating the negative affective cues associated with parenting (as in parental response to infant crying). The MeA has long been recognized as involved in sexual behavior in males and females. Indeed, the extended amygdala generally has been previously implicated in various aspects of affiliative behavior, including maintenance of physical proximity with adult conspecifics (Kling, 1972), maternal behavior (Fleming, Miceli, & Moretto, 1983), and facial recognition (Brothers, Ring, & Kling, 1990; Everitt, 1990). The MeA specifically has also been implicated in the control of social behavior (Fleming et al., 1980; Lehman & Winans, 1982), and there is accumulating evidence for the effect of the amygdala complex on parental behavior. It may be that the MeA (along with other associated regions) plays a different role at different points in development—for example, at the initiation of parenting compared to later parenthood.

In rodents, a key role of the MeA in the control of parenting (and possibly other social) behaviors is in the processing of olfactory cues (Lehman & Winans, 1982). The MeA receives direct input from the accessory olfactory bulb, which in turn receives input from the vomeronasal organ (a region of the olfactory system key to pheromone detection) and also from the main olfactory bulb through the cortical amygdaloid nucleus. Hence, the MeA may be especially central in mediating olfactory-cue-regulated affective changes to pups in the newly parturient animal. Nonpregnant, nulliparous female rats do not show maternal behavior unless sensitized by repeated pup exposure (Rosenblatt, 1967). If the females are rendered anosmic, this period is considerably shortened, suggesting that pup odors may actually inhibit the onset of maternal behavior (Fleming & Rosenblatt, 1974). Fleming and colleagues showed that virgin females with lesions to the MeA or to olfactory projections to the MeA have significantly shorter latencies to the onset of maternal behavior (Fleming et al., 1980; Fleming, Vaccarino, Tambosso, & Chee, 1979). Electrical stimulation of the MeA lengthens sensitization in estrous-cycling female rats (Morgan, Wachtus, Milgram, & Fleming, 1999). Hence the MeA may be responsible for an olfactory-based neophobia to pups, at least prior to exposure, and thus it may inhibit maternal activity on early exposure. Even in experienced animals, early MeA stimulation inhibits maternal behavior, as measured by the latency to retrieve or crouch over the pups (Morgan et al., 1999).

However, the initiation of maternal behavior, especially in the rodent model, involves a decrease in fearfulness and especially in the aversive response to pup-related stimuli, as well as an increase in maternal behaviors. The reduction in fearfulness may be in part mediated by a number of hormones related to pregnancy and parturition, including estrogen, progesterone, prolactin, and oxytocin (Numan & Numan, 1994). Once parental behavior is established, however, MeA lesions do not disrupt maternal behavior (Fleming, Gavarth, & Sarker, 1992; Kolunie & Stern, 1995); indeed, in some instances, MeA activity seems to facilitate parental behavior. Axon-sparing lesions of the corticomedial amygdala using the excitotoxin N-methyl-D,L-aspartic acid with a highly affiliative species, the prairie vole, resulted in males that showed significantly less contact with a familiar adult female and a pup than either males with lesions of the basolateral nucleus or controls did (Kirkpatrick, Carter, Newman, & Insel, 1994). A more selective, specific MeA lesion decreased paternal behavior but not contact with a familiar adult conspecific. In neither case were other behaviors affected, including exploratory behaviors or fearfulness in novel situations. Other investigators have reported that selective lesions of the basolateral nucleus of the amygdala do not have any impact on parental behavior (Numan, Numan, & English, 1993; Slotnick & Nigrosh, 1975).

However, the role of the MeA in social behavior cannot be considered in isolation. A major projection of the medial nucleus of the amygdala is to the hypothalamus and especially the MPOA (Numan & Numan, 1997). Indeed, the major efferents of the MeA are to both the anterior and ventromedial hypothalamus, the BNST, and the MPOA, with minor projections to other limbic areas, including the LS and the hippocampus—areas implicated in social and memory processes and in maternal behavior (Kalinichev, Rosenblatt, Nakabeppu, & Morrell, 2000; Numan, 2004; Numan & Numan, 1997; Sheehan, 2000). Pup exposure has been shown to increase the expression of Fos peptide in both the medial nucleus and amygdala of the MPOA in male prairie voles (Kirkpatrick & Insel, 1993) and in both virgin and experienced female rats (Morgan et al., 1999). The MeA also innervates other targets that have been implicated in female parental care. In the rat, both the peripeduncular nucleus and the VTA receive projections from cells in the MeA (deOlmos, Alheid, & Beltramino, 1985). The peripeduncular nucleus is implicated in maternal aggression (Factor, Mayer, & Rosenblatt, 1993), and the VTA is involved in several aspects of maternal behavior, especially pup retrieval (Hansen, Harthon, Wallin, Lofberg, & Svensson, 1991b).

The analysis of c-fos gene activation within neurons, as measured by the increased production of various Fos proteins, is an important advance in outlining the larger neural circuits that may be involved in either inhibiting or facilitating maternal behavior. Utilizing the neural c-fos response to pups in two groups of naïve females with no prior maternal experience with one group hormonally primed for maternal behaviors and pup care and the other not, Sheehan and colleagues identified putative inhibitory and excitatory regions within the forebrain (Sheehan, 2000; Sheehan et al., 2000). Inhibitory regions included the anterior hypothalamic nuclei (AHN), the ventral part of the LS, the dorsal premammillary nucleus (PM), and the parvocellular part of the PVN. The excitatory regions included the dorsal part of the MPOA, the ventral part of the BNST, and the intermediate part of the LS. Importantly, the MeA was activated regardless of either pup exposure or maternal behavior in the animals. Thus it may be that one circuit from the MeA to the AHN, dorsal PM, ventral LS, and parvocellular PVN is a "central aversion system" that is activated not only by novel olfactory stimuli from pups leading to maternal withdrawal, but more generally by a variety of stressful or threatening situations (Canteras, Chiavegatto, Ribeiro de Valle, & Swanson, 1997; Dielenberg, Hunt, & McGregor, 2001).

There are also projections from the AHN and ventromedial nucleus of the hypothalamus to the MPOA, which may also serve to actively inhibit approach to pups (Numan & Insel, 2003). However, when female rats are fully primed with maternal hormones, they are attracted to pup odors—the shift in behavior referred to earlier. It may be that the hormonal events of late pregnancy and parturition shut down the inhibitory MeA output system with projections to the AHN, while promoting activity in an MeA output system to the MPOA and ventral BNST that stimulates interest in and approach to pup olfactory cues. That is, one inhibitory MeA system is shut down, leaving active a facilitatory MeA-to-MPOA/ventral BNST system (Numan & Insel, 2003). It may also be that in nonmaternal or inexperienced mothers, input from the MeA to the AHN inhibits maternal behavior, whereas in experienced animals, olfactory cues activate an MeA-to-MPOA/ventral BNST circuit that facilitates approach and care. This area requires considerable study.

How this potential change in the balance of inhibitory and facilitative MeA to hypothalamic circuits occurs is not clear, but it may involve mechanisms of conditioned or associative learning (Numan & Sheehan, 1997) in addition to hormone regulation. Indeed, considerably less is known about the control of maternal/parental behavior or the circuitry of learned or established parental behavior after the initial period of hormone priming and exposure. We know that long-lasting parental experience involves activation of both the somatosensory and olfactory systems, and is mediated by some of the same neurochemical systems that underlie learning in other functional contexts (Morgan, Fleming, & Stern, 1992)—particularly the norepinephrine system, which is so key to learning in, for example, traumatic situations. With repeated pup exposure, there may be a change in neural structure–function in the amygdala consistent with learning an enhanced attraction to pups and diminished aversion to pup cues. Such a change in amygdala function based on exposure and learning has recently been reported in adults conditioned to odors as pups (Wilson & Sullivan, 1994).

Furthermore, as discussed above, established maternal behavior is associated with a general reduction in fearfulness and a concomitant increase in aggression, especially toward intruders posing potential danger to pups. Sensitization to pup cues and habituation to their withdrawal-eliciting properties on repeated exposure may be reflected in changes in the MeA as well as associated regions, especially hypothalamic connections. It may also be that true associative learning involving the pairing of conditioned and nonconditioned cues (along with the reinforcing effects of pup responsive behavior) plays a role in the maintenance of maternal behavior, and thus is especially relevant to the extended amygdaloid complex and the MeA. For example, there is a difference in Fos-Lir immunoreactivity between experienced and nonexperienced rat mothers in the MPOA, the basolateral nucleus of the amygdala, and the PFC (Fleming & Korsmit, 1996). How continued exposure to pups changes the balance of inhibitory versus facilitative activity in the MeA, and how reducing or blocking MeA activity in the context of established parenting influences maternal behavior, are open questions.

Also important to the relation between the MeA and parenting is the role of oxytocin in the amygdala generally and the MeA specifically. Oxytocin and a related neuropeptide, vasopressin, are intimately involved in the initiation of parenting behavior (Insel, 1997). Oxytocin is primarily synthesized in the magnocellular secretory neurons of two hypothalamic nuclei, the PVN and SON, which project to the pituitary. Oxytocin fibers from the PVN project to areas of the limbic system, including the amygdala, LS, and BNST, all of which are involved in regulating parental behavior (Sofroniew & Weindl, 1981). Oxytocin receptors are located in the central amygdala and MeA, oxytocin's action on the central nucleus of the amygdala produces an anxiolytic effect (Bale, Davis, Auger, Dorsa, & McCarthy, 2001). Furthermore, oxytocin influences olfactory-related processes in the MeA that are aspects of social recognition and also may be aspects of maternal aggression toward intruders (Ferguson, Aldag, Insel, & Young, 2001; Ferreira, Dahlof, & Hansen, 1987). It may be that the anxiolytic effect of oxytocin moderates the initial pup aversion as mediated through the MeA and is a part of the shift between initial fearfulness to pup cues to attraction and caring behavior—a hypothesis needing detailed study, especially in relation to the MeA and parenting.

We now turn to opiates and social behavior. The brain opioid theory of social attachment has accumulating empirical support. Lines of evidence indicate that (1) opioids diminish the reaction to social separation; (2) opioids are released during bouts of social contact; (3) opioids are rewarding and can induce odor and place preferences; and (4) low basal levels of opioids induce motivation to seek social contact. Like the neuropeptides oxytocin and vasopressin (Carter, 1998; Insel, 1997; Nelson & Panksepp, 1998), the endogenous opioid system may be shared by different forms of attachment. It has been hypothesized that a release of endogenous opioids mediates the rewarding properties of attachment, and that a reduction results in emotional distress and a need to seek and maintain proximity with the attachment object (Panksepp, 1981). Thus exogenous opiates such as morphine should reduce the motivation to seek social contact, and opiate blockers such as naltrexone should increase social contact—hypotheses demonstrated now by many studies across nonprimate mammalian species (Nelson & Panksepp, 1998; Panksepp, Nelson, & Siviy, 1994) in adult and offspring–parent attachment. For example, opiate agonists decrease separation induced distress among infant offspring (Carden, Barr, & Hofer, 1991), and these effects of endogenous opioids on infant-attachment-related behavior have been primarily linked to the mu receptors (Carden, Davachi, & Hofer, 1994). Furthermore, mice lacking the mu-opioid receptor gene show a marked reduction in attachment-related behaviors, including reduction in vocalization and preference for maternal cues (Moles, Kieffer, & D'Amato, 2004). In addition, several studies have indicated that endogenous opioids are released in response to somatosenory contact and milk transfer, both of special relevance to infant care and parent–infant attachment (Nelson & Panksepp, 1998). In situations of rough-and-tumble play, common to nearly all mammalian species, there is an increase in the release of endogenous opioids (Panksepp, 1981). Blocking this release reduces levels of play and physical contact among rats (Panksepp, Siviy, & Normansell, 1984).

Fewer studies have examined the effect of endogenous or exogenous opioids or opiate antagonists on maternal attachment behaviors toward infants though much more work has been done on infant-to-mother attachment (Nelson & Panksepp, 1998). The MPOA is rich in both fibers and receptors for opioid peptides (Simerly, Gorski, & Swanson, 1986), and the major receptor for beta-endorphin, the mu receptor, is found specifically in the MPOA (Hammer & Bridges, 1987). In research with lactating females, morphine injected directly into the MPOA reduced maternal behavior, including pup grouping and pup retrieval; this effect was blocked by naloxone (Rubin & Bridges, 1984), and the morphine effects appeared to be mediated by the mu receptor (Mann, Kinsley, & Bridges, 1991). In one study examining mother–infant separation and reunion in rhesus macaques, morphine decreased clinging with the infant upon reunion, whereas naltrexone increased clinging (Kalin, Shelton, & Lynn, 1995). Conversely, in the VTA (another region involved in parenting behavior), opioids appear to facilitate maternal responsiveness (Thompson & Kristal, 1996). It may be that maternal experience affects the balance between the inhibitory and facilitative aspects of the opioid system on maternal behavior, such that with experience, the inhibitory effect on the MPOA is decreased and the facilitative effect on the VTA is increased (Numan & Insel, 2003). Finally, there is considerable evidence that abuse of cocaine or opiates disrupts the neural mechanisms, especially those regulated by the MPOA, that contribute to maternal motivation and care (Bridges, 1996; Elliott, Lubin, Walker, & Johns, 2001; Thompson & Kristal, 1996).

Although there is much more work to be done, we may hypothesize that in situations of established parenting, a mother separated from her offspring experiences a reduction in endogenous opiates with an increase in her conditioned fear response to the distress calls of her infant, mediated in part through the amygdalar complex and a parallel drive to increase contact and increase endogenous opioids. Upon contact, there is an increase in endogenous opioids and a reduction in amygdala activity, as well as a reduction in anxiety. There may well be a parallel with oxytocin and the related anxiolytic properties (Mayes, 2006). We may also speculate that the link between primary and secondary motivational systems as outlined earlier is relevant to how primary dopaminergically regulated motivational behavior is modified and modulated by inhibitory systems in the amygdalar complex and by endogenous opioids, and vice versa (i.e., how dopaminergic systems may down-regulate aversive or negative affective states). It is important to underscore, though, that there is much to be worked out about the overlap between basic attachment- or reward-based circuits and those monitoring and modulating fear responses.

Social Relationships as Mediators of Reward and Punishment

The overlap between systems mediating reward- or motivation-driven behavior and attachment suggests that social relationships may be considered primary motivational cues, and in and of themselves may serve an emotion-regulating function. Because psychopathological perspectives may often inform models of normative development, we turn now to three examples in which the rewarding properties of attachments may be disrupted and thus lead to ineffective emotional regulation by salient social relationships. In the first instance—the pervasive effect of early life stress—we argue that the reward system itself is dysregulated and less sensitive to the intermittent rewarding properties of social relationships. The second instance—emotional and physical abuse in adulthood—further demonstrates the pervasive effects of early life stress, as well as how later abuse can further disrupt healthy attachment and eliminate the rewarding properties of intimate relationships. In the third instance—addiction—we suggest that drugs of abuse co-opt reward systems, so that these are no longer as responsive to the salient reward of attachments. We also explore the frequent overlap of these three instances, and consider how this co-occurrence serves to disrupt relationships. In all three instances, and particularly when they overlap, considering social relationships as primary motivators of behavior may have treatment implications.

Early Life Stress

Accumulating evidence from both preclinical and clinical laboratories indicates that early failures in parental care have a compromising and enduring impact on the stress-regulating capacities of offspring. In particular, the balance between the fear/stress response and reward systems is affected, so that offspring are less sensitive to reward and more vulnerable to stress. This increased vulnerability to stress and decreased sensitivity to reward have been linked to a range of psychopathologies throughout development, including substance use. Furthermore, early failures in parental care have been shown to negatively affect the parenting abilities of those offspring as adults (Harmer, Sanderson, & Mertin, 1999; Pajulo et al., 2001; Porter & Porter, 2004). Recent work in both preclinical and clinical settings is delineating a specific neural circuitry that is key to early parenting and attachment to offspring, and that overlaps considerably with mesolimbic dopaminergic reward systems and with stress response circuitry.

There has been considerable preclinical research characterizing key aspects of maternal behavior in animal models and linking individual variation in these behaviors to offspring development. In general, these findings suggest that maternal behavior in the days following birth serves to "program" the subsequent maternal behavior of the adult offspring, as well as to establish the pups' level of HPA responsiveness to stress (Denenberg, Rosenberg, Paschke, & Zarrow, 1969; Francis, Diorio, Liu, & Meaney, 1999; Levine, 1975). This complex programming also appears to influence aspects of learning and memory. Furthermore, many of the brain regions implicated in these experimental interventions are the same as those identified in the knockout gene and lesioning studies (see below).

Repeated handling of pups in conjunction with *brief* maternal separations induces more licking and grooming by the rat dams (Liu et al., 1997). As adults, the offspring of mothers that exhibit more licking and grooming of pups during the first 10 days of life show reduced plasma ACTH and corticosterone responses to acute restraint stress, as well as increased hippocampal GC receptor messenger RNA (mRNA) expression, and decreased levels of hypothalamic CRF mRNA (Liu et al., 1997; Plotsky & Meaney, 1993). Subsequent studies by the same group of investigators have shown that the offspring of these high-licking and high-grooming mothers also show reduced acoustic startle responses, as well as enhanced spatial learning and memory (Ladd et al., 2000; Liu, Diorio, Day, Francis, & Meaney, 2000). In contrast, repeated handling of pups in conjunction with *prolonged* maternal separations induces deranged maternal behavior, including a reduction in licking and grooming by the rat dams and reduced maternal aggression (Ladd et al., 2000). Similarly, the adult offspring show increased neuroendocrine responses to acute restraint stress and air puff startle, including elevated levels of PVN CRF mRNA and elevated plasma levels of ACTH and corticosterone (Ladd et al., 2000; Liu et al., 2000). These animals also show as an increased acoustic startle response and enhanced anxiety or fearfulness to novel environments (Ladd et al., 2000).

In other research, early adoption (3–6 hours after birth) was found to be associated with increased maternal licking behavior (Barbazanges et al., 1996), and to prevent the prolonged stress-induced secretion of corticosterone evident in early-separated offspring that were returned to the nest with their biological mother. Similarly, as adults the early-adopted pups demonstrated lower novelty-induced locomotion and better recognition performance in a Y-maze than the early-separated offspring did. However, later adoption at either 5 or 10 days resulted in a prolonged stress-induced corticosterone secretion, increased the locomotor response to novelty, and disrupted cognitive performance in the adult offspring. This has been further supported by work on maternal separation of mice, which suggests a role for nerve growth factor in mediating the effects of external manipulations on the developing brain (Cirulli, Berry, & Alleva, 2003). It has also been shown that the amount of licking and grooming a female pup receives in infancy is associated with how much licking and grooming she provides to her offspring as a new mother (Francis et al., 1999). Low-licking/low-grooming dams can be transformed into high-licking/high-grooming dams by handling. These changes are also passed on to the next generation; that is, the female offspring of the low-licking/low-grooming dams become high-licking/high-grooming mothers if they are either cross-fostered by high-licking/high-grooming dams or handled. The converse is also true; namely, the female offspring of the high-licking/high-grooming dams become low-licking/low-grooming mothers if they are cross-fostered by low-licking/low-grooming dams. These naturally occurring variations in licking, grooming, and arched-back nursing have also been associated with the development of individual differences in behavioral responses to novelty in adult offspring. Adult offspring of the mothers with low licking, grooming, and arched-back nursing show increased startle responses, decreased open-field exploration, and longer latencies to eat food provided in a novel environment (Francis et al., 1999).

Furthermore, Francis and coworkers have demonstrated that the influence of maternal care on the development of stress reactivity is mediated by changes in gene expression in regions of the brain that regulate stress responses. For example, adult offspring of dams high in licking, grooming, and arched-back nursing show increased hippocampal GC receptor mRNA expression, increased expression of N-methyl-D-aspartate receptor subunit and brain-derived neurotrophic factor mRNA, and increased cholinergic innervation of the hippocampus (Francis et al., 1999). In the amygdala, there are increased central benzodiazepine receptor levels in the central and basolateral nuclei; in the PVN, there is decreased CRF mRNA. These adult pups also show a number of changes in receptor density in the locus coeruleus, including increased alpha-2 adrenoreceptors, reduced GABA-A receptors, and decreased CRF receptors (Caldji, Francis, Sharma, Plotsky, & Meaney, 2000; Caldji et al., 1998). In another study, oxytocin receptor binding levels were examined in brain sections from animals high and low in licking, grooming, and arched-back nursing, sacrificed either as nonlactating virgins or during lactation (Francis, Champagne, & Meaney, 2000). Examination of the MPOA and the intermediate and ventral regions of the LS disclosed that oxytocin receptor levels were significantly higher in lactating females than in nonlactating females. Lactation-induced increases in oxytocin receptor binding were greater in high- than in low-licking/grooming/arched-back-nursing females in the BNST and ventral region of the septum. Francis and colleagues suggest, therefore, that variations in maternal behavior in the rat may be reflected in, and influenced by, differences in oxytocin receptor levels in the brain.

In sum, from the preclinical data, it appears that the nature of early caregiving experiences (which may also include environmental enrichment) can have potentially enduring consequences on individual differences in subsequent learning, novelty responsiveness, anxiety regulation, and patterns of stress response through specific neuropharmacological mechanisms (Weaver et al., 2004). Effects on these systems change whether salient social cues are either positively rewarding or negatively aversive and thus inhibiting. Parallel data from humans comes from studies of early child abuse and neglect, in which a range of findings suggest an enduring impact on stress regulatory systems, which may in turn have an impact on social relationships and can lead to a host of other negative consequences.

Recent studies suggest that chronic maltreatment seems to contribute to later dysregulated stress reactivity in a way that may parallel the preclinical models (Glaser, 2000; Watts-English, Fortson, Gibler, Hooper, & De Bellis, 2006). Early abuse affects at least three components of the stress response system: the HPA axis, the serotonergic system, and the sympathetic nervous system (Kaufman et al., 1997, 1998; Watts-English et al., 2006). One result of this dysregulated stress response is prolonged exposure to elevated catechoamines and adrenal steroids, which may in turn adversely affect brain development in early child-hood (Watts-English et al., 2006).

The structural brain differences associated with early maltreatment have been investi-gated in only a handful of studies. Maltreated children have been reported to show reduced brain volumes in a number of areas. Several groups have reported reduced volume in the middle portion of the corpus callosum (De Bellis et al., 1999; Teicher et al., 1997). De Bel-lis et al. (1999) also reported an overall 8% reduction in total brain volume among children with early maltreatment and posttraumatic stress disorder (PTSD), with greater effects being related to earlier and more prolonged abuse. In follow-up studies, De Bellis et al. (2002) have reported a range of structural differences in the frontal cortex between children with PTSD and sociodemographically matched controls; these differences include reductions in PFC, prefrontal cortical white matter, and right temporal lobe. How these structural differences are reflected functionally in children is less clear in the literature to date. However, in adults with a history of childhood abuse and PTSD, functional differences in processing emotion-ally valenced verbal material are associated with differences in PFC and anterior cingulate function (Bremner et al., 1999, 2003; Shin et al., 2001). As in preclinical models, it may be that early adversity or disrupted caregiving in humans influences both stress responsivity and the function of corticolimbic circuits key to reward, so that responses to salient emo-tional cues are either blunted or distorted.

The neurobiological effects of childhood maltreatment may also parallel clinical research suggesting the link between childhood maltreatment and a wide range of psychopathology (Famularo, Kinscherff, & Fenton, 1992). Children who have been maltreated are more likely to present with a range of psychiatric disorders, especially depressive and anxiety problems (Ethier, Couture, & Lacharite, 2004; Finzi, Ram, Har-Even, Shnit, & Weizman, 2001; Stern-berg et al., 1993; Toth & Cicchetti, 1996). During adolescence, young people with a history of physical maltreatment show higher levels of aggression, anxiety/depression, PTSD symp-toms, social problems, and social withdrawal than do their peers who have not experienced maltreatment (Lansford et al., 2002). Adults with a history of childhood maltreatment con-tinue to show increased psychiatric problems (Arnow, 2004), including personality disorders in general and antisocial personality disorder in particular (Johnson, Cohen, Brown, Smailes, & Bernstein, 1999; Luntz & Widom, 1994), with an increased likelihood of violence and criminality (Maxfield & Widom, 1996). Adults with a history of maltreatment are also at a greater risk of developing major depressive disorder (Widom, DuMont, & Czaja, 2007a), and are more likely to show symptoms of anxiety and related posttraumatic disorders (Kaplow & Widom, 2007; Widom, 1999). Moreover, a number of studies have demonstrated an asso-ciation between early maltreatment and later drug and alcohol misuse in adulthood (Lo & Cheng, 2007; Mullings, Hartley, & Marquart, 2004; Widom, White, Czaja, & Marmorstein, 2007b). It seems clear as well that these effects of early abuse may be amplified or reduced by genetic factors and later social context (Kim-Cohen et al., 2006; Moffitt, Caspi, & Rutter, 2005). For example, adolescents with the short allele of the serotonin transporter gene who have suffered early abuse are at greater risk for depression than those without the short allele. But if those same children are surrounded by greater social support, their risk for depression is reduced even in the presence of genetic risk (Kaufman et al., 2006). Taken together, pre-clinical and clinical studies demonstrate the long-lasting neurobiological and psychological effects of early maltreatment, which can serve to disrupt the capacity to form positive and healthy relationships and to be motivated by the rewarding aspects of social relationships.

Finally, personality variables may influence the link between childhood abuse and HIV risk behaviors, including substance use and risky sexual behavior. Specifically, sensation seeking and reward seeking on a behavioral task (Lejuez et al., 2002) mediated the relationship between abuse and the occurrence of HIV risk behavior in adolescents (Bornovalova, Gwadz, Kahler, Aklin, & Lejuez, 2008). Ongoing research appears to have replicated this finding in a younger, preadolescent sample, suggesting that these links are well developed by adolescence (MacPherson, Bornovalova, Weitzman, Wang, & Lejuez, 2008).

Abuse in Adulthood

In addition to early abuse and neglect, abuse later in life—emotional and physical interpersonal violence—can interfere with the ability to form healthy relationships and the motivation to achieve intimate relationships. Research has demonstrated that insecure attachment is both a precursor and an effect of emotional abuse. Those high in secure attachment are less likely to receive and inflict emotional abuse in relationships, while for those high in insecure attachment, receiving and inflicting emotional abuse are both more common (O'Hearn & Davis, 1997). Moreover, when one falls into a cycle of emotional abuse in adulthood relationships, one can no longer clearly view the rewards of healthy relationships or be motivated to achieve positive social relationships. Thus a cycle of interpersonal abuse can rob one of the inherent rewards of social relationships. Findings are similar with regard to physical violence in men: Adult attachment style has been shown to be related to frequency of violence. In one study, insecurely attached men with a high need for dominance in a relationship showed the highest levels of interpersonal violence in their relationships (Mauricio & Gormley, 2001). A common theoretical framework for explaining such findings is avoidance of intimacy; early abuse and a later cycle of violence can contribute to intimacy deficits, preventing individuals from achieving healthy, stable relationships (Bartholomew, 1990; Ward, Hudson, Marshall, & Siegert, 1995). Although early maltreatment clearly affects lifelong attachment capabilities, later-onset abuse can also have a significant impact on one's motivations to form relationships and to seek healthy rewards from these relationships.

Addiction, Reward, and Attachment

There is a considerable body of literature on addiction as a disorder of reward systems or reward-based motivational learning (Koob & Le Moal, 2001). In this section, we focus on how addiction interferes with social relationships by co-opting dopaminergically regulated reward systems, so that these systems are not readily available for other motivational contexts and especially for social relationships (Insel, 2003; Leckman & Mayes, 1998). Symptoms of dependence tend to be similar across the various drugs of abuse. These symptoms include a strong "craving" for the drug because of the highly reinforcing euphoriant effects that support habitual drug-seeking and drug-taking behaviors (Koob & Le Moal, 1997). Another common feature across addictions is the persistent and compulsive engagement in these unhealthy behaviors, despite serious adverse health, social, and legal consequences (Volkow & Wise, 2005). This persistent motivation to engage in addictive behaviors has been linked to two related concepts: first, a reduced ability to control the impulse to engage in these behaviors in the face of internal or external challenges or stress (Baumeister, 2003; Tice, Bratslavsky, & Baumeister, 2001); and second, the drive to feel good or feel better, in order to regain hedonic homeostasis (Koob et al., 2004; Sinha et al., 2005; Volkow & Li, 2005). While intoxicated, engaging in compulsive behavior to obtain their drug of abuse, or preoccupied about their next dose of the drug, addicted individuals neglect their personal relationships and their children's needs.

This pattern, fueled by an individual's addiction, can have serious ramifications for many of his or her close relationships. The effects are pervasive throughout all levels of family relationships—between spouses, parent and children, and siblings. They can severely disrupt healthy communication and positive, productive family functioning (Le Poire, 2003). The consequences of substance use are particularly striking in research examining negative outcomes related to relationships and marriage (Ripley, Cunion, & Noble, 2006). Research has demonstrated that higher levels of alcohol intoxication during marriage significantly predict later divorce, even after other factors are controlled for (Collins, Ellickson, & Klein, 2007). In addition to frequency of intoxication, discrepancy in partners' drinking patterns may contribute to marital dysfunction and unhealthy relationships (Mudar, Leonard, & Soltysinski, 2001). A substance-using partner, driven by the rewards of addiction rather than the perceived rewards of the relationship, is less motivated and less able to meet the other partner's expectations, fulfill social obligations, and embody his or her role as a spouse, partner, or parent.

This pattern of social neglect in those who use substances is apparent as early as adolescence. In a longitudinal sample of 12,686 adolescents from ages 14 to 21, cocaine and cannabis use was associated with more problematic interpersonal relationships and relationship conflicts (Macleod et al., 2004a). Many studies have demonstrated a relationship between adolescent substance use and problems with peers/siblings (Brook, Balka, & Whiteman, 1999), as well as higher rates of antisocial behavior (Fergusson, Horwood, & Swain-Campbell, 2002; Friedman, Kramer, Kreisher, & Granick, 1996; Macleod et al., 2004b). Higher rates of antisocial behavior in substance-using adolescents illustrate the initial contexts in which substance use disenables the rewarding properties of social relationships. Research also indicates a pervasive link between adolescent substance use and violent behaviors (Dornbusch, Lin, Munroe, & Bianchi, 1999; Friedman et al., 1996; White, Loeber, Stouthamer-Loeber, & Farrington, 1999). This evidence further suggests that substance-using individuals' disregard for others fuels involvement in violent behavior/criminal activity, whether as a means to obtain substances or as a consequence of being intoxicated or high. The motivations characteristic of healthy attachments are obscured in the lives of young people who use substances, thus making violent and antisocial behaviors seemingly more natural alternatives. This pattern continues throughout adulthood, as demonstrated by the higher rates of co-occurring substance use and interpersonal violence (Kilpatrick, Acierno, Resnick, Saunders, & Best, 1997; Testa, Livingston, & Leonard, 2003).

Maternal substance use is a unique demonstration of how addiction interferes with the motivational context of relationships. Primarily, we can see this manifested in mothers' decision in the first place to drink or use drugs during pregnancy, with the knowledge that prenatal substance use can severely disrupt a child's physical and cognitive development (Griffith, Azuma, & Chasnoff, 1994). After birth, parental substance use is associated with higher rates of neglect, child abuse, and poor parenting (Johnson & Leff, 1999; Nair, Schuler, Black, Kettinger, & Harrington, 2003). It can also lead to negative behavioral outcomes in children, such as attention-deficit/hyperactivity disorder and other externalizing disorders or symptoms (Barnow, Schuckit, Smith, Preuss, & Danko, 2002; Chronis et al., 2003), higher rates of psychopathology in general (Peleg-Oren & Teichman, 2006), and future engagement in risk behaviors (e.g., substance use, delinquency; Grekin, Brennan, & Hammen, 2005; Lam et al., 2007). With the exception of poverty, maternal substance abuse is the most common psychiatric or social problem involved when children are referred to the child welfare system because of suspected parental abuse or neglect (U.S. Department of Health and Human Services, 1999; Child Welfare League of America, 1998).

Observations of mother–child interactions involving mothers with histories of abuse and/ or dependence on illicit drugs (e.g., heroin and cocaine) have indicated poor sensitivity, unresponsiveness to children's emotional cues, and heightened physical provocation and intrusiveness (Burns, Chetnik, Burns, & Clark, 1997; Hans, Bernstein, & Henson, 1999). Research has consistently demonstrated that mothers with a substance use history may experience higher levels of infant parenting stress than nonusing mothers may (Kelley, 1992; Sheinkopf et al., 2006), and that a mother's level of drug use may serve to magnify the perceived burden of infant care (Porter & Porter, 2004). The characteristics of cocaine-exposed infants may serve to exacerbate the effect of addiction on disrupted attachment. Cocaine-exposed infants often demonstrate greater reactivity (e.g., excessive crying, irritability, jitteriness, tone abnormalities, and attention problems) than non-cocaine-exposed infants do (Lester et al., 2002), and these behaviors may be considered particularly distressing by substance-using parents (Nair et al., 2003; Porter & Porter, 2004). In addition to visible or audible cues that may appear burdensome to a parent, an absence of certain cues in cocaine-exposed infants (e.g., lethargy, unresponsiveness, lack of enjoyment during play) can lead to increased frustration and guilt in mothers (Nair et al., 2003; Porter & Porter, 2004). In other words, postnatal effects of drug exposure can interfere with an infant's ability to send healthy, positive cues to the mother that normally define the mother–infant attachment, thus feeding the guilt and decreased parenting confidence experienced by a substance-using mother (Porter & Porter, 2004; Suchman & Luthar, 2001).

Studies reporting drug-abusing mothers' views about parenting have indicated a lack of understanding about basic child development issues and ambivalent feelings about having and keeping children (Mayes, 2002; Murphy & Rosenbaum, 1999). Substance-abusing mothers have often been raised by parents with poor parenting skills, thus contributing to their lack of knowledge of infant characteristics and adaptive parenting techniques (Harmer et al., 1999; Pajulo et al., 2001; Porter & Porter, 2004). As a group, drug-dependent mothers fare worse than non-drug-dependent mothers on a wide range of parenting indices, and more frequently lose their children to foster care than non-drug-dependent mothers do (Chaffin, Kelleher, & Hollenberg, 1996; Mayes & Bornstein, 1996). Self-reported behaviors among drug-dependent mothers have also revealed harsh, threatening, overly involved, authoritarian parenting styles juxtaposed with permissiveness, neglect, poor involvement, low tolerance of child demands and misbehavior, and parent–child role reversals (Harmer et al., 1999; Mayes & Truman, 2002; Suchman & Luthar, 2000). Substance-using parents also less frequently seek external help or resources for their children; not only are they consumed with their own substance use, but they may fear losing their children as a consequence of their use (Davis, 1990; Pajulo et al., 2001; Suchman & Luthar, 2001).

Although these behaviors reflect in part the confluence of early abuse and neglect with concurrent psychopathology in the lives of substance-abusing parents, the behaviors also reflect the parents' difficulties in attending to the needs of their infants as well as to other relationships in their lives. Indeed, many addicted adults speak about their relationships with their drugs as the most central relationships of their lives (Hirschman, 1992). Hence, substance addiction affords an example of a type of psychopathology originating in alterations in mesolimbic dopaminergic reward systems. These value systems are essential to detecting or assigning salience to elements in the environment with adaptive significance (Edelman & Tononi, 1995), including food, sex, and (as we propose) relationships. Addiction co-opts or hijacks these endogenous value systems, which are usually activated in response to positive environmental opportunities; it thus numbs individuals to the emotional cues and needs of others (Friston & Tononi, 1996; Graybiel, Aosaki, Flaherty, & Kimura, 1994; Schultz, Dayan, & Montague, 1997). The co-opting of these reward systems has problematic consequences not only for addicted individuals themselves, but also severe implications for their children and other social relationship partners.

Addiction, Early Stress, and Adulthood Abuse

Early stress, later abuse, and addiction do not exist in isolation. Rather, research demonstrates that early stress and abuse are associated with later substance use, and that substance use is associated with increased abuse both of partners and children, particularly during periods of use (El-Bassel, Gilbert, Schilling, & Wada, 2000). Interpersonal violence in the relationships of substance-using individuals is a particularly salient illustration of our model. Research consistently demonstrates a vicious cycle, with substance use increasing the likelihood of future violence, and violence increasing the risk of later substance use (Kilpatrick et al., 1997). Illicit drug use in women has been shown to be associated with increased chances of experiencing interpersonal violence, both in ongoing relationships and in newly formed relationships (Testa et al., 2003). These findings do not pertain only to physical violence; research suggests that all types of abuse, including emotional abuse, are associated with higher levels of substance use. This link is particularly strong, however, for sexual and physical abuse (Moran, Vuchinich, & Hall, 2004). As discussed previously, insecure attachment has been shown to predict interpersonal violence and emotional abuse, and it has also been shown to be related to substance use (Golder, Gillmore, Spieker, & Morrison, 2005). Clearly, not only do addiction, early stress, and later abuse often coexist, but the overlap in neural circuitry among these conditions is perhaps most obvious when they coexist. We can view the scenario in which substance use, early stress, and later abuse interact as being the most likely to jeopardize the rewarding properties of social relationships.

Conclusions

Accumulating evidence suggests that there is a significant overlap between the cortical–striatal–thalamic–cortical systems responsive to reward and to social cues and attachment. The findings from studies of parents responding to their infants particularly underscore the primary rewarding salience of attachment relationships; they also show how, under conditions of optimal development, attachment relationships serve to modulate stress and fear, and negative affective states lead individuals to seek the comfort of others. However, under conditions of neglect and abuse, the balance between reward and stress regulation is distorted: The stress-reducing aspects of reward-motivated behavior are blunted, and social relationships are not sufficient to regulate negative affect. Other processes such as addiction, also based in mesolimbic reward systems, co-opt these systems, so that social relationships are not effectively salient and thus not instrumental in emotion regulation. Particularly in the context of coexisting addiction, early stress, and later abuse, we see an intensification of blunted reward sensitivity with regard to social relationships. Considering social relationships as primary rewarding motivators will expand the focus of research on reward system functions; it is especially relevant to how environmental context and early caretaking may influence the threshold or sensitivity of the neural systems subserving reward-motivated behaviors.

Acknowledgments

Linda C. Mayes's work is supported by National Institute on Drug Abuse (NIDA) Grant Nos. RO1-DA-06025, DA-017863, and KO5 DA020091, and by National Institute of Child Health and Human Development Grant No. RO1 HD044796. Her work is also supported by the Harris Foundation, the Donaghue Foundation, and the Pfeffer Foundation. Portions of this chapter draw extensively on related publication by Mayes and colleagues (Mayes, 2006; Mayes, Swain, & Leckman, 2005). C. W. Lejuez's work is supported by NIDA Grant Nos. RO1-DA-18647, RO1-DA-18703, and RO1-DA-19405 as Principal Investigator; R03-DA-23001 and R21-DA22741 as Co-Principal Investigator; and R36-DA-18506 and F31-DA23302 as faculty mentor.

References

Ainsworth, M. S., Blehar, M. C., Waters, E., & Wall, S. (1978). *Patterns of attachment: A psychological study of the Strange Situation.* Hillsdale, NJ: Erlbaum.

Arnow, B. A. (2004). Relationships between childhood maltreatment, adult health and psychiatric outcomes, and medical utilization. *Journal of Clinical Psychiatry, 65*(Suppl. 12), 10–15.

Baker, T. B., Piper, M. E., McCarthy, D. E., Majeskie, M. R., & Fiore, M. C. (2004). Addiction motivation reformulated: An affective processing model of negative reinforcement. *Psychological Review, 111*(1), 33–51.

Bale, T. L., Davis, A. M., Auger, A. P., Dorsa, D. M., & McCarthy, M. M. (2001). CNS region-specific oxytocin receptor expression: Importance in regulation of anxiety and sex behavior. *Journal of Neuroscience, 21,* 2546–2552.

Barbazanges, A., Vallee, M., Mayo, W., Day, J., Simon, H., Le Moal, M., et al. (1996). Early and later adoptions have different long-term effects on male rat offspring. *Journal of Neuroscience, 16,* 7783–7790.

Bardo, M. T., Donohew, R. L., & Harrington, N. G. (1996). Psychobiology of novelty seeking and drug seeking behavior. *Behavioural Brain Research, 77*(1), 23–43.

Barnow, S., Schuckit, M., Smith, T. L., Preuss, U., & Danko, G. (2002). The real relationship between the family density of alcoholism and externalizing symptoms among 146 children. *Alcohol and Alcoholism, 37*(4), 383–387.

Bartels, A., & Zeki, S. (2000). The neural basis of romantic love. *NeuroReport, 11*(17), 3829–3834.

Bartels, A., & Zeki, S. (2004). The neural correlates of maternal and romantic love. *NeuroImage, 21,* 1155–1166.

Bartholomew, K. (1990). Avoidance of intimacy: An attachment perspective. *Journal of Social and Personal Relationships, 7*(2), 147–178.

Baumann, M. H., Gendron, T. M., Becketts, K. M., Henningfield, J. E., Gorelick, D. A., & Rothman, R. B. (1995). Effects of intravenous cocaine on plasma cortisol and prolactin in human cocaine abusers. *Biological Psychiatry, 38*(11), 751–755.

Baumeister, R. F. (2003). Ego depletion and self-regulation failure: A resource model of self-control. *Alcohol: Clinical and Experimental Research, 27,* 281–284.

Bechara, A. (2001). Neurobiology of decision-making: Risk and reward. *Seminars in Clinical Neuropsychiatry, 6*(3), 205–216.

Bornovalova, M. A., Gwadz, M., Kahler, C. W., Aklin, W. M., & Lejuez, C. W. (2008). Sensation seeking and risk-taking propensity as mediators in the relationship between childhood abuse and HIV-related risk behavior. *Child Abuse and Neglect, 32*(1), 99–109.

Bowlby, J. (1973). *Attachment and loss: Vol. 2. Separation.* New York: Basic Books.

Bowlby, J. (1982). *Attachment and loss: Vol. 1. Attachment* (2nd ed.). New York: Basic Books. (Original work published 1969)

Breiter, H., Gollub, R. L., Weisskopf, R. M., Kennedy, D. N., Makris, N., Berke, J. D., et al. (1997). Acute effects of cocaine on human brain activity and emotion. *Neuron, 19*, 591–611.

Bremner, J., Narayan, M., Staib, L. H., Southwick, S. M., McGlashan, T., & Charney, D. S. (1999). Neural correlates of memories of childhood sexual abuse in women with and without posttraumatic stress disorder. *American Journal of Psychiatry, 156*(11), 1787–1795.

Bremner, J., Vythilingam, M., Vermetten, E., Southwick, S. M., McGlashan, T., Staib, L. H., et al. (2003). Neural correlates of declarative memory for emotionally valenced words in women with posttraumatic stress disorder related to early childhood sexual abuse. *Biological Psychiatry, 53*(10), 879–889.

Bretherton, I. (1985). Attachment theory: Retrospect and prospect. In I. Bretherton & E. Waters (Eds.), Growing points of attachment theory and research. *Monographs of the Society for Research in Child Development, 50*(1–2, Serial No. 209), 3–35.

Bretherton, I., & Munholland, K. A. (1999). Internal working models in attachment relationships: A construct revisited. In J. Cassidy & P. R. Shaver (Eds.), *Handbook of attachment: Theory, research, and clinical applications* (pp. 89–111). New York: Guilford Press.

Bridges, R. S. (1996). Biochemical basis of parental behavior in the rat. *Advances in the Study of Behavior, 25*, 215–242.

Bridges, R. S., Numan, M., Ronsheim, P. M., Mann, P. E., & Lupini, C. E. (1990). Central prolactin infusions stimulate maternal behavior in steroid-treated, nulliparous female rats. *Proceedings of the National Academy of Sciences USA, 87*, 8003–8007.

Brook, J. S., Balka, E. B., & Whiteman, M. (1999). The risks for late adolescence of early adolescent marijuana use. *American Journal of Public Health, 89*(10), 1549–1554.

Brothers, L., Ring, B., & Kling, A. (1990). Response of neurons in the macaque amygdala to complex social stimuli. *Behavioural Brain Research, 41*(3), 199–213.

Burns, K. A., Chetnik, L., Burns, W. J., & Clark, R. (1997). The early relationship of drug abusing mothers and their infants: An assessment at eight to twelve months of age. *Journal of Clinical Psychology, 53*, 279–287.

Caldji, C., Francis, D., Sharma, S., Plotsky, P. M., & Meaney, M. J. (2000). The effects of early rearing environment on the development of GABAA and central benzodiazepine receptor levels and novelty-induced fearfulness in the rat. *Neuropsychopharmacology, 22*, 219–229.

Caldji, C., Tannenbaum, B., Sharma, S., Francis, D., Plotsky, P. M., & Meaney, M. J. (1998). Maternal care during infancy regulates the development of neural systems mediating the expression of fearfulness in the rat. *Proceedings of the National Academy of Sciences USA, 95*, 5335–5340.

Canteras, N. S., Chiavegatto, S., Ribeiro de Valle, L. E., & Swanson, L. W. (1997). Severe reduction in rat defensive behavior to a predator by discrete hypothalamic chemical lesions. *Brain Research Bulletin, 44*, 297–305.

Carden, S. E., Barr, G. A., & Hofer, M. A. (1991). Differential effects of specific opioid receptor agonists on rat pup isolation calls. *Brain Research: Developmental Brain Research, 62*(1), 17–22.

Carden, S. E., Davachi, L., & Hofer, M. A. (1994). U50,488 increases ultrasonic vocalizations in 3-, 10-, and 18-day-old rat pups in isolation and the home cage. *Developmental Psychobiology, 27*(1), 65–83.

Carlson, E. A. (1998). A prospective longitudinal study of attachment disorganization/disorientation. *Child Development, 69*(4), 1107–1128.

Carter, C. S. (1998). Neuroendocrine perspectives on social attachment and love. *Psychoneuroendocrinology, 23*, 779–818.

Cassidy, J. (1999). The nature of the child's ties. In J. Cassidy & P. R. Shaver (Eds.), *Handbook of attachment: Theory, research, and clinical applications* (pp. 3–20). New York: Guilford Press.

Chaffin, M., Kelleher, K., & Hollenberg, J. (1996). Onset of physical abuse and neglect: Psychiatric, substance abuse, and social risk factors from prospective community data. *Child Abuse and Neglect, 20*, 191–203.

Chambers, R. A., & Self, D. W. (2002). Motivational responses to natural and drug rewards in rats with neonatal ventral hippocampal lesions: An animal model of dual diagnosis schizophrenia. *Neuropsychopharmacology, 27*, 889–905.

Chambers, R. A., Taylor, J. R., & Potenza, M. N. (2003). Developmental neurocircuitry of motivation in adolescence: A critical period of addiction vulnerability. *American Journal of Psychiatry, 160*, 1041–1052.

Child Welfare League of America. (1998). *Alcohol and other drug survey of state child welfare agencies*. Washington, DC: Author.

Chronis, A. M., Lahey, B. B., Pelham, W. E., Kipp, H. L., Baumann, B. L., & Lee, S. S. (2003). Psychopathology and substance abuse in parents of young children with attention-deficit/hyperactivity disorder. *Journal of the American Academy of Child and Adolescent Psychiatry, 42*(12), 1424–1432.

Cirulli, F., Berry, A., & Alleva, E. (2003). Early disruption of the mother–infant relationship: Effects on brain plasticity and implications for psychopathology. *Neuroscience and Biobehavioral Reviews, 27*(1–2), 73–82.

Collins, R. L., Ellickson, P. L., & Klein, D. J. (2007). The role of substance use in young adult divorce. *Addiction, 102*(5), 786–794.

Crittenden, P. M. (1992). Quality of attachment in the preschool years. *Development and Psychopathology, 4*, 209–241.

Dallman, M. F. (2005). Fast glucocorticoid actions on brain: Back to the future. *Frontiers in Neuroendocrinology, 26*(3–4), 103–108.

Davis, S. K. (1990). Chemical dependency in women: A description of its effects and outcome on adequate parenting. *Journal of Substance Abuse Treatment, 7*(4), 225–232.

De Bellis, M. D., Keshavan, M. S., Clark, D. B., Casey, B., Giedd, J. N., Boring, A. M., et al. (1999). Developmental traumatology: II. Brain development. *Biological Psychiatry, 45*(10), 1271–1284.

De Bellis, M. D., Keshavan, M. S., Shifflett, H., Iyengar, S., Beers, S. R., Hall, J., et al. (2002). Brain structures in pediatric maltreatment-related posttraumatic stress disorder: A sociodemographically matched study. *Biological Psychiatry, 52*(11), 1066–1078.

Dellu, F., Piazza, P., Mayo, W., Le Moal, M., & Simon, H. (1996). Novelty-seeking in rats: Biobehavioural characteristics and possible relationship with the sensation-seeking trait in man. *Neuropsychobiology, 34*(3), 136–145.

Denenberg, V. H., Rosenberg, K. M., Paschke, R., & Zarrow, M. X. (1969). Mice reared with rat aunts: Effects on plasma corticosterone and open-field activity. *Nature, 221*, 73–74.

deOlmos, J., Alheid, G. F., & Beltramino, C. A. (1985). Amygdala. In G. Paxinos (Ed.), *The rat nervous system: Vol. 1. Forebrain and midbrain* (pp. 223–334). Orlando, FL: Academic Press.

Dielenberg, R. A., Hunt, G. E., & McGregor, I. S. (2001). When a rat smells a cat: The distribution of Fos immunoreactivity in rat brain following exposure to a predatory odor. *Neuroscience, 104*, 1085–1097.

DiLeone, R., Georgescu, D., & Nestler, E. (2003). Lateral hypothalamic neuropeptides in reward and drug addiction. *Life Science, 73*(6), 759–768.

Donohew, L., Bardo, M. T., & Zimmerman, R. S. (2004). *Personality and risky behavior: Communication and prevention*. In R. M. Stelmack (Ed.), *On the psychobiology of personality: Essays in honor of Marvin Zuckerman* (pp. 223–245). Amsterdam: Elsevier.

Dornbusch, S. M., Lin, I. C., Munroe, P. T., & Bianchi, A. J. (1999). Adolescent polydrug use and violence in the United States. *International Journal of Adolescent Medicine and Health, 11*(3–4), 197–219.

Edelman, G. M., & Tononi, G. (1995). Neural Darwinism: The brain as a selectional system. In J. Cornwell (Ed.), *Nature's imagination: The frontiers of scientific vision* (pp. 78–100). New York: Oxford University Press.

El-Bassel, N., Gilbert, L., Schilling, R., & Wada, T. (2000). Drug abuse and partner violence among women in methadone treatment. *Journal of Family Violence, 15*(3), 209–228.

Elliott, J. C., Lubin, D. A., Walker, C. H., & Johns, J. M. (2001). Acute cocaine alters oxytocin levels in the medial preoptic area and amygdala in lactating rat dams: Implications for cocaine-induced changes in maternal behavior and maternal aggression. *Neuropeptides, 35,* 127–134.

Epel, E., Lapidus, R., McEwen, B., & Brownell, K. (2001). Stress may add bite to appetite in women: A laboratory study of stress-induced cortisol and eating behavior. *Psychoneuroendocrinology, 26*(1), 37–49.

Ethier, L. S., Couture, G., & Lacharite, C. (2004). Risk factors associated with the chronicity of high potential for child abuse and neglect. *Journal of Family Violence, 19*(1), 13–24.

Evans, C. S. (1997). Referential signalling. *Perspectives in Ethology, 12,* 99–143.

Everitt, B. J. (1990). Sexual motivation: A neural and behavioural analysis of the mechanisms underlying appetitive and copulatory responses of male rats. *Neuroscience and Biobehavioral Reviews, 14*(2), 217–232.

Everitt, B. J., & Wolf, M. E. (2002). Psychomotor stimulant addiction: A neural systems perspective. *Journal of Neuroscience, 22,* 3312–3320.

Eysenck, H. J. (1985). *Personality and individual differences: A natural science approach.* New York: Plenum Press.

Factor, E. M., Mayer, A. D., & Rosenblatt, J. S. (1993). Peripeduncular nucleus lesions in the rat: I. Effects on maternal aggression, lactation, and maternal behavior during pre- and postpartum periods. *Behavioral Neuroscience, 107,* 166–185.

Famularo, R., Kinscherff, R., & Fenton, T. (1992). Psychiatric diagnoses of maltreated children: Preliminary findings. *Journal of the American Academy of Child and Adolescent Psychiatry, 31*(5), 863–867.

Ferguson, J. N., Aldag, J. M., Insel, T. R., & Young, L. J. (2001). Oxytocin in the medial amygdala is essential for social recognition in the mouse. *Journal of Neuroscience, 21,* 8278–8285.

Fergusson, D. M., Horwood, L., & Swain-Campbell, N. (2002). Cannabis use and psychosocial adjustment in adolescence and young adulthood. *Addiction, 97*(9), 1123–1135.

Ferreira, A., Dahlof, L., & Hansen, S. (1987). Olfactory mechanisms in the control of maternal aggression, appetite, and fearfulness: Effects of lesions to olfactory receptors, mediodorsal thalamic nucleus, and insular prefrontal cortex. *Behavioral Neuroscience, 101,* 709–717.

Finzi, R., Ram, A., Har-Even, D., Shnit, D., & Weizman, A. (2001). Attachment styles and aggression in physically abused and neglected children. *Journal of Youth and Adolescence, 30*(6), 769–786.

Fleming, A. S., Gavarth, K., & Sarker, J. (1992). Effects of transections to the vomeronasal nerves or to the main olfactory bulbs on the initiation and long-term retention of maternal behavior in primiparous rats. *Behavioral and Neural Biology, 57,* 177–188.

Fleming, A. S., & Korsmit, M. (1996). Plasticity in the maternal circuit: Effects of maternal experience on Fos-Lir in hypothalamic, limbic, and cortical structures in the postpartum rat. *Behavioral Neuroscience, 110,* 567–582.

Fleming, A. S., Miceli, M., & Moretto, D. (1983). Lesions of the medial preoptic area prevent the facilitation of maternal behavior produced by amygdala lesions. *Physiology and Behavior, 31,* 503–510.

Fleming, A. S., O'Day, D. H., & Kraemer, G. W. (1999). Neurobiology of mother–infant interactions: Experience and central nervous system plasticity across development and generations. *Neuroscience and Biobehavioral Reviews, 23,* 673–685.

Fleming, A. S., & Rosenblatt, J. S. (1974). Olfactory regulation of maternal behavior in rats: II. Effects of peripherally induced anosmia and lesions of the lateral olfactory tract in pup-induced virgins. *Journal of Comparative Physiology and Psychology, 86,* 233–246.

Fleming, A. S., Steiner, M., & Corter, C. (1997). Cortisol, hedonics, and maternal responsiveness in human mothers. *Hormones and Behavior, 32*(2), 85–98.

Fleming, A. S., Vaccarino, F., & Luebke, C. (1980). Amygdaloid inhibition of maternal behavior in the nulliparous female rat. *Physiology and Behavior, 25,* 731–743.

Fleming, A. S., Vaccarino, F., Tambosso, L., & Chee, P. (1979). Vomeronssal and olfactory system modulation of modulation of maternal behavior in the rat. *Science, 203*, 372–374.

Fonagy, P., Gergely, G., Jurist, E. L., & Target, M. (2002). Affect regulation, mentalization, and development of the self." New York: Other Press.

Francis, D. D., Champagne, F. C., & Meaney, M. J. (2000). Variations in maternal behavior are associated with differences in oxytocin receptor levels in the rat. *Journal of Neuroendocrinology, 12*, 1145–1148.

Francis, D. D., Diorio, J., Liu, D., & Meaney, M. J. (1999). Non-genomic transmission across generations of maternal behavior and stress responses in the rat. *Science, 286*, 1155–1158.

Friedman, A. S., Kramer, S., Kreisher, C., & Granick, S. (1996). The relationships of substance use to illegal and violent behavior, in a community sample of young adult African American men and women (gender differences). *Journal of Substance Abuse, 8*, 379–402.

Friston, K. J., & Tononi, G. (1996). Characterizing the complexity of neuronal interactions. *Human Brain Mapping, 3*, 302–314.

Glaser, D. (2000). Child abuse and neglect and the brain: A review. *Journal of Child Psychology and Psychiatry, 41*(1), 97–116.

Goeders, N. E. (1997). A neuroendocrine role in cocaine reinforcement. *Psychoneuroendocrinology, 22*(4), 237–259.

Goeders, N. E. (2002). Stress and cocaine addiction. *Journal of Pharmacology and Experimental Therapeutics, 301*(3), 785–789.

Golder, S., Gillmore, M. R., Spieker, S., & Morrison, D. (2005). Substance use, related problem behaviors and adult attachment in a sample of high risk older adolescent women. *Journal of Child and Family Studies, 14*(2), 181–193.

Graybiel, A. M., Aosaki, T., Flaherty, A. W., & Kimura, M. (1994). The basal ganglia and adaptive motor control. *Science, 265*(6), 1826–1831.

Grekin, E. R., Brennan, P. A., & Hammen, C. (2005). Parental alcohol use disorders and child delinquency: The mediating effects of executive functioning and chronic family stress. *Journal of Studies on Alcohol, 66*(1), 14–22.

Griffith, D. R., Azuma, S. D., & Chasnoff, I. J. (1994). Three-year outcome of children exposed prenatally to drugs. *Journal of the American Academy of Child and Adolescent Psychiatry, 33*(1), 20–27.

Groenewegen, H. J., Mulder, A. B., Beijer, A. V., Wright, C. I., Lopes da Silva, F. H., & Pennartz, C. M. (1999). Hippocampal and amygdaloid interactions in the nucleus accumbens. *Psychobiology, 27*(2), 149–164.

Gurden, H., Tassin, J. P., & Jay, T. M. (1999). Integrity of mesocortical dopaminergic system is necessary for complete expression of *in vivo* hippocampal–prefrontal cortex long-term potentiation. *Neuroscience, 94*, 1019–1027.

Hammer, R. J., & Bridges, R. S. (1987). Preoptic area opioids opiate receptors increase during pregnancy and decrease during lactation. *Brain Research, 420*, 48–56.

Hans, S. L., Bernstein, V. J., & Henson, L. G. (1999). The role of psychopathology in the parenting of drug-dependent women. *Development and Psychopathology, 11*, 957–977.

Hansen, S., Harthon, C., Wallin, E., Lofberg, L., & Svensson, K. (1991a). Mesotelencephalic dopamine system and reproductive behavior in the female rat: Effects of ventral tegmental 6-hydroxydopamine lesions on maternal and sexual responsiveness. *Behavioral Neuroscience, 105*, 588–598.

Hansen, S., Harthon, C., Wallin, E., Lofberg, L., & Svensson, K. (1991b). The effects of 6-OHDA-induced dopamine depletions in the ventral or dorsal striatum on maternal and sexual behavior in the female rat. *Pharmacology, Biochemistry and Behavior, 39*, 71–77.

Harmer, A. L., Sanderson, J., & Mertin, P. (1999). Influence of negative childhood experiences on psychological functioning, social support, and parenting for mothers recovering from addiction. *Child Abuse and Neglect, 23*(5), 421–433.

Hazan, C., & Shaver, P. R. (1994). Attachment as an organizational framework for research on close relationships. *Psychological Inquiry, 5*(1), 1–22.

Heesch, C. M., Negus, B. H., Keffer, J. H., Snyder, R. W., II, Risser, R. C., & Eichhorn, E. J. (1995). Effects of cocaine on cortisol secretion in humans. *American Journal of the Medical Sciences, 310*(2), 61–64.

Hirschman, E. C. (1992). The consciousness of addiction: Toward a general theory of compulsive consumption. *Journal of Consumer Research, 19*(2), 155–179.

Horger, B. A., Lyasere, C. A., Berhow, M. T., Messer, C. J., Nestler, E. J., & Taylor, J. R. (1999). Enhancement of locomotor activity and conditioned reward to cocaine by brain-derived neurotrophic factor. *Journal of Neuroscience, 19*, 4110–4122.

Hyman, S. E., & Malenka, R. C. (2001). Addiction and the brain: The neurobiology of compulsion and its persistence. *Neuroscience, 2*, 695–703.

Insel, T. R. (1997). A neurobiological basis of social attachment. *American Journal of Psychiatry, 154*(6), 726–735.

Insel, T. R. (2003). Is social attachment an addictive disorder?" *Physiology and Behavior, 79*(3), 351–357.

Insel, T. R., & Harbaugh, C. R. (1989). Lesions of the hypothalamic paraventricular nucleus disrupt the initiation of maternal behavior. *Physiology and Behavior, 45*, 1033–1041.

Ito, R., Dalley, J. W., Robbins, T. W., & Everitt, B. J. (2002). Dopamine release in the dorsal striatum during cocaine-seeking behavior under the control of a drug-associated cue. *Journal of Neuroscience, 22*, 6247–6253.

Johnson, J. G., Cohen, P., Brown, J., Smailes, E., & Bernstein, D. P. (1999). Childhood maltreatment increases risk for personality disorders during early adulthood. *Archives of General Psychiatry, 56*(7), 600–606.

Johnson, J. L., & Leff, M. (1999). Children of substance abusers: Overview of research findings. *Pediatrics, 103*, 1085–1099.

Kalat, J. W. (2004). *Biological psychology.* Belmont, CA: Thomson/Wadsworth.

Kalin, N. H., Shelton, S. E., & Lynn, D. E. (1995). Opiate systems in mother and infant primates coordinate intimate contact during reunion. *Psychoneuroendocrinology, 20*, 735–742.

Kalinichev, M., Rosenblatt, J. S., Nakabeppu, Y., & Morrell, J. I. (2000). Induction of c-Fos-like and FosB-like immunoreactivity reveals forebrain neuronal populations involved differentially in pup-mediated maternal behavior in juvenile and adult rats. *Journal of Comparative Neurology, 416*, 45–78.

Kalivas, P. W. (1993). Neurotransmitter regulation of dopamine neurons in the ventral tegmental area. *Brain Research: Brain Research Reviews, 18*, 75–113.

Kalivas, P. W., Churchill, L., & Romanides, A. (1999). Involvement of the pallidal–thalamocortical circuit in adaptive behavior. In J. F. McGinty (Ed.), *Advancing from the ventral striatum to the extended amygdala: Implications for neuropsychiatry and drug use. In honor of Lennart Heimer* (Vol. 877, pp. 64–70). New York: New York Academy of Sciences.

Kalivas, P. W., & Duffy, P. (1989). Similar effects of daily cocaine and stress on mesocorticolimbic dopamine neurotransmission in the rat. *Biological Psychiatry, 25*(7), 913–928.

Kaplow, J. B., & Widom, C. S. (2007). Age of onset of child maltreatment predicts long-term mental health outcomes. *Journal of Abnormal Psychology, 116*(1), 176–187.

Kaufman, J., Birmaher, B., Perel, J., Dahl, R., Moreci, P., Nelson, B., et al. (1997). The corticotropin-releasing hormone challenge in depressed abused, depressed nonabused, and normal control children. *Biological Psychiatry, 42*(8), 669–679.

Kaufman, J., Birmaher, B., Perel, J., Dahl, R., Stull, S., Brent, D., et al. (1998). Serotonergic functioning in depressed abused children: Clinical and familial correlates. *Biological Psychiatry, 44*(10), 973–981.

Kaufman, J., Yang, B.-Z., Douglas-Palumberi, H., Grasso, D., Lipschitz, D., Houshyar, S., et al. (2006). Brain-derived neurotrophic factor-5-HHTLPR gene interactions and environmental modifiers of depression in children. *Biological Psychiatry, 59*(8), 673–680.

Kelley, S. J. (1992). Parenting stress and child maltreatment in drug-exposed children. *Child Abuse and Neglect, 16*(3), 317–328.

Kilpatrick, D. G., Acierno, R., Resnick, H. S., Saunders, B. E., & Best, C. L. (1997). A 2-year longitudinal analysis of the relationships between violent assault and substance use in women. *Journal of Consulting and Clinical Psychology, 65*, 834–847.

Kim-Cohen, J., Caspi, A., Taylor, A., Williams, B., Newcombe, R., Craig, I., et al. (2006). MAOA, maltreatment, and gene–environment interaction predicting children's mental health: New evidence and a meta-analysis. *Molecular Psychiatry, 11*(10), 903–913.

Kirkpatrick, B., Carter, C., Newman, S., & Insel, T. (1994). Axon sparing lesions of the medial nucleus of the amygdala decrease affiliative behaviors in the prairie vole (*Microtus ochrogaster*): Behavioral and anatomic specificity. *Behavioral Neuroscience, 108*, 501–513.

Kirkpatrick, B., & Insel, T. R. (1993). Fos immunoreactivity increases in the medial nucleus of the amygdala after pup exposure in prairie vole males. *Society for Neuroscience Abstracts*, No. 661.1.

Kling, A. (1972). Effects of amygdalectomy on social-affective behavior in non-human primates. In B. E. Eleftheriou (Ed.), *The neurobiology of the amygdala* (pp. 511–536). New York: Plenum Press.

Kobak, R. (1999). The emotional dynamics of disruptions in attachment relationships: Implications for theory, research, and clinical intervention. In J. Cassidy & P. R. Shaver (Eds.), *Handbook of attachment: Theory, research, and clinical applications* (pp. 21–43). New York: Guilford Press.

Kolomiets, B., Deniau, J. M., Mailly, P., Menetrey, A., & Thierry, A. M. (2001). Segregation and convergence of information flow through the cortico-subthalamic pathways. *Journal of Neuroscience, 21*, 5764–5772.

Kolunie, J. M., & Stern, J. M. (1995). Maternal aggression in rats: Effects of olfactory bulbectomy, ZnSO4-induced anosmia, and vomeronasal organ removal. *Hormones and Behavior, 29*, 492–518.

Koob, G. F., Ahmed, S. H., Boutrel, B., Chen, S. A., Kenny, P. J., Markou, A., et al. (2004). Neurobiological mechanisms in the transition from drug use to drug dependence. *Neuroscience and Biobehavioral Reviews, 27*(8), 739–749.

Koob, G. F., & Le Moal, M. (1997). Drug abuse: Hedonic homeostatic dysregulation. *Science, 278*(5335), 52–58.

Koob, G. F., & Le Moal, M. (2001). Drug addiction, dysregulation of reward, and allostasis. *Neuropsychopharmacology, 24*(2), 97–129.

Ladd, C. O., Huot, R. L., Thrivikraman, K. V., Nemeroff, C. B., Meaney, M. J., & Plotsky, P. M. (2000). Long-term behavioral and neuroendocrine adaptations to adverse early experience. *Progress in Brain Research, 122*, 81–103.

Lam, W. K., Cance, J. D., Eke, A. N., Fishbein, D. H., Hawkins, S. R., & Williams, J. (2007). Children of African-American mothers who use crack cocaine: Parenting influences on youth substance use. *Journal of Pediatric Psychology, 32*(8), 877–887.

Lansford, J. E., Dodge, K. A., Pettit, G. S., Bates, J. E., Crozier, J., & Kaplow, J. (2002). A 12-year prospective study of the long-term effects of early child physical maltreatment on psychological, behavioral, and academic problems in adolescence. *Archives of Pediatrics and Adolescent Medicine, 156*(8), 824–830.

Larsen, R. J., & Ketelaar, T. (1991). Personality and susceptibility to positive and negative emotional states. *Journal of Personality and Social Psychology, 61*(1), 132–140.

Leckman, J. F., & Herman, A. (2002). Maternal behavior and developmental psychopathology. *Biological Psychiatry, 51*, 27–43.

Leckman, J. F., & Mayes, L. C. (1998). Maladies of love: An evolutionary perspective on some forms of obsessive–compulsive disorder. In D. H. Hann, L. Huffman, I. Lederhendler, & D. Meinecke (Eds.), *Advancing research in developmental plasticity: Integrating the behavioral science and neuroscience of mental health* (pp. 134–152). Rockville, MD: National Institute of Mental Health.

Lehman, M. N., & Winans, S. S. (1982). Vomeronasal and olfactory pathways to the amygdala controlling male hamster sexual behavior: Autoradiographic and behavioral analyses. *Brain Research, 240*(1), 27–41.

Leibenluft, E., Gobbini, M. I., Harrison, T., & Haxby, J. V. (2004). Mothers' neural activation in response to pictures of their children and other children. *Biological Psychiatry, 56,* 225–232.

Lejuez, C., Read, J. P., Kahler, C. W., Richards, J. B., Ramsey, S. E., Stuart, G. L., et al. (2002). Evaluation of a behavioral measure of risk taking: The Balloon Analogue Risk Task (BART). *Journal of Experimental Psychology: Applied, 8*(2), 75–84.

Le Poire, B. A. (2003). The influence of drugs and alcohol on family communication: The effects that substance abuse has on family members and the effects that family members have on substance abuse. In A. L. Vangelisti (Ed.), *Handbook of family communication* (pp. 609–628). New York: Erlbaum.

Lester, B. M., Tronick, E. Z., LaGasse, L., Seifer, R., Bauer, C. R., & Shankaran, S. (2002). The maternal lifestyle study: Effects of substance exposure during pregnancy on neurodevelopmental outcome in 1-month old infants. *Pediatrics, 110,* 1182–1192.

Levine, S. (1975). Psychosocial factors in growth and development. In L. Levi (Ed.), *Society, stress and disease* (pp. 43–50). London: Oxford University Press.

Levy, A. D., Li, Q. A., Kerr, J. E., Rittenhouse, P. A., Milonas, G., Cabrera, T. M., et al. (1991). Cocaine-induced elevation of plasma adrenocorticotropin hormone and corticosterone is mediated by serotonergic neurons. *Journal of Pharmacology and Experimental Therapeutics, 259*(2), 495–500.

Lichter, D. G., & Cummings, J. L. (Eds.). (2001). *Frontal–subcortical circuits in psychiatric and neurological disorders.* New York: Guilford Press.

Liu, D., Diorio, J., Day, J. C., Francis, D. D., & Meaney, M. J. (2000). Maternal care, hippocampal synaptogenesis and cognitive development in rats. *Nature Neuroscience, 3,* 799–806.

Liu, D., Diorio, J., Tannenbaum, B., Caldji, C., Francis, D., Freedman, A., et al. (1997). Maternal care, hippocampal glucocorticoid receptors, and hypothalamic–pituitary–adrenal responses to stress. *Science, 277,* 1659–1662.

Ljungberg, T., Apicella, P., & Schultz, W. (1992). Responses of monkey dopamine neurons during learning of behavioral reactions. *Journal of Neurophysiology, 67,* 145–163.

Lo, C. C., & Cheng, T. C. (2007). The impact of childhood maltreatment on young adults' substance abuse. *American Journal of Drug and Alcohol Abuse, 33*(1), 139–146.

Lopez, J. F., Akil, H., & Watson, S. J. (1999). Neural circuits mediating stress. *Biological Psychiatry, 46,* 1461–1471.

Lorberbaum, J. P., Newman, J. D., Dubno, J. R., Horwitz, A. R., Nahas, Z., Teneback, C. C., et al. (1999). Feasibility of using fMRI to study mothers responding to infant cries. *Depression and Anxiety, 10*(3), 99–104.

Lorberbaum, J. P., Newman, J. D., Horwitz, A. R., Dubno, J. R., Lydiard, R. B., Hamner, M. B., et al. (2002). A potential role for thalamocingulate circuitry in human maternal behavior. *Biological Psychiatry, 51,* 431–445.

Luntz, B. K., & Widom, C. S. (1994). Antisocial personality disorder in abused and neglected children grown up. *American Journal of Psychiatry, 151*(5), 670–674.

MacLean, P. D. (1990). *The triune brain in evolution.* New York: Plenum Press.

Macleod, J., Oakes, R., Copello, A., Crome, I., Egger, M., Hickman, M., et al. (2004a). Psychological and social sequelae of cannabis and other illicit drug use by young people: A systematic review of longitudinal, general population studies. *Lancet, 363*(9421), 1579–1588.

Macleod, J., Oakes, R., Oppenkowski, T., Stokes-Lampard, H., Copello, A., Crome, I., et al. (2004b). How strong is the evidence that illicit drug use by young people is an important cause of psychological or social harm?: Methodological and policy implications of a systematic review of longitudinal, general population studies. *Drugs: Education, Prevention and Policy, 11*(4), 281–297.

MacPherson, L., Bornovalova, M. A., Weitzman, M., Wang, F., & Lejuez, C. W. (2008). *Sensation seeking and impulsivity mediate the relationship between reported emotional abuse and risk behavior engagement in early adolescents.* Unpublished manuscript.

Main, M. (2000). The organized categories of infant, child, and adult attachment: Flexible vs. inflexible attention under attachment-related stress. *Journal of the American Psychoanalytic Association, 48*, 1055–1096.

Main, M., Kaplan, N., & Cassidy, J. (1985). Security in infancy, childhood, and adulthood: A move to the level of representation. In I. Bretherton & E. Waters (Eds.), Growing points of attachment theory and research. *Monographs of the Society for Research in Child Development, 50*(1–2, Serial No. 209), 66–104.

Mann, P. E., Kinsley, C. H., & Bridges, R. S. (1991). Opioid receptor subtype involvement in maternal behavior in lactating rats. *Neuroendocrinology, 53*, 487–492.

Masterman, D. L., & Cummings, J. L. (1997). Frontal–subcortical circuits: The anatomical basis of executive, social and motivational behaviors. *Journal of Psychopharmacology, 11*, 107–114.

Mauricio, A., & Gormley, B. (2001). Male perpetration of physical violence against female partners. *Journal of Interpersonal Violence, 16*(10), 1066–1081.

Maxfield, M. G., & Widom, C. S. (1996). The cycle of violence: Revisited six years later. *Archives of Pediatrics and Adolescent Medicine, 150*, 300–395.

Mayes, L. C. (2002). A behavioral teratogenic model of the impact of prenatal cocaine exposure on arousal regulatory systems. *Neurotoxicology and Teratology, 24*(3), 385–395.

Mayes, L. C. (2006). Arousal regulation, emotional flexibility, medial amygdala function, and the impact of early experience. *Annals of the New York Academy of Sciences, 1094*, 178–192.

Mayes, L. C., & Bornstein, M. (1996). The context of development for young children from cocaine-abusing families. In P. Kato & T. Mann (Eds.), *Handbook of diversity issues in health psychology* (pp. 69–95). New York: Plenum Press.

Mayes, L. C., Swain, J., & Leckman, J. F. (2005). Parental attachment systems, neural circuits, genes, and experiential contributions to parental engagement. *Clinical Neuroscience Research, 4*, 301–313.

Mayes, L. C., & Truman, S. (2002). Substance abuse and parenting. In M. Bornstein (Ed.), *Handbook of parenting: Vol. 4. Social conditions and applied parenting* (pp. 329–359). Mahwah, NJ: Erlbaum.

Mello, N. K., & Mendelson, J. H. (1997). Cocaine's effects on neuroendocrine systems: Clinical and preclinical studies. *Pharmacology, Biochemistry and Behavior, 57*(3), 571–599.

Moffitt, T. E., Caspi, A., & Rutter, M. (2005). Strategy for investigating interactions between measured genes and measured environments. *Archives of General Psychiatry, 62*(5), 473–481.

Moldow, R., & Fischman, A. (1987). Cocaine induced secretion of ACTH, beta-endorphin, and corticosterone. *Peptides, 8*(5), 819–822.

Moles, A., Kieffer, B. L., & D'Amato, F. R. (2004). Deficit in attachment behavior in mice lacking the mu-opioid receptor gene. *Science, 304*, 1983–1986.

Moran, P. B., Vuchinich, S., & Hall, N. K. (2004). Associations between types of maltreatment and substance use during adolescence. *Child Abuse and Neglect, 28*(5), 565–574.

Morgan, H. D., Fleming, A. S., & Stern, J. M. (1992). Somatosensory control of the onset and retention of maternal responsiveness in primiparous Sprague–Dawley rats. *Physiology and Behavior, 51*, 541–555.

Morgan, H. D., Wachtus, J. A., Milgram, N. W., & Fleming, A. S. (1999). The long lasting effects of electrical stimulation of the medial preoptic area and medial amygdala on maternal behavior in female rats. *Behavioral Brain Research, 99*, 61–73.

Mudar, P., Leonard, K. E., & Soltysinski, K. (2001). Discrepant substance use and marital functioning in newlywed couples. *Journal of Consulting and Clinical Psychology, 69*(1), 130–134.

Mullings, J. L., Hartley, D. J., & Marquart, J. W. (2004). Exploring the relationship between alcohol use, childhood maltreatment, and treatment needs among female prisoners. *Substance Use and Misuse, 39*, 277–305.

Murphy, S., & Rosenbaum, M. (1999). *Pregnant women on drugs: Combating stereotypes and stigma.* New Brunswick, NJ: Rutgers University Press.

Nair, P., Schuler, M. E., Black, M. M., Kettinger, L., & Harrington, D. (2003). Cumulative environmental risk in substance abusing women: Early intervention, parenting stress, child abuse potential and child development. *Child Abuse and Neglect, 27*(9), 997–1017.

Nelson, E. E., & Panksepp, J. (1998). Brain substrates of infant–mother attachment: Contributions of opioids, oxytocin, and norepinephrine. *Neuroscience and Biobehavioral Reviews, 22*(3), 437–452.

Nestler, E. J., Barrot, M., & Self, D. W. (2001). AFosB: A sustained molecular switch for addiction. *Proceedings of the National Academy of Sciences USA, 98,* 11042–11046.

Nitschke, J. B., Nelson, E. E., Rusch, B. D., Fox, A. S., Oakes, T. R., & Davidson, R. J. (2004). Orbitofrontal cortex tracks positive mood in mothers viewing pictures of their newborn infants. *NeuroImage, 21,* 583–592.

Numan, M. (1994). Maternal behavior. In E. Knobil & J. F. Neill (Eds.), *The physiology of reproduction* (2nd ed., pp. 221–301). New York: Raven Press.

Numan, M. (2004). Maternal behaviors: Central integration or independent parallel circuits? Theoretical comment on Popeski and Woodside. *Behavioral Neuroscience, 118*(6), 1469–1472.

Numan, M., & Insel, T. R. (2003). *The neurobiology of parental behavior.* New York: Springer.

Numan, M., & Numan, M. J. (1994). Expression of Fos-like immunoreactivity in the preoptic area of maternally behaving virgin and postpartum rats. *Behavioral Neuroscience, 108,* 379–394.

Numan, M., & Numan, M. J. (1997). Projection sites of medial preoptic area and ventral bed nucleus of the stria terminalis neurons that express Fos during maternal behavior in female rats. *Journal of Neuroendocrinology, 9*(5), 369–384.

Numan, M., Numan, M. J., & English, J. B. (1993). Excitotoxic amino acid injections into the medial amygdala facilitate maternal behavior in virgin female rats. *Hormones and Behavior, 27,* 56–81.

Numan, M., Rosenblatt, J. S., & Kiminsaruk, B. R. (1997). Medial preoptic area and onset of maternal behavior in the rat. *Journal of Comparative Physiology and Psychology, 91,* 146–164.

Numan, M., & Sheehan, T. P. (1997). Neuroanatomical circuitry for mammalian maternal behavior. *Annals of the New York Academy of Sciences, 807,* 101–125.

O'Donnell, P., Greene, J., Pabello, N., Lewis, B. L., & Grace, A. A. (1999). Modulation of cell firing in the nucleus accumbens. *Annals of the New York Academy of Sciences, 877,* 157–175.

O'Hearn, R. E., & Davis, K. E. (1997). Women's experience of giving and receiving emotional abuse. *Journal of Interpersonal Violence, 12*(3), 375–391.

O'Malley, M. N., & Gillette, C. S. (1984). Exploring the relations between traits and emotions. *Journal of Personality, 52*(3), 274–284.

Pajulo, M., Savonlahti, E., Sourander, A., Ahlqvist, S., Helenius, H., & Piha, J. (2001). An early report on the mother–baby interactive capacity of substance-abusing mothers. *Journal of Substance Abuse Treatment, 20*(2), 143–151.

Panksepp, J. (1981). Brain opioids: A neurochemical substrate for narcotic and social dependence. In J. S. Cooper (Ed.), *Theory in psychopharmacology* (pp. 149–175). London: Academic Press.

Panksepp, J. (1996). Affective neuroscience: A paradigm to study the animate circuits for human emotions. In R. D. Kavanaugh, B. Zimmerberg, & S. Fein (Eds.), *Emotion: Interdisciplinary perspectives* (pp. 29–60). Hillsdale, NJ: Erlbaum.

Panksepp, J. (1998). *Affective neuroscience.* New York: Oxford University Press.

Panksepp, J., Nelson, E., & Siviy, S. (1994). Brain opioids and mother–infant social motivation. *Acta Paediatrica, 83*(Suppl. 397), 40–46.

Panksepp, J., Siviy, S., & Normansell, L. (1984). The psychobiology of play: Theoretical and methodological perspectives. *Neuroscience and Biobehavioral Reviews, 8*(4), 465–492.

Pecoraro, N., Gomez, F., & Dallman, M. F. (2005). Glucocorticoids dose-dependently remodel energy stores and amplify incentive relativity effects. *Psychoneuroendocrinology, 30*(9), 815–825.

Pecoraro, N., Reyes, F., Gomez, F., Bhargava, A., & Dallman, M. F. (2004). Chronic stress promotes palatable feeding, which reduces signs of stress: Feedforward and feedback effects of chronic stress. *Endocrinology, 145*(8), 3754–3762.

Pedersen, C. A. (1997). Oxytocin control of maternal behavior: Regulation by sex steroids and off-spring stimuli. *Annals of the New York Academy of Sciences, 807*, 126–145.

Pedersen, C. A., Caldwell, J. D., Walker, C., Ayers, G., & Mason, G. A. (1994). Oxytocin activates the postpartum onset of rat maternal behavior in the ventral tegmental and medial preoptic areas. *Behavioral Neuroscience, 108*, 1163–1171.

Pederson, C. A., & Prange, A. J. (1979). Induction of maternal behavior in virgin rats after intracerebroventricular administration of oxytocin. *Proceedings of the National Academy of Sciences USA, 76*, 6661–6665.

Peleg-Oren, N., & Teichman, M. (2006). Young children of parents with substance use disorders (SUD): A review of the literature and implications for social work practice. *Journal of Social Work Practice in the Addictions, 6*(1–2), 49–61.

Pennartz, C., Groenewegen, H. J., & Lopez da Silva, F. H. (1994). The nucleus accumbens as a complex of functionally distinct neuronal ensembles: An integration of behavioral, electrophysiological and anatomical data. *Progress in Neurobiology, 42*, 719–761.

Piazza, P. V., Barrot, M., Rougé-Pont, F., Marinelli, M., Maccari, S., Abrous, D. N., et al. (1996a). Suppression of glucocorticoid secretion and antipsychotic drugs have similar effects on the mesolimbic dopaminergic transmission. *Proceedings of the National Academy of Sciences USA, 93*(26), 15445–15450.

Piazza, P. V., & Le Moal, M. (1998). The role of stress in drug self-administration. *Trends in Pharmacological Sciences, 19*(2), 67–71.

Piazza, P. V., Rougé-Pont, F., Deroche, V., Maccari, S., Simon, H., & Le Moal, M. (1996b). Glucocorticoids have state-dependent stimulant effects on the mesencephalic dopaminergic transmission. *Proceedings of the National Academy of Sciences USA, 93*(16), 8716–8720.

Plotsky, P. M., & Meaney, M. J. (1993). Early, postnatal experience alters hypothalamic corticotropin-releasing factor (CRF) mRNA, median eminence CRF content and stress-induced release in adult rats. *Brain Research: Molecular Brain Research, 18*, 195–200.

Porter, L. S., & Porter, B. O. (2004). A blended infant massage-parenting enhancement program for recovering substance-abusing mothers. *Pediatric Nursing, 30*, 363–401.

Potenza, M. N. (2001). The neurobiology of pathological gambling. *Seminars in Clinical Neuropsychiatry, 6*, 217–226.

Ranote, S., Elliott, R., Abel, K. M., Mitchell, R., Deakin, J. F., & Appleby, L. (2004). The neural basis of maternal responsiveness to infants: An fMRI study. *NeuroReport, 15*(11), 1825–1829.

Ripley, J. S., Cunion, A., & Noble, N. (2006). Alcohol abuse in marriage and family contexts: Relational pathways to recovery. *Alcoholism Treatment Quarterly, 24*(1–2), 171–184.

Robinson, T. E., & Berridge, K. C. (1993). The neural basis of drug craving: An incentive-sensitization theory of addiction. *Brain Research: Brain Research Reviews, 18*(3), 247–291.

Rolls, E. T., & Treves, A. (1998). *Neural networks and brain function.* New York: Oxford University Press.

Rosenblatt, J. S. (1967). Nonhormonal basis of maternal behavior in the rat. *Science, 156*, 1512–1514.

Rubin, B. S., & Bridges, R. S. (1984). Disruption of ongoing maternal responsiveness by central administration of morphine sulfate. *Brain Research, 307*, 91–97.

Sano, M., Marder, K., & Dooneief, G. (1996). Basal ganglia diseases. In B. S. Fogel, R. B. Schiffer, & S. M. Rao (Eds.), *Neuropsychiatry* (pp. 805–834). Baltimore: Williams & Wilkins.

Schultz, W., Dayan P., & Montague, R. R. (1997). A neural substrate of prediction and reward. *Science, 275*, 1593–1599.

Seifritz, E., Esposito, F., Neuhoff, J. G., Lüthi, A., Mustovic, H., Dammann, G., et al. (2003). Differential sex-independent amygdala response to infant crying and laughing in parents versus nonparents. *Biological Psychiatry, 54*(12), 1367–1375.

Self, D. W. (1998). Neural substrates of drug craving and relapse in drug addiction. *Annals of Medicine, 30*(4), 379–389.

Self, D. W., & Nestler, E. J. (1998). Relapse to drug-seeking: Neural and molecular mechanisms. *Drug and Alcohol Dependence, 51*, 49–60.

Sheehan, T. P. (2000). *An investigation into the neural and hormonal inhibition of maternal behavior in rats.* Unpublished doctoral dissertation, Boston College.

Sheehan, T. P., Cirrito, J., Numan, M. J., & Numan, M. (2000). Using c-Fos immunocyto-chemistry to identify forebrain regions that may inhibit maternal behavior in rats. *Behavioral Neuroscience, 114*, 337–352.

Sheinkopf, S. J., Lester, B. M., LaGasse, L. L., Seifer, R., Bauer, C. R., Shankaran, S., et al. (2006). Interactions between maternal characteristics and neonatal behavior in the prediction of parenting stress and perception of infant temperament. *Journal of Pediatric Psychology, 31*(1), 27–40.

Shin, L. M., Whalen, P. J., Pitman, R. K., Bush, G., Macklin, M. L., Lasko, N. B., et al. (2001). An fMRI study of anterior cingulate function in posttraumatic stress disorder. *Biological Psychiatry, 50*(12), 932–942.

Simerly, R. B., Gorski, R. A., & Swanson, L. (1986). Neurotransmitter specificity of cells and fibers in the medial preoptic nucleus: An immunohistochemical study in the rat. *Journal of Comparative Neurology, 246*, 343–363.

Sinha, R. (2001). How does stress increase risk of drug abuse and relapse? *Psychopharmacology, 158*(4), 343–359.

Sinha, R., Lacadie, C., Skudlarski, P., Fulbright, R. K., Rounsaville, B. J., Kosten, T. R., et al. (2005). Neural activity associated with stress-induced cocaine craving: A functional magnetic resonance imaging study. *Psychopharmacology, 183*(2), 171–180.

Slotnick, B. M., & Nigrosh, B. J. (1975). Maternal behavior of mice with cingulate cortical, amygdala, or septal lesions. *Journal of Comparative and Physiological Psychology, 88*(1), 118–127.

Sofroniew, M. V., & Weindl, A. (1981). Central nervous system distribution of vasoressin, oxytocin, and neurophysin. In J. L. Martinez, R. A. Jensen, R. B. Messing, H. Rigter, & J. L. McGaugh (Eds.), *Endogenous peptides and learning and memory processes* (pp. 327–369). New York: Academic Press.

Sofuoglu, M., Nelson, D., Babb, D. A., & Hatsukami, D. K. (2001). Intravenous cocaine increases plasma epinephrine and norepinephrine in humans. *Pharmacology Biochemistry and Behavior, 68*(3), 455–459.

Solomon, R. (1977). An opponent-process theory of acquired motivation: The affective dynamics of addiction. In J. D. Maser & M. E. P. Seligman (Eds.), *Psychopathology: Experimental models* (pp. 67–103). San Francisco: Freeman.

Spear, L. (2000). The adolescent brain and age-related behavioral manifestations. *Neuroscience and Biobehavioral Reviews, 24*(4), 417–463.

Squire, S., & Stein, A. (2003). Functional MRI and parental responsiveness: A new avenue into parental psychopathology and early parent–child interactions? *British Journal of Psychiatry, 183*, 481–483.

Sroufe, L. A. (1983). Infant–caregiver attachment and patterns of adaption in preschool: The roots of maladaptation and competence. In M. Perlmutter (Ed.), *Minnesota Symposium on Child Psychology* (Vol. 13, pp. 41–83). Hillsdale, NJ: Erlbaum.

Sroufe, L. A. (1997). Psychopathology as an outcome of development. *Development and Psychopathology, 9*(2), 251–268.

Sroufe, L. A. (2000). Early relationships and the development of children. *Infant Mental Health Journal, 21*(1–2), 67–74.

Sroufe, L. A., Carlson, E., & Shulman, S. (1993). Individuals in relationships: Development from infancy through adolescence. In D. C. Funder, R. Parke, C. Tomlinson-Keesey, & K. Widaman (Eds.), *Studying lives through time: Approaches to personality and development* (pp. 315–342). Washington, DC: American Psychological Association.

Sroufe, L. A., & Waters, E. (1977). Attachment as an organizational construct. *Child Development, 48*, 1184–1199.

Stein, E. A., Pankiewicz, J., Harsch, H. H., Cho, J.-K., Fuller, S. A., Hoffman, R. G., et al. (1998).

Nicotine-induced limbic cortical activation in the human brain: A functional MRI study. *American Journal of Psychiatry, 155*(8), 1009–1015.

Sternberg, K. J., Lamb, M. E., Greenbaum, C., Cicchetti, D., Dawud, S., Cortes, R. M., et al. (1993). Effects of domestic violence on children's behavior problems and depression. *Developmental Psychology, 29*(1), 44–52.

Strafella, A. P., Paus, T., Barrett, J., & Dagher, A. (2001). Repetitive transcranial magnetic stimulation of the human prefrontal cortex induces dopamine release in the caudate nucleus. *Journal of Neuroscience, 21*, RC157, 1–4.

Strathearn, L., & McClure, S. M. (2002). A functional MRI study of maternal responses of infant facial cues. *Society for Neuroscience Abstracts*, Abstract Viewer No. 517.5.

Suchman, N. E., & Luthar, S. S. (2000). Maternal addiction, child maladjustment, and sociodemographic risks: Implications for parenting behaviors. *Addictions, 95*(9), 1417–1428.

Suchman, N. E., & Luthar, S. S. (2001). The mediating role of parenting stress in methadone-maintained mothers' parenting. *Parenting: Science and Practice, 1*(4), 285–315.

Swain, J. E., Leckman, J. F., Mayes, L. C., Feldman, R., Constable, R. T., & Schultz, R. T. (2004, April–May). *Neural substrates of human parent–infant attachment in the postpartum*. Paper presented at the 59th annual meeting of the Society of Biological Psychiatry, New York.

Swain, J. E., Leckman, J. F., Mayes, L. C., Feldman, R., Eicher, V., & Schultz, R. T. (2003, December). *Neural circuitry of human parent–infant attachment in the early postpartum*. Paper presented at the 42nd annual meeting of the American College of Neuropsychopharmacology, San Juan, Puerto Rico.

Swain, J. E., Lorberbaum, J. P., Kose, S., & Strathearn, L. (2007). Brain basis of early parent–infant interactions: Psychology, physiology, and *in vivo* functional neuroimaging studies. *Journal of Child Psychology and Psychiatry, 48*(3–4), 262–287.

Swanson, L. W. (2000). Cerebral hemisphere regulation of motivated behavior. *Brain Research, 886*, 113–164.

Teicher, M. H., Ito, Y., Glod, C. A., Andersen, S. L., Dumont, N., & Ackerman, E. (1997). Preliminary evidence for abnormal cortical development in physically and sexually abused children using EEG coherence and MRI. *Annals of the New York Academy of Sciences, 821*, 160–175.

Testa, M., Livingston, J. A., & Leonard, K. E. (2003). Women's substance use and experiences of intimate partner violence: A longitudinal investigation among a community sample. *Addictive Behaviors, 28*(9), 1649–1664.

Thierry, A., Tassin, J., Blanc, G., & Glowinski, J. (1976). Selective activation of the mesocortical DA system by stress. *Nature, 263*(5574), 242–244.

Thompson, A. C., & Kristal, M. B. (1996). Opioid stimulation in the ventral tegmental area facilitates the onset of maternal behavior in rats. *Brain Research, 743*, 184–201.

Tice, D., Bratslavsky, E., & Baumeister, R. (2001). Emotional distress regulation takes precedence over impulse control: If you feel bad, do it! *Journal of Personality and Social Psychology, 80*(1), 53–67.

Toth, S. L., & Cicchetti, D. (1996). Patterns of relatedness, depressive symptomatology, and perceived competence in maltreated children. *Journal of Consulting and Clinical Psychology, 64*(1), 32–41.

U.S. Department of Health and Human Services. (1999, April). *Blending perspectives and building common ground: A report to Congress on substance abuse and child protection*. Washington, DC: U.S. Government Printing Office.

van IJzendoorn, M. H. (1995). Adult attachment representations, parental responsiveness, and infant attachment: A meta-analysis on the predictive validity of the Adult Attachment Interview. *Psychological Bulletin, 117*(3), 387–403.

Van Leengoed, E., Kerker, E., & Swanson, H. H. (1987). Inhibition of postpartum maternal behavior in the rat by injecting an oxytocin antagonist into the cerebral ventricles. *Journal of Endocrinology, 112*, 275–282.

Volkow, N. D., & Fowler, J. S. (2000). Addiction, a disease of compulsion and drive: Involvement of the orbitofrontal cortex. *Cerebral Cortex, 10*(3), 318–325.

Volkow, N. D., & Li, T.-K. (2005). The neuroscience of addiction. *Nature Neuroscience, 8*(11), 1429–1430.

Volkow, N. D., & Wise, R. A. (2005). How can drug addiction help us understand obesity? *Nature Neuroscience, 8*(5), 555–560.

Waelti, P., Dickinson, A., & Schultz, W. (2001). Dopamine responses comply with basic assumptions of formal learning theory. *Nature, 412*, 43–48.

Ward, T., Hudson, S., Marshall, W. L., & Siegert, R. (1995). Attachment style and intimacy deficits in sexual offenders: A theoretical framework. *Sexual Abuse: Journal of Research and Treatment, 7*(4), 317–335.

Warren, S. L., Huston, L., Egeland, B., & Sroufe, L. A. (1997). Child and adolescent anxiety disorders and early attachment. *Journal of the American Academy of Child and Adolescent Psychiatry, 36*(5), 637–644.

Waters, E., Merrick, S., Treboux, D., Crowell, J., & Albersheim, L. (2000a). Attachment security in infancy and early adulthood: A twenty-year longitudinal study. *Child Development, 71*, 684–689.

Waters, E., Weinfield, N. S., & Hamilton, C. E. (2000b). The stability of attachment security from infancy to adolescence and early adulthood: General discussion. *Child Development, 71*(3), 703–706.

Watts-English, T., Fortson, B. L., Gibler, N., Hooper, S. R., & De Bellis, M. D. (2006). The psychobiology of maltreatment in childhood. *Journal of Social Issues, 62*(4), 717–736.

Weaver, I. C., Cervoni, N., Champagne, F. A., D'Alessio, A. C., Sharma, S., Seckl, J. R., et al. (2004). Epigenetic programming by maternal behavior. *Nature Neuroscience, 7*, 847–854.

Weiss, R. S. (1982). Attachment in adult life. In C. M. Parkes & J. Stevenson-Hinde (Eds.), *The place of attachment in human behavior* (pp. 171–184). New York: Basic Books.

White, H. R., Loeber, R., Stouthamer-Loeber, M., & Farrington, D. P. (1999). Developmental associations between substance use and violence. *Development and Psychopathology, 11*, 785–803.

Widom, C. S. (1999). Posttraumatic stress disorder in abused and neglected children grown up. *American Journal of Psychiatry, 156*(8), 1223–1229.

Widom, C. S., DuMont, K., & Czaja, S. J. (2007a). A prospective investigation of major depressive disorder and comorbidity in abused and neglected children grown up. *Archives of General Psychiatry, 64*(1), 49–56.

Widom, C. S., White, H. R., Czaja, S. J., & Marmorstein, N. R. (2007b). Long-term effects of child abuse and neglect on alcohol use and excessive drinking in middle adulthood. *Journal of Studies on Alcohol and Drugs, 68*(3), 317–326.

Wikler, A. (1965). Conditioning factors in opiate addictions and relapse. In D. M. Wilner & G. G. Kassebaum (Eds.), *Narcotics* (pp. 85–100). New York: McGraw-Hill.

Wilson, D. A., & Sullivan, R. M. (1994). Neurobiology of associative learning in the neonate: Early olfactory learning. *Behavioral and Neural Biology, 61*(1), 1–18.

Zuckerman, M., & Kuhlman, D. (2000). Personality and risk-taking: Common biosocial factors. *Journal of Personality, 68*(6), 999–1029.

CHAPTER 19

A Brain-Based Account of Developmental Changes in Social Decision Making

EVELINE A. CRONE
P. MICHIEL WESTENBERG

Among the most salient changes in childhood and adolescence are increases in the ability to control and regulate thoughts and actions, and in the use of this skill for social decision making. Despite these important changes, the neurobiological correlates of these changes are still poorly understood. What changes take place that result in increased consideration of others in childhood and adolescence? What causes children and adolescents to change their perceptions of fairness and trust? Which brain regions facilitate the development of social decision-making skills? Understanding how developmental changes in social decision making are initiated by neural changes will enhance our understanding of social development in terms of brain–behavior relations.

In this chapter, we review recent work on the functional development of brain regions that support key aspects of social decision making: cognitive control, future orientation, and perspective taking. Using this review, we hypothesize about how recent developments in the field of social-cognitive neuroscience can explain developmental changes in social decision making in childhood and adolescence. Importantly, we devote special attention to the need to characterize developmental changes in terms not only of chronological age changes, but also of psychosocial maturation levels. The chapter focuses on the changes that occur in late childhood and adolescence, which is a relatively unexplored developmental period, despite evidence that changes in brain organization continue until the early 20s. Our working hypothesis throughout the chapter is that developmental changes in social decision making during this age period are the result of an imbalance between emotion-inducing and emotion-regulating brain systems. We devote special attention to fractionation of the regulation system, with a special focus on the development of different regions within the prefrontal cortex (PFC).

Social Decision-Making Processes

Neuroscientific theories have suggested that social decision making by adults is the result of two interacting systems: an innate, emotion-inducing system, which activates primary emotions; and an acquired, emotion-regulating system, which is an evolutionarily younger system that distinguishes humans from animals (Adolphs, 2003; Gallagher & Frith, 2003). The development of social information processing has previously been interpreted in the context of a similar model, the social information-processing network (SIPN) model (Nelson, Leibenluft, McClure, & Pine, 2005). The SIPN model suggests that social information processing results from three interacting neural "nodes": a detection, affective, and regulation nodes. The detection node, which comprises the intraparietal sulcus, superior temporal sulcus, fusiform face area, and temporal and occipital regions, detects social properties of a stimulus and is functionally mature early in development. The affective node comprises the limbic area of the brain, including the amygdala, ventral striatum, hypothalamus, and orbital frontal cortex (OFC); it processes the emotional significance of a social stimulus, and influences the behavioral and emotional responses to social stimuli. This node is particularly active in early adolescence (see Ernst & Spear, Chapter 17, this volume, for further discussion). Finally, the cognitive regulation node, which consists of the PFC, is important for goal-directed behavior, impulse control, and theory of mind. This third node is thought to have the slowest developmental trajectory. Although these models provide an interesting starting point, the cognitive regulation system probably consists of several separate processes that influence social decision making in different ways. The different components of the regulation system constitute the focus of this chapter.

The cognitive regulation aspect of social decision making is a complex process that can be fractionated into several cognitive functions, including the abilities to (1) keep relevant information in an active state and exert goal-directed behavior (cognitive control), (2) anticipate future consequences on the basis of reward and punishment (future orientation), and (3) consider the thoughts and perspectives of other people (perspective taking). In the social decision-making literature, perspective taking is thought to be important when there is competition between the utility people derive from their own outcomes and from the outcomes of others (De Dreu, Lualhati, & McCusker, 1994; Handgraaf, Van Dijk, & De Cremer, 2003).

Developmental changes in cognitive control, future orientation, and perspective taking have been well documented in isolation, but the interactions of these systems in social decision making, as well as the neural substrates underlying these changes, remain poorly understood. Longitudinal research examining changes in brain structure over development within individuals has demonstrated that different brain structures develop at different rates, and these brain regions may contribute in different ways to separable social decision-making processes. For example, structural brain development studies have shown that cortical white matter increases approximately linearly with age throughout childhood and adolescence, and differs little across regions (Giedd, 2004; Gogtay et al., 2004; Sowell et al., 2004). In contrast, cortical gray matter, which reflects neuronal density and the number of connections between neurons, follows an inverted-U shape over development, peaking at different ages, depend-

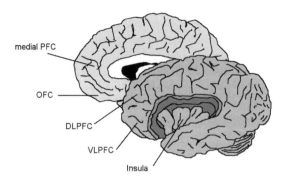

FIGURE 19.1. Brain regions implicated in social decision making. PFC, prefrontal cortex; OFC, orbital frontal cortex; DLPFC, dorsolateral PFC; VLPFC, ventrolateral PFC. See text for further details.

ing on the region. Therefore, gray matter loss is considered an index of the time course of maturation of a region (Sowell et al., 2004). It has been suggested that the nonlinearity of gray matter development reflects the synaptic reorganization that takes place during that period (Paus, 2005).

In comparison to what is known about structural brain development, much less is known about the *functional* development of different brain regions. Thus, despite the evidence for differential development of brain regions, it is currently unknown how the maturation of brain regions supports the development of social decision making. The functional development of a brain region is critically dependent on the network in which that region is engaged; that is, a specific brain region can function at an adult level in one task, but show an immature developmental pattern in another. Likewise, functional immaturity can be associated with reduced brain activation in a certain network, but with increased brain activation in another. Below, we review behavioral and functional magnetic resonance imaging (fMRI) studies that have focused specifically on the development of cognitive control, future orientation and perspective taking. Figure 19.1 presents the brain regions that are the focus of this review. We hypothesize that the subtle changes in social decision making over the course of childhood and adolescent development are associated with different developmental trajectories of separable regulatory systems involved in social cognition.

Cognitive Control and the Dorsolateral PFC Network

"Cognitive control" refers to the ability to control thoughts and actions in order to make them consistent with internal goals. A number of recent fMRI studies have demonstrated that the ability to exert cognitive control relies heavily on the PFC (Bunge, 2004; Miller & Cohen, 2001). Within the PFC, different subregions are thought to contribute to cognitive control processes in different ways, although working closely together.

Working Memory and Lateral PFC

One of the most studied components of cognitive control is "working memory," which refers to the ability to keep information in an active state while ignoring irrelevant information (Baddeley & Logie, 1999). The ability to keep information in working memory matures slowly in childhood and is often conceptualized as the driving force behind cognitive development (Case, 1992). Separate component processes of working memory have been associated with activation in different subregions within the lateral PFC (Figure 19.1). For example, the ability to keep information in an active state results in increased activation in ventrolateral (VL) PFC, such as when a person needs to rehearse a string of letters (Bunge, Kahn, Wallis, Miller, & Wagner, 2003). In contrast, when there is a need to manipulate information in working memory, or to select among several response alternatives, increased neural activation is seen in dorsolateral (DL) PFC (Smith & Jonides, 1999).

Development of the Working Memory Network

Neuroimaging studies of healthy children and adults have explored whether the behavioral changes in working memory maintenance and manipulation are associated with changes in lateral PFC subregions across childhood. Several studies that have examined the ability to maintain information in visuospatial working memory indicate that the most active regions in adults, including the lateral PFC and parietal cortex, are increasingly engaged over childhood (Klingberg, Forssberg, & Westerberg, 2002; Kwon, Reiss, & Menon, 2002).

In the developmental literature, it is well documented that the ability to maintain information in an active state matures late in childhood, but the developmental changes are much more dramatic when there is a need to manipulate or work with information in working memory (Diamond, 2002). In a recent study, Crone, Wendelken, Donohue, van Leijenhorst, and Bunge (2006b) asked children ages 8–12 years, adolescents ages 13–17 years, and young adults to perform a working memory task in which three objects had to be maintained in forward order, or had to be rearranged in backward order, during a delay period. As in prior studies, the need to maintain information in working memory in forward order was associated with increased VLPFC activation, and this activation pattern did not differ between the age groups. In contrast, when information had to be reordered in working memory in backward order, adults and adolescents also activated DLPFC, but this increased activation was not present in 8- to 12-year-old children. The number of correct responses in the backward manipulation task correlated with the activation in DLPFC, strengthening the hypothesis that DLPFC is important for improvements in working memory manipulation performance. Together, these data indicate that increased DLPFC activation over development is important for improvements in working memory maintenance and manipulation. Changes in lateral PFC activation have also been observed for other domains of cognitive control, such as response inhibition (Bunge, Dudukovic, Thomason, Vaidya, & Gabrieli, 2002; Durston et al., 2006) and task switching (Crone, Donohue, Honomichl, Wendelken, & Bunge, 2006a).

Together, these studies show that the development of separable cognitive control functions may be associated with different maturational trajectories of subregions within the lateral PFC. The differential engagement of specific PFC regions is consistent with the structural changes that are observed in development. Within the PFC, gray matter reduction is completed earlier for VLPFC than for DLPFC: Cortical gray matter loss continues until the early 20s for DLPFC, whereas VLPFC is mature in early adolescence (Giedd, 2004). The dynamics of gray matter increases and decreases have previously been associated with differences in intellectual ability (Shaw et al., 2006). As will become evident below, the DLPFC is one of the key brain regions in regulating social decisions; therefore, the changes observed in the recruitment of DLPFC may contribute to the differential pattern of social decisions observed in childhood and adolescence.

Future Orientation and the OFC Network

Future-oriented decision making involves the ability to choose between competing actions that are associated with uncertain benefits and penalties. Throughout childhood, we learn and develop the ability to make behavioral changes that are advantageous in the long run, based on sometimes incomplete or uncertain information about future benefits. Importantly, this ability matures slowly over childhood, and is one of the few cognitive functions that does not reach an adult level until late adolescence (Crone & van der Molen, 2004; Hooper, Luciana, Conklin, & Yarger, 2004; Overman et al., 2004). These changes are mainly reflected in a reward-oriented or risky response pattern in young children and adolescents, followed by more conservative decision making in adults (Blakemore & Choudhury, 2006; Boyer, 2006).

Whereas the lateral PFC is most likely to be involved in the direction of goal states, the OFC is thought to be directly involved in representing the affective states under uncertainty (Figure 19.1). The functions of the OFC have only recently been examined systematically in neuropsychological and neuroimaging research, inspired by the initial reports of Damasio and colleagues (see Damasio, 1994). Damasio's "somatic marker" hypothesis was specifically developed to explain the behaviors of patients with damage to the OFC region. Such patients have relatively intact intellectual and working memory functions, but suffer from poor daily-life decision making. Emotions play a critical role within this theory and are defined as "somatic states," referring to the musculoskeletal, visceral, and internal milieu components of the soma (Damasio, 1996). The somatic marker hypothesis suggests that the OFC plays a critical role in forming associations between somatic responses associated with previously learned outcomes of situations and the reinstatement of these somatic states when a similar decision has to be made. These associations potentially reactivate emotions by acting on the appropriate cortical or subcortical structures, and become highly relevant in situations where future outcomes cannot be easily predicted on the basis of logical cost–benefit comparisons.

The Iowa Gambling Task as an Index of OFC Functioning

A now classic task that was developed to examine the decision-making impairments in patients with OFC damage is the Iowa Gambling Task (IGT), which mimics real-life decisions in the way it allocates rewards and punishments in the context of uncertain outcomes (Bechara, Damasio, Damasio, & Anderson, 1994). This task (typically computerized) requires individuals to sample from four decks of cards, from which two can result in immediate high gain, whereas two others result in immediate low gain. The uncertainty in outcomes lies in the way delayed punishment is presented. The two decks that result in high gain are accompanied by large delayed punishment, from which one of the decks is accompanied by frequent, relatively small punishment (50%), whereas the other is accompanied by infrequent, relatively large punishment (10%). Similarly, from the decks that result in low immediate gain, one is accompanied by frequent but small delayed punishment (50%), and the other by infrequent but large delayed punishment (10%). The first two decks (A and B) are disadvantageous in the long run, because they result in a net loss, whereas the last two decks (C and D) are advantageous in the long run, because they result in net gain.

Bechara and colleagues have demonstrated that healthy individuals learn to adopt an advantageous response strategy during the course of the IGT, whereas patients with damage to the OFC keep selecting from the disadvantageous decks. Intriguingly, healthy individuals develop anticipatory skin conductance responses prior to selecting from disadvantageous decks, whereas this autonomic "warning signal" is absent in patients with OFC damage (Bechara, Damasio, Tranel, & Damasio, 1997; Bechara, Tranel, Damasio, & Damasio, 1996; Tomb, Hauser, Deldin, & Caramazza, 2002). These results have been interpreted as showing that patients with OFC damage have a "myopia for the future"; that is, they fail to anticipate the future outcomes of their decisions. Indeed, neuroimaging research has confirmed that the OFC is important for processing abstract reward and punishment in healthy adults (Breiter, Aharon, Kahneman, Dale, & Shizgal, 2001; Elliott, Newman, Longe, & Deakin, 2003; Knutson, Taylor, Kaufman, Peterson, & Glover, 2005; O'Doherty, Kringelbach, Rolls, Hornak, & Andrews, 2001; Rogers et al., 1999; Ursu & Carter, 2005). Subsequent studies have demonstrated that impaired performance is most evident for patients with right-hemisphere OFC damage (Clark, Manes, Antoun, Sahakian, & Robbins, 2003; Tranel, Bechara, & Denburg, 2002), but patients with damage to DLPFC and dorsomedial PFC (Fellows & Farah, 2005; Manes et al., 2002) perform disadvantageously on the IGT. Thus the IGT probably recruits several regions within the PFC that contribute to different aspects of the task (Dunn, Dalgleish, & Lawrence, 2006).

Development of Future-Oriented Decision Making

Several behavioral studies have examined developmental differences in performance on the IGT, and these studies have demonstrated that across childhood and adolescence, there is an increase in the rate with which participants learn to select from the advantageous decks (Blair, Colledge, & Mitchell, 2001; Crone, Jennings, & van der Molen, 2004; Hooper et al., 2004; Overman et al., 2004). Using age-appropriate tasks, researchers have demonstrated that developmental changes in future-oriented decision making also occur between ages 3 and 6 (Garon & Moore, 2004; Kerr & Zelazo, 2004). These studies show that future orientation is already sensitive to developmental changes in early childhood, and that it continues to mature until late adolescence.

To examine the developmental trajectory of future-oriented decision making in more detail, Crone and van der Molen (2004) had participants ages 6–25 years perform a developmentally appropriate analogue of the IGT, called The Hungry Donkey Task (HDT). The basic format of the IGT was retained, but card gambling was changed into a prosocial game inviting the player to assist a hungry donkey to win as many apples as possible. The data demonstrated that the children ages 6–9 years and 10–12 years had a strong bias toward disadvantageous choices, even though they switched decks immediately after receiving punishment, just as the adults did. Adolescents ages 13–15 years made more advantageous choices than younger children, but still made more disadvantageous choices than adults, suggesting that advantageous decision making does not reach adult levels until late adolescence. Hooper et al. (2004) and Overman et al. (2004) further demonstrated that developmental changes in IGT performance still continue during late adolescence. A comparison of the performance of children ages 11–13 years and adolescents ages 14–17 years revealed that both age groups learned to make advantageous choices over the course of the task, but that the learning curve was faster for the older group (although not yet at an adult level, in comparison to findings from other published studies). In addition to the IGT, Hooper et al. (2004) asked all participants to complete a go/no-go task indexing response inhibition, and a digit span task indexing working memory. Developmental differences were observed for all tasks, but hierarchical regressions did not support a specific relationship among the development of inhibition, working memory, and IGT performance. Thus these tasks may tap into specific cognitive processes with separate underlying neural structures.

In addition, Crone and van der Molen (2007) examined whether children's performance on an analogue of the IGT was also associated with reduced autonomic activity prior to advantageous choices, just as seen in patients with OFC damage. Participants ages 8–10, 12–14, and 16–18 years were asked to perform the HDT while heart rate and skin conductance level were recorded prior to the deck choices and following the delivery of reward and punishment (see Plate 19.1). Consistent with prior studies, 16- to 18-year-olds learned to select from the advantageous decks, whereas 8- to 10-year-olds and 12- to 14-year-olds kept selecting from the disadvantageous decks. Whereas all age groups showed an increase in skin conductance and a slowing of heart rate following the delivery of punishment, only 16- to 18-year-olds demonstrated differential skin conductance activity prior to selection from the decks. Contrary to adults, who showed increased skin conductance activity prior to disadvantageous decks, the 16- to 18-year-olds showed increased skin conductance activity prior to decks that could result in frequent relative to infrequent punishment, regardless of whether these choices were advantageous or disadvantageous. Together, these studies suggest that children ages 8–10 and 12–14 demonstrate a "myopia for the future," as patients with OFC damage do. However, punishment frequency may be a more dominant dimension in children's decision making than magnitude of punishment is (Dunn et al., 2006; Huizenga, Crone, & Jansen, 2007).

Development of Brain Regions Supporting Future-Oriented Decision Making

Two developmental fMRI studies have demonstrated that children and adolescents show greater OFC activation in the presence of reward and punishment than adults do. van Leijenhorst, Crone, and Bunge (2006) examined the brain regions contributing to decision making under uncertainty in a simple gambling task. This study indicated different developmental patterns for the decision phase and the feedback phase. Children ages 8–12 years and young adults were asked to make decisions where the probability or getting a reward was low (high risk) or high (low risk). In the high-risk condition, children showed greater activation in the anterior cingulate cortex (ACC) than did adults; the ACC is a brain region that works closely together with OFC and signals uncertainty (Cohen, Heller, & Ranganath, 2005; Krain et al., 2006). In contrast, following feedback indicating loss of money, both age groups showed increased activation in lateral OFC, but this increase was larger for children than for adults (Plate 19.2). These data suggest that the consequences of loss were more aversive for children than for adults.

The notion of protracted OFC development vis-à-vis decision making was further examined in an fMRI study including children, adolescents, and adults (Galvan et al., 2006). The researchers had participants within three age groups (7–11 years, 13–17 years, and 23–29 years) perform a delayed-response two-choice task, in which a cue indexed whether the response would be followed by a small, medium, or large reward. They demonstrated that increases in reward magnitude resulted in increased activity in the nucleus accumbens and OFC in all age groups. However, adolescents showed a larger increase in nucleus accumbens activation than children or adults did (Ernst et al., 2005; May et al., 2004), whereas both children and adolescents showed more activity in OFC than the adults did. Thus 7- to 11-year-old children were only deviant from adults in OFC activation, whereas 13- to 17-year-olds were deviant from adults in both OFC and nucleus accumbens activation. These results were interpreted in terms of protracted maturational changes in top-down control systems (OFC) relative to subcortical regions (nucleus accumbens) implicated in appetitive behaviors. Together, these behavioral and neuroimaging studies indicate that children and adolescents may be more driven by appetitive systems than by control systems, possibly leading to suboptimal choices in social decision-making tasks (see Ernst & Spear, Chapter 17, this volume for further discussion of the balance between appetitive and control systems).

Perspective Taking and the Medial PFC Network

To examine the brain regions that are important for consideration of others, neuroimaging studies in adults have made use of decision-making games that were initially developed in social psychology paradigms. In these studies, researchers examine the brain regions that are sensitive to "social" and "nonsocial" phases of the task. "Social" phases are those situations in which participants are required to take into account the perspective of other individuals, whereas "nonsocial" usually refers to a baseline condition of interacting with a computer. These studies have emphasized that brain regions that have previously been associated with primary emotions are also sensitive to social situations, such as rejection or mutual cooperation. In contrast, brain regions that have previously been implicated in cognitive control functions also regulate social emotions (such as trust) that guide behavior toward long-term goals.

Neural Substrates of Fairness

One example of such a social decision-making task is the "ultimatum bargaining game," often simply referred to as the "ultimatum game." This is a classic task for studying social decision making, because it requires communication between parties that involves a certain division of outcomes (Guth, Schmittberger, & Schwarze, 1982). In the standard game, two players once divide a certain amount of money between them. One player is the allocator and proposes a division of the money. The other player is the recipient and can either accept or reject the proposed division. If the recipient accepts, the money is divided as proposed. If the recipient rejects, both players get nothing. With its simple structure, the ultimatum game is an attractive tool for assessing the relative importance of brain regions subserving self-interest and perspective taking in social decision making.

Neuroscientific analysis of the classic ultimatum game demonstrated that both an emotion-inducing area in the PFC (the insula) and an emotion-regulating area in the PFC (the DLPFC) were active when participants experienced unfair decisions by their social partners (Sanfey, Rilling, Aronson, Nystrom, & Cohen, 2003). The insula has previously been implicated in primary emotions such as disgust, whereas the DLPFC has been implicated in goal-directed cognitive control (see our earlier discussion of DLPFC network). In the study reported by Sanfey et al., the insula was more active when recipients decided to reject unfair offers, reflecting the emotional goals of rejecting unfairness. In contrast, the DLPFC was active when recipients rejected *and* accepted offers, reflecting the cognitive goal of accumulating money. This difference was only observed when participants were playing with another person, and not when they were playing with a computer, emphasizing the social nature of this effect.

The importance of the DLPFC for the cognitive goal of accumulating money was confirmed in studies using transcranial magnetic stimulation, which showed that temporally altering neuron firing in this region resulted in an increase in acceptance of unfair offers (Knoch, Pascual-Leone, Meyer, Treyer, & Fehr, 2006). Interestingly, a study using a decision-making game that focuses on cooperation rather than rejection (the prisoner's dilemma game) demonstrated that the nucleus accumbens, a region in the limbic area traditionally associated with reward sensitivity, was more active following mutual collaboration (Rilling, Sanfey, Aronson, Nystrom, & Cohen, 2004). Thus social feelings of rejection and cooperation seem to depend on the same neural mechanisms as those that are sensitive to basic emotional signals of disgust and reward. Together, these studies demonstrate that social decision making is regulated by both emotion-inducing and emotion-regulating brain areas, which make independent contributions to the decision process.

Neural Substrates of Trust Decisions

As described above, the ultimatum game permits the study of fairness and self-interest in a simple experimental setting. However, in real life one often needs to make decisions about fairness, based on the trust that another person will reciprocate when given a fair offer. These situations put high demands on the ability to think about the other person's intentions (or to take another person's perspective). A typical task that captures the essentials of reciprocity is the "trust game" (e.g., Malhotra, 2004; Snijders, 1996; Snijders & Keren, 1999). In this game, two players are paired with each other as decision makers (player A and player B) over single or numerous trials. Player A makes the first decision, and player B observes player A's decision before making his or her own decision. More specifically, player A can choose between two options: a certain reward choice (the "no-trust" option), in which both players receive a small reward (e.g., both receive 45 euros), and an uncertain choice (the "trust" or "collaborative" option). When player A opts for the trust/collaborative option, player B can choose between two options. Either player B decides to reciprocate trust, in which case both receive a relatively large reward (e.g., player A receives 180 euros and player B receives 225 euros), or player B decides to exploit trust, in which case player A receives nothing and Player B receives a large reward (e.g., 405 euros). The trust game is an attractive task for capturing the risk to trust on the one hand, and reciprocity or exploitation on the other hand. Importantly, reciprocity is larger when a trusting player takes a large rather than a small risk. This pattern suggests that in order to generate positive reciprocal exchanges, or realize "positive reciprocity" (McCabe, Houser, Ryan, Smith, & Trouard, 2001), it is necessary to take a risk or be vulnerable in favor of greater postponed gains from mutual cooperation.

Neuroscientific studies using the trust game have demonstrated that in adult participants, the decision of player A to trust/collaborate is associated with increased activation in medial PFC when the other player is a human as opposed to a computer (McCabe et al., 2001). The medial PFC is a brain region that is active when individuals refer to others' states of mind as well as their own, or when there is a need to explain and predict the behaviors of others by attributing independent mental states to them, such as thoughts, beliefs, and desires (Adolphs, 2003). In the McCabe et al. study, the six subjects who scored highest on trust/cooperation showed significant increases in activation in medial PFC during human–human interactions when compared with human–computer interactions, whereas the six lowest-scoring participants did not show significant activation differences in medial PFC between the human and computer conditions. McCabe et al. conclude that "trust/cooperation requires an active convergence zone that binds joint attention to mutual gains with sufficient inhibition of immediate reward gratification to allow cooperative decisions" (2001, p. 11834). This finding is consistent with those of social neuroscience studies that have reported increased activation in medial PFC for social collaboration (Rilling et al., 2002), competition (Gallagher, Jack, Roepstorff, & Frith, 2002), and moral judgment (Greene, Nystrom, Engell, Darley, & Cohen, 2004).

Taken together, neuroscientific studies of social decision making demonstrate that the control functions described at the beginning of this chapter rely on different neural networks. First, positive and negative social emotions seem to draw upon basic affective neural mechanisms (such as the insula and the nucleus accumbens). Second, when individuals make decisions in which they take into account the larger cognitive goals of accumulating gain (even when this leads to, e.g., accepting unfairness), this process results in increased activation in the DLPFC, previously associated with cognitive control. Third, when considering intentions of other individuals, researchers report increased activation in the medial PFC, previously associated with the need to explain and predict the behavior of others.

Development of Social Decision Making

The neuroimaging studies described above provide a context for understanding behavioral differences in social decision-making tasks across development. Late childhood and adolescent development is typically characterized by increased self-consciousness, and adolescents become increasingly preoccupied with other people's concerns about their actions (Blakemore & Choudhury, 2006). Although theory-of-mind development has been well established early in childhood (Barresi & Moore, 1996; Perner, Leekam, & Wimmer, 1987), relatively few studies have examined perspective taking in adolescence. Blakemore and Choudhury (2006) recently reported that the ability to differentiate between first- and third-person perspective taking is still immature in adolescence (see Choudhury, Charman, & Blakemore, Chapter 9, this volume, for further discussion). These findings lead to the hypothesis that the ability to take another person's perspective and to differentiate between one's own concerns and those of others is still developing in adolescence.

Van Meel and Crone (2007) tested the development of concern for others in a behavioral study, by comparing developmental differences in response to fair and unfair offers, using an analogue of the ultimatum game that allowed repeated bargaining. Participants ages 10, 15, and 20 years played this game, with the counterplayer for each participant being an age-matched volunteer. Equality was most important in childhood: Young children were more likely to reject unfair offers than were older participants (see Figure 19.2). In contrast, adolescents and adults accepted more unfair offers in favor of personal gain. These findings are consistent with the assumption that the ability to weigh fairness against self-interest emerges in adolescence. Indeed, prior studies have demonstrated that young children operate according to principles of equality (i.e., they value fairness), whereas adolescents and adults accept unfairness by others when this leads to higher personal gain (self-interest) (Hoffman & Tee, 2006; Murninghan & Saxon, 1998). The developmental pattern leads to the hypothesis that especially young children are led more by activity in emotion-inducing areas for unfairness (insula), relative to emotion-regulating areas, which are important for personal gain (DLPFC).

Another phenomenon that has been observed in late childhood and adolescence concerns the shift from self-centered to collaborative decisions, which may bear a close similar-

(A)

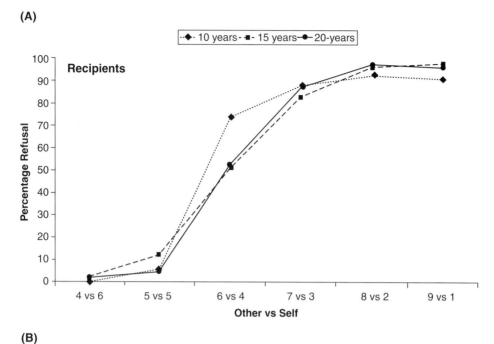

(B)

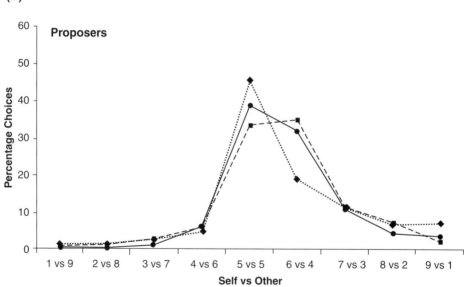

FIGURE 19.2. (A) Percentage refusal of offers in the ultimatum game for children, adolescents, and adults. Children refused more offers when the division was unfair. (B) Percentage of offers for different money splits in the ultimatum game. Children made more fair offers than adolescents and adults did.

ity to the act of reciprocal exchange, or the tendency for one person to repay what another has provided for him or her (Gouldner, 1960). Behavioral studies using the trust game demonstrated that the amount of trust given to another person increases between the ages of 8 and 22 (Sutter & Kocher, 2007). We (van den Bos, Westenberg, Van Dijk, & Crone, 2007) examined how trust, and the repayment of trust (reciprocity), are influenced by the risk that the trusting partner takes (i.e., "If I trust the other individual, will this lead to a large or a small loss if I am not repaid?"), and the benefit that the trusted partner gets from being trusted (i.e., "How much will I gain by being trusted relative to not being trusted?"). Both trust and reciprocity increased across adolescence, but trust had a steeper trajectory than reciprocity. Importantly, the age groups were differentially sensitive to the risk and benefit manipulations, in such a way that risk considerations developed earlier than benefit considerations. These developmental trends may indicate that that the ability to make inferences about other people's minds, purportedly regulated by the medial PFC, undergoes important changes in childhood and adolescence.

Although these behavioral studies provide only indirect indices of brain development, the differential pattern of choices suggests that the balance between emotion-inducing networks and emotion-regulating networks changes in development. There is a surprising lack of literature in the neuroimaging domain on developmental changes in concern for others and perspective taking. The recent imaging studies on perspective taking and social decision making in adults have suggested an important role for the medial PFC (Gallagher & Frith, 2003). Structurally, this region continues to undergo developmental changes well into adolescence or early adulthood (Sowell et al., 2004). Thus specific changes in brain regions subserving the ability to take another person's perspective may continue during childhood and adolescence. In the previous section, we have discussed tasks permitting the investigation of neural regions that are important for interactions between individuals (McCabe et al., 2001; Rilling et al., 2002; Sanfey et al., 2003). We are currently conducting neuroimaging studies using such tasks with adolescents, to understand changes in neural systems that are engaged during social decision-making processes.

Age-Independent Indices of Psychosocial Development: Moving beyond Age as a Proxy for Developmental Maturity

The investigation of brain-based aspects of cognitive and social development is invariably based on the comparison of different age groups. However, variability in developmental maturity *within* age groups—here referred to as "differential maturity"—may obscure differences *between* age groups. That is, age differences may not emerge, or may be very weak, because of large variability within each age cohort. The problem of differential maturity is aggravated when age groups consist of participants of varying ages (e.g., 13- to 17-year-olds). In short, the age-based approach may not yield the developmental effects that may be "out there."

For example, Eshel, Nelson, Blair, Pine, and Ernst (2007) did not observe the expected age difference in risk taking: Adolescents 9–17 years of age did *not* make more risky decisions than adults ages 20–40 years. As was expected, however, adolescents and adults differed in activation of the OFC and VLPFC, and activation of those brain areas was negatively related to risk taking. Given the lack of behavioral differences between adolescents and adults, however, one cannot be certain that the differences in brain activation reflect maturational differences. The difficulty of interpreting group differences in brain activation in terms of maturational differences is compounded when some of the findings are the opposite of what is expected. Similarly, Krain et al. (2006) examined developmental differences in the neural correlates of "intolerance of uncertainty" (IU) and decision making. Contrary to expectations, adolescents ages 13–17 years were more tolerant of ambiguity than were adults ages 19–36 years. This finding suggests that the adolescents were, on average, more mature than the adults. Yet the authors maintained that the differences found between adolescents and adults in the relationship between IU and ACC activity indicate that "mature brain circuits may develop compensatory mechanisms so that ACC activity ... is no longer associated with IU [in adults]" (Krain et al., 2006, p. 1028). However, given that adolescents scored lower on IU than the adults, it is not a foregone conclusion that the brain–IU relations discerned in the adult sample indicate greater maturity of the adult brain.

The problem of differential maturity is particularly pronounced during adolescence. Adolescence is a period of major change, and consequently studies often report high variability in adolescent samples (Luna et al., 2001). Steinberg (2005) suggests that this variability could be due to the fact that the adolescents within these samples are in different stages of development, because of individual variability in the timing of the hormonal and physical changes that take place during this period. As a result, age may be a particularly poor predictor of changes in adolescent decision making and its relationship to the developing brain. To reliably detect developmental changes within adolescence, as well as between adolescence and adulthood, researchers need to move beyond age as a proxy for developmental maturity. Cauffman and Steinberg (2000) studied the relationship among age, psychosocial maturity, and antisocial decision making in a sample of over 1000 adolescents and adults (grades 8, 10, and 12; college students "under 21" and "over 21"). "Psychosocial maturity" was defined as a composite of three components: responsibility, perspective, and temperance. The researchers found that age was a significant, but weak, predictor of antisocial decision making ($\beta = .08$, $p < .01$). However, when psychosocial maturity was also entered as a predictor in the regression analysis, the effect of age was no longer significant. The findings indicated that individual differences in psychosocial maturity were the more powerful correlates ($\beta = .52$, $p < .0001$). These findings demonstrate the advantage of using a direct measure of psychosocial maturity.

A limitation of most models of psychosocial maturity, however, is that they lack a developmental perspective. This makes it difficult to disentangle differential maturity from other individual differences (e.g., temperament, personality) and from gender differences (Cauffman & Steinberg [2000] reported substantial gender differences in antisocial decision making). For example, a 13-year-old female may display prosocial decision making because of developmental precocity, a conscientious and agreeable personality, or a mixture of both. Conversely, a 22-year-old male may display antisocial decision making because of developmental immaturity, a general lack of conscientiousness and agreeableness, or a combination of these. To investigate the development of social decision making and its relationship to the developing brain, it is imperative to classify such a 13-year-old as mature and such a 22-year-old as immature, independently of their age, gender, temperament, and personality characteristics.

An elegant solution is offered by the use of Loevinger's (1985, 1997) developmental model of personality. Within this model, personality growth is characterized as a series of changes in impulse control, interpersonal style, and conscious preoccupations. Developmental advances in these domains are depicted in terms of levels of development. The pace and extent of ego development within an individual depend upon many influences beyond the mere passage of time, such as environmental influences and hereditary factors (Allen, Hauser, Bell, & O' Connor, 1994; Newman, Tellegen, & Bouchard, 1998). Hence, within any age cohort there will be a range of ego development levels, reflecting individual differences in impulse control, time perspective, and perspective taking (Westenberg & Block, 1993). As such, Loevinger's model is a comprehensive model of psychosocial maturity (Westenberg, Hauser, & Cohn, 2004). A recent meta-analysis has validated the contention that this model is distinct from cognitive development and intelligence (Cohn & Westenberg, 2004). Recent studies from our own laboratory have shown that, especially in the adolescent period, maturation levels as indicated by Loevinger's model are much better predictors of changes in sensitivity to social and environmental factors than the specific ages of the adolescents are (Westenberg, Drewes, Goedhart, Siebelink, & Treffers, 2004).

Although a link between psychosocial maturity and brain maturation has been suggested (Steinberg, 2005), the evidence to date is only indirect and suggestive. However, the neurobiological evidence suggests that the brain systems involved in impulse control, time perspective, and perspective taking continue to mature into late adolescence. Hence it is likely that individual differences in psychosocial maturity are related to the timing and speed of brain maturation during adolescence. Loevinger's model seems particularly relevant for tracking brain-based aspects of psychosocial maturation during adolescence and into adulthood: (1) It represents the emotion-regulating components implicated in the development of social decision making; (2) the growth curve is steepest between late childhood and late adolescence, while leveling off during young adulthood (Cohn, 1998); and (3) it permits a research participant to be classified at a specific level of development, independently of the person's age or personality. The inclusion of such a developmental model will help to bring the maturation of social decision making into sharper focus. It might also help to answer the question whether all components of the emotion-regulating system (DLPFC, OFC, medial PFC) develop according to the same timetable, or whether these components display fractionated development.

Conclusions

During late childhood and adolescence, there are important changes in social awareness, which are probably associated with the late development of brain regions regulating social processes. Earlier models (e.g., the SIPN model; Nelson et al., 2005) have suggested that social decision making arises from the interaction between affective and regulation systems, which are reorganized in adolescence.

In this chapter, we propose that developmental changes in the regulation system should be fractionated into different control processes, which may have different developmental trajectories. We have highlighted three of these possible regulatory mechanisms: (1) cognitive control (regulated by the DLPFC network), (2) future orientation (regulated by the OFC network), and (3) perspective taking (regulated by the limbic–medial PFC network). The tight race between the development of emotion-inducing and emotion-regulating brain systems makes the adolescent period especially vulnerable to influences of environmental stress (Raine, 2002).

In future research, it will be important to understand the developmental trajectories of these different networks, and their *relative* contribution (i.e., how they work together) to making social judgments. It seems that new paradigms should be constructed allowing the assessment of the separate contributions of emotion-regulating versus emotion-inducing systems to the observed developmental change in social decision making.

Acknowledgments

Funding was provided by the Dutch Organization for Scientific Research (NWO-VENI/VIDI) to Eveline A. Crone. We thank Maurits van der Molen and Wouter van den Bos for helpful comments on a prior version of this chapter.

References

Adolphs, R. (2003). Cognitive neuroscience of human social behaviour. *Nature Reviews Neuroscience, 4*(3), 165–178.

Allen, J. P., Hauser, S. T., Bell, K. L., & O' Connor, T. G. (1994). Longitudinal assessment of autonomy and relatedness in adolescent family interactions as predictors of adolescent ego development and self esteem. *Child Development, 65,* 179–194.

Baddeley, A. D., & Logie, R. H. (1999). *Working memory: The multiple component model.* New York: Cambridge University Press.

Barresi, J., & Moore, C. (1996). Intentional relations and social understanding. *Behavioral and Brain Sciences, 19,* 107–154.

Bechara, A., Damasio, A. R., Damasio, H., & Anderson, S. W. (1994). Insensitivity to future consequences following damage to human prefrontal cortex. *Cognition, 50*(1–3), 7–15.

Bechara, A., Damasio, H., Tranel, D., & Damasio, A. R. (1997). Deciding advantageously before knowing the advantageous strategy. *Science, 275*(5304), 1293–1295.

Bechara, A., Tranel, D., Damasio, H., & Damasio, A. R. (1996). Failure to respond autonomically to anticipated future outcomes following damage to prefrontal cortex. *Cerebral Cortex, 6*(2), 215–225.

Blair, R. J., Colledge, E., & Mitchell, D. G. (2001). Somatic markers and response reversal: Is there orbitofrontal cortex dysfunction in boys with psychopathic tendencies? *Journal of Abnormal Child Psychology, 29*(6), 499–511.

Blakemore, S. J., & Choudhury, S. (2006). Development of the adolescent brain: Implications for executive function and social cognition. *Journal of Child Psychology and Psychiatry, 47,* 296–312.

Boyer, T. W. (2006). The development of risk-taking: A multi-perspective review. *Developmental Review, 26*(3), 291–345.

Breiter, H. C., Aharon, I., Kahneman, D., Dale, A., & Shizgal, P. (2001). Functional imaging of neural responses to expectancy and experience of monetary gains and losses. *Neuron, 30*(2), 619–639.

Bunge, S. A. (2004). How we use rules to select actions: A review of evidence from cognitive neuroscience. *Cognitive, Affective, and Behavioral Neuroscience, 4*(4), 564–579.

Bunge, S. A., Dudukovic, N. M., Thomason, M. E., Vaidya, C. J., & Gabrieli, J. D. (2002). Immature frontal lobe contributions to cognitive control in children: Evidence from fMRI. *Neuron, 33*(2), 301–311.

Bunge, S. A., Kahn, I., Wallis, J. D., Miller, E. K., & Wagner, A. D. (2003). Neural circuits subserving the retrieval and maintenance of abstract rules. *Journal of Neurophysiology, 90*(5), 3419–3428.

Case, R. (1992). *The mind's staircase: Exploring the conceptual underpinnings of children's thought and knowledge.* Hillsdale, NJ: Erlbaum.

Cauffman, E., & Steinberg, L. (2000). (Im)maturity of judgment in adolescence: Why adolescents may be less culpable than adults. *Behavioral Sciences and the Law, 18,* 741–760.

Clark, L., Manes, F., Antoun, N., Sahakian, B. J., & Robbins, T. W. (2003). The contributions of lesion laterality and lesion volume to decision-making impairment following frontal lobe damage. *Neuropsychologia, 41,* 1474–1483.

Cohen, M. X., Heller, A. S., & Ranganath, C. (2005). Functional connectivity with anterior cingulate and orbitofrontal cortices during decision-making. *Brain Research: Cognitive Brain Research, 23*(1), 61–70.

Cohn, L. D. (1998). *Age trends in personality development: A quantitative review.* Mahwah, NJ: Erlbaum.

Cohn, L. D., & Westenberg, P. M. (2004). Intelligence and maturity: Meta-analytic evidence for the incremental and discriminant validity of Loevinger's measure of ego development. *Journal of Personality and Social Psychology, 86,* 760–772.

Crone, E. A., Donohue, S. E., Honomichl, R., Wendelken, C., & Bunge, S. A. (2006a). Brain regions mediating flexible rule use during development. *Journal of Neuroscience, 26*(43), 11239–11247.

Crone, E. A., Jennings, J. R., & van der Molen, M. W. (2004). Developmental change in feedback processing as reflected by phasic heart rate changes. *Developmental Psychology, 40*(6), 1228–1238.

Crone, E. A., & van der Molen, M. W. (2004). Developmental changes in real life decision making: Performance on a gambling task previously shown to depend on the ventromedial prefrontal cortex. *Developmental Neuropsychology, 25*(3), 251–279.

Crone, E. A., & van der Molen, M. W. (2007). Development of decision-making in school-aged children and adolescents: Evidence from heart rate and skin conductance analysis. *Child Development, 78,* 1288–1301.

Crone, E. A., Wendelken, C., Donohue, S., van Leijenhorst, L., & Bunge, S. A. (2006b). Neurocognitive development of the ability to manipulate information in working memory. *Proceedings of the National Academy of Sciences USA, 103*(24), 9315–9320.

Damasio, A. R. (1994). *Descartes' error.* New York: Putnam.

Damasio, A. R. (1996). The somatic marker hypothesis and the possible functions of the prefrontal cortex. *Philosophical Transactions of the Royal Society of London, Series B, Biological Sciences, 351*(1346), 1413–1420.

De Dreu, C. K. W., Lualhati, J. C., & McCusker, C. M. (1994). Effects of gain–loss frames on satisfaction with self–other outcome differences. *European Journal of Social Psychology, 24,* 497–510.

Diamond, A. (2002). Normal development of prefrontal cortex from birth to young adulthood: Cognitive functions, anatomy and biochemistry. In D. T. Stuss & R. T. Knight (Eds.), *Principles of frontal lobe function* (pp. 466–503). London: Oxford University Press.

Dunn, B. D., Dalgleish, T., & Lawrence, A. D. (2006). The somatic marker hypothesis: A critical evaluation. *Neuroscience Biobehavioral Reviews, 30*(2), 239–271.

Durston, S., Davidson, M. C., Tottenham, N., Galvan, A., Spicer, J., Fossella, J. A., et al. (2006). A shift from diffuse to focal cortical activity with development. *Developmental Science, 9*(1), 1–8.

Elliott, R., Newman, J. L., Longe, O. A., & Deakin, J. F. (2003). Differential response patterns in the striatum and orbitofrontal cortex to financial reward in humans: A parametric functional magnetic resonance imaging study. *Journal of Neuroscience, 23*(1), 303–307.

Ernst, M., Nelson, E. E., Jazbec, S., McClure, E. B., Monk, C. S., Leibenluft, E., et al. (2005). Amygdala and nucleus accumbens in responses to receipt and omission of gains in adults and adolescents. *NeuroImage, 25*(4), 1279–1291.

Eshel, N., Nelson, E. E., Blair, R. J., Pine, D. S., & Ernst, M. (2007). Neural substrates of choice selection in adults and adolescents: Development of the ventrolateral prefrontal and anterior cingulate cortices. *Neuropsychologia, 45*(6), 1270–1279.

Fellows, L. K., & Farah, M. J. (2005). Different underlying impairments in decision-making following ventromedial and dorsolateral frontal lobe damage in humans. *Cerebral Cortex, 15*(1), 58–63.

Gallagher, H. L., & Frith, C. D. (2003). Functional imaging of 'theory of mind.' *Trends in Cognitive Sciences, 7,* 77–83.

Gallagher, H. L., Jack, A. I., Roepstorff, A., & Frith, C. D. (2002). Imaging the intentional stance in a competitive game. *NeuroImage, 16,* 814–821.

Galvan, A., Hare, T. A., Parra, C. E., Penn, J., Voss, H., Glover, G., et al. (2006). Earlier development of the accumbens relative to orbitofrontal cortex might underlie risk-taking behavior in adolescents. *Journal of Neuroscience, 26*(25), 6885–6892.

Garon, N., & Moore, C. (2004). Complex decision-making in early childhood. *Brain and Cognition, 55*(1), 158–170.

Giedd, J. N. (2004). Structural magnetic resonance imaging of the adolescent brain. *Annals of the New York Academy of Sciences, 1021,* 77–85.

Gogtay, N., Giedd, J. N., Lusk, L., Hayashi, K. M., Greenstein, D., Vaituzis, A. C., et al. (2004). Dynamic mapping of human cortical development during childhood through early adulthood. *Proceedings of the National Academy of Sciences USA, 101*(21), 8174–8179.

Gouldner, A. W. (1960). The norm of reciprocity: A preliminary statement. *American Sociological Review, 25,* 161–178.

Greene, J. D., Nystrom, L. E., Engell, A. D., Darley, J. M., & Cohen, J. D. (2004). The neural basis of cognitive conflict and control in moral judgment. *Neuron, 44,* 389–400.

Guth, W., Schmittberger, R., & Schwarze, B. (1982). An experimental analysis of ultimatum games. *Journal of Economic Behavior and Organization, 3,* 367–388.

Handgraaf, M. J. J., Van Dijk, E., & De Cremer, D. (2003). Social utility in ultimatum bargaining. *Social Justice Research, 16*(3), 263–283.

Hoffman, R., & Tee, J. Y. (2006). Adolescent–adult interactions and culture in the ultimatum game. *Journal of Economic Psychology, 27,* 98–116.

Hooper, C. J., Luciana, M., Conklin, H. M., & Yarger, R. S. (2004). Adolescents' performance on the Iowa Gambling Task: Implications for the development of decision making and ventromedial prefrontal cortex. *Developmental Psychology, 40*(6), 1148–1158.

Huizenga, H., Crone, E. A., & Jansen, B. (2007). Decision making in healthy children, adolescents, and adults explained by the use of increasingly complex proportional reasoning rules. *Developmental Science, 10*(6), 814–825.

Kerr, A., & Zelazo, P. D. (2004). Development of "hot" executive function: The Children's Gambling Task. *Brain and Cognition, 55*(1), 148–157.

Klingberg, T., Forssberg, H., & Westerberg, H. (2002). Increased brain activity in frontal and parietal cortex underlies the development of visuospatial working memory capacity during childhood. *Journal of Cognitive Neuroscience, 14*(1), 1–10.

Knoch, D., Pascual-Leone, A., Meyer, K., Treyer, V., & Fehr, E. (2006). Diminishing reciprocal fairness by disrupting the right prefrontal cortex. *Science, 314,* 829–832.

Knutson, B., Taylor, J., Kaufman, M., Peterson, R., & Glover, G. (2005). Distributed neural representation of expected value. *Journal of Neuroscience, 25*(19), 4806–4812.

Krain, A. L., Hefton, S., Pine, D. S., Ernst, M., Xavier Castellanos, F., Klein, R. G., et al. (2006). An fMRI examination of developmental differences in the neural correlates of uncertainty and decision-making. *Journal of Child Psychology and Psychiatry, 47*(10), 1023–1030.

Kwon, H., Reiss, A. L., & Menon, V. (2002). Neural basis of protracted developmental changes in visuo-spatial working memory. *Proceedings of the National Academy of Sciences USA, 99,* 13336–13341.

Loevinger, J. (1985). Revision of the sentence completion test for ego development. *Journal of Personality and Social Psychology, 48,* 420–427.

Loevinger, J. (1997). *Stages of personality development.* San Diego, CA: Academic Press.

Luna, B., Thulborn, K. R., Munoz, D. P., Merriam, E. P., Garver, K. E., Minshew, N. J., et al. (2001). Maturation of widely distributed brain function subserves cognitive development. *NeuroImage, 13*(5), 786–793.

Malhotra, D. (2004). Trust and reciprocity decisions: The different perspectives of trusters and trusted parties. *Organizational Behavior and Human Decision Processes, 94,* 61–73.

Manes, F., Sahakian, B. J., Clark, L., Rogers, R. D., Antoun, N., Aitken, M., et al. (2002). Decision-making processes following damage to the prefrontal cortex. *Brain, 125*, 624–639.

May, J. C., Delgado, M. R., Dahl, R. E., Stenger, V. A., Ryan, N. D., Fiez, J. A., et al. (2004). Event-related functional magnetic resonance imaging of reward-related brain circuitry in children and adolescents. *Biological Psychiatry, 55*(4), 359–366.

McCabe, K., Houser, D., Ryan, L., Smith, V., & Trouard, T. (2001). A functional imaging study of cooperation in two-person reciprocal exchange. *Proceedings of the National Academy of Sciences USA, 98*, 11832–11835.

Miller, E. K., & Cohen, J. D. (2001). An integrative theory of prefrontal cortex function. *Annual Review of Neuroscience, 24*, 167–202.

Murninghan, J. K., & Saxon, M. S. (1998). Ultimatum bargaining by children and adolescents. *Journal of Economic Psychology, 19*, 415–445.

Nelson, E., Leibenluft, E., McClure, E. B., & Pine, D. S. (2005). The social re-orientation of adolescence: A neuroscience perspective on the process and its relation to psychopathology. *Psychological Medicine, 35*, 163–174.

Newman, D. L., Tellegen, A., & Bouchard, T. J., Jr. (1998). Individual differences in adult ego development: Sources of influence in twins reared apart. *Journal of Personality and Social Psychology, 74*, 985–995.

O'Doherty, J., Kringelbach, M. L., Rolls, E. T., Hornak, J., & Andrews, C. (2001). Abstract reward and punishment representations in human orbitofrontal cortex. *Nature Neuroscience, 4*(1), 95–102.

Overman, W. H., Frassrand, K., Ansel, S., Trawalter, S., Bies, B., & Redmond, A. (2004). Performance on the Iowa Card Task by adolescents and adults. *Neuropsychologia, 42*(13), 1838–1851.

Paus, T. (2005). Mapping brain maturation and cognitive development during adolescence. *Trends in Cognitive Sciences, 9*(2), 60–68.

Perner, J., Leekam, S., & Wimmer, H. (1987). Three-year-olds' difficulty with false belief: The case for a conceptual deficit. *British Journal of Developmental Psychology, 5*, 125–137.

Raine, A. (2002). Biosocial studies of antisocial and violent behavior in children and adults: A review. *Journal of Abnormal Child Psychology, 30*(4), 311–326.

Rilling, J. K., Gutman, D. A., Zeh, T. R., Pagnoni, G., Berns, G. S., & Kilts, C. D. (2002). A neural basis for social cooperation. *Neuron, 35*, 395–405.

Rilling, J. K., Sanfey, A. G., Aronson, J. A., Nystrom, L. E., & Cohen, J. D. (2004). The neural correlates of theory of mind within interpersonal interactions. *NeuroImage, 22*, 1694–1703.

Rogers, R. D., Owen, A. M., Middleton, H. C., Williams, E. J., Pickard, J. D., Sahakian, B. J., et al. (1999). Choosing between small, likely rewards and large, unlikely rewards activates inferior and orbital prefrontal cortex. *Journal of Neuroscience, 19*(20), 9029–9038.

Sanfey, A. G., Rilling, J. K., Aronson, J. A., Nystrom, L. E., & Cohen, J. D. (2003). The neural basis of economic decision-making in the ultimatum game. *Science, 300*, 1755–1758.

Shaw, P., Greenstein, D., Lerch, J., Clasen, L., Lenroot, R., Gogtay, N., et al. (2006). Intellectual ability and cortical development in children and adolescents. *Nature, 440*(7084), 676–679.

Smith, E. E., & Jonides, J. (1999). Storage and executive processes in the frontal lobes. *Science, 283*(5408), 1657–1661.

Snijders, C. (1996). *Trust and commitments.* Amsterdam: Thesis Publishers.

Snijders, C., & Keren, G. (1999). Determinants of trust. In D. V. Budescu & I. Erev (Eds.), *Games and human behavior: Essays in honor of Amnon Rapoport* (pp. 355–385). Mahwah, NJ: Erlbaum.

Sowell, E. R., Thompson, P. M., Leonard, C. M., Welcome, S. E., Kan, E., & Toga, A. W. (2004). Longitudinal mapping of cortical thickness and brain growth in normal children. *Journal of Neuroscience, 24*(38), 8223–8231.

Steinberg, L. (2005). Cognitive and affective development in adolescence. *Trends in Cognitive Sciences, 9*, 69–74.

Sutter, M., & Kocher, M. G. (2007). Trust and trustworthiness across different age groups. *Games and Economic Behavior, 59*(2), 364–382.

Tomb, I., Hauser, M., Deldin, P., & Caramazza, A. (2002). Do somatic markers mediate decisions on the gambling task? *Nature Neuroscience, 5*(11), 1103–1104 (author reply, 1104).

Tranel, D., Bechara, A., & Denburg, N. L. (2002). Asymmetric functional roles of right and left ventromedial prefrontal cortices in social conduct, decision-making, and emotional processing. *Cortex, 38,* 589–612.

Ursu, S., & Carter, C. S. (2005). Outcome representations, counterfactual comparison and the human orbitfrontal cortex: Implications for neuroimaging studies of decision-making. *Brain Research: Cognitive Brain Research, 23*(1), 51–60.

van den Bos, W., Westenberg, P. M., Van Dijk, E., & Crone, E. A. (2007, June). *Development of trust in adolescence.* Poster presented at the Jean Piaget Society Conference, Amsterdam.

van Leijenhorst, L., Crone, E. A., & Bunge, S. A. (2006). Neural correlates of developmental differences in risk estimation and feedback processing. *Neuropsychologia, 44*(11), 2158–2170.

Van Meel, C. S., & Crone, E. A. (2007). *Age differences in ultimatum bargaining.* Unpublished manuscript.

Westenberg, P. M., & Block, J. (1993). Ego development and individual differences in personality. *Journal of Personality and Social Psychology, 65,* 792–800.

Westenberg, P. M., Drewes, M. J., Goedhart, A. W., Siebelink, B. M., & Treffers, P. D. A. (2004). A developmental analysis of self reported fears in late childhood through mid adolescence: Social-evaluative fears on the rise? *Journal of Child Psychology and Psychiatry, 45,* 481–495.

Westenberg, P. M., Hauser, S. T., & Cohn, L. D. (2004). Sentence completion measurement of psychosocial maturity. In M. Hersen (Series Ed.) & M. J. Hilsenroth & D. L. Segal (Vol. Eds.), *Comprehensive handbook of psychological assessment: Vol. 2. Personality assessment* (pp. 595–616). Hoboken, NJ: Wiley.

PART VI

PERSPECTIVES ON PSYCHOPATHOLOGY

CHAPTER 20

A Social Neuroscience Approach
to Adolescent Depression

DANIEL S. PINE

Since Aristotle, philosophers and scientists have marveled at the changes of adolescence. Now breakthroughs in neuroscience provide an opportunity to observe adolescence through a new lens. This period signals the emergence of dramatic improvements in diverse capabilities. Reaction times become faster; muscles become stronger; capacity for reflection becomes greater; and working memory capacity becomes larger. In fact, one would be hard pressed to find a physical or mental function that fails to improve during adolescence.

Paradoxically, adolescence is also a tragically cruel time of life: Although it is associated with remarkably improved prowess, it is also characterized by abrupt increases in morbidity and mortality. For example, as shown in Figure 20.1, the number of deaths among 15- to 24-year-olds is nearly five times that experienced by 5- to 14-year-olds (National Center for Health Statistics, 2005). How can we understand this paradox of rising mortality in the face of peak physical function?

Indeed, adolescence is not only unique in its juxtaposition of prime physiological function and increased mortality; it is also unique in the factors contributing to mortality. Adolescents die from fundamentally different things than either children or adults do. In 2003, 73% of all adolescent deaths in the United States were attributable to three causes: injuries, homicide, and suicide. In all other decades of life, these causes together accounted for fewer than 50% of deaths (National Center for Health Statistics, 2005). Thus only in adolescence are the causes of mortality overwhelmingly behavioral. Given this fact, neuroscience may provide particularly important insights into contributors to adolescent mortality.

Among the three major causes of adolescent mortality, we know the most about suicide. Approximately 90% of adolescents who commit suicide suffer from psychopathology—illness that is typically undiagnosed and untreated (Shaffer et al., 1996). The association between

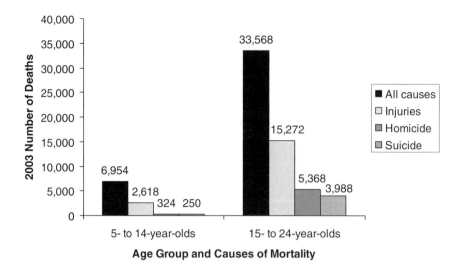

FIGURE 20.1. The number of deaths from all causes and three specific causes in two age groups. The data reveal the marked increase in death between the age bands of 5–14 years and 15–24 years, with behavioral causes of death making a major contribution to this increase. Data from National Center for Health Statistics (2005).

major depressive disorder (MDD) and suicide has been emphasized in recent scientific publications and the mass media. Indeed, MDD exhibits robust increases in prevalence during adolescence and has indisputably strong associations with adolescent suicide (Shaffer et al., 1996). As a result, MDD and other mood disorders show a particularly strong association with adolescent morbidity and mortality. The current chapter focuses on MDD. Nevertheless, many other psychopathologies also dramatically increase during adolescence, including some forms of anxiety and substance use disorders. Each of these is associated with disturbances in emotion regulation; their upsurge thus supports long-held notions of tight relationships between adolescent development and changes in emotional function. Moreover, each shows strong associations with suicide. Clearly, then, understanding the emergence of adolescent psychopathology requires research in neuroscience focused on this critical period. Given the role of emotion in adolescent psychopathology, a particularly important focus is on the relationship between emotional processes and brain function.

Despite the importance of studying emotion in a range of disorders, the current chapter explores neural processes associated with one specific form of emotional psychopathology, MDD, which becomes prominent during adolescence. The focus on this specific condition permits a more in-depth review to be provided; this review also illustrates in some detail an approach that may also prove useful in other conditions beyond MDD. The chapter reviews data on MDD in three stages. The initial section reviews clinical features of various emotional psychopathologies manifested during adolescence before narrowing the focus to pediatric MDD. As an aside, because pediatric MDD occurs predominantly among adolescents, most studies on pediatric MDD focus on adolescents. For the current review, the term "adolescent" is used to describe research focused only on adolescents, whereas the term "pediatric" is used to refer to research focused on both children and adolescents, with the understanding that the overwhelming majority of research on pediatric MDD has predominantly examined adolescents. The section ends with a framework for integrating clinical understandings with research in neuroscience.

As articulated in the second section, research on information processing provides an outstanding avenue for achieving such integration. Accordingly, this second section delineates the various information-processing functions that appear perturbed in pediatric MDD. Finally, the third section delineates the neural circuits implicated in these information-processing functions. This delineation focuses in most detail on results from brain imaging studies in human adolescents.

Clinical Features

A relatively clear consensus has emerged concerning the association between adolescence and changes in emotional processes (Steinberg et al., 2006). Adolescence is uniformly recognized as a period of life where dramatic changes in emotional processes emerge. Because some marked degree of change is consistently observed across cultures and even across mammalian species, these changes are generally agreed to reflect developmental adaptations in core underlying neural circuitry (Spear, 2000). However, debate continues about the most appropriate conceptualization of these changes in humans.

The current chapter focuses on adolescent changes in emotional processes viewed from the perspective of clinical psychopathology. Even from this relatively narrow perspective, a strong consensus has yet to emerge on the most appropriate conceptualization of emotional changes emerging in adolescence. Debate focuses on two specific aspects of emotional psychopathology during adolescence: (1) the degree to which emotional psychopathology reflects changes in a collection of discrete syndromes; and (2) the degree to which these syndromes can best be characterized as categorical entities, as opposed to extremes of continuously distributed dimensions.

Categories of Emotional Psychopathology

The most widely used system for categorizing adolescent psychopathology relies on the fourth edition, text revision, of the *Diagnostic and Statistical Manual of Mental Disorders* (DSM-IV-TR; American Psychiatric Association, 2000). This system divides emotional psychopathologies manifested in adolescence into two broad categories, each of which is further categorized into narrower conditions. These two broad categories are the anxiety and mood disorders. The anxiety disorders include conditions associated with broad patterns of fear or worry that interfere with functioning or that cause high degrees of distress. Anxiety about social circumstances appears particularly salient for adolescents, as reflected in particularly strong elevations in social anxiety disorder during this period (Pine & Klein, 2008). Because data on the neuroscience of adolescent anxiety disorders have been recently reviewed (Pine, 2003, 2007), the current chapter does not focus on these conditions. Rather, the anxiety data are reviewed only when these data address questions concerning the degree to which findings in MDD show similarities and differences with findings in anxiety disorders.

Mood disorders represent the second category of emotional psychopathologies that are highly prevalent during adolescence. Mood disorders in DSM-IV-TR are divided into two subclasses of syndromes. One category, bipolar disorders, involves episodes of both mania and major depression. Data on the neuroscience and clinical correlates of adolescent bipolar disorders have also been recently reviewed (Dickstein & Leibenluft, 2006; Leibenluft, Charney, & Pine, 2003). As a result, much as for the data on pediatric anxiety, the current chapter only considers data in bipolar disorder when such data address questions concerning the degree to which findings in MDD show similarities and differences with findings in bipolar disorder. The second category of mood disorders involves episodes of major depression, as defined for MDD, the focus of the current chapter. To meet criteria for MDD, an adolescent must exhibit a persistent alteration in mood, characterized by pervasive sadness, irritability, or a loss of interest in activities that usually are experienced as pleasurable. To be considered "pervasive," this disturbance must be present every day for the entire day for a period of at least 2 weeks. Moreover, the disturbance must be associated with at least five of nine criteria, including changes in cognitive and vegetative functions.

DSM-IV-TR provides clear criteria for distinguishing MDD from bipolar disorder and from anxiety disorders. Nevertheless, the degree to which these criteria reflect valid distinctions among forms of adolescent psychopathology continues to be debated. Probably the strongest evidence against validity emerges from data on comorbidity. That is, adolescents with MDD typically also present with prominent features of other psychopathologies, including the anxiety disorders (Angold, Costello, & Erkanli, 1999a). Moreover, longitudinal data note relatively strong associations between anxiety and MDD in adolescents followed into adulthood (Pine, Cohen, Gurley, Brook, & Ma, 1998; Beesdo et al., 2007). These data have led some to question the validity of the distinction between MDD and anxiety or other emotional psychopathologies. Nevertheless, some data reviewed below do reveal meaningful differences between MDD and other psychiatric conditions in adolescence. Moreover, adolescents with MDD can be clearly differentiated on the basis of their presenting symptoms from adolescents with either bipolar disorder or anxiety disorders, which provides clinicians a useful framework for communicating about patients. Thus DSM-IV-TR remains a valuable tool that will provide a useful basis for future classification schemes that might be more closely based on neuroscience understandings of pathophysiology.

Dimensions versus Categories

The second major area of debate concerning adolescent psychopathology focuses on the degree to which mental health problems in adolescents can be conceptualized as categorical or continuous entities. DSM-IV-TR adopts a clear categorical view of MDD and other syndromes. The strongest rationale for this approach actually emerges from practical as opposed to scientific considerations. In the face of limited resources, clinicians are forced to decide which adolescents should receive prioritization for clinical services. A nosology adopting a categorical view of behavior assists in these decisions: Children suffering from "disorders" should receive priority. In the case of MDD, the decision as to whether or not an adolescent meets criteria for MDD, as opposed to subclinical forms of depression, rests on two features. First, the depressive symptoms must cause a "clinically significant" degree of impairment, meaning that they interfere with the adolescent's ability to function. Second, the symptoms must be pervasive, in that they occur consistently over time, and they must be associated with the requisite constellations of associated features.

Although this focus on MDD as a categorical entity carries clear advantages from a service allocation perspective, epidemiological data actually suggest problems with this approach. (Parenthetically, similar concerns apply for research on anxiety disorders, and such concerns may also apply for research on bipolar disorder, although insufficient research exists to draw definitive conclusions in this final area.) For MDD, categorical diagnosis clearly carries a high risk for future psychopathology, including MDD. Nevertheless, adolescents who present with subclinical forms of depression also face very high risk for MDD during adulthood (Pine et al., 1998; Pine, Cohen, Cohen, & Brook, 1999). Moreover, from the perspective of genetics and familial risk, data provide stronger support for MDD as the extreme of a continuously distributed trait, as opposed to a categorical diagnostic entity (Kendler & Greenspan, 2006; Kendler, Hettema, Butera, Garnder, & Prescott, 2003). These data suggest that although categorical perspectives may be useful for resource allocation purposes, a continuous view of MDD may be more fully justified by the scientific data. Nevertheless, until a uniformly agreed-upon alternative conceptualization is forthcoming, the DSM-IV-TR view of MDD provides a useful definition for summarizing available literature.

MDD, Development, and Gender

MDD can be conceptualized as a classic developmental disorder. This characterization reflects the prominent changes in prevalence and demography across development. The rate of MDD is very low prior to puberty, with prevalence rarely appearing higher than a few percent. Moreover, rates of MDD appear approximately equal in boys and girls during this period. In early adolescence, however, rates of MDD show marked increases (reaching lifetime rates in the 10–15% range), and this increase arises largely from particularly marked increases in females, with much weaker increases in males. These gender-specific changes in MDD rates are paralleled by changes in hormone levels. Thus girls of the same age who differ in hormonally defined pubertal status show differential risks for MDD, with the more mature girls exhibiting the higher risk for MDD (Angold, Costello, Erkanli, & Worthman, 1999b). Nevertheless, for a few reasons, these associations between hormonal changes and rising risks of MDD in females are unlikely to reflect direct effects of hormones on brain functions associated with MDD. The magnitude of these associations is relatively weak, which is inconsistent with strong, direct hormonal effects on risk for MDD. Moreover, changes in hormonal status seem to interact with psychosocial stress, in that pubertal girls who also are exposed to stress appear to face a particularly high risk for MDD (Hankin & Abramson, 2001; Pine, Cohen, Johnson, & Brooks, 2002). This risk is elevated even in relation to that for prepubertal girls with high stress exposure or for pubertal girls with low stress exposure. Thus MDD represents a condition that emerges when developmental changes create a vulnerability to certain forms of stress.

The variety of stress that predicts high risk for MDD implicates social processes in the syndrome. MDD is particularly likely to emerge following stressors that involve changes in the social milieu (Hankin & Abramson, 2001; Nelson, Leibenluft, McClure, & Pine, 2005). Similar associations with social stressors occur for adolescent suicide, which is often associated with MDD (Gould, Fisher, Parides, Flory, & Shaffer, 1996). These data suggest that MDD and suicidal behavior can result from differences in social perception among at-risk and low-risk adolescents. Adolescents at risk for MDD may exhibit subtle differences in their responses to social signals from peers (Dodge, 1993; Rudolph & Conley, 2005). For example, adolescents at risk for MDD may be more likely to detect signs of social disapproval or to respond with mood deterioration following the detection of these signs. In the face of social stress, those adolescents with perturbed social processing may be most likely to develop MDD. These differences may also relate to gender differences in the risk for MDD, in that females may be more likely than males to show patterns of social information processing that place them at risk for MDD (Hankin & Abramson, 2001). From this perspective, developmental changes are hypothesized to account for increasing MDD risk by altering underlying neural circuitry associated with social information processing (Nelson et al., 2005).

Social processes associated with gender-related differences in risk for MDD emerge most prominently during adolescence. However, other processes associated with the gender-related differences in risk for MDD manifest themselves prior to puberty. Specifically, whereas rates of MDD are higher in girls than boys only after puberty, rates of anxiety are higher in girls than boys long before puberty. This includes both rates of overt anxiety disorders and scores on anxiety symptom scales among nonaffected adolescents. Moreover, childhood anxiety emerges as a particularly robust predictor of adolescent and adult MDD (Hankin & Abramson, 2001; Pine et al., 1998). Thus children with anxiety, who face a high risk for MDD, show abnormalities even before puberty in the degree to which they process threat. These prepubertal differences in threat processing may represent an early manifestation of the female-specific increased risk for MDD. Because anxiety disorders are conceptualized as perturbations in threat processing, these data suggest that risk for MDD is reflected in perturbed information processing for threats.

A Neuroscience Framework for Adolescent MDD

Debate concerning the validity of current categorical definitions of MDD has fueled efforts to develop alternative frameworks for conceptualizing MDD. Research on both social information processing and threat processing emphasizes the importance of developing a framework focusing on information-processing functions. Accordingly, information processing lies at the heart of the current framework. Figure 20.2 presents a pictorial illustration of this framework. Causal processes related to the pathophysiology of MDD begin from the left side of the figure. As such, this framework recognizes the role of both genes and the environment, through main effects and interactions, in contributing to MDD (Gross & Hen, 2004). The interactions do not directly cause MDD, but they influence risk by altering functioning in a specific neural circuit.

Perturbed functioning in this neural circuit manifests itself as perturbations in information processing, as illustrated in the middle box of this figure. In turn, clinical phenotypes are associated with these perturbations in information processing, as indicated in the box on the right. As noted above, DSM-IV-TR emphasizes the categorical perspective of the MDD phenotype, but conceptualizations of the phenotypes are likely to change as relationships with information processing and brain function are examined. Some perturbations ultimately may map closely onto diagnosis, as currently conceptualized in DSM-IV-TR. Other perturbations, however, may map onto dimensional constructs, such as one or another depressive symptom (e.g., anhedonia) or personality construct (e.g., neuroticism). Figure 20.2 lists some potentially interesting phenotypes. Finally, some information-processing perturbations may map more closely onto behaviors associated with risk for MDD, as opposed to overt expressions of MDD or other clinical phenotypes. For example, behavioral inhibition is a temperament type manifested within the first years of life that has been linked both to perturbed social appraisal and to perturbed threat processing. This temperament type also predicts risk for MDD (Caspi, Moffitt, Newman, & Silva, 1996), although most research emphasizes its connections with anxiety disorders. Thus elucidating the pathophysiology of MDD and related conditions will require research that ties functioning within a specific neural circuit

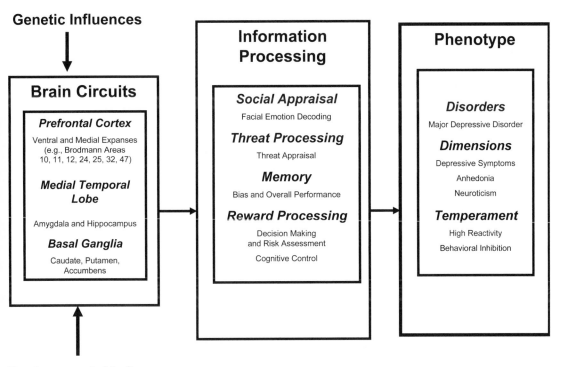

FIGURE 20.2. An illustrative framework for conceptualizing the relationship among brain circuits, information processing, and phenotypes related to pediatric MDD.

to specific information-processing functions, which in turn must be linked to specific sets of phenotypic expressions. The next two sections of this chapter illustrate the manner in which available research links perturbed information processing and perturbed neural circuit function specifically to MDD.

The framework illustrated in Figure 20.2 depicts connections among genes, environmental factors, brain function, information processing, and clinical states. As such, this framework adopts a multitiered approach, recognizing the many levels at which individual differences in behavior and in psychology are manifested. A major issue not illustrated in Figure 20.2 concerns the manner in which each component of the figure changes with development, as well as the manner in which relationships among these components change. Very few studies have begun to map relationships among brain function, information-processing functions, and depressive symptoms in child and adolescent populations at isolated stages of development, let alone across different stages. The current chapter reviews data from the few available studies in this area, which are predominantly cross-sectional. As these data begin to generate consistent conclusions and begin to spawn longitudinal research, refinements in the framework depicted in Figure 20.2 will need to document the manner in which relationships among factors in the figure change during development. For example, this framework will need to consider the manner in which neural circuits and associated information-processing functions implicated in pediatric anxiety relate to the circuits and information-processing functions implicated in both pediatric and adult MDD.

Information Processing

Figure 20.2 lists four information-processing constructs that are linked in at least some literature both to one or another measure of pediatric MDD and to functioning within brain structures listed in the figure. Because of space limitations, the current section only broadly summarizes the nature of findings in each of these four areas and describes a few key representative studies. After summarizing this work, the section concludes by describing potential hormonal and neurochemical modulators of these information-processing functions. Data from brain imaging studies of MDD are reviewed in the next section.

Social Information Processing

Many clinically focused reports have examined social functioning in adolescent MDD, noting relationships between symptoms of MDD and problems in social relationships. However, relatively few studies have relied on an information-processing approach to delineate the nature of perturbed social information processing in MDD. Probably the most extensive research in this area relies on tests of facial emotion processing (see de Haan & Matheson, Chapter 6, this volume, for further discussion of facial emotion processing). Specifically, studies of facial emotion processing in MDD focus on the accuracy with which individuals appropriately label or classify photographs of prototypical facial emotion displays. From this perspective, data in adult MDD document perturbed facial emotion processing, manifested as a biased tendency to rate facial emotions as more negative than healthy adults rate them (Gur et al., 1992). Fewer studies have examined the relationship between pediatric MDD and this aspect of facial emotion processing. Nevertheless, the few available studies in this area do find signs of bias in adolescent MDD similar to those reported in adult MDD. Namely, this work finds that symptoms of MDD in children and adolescents predict a tendency to misclassify facial emotion displays depicted in photographs (Arsenio, Sesin, & Siegel, 2004; Lenti, Giacobbe, & Pegna, 2000). Moreover, at least one study found abnormalities specifically in the processing of fear faces (Lenti et al., 2000). As reviewed below, other work links MDD to the amygdala, a brain region previously linked specifically to the processing of fear faces. Thus findings noting deficient ability to appropriately classify facial displays of fear in MDD are consistent with broader literatures documenting relationships among anxiety, MDD, processing of fear faces, and the neural circuitry engaged during the viewing of fearful facial expressions. Of note, data in bipolar disorder also implicate perturbed facial emotion processing in both adult and pediatric forms of the illness (McClure et al., 2005). When results are compared across the various studies, preliminary evidence suggests that MDD and bipolar disorder show biases in the manner in which they rate different subsets of facial emotion classes. Nevertheless, few data are available on this issue, and future research may show that deficits in facial emotion classification exhibit associations with mood disorders as a broad class, as opposed to MDD or bipolar disorder specifically.

   Deficits in facial emotion identification are thought to reflect broader tendencies to misread social cues in MDD (Dodge, 1993). Theoretical perspectives suggest that these deficits actually contribute to pathophysiology by placing children at risk for MDD. The few available studies examining the relationships between familial risk for MDD and perturbed facial emotion processing provide mixed support for such perspectives (Arsenio et al., 2004; Pine et al., 2005). Other work suggests that perturbed facial emotion processing is a result of adverse social experiences, such as abusive parenting (Pollak, Cicchetti, Hornung, & Reed, 2000).

Regardless of the mechanism, the importance of work in this area relates to the strong literature delineating neural correlates of facial emotion processing, focusing on the amygdala and other structures implicated in MDD (Haxby, Hoffman, & Gobbini, 2002). This provides a strong rationale for using such processing as a paradigm in brain imaging studies of MDD.

Threat Processing

As noted above, clinical studies implicate abnormal threat processing in MDD. These studies show that childhood clinical problems with anxiety are associated with parental histories of MDD, and with risk for MDD in these children themselves as they mature toward adulthood. Such clinical findings generate interest in the nature of threat-processing profiles on experimental paradigms among children and adolescents with MDD. Such experimental paradigms examine diverse aspects of threat processing, including attention allocation, startle potentiation, verbal threat classification, and measures of experimental avoidance. Moreover, work on the neural correlates of MDD in adults generates further interest in the relationship between threat processing and pediatric MDD, given the overlap in the neural correlates of MDD and threat processing (Drevets, 2003; Sheline, 2003). Nevertheless, findings on threat-processing profiles among children and adolescents studied in the laboratory provide relatively weak evidence of perturbed threat processing in MDD. As reviewed elsewhere (Pine, 2007), a wealth of work documents two sets of threat-processing perturbations in pediatric anxiety disorders: perturbed attention allocation to threats, and a biased tendency to appraise one or another situation as threatening. Neither perturbation occurs consistently in studies of pediatric MDD, suggesting dissociable threat-processing perturbations in pediatric anxiety relative to MDD.

This work also notes that relatively specific threat appraisal biases emerge in specific anxiety disorders. It is possible that pediatric MDD is associated with a particular form of threat-processing abnormality that differs from that found in pediatric anxiety disorders. Consistent with some abnormality in threat processing, recent work suggests that offspring of parents with MDD show a heightened physiological and anxiety symptom response during experiments involving expectation of an unpleasant air blast or exposure to darkness (Grillon et al., 2005).

Memory

Perturbations in memory are among the most consistent correlates of adult MDD. Two forms of memory perturbation in MDD emerge with some consistency. First, MDD in adulthood is characterized by an overall reduction in performance on standard tests of declarative memory; however, the most consistent evidence of these abnormalities emerges in studies among relatively older depressed individuals (Zakzanis, Leach, & Kaplan, 1998). Second, MDD in adulthood is characterized by a biased retention of emotionally negative items on memory tests that involve encoding of both neutral and emotionally evocative stimuli (Hertel, 2004). These data in adults note a double dissociation with adult anxiety, which involves attention but not memory biases, unlike MDD, which involves memory but not attention biases (Macleod & Matthews, 2004).

Data on emotional but not neutral memory in pediatric MDD generally support the data in adult MDD. In terms of overall reductions in memory on standardized tests using neutral stimuli, the data do implicate mnemonic perturbations in pediatric anxiety (Vasa et al., 2007).

However, adolescent MDD is associated with normal performance on standard memory tests using neutral material, consistent with data in adults, which only find abnormalities among relatively older subjects (Frost, Moffitt, & McGee, 1989). In terms of memory for emotional material, findings in pediatric MDD generally confirm findings in adult MDD, where consistent evidence of biased recall emerges among youth as it does among adults (Neshat-Doost, Taghavi, Moradi, Yule, & Dalgleish, 1998; Pine et al., 2004).

Reward Processing, Decision Making, and Cognitive Control

As it has for the circuitry involved in facial emotion processing, considerable work in basic cognitive neuroscience has delineated the neural circuitry engaged in reward processing, decision making, and cognitive control (Ernst, Pine, & Hardin, 2006). This has stimulated efforts to delineate the relationships between pediatric MDD and these constructs (Forbes & Dahl, 2005). Reward processing consists of multiple cognitive subprocesses associated with evaluating behavioral choice, weighing alternatives, reacting to feedback, and using this feedback to inform subsequent behavior. Work on the manner in which pediatric MDD relates to each of these subprocesses has only just begun; however, at this writing, pediatric MDD has been linked to perturbations in each. In research on the initial process of decision making, emotional information has been shown to influence allocation of attention during decision making to a greater extent in adolescents with MDD than in those without MDD (Kyte, Goodyer, & Sahakian, 2005; Ladouceur et al., 2006). In research on implementation of behavioral choice, pediatric MDD also has been linked to alterations in relevant processes. For example, these alterations can be manifested as a failure to chose the most rewarding options on tasks where subjects must weigh the probability and magnitude of potential rewards associated with distinct choices (Forbes, Shaw, & Dahl, 2007). Finally, pediatric MDD has been linked to perturbed regulatory responses either immediately following motor execution or following feedback concerning choice behavior (Jazbec, McClure, Hardin, Pine, & Ernst, 2005; Guyer et al., 2006).

One set of questions emerging from these data concerns the degree to which impairments in decision making and cognitive control emerge on tasks that present only neutral but not emotional information. Distinct neural circuitry is engaged in the process of decision making on tasks that involve emotional as opposed to nonemotional material. Although relatively few studies in pediatric MDD have contrasted decision-making skill across emotional and nonemotional contexts, signs of perturbed executive control emerge most consistently on tasks that involve an emotional component (Kyte et al., 2005; Ladouceur et al., 2006).

A second set of questions concerns the degree to which decision-making deficits reflect risk factors for MDD as opposed to downstream effects of living with MDD. From a theoretical perspective, deficient reward-processing and decision-making skills are viewed as risk factors for pediatric MDD, rather than as consequences of the condition (Forbes et al., 2007). Emerging data have begun to support this contention. Thus children born to parents with MDD exhibit perturbations on decision-making tasks similar to those found in children and adolescents affected by MDD. Specifically, risk for MDD has been linked to inefficient use of emotion regulation strategies during the process of decision making, as well as to more intense negative affect following errant decisions (Murray, Woolgar, Cooper, & Hipwell, 2001; Silk, Shaw, Skuban, Oland, & Kovacs, 2006). Similarly, deficient decision-making skill relates to both current and future MDD symptoms in youth (Forbes et al., 2007).

Modulation of Information Processing

Figure 20.2 suggests that information-processing perturbations result from genetic and environmental influences on neural architecture. Understanding the mechanisms by which these influences emerge may provide crucial insights for both therapeutics and prediction of risk. Two sets of chemical systems have been implicated as possible mediators in the link that runs from genes and the environment through the brain to shape information processing. These comprise systems involved in regulation of serotonin (5-HT) and the hypothalamic–pituitary–adrenal (HPA) axis.

A wealth of research implicates the 5-HT system both in adult MDD and in developmental processes associated with the condition (Davidson et al., 2002; Gross & Hen, 2004). Genetically mediated perturbations in 5-HT are thought to place individuals at risk for MDD, with MDD ultimately manifesting itself in individuals with 5-HT abnormalities who also are exposed to stress (Caspi & Moffitt, 2006). Some data also suggest that early life stress can alter 5-HT function, such that environmental effects may modulate the 5-HT system with reverberating effects on brain function (Gross & Hen, 2004; Pine et al., 1997). These associations among 5-HT, brain function, and risk for MDD are thought to emerge from 5-HT's effects on information-processing functions implicated in MDD. Available work in adults provides some support for these views (Marsh et al., 2006). However, most work in this area examines adults (Davidson et al., 2002), and far more work is needed focusing specifically on the relationships in children and adolescents among 5HT, depressive symptoms, and information processing.

Much like work linking the 5-HT system to adult MDD, considerable work implicates HPA axis dysfunction to risk for MDD in adults. In general, this work documents signs of aberrant negative feedback in adult MDD, which suggests abnormal stress regulation capacity in the condition (Davidson et al., 2002). As work on the 5-HT system does, basic science work documents robust developmental influences on HPA axis function, such that early life experiences with stress exert robust, long-term influences on the axis (Gross & Hen, 2004). This has led some to suggest that HPA axis abnormalities in adult MDD reflect stress-related experiential influences on the axis during development. Nevertheless, data linking the HPA axis to either ongoing MDD or risk for MDD in children and adolescents appear inconsistent (Terleph et al., 2006; Young, Vazquez, Jiang, & Pfeffer, 2006). Moreover, relatively little work has examined associations between HPA axis activity and information processing. Nevertheless, some evidence links the HPA axis to abnormal memory performance in adults and abnormal affective responses in youth (Terleph et al., 2006).

Neural Processes

The final major section of this chapter reviews the manner in which perturbed neural function has been linked to pediatric MDD. The section begins by delineating a set of neural circuits implicated in MDD. This delineation follows largely from data in adult MDD and in work on neural responses in rodents and nonhuman primates to various forms of stress, threats, and rewards. Next, the section summarizes literature from brain imaging studies in pediatric MDD and related conditions. This includes a review of data from structural magnetic resonance imaging (MRI) studies, functional MRI (fMRI) work on information-processing constructs discussed in the preceding section, and studies relying on other techniques, including quantitative electroencephalography (qEEG).

Neural Circuitry

Considerable work in rodents and nonhuman primates delineates neural circuitry engaged by various emotionally salient events. This work notes many parallels between the circuitry associated with adult MDD and that associated with emotional processes in rodents and nonhuman primates (Davidson et al., 2002; Gross & Hen, 2004; Nelson et al., 2005; Sheline, 2003). Although many details remain to be clarified, Plate 20.1 summarizes core components of this circuitry, which can be roughly parsed into three interacting components.

First, considerable work both in animal models and in adults implicates perturbed medial temporal lobe (MTL) structure and function in MDD. Plate 20.1 illustrates the two most important MTL components: the amygdala and hippocampus. In general, the amygdala serves to mediate psychological processes in which the emotional salience of stimuli or events affects an evaluation of overall salience. These processes are engaged in diverse circumstances requiring modulations of attention or learning and memory, and consideration of events with both positive and negative valence (Baxter & Murray, 2002; Blair, Mitchell, & Blair, 2005; Davis & Whalen, 2001; Davidson et al., 2002). This suggests that one vital role played by the amygdala is to regulate attention during processing of emotional stimuli to facilitate other psychological processes, such as decision making and memory formation. The amygdala exhibits particularly robust anatomical and functional connections with the hippocampus, the other key component of the MTL. Memory represents the psychological process most consistently tied to hippocampal function, and hippocampal–amygdala interactions are thought to underlie aspects of emotional memory formation. The structure also has been implicated in various other emotional processes, including the response to antidepressant medications (Santarelli et al., 2003).

Second, other work implicates the basal ganglia in affective modulation of behavior. Much of this work focuses on one particular subcomponent of the basal ganglia, the nucleus accumbens, as illustrated in Plate 20.1. Historically, this set of structures has been most consistently implicated in aspects of reward processing and decision making (Nestler & Carlezon, 2006). Recent work also suggests a broader role in error monitoring and related processes engaged during early stages of learning (Miller & Cohen, 2001). The consistent evidence implicating reward in MDD has led to considerable focus on the role of this structure in MDD (Bonelli, Kapfhammer, Pillay, & Yurgelun-Todd, 2006).

Third, considerable work focused on the role played by the prefrontal cortex (PFC) in MDD. Although some controversy persists concerning the most appropriate manner in which to parse PFC functions, most current perspectives view this structure as possessing three interrelated functional units: a dorsal, ventral, and medial PFC region, each of which can be further divided (Miller & Cohen, 2001). As a whole, the PFC has been reliably implicated in complex aspects of motor planning and executive function, with distinct features of such psychological processes being assigned to one or another specific subregion. For mediation of emotional processes, lesion studies most consistently implicate the ventral PFC, including both its medial and lateral components. These regions encompass a range of Brodmann areas (BAs), including BAs 10, 11, 12, 24, 25, 32, and 47. These regions show rich anatomical connections with the amygdala, leading to strong interactions between the amygdala and ventral PFC, as suggested by the arrows in Plate 20.1. These interactions are thought to facilitate emotion regulation processes, such as extinction or other forms of contextual modulation of emotional processes (Miller & Cohen, 2001; Davidson, 2002). The dorsal PFC shows no such rich anatomical connectivity with the amygdala, and this region has been most consistently implicated in nonemotional planning processes, such as spatial working memory. Nevertheless, strong anatomical connections between the ventral and dorsal PFC provide the substrate through which processes such as planning or working memory can affect and be affected by processes instantiated in the ventral PFC.

Morphometry

Most MRI studies of the relationship between brain structure and MDD have examined components of the neural circuitry described in the preceding section. Most of this work has examined brain structure in adult patients with MDD, though emerging studies are attempting to extend this work to pediatric MDD. Probably the strongest findings in adult MDD note reduced hippocampal volume in the condition, with enough findings emerging to support a quantitative meta-analysis (Videbech & Ravnkilde, 2004). Data for the amygdala appear more variable, with studies reporting both increases and decreases in amygdala volume among adults with MDD (Hastings, Parsey, Oquendo, Aragno, & Mann, 2004; Lange & Irle, 2004). In pediatric MDD, however, very few studies have considered these issues. One paper did report reduced amygdala but not hippocampal volume in pediatric MDD (Rosso et al., 2005), whereas a second found no differences in either structure (MacMillan et al., 2003), and a third found reduced volume only in the hippocampus (MacMaster & Kusumakar, 2004). Similarly, one study in pediatric anxiety found reduced amygdala volume and another found increased amygdala volume, with neither study finding any evidence of changes in the hippocampus (see Pine, 2007, for a review).

In studies of other brain regions, data for basal ganglia volumes appear quite inconsistent in adults, let alone childhood MDD (Bonelli et al., 2006). The PFC represents the only other brain region where evidence consistently implicates perturbed brain structure in MDD. Among adults with some forms of early-onset, highly familial MDD, reductions in ventral medial aspects of the PFC emerge as a reasonably consistent finding, even among young adults (Botteron, Raichle, Drevets, Heath, & Todd, 2002; Hastings et al., 2004; Sheline, 2003). However, the few available studies in pediatric MDD have not examined structural integrity of this specific region; rather, they report findings that generally have not emerged in studies among adults (Nolan et al., 2002).

Finally, the degree to which any of these structural abnormalities may represent risk factors for MDD, as opposed to downstream complications of suffering with the condition, remains unclear. Evidence exists to support both possibilities. In terms of perspectives favoring the view of morphometric findings as risk factors, a genetic polymorphism of the 5-HT reuptake transporter implicated in MDD also predicts amygdala and PFC volumetric abnormalities. On the other hand, in terms of perspectives favoring a view focusing on downstream complications, hippocampal abnormalities appear most strongly related to long-term, chronic MDD (Sheline, 2003).

Facial Emotion Processing and Memory Encoding

Evidence of social dysfunction in MDD, combined with signs of perturbed facial emotion processing in the disorder, generates considerable interest in the neural circuitry of facial emotion-processing deficits in pediatric MDD. Moreover, facial emotion processing engages the amygdala in healthy adults, generating further interest in neural correlates of facial emotion processing in pediatric MDD. Two studies have considered this issue.

First, one study compared amygdala activation in five adolescents with MDD, five with anxiety, and five with no psychopathology, noting signs of reduced amygdala activation in the subjects with MDD (Thomas et al., 2001). Of note, this study was based on a block design fMRI paradigm with passive viewing of the facial emotion stimuli. This leaves unanswered questions about the specific psychological processes engaged during face viewing.

Second, in the only other fMRI study of pediatric MDD, *enhanced* amygdala activation was detected (Roberson-Nay et al., 2006). Features of this second study may provide insights on factors contributing to the discrepant findings. Plate 20.2 illustrates the nature of the design and the findings in this second study. Unlike the first study, this report relied on an event-related paradigm during the memory encoding of evocative faces. This method allowed the parsing of brain activation data into scans collected during the viewing of faces correctly recognized after scanning, as opposed to scans collected during the viewing of faces not correctly recognized. As shown in Plate 20.2, evidence of enhanced amygdala response to facial emotion displays only emerged for faces that were successfully encoded into memory. For unsuccessfully encoded faces, the opposite trend emerged. Finally, these findings implicating enhanced amygdala response in MDD during memory encoding contrast with findings in pediatric bipolar and anxiety disorders, where enhanced amygdala response occurs under distinct conditions (Rich et al., 2006; McClure et al., 2007).

Reward Processing

Findings of considerable interest have also emerged from both basic science research and data on reward processing in the laboratory. Only one study has examined the association between pediatric MDD and neural circuitry function engaged during reward processing (Forbes et al., 2006a). Consistent with data in adult MDD (Epstein et al., 2006), this study found evidence of reduced striatal engagement to reward stimuli in pediatric MDD. Of note, much like the study of facial emotion processing described above, this report relied on an event-related design that allowed the investigators to isolate the specific aspects of reward processing associated with deficient striatal engagement. Moreover, these data support suggestions of a distinction between pediatric MDD and anxiety, where evidence of enhanced as opposed to reduced striatal engagement to rewarding events emerges (Pine, 2007).

Other Brain Imaging Findings

Clearly, research on brain morphometry and fMRI studies of facial emotion processing or reward represent research areas worthy of considerable attention. Nevertheless, a few other findings in various aspects of brain imaging also merit consideration. Each of these areas is only mentioned briefly, either because limited imaging research has examined patients with MDD or because alternative techniques are likely to be used to extend these findings in the future.

In regard to areas with limited previous research, considerable work has examined aspects of cognitive control, but no fMRI studies have employed cognitive control paradigms in MDD. Moreover, given data implicating perturbed emotion regulation ability in MDD, studies are needed examining the brain regions engaged by cognitive control experiments in pediatric MDD. Similarly, work is only beginning that uses magnetic resonance spectroscopy (MRS), a technique that noninvasively generates estimates of chemical concentrations in specific brain regions. The few available MRS studies in pediatric MDD note perturbations in PFC components required for cognitive control; these findings further emphasize the need for other imaging work focused on this region (Caetano et al., 2005; Rosenberg et al., 2005).

In contrast, qEEG represents a research area where alternative techniques might prove useful in attempting to extend recent findings on pediatric MDD. qEEG provides an easily acquired quantitative index of neural electrical activity. The main disadvantage of the technique is its limited spatial resolution. Nevertheless, the ease with which data can be acquired facilitates work with relatively large samples in many laboratories. The main findings in qEEG studies of pediatric MDD demonstrate abnormalities in the degree to which qEEG power varies across the two brain hemispheres. Two main patterns of aberrant laterality have been associated with pediatric MDD—both in children or adolescents with ongoing MDD, and in those at high risk for the condition because of either parental history or early childhood temperament. One profile involves a pattern of enhanced brain activity over right frontal regions, relative to left frontal regions (Davidson et al., 2002; Forbes et al., 2006b). The other profile involves a pattern of enhanced brain activity over left posterior regions (Kentgen et al., 2000; Bruder et al., 2005). For both profiles, the data are relatively well replicated, and are provocative in their suggestion that perturbed laterality represents a marker not only of ongoing MDD but of underlying risk. The main limitation of this work concerns the unclear relationship between perturbations in qEEG profiles and functioning within specific brain regions, as reflected in information-processing profiles. Future work might probe functioning in specific brain regions that typically become engaged differentially across the two hemispheres by cognitive processes implicated in MDD.

Conclusions

The current chapter focuses on three aspects of research linking neuroscience to developmental psychopathology. In the first section, the chapter reviews conceptualizations of relevant phenotypes, with a focus on relating phenotype to perturbations in neural systems through a focus on information processing. The second section summarizes data from a set of information-processing functions as they relate to pediatric MDD. Finally, the third section delineates the specific neural circuitry implicated in adult MDD and reviews neuroimaging data examining findings in pediatric MDD.

This review clearly illustrates the considerable progress that has been made in research over the past 20 years on the development and pathophysiology of MDD. Such research demonstrates the strong developmental nature of MDD, now conceptualized as a condition that typically has its roots in childhood and adolescence. Moreover, such research suggests that a focus on information processing in future work may precisely elucidate the developmental events that underlie the associations between behaviors manifested during childhood and risk for MDD in adulthood. Specifically, considerable work is needed demonstrating the manner in which information-processing functions show cross-sectional and longitudinal associations, both with symptoms of MDD and with perturbed brain function. Such research holds the promise of ultimately mapping the neurodevelopmental unfolding of MDD.

References

American Psychiatric Association. (2000). *Diagnostic and statistical manual of mental disorders* (4th ed., text rev.). Washington, DC: Author.

Angold, A., Costello, E. J., & Erkanli, A. (1999a). Comorbidity. *Journal of Child Psychology and Psychiatry, 40*(1), 57–87.

Angold, A., Costello, E. J., Erkanli, A., & Worthman, C. M. (1999b). Pubertal changes in hormone levels and depression in girls. *Psychological Medicine, 29*(5), 1043–1053.

Arsenio, W. F., Sesin, M., & Siegel, L. (2004). Emotion-related abilities and depressive symptoms in Latina mothers and their children. *Development and Psychopathology, 16*(1), 95–112.

Baxter, M. G., & Murray, E. A. (2002). The amygdala and reward. *Nature Reviews Neuroscience, 3*(7), 563–573.

Beesdo, K., Bittner, A., Pine, D. S., Stein, M. B., Hofler, M., Lieb, R., et al. (2007). Incidence of social anxiety disorder and the consistent risk for secondary depression in the first three decades of life. *Archives of General Psychiatry, 64*(8), 903–912.

Blair, J., Mitchell, D., & Blair, K. (2005). *The psychopath: Emotion and the brain.* Oxford: Blackwell.

Bonelli, R. M., Kapfhammer, H. P., Pillay, S. S., & Yurgelun-Todd, D. A. (2006). Basal ganglia volumetric studies in affective disorder: What did we learn in the last 15 years? *Journal of Neural Transmission, 113*(2), 255–268.

Botteron, K. N., Raichle, M. E., Drevets, W. C., Heath, A. C., & Todd, R. C. (2002). Volumetric reduction in left subgenual prefrontal cortex in early onset depression. *Biological Psychiatry, 51*(4), 342–344.

Bruder, G. E., Tenke, C. E., Warner, V., Noumura, Y., Grillon, C., Hille, J., et al. (2005). Electroencephalographic measures of regional hemispheric activity in offspring at risk for depressive disorders. *Biological Psychiatry, 57*(4), 328–335.

Caetano, S. C., Fonseca, M., Olvera, R. L., Nicoletti, M., Hatch, J. P., Stanley, J. A., et al. (2005). Proton spectroscopy study of the left dorsolateral prefrontal cortex in pediatric depressed patients. *Neuroscience Letters, 384*(3), 321–326.

Caspi, A., & Moffitt, T. E. (2006). Gene–environment interactions in psychiatry: Joining forces with neuroscience. *Nature Reviews Neuroscience, 7*(7), 583–590.

Caspi, A., Moffitt, T. E., Newman, D. L., & Silva, P. A. (1996). Behavioral observations at age 3 years predict adult psychiatric disorders: Longitudinal evidence from a birth cohort. *Archives of General Psychiatry, 53*(11), 1033–1039.

Davidson, R. J. (2002). Anxiety and affective style: Role of prefrontal cortex and amygdala. *Biological Psychiatry, 51*(1), 68–80.

Davidson, R. J., Lewis, D. A., Alloy, L. B., Amaral, D. G., Bush, G., Cohen, J. D., et al. (2002). Neural and behavioral substrates of mood and mood regulation. *Biological Psychiatry, 52*(6), 478–502.

Davis, M., & Whalen, P. J. (2001). The amygdala: Vigilance and emotion. *Molecular Psychiatry, 6*(1), 13–34.

Dickstein, D. P., & Leibenluft, E. (2006). Emotion regulation in children and adolescents: Boundaries between normalcy and bipolar disorder. *Development and Psychopathology, 18*(4), 1105–1131.

Dodge, K. A. (1993). Social-cognitive mechanisms in the development of conduct disorder and depression. *Annual Review of Psychology, 44*, 559–584.

Drevets, W. C. (2003). Neuroimaging abnormalities in the amygdala in mood disorders. *Annals of the New York Academy of Sciences, 985*, 420–444.

Epstein, J., Pan, H., Kocsis, J. H., Yang, Y., Butler, T., Chusid, J., et al. (2006). Lack of ventral striatal response to positive stimuli in depressed versus normal subjects. *American Journal of Psychiatry, 163*(10), 1784–1790.

Ernst, M., Pine, D. S., & Hardin, M. (2006). Triadic model of the neurobiology of motivated behavior in adolescence. *Psychological Medicine, 36*(3), 299–312.

Forbes, E. E., & Dahl, R. E. (2005). Neural systems of positive affect: Relevance to understanding child and adolescent depression? *Development and Psychopathology, 17*(3), 827–850.

Forbes, E. E., May, J. C., Siegle, G. J., Ladouceur, C. D., Ryan, N. D., Carter, C. S., et al. (2006a). Reward-related decision-making in pediatric major depressive disorder: An fMRI study. *Journal of Child Psychology and Psychiatry, 47*(10), 1031–1040.

Forbes, E. E., Shaw, D. S., & Dahl, R. E. (2007). Alterations in reward-related decision making in boys with recent and future depression. *Biological Psychiatry, 61*(5), 633–639.

Forbes, E. E., Shaw, D. S., Fox, N. A., Cohn, J. F., Silks, J. S., & Kovacs, M. (2006b). Maternal depression, child frontal asymmetry, and child affective behavior as factors in child behavior problems. *Journal of Child Psychology and Psychiatry, 47*(1), 79–87.

Frost, L. A., Moffitt, T. E., & McGee, R. (1989). Neuropsychological correlates of psychopathology in an unselected cohort of young adolescents. *Journal of Abnormal Psychology, 98*(3), 307–313.

Gould, M. S., Fisher, P., Parides, M., Flory, M., & Shaffer, D. (1996). Psychosocial risk factors of child and adolescent completed suicide. *Archives of General Psychiatry, 53*(12), 1155–1162.

Grillon, C., Warner, V., Hille, J., Merikangas, K. R., Bruder, G. E., Tenke, C. E., et al. (2005). Families at high and low risk for depression: A three-generation startle study. *Biological Psychiatry, 57*(9), 953–960.

Gross, C., & Hen, R. (2004). The developmental origins of anxiety. *Nature Reviews Neuroscience, 5*(7), 545–552.

Gur, R. C., Erwin, R. J., Gur, E. E., Zwil, A. S., Heimberg, C., & Kraemer, H. C. (1992). Facial emotion discrimination: II. Behavioral findings in depression. *Psychiatry Research, 42*(3), 241–251.

Guyer, A. E., Kaufman, J., Hodgdon, H. B., Masten, C. L., Jazbec, S., Pine, D. S., et al. (2006). Behavioral alterations in reward system function: The role of childhood maltreatment and psychopathology *Journal of the American Academy of Child and Adolescent Psychiatry, 45*(9), 1059–1067.

Hankin, B. L., & Abramson, L. Y. (2001). Development of gender differences in depression: An elaborated cognitive vulnerability–transactional stress theory. *Psychological Bulletin, 127*(6), 773–796.

Hastings, R. S., Parsey, R. V., Oquendo, M. A., Arango, V., & Mann, J. (2004). Volumetric analysis of the prefrontal cortex, amygdala, and hippocampus in major depression. *Neuropsychopharmacology, 29*(5), 952–959.

Haxby, J. V., Hoffman, E. A., & Gobbini, M. I. (2002). Human neural systems for face recognition and social communication. *Biological Psychiatry, 51*(1), 59–67.

Hertel, P. (2004). Memory for emotional and nonemotional events in depression: A question of habit? In J. P. Reisberg (Ed.), *Memory and emotion* (pp. 186–216). New York: Oxford University Press.

Jazbec, S., McClure, E., Hardin, M., Pine, D. S., & Ernst, M. (2005). Cognitive control under contingencies in anxious and depressed adolescents: An antisaccade task. *Biological Psychiatry, 58*(8), 632–639.

Kendler, K. S., & Greenspan, R. J. (2006). The nature of genetic influences on behavior: Lessons from "simpler" organisms. *American Journal of Psychiatry, 163*(10), 1683–1694.

Kendler, K. S., Hettema, J. M., Butera, F., Gardner, C. O., & Prescott, C. A. (2003). Life event dimensions of loss, humiliation, entrapment, and danger in the prediction of onsets of major depression and generalized anxiety. *Archives of General Psychiatry, 60*(8), 789–796.

Kentgen, L. M., Tenke, C. E., Pine, D. S., Fong, R., Klein, R. G., & Bruder, G. E. (2000). Electroencephalographic asymmetries in adolescents with major depression: Influence of comorbidity with anxiety disorders. *Journal of Abnormal Psychology, 109*(4), 797–802.

Kyte, Z. A., Goodyer, I. M., & Sahakian, B. J. (2005). Selected executive skills in adolescents with recent first episode major depression. *Journal of Child Psychology and Psychiatry, 46*(9), 995–1005.

Ladouceur, C. D., Dahl, R. E., Williamson, D. E., Birmaher, B., Axelson, D. A., Ryan, N. D., et al. (2006). Processing emotional facial expressions influences performance on a go/nogo task in pediatric anxiety and depression. *Journal of Child Psychology and Psychiatry, 47*(11), 1107–1115.

Lange, C., & Irle, E. (2004). Enlarged amygdala volume and reduced hippocampal volume in young women with major depression. *Psychological Medicine, 34*(6), 1059–1064.

Leibenluft, E., Charney, D. S., & Pine, D. S. (2003). Researching the pathophysiology of pediatric bipolar disorder. *Biological Psychiatry, 53*(11), 1009–1020.

Lenti, C., Giacobbe, A., & Pegna, C. (2000). Recognition of emotional facial expressions in depressed children and adolescents. *Perceptual and Motor Skills, 91*(1), 227–236.

Macleod, C., & Mathews, A. (2004). Selective memory effects in anxiety disorders: An overview of research findings and their implications. In J. P. Reisberg (Ed.), *Memory and emotion* (pp. 155–185). New York: Oxford University Press.

MacMaster, F. P., & Kusumakar, V. (2004). Hippocampal volume in early onset depression. *BMC Medical Education, 2*, 2.

MacMillan, S., Szeszko, P. R., Moore, G. J., Madden, R., Lorch, E., Ivey, J., et al. (2003). Increased amygdala:hippocampal volume ratios associated with severity of anxiety in pediatric major depression. *Journal of Child and Adolescent Psychopharmacology, 13*(1), 65–73.

Marsh, A. A., Finger, E. C., Buzas, B., Soliman, N., Richell, R. A., Vythilingham, M., et al. (2006). Impaired recognition of fear facial expressions in 5-HTTLPR S-polymorphism carriers following tryptophan depletion. *Psychopharmacology (Berlin), 189*(3), 387–394.

McClure, E. B., Monk, C. S., Nelson, E. E., Parish, J. M., Adler, A., Blaire, R. J., et al. (2007). Abnormal attention modulation of fear circuit function in pediatric generalized anxiety disorder. *Archives of General Psychiatry, 64*(1), 97–106.

McClure, E. B., Treland, J. E., Snow, J., Schmajuk, M., Dickestein, D. P., Towbin, K. E., et al. (2005). Deficits in social cognition and response flexibility in pediatric bipolar disorder. *American Journal of Psychiatry, 162*(9), 1644–1651.

Miller, E. K., & Cohen, J, D. (2001). An integrative theory of prefrontal cortex function. *Annual Review of Neuroscience, 24*, 167–202.

Murray, L., Woolgar, M., Cooper, P., & Hipwell, A. (2001). Cognitive vulnerability to depression in 5-year-old children of depressed mothers. *Journal of Child Psychology and Psychiatry, 42*(7), 891–899.

National Center for Health Statistics. (2005). *Health, United States, 2005, with chartbook on trends in the health of Americans.* Hyattsville, MD: Author.

Nelson, E. E., Leibenluft, E., McClure, E. B., & Pine, D. S. (2005). The social re-orientation of adolescence: A neuroscience perspective on the process and its relation to psychopathology. *Psychological Medicine, 35*(2), 163–174.

Neshat-Doost, H. T., Taghavi, M. R., Moradi, A. R., Yule, W., & Dalgleish, T. (1998). Memory for emotional trait adjectives in clinically depressed youth. *Journal of Abnormal Psychology, 107*(4), 642–650.

Nestler, E. J., & Carlezon, W. A., Jr. (2006). The mesolimbic dopamine reward circuit in depression. *Biological Psychiatry, 59*(12), 1151–1159.

Nolan, C. L., Moore, G. J., Madden, R., Farchione, T., Bartoi, M., Lorch, E., et al. (2002). Prefrontal cortical volume in childhood-onset major depression: Preliminary findings. *Archives of General Psychiatry, 59*(2), 173–179.

Pine, D. S. (2003). Developmental psychobiology and response to threats: Relevance to trauma in children and adolescents. *Biological Psychiatry, 53*(9), 796–808.

Pine, D. S. (2007). Research review: A neuroscience framework for pediatric anxiety disorders. *Journal of Child Psychology and Psychiatry, 48*(7), 631–648.

Pine, D. S., Cohen, E., Cohen, P., & Brook, J. (1999). Adolescent depressive symptoms as predictors of adult depression: Moodiness or mood disorder? *American Journal of Psychiatry, 156*(1), 133–135.

Pine, D. S., Cohen, P., Gurley, D., Brook, J., & Ma, Y. (1998). The risk for early-adulthood anxiety and depressive disorders in adolescents with anxiety and depressive disorders. *Archives of General Psychiatry, 55*(1), 56–64.

Pine, D. S., Cohen, P., Johnson, J. G., & Brooks, J. S. (2002). Adolescent life events as predictors of adult depression. *Journal of Affective Disorders, 68*(1), 49–57.

Pine, D. S., Coplan, J. D., Wasserman, G. A., Miller, L. S., Fried, J. E., Davies, M., et al. (1997). Neuroendocrine response to fenfluramine challenge in boys: Associations with aggressive behavior and adverse rearing. *Archives of General Psychiatry, 54*(9), 839–846.

Pine, D. S., & Klein, R. G. (2008). Anxiety disorders. In M. Rutter, D. Bishop, D. S. Pine, S. Scott, J. Stevenson, E. Taylor, & A. Thapar (Eds.), *Rutter's child and adolescent psychiatry* (5th ed., pp. 628–669), Oxford: Blackwell.

Pine, D. S., Klein, R. G., Roberson-Nay, R., Mannuzza, S., Moulton, J. L., Woldehawariat, G., et al. (2005). Face emotion processing and risk for panic disorder in youth. *Journal of the American Academy of Child and Adolescent Psychiatry, 44*, 664–672.

Pine, D. S., Lissek, S., Klein, R. G., Mannuzza, S., Moulton, J. L., 3rd, Guardino, M., et al. (2004). Face-memory and emotion: Associations with major depression in children and adolescents. *Journal of Child Psychology and Psychiatry, 45*(7), 1199–1208.

Pollak, S. D., Cicchetti, D., Hornung, K., & Reed, A. (2000). Recognizing emotion in faces: Developmental effects of child abuse and neglect. *Developmental Psychology, 36*(5), 679–688.

Rich, B. A., Vinton, D. T., Roberson-Nay, R., Hommer, R. E., Berghorst, L. H., McClure, E. B., et al. (2006). Limbic hyperactivation during processing of neutral facial expressions in children with bipolar disorder. *Proceedings of the National Academy of Sciences USA, 103*(23), 8900–8905.

Roberson-Nay, R., McClure, E. B., Monk, C. S., Nelson, E. E., Guyer, A. E., Fromm, S. J., et al. (2006). Increased amygdala activity during successful memory encoding in adolescent major depressive disorder: An FMRI study. *Biological Psychiatry, 60*(9), 966–973.

Rosenberg, D. R., Macmaster, F. P., Mirza, Y., Smith, J. M., Easter, P. C., Banerjee, S. P., et al. (2005). Reduced anterior cingulate glutamate in pediatric major depression: A magnetic resonance spectroscopy study. *Biological Psychiatry, 58*(9), 700–704.

Rosso, I. M., Cintron, C. M., Steingard, R. J., Renshaw, P. F., Young, A. D., & Yurgelun-Todd, D. A. (2005). Amygdala and hippocampus volumes in pediatric major depression. *Biological Psychiatry, 57*(1), 21–26.

Rudolph, K. D., & Conley, C. S. (2005). The socioemotional costs and benefits of social-evaluative concerns: Do girls care too much? *Journal of Personality, 73*(1), 115–137.

Santarelli, L., Saxe, M., Gross, C., Surget, A., Battaglia, F., Dulawa, S., et al. (2003). Requirement of hippocampal neurogenesis for the behavioral effects of antidepressants. *Science, 301*(5634), 805–809.

Shaffer, D., Gould, M. S., Fisher, P., Trautman, P., Moreau, D., Kleinman, M., et al. (1996). Psychiatric diagnosis in child and adolescent suicide. *Archives of General Psychiatry, 53*(4), 339–348.

Sheline, Y. I. (2003). Neuroimaging studies of mood disorder effects on the brain. *Biological Psychiatry, 54*(3), 338–352.

Silk, J. S., Shaw, D. S., Skuban, E. M., Oland, A. A., & Kovacs, M. (2006). Emotion regulation strategies in offspring of childhood-onset depressed mothers. *Journal of Child Psychology and Psychiatry, 47*(1), 69–78.

Spear, L. P. (2000). The adolescent brain and age-related behavioral manifestations. *Neuroscience and Biobehavioral Reviews, 24*(4), 417–463.

Steinberg, L., Dahl, R., Keating, D., Kupfer, D. J., Masten, A. S., & Pine, D. S. (2006). The study of developmental psychopathology in adolescence: Integrating affective neuroscience with the study of context. In D. Cicchetti & D. J. Cohen (Eds.), *Developmental psychopathology: Vol. 2. Developmental neuroscience* (2nd ed., pp. 710–741). Hoboken, NJ: Wiley.

Terleph, T. A., Klein, R. G., Roberson-Nay, R., Mannuzza, S., Moulton, J. L., 3rd, Woldehawariat, G., et al. (2006). Stress responsivity and HPA axis activity in juveniles: Results from a home-based CO_2 inhalation study. *American Journal of Psychiatry, 163*(4), 738–740.

Thomas, K. M., Drevets, W. C., Dahl, R. E., Ryan, N. D., Birmaher, B., Eccard, C. H., et al. (2001).

Amygdala response to fearful faces in anxious and depressed children. *Archives of General Psychiatry, 58*(11), 1057–1063.

Vasa, R. A., Roberson-Nay, R., Klein, R. G., Mannuzza, S., Moulton, J. L., Guardino, M., et al. (2007). Memory deficits in children with and at risk for anxiety disorders. *Depression and Anxiety, 24*(2), 85–94.

Videbech, P., & Ravnkilde, B. (2004). Hippocampal volume and depression: A meta-analysis of MRI studies. *American Journal of Psychiatry, 161*(11), 1957–1966.

Young, E. A., Vazquez, D., Jiang, H., & Pfeffer, C. R. (2006). Saliva cortisol and response to dexamethasone in children of depressed parents. *Biological Psychiatry, 60*(8), 831–836.

Zakzanis, K. K., Leach, L., & Kaplan, E. (1998). On the nature and pattern of neurocognitive function in major depressive disorder. *Neuropsychiatry, Neuropsychology and Behavioral Neurology, 11*(3), 111–119.

CHAPTER 21

The Development and Neural Bases of Psychopathy

ROBERT JAMES RICHARD BLAIR
ELIZABETH FINGER
ABIGAIL MARSH

In this chapter, we address three main issues, the first being the nature of psychopathy. Specifically, we describe the defining features of the disorder and indicate how the construct differs from the of *Diagnostic and Statistical Manual of Mental Disorders*, fourth edition, text revision (DSM-IV-TR; American Psychiatric Association, 2000) diagnoses of antisocial personality disorder (APD) and the pediatric counterpart to this disorder, conduct disorder (CD). Second, we address the origins of psychopathy, giving particular consideration to evidence regarding whether psychopathy has a genetic or social basis. Third, we consider the neurocognitive basis of psychopathy, describing both areas of relative consensus and points of contention within the literature.

What Is Psychopathy?

Cleckley (1976) provided the foundation for the modern conceptualization of psychopathy in his book *The Mask of Sanity*. His description was used by Robert Hare to develop the original Psychopathy Checklist (PCL; Hare, 1980), which was first revised in 1991 (the PCL-R; Hare, 1991) and then again in 2003 (Hare, 2003). The PCL-R is an empirically determined, formalized tool for the assessment of psychopathy in adults. Comparable measures have been developed for assessment of children and adolescents—the PCL: Youth Version (Forth, Kosson, & Hare, 2004; Kosson, Cyterski, Steuerwald, Neumann, & Walker-Matthews, 2002) and the Antisocial Process Screening Device (Frick & Hare, 2001).

Psychopathy involves two core components: emotional dysfunction and antisocial behavior. These components have been identified through factor analysis (Frick, 1995; Frick, O'Brien, Wootton, & McBurnett, 1994; Harpur, Hakstian, & Hare, 1988; Hart, Forth, & Hare, 1990; Hobson & Shine, 1998). The emotional dysfunction involves "callous and unemotional" (CU) traits, such as reduced guilt and empathy, as well as reduced attachment to significant others. The antisocial behavior component involves a predisposition to antisocial behavior from an early age. Some recent work has challenged the two-factor description of psychopathy in favor of three-factor (Cooke & Michie, 2001; Frick & Hare, 2001) or four-factor (Williams, Paulhus, & Hare, 2007) solutions. However, it should be noted that both three- and four-factor solutions effectively involve the subdivision of the original emotional dysfunction into two subcategories: the CU dimension (concentrating on the lack of guilt, empathy, or attachment to significant others) and a narcissism dimension (concentrating on grandiose feelings of self-worth) (Cooke & Michie, 2001; Frick & Hare, 2001).

Given its reliance on antisocial behavior, the construct of psychopathy overlaps with the DSM-IV-TR diagnoses of both childhood CD and adult APD. However, only a subset of these diagnosed with either CD or APD meet criteria for psychopathy. In adult forensic samples, only approximately 25% of individuals who warrant a diagnosis of APD meet criteria for psychopathy (Hare, 2003). The major difference between psychopathy and the DSM-IV-TR diagnoses of CD and APD is that the latter focus on the antisocial behavior and do not consider its potential etiology. In contrast, the construct of psychopathy emphasizes the specific form of emotional dysfunction thought to underlie the emergence of the antisocial behavior (Blair, Mitchell, & Blair, 2005).

The failure of the DSM-IV-TR to consider the causes of CD and APD is regrettable. Many different factors may increase an individual's risk for displaying antisocial behaviors such as aggression (see Blair et al., 2005); as a result, individuals with wholly different pathophysiologies are being given these diagnoses. Thus the antisocial behavior of some children with CD and adults with APD is associated with the reduced emotional responsiveness of psychopathy, whereas that of others is associated with exaggerated emotional responsiveness or otherwise dysregulated emotional responding.

Its focus on etiology renders psychopathy a more useful classification. For example, studies have shown the predictive power of scores on the PCL-R with respect to recidivism (Hare, Clark, Grann, & Thornton, 2000; Hart, Kropp, & Hare, 1988; Kawasaki et al., 2001). Variance among individuals in CU traits (the affective component of psychopathy) predicts receptivity to parenting strategies, with heightened levels of CU traits being associated with less receptivity to standard parenting strategies (Oxford, Cavell, & Hughes, 2003; Wootton, Frick, Shelton, & Silverthorn, 1997). Children's CU traits also predict their responsiveness to behavioral parent-training intervention (Hawes & Dadds, 2005).

A defining behavioral feature of psychopathy is an increased risk for "instrumental" aggression (Cornell et al., 1996; Williamson, Hare, & Wong, 1987). Instrumental aggression (also referred to as "proactive aggression") is purposeful and goal-directed; the aggression is used instrumentally to achieve a specific desired goal (Berkowitz, 1993). This contrasts with "reactive" aggression (also referred to as "affective" or "impulsive" aggression), which is not goal-directed, but is triggered by a frustrating or threatening event and is frequently associated with anger. Many conditions increase the risk for reactive aggression—for example, lesions including the orbital frontal cortex (OFC) (Anderson, Bechara, Damasio, Tranel, & Damasio, 1999; Grafman, Schwab, Warden, Pridgen, & Brown, 1996), childhood bipolar disorder (Leibenluft, Blair, Charney, & Pine, 2003), posttraumatic stress disorder (Calhoun et al., 2002), intermittent explosive disorder (Coccaro, 1998), or borderline personality disorder (Siever, Torgersen, Gunderson, Livesley, & Kendler, 2002). The only condition known to increase the risk for instrumental (and reactive) aggression is psychopathy (Cornell et al., 1996; Williamson et al., 1987).

What Are the Origins of Psychopathy?

Genetic Factors

Many studies have implied that antisocial behavior is heritable (for a review, see Rhee & Waldman, 2002). It is improbable that genes code for antisocial behaviors directly, however. Instead, genes may influence those systems that increase an individual's risk for antisocial behavior.

In the section above, we have described psychopathy as an emotional disorder marked by reduced guilt and empathy, and associated with an increased risk for antisocial behavior. Is there a genetic basis to psychopathy? Several recent studies have investigated this issue and concluded that it is heritable and under moderate environmental influence (Blonigen, Hicks, Krueger, Patrick, & Iacono, 2005; Viding, Blair, Moffitt, & Plomin, 2005). Moreover, whereas antisocial behavior in the presence of elevated levels of CU traits is strongly heritable (81%), heritability of antisocial behavior without CU traits is only moderate (30%) (Viding et al., 2005).

Although behavioral genetic work indicates a genetic contribution to the development of psychopathy, the molecular genetics of psychopathy are currently unknown. Suggestive data are provided by recent investigations of the influence of specific genetic polymorphisms on neural and behavioral responding. These specific genetic polymorphisms appear to have implications for the responsiveness of the amygdala, a structure that appears dysfunctional in psychopathy (see below). For example, several studies have reported that individuals who are homozygous for the long allele of the serotonin transporter gene promoter (5-HTTLPR) show significantly reduced amygdala responding to emotional expressions, relative to those who have the short-form polymorphism of 5-HTTLPR (Hariri et al., 2002). In addition, such individuals show behavioral impairment on some emotional learning tasks reliant on the amygdala (Finger et al., 2007). This is not to suggest that long-allele homozygosity causes psychopathy, but rather that the serotonin transporter gene may be one of several genes that have polymorphisms associated with decreased emotional and amygdala responsiveness. The basic genetic risk for psychopathy may emerge in individuals who have several independent polymorphisms that predispose them toward reduced emotional and amygdala responsiveness.

Environmental Factors

An increased risk for aggression, and potentially for psychopathy, has been associated with a variety of environmental causes. In this section we will consider one environmental cause—exposure to extreme threats.

An individual is at increased risk for aggression following exposure to extreme threat, whether it is violence in the home or neighborhood (Miller, Wasserman, Neugebauer, Gorman-Smith, & Kamboukos, 1999; Schwab-Stone et al., 1999), or physical or sexual abuse (Dodge, Pettit, Bates, & Valente, 1995; Farrington & Loeber, 2000). However, these data do not necessarily imply that this environmental factor increases the risk for the development of psychopathy.

Considerable animal work has shown that prolonged threats and stress lead to long term potentiation of the neural and neurochemical systems that respond to threat; in other words, the individual becomes more responsive to aversive stimuli (Bremner & Vermetten, 2001; King, 1999). Moreover, traumatic exposure in humans, such as exposure to violence in the home or neighborhood, increases the risk for mood and anxiety disorders, which are typically associated with increased emotional responsiveness rather than with the decreased emotional responsiveness seen in psychopathy (Charney, 2003; Gorman-Smith & Tolan, 1998; Schwab-Stone et al., 1999).

Because exposure to extreme threats is associated with an increase in emotional responsiveness rather than a decrease, it is unlikely that it causes psychopathy to develop. Why then does it increase the risk for aggression? The answer to this question probably relates to the type of aggression being displayed. For example, exposure to threat as a consequence of abusive parenting or other trauma is associated with an increased risk for *reactive* aggression (Dodge et al., 1995), such that the individual is more likely to respond with aggression when provoked by threatening or frustrating events.

The mammalian response to threat is gradated. At low levels of danger from a distant threat, animals tend to freeze. As the danger level increases and the threat draws closer, animals attempt to escape. At the highest level of danger, when the threat is very close and escape is no longer possible, reactive aggression is displayed (Blanchard, Blanchard, & Takahashi, 1977). This gradated response is mediated by a basic threat system that runs from medial amygdaloidal areas downward, largely via the stria terminalis to the medial hypothalamus, and from there to the dorsal half of the periaqueductal gray (Gregg & Siegel, 2001; Panksepp, 1998). This system is regulated by OFC, medial frontal cortex, and ventrolateral frontal cortex (Blair, 2004).

The responsiveness of the neural threat system can be potentiated by prior exposure to threatening or stressful events (Charney, 2003), and the behavioral response to threat (whether freezing, fleeing, or fighting) is determined by the responsiveness of the system. If the system has been potentiated by prior threat exposure, an individual will escalate his or her response to the threat more rapidly than another individual who has not had that prior exposure. In other words, we believe that exposure to extreme threat or stress increases the risk for *reactive* aggression, because it increases the underlying responsiveness of the system and therefore makes the individual more likely to escalate his or her response to threats.

What Is the Neurocognitive Basis of Psychopathy?

Various accounts have been offered to explain psychopathy. Some exist primarily at the neural level (Gorenstein, 1982; Kiehl, 2006), and others primarily at the cognitive level (Blair, 1995; Frick & Marsee, 2006; Hartung, Milich, Lynam, & Martin, 2002; Lykken, 1995); still others are cognitive neuroscience positions, which consider the cognitive functions mediated by particular neural regions (Blair, 2001; Damasio, 1994; Patrick, 1994; Raine, 2002). The accounts have included the frontal lobe hypothesis (Gorenstein, 1982), the somatic marker hypothesis (Damasio, 1994), the amygdala-based hypotheses (Blair, 2001; Patrick, 1994), fear dysfunction accounts (Frick & Marsee, 2006; Lykken, 1995; Patrick, 1994), the paralimbic hypothesis (Kiehl, 2006), the response set modulation hypothesis, models based on impulse control problems (Hartung et al., 2002), and the violence inhibition mechanism and integrated emotion systems accounts (Blair, 2001, 2006).

Some positions can be considered to have been superseded (Blair, 1995; Gorenstein, 1982; Raine, 2002). That is, some early positions appeared to suggest that there is general frontal lobe impairment in psychopathy (Gorenstein, 1982; cf. Raine, 2002). However, because the frontal lobes account for nearly 50% of the total cerebral cortex, such positions were highly underspecified. Moreover, neuropsychological work demonstrated that functions mediated by the dorsolateral and dorsomedial frontal cortex are intact in individuals with psychopathy, and that only functions mediated by the OFC, ventromedial frontal cortex, and possibly inferior frontal cortex appear compromised (Blair et al., 2006b; LaPierre, Braun, & Hodgins, 1995; Mitchell, Colledge, Leonard, & Blair, 2002). For a review of this literature and criticisms of these positions, see Blair et al. (2005).

Blair's (1995) original violence inhibition mechanism model faced difficulty accounting for data indicating more general problems in individuals with psychopathy in the processing of punishment information (e.g., Patrick, 1994). This has led to a revision and expansion of the model at both the cognitive and neural levels (i.e., the integrated emotion systems model; Blair, 2004, 2006).

Although some accounts have been superseded, exciting progress has been made in the field of psychopathy over the past 10 years. This progress has permitted refinements to be made to the frontal lobe and violence inhibition mechanism positions. A specific form of frontal lobe dysfunction does appear to be a main component of the pathophysiology of psychopathy. An emphasis on the importance of an appropriate response to the distress of others (as in the violence inhibition mechanism model) remains an important part of a developmental account of the disorder. Moreover, the general refinement and increased specification of positions such as these has resulted in increased consensus among the various positions emerging. We briefly describe the points of (rough) consensus below.

Consensus at the Neural Level

1. *At the anatomical level, amygdala dysfunction is a core feature of psychopathy.* This position was first proposed by Patrick (1994) and has been considerably extended and revised by Blair (e.g., 2001, 2006). The amygdala is one of the systems considered impaired within Kiehl's (2006) paralimbic hypothesis, and although neither Lykken nor Frick makes extensive reference to the structure, the impairments described by both are highly consistent with an amygdalocentric view (Frick & Marsee, 2006; Lykken, 1995).

Core deficits seen in psychopathy are in aversive conditioning (Birbaumer et al., 2005; Lykken, 1957), the augmentation of the startle reflex by visual threat primes (Levenston, Patrick, Bradley, & Lang, 2000; Patrick, Bradley, & Lang, 1993), passive avoidance learning (Newman & Kosson, 1986), and the recognition of fearful expressions (Blair, Colledge, Murray, & Mitchell, 2001). These impairments are all seen following amygdala lesions (for a review of the literature, see Blair, 2006). Moreover, recent neuroimaging work with adult forensic populations has consistently indicated that individuals with psychopathy show reduced amygdala responses to emotional words in the context of emotional memory paradigms (Kiehl et al., 2001) and during aversive conditioning (Birbaumer et al., 2005). Work with subclinical populations has similarly found that individuals with psychopathic traits show reduced amygdala responses to emotional expressions (Gordon, Baird, & End, 2004) and less amygdala differentiation in responding when making the choice to cooperate versus defect in a prisoner's dilemma paradigm (Rilling et al., 2007).

One important position that does not regard amygdala dysfunction as important for the development of psychopathy is the somatic marker hypothesis of Damasio (1994). Damasio (1994) and colleagues propose that psychopathy may be a developmental consequence of early damage to the OFC. However, this view has been extensively criticized (Blair, 2004; Blair & Cipolotti, 2000). Importantly, none of the core impairments seen in psychopathy are found following OFC lesions. They are, however, found following lesions of the amygdala (for extended discussions of these data, see Blair, 2004, 2006).

2. *It is necessary to consider OFC functioning when considering psychopathy.* While Damasio's viewpoint that psychopathy is a developmental consequence of early OFC damage is untenable, given data that the core impairments seen in psychopathy are not found following OFC lesions (see Blair, 2004, 2006), this does not imply that OFC activity is irrelevant to the understanding of psychopathy. Both Blair (2004) and Kiehl (2006) consider atypical OFC activity to contribute to the pathophysiology of psychopathy. However, other dominant theoretical viewpoints remain unspecified at the anatomical level (Frick & Marsee, 2006; Lykken, 1995) or have remained agnostic (Patrick, 1994).

Kiehl's (2006) paralimbic hypothesis is primarily a model at the neural level. There is considerable discussion of structures that may be impaired, but rather less consideration of their functional contribution. In contrast, Blair's cognitive neuroscience position, the integrated emotion systems model (Blair, 2004, 2006), follows Gallagher, Schoenbaum, and their colleagues regarding the function of the OFC and its relationship with the amygdala. The basic suggestion is that the amygdala feeds forward reinforcement information associated with stimuli to medial OFC, which then represents this outcome information, allowing decision making in regard to behavior such as approach and avoidance (Gallagher, McMahan, & Schoenbaum, 1999; Schoenbaum, Nugent, Saddoris, & Setlow, 2002; Schoenbaum, Setlow, Saddoris, & Gallagher, 2003). This suggests that even if there is not primary pathology within the OFC, individuals with psychopathy should show anomalous medial OFC activity in the context of tasks that activate the amygdala. This is exactly what is seen. In Kiehl et al.'s (2001) study of emotional memory, individuals with psychopathy showed not only reduced amygdala responses to the emotional words, but also reduced rostral anterior cingulate cortex/medial OFC activation. In addition to showing reduced amygdala activity, individuals with psychopathy also exhibited medial OFC activity during aversive conditioning (Birbaumer et al., 2005), as well as less medial OFC differentiation in responding when making cooperation versus defection choices in the prisoner's dilemma paradigm (Rilling et al., 2007).

• *Point of contention: The functions of many neural systems are compromised in psychopathy.* According to Kiehl's (2006) paralimbic hypothesis, individuals with psychopathy face disruption of the neural systems that make up what he terms the "paralimbic system" (i.e., the OFC, insula, anterior and posterior cingulate cortex, amygdala, parahippocampal gyrus, and anterior superior temporal gyrus). This is in notable contrast to Blair's position, which states that although the functioning of regions beyond the amygdala and medial OFC may be anomalous under certain conditions, this is due to deficient input from the amygdala and medial OFC rather than to direct dysfunction of the systems themselves.

In some respects, Kiehl's position is likely to be correct. As noted above, data are emerging to indicate a strong genetic contribution to psychopathy (Blonigen et al., 2005; Viding et al., 2005). It is likely that any genetic contribution will affect the development of structures beyond the amygdala and OFC.

In other respect, Kiehl's position is clearly incorrect. The position uses a neurological approach to understand a psychiatric condition. Kiehl's position implies that the functions of all the regions making up the paralimbic system are compromised under a wide range of circumstances. But available data suggest that this is not the case. For example, some social cognition functions reliant on amygdala function, such as the making of trustworthiness judgments or mental state judgments from eye information, appear spared in psychopathy (Richell et al., 2003; Richell et al., 2005). Similarly, the crucial role of dorsal anterior cingulate cortex in the resolution of response conflict appears spared in psychopathy (Blair et al., 2006b; Hiatt, Schmitt, & Newman, 2004). The problem with Kiehl's position is its neurological approach to this psychiatric condition (i.e., a reference specifically to dysfunctional neural structures). It is unlikely that psychiatric conditions should be explained with reference to dysfunction within particular neural structures, rather than dysfunction in functional systems that rely on multiple structures (and where other functional systems reliant on the same neural structures may be intact).

Consensus at the Cognitive Level

3. *Psychopathy is an emotional disorder.* This is part of the clinical description, and almost all positions agree on this point (Blair, 2001, 2006; Damasio, 1994; Frick & Marsee, 2006; Kiehl, 2006; Lykken, 1995; Patrick, 1994).

4. *Psychopathy reflects impaired processing of reinforcement information.* Almost all positions also agree on this general statement (Blair, 2001, 2006; Damasio, 1994; Frick & Marsee, 2006; Kiehl, 2006; Lykken, 1995; Patrick, 1994). Some of the very earliest positions on psychopathy were variants of this type of position (Hare, 1970; Lykken, 1957).

There are some disagreements in the details, however. The first is whether the impairment in reinforcement processing reflects impaired *learning* on the basis of reinforcement information or the impaired *use* of reinforcement information. Most positions on psychopathy have assumed that the problem reflects an inability to learn from reinforcement information (Blair, 2001, 2006; Frick & Marsee, 2006; Kiehl, 2006; Lykken, 1995; Patrick, 1994). However, Damasio's (1994) viewpoint on psychopathy assumes that it reflects the impaired use of prior learning. Within his model, the OFC is described as responding to somatic marker information to allow decision making. Certain data strongly refute this position, however. For example, individuals with psychopathy show impairment in aversive conditioning (Birbaumer et al., 2005; Lykken, 1957). According to the Damasio model, this process does not require somatic marker information, but rather allows the formation of somatic markers that the OFC can process (Bechara, Damasio, Damasio, & Lee, 1999). Indeed, the data on aversive conditioning in individuals with psychopathy formed the basis of some of the earliest criticism of the somatic marker hypothesis of psychopathy (Blair, 2001).

The second disagreement pertains to whether psychopathy is characterized by dysfunction in processing punishment information. This view is shared by most of the models (Frick & Marsee, 2006; Kiehl, 2006; Lykken, 1995; Patrick, 1994). However, it faces two serious challenges:

a. There are many forms of punishment processing (i.e., partially independent neurocognitive systems process punishment information), and not all are equivalently impaired in psychopathy. For example, social threats such as another's angry expression are arguably processed by different systems from those that respond to fearful expressions (Blair, 2004). Individuals with psychopathy show no impairment in the processing of angry expressions, but considerable difficulty with processing fearful expressions (Blair et al., 2004a).

Similarly, the use of punishment in stimulus–reinforcement learning (i.e., the ability to form stimulus–punishment associations) is partially separable from the ability to use punishment information in the context of stimulus–response associations (Baxter & Murray, 2002). For example, in the classic passive avoidance learning paradigm, the individual learns that responding to some stimuli gives rise to reward, while responding to others gives rise to punishment. This paradigm can be solved through stimulus–reinforcement associations: Stimulus–reward associations guide the individual toward the good stimuli, while stimulus–punishment associations guide the individual away from the bad stimuli. Individuals with psychopathy show significant impairment in passive avoidance learning and other stimulus–reinforcement learning paradigms (Blair, Leonard, Morton, & Blair, 2006a; Blair et al., 2004b; Newman & Kosson, 1986). However, a modification of the passive learning paradigm can change it to a stimulus–*response* learning task. This occurs if the individual learns that responding to some stimuli gives rise to reward, while *not* responding to gives rise to reward (an alternative version of this modification involves punishment for responding to some and punishment for *not* responding to others). In these variants, there are no good or bad stimuli; whether a stimulus is good (or bad) depends on how the individual responds to that stimulus. Individuals with psychopathy do not show significant impairment on such conditional learning variants of passive avoidance paradigms or on other stimulus–response paradigms, such as object discrimination (Mitchell et al., 2002, 2006a; Newman, Patterson, Howland, & Nichols, 1990);

b. Individuals with psychopathy are impaired in the processing not only of punishment information, but also of reward information (Blair et al., 2006a; Mitchell, Richell, Leonard, & Blair, 2006b; Verona, Patrick, Curtin, Bradley, & Lang, 2004). For example, individuals with psychopathy show impairment in deciding between two objects associated with different levels of reward on the differential reward punishment task (albeit less impairment than when deciding between two objects associated with different levels of punishment) (Blair et al., 2006a). Similarly, individuals with psychopathy show less interference by positive as well as negative emotional distractors on an emotional interruption task (Mitchell et al., 2006b). In addition, individuals with psychopathy show reduced autonomic activity to positive as well as negative emotionally evocative noises (Verona et al., 2004).

One way to reconcile these potentially contradictory findings (not all forms of punishment processing are impaired, and not only punishment processing but also reward processing is impaired) is by reference to the neural level (Blair, 2006). The amygdala is less involved in processing some threats than others. For example, the amygdala shows considerably less responsiveness to angry expressions than to fearful expressions (Whalen, Shin, McInerney, & Rauch, 1998). Similarly, the amygdala is not necessary for the processing of punishment information in the context of stimulus–response tasks, though it is necessary for the formation and use of stimulus–reinforcement associations (Baxter & Murray, 2002). Finally, the amygdala is involved in the processing not only of punishment information, but also of reward information (Baxter & Murray, 2002). In short, the data suggest that psychopathy should be characterized not as an impairment in punishment processing, but as an impairment in those forms of punishment processing mediated by the amygdala.

5. *The impairment in the processing of reinforcement, particularly the reinforcement of others' distress, underlies the difficulties in the socialization of individuals with psychopathy.* Almost all positions agree that the impairment seen in psychopathy interferes with their socialization (Blair, 2001, 2006; Frick & Marsee, 2006; Lykken, 1995; Patrick, 1994). Direct evidence that psychopathy is associated with disruption in the ability to be socialized was first obtained by Paul Frick and colleagues (Wootton, Frick, Shelton, & Silverthorn, 1997). They showed that although parenting type has a significant impact on healthy children (who show increased conduct problems following ineffective parenting), it does not have a significant impact on children who show high levels of the CU traits that constitute the emotional component of psychopathy. This result has been subsequently replicated (Oxford et al., 2003).

Very early viewpoints on psychopathy considered that socialization would be best achieved through the use of punishment-based learning strategies; these positions were based on the idea that psychopathy is due to reduced sensitivity to punishment (Trasler, 1978). The problem with these views is that punishment-based socialization strategies are not particularly effective with respect to socialization (Brody & Shaffer, 1982). This is not surprising, because conditioning theory suggests that the stimulus associated with punishment will be the one that best predicts the punishment. This is unlikely to be the transgression committed at variable lengths in the past, and far more likely to be the individual who actually delivers the punishment.

More recent work has stressed the importance of empathic responses. Work has shown that empathy induction techniques, such as focusing the transgressor's attention on the victim's distress, are effective in socialization (Brody & Shaffer, 1982; Hoffman, 1970). There are considerable data suggesting that the processing of the victim's distress is impaired in individuals with psychopathy (Aniskiewicz, 1979; Blair, Jones, Clark, & Smith, 1997; Blair et al., 2004a; Dolan & Fullam, 2006).

Views That Disagree with the Consensus

Although there is considerable consensus regarding the five points raised above, two positions do not share this consensus: Newman's response set modulation hypothesis and Lynam's inhibition account (Hartung et al., 2002; Whiteside & Lynam, 2001). Both views have their origins in Gray's (1971) behavior inhibition system model, despite the attacks on this model at both the anatomical level (functions ascribed to the septum and hippocampus have been shown to be reliant on the amygdala; LeDoux, 2000) and the cognitive level (functions ascribed to a unitary system, the behavior inhibition system, have been shown to be dissociable; Blair, 2004).

Response set modulation has been defined as the rapid and relatively automatic (i.e., non-effortful or involuntary) shift of attention from the effortful organization and implementation of goal-directed behavior to its evaluation (Newman, 1998; Patterson & Newman, 1993). The impulsivity associated with psychopathy (Miller, Flory, Lynam, & Leukefeld, 2003; Whiteside & Lynam, 2001) has been conceptualized as a lack of premeditation and perseverance. Within this account, this lack of premeditation reflects the inability to inhibit previously rewarded behavior when presented with changing contingencies (Whiteside & Lynam, 2001), whereas the lack of perseverance may be related to disorders that involve the inability to ignore distracting stimuli or to remain focused on a particular task (Whiteside & Lynam, 2001).

Both the response set modulation and impulsivity positions disagree with all the points of consensus outlined above. Neither regards the amygdala or OFC as central to the pathophysiology of psychopathy. Moreover, although both positions acknowledge the existence of emotional deficits in individuals with psychopathy, they do not consider psychopathy to be a primarily emotional disorder. Instead, they consider emotional deficits to be secondary to problems in response set modulation or impulsivity. In addition, they do not consider problems in processing reinforcement, social or otherwise, to be central to the disorder or to explain the difficulties with socialization. Rather, both consider the problem primarily one of attention: Either the individual attends primarily to reward information and cannot shift attention to punishment information (i.e., is incapable of response set modulation; Newman, 1998); or, alternatively, the individual is not attending at all but has been distracted (i.e., exhibits impulsivity; Miller et al., 2003).

It should be noted that such domain-general accounts face considerable difficulties with the psychopathy literature, however. For example, these positions should predict that individuals with psychopathy would be impaired on a broad range of tasks, because many tasks can be considered to involve response set modulation or attention—for example, the intradimensional/extradimensional (ID/ED) and spatial alteration/object alteration (SA/OA) tasks. In these tasks, there are two principal measures. The first of these is the number of response reversal errors/object reversals; these errors occur when subjects fail to change their response from one object to another following a change in reinforcement contingency. The second measure is the number of ED errors/spatial reversals; these errors occur when subject fail to change their response from one semantic category to another (shapes to lines; ED) or from one spatial location to another (SA).

Response/object reversal, ED shifting, and SA all appear to require the inhibition of a previously rewarded behavior/response modulation. Thus, according to an inhibition or response set modulation account, individuals with psychopathy should show impairment not only in response/object reversal, but also in ED shifting and SA. Yet they do not. While individuals with psychopathy do show impairment in response/object reversal, they show no significant difficulty with ED shifting/SA (Blair et al., 2006a; Mitchell et al., 2002). This would be more problematic if the consensus view also had difficulty explaining these data. However, it does not. The functional integrity of the OFC and ventromedial frontal cortex is necessary for successful performance on response reversal and OA tasks (Dias, Robbins, & Roberts, 1996; Freedman, Black, Ebert, & Binns, 1998). As we have argued above, individuals with psychopathy show OFC and ventromedial frontal cortex dysfunction; thus their impairment on response reversal and OA tasks is to be expected. The functional integrity of dorsolateral frontal cortex is necessary for successful performance on ED shifting and SA tasks (Dias et al., 1996; Freedman et al., 1998). As we have argued above, individuals with psychopathy do not appear to have dorsolateral frontal cortex dysfunction; thus their lack of impairment on ED shifting and SA tasks is to be expected.

Conclusions

Individuals with psychopathy represent a subset of individuals with CD/APD whose pathological development is related to a specific form of emotional dysfunction: CU traits. Behaviorally, individuals with psychopathy are marked by an increased risk for reactive *and* instrumental aggression. The increased risk for instrumental aggression is noteworthy because it is not seen in any other psychiatric condition, and its presence suggests that the pathology of psychopathy interferes with appropriate socialization. The psychopathy construct has been extremely successful in predicting long-term prognosis, particularly in adult samples. But perhaps the greatest importance of this construct for child psychiatrists is that it identifies a sample of children who are unlikely to benefit from standard treatment strategies for children with CD or oppositional defiant disorder (Hawes & Dadds, 2005).

Behavioral genetic work indicates that CU traits are strongly heritable, and that the antisocial behavior of children with these traits is under considerable genetic influence. However, the corresponding molecular work remains in its infancy. Social factors clearly influence the behavioral manifestation of CU traits. They are likely to have a considerable influence on an individual's motives, such as whether he or she needs to offend in order to gain resources or whether a range of alternative strategies is available.

With the exception of the response set modulation and impulsivity positions, a relatively good consensus exists within the field with respect to several main issues. Given this consensus, it is unsurprising that both the response set modulation and impulsivity positions face considerable difficulty accounting for much of the literature. Most positions agree, or at least do not explicitly disagree, on the involvement of the amygdala and OFC in the pathophysiology of psychopathy. Similarly, all positions agree that the disorder is an emotional disorder, and that problems in the processing of social and nonsocial reinforcement are central to the disorder. These emotional impairments are thought to interfere with socialization and to allow the development of the disorder.

There are, of course, points of disagreement within this consensus. A main one concerns the extent of the neural dysfunction, with Kiehl proposing that an extensive array of systems may be impaired in the disorder, and Blair arguing for a far more limited number (the amygdala and ventromedial prefrontal cortex). Moreover, according to Blair, even within those neural regions, the impairment may be confined to some specific functional capacities but leave others intact. It is to be hoped that the triangulation of neuroimaging and neuropsychological data will be able to settle this point of contention.

A secondary point of disagreement concerns the nature of the impairment in reinforcement processing. Many positions have suggested that the impairment is in the processing of punishment information. However, there is growing evidence that this is an inaccurate characterization. The use of punishment information in the context of learning stimulus–response associations appears intact in individuals with psychopathy. Moreover, individuals with psychopathy appear to show problems in using reward information in some contexts. It is possible that the impairment in psychopathy is in the use of reinforcement information in the context of stimulus–reinforcement association formation. A particularly important reinforcement with respect to socialization is, of course, the distress of victims.

Importantly, this growing consensus has useful clinical implications. If psychopathy is understood as a form of emotional disorder, albeit one marked by reduced rather than increased emotional responsiveness, it is possible to consider potential treatment interventions. Many emotional disorders are being treated relatively successfully. These treatments usually involve the suppression of emotional responding; however, such treatments provide interesting suggestions with respect to psychopathy. For example, agents that reduce noradrenergic and, consequently, amygdala activity appear useful in treating posttraumatic stress disorder (Strawn & Geracioti, 2008). Perhaps psychopathy could be treated by agents that increase noradrenergic activity. Given that psychopathy is currently regarded as untreatable, such possibilities provide hope for the future.

Acknowledgments

This research was supported by grants from the Intramural Research Program of the National Institute of Mental Health (to Robert James Richard Blair) and from the U.K. Department of Health and Medical Research Council (to Elizabeth Finger).

References

American Psychiatric Association. (2000). *Diagnostic and statistical manual of mental disorders* (4th ed., text rev.). Washington, DC: Author.

Anderson, S. W., Bechara, A., Damasio, H., Tranel, D., & Damasio, A. R. (1999). Impairment of social and moral behaviour related to early damage in human prefrontal cortex. *Nature Neuroscience, 2*, 1032–1037.

Aniskiewicz, A. S. (1979). Autonomic components of vicarious conditioning and psychopathy. *Journal of Clinical Psychology, 35*, 60–67.

Baxter, M. G., & Murray, E. A. (2002). The amygdala and reward. *Nature Reviews Neuroscience, 3*(7), 563–573.

Bechara, A., Damasio, H., Damasio, A. R., & Lee, G. P. (1999). Different contributions of the human amygdala and ventromedial prefrontal cortex to decision-making. *Journal of Neuroscience, 19*, 5473–5481.

Berkowitz, L. (1993). *Aggression: Its causes, consequences, and control.* Philadelphia: Temple University Press.

Birbaumer, N., Veit, R., Lotze, M., Erb, M., Hermann, C., Grodd, W., et al. (2005). Deficient fear conditioning in psychopathy: A functional magnetic resonance imaging study. *Archives of General Psychiatry, 62*(7), 799–805.

Blair, K. S., Leonard, A., Morton, J., & Blair, R. J. R. (2006a). Impaired decision making on the basis of both reward and punishment information in individuals with psychopathy. *Personality and Individual Differences, 41*, 155–165.

Blair, K. S., Newman, C., Mitchell, D. G., Richell, R. A., Leonard, A., Morton, J., et al. (2006b). Differentiating among prefrontal substrates in psychopathy: Neuropsychological test findings. *Neuropsychology, 20*(2), 153–165.

Blair, R. J. R. (1995). A cognitive developmental approach to morality: Investigating the psychopath. *Cognition, 57*, 1–29.

Blair, R. J. R. (2001). Neuro-cognitive models of aggression, the antisocial personality disorders and psychopathy. *Journal of Neurology, Neurosurgery and Psychiatry, 71*, 727–731.

Blair, R. J. R. (2004). The roles of orbital frontal cortex in the modulation of antisocial behavior. *Brain and Cognition, 55*(1), 198–208.

Blair, R. J. R. (2006). The emergence of psychopathy: Implications for the neuropsychological approach to developmental disorders. *Cognition, 101*, 414–442.

Blair, R. J. R., & Cipolotti, L. (2000). Impaired social response reversal: A case of "acquired sociopathy." *Brain, 123*, 1122–1141.

Blair, R. J. R., Colledge, E., Murray, L., & Mitchell, D. G. (2001). A selective impairment in the processing of sad and fearful expressions in children with psychopathic tendencies. *Journal of Abnormal Child Psychology, 29*(6), 491–498.

Blair, R. J. R., Jones, L., Clark, F., & Smith, M. (1997). The psychopathic individual: A lack of responsiveness to distress cues? *Psychophysiology, 34*, 192–198.

Blair, R. J. R., Mitchell, D. G. V., & Blair, K. S. (2005). *The psychopath: Emotion and the brain.* Oxford: Blackwell.

Blair, R. J. R., Mitchell, D. G. V., Colledge, E., Leonard, R. A., Shine, J. H., Murray, L. K., et al.

(2004a). Reduced sensitivity to others' fearful expressions in psychopathic individuals. *Personality and Individual Differences, 37,* 1111–1121.

Blair, R. J. R., Mitchell, D. G. V., Leonard, A., Budhani, S., Peschardt, K. S., & Newman, C. (2004b). Passive avoidance learning in individuals with psychopathy: Modulation by reward but not by punishment. *Personality and Individual Differences, 37,* 1179–1192.

Blanchard, R. J., Blanchard, D. C., & Takahashi, L. K. (1977). Attack and defensive behaviour in the albino rat. *Animal Behaviour, 25,* 197–224.

Blonigen, D. M., Hicks, B. M., Krueger, R. F., Patrick, C. J., & Iacono, W. G. (2005). Psychopathic personality traits: Heritability and genetic overlap with internalizing and externalizing psychopathology. *Psychological Medicine, 35,* 637–648.

Bremner, J. D., & Vermetten, E. (2001). Stress and development: Behavioral and biological consequences. *Development and Psychopathology, 13,* 473–489.

Brody, G. H., & Shaffer, D. R. (1982). Contributions of parents and peers to children's moral socialisation. *Developmental Review, 2,* 31–75.

Calhoun, P. S., Beckham, J. C., Feldman, M. E., Barefoot, J. C., Haney, T., & Boswort, H. B. (2002). Partners' ratings of combat veterans' anger. *Journal of Traumatic Stress, 15*(2), 133–136.

Charney, D. S. (2003). Neuroanatomical circuits modulating fear and anxiety behaviors. *Acta Psychiatrica Scandinavica, 108*(Suppl. 417), 38–50.

Cleckley, H. M. (1976). *The mask of sanity* (5th ed.). St. Louis, MO: Mosby.

Coccaro, E. F. (1998). Impulsive aggression: A behavior in search of clinical definition. *Harvard Review of Psychiatry, 5*(6), 336–339.

Cooke, D. J., & Michie, C. (2001). Refining the construct of psychopathy: Towards a hierarchical model. *Psychological Assessment, 13*(2), 171–188.

Cornell, D. G., Warren, J., Hawk, G., Stafford, E., Oram, G., & Pine, D. (1996). Psychopathy in instrumental and reactive violent offenders. *Journal of Consulting and Clinical Psychology, 64,* 783–790.

Damasio, A. R. (1994). *Descartes' error: Emotion, rationality and the human brain.* New York: Putnam.

Dias, R., Robbins, T. W., & Roberts, A. C. (1996). Dissociation in prefrontal cortex of affective and attentional shifts. *Nature, 380,* 69–72.

Dodge, K. A., Pettit, G. S., Bates, J. E., & Valente, E. (1995). Social information-processing patterns partially mediate the effect of early physical abuse on later conduct problems. *Journal of Abnormal Psychology, 104,* 632–643.

Dolan, M., & Fullam, R. (2006). Face affect recognition deficits in personality-disordered offenders: Association with psychopathy. *Psychological Medicine, 36*(11), 1563–1569.

Farrington, D. P., & Loeber, R. (2000). Epidemiology of juvenile violence. *Child and Adolescent Psychiatric Clinics of North America, 9*(4), 733–748.

Finger, E. C., Marsh, A. A., Buzas, B., Kamel, N., Rhodes, R., Vythilingham, M., et al. (2007). The impact of tryptophan depletion and 5-HTTLPR genotype on passive avoidance and response reversal instrumental learning tasks. *Neuropsychopharmacology, 32*(1), 206–215.

Forth, A. E., Kosson, D. S., & Hare, R. D. (2004). *The Psychopathy Checklist: Youth Version.* Toronto: Multi-Health Systems.

Freedman, M., Black, S., Ebert, P., & Binns, M. (1998). Orbitofrontal function, object alternation and perseveration. *Cerebral Cortex, 8*(1), 18–27.

Frick, P. J. (1995). Callous–unemotional traits and conduct problems: A two-factor model of psychopathy in children. *Issues in Criminological and Legal Psychology, 24,* 47–51.

Frick, P. J., & Hare, R. D. (2001). *The Antisocial Process Screening Device.* Toronto: Multi-Health Systems.

Frick, P. J., & Marsee, M. A. (2006). Psychopathy and developmental pathways to antisocial behavior in youth. In C. J. Patrick (Ed.), *Handbook of psychopathy* (pp. 353–374). New York: Guilford Press.

Frick, P. J., O'Brien, B. S., Wootton, J. M., & McBurnett, K. (1994). Psychopathy and conduct problems in children. *Journal of Abnormal Psychology, 103*, 700–707.

Gallagher, M., McMahan, R. W., & Schoenbaum, G. (1999). Orbitofrontal cortex and representation of incentive value in associative learning. *Journal of Neuroscience, 19*, 6610–6614.

Gordon, H. L., Baird, A. A., & End, A. (2004). Functional differences among those high and low on a trait measure of psychopathy. *Biological Psychiatry, 56*(7), 516–521.

Gorenstein, E. E. (1982). Frontal lobe functions in psychopaths. *Journal of Abnormal Psychology, 91*, 368–379.

Gorman-Smith, D., & Tolan, P. (1998). The role of exposure to community violence and developmental problems among inner-city youth. *Development and Psychopathology, 10*, 101–116.

Grafman, J., Schwab, K., Warden, D., Pridgen, B. S., & Brown, H. R. (1996). Frontal lobe injuries, violence, and aggression: A report of the Vietnam head injury study. *Neurology, 46*, 1231–1238.

Gray, J. A. (1971). *The psychology of fear and stress.* London: Weidenfeld & Nicolson.

Gregg, T. R., & Siegel, A. (2001). Brain structures and neurotransmitters regulating aggression in cats: Implications for human aggression. *Progress in Neuropsychopharmacology and Biological Psychiatry, 25*(1), 91–140.

Hare, R. D. (1970). *Psychopathy: Theory and research.* New York: Wiley.

Hare, R. D. (1980). A research scale for the assessment of psychopathy in criminal populations. *Personality and Individual Differences, 1*, 111–119.

Hare, R. D. (1991). *The Hare Psychopathy Checklist—Revised.* Toronto: Multi-Health Systems.

Hare, R. D. (2003). *Hare Psychopathy Checklist—Revised* (2nd ed.). Toronto: Multi-Health Systems.

Hare, R. D., Clark, D., Grann, M., & Thornton, D. (2000). Psychopathy and the predictive validity of the PCL-R: An international perspective. *Behavioral Sciences and the Law, 18*, 623–645.

Hariri, A. R., Mattay, V. S., Tessitore, A., Kolachana, B., Fera, F., Goldman, D., et al. (2002). Serotonin transporter genetic variation and the response of the human amygdala. *Science, 297*(5580), 400–403.

Harpur, T. J., Hakstian, A. R., & Hare, R. D. (1988). The factor structure of the Psychopathy Checklist. *Journal of Consulting and Clinical Psychology, 56*, 741–747.

Hart, S. D., Forth, A. E., & Hare, R. D. (1990). Performance of criminal psychopaths on selected neuropsychological tests. *Journal of Abnormal Psychology, 99*, 374–379.

Hart, S., Kropp, P. R., & Hare, R. D. (1988). Performance of male psychopaths following conditional release from prison. *Journal of Consulting and Clinical Psychology, 56*, 227–232.

Hartung, C. M., Milich, R., Lynam, D. R., & Martin, C. A. (2002). Understanding the relations among gender, disinhibition, and disruptive behavior in adolescents. *Journal of Abnormal Psychology, 111*(4), 659–664.

Hawes, D. J., & Dadds, M. R. (2005). The treatment of conduct problems in children with callous–unemotional traits. *Journal of Consulting and Clinical Psychology, 73*(4), 737–741.

Hiatt, K. D., Schmitt, W. A., & Newman, J. P. (2004). Stroop tasks reveal abnormal selective attention among psychopathic offenders. *Neuropsychology, 18*, 50–59.

Hobson, J., & Shine, J. (1998). Measurement of psychopathy in a UK prison population referred for long-term psychotherapy. *British Journal of Criminology, 38*, 504–515.

Hoffman, M. L. (1970). Conscience, personality and socialization techniques. *Human Development, 13*, 90–126.

Kawasaki, H., Kaufman, O., Damasio, H., Damasio, A. R., Granner, M., Bakken, H., et al. (2001). Single-neuron responses to emotional visual stimuli recorded in human ventral prefrontal cortex. *Nature Neuroscience, 4*(1), 15–16.

Kiehl, K. A. (2006). A cognitive neuroscience perspective on psychopathy: Evidence for paralimbic system dysfunction. *Psychiatry Research, 142*(2–3), 107–128.

Kiehl, K. A., Smith, A. M., Hare, R. D., Mendrek, A., Forster, B. B., Brink, J., et al. (2001). Limbic abnormalities in affective processing by criminal psychopaths as revealed by functional magnetic resonance imaging. *Biological Psychiatry, 50*, 677–684.

King, S. M. (1999). Escape-related behaviours in an unstable, elevated and exposed environment: II. Long-term sensitization after repetitive electrical stimulation of the rodent midbrain defence system. *Behavioural Brain Research, 98*(1), 127–142.

Kosson, D. S., Cyterski, T. D., Steuerwald, B. L., Neumann, C. S., & Walker-Matthews, S. (2002). The reliability and validity of the Psychopathy Checklist: Youth Version (PCL:YV) in nonincarcerated adolescent males. *Psychological Assessment, 14*(1), 97–109.

LaPierre, D., Braun, C. M. J., & Hodgins, S. (1995). Ventral frontal deficits in psychopathy: Neuropsychological test findings. *Neuropsychologia, 33*, 139–151.

LeDoux, J. E. (2000). Emotion circuits in the brain. *Annual Review of Neuroscience, 23*, 155–184.

Leibenluft, E., Blair, R. J., Charney, D. S., & Pine, D. S. (2003). Irritability in pediatric mania and other childhood psychopathology. *Annals of the New York Academy of Sciences, 1008*, 201–218.

Levenston, G. K., Patrick, C. J., Bradley, M. M., & Lang, P. J. (2000). The psychopath as observer: Emotion and attention in picture processing. *Journal of Abnormal Psychology, 109*, 373–386.

Lykken, D. T. (1957). A study of anxiety in the sociopathic personality. *Journal of Abnormal and Social Psychology, 55*, 6–10.

Lykken, D. T. (1995). *The antisocial personalities*. Hillsdale, NJ: Erlbaum.

Miller, J. D., Flory, K., Lynam, D. R., & Leukefeld, C. (2003). A test of the four-factor model of impulsivity-related traits. *Personality and Individual Differences, 34*, 1403–1418.

Miller, L. S., Wasserman, G. A., Neugebauer, R., Gorman-Smith, D., & Kamboukos, D. (1999). Witnessed community violence and antisocial behavior in high-risk, urban boys. *Journal of Clinical Child Psychology, 28*(1), 2–11.

Mitchell, D. G., Colledge, E., Leonard, A., & Blair, R. J. R. (2002). Risky decisions and response reversal: Is there evidence of orbitofrontal cortex dysfunction in psychopathic individuals? *Neuropsychologia, 40*, 2013–2022.

Mitchell, D. G., Fine, C., Richell, R. A., Newman, C., Lumsden, J., Blair, K. S., et al. (2006a). Instrumental learning and relearning in individuals with psychopathy and in patients with lesions involving the amygdala or orbitofrontal cortex. *Neuropsychology, 20*(3), 280–289.

Mitchell, D. G., Richell, R. A., Leonard, A., & Blair, R. J. R. (2006b). Emotion at the expense of cognition: Psychopathic individuals outperform controls on an operant response task. *Journal of Abnormal Psychology, 115*(3), 559–566.

Newman, J. P. (1998). Psychopathic behaviour: An information processing perspective. In D. J. Cooke, A. E. Forth, & R. D. Hare (Eds.), *Psychopathy: Theory, research and implications for society* (pp. 81–105). Dordrecht, The Netherlands: Kluwer Academic.

Newman, J. P., & Kosson, D. S. (1986). Passive avoidance learning in psychopathic and nonpsychopathic offenders. *Journal of Abnormal Psychology, 95*, 252–256.

Newman, J. P., Patterson, C. M., Howland, E. W., & Nichols, S. L. (1990). Passive avoidance in psychopaths: The effects of reward. *Personality and Individual Differences, 11*, 1101–1114.

Oxford, M., Cavell, T. A., & Hughes, J. N. (2003). Callous–unemotional traits moderate the relation between ineffective parenting and child externalizing problems: A partial replication and extension. *Journal of Clinical Child and Adolescent Psychology, 32*, 577–585.

Panksepp, J. (1998). *Affective neuroscience: The foundations of human and animal emotions*. New York: Oxford University Press.

Patrick, C. J. (1994). Emotion and psychopathy: Startling new insights. *Psychophysiology, 31*, 319–330.

Patrick, C. J., Bradley, M. M., & Lang, P. J. (1993). Emotion in the criminal psychopath: Startle reflex modulation. *Journal of Abnormal Psychology, 102*, 82–92.

Patterson, C. M., & Newman, J. P. (1993). Reflectivity and learning from aversive events: Toward a psychological mechanism for the syndromes of disinhibition. *Psychological Review, 100*, 716–736.

Raine, A. (2002). Annotation: The role of prefrontal deficits, low autonomic arousal, and early health factors in the development of antisocial and aggressive behavior in children. *Journal of Child Psychology and Psychiatry, 43*(4), 417–434.

Rhee, S. H., & Waldman, I. D. (2002). Genetic and environmental influences on antisocial behavior: A meta-analysis of twin and adoption studies. *Psychological Bulletin, 128*(3), 490–529.

Richell, R. A., Mitchell, D. G., Newman, C., Leonard, A., Baron-Cohen, S., & Blair, R. J. (2003). Theory of mind and psychopathy: Can psychopathic individuals read the 'language of the eyes'? *Neuropsychologia, 41*(5), 523–526.

Richell, R. A., Mitchell, D. G. V., Peschardt, K. S., Winston, J. S., Leonard, A., Dolan, R. J., et al. (2005). Trust and distrust: The perception of trustworthiness of faces in psychopathic and non-psychopathic offenders. *Personality and Individual Differences, 38,* 1735–1744.

Rilling, J. K., Glenn, A. L., Jairam, M. R., Pagnoni, G., Goldsmith, D. R., Elfenbein, H. A., et al. (2007). Neural correlates of social cooperation and non-cooperation as a function of psychopathy. *Biological Psychiatry, 61*(11), 1260–1271.

Schoenbaum, G., Nugent, S. L., Saddoris, M. P., & Setlow, B. (2002). Orbitofrontal lesions in rats impair reversal but not acquisition of go, no-go odor discriminations. *NeuroReport, 13*(6), 885–890.

Schoenbaum, G., Setlow, B., Saddoris, M. P., & Gallagher, M. (2003). Encoding predicted outcome and acquired value in orbitofrontal cortex during cue sampling depends upon input from baso-lateral amygdala. *Neuron, 39*(5), 855–867.

Schwab-Stone, M., Chen, C., Greenberger, E., Silver, D., Lichtman, J., & Voyce, C. (1999). No safe haven: II. The effects of violence exposure on urban youth. *Journal of the American Academy of Child and Adolescent Psychiatry, 38*(4), 359–367.

Siever, L. J., Torgersen, S., Gunderson, J. G., Livesley, W. J., & Kendler, K. S. (2002). The border-line diagnosis: III. Identifying endophenotypes for genetic studies. *Biological Psychiatry, 51*(12), 964–968.

Strawn, J. R., & Geracioti, T. D., Jr. (2008). Noradrenergic dysfunction and the psychopharmacology of posttraumatic stress disorder. *Depression and Anxiety, 25*(3), 260–271.

Trasler, G. (1978). Relations between psychopathy and persistent criminality: Methodological and theoretical issues. In R. D. Hare & D. Schalling (Eds.), *Psychopathic behaviour: Approaches to research.* Chichester, UK: Wiley.

Verona, E., Patrick, C. J., Curtin, J. J., Bradley, M. M., & Lang, P. J. (2004). Psychopathy and physi-ological response to emotionally evocative sounds. *Journal of Abnormal Psychology, 113*(1), 99–108.

Viding, E., Blair, R. J. R., Moffitt, T. E., & Plomin, R. (2005). Evidence for substantial genetic risk for psychopathy in 7-year-olds. *Journal of Child Psychology and Psychiatry, 46,* 592–597.

Whalen, P. J., Shin, L. M., McInerney, S. C., & Rauch, S. L. (1998). Greater fMRI activation to fearful vs. angry expressions in the amygdaloid region. *Society for Neuroscience Abstracts, 24,* 692.

Whiteside, S. P., & Lynam, D. R. (2001). The five factor model and impulsivity: Using a structural model of personality to understand impulsivity. *Personality and Individual Differences, 30,* 669–689.

Williams, K. M., Paulhus, D. L., & Hare, R. D. (2007). Capturing the four-factor structure of psy-chopathy in college students via self-report. *Journal of Personality Assessment, 88*(2), 205–219.

Williamson, S., Hare, R. D., & Wong, S. (1987). Violence: Criminal psychopaths and their victims. *Canadian Journal of Behavioural Science, 19,* 454–462.

Wootton, J. M., Frick, P. J., Shelton, K. K., & Silverthorn, P. (1997). Ineffective parenting and child-hood conduct problems: The moderating role of callous–unemotional traits. *Journal of Consult-ing and Clinical Psychology, 65,* 292–300.

CHAPTER 22

Autism

Risk Factors, Risk Processes, and Outcome

GERALDINE DAWSON
LINDSEY STERLING
SUSAN FAJA

Autism is a developmental disorder characterized by impairments in social and communication behavior, and by a restricted range of activities and interests. Current estimates indicate that the full range of autism spectrum disorders (ASD) affects approximately 1 in 150 persons—a prevalence rate higher than that of type 1 diabetes, blindness, Down syndrome, childhood cancer, or cystic fibrosis (Kuehn, 2007). Advances in the fields of cognitive and affective developmental neuroscience, developmental psychopathology, neurobiology, and genetics hold promise for discovering the causes of autism, as well as effective prevention and treatment approaches. In this chapter, we review current findings regarding early brain and behavioral development in autism, and provide a perspective that offers hope for improved outcomes through early identification and intervention.

We propose a developmental model of risk factors, risk processes, and symptom emergence, illustrated in Figure 22.1. This model posits that measurable *risk factors*, including both genetic and environmental factors, have the potential to allow very early identification of infants at increased risk for developing autism symptoms. Identification of such risk factors is a focus of current research in the field. It is hypothesized that these early autism risk factors (genetic and environmental) lead to abnormalities in brain development, altering a child's pattern of responses to and perception of his or her environment. These altered interactions between child and environment, or *risk processes*, are hypothesized to disrupt critical input influencing the development, specialization, and integration of brain circuitry during early sensitive periods, thus mediating the effects of risk factors on outcome. Through this mediational process, early susceptibilities contribute to *outcome*, the development of

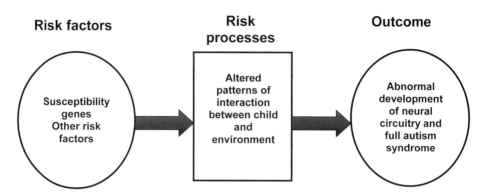

FIGURE 22.1. Experience-based risk processes in autism.

the full autism syndrome. Thus there is not a one-to-one correspondence between genetic or environmental factors and the development of autism. Rather, the possible developmental pathways for an individual child will differ as a function of the interaction between early risk factors and the context in which the child develops. Although the developmental pathway can always deviate, canalization constrains the magnitude and quality of change. Thus "the longer an individual continues along a maladaptive ontogenetic pathway, the more difficult it is to reclaim a normal developmental trajectory" (Cicchetti & Cohen, 1995, p. 7). As such, this developmental model further posits that the earlier the detection of risk factors and initiation of intervention, the greater the potential for intervention to alter the abnormal developmental trajectory of individuals with autism and to help guide brain and behavioral development back toward a normal pathway (and, in some cases, to prevent the full syndrome of autism from becoming manifest).

We begin by briefly reviewing findings on genetic risk factors in autism. This is followed by a review of the earliest indications of abnormal behavioral and brain development in autism, and a discussion of some of the models that have been proposed to account for these abnormalities. Finally, we discuss how early intervention might alter the course of early brain and behavioral development in autism.

Evidence of Genetic Risk Factors in Autism

Evidence from twin studies provides strong support for the contribution of genetic risk factors in the development of autism. Concordance rates for autism for monozygotic (MZ) twins have ranged from 69% to 95% (Bailey et al., 1995; Folstein & Rutter, 1977a, 1977b; Ritvo, Freeman, Mason-Brothers, Mo, & Ritvo, 1985; Steffenburg et al., 1989), whereas for dizygotic twins (DZ) from the same studies, concordance rates were 0–24%. When a broader autism phenotype (e.g., language and/or social impairment) was used, concordance ranges increased for both MZ (88–91%) and DZ (9–30%) twins (Bailey et al., 1995; Folstein & Rutter, 1977b; Steffenburg et al., 1989).

Sibling risk rates for autism, which range from 2.8% to 7.0%, are higher than rates found in control families (August, Stewart, & Tsai, 1981; Bailey, Phillips, & Rutter, 1996; Smalley, Asarnow, & Spence, 1988). Furthermore, siblings who do not meet criteria for autism nevertheless show an elevated rate of autism-related symptoms, including social dysfunction and isolation (Piven et al., 1990). A number of studies have indicated that unaffected relatives exhibit elevated rates of autism-related symptoms (Bailey et al., 1995, 1996; Baker, Piven, Schwartz, & Patil, 1994; Folstein & Rutter, 1977b; Landa, Folstein, & Isaacs, 1991; Landa et al., 1992; Narayan, Moyes, & Wolff, 1990; Wolff, Naravan, & Moyes, 1988).

Although the mode of inheritance in autism is not completely understood, most cases of autism appear to be caused by multiple genes interacting epistatically, perhaps with 2–10 contributing loci (Pickles et al., 1995). The hypothesis that multiple autism susceptibility genes must be present for autism to occur is based on the rapid falloff in risk rates with decreasing genetic relatedness. Concordance rates for MZ twins is much higher than the concordance rates for DZ twins or the risk to siblings of autistic subjects (Bailey et al., 1995; Folstein & Rutter, 1977a, 1977b; Ritvo et al., 1985; Steffenburg et al., 1989), and the rate of autism found in relatives of probands who are more distant than siblings is extremely low (Delong & Dwyer, 1988).

Many genome-wide linkage studies of autism have been conducted (Auranen et al., 2002; Barrett et al., 1999; Buxbaum et al., 2001; Cantor et al., 2005; International Molecular Genetic Study of Autism Consortium, 2001; Lamb et al., 2005; Liu et al., 2001; McCauley et al., 2005; Philippe et al., 1999; Risch et al., 1999; Schellenberg et al., 2006; Shao et al., 2002; Stone et al., 2004; Yonan et al., 2003), and over 100 genes have been tested as candidates for autism susceptibility loci. Some of these candidate genes offer promise. An association between autism and Engrailed 2 (EN-2), located on chromosome 7, has been reported. EN-2 is expressed primarily in the cerebellum during nervous system development and is critical for cerebellar development in mice (Cheh et al., 2006; Millen, Wurst, Herrup, & Joyner, 1994). Abnormalities in cerebellar development have been consistently found in individuals with autism, including reduced Purkinje cells in the cerebellar cortex (Bailey et al., 1998; Courchesne, 1997, 2004; Kemper & Bauman, 1998; Ritvo et al., 1986). EN-2 knockout mice have a reduction in Purkinje cells and a decreased size of the cerebellar lobes (Kuemerle, Zanjani, Joyner, & Herrup, 1997; Millen et al., 1994); they also display a number of autism-like behaviors, including reduced social play, failure to differentiate responses to intruders from those to conspecifics, and an increase in repetitive behavior (Cheh et al., 2006).

The serotonin transporter gene, SLC6A4, has been tested as a candidate gene, yielding positive but not entirely consistent results (reviewed in Devlin et al., 2005). Elevated levels of platelet serotonin (5-HT) have been reported in autism (Rolf, Haarmann, Grotemeyer, & Kehrer, 1993). Selective 5-HT reuptake inhibitors are frequently used in the psychopharmacological treatment of autism. Serotonin facilitates neuronal development; modulates sensory input and arousal; and affects behaviors such as sleep, mood, aggression, impulsivity, and affiliation (Lucki, 1998). It also innervates the limbic regions involved in emotional expression and social behavior. Evidence for involvement of the serotonin transporter gene in autism has been demonstrated in two studies. In the first study, using a large, independent family-based sample (390 families, 1528 individuals), Devlin et al. (2005) investigated the impact of alleles at the SLCGA4 promoter region (5-HTTLPR) and three other polymorphisms within SLC6A4. Devlin and colleagues found an excess transmission of the short allele of 5-HTTLPR both for individuals with the narrow diagnosis of autism ($p = .035$) and for those with the broader diagnosis of ASD ($p = .007$). In a second study, investigators examined relations between variability in 5-HTTLPR and early abnormalities in brain growth in a sample of children with autism. Enlargement of the cerebrum, especially in early life, is a highly replicated biological finding in autism (discussed in more detail below).

Findings indicated that the short allele was strongly associated with increased cortical gray matter when samples of magnetic resonance imaging (MRI) data were combined with DNA collected on 2- to 4-year-old males with ASD (Wassink et al., 2007). These findings are the first to establish a direct association between a genetic marker and abnormalities in brain development in autism.

Several single-gene disorders have been associated with increased risk for autism or expression of an autism-like phenotype, including fragile X syndrome, Rett syndrome, Angelman syndrome, tuberous sclerosis, and neurofibromatosis (see Veenstra-VanderWeele & Cook, 2004, for a review). The 15q11–q13 region associated with Angelman syndrome codes for subunits of the gamma-aminobutyric acid-A (GABA-A) receptor. GABAergic interneurons have a role in establishing the architecture of cortical columns (DeFelipe, Hendry, Hashikawa, Molinari, & Jones, 1990; Peters & Sethares, 1997). The increased prevalence of epilepsy in individuals with autism and 15q11–q13 duplications is consistent with the involvement of GABA. Hippocampal GABA receptor binding in autism is abnormally low (Blatt et al., 2001), as are platelet GABA levels (Rolf et al., 1993).

Based on linkage results showing consistent findings on 7q, Levitt and colleagues (Campbell et al., 2006) analyzed the gene encoding the methionine (MET) receptor tyrosine kinase and showed a genetic association between the C allele in the promoter region of the MET gene. MET signaling is involved in neocortical and cerebellar development, immune function, and gastrointestinal repair. Another study (Szatmari et al., 2007), involving collaboration among 50 institutions that pooled genetic data from 1200 multiplex families, found evidence that autism was associated with neurexin 1. Neurexin 1 is part of a family of genes affecting the neurotransmitter glutamate, which is involved in both synaptogenesis and learning.

In summary, there is strong evidence for genetic influences in autism; however, the role of susceptibility genes in autism is complex. Evidence thus far indicates that the interaction of multiple genes influences gene expression or encodes functional changes in proteins that are part of complex regulatory networks, thereby increasing susceptibility to the development of autism. Furthermore, environmental factors affect the expression of many genes, offering hope that early intervention can alter genetic expression, brain development, and behavioral outcome. In the next section, we provide a brief overview of what is currently known about the early behavioral manifestations of autism in infancy.

Early Behavioral Development in Autism

Behavioral Symptoms Apparent in Infancy

Home videotape studies suggest that many infants at risk for autism show very few, if any, behavioral symptoms at 6 months; however, by 12 months, core autism symptoms are apparent. One study found that at 6 months, infants who later developed autism were distinguished from typically developing children by slightly lower frequency of attempts to seek physical contact, vocalizations, looking at faces, and smiling at others (Maestro et al., 2002). Failure to respond to name has been documented as a symptom by 8–10 months (Werner, Dawson, Osterling, & Dinno, 2000), and is a feature that distinguishes infants later diagnosed with autism from typical infants by 12 months (Baranek, 1999; Osterling & Dawson, 1994; Osterling, Dawson, & Munson, 2002). Other behaviors that distinguish children with autism from typically developing children by 12 months include less time spent looking at the faces of others, as well as fewer instances of showing things to others and of pointing to request or to share interest (Adrien et al., 1993; Maestro et al., 2002; Osterling & Dawson, 1994; Osterling et al., 2002; Werner & Dawson, 2005).

Robust differences in social communication skills—including sharing experiences, interests, and attention with others by pointing or showing objects—emerge by the second year of life (Maestro et al., 2001; Werner & Dawson, 2005). Moreover, the language development of children with autism diverges from the normal course by age 2, as reflected by reduced ability to follow verbal instructions, babble or make complex vocalizations, vocally imitate, or use single words and phrases (Mars, Mauk, & Dowrick, 1998; Werner & Dawson, 2005). Not looking at people and not responding to their names when called continue to distinguish children with autism beyond their first birthday as well (Mars et al., 1998; Werner & Dawson, 2005).

Prospective studies of infants at risk for autism (most commonly infants with an older, affected sibling) have also greatly informed the understanding of early emergence of symptoms. In one such study, high-risk and low-risk infant siblings were assessed with the Autism Observation Scale for Infants (Bryson, Zwaigenbaum, McDermott, Rombough, & Brian, 2008), which measures visual attention, response to name, response to a brief still face, anticipatory responses, imitation, social babbling, eye contact and social smiling, reactivity, affect, ease of transitioning, and atypical motor and sensory behaviors (Zwaigenbaum et al., 2005). Although these markers did not distinguish groups at 6 months of age on the basis of their diagnostic classification at 24 months, a subset of the children who were later diagnosed exhibited impairments in responding to name or unusual sensory behaviors. By 12 months, groups could be distinguished significantly by the presence of markers. Specifically, the most predictive markers included atypical eye contact, visual tracking, disengaging visual attention, orienting to name, imitation, social smiling, reactivity, social interest, and sensory-oriented behaviors. Landa and Garrett-Mayer (2006) reported prospective longitudinal data following the cognitive development of high-risk infant siblings who later developed ASD, compared with language-delayed and unaffected groups. By 14 months, the children who received ASD differed from the unaffected groups in gross and fine motor abilities, receptive and expressive language, and overall intelligence on the Mullen Scales of Early Learning (Mullen, 1995).

Social Behavior Abnormalities in the Toddler–Preschool Period

At least five domains of social behavior are affected in autism by the toddler–preschool period. These domains are social orienting, attention to emotional cues, joint attention, motor imitation, and face processing. Each of these is discussed briefly.

Dawson and her coworkers coined the term "social orienting impairment" to refer to the failure of young children with autism to orient spontaneously to naturally occurring social stimuli in their environment (Dawson, Meltzoff, Osterling, & Brown, 1998a). Compared to children with mental retardation without autism and typically developing children, children with autism more frequently fail to orient to both social and nonsocial stimuli, but this failure is much more extreme for social stimuli (Dawson, Meltzoff, Osterling, Rinaldi, & Brown, 1998c; Dawson et al., 2004a). Typical infants, on the other hand, show remarkable sensitivity to social stimuli very early in life (Rochat, 1999). Neonates are naturally attracted to people, including human sounds, human movements, and features of the human face (Maurer & Salapatek, 1976; Morton & Johnson, 1991). This inherent sensitivity and attention to the social world is reflexive rather than voluntary, and it probably provides the foundation for the acquisition of subsequent social behaviors (Rochat & Striano, 1999). As mentioned above, failure to orient to social stimuli is one of the earliest and most basic social impairments in autism; this almost certainly contributes to the later social and communicative impairments observed in the disorder (Dawson, Meltzoff, Osterling, & Rinaldi, 1998b; Mundy & Neal, 2001).

The second domain of impairment, "joint attention," refers to the ability to coordinate attention between interactive social partners with respect to objects or events in order to share an awareness of the objects or events (Mundy, Sigman, Ungerer, & Sherman, 1986). Joint attention behaviors include sharing attention (e.g., through the use of alternating eye gaze), following the attention of another (e.g., following eye gaze or a point), and directing the attention of another (see Carver & Cornew, Chapter 7, this volume, for further discussion of joint attention). Typically developing infants generally demonstrate all of these skills by 12 months of age (Carpenter, Nagell, & Tomasello, 1998). Joint attention ability has been unequivocally established as an early-emerging and fundamental impairment in autism, present by 1 year of age and incorporated into the diagnostic criteria for autistic disorder (American Psychiatric Association, 2000; Mundy et al., 1986). Joint attention skills have also been found to be good predictors of both concurrent and future language abilities in children with autism (Dawson et al., 2004a).

The third core domain of impairment is a failure to attend and respond appropriately to others' emotions. Typically developing infants show great sensitivity to the emotions displayed by others. Within the first 6 months of life, infants begin to attend to the affective displays of others (Trevarthen, 1979) and differentially respond to faces showing different emotions (e.g., neutral, happy, sad). By 9–12 months of age, social referencing is established, whereby children seek emotional information from an adult's face when presented with a stimulus of uncertain valence (Feinman, 1982; Moore & Corkum, 1994). Children begin to respond to another person's distress affectively and prosocially by helping, comforting, and sharing by 2 years of age (Rheingold, Hay, & West, 1976; Zahn-Waxler & Radke-Yarrow, 1990).

Many children with autism demonstrate a lack of sensitivity to others' emotional states. Studies have shown that when adults displayed facial expressions of distress, children with autism looked less at the adults, and showed less concern than did children with either mental retardation or typical development (Bacon, Fein, Morris, Waterhouse, & Allen, 1998; Charman et al., 1998; Dawson et al., 1998b, 2004a; Sigman, Kasari, Kwon, & Yirmiya, 1992). To examine whether young children with autism responded differentially to distinct emotional facial expressions (neutral vs. fearful), Dawson, Webb, Carver, Panagiotides, and McPartland (2004b) measured brain response via event-related potentials (ERPs). Differential ERPs to facial expressions depicting distinct emotions have previously been shown in adults (Eimer & Holmes, 2002), in children (de Haan, Nelson, Gunnar, & Tout, 1998), and even in infants as young as 7 months of age (Nelson & de Haan, 1996). Compared to typically developing children, 3- to 4-year-old children with autism exhibited significantly slower early (N300) brain responses to the facial expression of fear. Children with autism also failed to show the typical children's larger-amplitude negative slow-wave responses to the fearful face. The delayed response to the fearful face suggests that information-processing speed is compromised. Moreover, the children with autism displayed aberrant ERP scalp topography in response to the fearful face, suggesting a failure of cortical specialization or atypical recruitment of cortical areas. Finally, individual differences in N300 latency to the fearful face were related to performance on behavioral tasks requiring social attention.

The fourth domain of impairment is in the ability to imitate others. In normal development, imitation ability develops rapidly; by 1 year of age, infants are able to imitate actions on objects, as well as such gestures as waving. The importance of imitation in social develop-ment has long been recognized (see Decety & Meyer, Chapter 8, and Myowa-Yamakoshi & Tomanga, Chapter 11, this volume, for further discussions of imitation). The ability to imitate facilitates a child's social connectedness with others and the capacity to differentiate self from others (Meltzoff & Gopnik, 1993; Eckerman, Davis, & Didow, 1989; Uzgiris, 1981; Nadel, Guerini, Peze, & Rivet, 1999; Trevarthen, Kokkinaki, & Fiamenghi, 1999). Through imitation, children begin to perceive and understand others' intentions and goals (Uzgiris, 1999; Kugiumutzakis, 1999). This probably serves as a precursor for the development of a theory of mind (Meltzoff & Gopnik, 1993; Rogers & Pennington, 1991). Imitation is also an important component of symbolic play (Piaget, 1962), peer relationships (Trevarthen et al., 1999), language (Avikainen, Wohlschlager, Liuhanen, Hanninen, & Hari, 2003; Charman et al., 2003), and emotional sharing (Hatfield, Cacioppo, & Rapson, 1994). Children with autism have early core impairments in the ability to imitate others spontaneously, especially in social play contexts (Dawson & Adams, 1984; Dawson & Lewy, 1989; Rogers, Bennetto, McEvoy, & Pennington, 1996; Rogers & Pennington, 1991; Williams, Whiten, Suddendorf, & Perrett, 2001). Failure to imitate is specific to autism, and discriminates toddlers with autism from those with mental retardation or a communication disorder (Stone, Lemanek, Fishel, Fer-nandez, & Altemeier, 1990; Stone, Ousley, & Littleford, 1997). Functional MRI (fMRI) and EEG studies of mirror neuron activities, which are activated while observing and perform-ing an action, have documented deficits in this neural system in autism during imitation tasks (Dapretto et al., 2006; Bernier, Dawson, Webb, & Murias, 2007; see also Decety & Meyer, Chapter 8, this volume).

Finally, the fifth domain of impairment is in face processing. During typical develop-ment, face recognition ability is present very early in life (see Pascalis, Kelly, & Schwarzer, Chapter 4, this volume). The neonate's capacity for very rapid face recognition and a visual preference for faces are evident at birth (Goren, Sarty, & Wu, 1975; Walton & Bower, 1993). Infants as young as 4 months recognize upright faces better than upside-down faces (Fagan, 1972). By 6 months, infants show differential ERPs to familiar versus unfamiliar faces (de Haan & Nelson, 1997, 1999). Face-processing abilities in individuals with autism have con-sistently been shown to be impaired. Using ERPs, Dawson et al. (2002a) examined face rec-ognition abilities in 3- to 4-year-old children with autism and typical development. High-density ERPs were recorded from the children while they watched images of familiar (their mothers) and unfamiliar (another female) faces, and familiar (a favorite toy) and unfamiliar (a novel toy) objects. The typical children demonstrated increased amplitude when presented with novel faces and novel objects for two ERP components. Children with autism failed to show differential ERP responses to novel versus familiar faces, but did show amplitude dif-ferences when viewing objects. These findings indicated that face recognition impairments are manifested as early as 3 years of age in children with autism.

Early Behavioral Development: Summary

In summary, it appears that symptoms of autism are mild and not readily apparent by 6 months of age; however, by 8–12 months of age, several behavioral symptoms associated with autism can be observed. By the toddler–preschool period, social impairments in social orient-ing, joint attention, responses to emotion, imitation, and face processing are readily apparent. These social impairments, in combination with delayed and atypical language development and with a restricted range of activities and repetitive behaviors, constitute the syndrome of autism in early childhood. At the same time that the early symptoms of autism are emerging during the first year of life, studies suggest that abnormalities in brain development are also apparent. In the next section, we provide a review of evidence for early abnormal brain development in children with autism.

Evidence of Early Abnormal Brain Development in Autism

Whereas several behavioral symptoms of autism are apparent during the first year of life, few early biological risk indices have yet been identified. One risk index is an atypical pattern of growth in head circumference (HC), characterized by small to normal head size at birth, followed by an accelerated pattern of growth that appears to begin at about 4 months of age (Courchesne & Pierce, 2005; Redcay & Courchesne, 2005; Gillberg & de Souza, 2002). Courchesne, Carper and Akshoomoff (2003) reported an increase in HC of 1.67 standard deviation (*SD*) between birth and 6–14 months. In a meta-analysis using HC (converted to brain volume), brain volume measured from MRI, and brain weight from autopsy studies, Redcay and Courchesne (2005) found that brain size changed from 13% smaller than controls at birth to 10% greater than controls at 1 year, and only 2% greater by adolescence.

Dawson et al. (2007) examined HC growth longitudinally in 28 children with ASD from birth through 36 months of age, replicating earlier findings of atypical head growth. Whereas children with ASD, on average, did not have significantly larger HC at birth, HC was nearly 1 *SD* larger than the national Centers for Disease Control and Prevention norms by 1 year of age. In comparison, the rate of growth in HC after 12 months was not significantly different from that in the normative sample. In other words, these longitudinal data suggest that rate of HC growth decelerates in infants with autism after 12 months of age, relative to the rate from birth to 12 months of age, and that the early period of exceptionally rapid head growth is restricted to the first year of life. Thus the period of accelerated head growth appears to precede and then coincide with the onset of autism behavioral symptoms. Interestingly (and perhaps coincidentally), the period after 12 months of age, marked by deceleration of head growth rate, appears to be associated with a slowing in acquisition or actual loss in skills in infants with autism (Dawson et al., 2007).

Structural Brain Imaging in Young Children with Autism

MRI studies provide evidence consistent with the results of HC studies. For example, Sparks et al. (2002) showed that 3- to 4-year-olds with ASD had significantly larger total cerebral volume than that of controls. In another study of 2- to 4-year-olds with autism, 90% of the children with autism were found to have MRI-based brain volumes larger than normal (Courchesne et al., 2001). This abnormal brain growth appears to be due primarily to excessive enlargement of cerebellar and cerebral white matter and cerebral gray matter. At 2–4 years old, children with autism show an anterior-to-posterior gradient of overgrowth, with the frontal lobe being the largest (Courchesne et al., 2001). Alterations in early cellular developmental processes, such as failure of apoptosis or synaptic pruning, have been hypothesized to account for early cerebral enlargement, but longitudinal magnetic resonance spectroscopic (MRS) findings of brain chemical changes in children with ASD do not support such mechanisms.

MRI studies have revealed several other abnormalities, especially of the medial temporal lobe (MTL) structures, and particularly the amygdala. These structures have been strongly implicated in autism-related symptom expression (Munson et al., 2006). Munson et al. (2006) reported amygdala enlargement (relative to total brain volume) in 3- to 4-year-old children with autism; larger amygdala volume was associated with a more severe course during the preschool period. Imaging studies have identified volumetric amygdalar abnormalities, which, similar to cerebral volume, vary as a function of age (e.g., Aylward et al., 1999; Schumann et al., 2004). Reviews examining structural and functional imaging findings in autism (Baron-Cohen et al., 2000; Schultz, 2005) have highlighted the potential role of early amygdalar developmental deficits in affecting social perception and social behavior. Autopsy studies of autism (Pickett & London, 2005) have implicated the amygdala by identifying histopathological features reflecting reduced numbers of neurons, particularly in adults (Schumann & Amaral, 2006), or reduced cell size and increased neuronal cell-packing density (Bauman & Kemper, 1985, 2005). From their postmortem studies of the amygdala, Schumann and Amaral (2006) have identified the lateral nucleus as having accentuated pathological features.

Chemical Brain Imaging Studies in Young Children with Autism

Whereas many studies have utilized MRI to investigate autism, relatively few have used brain proton MRS ([1H] MRS), a noninvasive method for characterizing tissue-based chemistry and cellular features *in vivo*. Although MRI is sensitive to changes in tissue water characteristics and defining structure at a macroscopic level, it is insensitive to much of cellular level organization. In this regard, [1H] MRS has been used to detect abnormalities in brain regions that appear normal in MRI, as well as to elucidate pathology underlying MRI-visible abnormalities. In brain tissue, the MRI-visible concentrations and mobility of several low-molecular-weight chemicals can be measured as spectral peaks, including N-acetyl aspartate (NAA), creatine, choline, and myo-inositol. Glutamate and glutamine have complicated peak shapes and resonate at similar spectral locations, resulting in a common description of the combined peaks (referred to as "GLX"). Found only in the nervous system, NAA appears to be a sensitive marker for neuronal integrity or neuronal–glial homeostasis.

A MRS study of 3- to 4-year-old children with ASD revealed regional and global decreases in NAA, as well as lower levels of other chemicals and prolonged chemical T2 relaxation times (Friedman et al., 2003). Analyses further demonstrated a predominant gray matter tissue distribution of these chemical abnormalities (Friedman et al., 2006). New evidence from longitudinal findings suggests that this pattern of chemical alterations in ASD generally persists between 3–4 and 6–7 years of age (S. Dager, August 1, 2006, personal communication). These findings have implications for understanding the mechanism for abnormal brain growth in autism, predominantly affecting gray matter at an early age. These alterations may reflect reduced synapse density, an artifact of migratory/apoptotic abnormalities (Fatemi & Halt, 2001), column density/packing abnormalities (Casanova, 2004), and/or active processes such as reactive gliosis and edema (Vargas, Nascimbene, Krishnan, Zimmerman, & Pardo, 2005).

Interestingly, in the same longitudinal sample mentioned above, there were new findings of elevated gray matter GLX at age 6–7 years (S. Dager, August 1, 2006, personal communication), not seen at the earlier age point. Theories have been put forth that autism is associated with reduced inhibitory balance (Belmonte et al., 2004) associated with atypical levels of glutamate. Research has also found elevated glutamate levels in association with temporal lobe epilepsy foci (Petroff, Errante, Rothman, Kim, & Spencer, 2002). Autism is associated with a high risk for seizures, especially of the temporal lobe.

Abnormal Brain Development: Summary

In summary, autism is associated with early abnormalities in brain development. Autism is not a static brain disorder, but rather is characterized by dynamic postnatal changes in the brain and behavior. To return to the model presented at the beginning of this chapter, various early genetic *risk factors* lower the threshold of vulnerability to suboptimal neuronal processes. It is likely that gene–environment and brain–environment interactions constitute the additional *risk processes* that contribute to the eventual *outcome* (i.e., manifestation of the full blown autism syndrome). Environmental contributions to risk processes probably include both biological and experiential factors. We next discuss how early experiential factors—namely, altered patterns of interaction between the child and his or her social environment—constitute one type of risk process associated with the development of autism.

Emergence of Social Brain Circuitry in Autism: Altered Patterns of Social Interaction as Risk Processes

We begin by describing a model of the emergence of social brain circuitry in the first year of life and discussing how the trajectory of normal development of this circuitry is altered in autism. As described above, early deviations from normal brain development result in a failure of normal social and communicative development in autism that is apparent early in life. Impairments in social orienting, joint attention, responses to emotions, imitation, and face processing are evident by toddlerhood. To help explain this wide range of impairments, all of which involve reduced attention to social input, Dawson and others have proposed the "social motivation hypothesis." This hypothesis posits that some of the social impairments evident in autism are not fundamental, but rather are secondary to a primary impairment in social motivation or affective tagging of socially relevant stimuli (Dawson et al., 2002a, 2005; Grelotti, Gauthier, & Schultz, 2002; Schultz, 2005; Waterhouse, Fein, & Modahl, 1996). Findings from empirical studies have indicated that young children with autism are less likely to smile when looking at their mothers during social interaction (Dawson, Hill, Galpert, Spencer, & Watson, 1990), or to express positive emotion during joint attention episodes (Kasari, Sigman, Mundy, & Yirmiya, 1990). In addition, young children with autism fail to show normal preferences for social-linguistic stimuli (Klin, 1991; Kuhl, Coffey-Corina, Padden, & Dawson, 2004). Evidence from the Broader Phenotype Autism Symptom Scale suggests that the social motivation trait may have a genetic basis in autism, because social motivation appears heritable among affected children and family members of multiplex autism families (Sung et al., 2005).

Consistent with the social motivation hypothesis, reduced social motivation results in less attention to social stimuli, such as faces, the human voice, and hand gestures. Previously, Dawson et al. (2002a) suggested that impaired social motivation in autism is related to a difficulty in forming representations of the reward value of social stimuli. The dopamine system is one of the primary neural systems involved in processing reward information (Schultz, 1998), including social rewards such as eye contact (Kampe, Frith, Dolan, & Frith, 2001). Dopaminergic projections to the striatum and frontal cortex, particularly the orbital frontal cortex, are crucial in mediating the effects of reward on approach behavior (Gingrich, Liu, Cascio, Wang, & Insel, 2000). Input from basolateral amygdala facilitates formation of representations of reward value in the orbital frontal cortex (Schoenbaum, Setlow, Saddoris, &

Gallagher, 2003). Dawson et al. (2002b) reported that severity of joint attention impairments in young children with autism strongly correlates with performance on neurocognitive tasks that recruit the MTL–orbital frontal circuit. The social motivation impairments found in autism may be due in part to early dysfunction of the dopamine reward system, especially in social contexts.

Oxytocin and Its Relation to the Dopamine Reward System

Waterhouse et al. (1996) hypothesized that atypical oxytocin system function reduces social bonding and affiliation in autism. The work of Insel and colleagues allows further speculation regarding the role of neuropeptides in autism and the potential genetic basis for impairments in social motivation (see Insel & Fernald, 2004, for a full discussion of this hypothesis; see chapters in Part IV of this volume for further discussions of the role of neuropeptides in social relationships). Insel (1997) has discussed the role of peptides—specifically, oxytocin and vasopressin—in the modulation of the dopamine reward circuit in social contexts (Pedersen, Caldwell, Walker, Ayers, & Mason, 1994). These peptides are particularly important for linking social signals to reinforcement pathways. Several animal studies have demonstrated that vasopressin and oxytocin are critical for facilitating "social memory." For example, oxytocin knockout mice show a profound and specific deficit in social memory (Ferguson, Young, Hearn, Insel, & Winslow, 2000; Ferguson, Young, & Insel, 2002; Nishimori et al., 1996), supporting the notion that social memory has a neural basis that is distinct from other forms of memory. Both oxytocin and vasopressin appear to play a role in a range of social behaviors, including social affiliation (Witt, Winslow, & Insel, 1992), maternal behavior (Pedersen et al., 1994), and social attachment (Insel & Hulihan, 1995; Winslow, Hasting, Carter, Harbaugh, & Insel, 1993).

Oxytocin and vasopression may affect social behavior via their influence on the mesocorticolimbic dopamine circuit. Specifically, Insel and Fernald (2004) suggest that a circuit linking the anterior hypothalamus to the ventral tegmental area and the nucleus accumbens may be especially important for mediating sensitivity to social reward in the context of social interaction. Like Insel, O'Brien, and Leckman (1999), Dawson and colleagues have speculated that autism involves abnormalities in oxytocin and vasopressin, which have an impact on functioning of the dopamine reward pathway, specifically in the context of social interactions. Several studies have found abnormalities in oxytocin and vasopressin in autism. Specifically, plasma concentration of oxytocin is reduced in children with autism (Modahl et al., 1998). In addition, Kim et al. (2002) found nominally significant transmission disequilibrium between an AVPR1A microsatellite and the presence of autism. AVPR1A is a V1a receptor in the brain that mediates vasopressin activity. Studies have also found an association between the oxytocin receptor gene (OXTR) and autism (Jacob et al., 2007; Wu et al., 2005). Moreover, research has demonstrated that intravenous oxytocin administration can reduce repetitive behavior (Hollander et al., 2003) and increase comprehension of affective meaning (Hollander et al., 2007) in individuals with ASD.

Emergence of Social Brain Circuitry in the First Year of Life

Dawson et al. (2005) have posited a developmental model, emphasizing the importance of the reward system, for the normal emergence of social brain circuitry during early infancy. The model, drawing upon the work of Insel and colleagues, posits that oxytocin modulates the dopamine reward system and is important for shaping the infant's early preference for and attention to social stimuli. As noted earlier, typically developing infants are particularly drawn to people, especially to the sounds, movements, and features of the human face (Maurer & Salapatek, 1976; Morton & Johnson, 1991). Infants as young as 5 months of age are sensitive to even very small deviations in eye gaze during social interactions with adults, smiling and attending less when eyes are averted (Symons, Hains, & Muir, 1998). In these earliest stages, an infant's orienting behavior is involuntary rather than intentional. Later-emerging aspects of social cognition likely depend on this very early propensity to devote particular attention to faces (Rochat & Striano, 1999). In typical development, infants 5–7 months of age turn their heads when their names are called, and match the direction of their mothers' head turns to a visible target as early as 6 months of age (Morales, Mundy, & Rojas, 1998).

By 7 months of age, the typical infant spontaneously and intentionally orients to naturally occurring social stimuli in his or her environment. Dawson et al. (2002a) hypothesize that the anticipation of pleasure (reward) associated with social stimuli drives infants to orient to such stimuli. This type of interaction involves activation of the reward circuit, including parts of the prefrontal regions (e.g., the orbital frontal cortex) that are involved in forming reward representations. As infants gain experience with faces and voices in the context of social interactions, cortical specialization for faces, speech, and other types of social stimuli develops. With increasing experience, areas of the brain specialized for the processing of social stimuli, such as the fusiform gyrus and superior temporal sulcus, become increasingly integrated with other regions, such as those involved in reward (e.g., amygdala) and actions and attention (cerebellum, prefrontal/cingulate cortex). This leads to the development of a more complex social brain circuitry system supporting more complex behaviors, including disengagement of attention, joint attention, intentional communication, and delayed imitation.

Implications for Autism

One of the earliest symptoms of autism is a lack of "social orienting" (Dawson et al., 1998a, 2002a). This reduced attention to social stimuli may result in a failure to develop expertise in face and language processing (Dawson et al., 2005; Grelotti et al., 2002). Because cortical specialization depends on experience (Nelson, 2001), reduced attention to people, including their faces, gestures, and speech, would lead to a failure of specialization of regions that typically mediate social cognition. This would be reflected in decreased cortical specialization and abnormal brain circuitry for social cognition, resulting in slower information-processing speed when a child is presented with social information (e.g., faces).

A simple lack of exposure to social information is not responsible for the abnormal trajectory for brain development in autism. Like typically developing infants, infants with autism receive ample social input during development; they are held, talked to, and fed by their parents during face-to-face interactions. Despite exposure to social information, an infant with autism may not actively attend to another person's face or voice, or perceive social information within a larger social/affective context, if such interactions are not inherently interesting or rewarding to the infant. Recent research by Kuhl and colleagues (Kuhl, Tsao, & Liu, 2003; Kuhl, 2007) has demonstrated that simple exposure to language does not necessarily facilitate the development of brain circuitry specialized for language. Rather, the infant must experience language within a social interactive context for speech perception to develop. Furthermore, Kuhl et al. (2004) found that listening preferences in 3- to 4-year-old children with autism differed dramatically from those of typically developing children: Children with autism preferred listening to mechanical-sounding auditory signals rather than human speech ("motherese"). This preference was significantly correlated with lower language ability, more severe autism symptoms, and abnormal ERPs to speech sounds. We hypothesize that a failure to affectively tag social stimuli as relevant prevents proper attention to such stimuli, thus impeding cortical specialization for face and language brain regions. Consequently, increasingly complex behaviors requiring integration of social stimuli with intentional movements and attention, such as joint attention, fail to emerge.

Fortunately, early interventions appropriate for young children with autism, and targeting the domains of social functioning, are becoming increasingly available. Evidence suggests that early intervention can alter the developmental trajectory of many young children with autism, and potentially redirect the course of development toward a more normative trajectory. To provide a better understanding of neuroplasticity in autism, we next briefly review the animal literature on the role of environmental stimulation in altering early behavior and brain development.

Early Environmental Enrichment and Brain Plasticity

Brain and behavioral plasticity resulting from environmental manipulation with animals, including animal models of genetic, developmental, and degenerative disorders, has been reported in recent years (see Lewis, 2004, and Nithianantharajah & Hannan, 2006, for excellent reviews of the enrichment literature).

Environmental enrichment has produced behavioral changes such as improved learning and memory, increased exploration, and decreased fearful responding to novelty (e.g., Benaroya-Milshtein et al., 2004; Duffy, Craddock, Abel, & Nguyen, 2001; Escorihuela, Tobena, & Fernandez-Teruel, 1995; Schrijver, Bahr, Weiss, & Wurbel, 2002; Wong & Jamieson, 1968). Enrichment can also counteract the effects of earlier environmental stressors; for example, it can bring about a reduction of exaggerated stress responses in prematurely weaned pups (Bredy, Humpartzoomian, Cain, & Meaney, 2003; Francis, Diorio, Plotsky, & Meany, 2002).

In addition to changes in behavior, the environment directly affects brain development and neural plasticity. Tissue changes include the weight and thickness of the cortex, the density or affinity of neurotransmitter receptors, and increased numbers of synapses and density of dendritic branching (Bredy et al., 2003; Diamond, Rosenzweig, Bennett, Linder, & Lyon, 1972). These changes at the synapse, as well as increases in the number of neurons in regions such as the hippocampus, have been induced in adult animals (Greenough, Volkmar, & Juraska, 1973; Kempermann, Kuhn, & Gage, 1997). Molecular changes also result from enrichment such as modulation of the genetic expression of neurotransmitter pathways, differential transcription of neurotransmitter-related target genes, and increased neurotrophic factors (Pham, Winblad, Granholm, & Mohammed, 2002; Rampon et al., 2000). Long-term potentiation of synapses via increased excitatory responses results from enrichment and may be a cellular representation of memory (e.g., Foster, Gagne, & Massicotte, 1996). Finally, during development, enrichment may be a neuroprotective factor responsible for inhibition of spontaneous apopotosis (Young, Lawlor, Leone, Dragunow, & During, 1999).

The effects of enrichment on attenuating or reversing symptoms in animal models where there is an initial vulnerability due to neural insult or deprivation is particularly pertinent to autism. For example, in strains of mice bred for high- or low-avoidance behaviors, enrichment increased incentive seeking in both high- and low-avoidance lines (e.g., Fernandez-Teruel et al., 2002). In deer mice, restricted, repetitive motor behaviors, similar to those observed in individuals with autism, develop as a result of standard housing conditions. However, if exposed to enriched rather than standard environments early in development, the mice do not develop motor stereotypies, suggesting a critical period for the emergence of such behaviors (e.g., Powell, Newman, McDonald, Bugenhagen, & Lewis, 2000; Turner, Lewis, & King, 2003; Turner, Yang, & Lewis, 2002). In a rat model of autism, enrichment reversed most behaviors associated with the model, including the frequency of social behavior and latency to social exploration, sensitivity to sensory input, and anxious behavior during learning tasks (Schneider, Turczak, & Przewlocki, 2006). The role of enrichment in animal models of genetic disorders affecting brain development has also been investigated. FMR1 knockout mice are commonly used to model fragile X syndrome and exhibit cognitive and brain anomalies associated with the disorder. Enrichment has an impact on exploratory behavior, dendritic branching, the number of dendritic spines, and expression of glutamate signaling, although it does not appear to affect the protein directly implicated in the genetic mutation (Restivo et al., 2005). In the Ts65Dn mouse model of Down syndrome, enrichment leads to improved learning in females but diminished performance in males during learning tasks (Martinez-Cue et al., 2002).

Whereas the evidence of neural plasticity in adulthood provides hope for continual reduction of symptom expression in autism throughout the lifespan, evidence of a critical period for development of motor stereotypy in deer mice suggests that the timing of intervention may be more important for some symptom domains than for others. Future research should investigate when brain and behavioral plasticity occurs, as well as the necessary intervention duration to produce change, in order to identify individuals most likely to respond to a particular intervention.

Conclusions

We have argued that altered patterns of social interaction constitute a risk process in autism, further amplifying early deviances in brain development by reducing critical input to the developing brain. We have also reviewed evidence from animal studies suggesting that enrichment of the early environment can mitigate the effects of early disease and neural insult. In addition, there is evidence to suggest that early, intensive behavioral interventions are effective for improving outcome in children with autism, particularly when initiated during the infant–toddler period. Interventions focused on promoting social engagement and social attention can potentially alter the development of brain circuitry underlying social, language, and cognitive behavior in children with autism. Studies of early, intensive behavioral intervention have been shown to have a significant impact on outcome in autism, including significant gains in IQ, language, and educational placement (see Faja & Dawson, 2006, for a review). The hope is that very early intervention will result in not only significant improvements in behavioral outcome, but also changes in neural responses to social and linguistic stimuli (Dawson, 2008; Smith, Rogers, & Dawson, 2007).

In conclusion, although the complexity of autism—in terms of both its etiology and its heterogeneity of symptom expression—poses significant challenges, research focused on identifying autism susceptibility indices, early identification, and early intervention offers real hope for the future.

Acknowledgments

The writing of this chapter was funded by grants from the National Institute of Child Health and Human Development (Nos. U19HD34565, P50HD066782, and R01HD-55741) and the National Institute of Mental Health (No. U54MH066399).

References

Adrien, J. L., Lenoir, P., Martineau, J., Perrot, A., Hameury, L., Larmande, C., et al. (1993). Blind ratings of early symptoms of autism based upon family home movies. *Journal of the American Academy of Child and Adolescent Psychiatry, 32*, 617–626.

American Psychiatric Association. (2000). *Diagnostic and statistical manual of mental disorders* (4th ed., text rev.). Washington, DC: Author.

August, G. J., Stewart, M. A., & Tsai, L. (1981). The incidence of cognitive disabilities in the siblings of autistic children. *British Journal of Psychiatry, 138*, 416–422.

Auranen, M., Vanhala, R., Varilo, T., Ayers, K., Kempas, E., Ylisaukko-oja, T., et al. (2002). A genome-wide screen for autism-spectrum disorders: Evidence for a major susceptibility locus on chromosome 3q25–27. *American Journal of Human Genetics, 71*, 777–790.

Avikainen, S., Wohlschlager, A., Liuhanen, S., Hanninen, R., & Hari, R. (2003). Impaired mirror-image imitation in Asperger and high-functioning autistic subjects. *Current Biology, 13*(4), 339–341.

Aylward, E. H., Minshew, N. J., Goldstein, G., Honeycutt, N. A., Augustine, A. M., Yates, K. O., et al. (1999). MRI volumes of amygdala and hippocampus in non-mentally retarded autistic adolescents and adults. *Neurology, 53*, 2145–2150.

Bacon, A. L., Fein, D., Morris, R., Waterhouse, L., & Allen, D. (1998). The responses of autistic children to the distress of others. *Journal of Autism and Developmental Disorders, 28*, 129–141.

Bailey, A., Le Couteur, A., Gottesman, I., Bolton, P., Simonoff, E., Yuzda, E., et al. (1995). Autism as a strongly genetic disorder: Evidence from a British twin study. *Psychological Medicine, 25*, 63–77.

Bailey, A., Luthert, P., Dean, A., Harding, B., Janota, I., Montgomery, M., et al. (1998). A clinicopathological study of autism. *Brain, 121*(Pt. 5), 889–905.

Bailey, A., Phillips, W., & Rutter, M. (1996). Autism: Towards an integration of clinical, genetic, neuropsychological, and neurobiological perspectives. *Journal of Child Psychology and Psychiatry, 37*, 89–126.

Baker, P., Piven, J., Schwartz, S., & Patil, S. (1994). Brief report: Duplication of chromosome 15q11–13 in two individuals with autistic disorder. *Journal of Autism and Developmental Disorders, 24*, 529–535.

Baranek, G. T. (1999). Autism during infancy: A retrospective video analysis of sensory–motor and social behaviours at 9–12 months of age. *Journal of Autism and Developmental Disorders, 29*, 213–224.

Baron-Cohen, S., Ring, H. A., Bullmore, E. T., Wheelwright, S., Ashwin, C., & Williams, S. C. (2000). The amygdala theory of autism. *Neuroscience and Biobehavioral Reviews, 24*(3), 355–364.

Barrett, S., Beck, J. C., Bernier, R., Bisson, E., Braun, T. A., Casavant, T. L., et al. (1999). An autosomal genomic screen for autism. *American Journal of Medical Genetics, 88*, 609–615.

Bauman, M., & Kemper, T. L. (1985). Histoanatomic observations of the brain in early infantile autism. *Neurology, 35*(6), 866–874.

Bauman, M., & Kemper, T. L. (2005). Neuroanatomic observations of the brain in autism: A review and future directions. *International Journal of Developmental Neuroscience, 23*, 183–187.

Belmonte, M. K., Cook, E. H., Anderson, G. M., Rubenstein, J. L. R., Greenough, W. R., Beckel-Mitchener, A., et al. (2004). Autism as a disorder of neural information processing: Directions for research and targets for therapy. *Molecular Psychiatry, 9*, 646–663.

Benaroya-Milshtein, N., Hollander, N., Apter, A., Kukulansky, T., Raz, N., Wilf, A., et al. (2004). Environmental enrichment in mice decreases anxiety, attenuates stress responses and enhances natural killer cell activity. *European Journal of Neuroscience, 20*, 1341–1347.

Bernier, R., Dawson, G., Webb, S., & Murias, M. (2007). EEG mu rhythm and imitation impairments in individuals with autism spectrum disorder. *Brain and Cognition, 64*(3), 228–237.

Blatt, G. J., Fitzgerald, C. M., Guptill, J. T., Booker, A. B., Kemper, T. K., & Bauman, M. L. (2001). Density and distribution of hippocampal neurotransmitter receptors in autism: An autoradiographic study. *Journal of Autism and Developmental Disorders, 31*, 537–543.

Bredy, T. W., Humpartzoomian, R. A., Cain, D. P., & Meaney, M. J. (2003). Partial reversal of the effect of maternal care on cognitive function through environmental enrichment. *Neuroscience, 118*, 571–576.

Bryson, S. E., Zwaigenbaum, L., McDermott, C., Rombough, V., & Brian, J. (2008). The Autism Observation Scale for Infants: Scale development and reliability data. *Journal of Autism and Developmental Disorders, 38*(4), 731–738.

Buxbaum, J. D., Silverman, J. M., Smith, C. J., Kilifarski, M., Reichert, J., Hollander, E., et al. (2001). Evidence for a susceptibility gene for autism on chromosome 2 and for genetic heterogeneity. *American Journal of Human Genetics, 68*, 1514–1520.

Campbell, D. B., Sutcliffe, J. S., Ebert, P. J., Militerni, R., Bravaccio, C., Trillo, S., et al. (2006). A genetic variant that disrupts MET transcription is associated with autism. *Proceedings of the National Academy of Sciences USA, 103*, 16834–16839.

Cantor, R. M., Kono, N., Duvall, J. A., AlvarezRetuerto, A., Stone, J. L., Alarcon, M., et al. (2005). Replication of autism linkage: Fine-mapping peak at 17q21. *American Journal of Human Genetics, 76*, 1050–1056.

Carpenter, M., Nagell, K., & Tomasello, M. (1998). Social cognition, joint attention, and communicative competence from 9 to 15 months of age. *Monographs of the Society for Research in Child Development, 63*(4, Serial No. 255), 1–143.

Casanova, M. F. (2004). White matter volume increase and minicolumns in autism. *Annals of Neurology, 56*, 453.

Charman, T., Baron-Cohen, S., Swettenham, J., Baird, G., Drew, A., & Cox, A. (2003). Predicting language outcome in infants with autism and pervasive developmental disorder. *International Journal of Language and Communication Disorders, 38*, 265–285.

Charman, T., Swettenham, J., Baron-Cohen, S., Cox, A., Baird, G., & Drew, A. (1998). An experimental investigation of social-cognitive abilities in infants with autism: Clinical implications. *Infant Mental Health Journal, 19*, 260–275.

Cheh, M. A., Millonig, J. H., Roselli, L. M., Ming, X., Jacobsen, E., Kamdar, S., et al. (2006). En2 knockout mice display neurobehavioral and neurochemical alterations relevant to autism spectrum disorder. *Brain Research, 1116*(1), 166–176.

Cicchetti, D., & Cohen, D. J. (1995). Perspectives on developmental psychopathology. In D. Cicchetti & D. J. Cohen (Eds.) *Developmental psychopathology: Vol. 1. Theory and methods* (pp. 3–22). New York: Wiley.

Courchesne, E. (1997). Brainstem, cerebellar and limbic neuroanatomical abnormalities in autism. *Current Opinion in Neurobiology 7*, 269–278.

Courchesne, E. (2004). Brain development in autism: Early overgrowth followed by premature arrest of growth. *Mental Retardation and Developmental Disabilities Research Reviews, 10*, 106–111.

Courchesne, E., Carper, R., & Akshoomoff, N. (2003). Evidence of brain overgrowth in the first year of life in autism. *Journal of the American Medical Association, 290*, 337–344.

Courchesne, E., Karns, C., Davis, H. R., Ziccardi, R., Carper, R., Tigue, Z., et al. (2001). Unusual brain growth patterns in early life in patients with autistic disorder: An MRI study. *Neurology, 57*, 245–254.

Courchesne, E., & Pierce, K. (2005). Brain overgrowth in autism during a critical time in development: Implications for frontal pyramidal neuron and interneuron development and connectivity. *International Journal of Developmental Neuroscience, 23*, 153–170.

Dapretto, M., Davies, M. S., Pfeifer, J. F., Scot, A. A., Sigman, M., Bookheimer, S. Y., et al. (2006). Understanding emotions in others: Mirror neuron dysfunction in children with autism spectrum disorders. *Nature Neuroscience, 9,* 28–30.

Dawson, G. (2008). Early behavioral intervention, brain plasticity, and the prevention of autism. *Development and Psychopathology, 20,* 775–803.

Dawson, G., & Adams, A. (1984). Imitation and social responsiveness in autistic children. *Journal of Abnormal Child Psychology, 12,* 209–225.

Dawson, G., Carver, L., Meltzoff, A. N., Panagiotides, H., McPartland, J., & Webb, S. J. (2002a). Neural correlates of face and object recognition in young children with autism spectrum disorder, developmental delay, and typical development. *Child Development, 73,* 700–717.

Dawson, G., Hill, D., Galpert, L., Spencer, A., & Watson, L. (1990). Affective exchanges between young autistic children and their mothers. *Journal of Abnormal Child Psychology, 18,* 335–345.

Dawson, G., & Lewy, A. (1989). Arousal, attention, and the socioemotional impairments of individuals with autism. In G. Dawson (Ed.), *Autism: Nature, diagnosis, and treatment* (pp. 49–74). New York: Guilford Press.

Dawson, G., Meltzoff, A. N., Osterling, J., & Brown, E. (1998a). Children with autism fail to orient to naturally occurring social stimuli. *Journal of Autism and Developmental Disorders, 28,* 479–485.

Dawson, G., Meltzoff, A. N., Osterling, J., & Rinaldi, J. (1998b). Neuropsychological correlates of early symptoms of autism. *Child Development, 69,* 1276–1285.

Dawson, G., Meltzoff, A. N., Osterling, J., Rinaldi, J., & Brown, E. (1998c). Children with autism fail to orient to naturally occurring social stimuli. *Journal of Autism and Developmental Disorders, 28,* 479–485.

Dawson, G., Munson, J., Estes, A., Osterling, J., McPartland, J., Toth, K., et al. (2002b). Neurocognitive function and joint attention ability in young children with autism spectrum disorder versus developmental delay. *Child Development, 73,* 345–358.

Dawson, G., Munson, J., Webb, S. J., Nalty, T., Abbott, R., & Toth, K. (2007). Rate of head growth decelerates and symptoms worsen in the second year of life in autism. *Biological Psychiatry, 61,* 458–464.

Dawson, G., Toth, K., Abbott, R., Osterling, J., Munson, J., Estes, A., et al. (2004a). Early social attention impairments in autism: Social orienting, joint attention, and attention to distress. *Developmental Psychology, 40,* 271–283.

Dawson, G., Webb, S., Carver, L., Panagiotides, H., & McPartland, J. (2004b). Young children with autism show atypical brain responses to fearful versus neutral facial expressions. *Developmental Science, 7,* 340–359.

Dawson, G., Webb, S. J., Wijsman, E., Schellenberg, G., Estes, A., Munson, J., et al. (2005). Neurocognitive and electrophysiological evidence of altered face processing in parents of children with autism: Implications for a model of abnormal development of social brain circuitry in autism. *Development and Psychopathology, 17,* 679–697.

de Haan, M., & Nelson, C. A. (1997). Recognition of the mother's face by 6-month-old infants: A neurobehavioral study. *Child Development, 68,* 187–210.

de Haan, M., & Nelson, C. A. (1999). Brain activity differentiates face and object processing in 6-month-old infants. *Developmental Psychology, 35,* 1113–1121.

de Haan, M., Nelson, C. A., Gunnar, M. R., & Tout, K. A. (1998). Hemispheric differences in brain activity related to the recognition of emotional expressions by 5-year-old children. *Developmental Neuropsychology, 14,* 495–518.

DeFelipe, J., Hendry, S. H., Hashikawa, T., Molinari, M., & Jones, E. G. (1990). A microcolumnar structure of monkey cerebral cortex revealed by immunocytochemical studies of double bouquet cell axons. *Neuroscience, 23,* 622–631.

Delong, G. R., & Dwyer, J. T. (1988). Correlation of family history with specific autistic subgroups:

Asperger's and bipolar affective disease. *Journal of Autism and Developmental Disorders, 18,* 593–600.

Devlin, B., Cook, E. H., Jr., Coon, H., Dawson, G., Grigorenko, E. L., McMahon, W., et al. (2005). Autism and the serotonin transporter: The long and short of it. *Molecular Psychiatry, 10,* 1110–1116.

Diamond, M. C., Rosenzweig, M. R., Bennett, E. L., Lindner, B., & Lyon, L. (1972). Effects of environmental enrichment and improverishment on rat cerebral cortex. *Journal of Neurobiology, 3,* 47–64.

Duffy, S. N., Craddock, K. J., Abel, T., & Nguyen, P. V. (2001). Environmental enrichment modifies the PKA-dependence of hippocampal LTP and improves hippocampus-dependent memory. *Learning and Memory, 8,* 26–34.

Eckerman, C. O., Davis, C. C., & Didow, S. M. (1989). Toddlers' emerging ways of achieving social coordinations with a peer. *Child Development, 60,* 440–453.

Eimer, M., & Holmes, A. (2002). An ERP study on the time course of emotional face processing. *NeuroReport, 13,* 427–431.

Escorihuela, R. M., Tobena, A., & Fernandez-Teruel, A. (1995). Environmental enrichment and postnatal handling prevent spatial learning deficits in aged hypoemotional (Roman high-avoidance) and hyperemotional (Roman low-avoidance) rats. *Learning and Memory, 2,* 40–48.

Fagan, J. (1972). Infants' recognition memory for face. *Journal of Experimental Child Psychology, 14,* 453–476.

Faja, S., & Dawson, G. (2006). Early intervention for autism. In J. Luby (Ed.), *Handbook of preschool mental health: Development, disorders, and treatment* (pp. 388–416). New York: Guilford Press.

Fatemi, S. H., & Halt, A. R. (2001). Altered levels of bcl2 and p53 proteins in parietal cortex reflect deranged apoptotic regulation in autism. *Synapse, 42,* 281–284.

Feinman, S. (1982). Social referencing in infancy. *Merrill–Palmer Quarterly, 28,* 445–470.

Ferguson, J., Young, H., Hearn, E., Insel, T. R., & Winslow, J. (2000). Social amnesia in mice lacking the oxytocin gene. *Nature Genetics, 25,* 284–288.

Ferguson, J., Young, H., & Insel, T. R. (2002). The neuroendocrine basis of social recognition. *Frontiers in Neuroendocrinology, 23,* 200–224.

Fernandez-Teruel, A., Driscoll, P., Gil, L., Aguilar, R., Tobena, A., & Escorihuela, R. M. (2002). Enduring effects of environmental enrichment on novelty seeking, saccharin and ethanol intake in two rat lines (RHA/Verh and RLA/Verh) differing in incentive-seeking behavior. *Pharmacology, Biochemistry and Behavior, 73,* 225–231.

Folstein, S., & Rutter, M. (1977a). Genetic influences and infantile autism. *Nature, 265,* 726–728.

Folstein, S., & Rutter, M. (1977b). Infantile autism: A genetic study of 21 twin pairs. *Journal of Child Psychology and Psychiatry, 18,* 297–321.

Foster, T. C., Gagne, J., & Massicotte, G. (1996). Mechanism of altered synaptic strength due to experience: Relation to long-term potentiation. *Brain Research, 736,* 243–250.

Francis, D. D., Diorio, J., Plotsky, P. M., & Meaney, M. J. (2002). Environmental enrichment reverses the effects of maternal separation on stress reactivity. *Journal of Neuroscience, 22,* 7840–7843.

Friedman, S. D., Shaw, D. W., Artru, A. A., Dawson, G., Petropoulos, H., & Dager, S. R. (2006). Gray and white matter brain chemistry in young children with autism. *Archives of General Psychiatry, 63,* 786–794.

Friedman, S. D., Shaw, D. W., Artru, A. A., Richards, T. L., Gardner, J., Dawson, G., et al. (2003). Regional brain chemical alterations in young children with autism spectrum disorder. *Neurology, 60,* 100–107.

Gillberg, C., & de Souza, L. (2002). Head circumference in autism, Asperger syndrome, and ADHD: A comparative study. *Developmental Medicine and Child Neurology, 44,* 296–300.

Gingrich, B., Liu, Y., Cascio, C., Wang, Z., & Insel, T. R. (2000). Dopamine D2 receptors in the nucleus accumbens are important for social attachment in female prairie voles. *Behavioral Neuroscience, 114,* 173–183.

Goren, C., Sarty, M., & Wu, P. (1975). Visual following and pattern discrimination of face-like stimuli by newborn infants. *Pediatrics, 56,* 544–549.

Greenough, W. T., Volkmar, F. R., & Juraska, J. M. (1973). Effects of rearing complexity on dendritic branching in frontotemporal and temporal cortex of the rat. *Experimental Neurology, 41,* 371–378.

Grelotti, D., Gauthier, I., & Schultz, R. (2002). Social interest and the development of cortical face specialization: What autism teaches us about face processing. *Developmental Psychobiology, 40,* 213–225.

Hatfield, E., Cacioppo, J. T., & Rapson, R. L. (1994). *Emotional contagion.* Cambridge, UK: Cambridge University Press.

Hollander, E., Bartz, J., Chaplin, W., Phillips, A., Sumner, J., Soorya, L., et al. (2007). Oxytocin increases retention of social cognition in autism. *Biological Psychiatry, 61,* 498–503.

Hollander, E., Novotny, S., Hanratty, M., Yaffe, R., DeCaria, C. M. Aronowitz, B. R., et al. (2003). Oxytocin infusion reduces repetitive behavior in adults with autistic and Asperger's disorders. *Neuropsychopharmacology, 28,* 193–198.

Insel, T. R. (1997). A neurobiological basis of social attachment. *American Journal of Psychiatry, 154,* 726–735.

Insel, T. R., & Fernald, R. D. (2004). How the brain processes social information: Searching for the social brain. *Annual Review of Neuroscience, 27,* 697–722.

Insel, T. R., & Hulihan, T. J. (1995). A gender-specific mechanism for pair bonding: Oxytocin and partner preference formation in monogamous voles. *Behavioral Neuroscience, 109,* 782–789.

Insel, T. R., O'Brien, D. J., & Leckman, J. F. (1999). Oxytocin, vasopressin, and autism: Is there a connection? *Biological Psychiatry, 45,* 145–157.

International Molecular Genetic Study of Autism Consortium. (2001). A genomewide screen for autism: Strong evidence for linkage to chromosomes 2q, 7q, and 16p. *American Journal of Human Genetics, 69,* 570–581.

Jacob, S., Brune, C. W. Carter, C. S., Leventhal, B. L., Lord, C., & Cook, E. H. (2007). Association of the oxytocin receptor gene (OXTR) in Caucasian children and adolescents with autism. *Neuroscience Letters, 417,* 6–9.

Kampe, K., Frith, C., Dolan, R., & Frith, U. (2001). Psychology: Reward value of attraction and gaze. *Nature, 413,* 589.

Kasari, C., Sigman, M., Mundy, P., & Yirmiya, N. (1990). Affective sharing in the context of joint attention interactions of normal, autistic, and mentally retarded children. *Journal of Autism and Developmental Disorders, 20,* 87–100.

Kemper, T. L., & Bauman, M. (1998). Neuropathology of infantile autism. *Journal of Neuropathology and Experimental Neurology, 57,* 645–652.

Kempermann, G., Kuhn, H. G., & Gage, F. H. (1997). More hippocampal neurons in adult mice living in an enriched environment. *Nature, 386,* 493–495.

Kim, S. J., Young, L. J., Gonen, D., Veenstra-VanderWeele, J., Courchesne, R., Courchesne, E., et al. (2002). Transmission disequilibrium testing of arginine vasopressin receptor 1A (AVPR1A) polymorphisms in autism. *Molecular Psychiatry, 7,* 503–507.

Klin, A. (1991). Young autistic children's listening preferences in regard to speech: A possible characterization of the symptom of social withdrawal. *Journal of Autism and Developmental Disorder, 21,* 29–42.

Kuehn, B. M. (2007). CDC: Autism spectrum disorders common. *Journal of the American Medical Association, 297,* 940.

Kuemerle, B., Zanjani, H., Joyner, A., & Herrup, K. (1997). Pattern deformities and cell loss in Engrailed-2 mutant mice suggest two separate patterning events during cerebellar development. *Journal of Neuroscience, 17,* 7881–7889.

Kugiumutzakis, G. (1999). Genesis and development of early infant mimesis to facial and vocal models. In J. Nadel & G. Butterworth (Eds.), *Imitation in infancy* (pp. 36–59). New York: Cambridge University Press.

Kuhl, P. (2007). Is speech learning "gated" by the social brain? *Developmental Science, 10,* 110–120.

Kuhl, P., Coffey-Corina, S., Padden, D., & Dawson, G. (2004). Links between social and linguistic processing of speech in preschool children with autism: Behavioral and electrophysiological measures. *Developmental Science, 7,* 19–30.

Kuhl, P., Tsao, F., & Liu, H. (2003). Foreign-language experience in infancy: Effects of short-term exposure and social interaction on phonetic learning. *Proceedings of the National Academy of Sciences USA, 100,* 9096–9101.

Lamb, J. A., Barnby, G., Bonora, E., Sykes, N., Bacchelli, E., Blasi, F., et al. (2005). Analysis of IMG-SAC autism susceptibility loci: Evidence for sex limited and parent of origin specific effects. *Journal of Medical Genetics, 42,* 132–137.

Landa, R., Folstein, S. E., & Isaacs, C. (1991). Spontaneous narrative-discourse performance of parents of autistic individuals. *Journal of Speech and Hearing Research, 34,* 1339–1345.

Landa, R., & Garrett-Mayer, E. (2006). Development in infants with autism spectrum disorders: A prospective study. *Journal of Child Psychology and Psychiatry, 47,* 629–638.

Landa, R., Piven, J., Wzorek, M. M., Gayle, J. O., Chase, G. A., & Folstein, S. E. (1992). Social language use in parents of autistic individuals. *Psychological Medicine, 22,* 245–254.

Lewis, M. H. (2004). Environmental complexity and central nervous system development and function. *Mental Retardation and Developmental Disabilities Research Reviews, 10,* 91–95.

Liu, J. J., Nyholt, D. R., Magnussen, P., Parano, E., Pavone, P., Geschwind, D., et al. (2001). A genomewide screen for autism susceptibility loci. *American Journal of Human Genetics, 69,* 327–340.

Lucki, I. (1998). The spectrum of behaviors influence by serotonin. *Biological Psychiatry, 44,* 151–162.

Maestro, S., Muratori, F., Barbieri, F., Casella, C., Cattaneo, V., Cavallaro, M. C., et al. (2001). Early behavioral development in autistic children: The first two years of life through home movies. *Psychopathology, 34,* 147–152.

Maestro, S., Muratori, F., Cavallaro, M. C., Pei, F., Stern, D., Golse, B., et al. (2002). Attentional skills during the first 6 months of age in autism spectrum disorder. *Journal of the American Academy of Child and Adolescent Psychiatry, 41,* 1239–1245.

Mars, A. E., Mauk, J. E., & Dowrick, P. W. (1998). Symptoms of pervasive developmental disorders as observed in prediagnostic home videos of infants and toddlers. *Journal of Pediatrics, 132,* 500–504.

Martinez-Cue, C., Baamonde, C., Lumbreras, M., Paz, J., Davisson, M. T., Schmidt, C., et al. (2002). Differential effects of environmental enrichment on behavior and learning of male and female Ts65Dn mice, a model for Down syndrome. *Behavioural Brain Research, 134,* 185–200.

Maurer, D., & Salapatek, P. (1976). Developmental changes in the scanning of faces by young infants. *Child Development, 47,* 523–527.

McCauley, J. L., Li, C., Jiang, L., Olson, L. M., Crockett, G., Gainer, K., et al. (2005). Genome-wide and ordered-subset linkage analyses provide support for autism loci on 17q and 19p with evidence of phenotypic and interlocus genetic correlates. *BMC Medical Genetics, 6,* 1.

Meltzoff, A. N., & Gopnik, A. (1993). The role of imitation in understanding persons and developing a theory of mind. In S. Baron-Cohen, H. Tager-Flusberg, & D. J. Cohen (Eds.), *Understanding other minds: Perspectives from autism* (pp. 335–366). Oxford: Oxford University Press.

Millen, K. J., Wurst, W. W., Herrup, K., & Joyner, A. L. (1994). Abnormal embryonic cerebellar development and patterning of postnatal foliation in two mouse Engrailed-2 mutants. *Development, 120,* 695–706.

Modahl, C., Green, L., Fein, D., Morris, M., Waterhouse, L., Feinstein, C., et al. (1998). Plasma oxytocin levels in autistic children. *Biological Psychiatry, 43,* 270–277.

Moore, C., & Corkum, V. (1994). Social understanding at the end of the first year of life. *Developmental Review, 14,* 349–372.

Morales, M., Mundy, P., & Rojas, J. (1998). Brief report: Following the direction of gaze and language development in 6-month-olds. *Infant Behavior and Development, 21,* 373–377.

Morton, J., & Johnson, M. H. (1991). CONSPEC and CONLERN: A two-process theory of infant face recognition. *Psychological Review, 2,* 164–181.

Mullen, E. M. (1995). *Mullen Scales of Early Learning.* Circle Pines, MN: American Guidance Service.

Mundy, P., & Neal, R. (2001). Neural plasticity, joint attention and a transactional social-orienting model of autism. In L. Glidden (Ed.), *International review of research in mental retardation: Vol. 23. Autism* (pp. 139–168). San Diego, CA: Academic Press.

Mundy, P., Sigman, M., Ungerer, J., & Sherman, T. (1986). Defining the social deficits of autism: The contribution of nonverbal communication measures. *Journal of Child Psychology and Psychiatry, 27,* 657–669.

Munson, J., Dawson, G., Abbott, R., Faja, S., Webb, S. J., Friedman, S. D., et al. (2006). Amygdalar volume and behavioral development in autism. *Archives of General Psychiatry, 63,* 686–693.

Nadel, J., Guerini, C., Peze, A., & Rivet, C. (1999). The evolving nature of imitation as a format for communication. In J. Nadel & G. Butterworth (Eds.), *Imitation in infancy* (pp. 209–234). New York: Cambridge University Press.

Narayan, S., Moyes, B., & Wolff, S. (1990). Family characteristics of autistic children: A further report. *Journal of Autism and Developmental Disorders, 20,* 557–559.

Nelson, C. A. (2001). The development and neural bases of face recognition. *Infant and Child Development, 10,* 3–18.

Nelson, C. A., & de Haan, M. (1996). Neural correlates of infants' visual responsiveness to facial expressions of emotion. *Developmental Psychobiology, 29,* 577–595.

Nishimori, K., Young, L. J., Guo, Q., Wang, Z., Insel, T. R., & Matzuk, M. M. (1996). Oxytocin is required for nursing but is not essential for parturition or reproductive behavior. *Proceedings of the National Academy of Sciences USA, 93,* 11699–11704.

Nithianantharajah, J., & Hannan, A. J. (2006). Enriched environments, experience-dependent plasticity and disorders of the nervous system. *Nature Reviews Neuroscience, 7,* 697–709.

Osterling, J. A., & Dawson, G. (1994). Early recognition of children with autism: A study of first birthday home video tapes. *Journal of Autism and Developmental Disorders, 24,* 247–257.

Osterling, J. A., Dawson, G., & Munson, J. A. (2002). Early recognition of 1-year old infants with autism spectrum disorder versus mental retardation. *Development and Psychopathology, 14,* 239–251.

Pedersen, C. A., Caldwell, J. O., Walker, C., Ayers, G., & Mason, G. A. (1994). Oxytocin activates the postpartum onset of rat maternal behavior in the ventral tegmental and medial preoptic areas. *Behavioral Neuroscience, 108,* 1163–1171.

Peters, A., & Sethares, G. (1997). The organization of double bouquet cells in monkey striate cortex. *Journal of Neurocytology, 26,* 7779–7797.

Petroff, O. A., Errante, L. D., Rothman, D. L., Kim, J. H., & Spencer, D. D. (2002). Glutamate–glutamine cycling in the epileptic human hippocampus. *Epilepsia, 43,* 703–710.

Pham, T. M., Winblad, B., Granholm, A., & Mohammed, A. H. (2002). Environmental influences on brain neurotrophins in rats. *Pharmacology, Biochemistry and Behavior, 73,* 167–175.

Philippe, A., Martinez, M., Guilloudbataille, M., Gillberg, C., Rastam, M., Sponheim, E., et al. (1999). Genome-wide scan for autism susceptibility genes. *Human Molecular Genetics, 8,* 805–812.

Piaget, J. (1962). *Play, dreams and imitation in childhood.* New York: Norton.

Pickett, J., & London, E. (2005). The neuropathology of autism: A review. *Journal of Neuropathology and Experimental Neurology, 64,* 925–935.

Pickles, A., Bolton, P., Macdonald, H., Bailey, A., Le Couteur, A., Sim, C.-H., et al. (1995). Latent-class analysis of recurrence risks for complex phenotypes with selection and measurement error: A twin and family history study of autism. *American Journal of Human Genetics, 57,* 717–726.

Piven, J., Gayle, J., Chase, G. A., Fink, B., Landa, R., Wzorek, M. M., et al. (1990). A family history study of neuropsychiatric disorders in the adult siblings of autistic individuals. *Journal of the American Academy of Children and Adolescent Psychiatry, 29,* 177–183.

Powell, S. B., Newman, H. A., McDonald, T. A., Bugenhagen, P., & Lewis, M. H. (2000). Development of spontaneous stereotyped behavior in deer mice: Effects of early and late exposure to a more complex environment. *Developmental Psychobiology, 37,* 100–108.

Rampon, C., Jiang, C. H., Dong, H., Tang, Y. P., Lockhart, D. J., Schultz, P. G., et al. (2000). Effects of environmental enrichment on gene expression in the brain. *Proceedings of the National Academy of Sciences USA, 97,* 12880–12884.

Redcay, E., & Courchesne, E. (2005). When is the brain enlarged in autism?: A meta-analysis of all brain size reports. *Biological Psychiatry, 58,* 1–9.

Restivo, L., Ferrari, F., Passino, E., Sgobio, C., Bock, J., Oostra, B. A., et al. (2005). Enriched environment promotes behavioral and morphological recovery in a mouse model for the fragile X syndrome. *Proceedings of the National Academy of Sciences USA, 102,* 11557–11562.

Rheingold, H. L., Hay, D. F., & West, M. J. (1976). Sharing in the second year of life. *Child Development, 47,* 1148–1158.

Risch, N., Spiker, D., Lotspeich, L., Nouri, N., Hinds, D., Hallmayer, J., et al. (1999). A genomic screen of autism: Evidence for a multilocus etiology. *American Journal of Human Genetics, 65,* 493–507.

Ritvo, E. R., Freeman, B. J., Mason-Brothers, A., Mo, A., & Ritvo, A. M. (1985). Concordance for the syndrome of autism in 40 pairs of afflicted twins. *American Journal of Psychiatry, 142,* 74–77.

Ritvo, E. R., Freeman, B. J., Scheibel, A. B., Duong, T., Robinson, H., Guthrie, D., et al. (1986). Lower Purkinje cell counts in the cerebella of four autistic subjects: Initial findings of the UCLA–NSAC autopsy research report. *American Journal of Psychiatry, 143,* 862–866.

Rochat, P. (Ed.). (1999). *Early social cognition: Understanding others in the first months of life.* Mahwah, NJ: Erlbaum.

Rochat, P., & Striano, T. (1999). Social cognitive development in the first year. In P. Rochat (Ed.), *Early social cognition: Understanding others in the first months of life* (pp. 3–34). Mahwah, NJ: Erlbaum.

Rogers, S. J., Bennetto, L., McEvoy, R., & Pennington, B. F. (1996). Imitation and pantomime in high-functioning adolescents with autism spectrum disorders. *Child Development, 67,* 2060–2073.

Rogers, S. J., & Pennington, B. F. (1991). A theoretical approach to the deficits in infantile autism. *Developmental and Psychopathology, 3,* 137–162.

Rolf, L. H., Haarmann, F. Y., Grotemeyer, K. H., & Kehrer, H. (1993). Serotonin and amino acid content in platelets of autistic children. *Acta Psychiatrica Scandinavica, 87,* 312–316.

Schellenberg, G., Dawson, G., Sung, Y. J., Estes, A., Munson, J., Rosenthal, E., et al. (2006). Evidence for multiple loci from a genome scan of autism kindred: A CPEA study. *Molecular Psychiatry, 11,* 1049–1060.

Schneider, T., Turczak, J., & Przewlocki, R. (2006). Environmental enrichment reverses behavioral alterations in rats prenatally exposed to valproic acid: Issues for a therapeutic approach in autism. *Neuropsychopharmacology, 31,* 36–46.

Schoenbaum, G., Setlow, B., Saddoris, M. P., & Gallagher, M. (2003). Encoding predicted outcome and acquired value in orbitofrontal cortex during cue sampling depends upon input from basolateral amygdala. *Neuron, 39,* 731–733.

Schrijver, N. C., Bahr, N. I., Weiss, I. C., & Wurbel, H. (2002). Dissociable effects of isolation rearing and environmental enrichment on exploration, spatial learning and HPA activity in adult rats. *Pharmacology, Biochemistry and Behavior, 73,* 209–224.

Schultz, R. T. (2005). Developmental deficits in social perception in autism: The role of the amygdala and fusiform face area. *International Journal of Developmental Neuroscience, 23,* 125–141.

Schultz, W. (1998). Predictive reward signal of dopamine neurons. *Journal of Neurophysiology, 80,* 1–27.

Schumann, C. M., & Amaral, D. G. (2006). Stereological analysis of amygdala neuron number in autism. *Journal of Neuroscience, 26,* 7674–7679.

Schumann, C. M., Hamstra, J., Goodlin-Jones, B. L., Lotspeich, L. J., Kwon, H., Buonocore, M. H., et al. (2004). The amygdala is enlarged in children but not adolescents with autism; the hippocampus is enlarged at all ages. *Journal of Neuroscience, 24,* 6392–6401.

Shao, Y. J., Wolpert, C. M., Raiford, K. L., Menold, M. M., Donnelly, S. L., Ravan, S. A., et al. (2002). Genomic screen and follow-up analysis for autistic disorder. *American Journal of Medical Genetics, 114,* 99–105.

Sigman, M., Kasari, C., Kwon, J., & Yirmiya, N. (1992). Responses to the negative emotions of others by autistic, mentally retarded, and normal children. *Child Development, 63,* 796–807.

Smalley, S. L., Asarnow, R. F., & Spence, A. (1988). Autism and genetics. *Archives of General Psychiatry, 45,* 953–961.

Smith, M., Rogers, S., & Dawson, G. (2007). *The Early Start Denver Model: A comprehensive early intervention approach for toddlers with autism* (3rd ed., J. S. Handleman & S. L. Harris, Eds.). Austin, TX: PRO-ED.

Sparks, B. F., Friedman, S. D., Shaw, D. W., Aylward, E. H., Echelard, D., Artru, A. A., et al. (2002). Brain structural abnormalities in young children with autism spectrum disorder. *Neurology, 59,* 184–192.

Steffenburg, S., Gillberg, C., Hellgren, L., Andersson, L., Gillberg, I., Jakobsson, G., et al. (1989). A twin study of autism in Denmark, Finland, Iceland, Norway, and Sweden. *Journal of Child Psychology and Psychiatry, 30,* 405–416.

Stone, J. L., Merriman, B., Cantor, R. M., Yonan, A. L., Gilliam, T. C., Geschwind, D. H., et al. (2004). Evidence for sex-specific risk alleles in autism spectrum disorder. *American Journal of Human Genetics, 75,* 1117–1123.

Stone, W., Lemanek, K., Fishel, P., Fernandez, M., & Altemeier, W. (1990). Play and imitation skills in the diagnosis of autism in young children. *Pediatrics, 86,* 267–272.

Stone, W. L., Ousley, O. Y., & Littleford, C. D. (1997). Motor imitation in young children with autism: What's the object? *Journal of Abnormal Child Psychology, 25,* 475–485.

Sung, Y. J., Dawson, G., Munson, J., Estes, A., Schellenberg, G. D., & Wijsman, E. M. (2005). Genetic investigation of quantitative traits related to autism: Use of multivariate polygenic models with ascertainment adjustment. *American Journal of Human Genetics, 76,* 68–81.

Symons, L., Hains, S., & Muir, S. (1998). Look at me: 5-month-old infants' sensitivity to very small deviations in eye-gaze during social interactions. *Infant Behavior and Development, 21,* 531–536.

Szatmari, P., Paterson, A. D., Zwaigenbaum, L., Roberts, W., Brian, J., Liu, X. Q., et al. (2007). Mapping autism risk loci using genetic linkage and chromosomal rearrangements. *Nature Genetics, 39,* 319–328.

Trevarthen, C. (1979). Communication and cooperation in early infancy: A description of primary intersubjectivity. In M. Bullowa (Ed.), *Before speech: The beginnings of interpersonal communication.* Cambridge, UK: Cambridge University Press.

Trevarthen, C., Kokkinaki, T., & Fiamenghi, G., Jr. (1999). What infants' imitations communicate: With mothers, with fathers, and with peers. In J. Nadel & G. Butterworth (Eds.), *Imitation in infancy* (pp. 127–185). New York: Cambridge University Press.

Turner, C. A., Lewis, M. H., & King, M. A. (2003). Environmental enrichment: Effects on stereotyped behavior and dendritic morphology. *Developmental Psychobiology, 43,* 20–27.

Turner, C. A., Yang, M. C., & Lewis, M. H. (2002). Environmental enrichment: Effects on stereotyped behavior and regional neuronal metabolic activity. *Brain Research, 938,* 15–21.

Uzgiris, E. E. (1981). Probing immune reactions by laser light scattering spectroscopy. *Methods in Enzymology, 74*(Pt. C), 177–198.

Uzgiris, I. (1999). Imitation as activity: Its developmental aspect. In J. Nadel & G. Butterworth (Eds.), *Imitation in infancy* (pp. 209–234). New York: Cambridge University Press.

Vargas, D. L., Nascimbene, C., Krishnan, C., Zimmerman, A. W., & Pardo, C. A. (2005). Neuroglial activation and neuroinflammation in the brain of patients with autism. *Annals of Neurology, 57,* 67–81.

Veenstra-VanderWeele, J., & Cook, E. H., Jr. (2004). Molecular genetics of autism spectrum disorder. *Molecular Psychiatry, 9*, 819–832.

Walton, G. E., & Bower, T. G. (1993). Amodal representations of speech in infants. *Infant Behavior and Development, 16*, 233–243.

Wassink, T. H., Hazlett, H., Mosconi, M., Epping, E., Arndt, S., Schellenberg, G., et al. (2007). Cerebral cortical gray matter overgrowth and functional variation of the serotonin transporter gene in autism. *Archives of General Psychiatry, 64*(6), 709–717.

Waterhouse, L., Fein, D., & Modahl, C. (1996). Neurofunctional mechanisms in autism. *Psychological Review, 103*, 457–489.

Werner, E., & Dawson, G. (2005). Validation of the phenomenon of autistic regression using home videotapes. *Archives of General Psychiatry, 62*, 889–895.

Werner, E., Dawson, G., Osterling, J., & Dinno, N. (2000). Brief report: Recognition of autism spectrum disorder before one year of age: A retrospective study based on home videotapes. *Journal of Autism and Developmental Disorders, 30*, 157–162.

Williams, J. H., Whiten, A., Suddendorf, T., & Perrett, D. I. (2001). Imitation, mirror neurons and autism. *Neuroscience and Biobehavioral Review, 25*, 287–295.

Winslow, J., Hastings, N., Carter, C. S., Harbaugh, C. R., & Insel, T. R. (1993). A role for central vasopressin in pair bonding in monogamous prairie voles. *Nature, 365*, 545–548.

Witt, D. M., Winslow, J. T., & Insel, T. R. (1992). Enhanced social interactions in rats following chronic, centrally infused oxytocin. *Pharmacology, Biochemistry and Behavior, 43*, 855–861.

Wolff, S., Narayan, S., & Moyes, B. (1988). Personality characteristics of parents of autistic children: A controlled study. *Journal of Child Psychiatry, 29*, 143–153.

Wong, R., & Jamieson, J. L. (1968). Infantile handling and the facilitation of discrimination and reversal learning. *Quarterly Journal of Experimental Psychology, 20*, 197–199.

Wu, S., Jia, M., Ruan, Y., Liu, J., Guo, Y., Chuang, M., et al. (2005). Positive association of the oxytocin receptor gene (OXTR) with autism in the Chinese Han population. *Biological Psychiatry, 58*, 74–77.

Yonan, A. L., Alarcon, M., Cheng, R., Magnusson, P. K. E., Spence, S. J., Palmer, A. A., et al. (2003). A genomewide screen of 345 families for autism-susceptibility loci. *American Journal of Human Genetics, 73*, 886–897.

Young, D., Lawlor, P. A., Leone, P., Dragunow, M., & During, M. L. (1999). Environmental enrichment inhibits spontaneous apoptosis, prevents seizures and is neuroprotective. *Nature Medicine, 5*, 448–453.

Zahn-Waxler, C., & Radke-Yarrow, M. (1990). The origins of empathic concern. *Motivation and Emotion, 14*, 107–130.

Zwaigenbaum, L., Bryson, S., Rogers, T., Roberts, W., Brian, J., & Szatmari, P. (2005). Behavioral manifestations of autism in the first year of life. *International Journal of Developmental Neuroscience, 23*, 143–152.

CHAPTER 23

Social and Genetic Aspects of Turner, Williams–Beuren, and Fragile X Syndromes

DAVID SKUSE
LOUISE GALLAGHER

The genetic influences on the development of social-emotional processes are readily apparent in the context of genetic disorders associated with a specific behavioral phenotype. The term "behavioral phenotype" or "cognitive phenotype" is used to describe a typical style of behavior or cognition that occurs in association with developmental disorders in which a genetic anomaly has been demonstrated. Behavioral or cognitive phenotypes reflect the maldevelopment of dysfunctional neural systems. Some such characteristics are common to many disorders, such as inattention. Other characteristics may be more marked, in association with specific genetic etiologies. Cognitive deficits may include problems in visuospatial skills, language, social understanding, or mathematical abilities. Behavioral adjustment may be disrupted by self-injury, hyperphagia, or impulsivity. Occasionally a particular genetic mutation or deletion is associated with a behavioral phenotype that looks like a known disease, such as autism or schizophrenia. The finding does not imply that the genetic locus has any direct part to play in idiopathic cases of the syndrome. Major psychiatric disorders have a multigenic etiology; complex mental disorders are not due to mutations in single genes.

Behavioral and cognitive phenotypes may arise as a result of any one of a number of possible mechanisms. The simplest genetic aberration is a "causative mutation." Here a mutation in a specific gene effects a change in the nucleotide sequence and a corresponding alteration in gene product that affects the phenotype. Such a mutation must be both necessary and sufficient to effect the change. Mendelian inheritance occurs where a causative mutation is transmitted from one generation to the next. Inheritance in this form may be dominant or recessive; in dominant inheritance, a single mutation is sufficient to cause the phenotype, and this occurs in heterozygotes for the mutation. Recessive inheritance occurs where there is homozygosity for the mutation. Since this mechanism is conceptually simple, causative genes are easily identified. However phenotypic manifestations are still variable and multisystemic, since mutations of single genes may vary in size and position. Moreover, genetic modifiers are recognized as contributing to the variation in the phenotype through differences between individuals in the genetic background and the influence of environmental factors. Although little is known about the role of genetic modifiers in disease, they may directly influence metabolic and cellular pathways involved in the primary pathophysiology of a susceptibility gene (Haston & Hudson, 2005). Or they may act in secondary pathways that influence the course of the disorder, especially if it is neurodevelopmental in origin.

Mendelian inheritance is unlikely to underlie complex disorders, such as psychiatric conditions. The phenotypic outcome of a complex disorder or trait is a result of several interacting "polymorphisms" (variants in the genome that may influence gene expression). A single variant may be necessary but not sufficient to produce a phenotype. Others may be neither necessary nor sufficient in their own right and operate epistatically. Environmental factors also have a crucial part to play, influencing such factors as the phenotype's timing of onset and its severity.

Abnormalities in chromosomal structure or in the total complement of chromosomal material may also influence behavioral phenotypes. Chromosomal abnormalities can be numerical or structural. Numerical abnormalities are associated with either the loss of one of a pair of chromosomes, or with the aberrant formation of more than one copy of a chromosome (as in trisomy 21, 45,X, or 47,XXY). Structural anomalies usually involve microdeletions of a few thousand nucleotide bases, or (more rarely) the loss of a substantial part of a chromosome. Anomalies may arise during meiosis, at about the time of fertilization, or after cell division begins in early embryonic development. In the first situation, the abnormality is represented in every somatic cell; later-arising anomalies are present in a proportion of developing cells, and this condition is termed "mosaicism." In mosaicism, the distribution of affected cells is clonal, reflecting the original population of normal or abnormal cells from which they were derived. Abnormal cells may be selected against in certain tissues where they are not capable of adequately contributing to the development of the tissue. Examples of disorders caused by microdeletions include 22q11.2 deletion syndrome, Smith–Magenis syndrome, and Williams–Beuren syndrome. These are rarely mosaic. They are also termed "contiguous gene syndromes," as there are a number of genes affected by the microdeletion, many of which may be implicated in the clinical phenotype. A microdeletion may render the individual haploinsufficient for a deleted gene. In this case, both copies of the gene are required for the gene product to function, and therefore normal functioning is interrupted. More recently, it has been recognized that submicroscopic deletions undetectable by traditional methods of chromosomal analysis are also present in the genome and may contribute to disorder. These are known as "copy number variations" (CNVs) and may take the form of deletions, insertions, inversions, and multiple copies of genomic DNA (Redon, Ishikawa, et al., 2006). The significance of CNVs is still being determined. It appears that de novo events are likely to be more clinically significant. There is hope that structural variation may shed new light on complex diseases, such as neurodevelopmental psychiatric disorders (e.g., autism) in which the phenotype is due to the interaction of many different genes (Check, 2005).

Some dominant conditions result in phenotypes that become more severe in successive generations—a mechanism known as "anticipation." Trinucleotide repeat expansions have been proposed as the biological basis of this phenomenon (Pearson, Nichol Edamura, et al.,

2005). These are triplets of the nucleotides cytosine (C), adenine (A), guanine (G), or thymine (T) (e.g., CAG, CGG, CTG), which may occur within genes, or in regions of DNA between genes. Expansions occur in the course of DNA replication, with repeats growing in length in successive generations, and implicate germline instability. There are proportionately more neurological diseases caused by expansion of CTG, CGG, CAG, or GAA repeats. Considerable phenotypic heterogeneity may be associated with trinucleotide repeats related to variability in the length of the repeat sequences, and the degree to which transcription and gene function are disrupted. Characteristic cognitive or behavioral phenotypes are recognized in association with trinucleotide repeat expansions (e.g., fragile X syndrome and Huntington disease). Fragile X syndrome is caused by disruption to the FMR1 gene on the X chromosome and is discussed further below.

Other mechanisms that influence behavioral phenotypes include both epigenetics and imprinting. "Epigenetics" refers to such processes as methylation and chromatin remodeling, which lead to the silencing of genes, on one or the other copy, early in development. Epigenetic mechanisms may play a role in the disruption of patterns of DNA methylation in the embryo, leading to developmental alteration in the offspring, with persistent changes in the germline. It has been suggested that the significant gender differences in the relative risk of many psychiatric disorders (e.g., attention-deficit/hyperactivity disorder [ADHD], autism, eating disorders) could imply that phenotypic manifestations are modulated by sex steroids (Crews & McLachlan, 2006). It has been shown in rats that the quality of maternal care in the early postnatal period has an epigenetic effect on genes concerned with neuronal growth in the hippocampus (Szyf, Weaver, et al., 2005). Many genes are potentially regulated by maternal care, and they can influence the behavior of that offspring in adulthood (Weaver, Champagne, et al., 2005). Epigenetic influences on risk of mental illness have not been widely investigated, but accumulating evidence in human as well as animal models implies specific mechanisms of dysfunction (Persico & Bourgeron, 2006; Rutter, Moffitt, et al., 2006).

The phenomenon of "genomic imprinting" refers to differences in gene activity related to maternal or paternal inheritance where the gene in question is expressed from only one of a pair of alleles. Imprinting may only be detected in certain tissues or at certain periods of development, and does not alter the underlying DNA sequence. The existence of imprinted genes may be revealed when the only active copy is mutated and ceases to function, or is lost due to a structural abnormality of the chromosome. Occasionally the chromosome, or a region of the chromosome bearing the gene, undergoes "uniparental disomy": Both copies of the gene (and part of the surrounding genome) are inherited from just one parent. The clinical phenotypes resulting from abnormalities of imprinted loci tend to be severe. Most imprinted genes have been described on autosomes, but Skuse, James, et al. (1997) proposed that imprinted genes on the X chromosome could be associated with sexually dimorphic characteristics. Evidence exists for imprinting on the mouse X chromosome associated with a sexually dimorphic behavioral characteristic—anxiety in a novel and potentially threatening situation, which affects males to a greater degree than females (Davies, Isles, et al., 2005).

Turner Syndrome

Turner syndrome (TS) was first described by the endocrinologist Henry Turner in 1938, but it was not until 20 years later that the genetic basis of the syndrome was discovered. About 50% of clinically identified cases of TS are associated with a single X chromosome (X monosomy), and affected females therefore have 45 rather than the usual 46 intact chromosomes (Jacobs, Dalton, et al., 1997). In the majority, there is loss of the paternal sex chromosome, and so the single normal X is maternal in origin. In the other 50%, there is more than one obvious cell line, containing either the full complement of 46 chromosomes or some other complex variation. This condition is known as "Turner mosaicism," and these females may have a more mild or a more severe phenotype, depending on the nature of the genetic abnormality. The prevalence of TS in all its manifestations is 4 per 10,000 live female births. Although this is the most common chromosomal aneuploidy, the vast majority of affected conceptuses (95% or so) are spontaneously aborted.

A striking physical characteristic of TS is include short stature. This is associated with the deletion of the distal portion of the short arm of the X chromosome, which contains a gene (the short stature homeobox gene, or SHOX) that is important for normal growth in stature and is required in two working copies by both males (it is on the Y chromosome too) and females (Rao, Weiss, et al., 1997). TS is also usually associated with degeneration of the ovaries soon after birth ("streak ovaries"), so secondary sexual characteristics do not develop without estrogen replacement therapy. Verbal intelligence is normal in X monosomy, but there are associated deficits in nonverbal skills in general and arithmetical abilities in particular. The condition is associated with significant autistic traits, mainly in the domains of social reciprocity and communication, in the presence of good formal language skills and normal verbal intelligence.

Genetic Anomaly

In humans, partial or complete loss of one of the sex chromosomes—either the second X chromosome or the Y chromosome—results in TS. It is associated with a phenotype because of two main influences. First, there is haploinsufficiency for genes that are normally expressed from both X chromosomes in females. They fall into two classes: genes in the pseudoautosomal regions (PAR1 and PAR2) at the tips of the long and short arms of the X chromosome, which have complete homology with the equivalent regions on the Y chromosome. There are also many genes lying outside the PAR that escape X inactivation (Carrel & Willard, 2005). These are presumably needed in two copies for normal female development. Second, hormonal factors are likely to play a contributory role, because noninactivated genes on the X chromosome contribute to the development and maintenance of ovarian tissues. Early degeneration of the ovaries leads to estrogen insufficiency. Occasionally, milder cases are not detected until adulthood, but these are likely to be cases of mosaicism rather than pure X monosomy. This additional cell line may contain a normal 46,XX karyotype (in which case the phenotypic features of the syndrome are ameliorated); some structural anomaly of the X chromosome; or, rarely, a partial Y chromosome (lacking critical elements essential for the development of the male phenotype).

Because males invariably inherit their single X chromosome from their mothers, X-linked imprinted genes could theoretically have sexually dimorphic expression. This may arise because expression is exclusively from the paternally inherited X chromosome (and thus only in females). Alternatively, expression could be exclusively from the maternally inherited X chromosome and would be sexually dimorphic if the gene concerned was subject to X inactivation (Davies et al., 2005). Skuse et al. (1997) proposed, from a study comparing X-monosomic females whose single X was either maternal or paternal in origin, that a paternally expressed allele may be associated with enhanced social-cognitive abilities in normal females relative to males. X-linked imprinting could protect normal females from deleterious allelic variants of autosomal genes that influence the functions of the social brain. Despite considerable work over the last 15 years or so, the only incontrovertible gene that has been found to contribute to the TS phenotype is SHOX, although recent work with a mouse model of the syndrome has identified a paternally imprinted, maternally expressed gene as having a circumscribed impact on the cognitive-behavioral phenotype (Davies et al., 2005).

Physical Phenotype

There are many possible physical characteristics of TS, but none is invariable. If the condition is not detected at birth (usually suspected because of a transient edema that gives a "Michelin Man" appearance, but that rapidly clears), the diagnosis may not be made until middle childhood. The usual reason for ascertainment is growth delay. Specific characteristics include a narrow, high-arched palate, which (along with oromotor immaturity) is associated with severe feeding difficulties in infancy. The mouth may also be characterized by retrognathia (receding lower jaw). There are low-set ears, a low hairline, and occasionally a webbed neck, but this latter feature is much rarer than textbook descriptions would suggest. The eyes may have a strabismus and slight ptosis. The chest is typically broad, with widely spaced breasts, and a wide carrying angle at the elbow is described. Some form of cardiac abnormality occurs in approximately one-third of patients with TS. Problems are primarily left-sided and may include coarctation of the aorta and bicuspid aortic valve. There is a tendency toward hypertension. Other common problems include renal anomalies (30%), associated with an increased risk of urinary tract infections and hypertension. Hypothyroidism is associated with an autoimmune thyroiditis. One of the most serious, but often overlooked, complications is otitis media, exacerbated by anatomical anomalies of the Eustachian tube. This is extremely common in girls with TS, particularly in infancy and early childhood. The majority (50–90%) of women with TS also develop early sensorineural (nerve) hearing loss and may require hearing aids earlier than the general population.

Because of the small stature, which is almost invariably relative to the height of a child's mother in particular, treatment with human growth hormone is increasingly used. The average stature after treatment is increased by a few centimeters, although the condition is not associated with growth hormone insufficiency, and high doses are needed to achieve any significant benefit. There is no evidence that the many psychological problems associated with this condition are due to the stress of having an excessively short stature, although many pediatricians still believe that to be the case.

Behavioral Phenotype

Females with TS are typically reasonably well adjusted socially until adolescence. This is a time of difficulty for a number of different reasons. First, there is the absence of a normal growth spurt, and secondary sexual characteristics may be delayed deliberately by endocrinological management in order to maximize adult height. Second, because the syndrome is associated with specific deficits in social-cognitive competence, girls frequently have problems forming and maintaining peer relationships, and these become all the more complex in adolescence. As adults, many women experience difficulties with social and partner relationships, as well as the ability to fit into a work environment with other adults. Consequently, many choose to work primarily with children, and a disproportionate number find employment in nurseries or child care centers. The social deficits of TS are very striking, but they may not be immediately obvious to pediatricians or endocrinologists, who are primarily responsible for the care of people with the syndrome. Social deficits are probably more marked in childhood among girls who inherited their single X chromosome from their mothers than those who inherited it from their fathers (Skuse et al., 1997), suggesting that social adjustment could be influenced by an imprinted X-linked genetic mechanism.

Cognitive Phenotype

In terms of cognitive development, girls with TS have normal verbal intelligence, but they are deficient in visuospatial skills (such as the ability to complete a jigsaw puzzle). Deficits in nonverbal IQ occur in about 80% of cases of X monosomy. These are typified by poor-quality copying of complex designs and poor visuospatial memory. Individuals with the syndrome also perform poorly on a speeded peg-moving task (a motor task with visuospatial demands) (Bishop, Canning, et al., 2000). They also usually have difficulties in arithmetical abilities and may lack even a basic concept of numbers, implying that dosage-sensitive X-linked genes are also involved in numerical cognitive skills and spatial intelligence.

Together with the behavioural difficulties in social adjustment, females with X monosomy show deficits in social-perceptual processing. They have difficulties recognizing faces and facial expressions of emotion (Lawrence, Kuntsi, et al., 2003b). They are poor at judging direction of gaze, especially when a pictured face is looking at them (Elgar, Campbell, et al., 2002). In some respects, these social-perceptual deficits are reminiscent of those experienced by individuals with an autism spectrum disorder. The risk of autism may be increased 500-fold in women with X monosomy, compared to women of the normal karyotype (Creswell & Skuse, 1999). Autism is also characterized by a selective deficit in "theory of mind"—that is, the ability to attribute independent mental states in order to explain and predict behavior. One commonly used test of theory-of-mind skills in adults employs photographs of the eye region of faces and requires participants to choose a word (such as *alarmed, hostile, shy,* or *anxious*) to best describe what the person in the picture is thinking or feeling (Baron-Cohen, Wheelwright, et al., 2001). Women with X monosomy are severely impaired on this task (Lawrence et al., 2003b). They are also poor at attributing mental states to animated shapes: They produce few descriptions of thoughts, feelings, or mental states (intention scores) when attempting to describe animations that elicit such descriptions in normal populations (Frith & Frith, 2003). This finding is consistent with the observation that social-cognitive anomalies in TS extend to deficits in mentalizing skills. It is not explained by the relative weakness in visuospatial abilities.

To what extent are the deficits reported here in line with those reported for people with autism? The deficit in "reading the mind from the eyes," a mentalizing task dependent on face processing, is more severe in women with TS than in people with autism, while the deficit in processing "fear" from faces appears to be at least as profound in women with TS as in high-functioning people with autism (Lawrence, Campbell, et al., 2003a; Lawrence et al., 2003b). These claims of relative difference must, however, be taken as indicative rather than authoritative. Performance on the tasks is sensitive to verbal and nonverbal intelligence, and shows age-related differences (Campbell, Lawrence, et al., 2006).

Neural Phenotype

Functional imaging has confirmed that the (medial) orbital frontal cortex and superior temporal sulcus (STS) (Castelli, Frith, et al., 2002; Sabbagh, Moulson, et al., 2004) are involved in mentalizing. Activation in regions adjacent to the amygdala (anterior medial temporal cortex) has also been reported for the animations task under theory-of-mind presentation conditions. Cortical structural imaging in women with X monosomy has indicated abnormalities in each of these specific neural regions. There are abnormalities of the STS together with abnormal connectivity, especially between anterior and posterior temporal regions (Molko, Cachia, et al., 2004). Volumetric gray matter differences are found in amygdala and orbital frontal cortex (Good et al., 2003; Kesler, Garrett, et al., 2004). Functional anomalies in the amygdala have also been reported in relation to facial expression processing (Skuse, Morris, et al., 2005b). All these findings suggest that the anomalous processing of theory-of-mind measures may reflect anomalous cortical structure and function of the "social brain" in women with TS. There is also evidence supporting the theory that X-linked imprinted genes may play a role in the X-monosomic social phenotype. X-linked imprinting appears to influence the volume of the superior temporal gyrus, as well as those of occipital white matter and cerebellar gray matter. Cutter, Daly, et al. (2006) employed magnetic resonance imaging and proton magnetic resonance spectroscopy to investigate brain anatomy and metabolism in X monosomy, using both a hand-traced region-of-interest approach and voxel-based morphometry. Women with $45,X_m$ (i.e., X monosomy with maternal X chromosome) were shown to possess a significantly larger adjusted right hippocampal volume than subjects with $45,X_p$ (X monosomy with paternal X chromosome). This result may explain a prior finding that females with $45,X_p$ have poorer visual memory than those with $45,X_m$, despite their better social adjustment. Females with $45,X_m$ had significantly smaller volumes of caudate nucleus and thalamus than those with a single paternal X chromosome had. Dysfunction of the caudate nucleus could lead to abnormal executive functioning, with impaired working memory, planning ability, set shifting, and social cooperation. Maternally expressed X-linked genes may therefore influence hippocampal development, and paternally expressed genes may influence the normal development of the caudate nucleus and thalamus in females.

Genotype–Phenotype Relationships

Many phenotypic features of TS are due to haploinsufficiency for a gene, or several genes, that are expressed from both X chromosomes in normal females (i.e., are not subject to X inactivation). We know that a great many more genes escape X inactivation than have homologues on the Y chromosome, for reasons that are unclear but may have something to do with sexual dimorphism (Skuse, 2006). With molecular methods, it is possible to map the breakpoints in structural anomalies of the X chromosome and in samples of isochromosomes, as well as partial deletions of the short arm of the X chromosome, and also ring chromosomes. Since the publication of the genetic sequence of the X chromosome (Ross, Grafham, et al., 2005), it should be possible to map susceptibility genes more accurately than was formerly possible. There is a clear need to collect more detailed information on the cognitive and behavioral phenotypes of these structural anomalies, and to correlate systematic variations in those phenotypes with the nature and extent of chromosomal material that has been lost. This technique is known as "deletion mapping," and was successfully applied in the mapping of the only "TS" gene identified to date (SHOX), which influences growth in stature. The technique is also potentially applicable to phenotypes such as cognition (Good et al., 2003).

Williams–Beuren Syndrome

Williams–Beuren syndrome (WBS) is a rare sporadic disorder, with an incidence of about 1 in 7500 live births (Stromme, Bjornstad, et al., 2002). The condition is associated with a distinctive cognitive and behavioral profile. Affected individuals have a superficial facility with language, as well as sociability and good facial recognition; however, they usually possess very poor visuospatial skills, owing to their propensity to see objects as sets of parts rather than as coherent wholes.

Genetic Anomaly

In 95% of individuals with clinical features of WBS, there is a submicroscopic deletion of 1.55 megabase (Mb), and in 5%, there is a deletion of 1.84 megabase (Mb) in one copy of chromosome 7 on the long arm at 7q11.2 (Bayes, Magano, et al., 2003). The region of chromosome 7 most commonly deleted contains a number of genes that may have some relevance to the phenotype. These include the following:

1. Elastin (ELN), responsible for the connective tissue problems associated with the syndrome, including the cardiac anomalies found in a majority.
2. LIM kinase 1 (LIMK1), expressed in the brain, and linked to the visuospatial deficits of the syndrome by several authors (Hoogenraad, Akhmanova, et al., 2004); however, findings have not been consistent, and some individuals with intact LIMK1 nevertheless have the classic phenotype.
3. General transcription factor II (GTF2I), which has been linked to the mental retardation associated with WBS.
4. Syntaxin1A (STX1A), a transcription factor involved in neurotransmitter release).
5. Bromodomain adjacent to leucine zipper 1B (BAZ1B), which may have metabolic influences.
6. Cytoplasmic linker 2 (CYLN2), which is strongly expressed in the brain, and may be associated with cerebellar abnormalities (Hoogenraad et al., 2004).
7. GTF2IRD1, another transcription factor possibly implicated in the craniofacial features of the syndrome, as well as in the visuospatial anomalies (Tassabehji, Hammond, et al., 2005).
8. Neutrophil cystolic factor 1 (NCF1), which could be associated with risk of hypertension.

In total, at least 20 genes may be deleted, including a number of highly repetitive DNA sequences. The deletion arises because occasionally recombination occurs between misaligned repeat sequences on either side of the critical region during gamete formation. Inversion of the segment of chromosome 7 occurs in about 7% of the general population, and in up to one-third of the parents of probands who transmitted the deleted chromosome 7. Although the extent of the deletion is variable, the classic WBS phenotype is always associated with deletion of ELN, and usually of LIMK1.

Physical Phenotype

WBS is associated with a range of physical features, none of which is invariable and none of which is pathognomonic. The facial features are usually distinctive, with a small jaw, broad brow, short nose, long philtrum between nose and upper lip, wide mouth, and full lips. Children have full cheeks, periorbital puffiness, and a short, upturned nose. The facial phenotype changes between childhood and adulthood, becoming more angular and appearing thinner. There are usually cardiovascular complications due to the ELN deletion, and supravalvular aortic stenosis occurs in 75% of individuals with WBS. Puberty is often early, and there can be endocrinological anomalies leading to hypothyroidism, diabetes, and hypercalcemia. Growth is impaired throughout life, and most adults with WBS will be in the lowest 3% of the population for stature.

Cognitive and Behavioral Phenotype

The issue of intelligence in WBS has recently been the subject of debate (Meyer-Lindenberg, Mervis, et al., 2006). Due to the considerable discrepancy between verbal and nonverbal abilities, the computation of an overall or full scale IQ is inappropriate. The mean verbal IQ is about 70, as is the score for nonverbal reasoning. The standard deviation is similar to that for the general population; therefore, the mean level of ability is about 2 standard deviations below population norms. On tests that include measures of visuospatial skills, such as the Differential Ability Scales—School Age Version (Elliott, 1990), the mean level of ability is lower (about 2.5 standard deviations below population norms).

On the other hand, persons with WBS have relatively well-maintained verbal short-term memory and verbal fluency, although their verbal skills are qualitatively abnormal. It is important to note that claims are made about the strengths of children with WBS that later research shows are based on massive ascertainment bias (a situation that applies to many other neurodevelopmental syndromes). Language acquisition in the syndrome is delayed, and the progression of language development may follow a path that is different from normal (Scerif & Karmiloff-Smith, 2005). Grammatical abilities are consonant with verbal IQ, and there is no evidence for the use of unusual and sophisticated language, as was once claimed. In general, pragmatic skills (language used for social communication) are weak.

The claim that children with WBS have unusual musical talents, quite out of line with their general abilities, is not borne out on closer examination. Although anecdotal accounts suggest that they may have an undue interest in music, compared with typically developing children of similar mental age, this observation has not been tested systematically.

Children with WBS do not appear to have social deficits of an autistic character, although no study using standardized interview or observational measures of autistic symptomatology has yet been done. It is often said that their social interactions are relatively unimpaired. They are typically outgoing and gregarious. There is sometimes good eye contact, although this is debatably the exception rather than the rule, and they possess good empathic skills from infancy onward. They are interested in faces, although their face processing is abnormal: Their technique for face recognition memory is based on the sum of parts, rather than on a gestalt of the face as a whole. It would be a mistake to assume that their social interactions are normal. Rather, the abnormality is of a different character from that found in conditions where the impairment to social skills is autism-like. Children with WBS are often strikingly disinhibited and relate indiscriminately to unfamiliar children and adults, which may put them at social risk. The social disinhibition found in association with this syndrome, as occurs in many cases of ADHD of unknown origin, is correlated with poor performance on standard tests of attention. Tager-Flusberg and Sullivan (2000) found that whereas people with the syndrome possess superficially good social skills, they have deficits in specific social-cognitive abilities. These deficits include theory of mind and the ability to accurately interpret other people's facial expressions, especially negative facial expressions. Their lack of ability to perceive a face signalling "threat" is common to a range of neurodevelopmental disorders, and may indicate anomalous amygdalocortical connectivity.

Neural Phenotype

The brain phenotype of WBS has been described in terms of both structure and function. Postmortem studies have found reduced brain volume, Chiari malformations (congenital abnormalities of the posterior fossa, causing protrusion of the cerebellum through the foramen magnum into the spinal canal), and abnormalities of the corpus callosum (more convex than that of normal control subjects and overall smaller in volume, particularly in the splenium and in the caudal part of the callosal body). *In vivo* studies of brain structure have used automated methods, such as voxel-based morphometric comparisons of the brains of persons with WBS and normal controls. Because of the distinctively different shape of the brain in WBS, there are potential technical pitfalls in so doing (Eckert, Galaburda, et al., 2006). Nevertheless, there have been remarkably consistent findings of reduced posterior parietal gray matter compared to typical controls, although inconsistent findings have been reported for hypothalamus and orbital frontal gray matter regions. The most consistent abnormalities are found in terms of symmetrical gray matter volume reductions in the intraparietal sulcus (associated with abnormal white matter integrity), around the third ventricle, and in the orbital frontal cortex (Meyer-Lindenberg et al., 2006).

Functional anomalies in visual processing are to be expected, given the characteristic problems in visuospatial tasks such as drawing, jigsaws, Legos, and other forms of pattern construction. Meyer-Lindenberg, Kohn, et al. (2004) tested the hypothesis that the main functional anomaly would be in the dorsal visual processing pathway that lies between the primary visual cortex and higher cortical processing areas (including the intraparietal sulcus). They found that the pathway was hypofunctional, which could be a link to a recent finding that one of the genes deleted from the 7q11.2 critical region (GTF2IRD1) is inadequately expressed in the peripheral visual cortex and superior parietal regions.

There is hyporesponsivity of the amygdala to threatening faces in WBS (Meyer-Lindenberg, Hariri, et al., 2005). This is superficially similar to the case in autism, where the most consistent neural abnormality is a failure of amygdalocortical connectivity in response to socially relevant cues. The abnormal amygdala response in autism could be due to a failure to make adequate eye contact with the facial stimuli, but this is less likely to be the explanation in WBS. It seems instead that there is likely to be an anomaly of amygdala–orbital frontal connectivity, which results in a failure to link sensory representations of socially relevant stimuli (in this case, threats) with social judgments made about them. If the threat stimulus signaled by the amygdala is not associated in the orbital frontal cortex with a cortical representation, warning the individual about the potential consequences of continued interaction, disinhibited behavior could evolve independently of social learning.

Genotype–Phenotype Relationships

At this stage, there is no incontestable link between any single gene in the critical region and an aspect of the cognitive or behavioral phenotype of WBS. Some progress has been made by studying individuals who are lacking a smaller critical region than is usual (thereby apparently ruling out a few genes), and by constructing animal models in which putative contributory genes have been knocked out. The greatest interest to date has focused on genes LIMK1, GTF2IRD1, and GTF2I, together with CYLN2. All are thought to play some role in the visuoconstructional deficits associated with the syndrome. Although there is no doubt that ELN plays an important role in many features of WBS, this gene is not expressed in the brain, and therefore it is not a candidate for these aspects of the phenotype.

Fragile X Syndrome

Fragile X syndrome (FXS) is the most common cause of inherited mental retardation, occurring in approximately 1 in 4000 males (Pembrey, Barnicoat, et al., 2001). Because it is an X-linked syndrome, only a partial phenotype is seen in females, with a prevalence of 1 in 8000. Most mothers of boys with the syndrome have no knowledge of their potential risk for delivering an affected child, and no reason to be investigated for behavioral anomalies or cognitive deficits. Affected males have fairly distinctive physical features and behavioral characteristics. The most striking deficit is moderate to severe intellectual impairment, with most affected males having IQs under 50.

Genetic Anomaly

FXS is caused by the deficiency or absence of the fragile X mental retardation protein 1 (FMR1). In well over 99% of cases, the molecular defect is an expansion of CGG repeats in the 5' untranslated region of the FMR1 gene, which lies on the long arm of the X chromosome at Xq27.3. Large expansions in this region, known as "full mutations" (more than 200 repeats), inhibit transcription and lead to silencing of the gene (Musci & Caughey, 2005). The FMR1 gene is strongly expressed in the brain, where it binds RNA and acts as a translational suppressor of dendritic messenger RNAs that code for a variety of proteins, many of which are involved in signal transduction (Ennis, Murray, et al., 2006). The distribution of allele length in the general population is usually about 30 repeats. Expansions to greater degrees, falling short of full mutations, are known as "premutations" (55–200 repeats). Elderly male carriers of premutation alleles (maternal grandfathers, typically) have recently been recognized to have an associated fragile X tremor/ataxia syndrome (Hagerman & Hagerman, 2004). This premutation syndrome is rare among female carriers for reasons that are unknown, but affected females have a higher-than-average risk of premature ovarian failure, with a menopause at less than 40 years of age. Carrier frequencies are given as 1:259 females and 1:810 males. Premutations greater than 90 repeats expand to full mutations when they are transmitted to the next generation by females; they are intrinsically unstable. In the intermediate range of allele expansion (40–60 repeats), there is also increased risk of expansion, but rarely to the full mutation in one generation (Ennis et al., 2006). Intermediate alleles are probably associated with an increased risk of special educational needs in boys, but the clinical implications of this weak effect are not great (Youings, Murray, et al., 2000).

Physical Phenotype

The classical clinical features of FXS include a long, narrow face, with a large head and prominent ears; narrow palpebral fissures, with eyes closely set together; a high-arched palate; and a protruding jaw. There are also joint and skeletal features, including hypotonia (lax joints), flat feet, and hyperextensible thumb and fingers. Testicular volume is exceptionally large during and after puberty in the majority. There are similar, but milder, physical phenotypes in boys with intermediate mutations or premutations (Aziz, Stathopulu, et al., 2003).

Cognitive and Behavioral Phenotype

Evidence is emerging to suggest that FXS, like WBS, is associated with a complex pattern of cognitive disabilities, with some preserved areas and typical cognitive and behavioral characteristics. Despite global (verbal and nonverbal) mental retardation in most cases, there is relative sparing of skills that require face and emotion processing, as well as theory of mind (Cornish, Turk, et al., 2004). This is associated with poor sequential online processing, short-term visuospatial memory deficits, impaired motor planning, and a repetitive and impulsive style of interaction (Cornish et al., 2004). There is said to be a characteristic form of repetitive and impulsive speech (e.g., "Ca-Ca-Ca-Ca-Can I have a biscuit?"). Other language features have much in common with autistic styles of conversation, including concrete understanding and use of language, and turn-taking difficulties.

The behavioral style of a child with FXS is characterized by more inattentiveness, restlessness, and fidgetiness than would be found in unaffected children of equivalent intellectual level. In the small number of affected males with a normal IQ (possibly due to cellular mosaicism for the full mutation, which is often associated with premutation in a proportion of cells), there may be such psychiatric problems as anxiety, mood lability, social communication problems, and selective mutism (Hagerman, Ono, et al., 2005).

The association between FXS and autism has intrigued researchers and clinicians for many years. Various behaviors in FXS are reminiscent of autistic behaviors, including social anxiety and withdrawal, hyperarousal, peer relationship problems, stereotypic behaviors, gaze aversion, and other anomalies of social reciprocity and language of an autistic type (Hessl, Glaser, et al., 2006). There is some evidence that an increased risk could be associated with the premutation as well as the full mutation (Hagerman et al., 2005). It is rare to find the mutation among children with autism and a normal-range IQ. Just over 2% of children with autism overall will have the FMR1 full mutation (Reddy, 2005).

Neural Phenotype

Studies using structural neuroimaging of individuals with FXS have revealed that specific neuroanatomical anomalies are associated with the full mutation (Hessl, Rivera, et al., 2004). These include enlargement of the hippocampus, amygdala, caudate nucleus, and thalamus. The cerebellar vermis and superior temporal gyrus are reduced in size. Normally, these regions are recruited during social engagement and the processing of social information, including interpreting the direction of another's gaze. Some girls and boys with the full FMR1 mutation find direct gaze aversive, and it is associated with increased stress reflected in cortisol reactivity. Others with the genetic anomaly may avoid eye contact because they lack social interest. Females with the FMR1 premutation (or full mutation) do not adequately recruit brain regions that are associated with certain specific cognitive tasks, such as arithmetic processing, and their arithmetic abilities may be relatively more impaired than their reading or spelling skills (Lachiewicz, Dawson, et al., 2006). This finding has not been replicated in all studies (Fisch, 2006). Executive function skills, such as planning, attention, and set shifting, are also impaired in females with the premutation (or full mutation), and these impairments seem to worsen with chronological age.

Conclusions

There exist several well-described genetic syndromes that are associated to a greater or lesser extent with specific behavioral, cognitive, and neural processes. In this chapter, we have discussed three examples of such syndromes: TS, WBS, and FXS, which are associated with sex chromosome aneuploidy, submicroscopic deletions, and a trinucelotide repeat expansion, respectively. (Characteristics of these syndromes are summarized in Table 23.1.) The study of neural and cognitive processes and behavioral phenotypes associated with genetic anomalies is interesting for a variety of reasons. Compared with the study of complex genetic conditions such as psychiatric disorders, the study of behavioral phenotypes arguably offers a more productive way to study relationships between genetic anomalies and behavioral, cognitive, and neural processes. The relative advantage of this approach is that it reduces the genes of interest to a narrower range, compared with the large number of loci implicated in such disorders as schizophrenia, autism, or ADHD. One application of this approach might be to use evidence from genetic syndromes to identify potential candidate genes for association testing in idiopathic disorders where similar cognitive or behavioral characteristics are observed. Gene identification offers the opportunity to further investigate the molecular biology of the gene product and the potential role it plays in neural and cognitive processes more generally in the population. It is potentially more fruitful to investigate the cognitive outcome of disrupted neural processes as a result of identified genetic anomalies than to seek to find the underlying but unknown genetic anomalies among individuals with heritable cognitive or behavioral deficits. In psychiatry, genetic influences on "endophenotypes" or "intermediate phenotypes" (measurable components unseen by the unaided eye along the pathway between genes and distal phenotype; Gottesman & Gould, 2003) could be easier to identify than the role played by normal genetic variation on individual differences at the level of symptomatology (Gottesman & Hanson, 2005; Prathikanti & Weinberger, 2005).

The possibility that we could derive a rational system for parsing psychiatric disorders, by means of identifying the specific genes that increase susceptibility to them, is an idea Kendler (2006) discussed with considerable scepticism. Some of the potential problems are exemplified by examining the behavioral and cognitive phenotypes of disorders caused by known genetic disorders. First, single-gene anomalies (such as those causing FXS or Rett syndrome) are not invariably associated with characteristic cognitive or behavioral dysfunction. Mutations of the relevant FMR1 or MECP2 (methyl CpG binding protein 2) genes are found in people without the classical symptomatic phenotype. There are other people who have the symptomatic profile characteristic of these disorders but in whom no genetic anomaly can be found. Second, it has proved extraordinarily difficult to identify a clear relationship between the influence of any single gene (out of several potential candidates) and the cognitive or behavioral phenotype in conditions such as velocardiofacial syndrome or

TABLE 23.1. Overview of Behavioral, Cognitive, and Neural Phenotypes Associated with the Genetic Anomalies Arising in Turner Syndrome (TS), Williams–Beuren Syndrome (WBS), and Fragile X Syndrome (FXS)

	TS	WBS	FXS
Genetic anomaly	45,X, 45,X/46,XX, SHOX gene.	Submicroscopic deletion of 7q11.2. Implicated genes: ELN, LIMK1.	CGG repeat expansion (>200) in FMR1.
Physical phenotype	Short stature; streak ovaries; narrow, high-arched palate; receding lower jaw; low-set ears; low hairline; webbed neck; strabismus; ptosis; inner-ear anomalies; broad chest with widely spaced nipples; wide carrying angle; cardiac and renal anomalies; hypertension.	Distinctive facial features—small jaw, broad brow, short nose, long philtrum, wide mouth, full lips; cardiovascular complications/supravalvular aortic stenosis; early puberty; endocrinological anomalies—hypothyroidism, diabetes, hypercalcemia; short stature.	Long, narrow face; large head; prominent ears; close-set eyes; narrow palpebral fissures; high-arched palate; protruding jaw; hypotonia; flat feet; hyperextensible joints; macroorchidism.
Behavioral phenotype	Social deficits, more marked in 45,X$_m$ carriers; autistic disorder.	Sociable; socially disinhibited at times.	Repetitive and impulsive speech; social deficits; inattentiveness; restlessness; autistic features (social anxiety, withdrawal, peer difficulties stereotypies, gaze aversion); autistic disorder.
Cognitive phenotype	Mild learning disability; nonverbal deficits; poor visuospatial memory; facial emotion-processing deficits; deficits on mentalizing tasks.	Mild–moderate learning disability; marked nonverbal deficits; good verbal short-term memory and verbal fluency; delayed language; abnormal face processing; impaired facial emotion processing (negative valence); poor attention.	Moderate learning disability; visuospatial memory deficits; impaired motor planning; arithmetic deficits and executive function deficits in females with premutation and full mutation.
Neural phenotype	*Structural*: Abnormalities in STS; volumetric gray matter differences in amygdala and OFC. 45,X$_m$: Larger right hippocampal volume; smaller volume of caudate nucleus/thalamus.	*Structural*: Reduced brain volume; Chiari malformations; anomalies of corpus callosum; reduced post. parietal gray matter (intraparietal sulcus, third ventricle, OFC).	*Structural*: Enlarged hippocampus, amygdala, caudate, thalamus; reduced cerebellar vermis, STG.
	Functional: Abnormal connectivity between STS and ant. and post. temporal regions; functional anomalies in amygdala during face-processing tasks.	*Functional*: Hypofunction of dorsal visual processing pathway; hyporesponsivity of amygdala to threatening faces.	.

Note. STS, superior temporal sulcus; STG, superior temporal gyrus; OFC, orbital frontal cortex; 45,X, annotation X monosomy; 45,X/46,XX, annotation Turner mosaicism; 45,X$_m$, X monosomy with maternal X chromosome; SHOX, short stature homeobox gene; ELN, elastin gene; LIMK1, LIM domain kinase 1 gene; FMR1, fragile X mental retardation protein 1 gene.

WBS. For example, in velocardiofacial syndrome the "risk region" of the genome is clearly defined by a microdeletion and every nucleotide/gene in that region has been mapped. The risk of developing schizophrenia in this condition is very high, in the order of 25% and 1% of the schizophrenic population having the anomaly. Yet the causative mechanism has eluded discovery and no single genetic risk factor has yet been identified, despite over a decade of searching for it.

Within the last few years, one of the most striking discoveries in behavioral phenotype research is how many syndromes are associated with some features of autism. This is also reflected in the disorders discussed here. This lack of specificity could reflect, in part, the fact that many behavioral phenotypes are associated with moderate to severe mental retardation. In each case, the discovery of what appeared to be a clue to the etiology of "autism" was greeted with great excitement and sponsored new research efforts. It was thought that a greater understanding of the pathophysiology of disorders associated with an "autism-like" phenotype, such as tuberous sclerosis or Rett syndrome, might offer insights into the pathophysiology of autism itself. But a similar risk of autism is observed in FXS, velocardiofacial syndrome, Prader–Willi syndrome, TS, and even Duchenne muscular dystrophy (in each case, the proportion affected seems to be about 30%). These discoveries must call into question the hypothesis that there is a specific genetic susceptibility (in each case) in common with idiopathic autistic disorders. In all the examples given above, formal evaluation of autistic symptoms via standardized measures demonstrates an enormously increased risk, compared with the accepted population prevalence for autism spectrum disorders of 0.6%. However, it is also important to consider that the social and communication deficits and repetitive behaviors that represent the triad of impairments in autism are increasingly recognized as traits quantitatively distributed in the population (Skuse, Mandy, et al., 2005a; Ronald, Happé, et al., 2006). Thus autism may have more in common with ADHD than was previously thought. It might be reasonable to assume that traits associated with normal functioning, such as the ability to attend, concentrate, and interact socially, may be impaired in the presence of widespread disruption to neural processes.

Accordingly, in the search for genes that predispose to the development of child psychiatric disorders, it may not be efficient (and indeed is arguably conceptually muddled) to seek the genetic components of relatively amorphous diseases, still less "comorbid" disorders. As Kendler (2006) reflects, it is highly unlikely that we will discover any direct relationship between gene variation and conventional categories of psychiatric disorder. Rather, we should look for how genetic variation influences neural functions that may be common to a number of diseases or disorders. Creating a heuristic cognitive model of developmental disorders will in due course lead to the integration of disciplines, and this is already happening in child psychiatry. We need to learn how genes control neurobiological development, and in so doing, increasing use will be made of animal models (Gould & Gottesman, 2006). Studies of the genetic influences upon specific neural processes could reveal "double dissociations," analogous to the procedure used by neuropsychologists to contrast cognitive profiles in groups with distinct brain lesions. In other words, through the appropriate study of candidate gene effects upon neurocognitive phenotypes, we could discover systems that are influenced by relatively independent genetic mechanisms. Were it possible to generalize this technique, with the discovery of other genes for specific neural functions (which may in turn have neurobehavioral consequences), a more rational and scientific basis for the classification and treatment of child psychiatric disorders could be developed.

References

Aziz, M., Stathopulu, E., et al. (2003). Clinical features of boys with fragile X premutations and intermediate alleles. *American Journal of Medical Genetics: Part B. Neuropsychiatric Genetics, 121*(1), 119–127.

Baron-Cohen, S., Wheelwright, S., et al. (2001). The "Reading the Mind in the Eyes" Test, revised version: A study with normal adults, and adults with Asperger syndrome or high-functioning autism. *Journal of Child Psychology and Psychiatry, 42*(2), 241–251.

Bayes, M., Magano, L. F., et al. (2003). Mutational mechanisms of Williams–Beuren syndrome deletions. *American Journal of Human Genetics, 73*(1), 131–151.

Bishop, D. V., Canning, E., et al. (2000). Distinctive patterns of memory function in subgroups of females with Turner syndrome: Evidence for imprinted loci on the X-chromosome affecting neurodevelopment. *Neuropsychologia, 38*(5), 712–721.

Campbell, R., Lawrence, K., et al. (2006). Meanings in motion and faces: Developmental associations between the processing of intention from geometrical animations and gaze detection accuracy. *Development and Psychopathology, 18*(1), 99–118.

Carrel, L., & Willard, H. F. (2005). X-inactivation profile reveals extensive variability in X-linked gene expression in females. *Nature, 434*(7031), 400–404.

Castelli, F., Frith, C., et al. (2002). Autism, Asperger syndrome and brain mechanisms for the attribution of mental states to animated shapes. *Brain, 125*(Pt. 8), 1839–1849.

Check, E. (2005). Human genome: Patchwork people. *Nature, 437*(7062), 1084–1086.

Cornish, K. M., Turk, J., et al. (2004). Annotation: Deconstructing the attention deficit in fragile X syndrome: A developmental neuropsychological approach. *Journal of Child Psychology and Psychiatry, 45*(6), 1042–1053.

Creswell C., & Skuse, D. H. (1999). Autism in association with Turner syndrome: Genetic implications for male vulnerability to pervasive developmental disorders. *Neurocase, 5*, 101–108.

Crews, D., & McLachlan, J. A. (2006). Epigenetics, evolution, endocrine disruption, health, and disease. *Endocrinology, 147*(6, Suppl.), S4–S10.

Cutter, W. J., Daly, E. M., et al. (2006). Influence of X chromosome and hormones on human brain development: A magnetic resonance imaging and proton magnetic resonance spectroscopy study of Turner syndrome. *Biological Psychiatry, 59*(3), 273–283.

Davies, W., Isles, A., et al. (2005). Xlr3b is a new imprinted candidate for X-linked parent-of-origin effects on cognitive function in mice. *Nature Genetics, 37*(6), 625–629.

Eckert, M. A., Galaburda, A. M., et al. (2006). Anomalous sylvian fissure morphology in Williams syndrome. *NeuroImage, 33*(1), 39–45.

Elgar, K., Campbell, R., et al. (2002). Are you looking at me?: Accuracy in processing line-of-sight in Turner syndrome. *Proceedings of the Royal Society of London, Series B, Biological Sciences, 269*(1508), 2415–2422.

Elliott, C. D. (1990). The nature and structure of children's abilities: Evidence from the Differential Ability Scales. *Journal of Psychoeducational Assessment, 8*, 376–390.

Ennis, S., Murray, A., et al. (2006). An investigation of FRAXA intermediate allele phenotype in a longitudinal sample. *Annals of Human Genetics, 70*(Pt. 2), 170–180.

Fisch, G. S. (2006). Cognitive-behavioral profiles of females with the fragile X mutation. *American Journal of Medical Genetics: Part A, 140*(7), 673–677.

Frith, U., & Frith, C. D. (2003). Development and neurophysiology of mentalizing. *Philosophical Transactions of the Royal Society of London, Series B, Biological Sciences, 358*(1431), 459–473.

Good, C. D., Lawrence, K., et al. (2003). Dosage-sensitive X-linked locus influences the development of amygdala and orbitofrontal cortex, and fear recognition in humans. *Brain, 126*(Pt. 11), 2431–2446.

Gottesman, I. I., & Gould, T. D. (2003). The endophenotype concept in psychiatry: Etymology and strategic intentions. *American Journal of Psychiatry, 160*(4), 636–645.

Gottesman, I. I., & Hanson, D. R. (2005). Human development: Biological and genetic processes. *Annual Review of Psychology, 56*, 263–286.

Gould, T. D., & Gottesman, I. I. (2006). Psychiatric endophenotypes and the development of valid animal models. *Genes, Brain and Behavior, 5*(2), 113–119.

Hagerman, P. J., & Hagerman, R. J. (2004). Fragile X-associated tremor/ataxia syndrome (FXTAS). *Mental Retardation and Developmental Disabilities Research Reviews, 10*(1), 25–30.

Hagerman, R. J., Ono, M. Y., et al. (2005). Recent advances in fragile X: A model for autism and neurodegeneration. *Current Opinion in Psychiatry, 18*(5), 490–496.

Haston, C. K., & Hudson, T. J. (2005). Finding genetic modifiers of cystic fibrosis. *New England Journal of Medicine, 353*(14), 1509–1511.

Hessl, D., Glaser, B., et al. (2006). Social behavior and cortisol reactivity in children with fragile X syndrome. *Journal of Child Psychology and Psychiatry, 47*(6), 602–610.

Hessl, D., Rivera, S. M., et al. (2004). The neuroanatomy and neuroendocrinology of fragile X syndrome. *Mental Retardation and Developmental Disabilities Research Reviews, 10*(1), 17–24.

Hoogenraad, C. C., Akhmanova, A., et al. (2004). LIMK1 and CLIP-115: Linking cytoskeletal defects to Williams syndrome. *Bioessays, 26*(2), 141–150.

Jacobs, P., Dalton, P., et al. (1997). Turner syndrome: A cytogenetic and molecular study. *Annals of Human Genetics, 61*(Pt. 6), 471–483.

Kendler, K. S. (2006). Reflections on the relationship between psychiatric genetics and psychiatric nosology. *American Journal of Psychiatry, 163*(7), 1138–1146.

Kesler, S. R., Garrett, A., et al. (2004). Amygdala and hippocampal volumes in Turner syndrome: A high-resolution MRI study of X-monosomy. *Neuropsychologia, 42*(14), 1971–1978.

Lachiewicz, A. M., Dawson, D. V., et al. (2006). Arithmetic difficulties in females with the fragile X premutation. *American Journal of Medical Genetics: Part A, 140*(7), 665–672.

Lawrence, K., Campbell, R., et al. (2003a). Interpreting gaze in Turner syndrome: Impaired sensitivity to intention and emotion, but preservation of social cueing. *Neuropsychologia, 41*(8), 894–905.

Lawrence, K., Kuntsi, J., et al. (2003b). Face and emotion recognition deficits in Turner syndrome: A possible role for X-linked genes in amygdala development. *Neuropsychology, 17*(1), 39–49.

Meyer-Lindenberg, A., Hariri, A. R., et al. (2005). Neural correlates of genetically abnormal social cognition in Williams syndrome. *Nature Neuroscience, 8*(8), 991–993.

Meyer-Lindenberg, A., Kohn, P., et al. (2004). Neural basis of genetically determined visuospatial construction deficit in Williams syndrome. *Neuron, 43*(5), 623–631.

Meyer-Lindenberg, A., Mervis, C. B., et al. (2006). Neural mechanisms in Williams syndrome: A unique window to genetic influences on cognition and behaviour. *Nature Reviews Neuroscience, 7*(5), 380–393.

Molko, N., Cachia, A., et al. (2004). Brain anatomy in Turner syndrome: Evidence for impaired social and spatial–numerical networks. *Cerebral Cortex, 14*(8), 840–850.

Musci, T. J., & Caughey, A. B. (2005). Cost-effectiveness analysis of prenatal population-based fragile X carrier screening. *American Journal of Obstetrics and Gynecology, 192*(6), 1905–1912 (discussion, 1912–1915).

Pearson, C. E., Nichol Edamura, K., et al. (2005). Repeat instability: Mechanisms of dynamic mutations. *Nature Reviews Genetics, 6*(10), 729–742.

Pembrey, M. E., Barnicoat, A. J., et al. (2001). An assessment of screening strategies for fragile X syndrome in the UK. *Health Technology Assessment, 5*(7), 1–95.

Persico, A. M., & Bourgeron, T. (2006). Searching for ways out of the autism maze: Genetic, epigenetic and environmental clues. *Trends in Neurosciences, 29*(7), 349–358.

Prathikanti, S., & Weinberger, D. R. (2005). Psychiatric genetics—the new era: Genetic research and some clinical implications. *British Medical Bulletin, 73–74*, 107–122.

Rao, E., Weiss, B., et al. (1997). Pseudoautosomal deletions encompassing a novel homeobox gene cause growth failure in idiopathic short stature and Turner syndrome. *Nature Genetics, 16*(1), 54–63.

Reddy, K. S. (2005). Cytogenetic abnormalities and fragile-X syndrome in autism spectrum disorder. *BMC Medical Genetics, 6,* 3.

Redon, R., Ishikawa, S., et al. (2006). Global variation in copy number in the human genome. *Nature, 444*(7118), 444–454.

Ronald, A., Happé, F., et al. (2006). Genetic heterogeneity between the three components of the autism spectrum: A twin study. *Journal of the American Academy of Child and Adolescent Psychiatry, 45*(6), 691–699.

Ross, M. T., Grafham, D. V., et al. (2005). The DNA sequence of the human X chromosome. *Nature, 434*(7031), 325–337.

Rutter, M., Moffitt, T. E., et al. (2006). Gene–environment interplay and psychopathology: Multiple varieties but real effects. *Journal of Child Psychology and Psychiatry, 47*(3–4), 226–261.

Sabbagh, M. A., Moulson, M. C., et al. (2004). Neural correlates of mental state decoding in human adults: An event-related potential study. *Journal of Cognitive Neuroscience, 16*(3), 415–426.

Scerif, G., & Karmiloff-Smith, A. (2005). The dawn of cognitive genetics? Crucial developmental caveats. *Trends in Cognitive Sciences, 9*(3), 126–135.

Skuse, D. (2006). Genetic influences on the neural basis of social cognition. *Philosophical Transactions of the Royal Society of London, Series B, Biological Sciences, 361*(1476), 2129–2141.

Skuse, D. H., James, R. S., et al. (1997). Evidence from Turner's syndrome of an imprinted X-linked locus affecting cognitive function. *Nature, 387*(6634), 705–708.

Skuse, D. H., Mandy, W. P., et al. (2005a). Measuring autistic traits: Heritability, reliability and validity of the Social and Communication Disorders Checklist. *British Journal of Psychiatry, 187,* 568–572.

Skuse, D. H., Morris, J. S., et al. (2005b). Functional dissociation of amygdala-modulated arousal and cognitive appraisal, in Turner syndrome. *Brain, 128*(Pt. 9), 2084–2096.

Stromme, P., Bjornstad, P. G., et al. (2002). Prevalence estimation of Williams syndrome. *Journal of Child Neurology, 17*(4), 269–271.

Szyf, M., Weaver, I. C., et al. (2005). Maternal programming of steroid receptor expression and phenotype through DNA methylation in the rat. *Frontiers in Neuroendocrinology, 26*(3–4), 139–162.

Tager-Flusberg, H., & Sullivan, K. (2000). A componential view of theory of mind: Evidence from Williams syndrome. *Cognition, 76*(1), 59–90.

Tassabehji, M., Hammond, P., et al. (2005). GTF2IRD1 in craniofacial development of humans and mice. *Science, 310*(5751), 1184–1187.

Weaver, I. C., Champagne, F. A., et al. (2005). Reversal of maternal programming of stress responses in adult offspring through methyl supplementation: Altering epigenetic marking later in life. *Journal of Neuroscience, 25*(47), 11045–11054.

Youings, S. A., Murray, A., et al. (2000). FRAXA and FRAXE: The results of a five year survey. *Journal of Medical Genetics, 37*(6), 415–421.

CHAPTER 24

The Effects
of Early Institutionalization
on Social Behavior
and Underlying Neural Correlates

BETHANY C. REEB
NATHAN A. FOX
CHARLES A. NELSON
CHARLES H. ZEANAH

Since the early investigations of Goldfarb (1945), Spitz (1946), Bowlby (1951), and Provence and Lipton (1962), it has been well known that children reared in institutions develop atypical social behavior and exhibit significant delays in cognitive and motor development. Recent findings confirm these earlier reports (for reviews, see Johnson, 2000; Rutter, 2006). An equally well-documented finding is that after children are removed from institutional settings, they show significant gains in all domains (Johnson, 2002, Rutter, 2006). Nonetheless, deficits in social functioning frequently persist for a number of years after a child is either fostered or adopted (Rutter, 2006; Rutter et al., 2007).

Despite over 50 years of research examining the effects of early institutionalization on child development, researchers have only recently begun to investigate the underlying neurobiology that may be affected by such early-life events. Identification of these neural correlates may give us some insight into which mechanisms underlie the observed social deficits, as well as why these deficits persist several years after adoption. In this chapter, we review research that has investigated the effects of early institutionalization on both development of social behavior and its underlying neural mechanisms. In addition, we discuss findings from research using nonhuman primates and rodents, which suggest a possible mediating role of amygdala development in the development of institutionalization-induced social dysfunction.

Effects of Early Institutionalization on Social Behavior

Children who have experienced early institutionalization display abnormal socioemotional behavior. These abnormalities include decreased functional and symbolic play behavior (Kaler & Freeman, 1994; Kreppner, O'Connor, Dunn, Andersen-Wood, & English and Romanian Adoptees [ERA] Study Team, 1999) and increased aggression (Smyke, Dumitrescu, & Zeanah, 2002). However, the most stable and persistent finding in children who have experienced early institutionalization is their display of social disinhibition, also termed "indiscriminate sociability," "indiscriminate friendliness," or "disinhibited attachment disorder." This pattern of behavior has been characterized by an undiscriminating social approach to others, lack of awareness of social boundaries, and difficulty in identifying or responding to social cues about what is socially appropriate when engaging with other people (O'Connor et al., 2003; O'Connor, Rutter, & ERA Study Team, 2000). Early studies conducted by Tizard and colleagues were among the first to demonstrate that children who remained in an institution for at least the first 2 years of life displayed "overfriendly" behaviors toward strangers during early childhood (Tizard & Rees, 1975), and that this behavior persisted into late childhood (Tizard & Hodges, 1978) and early adolescence (Hodges & Tizard, 1989). Compared to more recent studies of institutions in Romania, where children were confined to cots for much of the day and had very few toys and interactions with caregivers, the children in Tizard's studies were raised in relatively high-quality institutions, where there was access to toys and where a low staff-to-child ratio allowed for a good deal of child–caregiver interaction. However, caregivers rotated shifts often and were instructed not to form relationships with the children, thus limiting the opportunity for the development of selective child–caregiver attachments. Tizard and colleagues found that, despite these children being raised in such high-quality institutions with plenty of mental and physical stimulation, overfriendliness toward strangers still emerged; this suggests that the development of normal social behavior is not dependent on the physical quality of the environment, but upon the presence of a loving and stable caregiver (Tizard & Rees, 1975). More recent studies have confirmed these findings (Roy, Rutter, & Pickles, 2004; Smyke et al., 2002).

Since Tizard's initial studies, several other studies have replicated the finding that institutionalized children are at significantly greater risk of displaying disinhibited attachment behaviors than are typically developing children (O'Connor et al., 2000, 2003; Rutter et al., 2007; Rutter, Kreppner, & O'Connor, 2001; Vorria, Rutter, Pickles, Wolkind, & Hobshaum, 1998; Zeanah, Smyke, & Dumitrescu, 2002; Zeanah, Smyke, Koga, Carlson, & Bucharest Early Intervention Project [BEIP] Core Group, 2005). However, in any given population of previously institutionalized children, heterogeneity exists: Some children display disinhibited attachment behaviors, while others may not (Rutter et al., 2001). This heterogeneity in social behavior may be due to heterogeneity in the time at which children are removed from the institutional environment. Selective attachments appear to form between the ages of 6 and 9 months in typically developing children (Cassidy & Shaver, 1999). Perhaps a sensitive period exists in which there is increased plasticity in the brain to optimize the formation of such attachments. This period of neural plasticity is likely to occur prior to the appearance of attachment behaviors at 6–9 months of age. Therefore, if sensitive, responsive, and consistent caregivers are available during this proposed sensitive period, there is a greater likelihood that normal attachments will develop. However, children who do not have such caregivers may develop abnormal attachments if the typical care is not received during this period of time. Thus it is possible that if an institutionalized child is removed from the deprived institutional environment and placed into a home with caring and responsive parents during this sensitive period, he or she will develop normal attachment patterns. In contrast, a child who continues to remain in an institution past the offset of this sensitive period without stable and responsive attachment figures is more likely to develop attachment disorders.

Indeed, some researchers have investigated the link between age of removal from an institution and later development of social dysfunction. Chisholm (1998) examined attachment behaviors in children from Romanian institutions at approximately 4.5 years of age. Compared to Canadian-born control children, the institutionalized children displayed increased disinhibited attachment and indiscriminate friendliness. Interestingly, these children were able to form attachment relationships with their adoptive caregivers. However, there was a significant increase in the incidence of insecure attachment patterns among the previously institutionalized children. When the group of previously institutionalized children was divided into those who were adopted from the orphanage before 4 months of age and those who were adopted after 8 months of age, only the children adopted after 8 months of age showed a significant increase in indiscriminate friendliness as well as in insecure attachment patterns. Children adopted before 4 months of age did not differ from the Canadian-born controls in either of these measures. These results suggest that children adopted prior to the end of the proposed sensitive period for attachment may not be as susceptible to deprivation-induced changes in social behavior, whereas those adopted after this period may be at higher risk.

In the ERA Study conducted by Rutter and colleagues, disinhibited attachment behavior was examined in children adopted from Romanian institutions when they were 4–11 years of age (O'Connor, Bredenkamp, Rutter, & ERA Study Team, 1999; O'Connor et al., 2000, 2003; Rutter et al., 2007). The researchers found that at 4 years of age, children who were institutionalized for 7–24 months exhibited significantly higher levels of attachment disturbances than did institutionalized children adopted before 6 months of age from a U.K. institution, as well as those who were adopted before 6 months of age from a Romanian institution (O'Connor et al., 1999). When child–parent attachment was assessed with a home-based version of the Strange Situation procedure, institutionalized children who were adopted between 6 and 24 months of age were significantly more likely to display either organized but insecure or disorganized patterns of attachment, while children adopted prior to 6 months of age were more likely to display a secure attachment pattern (O'Connor et al., 2003). In addition, increased levels of indiscriminate friendliness were related to an increased likelihood of displaying organized/insecure or disorganized attachment patterns.

Rutter and his colleagues conducted follow-up studies on this same group of children at 6 and 11 years of age. At 6 years of age, children showed some decrease in attachment disturbances, but children who were adopted between 6 and 24 months of age had a greater likelihood of displaying severe attachment disorder signs than those adopted before 6 months had (O'Connor et al., 2000). In addition, those who were adopted after 24 months of age showed another 30% increase in severe signs above children adopted prior to 24 months of age. At age 11, patterns of disinhibited attachment persisted among institutionalized children, especially those who lived in the institution for at least 6 months (Rutter et al., 2007). These children were also more likely to display socially inappropriate physical contact, lack of social reserve, verbal and social violation of conventional boundaries, and a high level of spontaneous comments. From both the initial and follow-up studies of this cohort of institutionalized children, it appears that children who were adopted before the age of 6 months were significantly less likely to develop attachment problems, thus providing additional evidence of the possibility that a sensitive period for development of attachment behavior exists and that this period may close at approximately 6 months of age.

Recent research from the BEIP suggests that the sensitive period's window may be extended to 22 months of age (Zeanah & Smyke, 2007). In this study, it was found that institutionalized children who were placed into foster care before 22 months of age were significantly less likely to develop disinhibited attachment than were either children who remained in the institution or those who were placed into foster care after 22 months of age. In addition, the children placed into foster care before 22 months did not differ from community children in measures of disinhibited attachment. Therefore, it may be that the majority of neural connections associated with the development of selective attachments are made prior to 6 months of age, but that some flexibility in the neural system still exists (such that if stable caregiving occurs between 6 and 22 months, the behavioral system may still develop normally).

Why would patterns of disinhibited attachment develop in children who have been institutionalized? It has been suggested that this behavior is adaptive for the institutional setting (Chisholm, 1998; Rutter, O'Connor, & ERA Study Team, 2004; Smyke et al., 2002), and that experience-adaptive developmental programming occurs in the brain during a particular sensitive period (Rutter et al., 2004). During sensitive periods, the brain expects input of a certain type to develop the foundational neural structure that will underlie a particular competency or skill (Knudsen, 2004). Therefore, a young child's brain may expect protection and responsive caregiving from individuals in his or her environment, to promote the formation of an adaptive attachment bond and facilitate the organization of adaptive attachment behaviors. A child living in an institutional environment will learn that there is no single caregiver or set of caregivers who can reliably provide a sense of security and be responsive to the child's distress. Therefore, such a child will develop an adaptive response in which he or she engages any adult who may provide comfort when exposed to a novel or unfamiliar situation. In this sense, disinhibited or indiscriminate friendliness may be viewed as adaptive.

Effects of Early Institutionalization on Neuroendocrine Function

Research investigating the effects of early deprivation on nonhuman primates and rodents has shown that early deprivation permanently alters the stress response system (for a review, see Sanchez, Ladd, & Plotsky, 2001). Given these findings in animal studies, it is likely that hypothalamic–pituitary–adrenal (HPA) function is also altered in children who have experienced institutionalization. To date, few studies have investigated the effects of early institutionalization on neuroendocrine function. One study by Carlson and Earls (1997) compared the cortisol levels of Romanian orphans to those of never-institutionalized, home-reared Romanian children. Between the ages of 2 and 9 months, the institutionalized children were divided into two groups; one group was randomly assigned to receive social enrichment with a low caretaker-to-child ratio (4:1), and the other group was left in the standard, socially deprived condition of a high caretaker-to-child ratio (20:1). The socially enriched group remained in the enriched environment for 13 months before being placed back into the standard orphanage conditions.

When the children were 2 years old (5–6 months after enrichment), morning, noon, and evening salivary cortisol levels were measured over 2 consecutive days. Institutionalized and home-reared children were found to differ significantly in their HPA diurnal profile. Home-reared children displayed typical rhythm in cortisol production with elevated cortisol in the morning and decreased levels as the day progressed. In contrast, the institutionalized children (enriched and deprived combined) displayed significantly lower morning cortisol values, but showed similar decreases in noon and evening cortisol levels, suggesting that early institutionalization led to an overall blunting of the diurnal pattern. In addition, when the two institutionalized groups were examined separately, the consistently deprived children displayed elevated noon cortisol levels, while the previously enriched group did not differ from the home-reared children; this finding suggests that consistent deprivation may have led to greater abnormality in the HPA response, and that intervention may have alleviated some of these effects. Similar blunting effects of institutionalization on morning cortisol levels were also found in a population of internationally adopted children within the first month after adoption (Bruce, Kroupina, Parker, & Gunnar, 2000).

Additional long-term effects of early institutionalization on neuroendocrine function were reported by Gunnar, Morison, Chisholm, and Schuder (2001). In this study, institutionalized Romanian children adopted into Canada were compared with never-institutionalized Canadian-born children. The Romanian orphans were divided into early- and late-adopted groups; the early-adopted group was adopted before 4 months of age, whereas the late-adopted group was adopted at 8 months of age or later. At the time of testing, all children had lived in their adoptive homes for at least 6 years. Morning, noon, and evening saliva samples were collected on 3 consecutive days. In contrast to Carlson and Earl's (1997) findings, Gunnar et al. found that rather than displaying hyporesponsiveness in HPA function, the late-adopted group had significantly higher average daily cortisol levels than those of the early-adopted and never-institutionalized children, with the greatest difference found in the morning. The early-adopted and never-institutionalized children did not differ significantly from one another in their daytime pattern of cortisol response. In addition, relations between time spent in the institution and cortisol levels suggested that longer durations of institutional care were related to higher evening cortisol levels. The highest increases in cortisol levels were found in children who were placed with an adoptive family after 24 months of age, whereas children placed with an adoptive family before 24 months of age were in the normal range of evening levels; these results suggest that perhaps the first 2 years of life may be a sensitive period for the development of normal neuroendocrine function.

It should be noted that the results and methods of collection in Gunnar's study differed from those reported by Carlson and Earls (1997). The cortisol measures in that study were taken from children while they were still in the institution. In contrast, in Gunnar and colleagues' study, children had been out of the institution for at least 6 years. These results suggest that when children were removed from the institutional environment, they displayed some recovery of HPA function (normal daily rhythmic patterns). However, because the sample of institutionalized children was not randomly assigned, it may be that these findings are the results of selection bias; that is, parents in Canada may have adopted children with fewer mental or health problems than the children Carlson and Earls tested in the Romanian institution. These children would not necessarily be a representative sample of institutionalized children. Nevertheless, the data suggest that these children still displayed overall heightened responsivity, compared to those who were never institutionalized. Taken together, both studies suggest that early social deprivation can lead to a general dysregulation of the neuroendocrine stress response system. Similar to the persistence in social dysfunction, there is also a parallel persistence of dysfunction in the stress response system that can last for at least 6 years after adoption.

Early Institutionalization's Effects on Electroencephalography

Neural activity may be noninvasively measured through collection of electroencephalographic (EEG) activity and event-related potentials (ERPs). These methods have been used to further examine the effects of early institutionalization on neural mechanisms that may be related to early social development. (Research using ERPs is described in the next section.) In the first study examining the effects of early institutionalization on EEG activity, Marshall, Fox, and the BEIP Core Group (2004) reported that institutionalized Romanian children displayed a significant increase in low-frequency band activity (theta power), as well as a significant decrease in high-frequency band activity (alpha and beta power), compared to age-matched never-institutionalized Romanian children. Furthermore, there was regional specificity for this increase in low-frequency activity and decrease in high-frequency activity: Group differences in theta power were primarily found in the posterior scalp regions, while group differences in alpha and beta power were found in the frontal and temporal regions. Similar patterns of EEG power were obtained in a comparison of internationally adopted postinstitutionalized children to age-matched never-institutionalized children (Tarullo, Chatham, & Gunnar, 2007). In addition, similar patterns of activity have been reported in children with disorders such as autism (Dawson, Klinger, Panagiotides, Lewy, & Castelloe, 1995) and attention-deficit/hyperactivity disorder (Harmony, Marosi, Diaz de Leon, Becker, & Fernandez, 1990)—disorders known to be prevalent in institutionalized children (Kreppner, O'Connor, & Rutter, 2001; Rutter et al., 1999; Vorria et al., 1998). Marshall et al. (2004) suggested that these patterns of EEG may be due to either a maturational lag in nervous system development or a general hypoarousal due to a significant decrease in stimulation; however, the data are not suitable to determine which model is the better fit.

A follow-up study by Marshall, Reeb, Fox, and the BEIP Core Group (2008) of the same group of institutionalized children investigated the effects of foster care intervention on the development of EEG patterns at 42 months of age. Through random assignment, half of the institutionalized children were placed into foster care, while the other half remained in the institution[1] (Zeanah et al., 2003). EEG was reassessed when all the children were 42 months of age. EEG power did not differ significantly between the institutionalized and foster care groups in any frequency band. However, this lack of difference may be due to the variability in the age at which children were placed into foster care intervention, since it is likely that earlier-placed children will show greater differences in EEG than later-placed children when compared to institutionalized children. As predicted, when the relation between age of placement and EEG was analyzed among the foster care group, it was found that children placed into foster care at younger ages displayed increased EEG power, but only in alpha

---

[1]Due to a policy of noninterference, some children in the group assigned to remain in the institution may have been placed into government foster care or placed back into the care of their biological parents.

band activity (Marshall et al., 2008). These results suggest that institutionalized children who experience early-life foster care intervention display some recovery of EEG power compared to those who remain in the institution, and that this recovery is likely to depend upon the age at which children are placed into foster care (with earlier placement leading to greater recovery).

Marshall et al. (2008) additionally investigated the effects of foster care intervention on measures of EEG coherence. "EEG coherence" is a statistical measure of phase consistency between two time series, with high coherence between two electrode sites thought to reflect a high degree of synchronization between two cortical areas. It has been proposed that decreased coherence represents increased spatial differentiation and thus increased complexity of the brain, which in turn is related to increased speed and efficiency of information processing (Thatcher, 1992). As was found for EEG power, the foster care and institutionalized groups did not differ in their overall EEG coherence. However, age of placement was associated with a decrease in short-distance EEG coherence, specifically between frontal and temporal regions, among the children placed into foster care. These results suggest that differentiation of the neural circuits of the brain as measured by EEG coherence is dependent upon the age at which children are removed from institutionalization, with the more mature pattern associated with earlier removal.

Early Institutionalization's Effects on ERPs

All studies that have been described thus far demonstrated the effects of institutionalization on the stress response system and EEG measures—neural correlates that may subsequently affect social responsivity. Only a few studies have directly examined the neural processing of social information by using facial stimuli (Moulson, Fox, Zeanah, & Nelson, in press; Parker, Nelson, & BEIP Core Group, 2005a, 2005b). In one study, ERPs were recorded while institutionalized Romanian children (ranging from 5 to 31 months of age) or never-institutionalized Romanian children viewed pictures of either their caregivers' faces or a stranger's face (Parker et al., 2005a). For the caregivers' faces, institutionalized children were shown pictures of their preferred caregivers, while never-institutionalized children were shown pictures of their mothers. The institutionalized children displayed attenuation in their N170, negative component (Nc), and positive slow-wave (PSW) component, and increases in their P250 component, compared to the never-institutionalized group. Both groups of children had a normative increase in the Nc amplitude to the face of a stranger compared to their caregivers' faces. In contrast, the institutionalized children displayed an abnormal pattern for the PSW component: They showed a tendency toward a decrease in amplitude to the stranger's face compared to their familiar caregivers' faces, while the never-institutionalized children showed an opposite pattern. Age at testing appears to affect response to novel and familiar faces, in which greater Nc amplitude to the caregiver is typically found during infancy (de Haan & Nelson, 1997), whereas greater Nc amplitude to the stranger is apparent during toddlerhood (Carver et al., 2003; Dawson et al., 2002). Similar patterns were found when a younger (age range = 7–23.4 months) and older group (23.5–31.6 months) of children were compared; however, there were no differences in either the younger or older group between institutionalized and never-institutionalized children, suggesting that they have a similar developmental pattern. However, it should be pointed out that the variability in age in both the younger and older groups, particularly in the younger group (which spanned across a 16-month time period),

may have obscured any differences in development patterns. Therefore, effects of early institutionalization on caregiver–stranger facial discrimination cannot be ruled out entirely.

In a second study, ERPs were collected while the same group of institutionalized children
or never-institutionalized children viewed faces that displayed different emotional expressions—happy, sad, angry, and fearful. Similar to the results presented above, the institutionalized children showed overall attenuated amplitude of the N170, Nc, and PSW components,
but significantly larger P250 amplitude, suggesting that institutionalization affected the overall processing of facial information. In addition, there was some evidence that the institutionalized children processed expressions of fear differently than the never-institutionalized
children did. Parker et al. (2005b) speculated that the institutionalized children might have
displayed such overactivity in response to fearful expressions because of hyperactivity in
the amygdala. The amygdala has been shown to play a significant role in the recognition of
fear in both adults (Adolphs, Tranel, Damasio, & Damasio, 1994) and children (Thomas et
al., 2001). These findings were recently replicated and expanded to show that institutionalized children additionally displayed smaller amplitudes for the P1 and P400 components
than the never-institutionalized group did, and that the latencies for these components were
longer among the institutionalized children than among the never-institutionalized children,
indicating slower information processing in children with a history of institutionalization.
Finally, never-institutionalized children showed right-hemispheric specialization for faces,
whereas the institutionalized children did not.

After these initial baseline measures were collected, the institutionalized children
were randomly assigned to be placed into foster care or to remain in the institution. At 42
months of age, ERPs were once again collected during presentation of emotion face stimuli
(Moulson et al., in press). It was found that P1 and P400 amplitudes of children in the foster
care group displayed amplitudes midway between those of the institutionalized and never-
institutionalized children, suggesting that foster care may have been normalizing these components. Interestingly, neither age at placement in foster care nor duration of time in foster
care contributed to this outcome. Although these results are very promising, in that some
neural functions may show some recovery after children are removed from the institutional
environment, additional assessments are still needed to evaluate whether full recovery of
processing is possible over longer periods of intervention.

These observed effects of early institutionalization on the neural processing of faces are
likely to underlie the early deprivation deficits observed in emotion identification and understanding (Fries & Pollak, 2004; Sloutsky, 1997). This overall attenuation in neural activity
observed in the institutionalized group is also consistent with the "hypoactivation" observed
in EEG power (Marshall et al., 2004). In general, institutionalization leads to decreased neural activation during both states of rest and processing of social stimuli, but these effects can
be somewhat ameliorated by foster care placement. Although these studies are informative
in the examination of the effects of early institutionalization on the brain, they only measure
activity at the scalp.

Effects of Institutionalization
on Brain Metabolism and Connectivity

Studies conducted by Chugani and colleagues (Chugani et al., 2001; Eluvathingal et al.,
2006) give some insight into which underlying brain structures are affected by early insti-

tutionalization. These researchers examined a small sample of previously institutionalized children, studying glucose metabolism in the brain (Chugani et al., 2001) as well as white matter connectivity (Eluvathingal et al., 2006). Positron emission tomography (PET) was used to measure brain metabolism in children who were adopted from Romanian institutions into the United States between the ages of 16 and 90 months. Compared to normal adults and a set of controls that included age-matched never-institutionalized children with one epileptic hemisphere and one nonepileptic hemisphere, institutionalized children displayed significant decreases in metabolism in the orbital frontal gyrus, infralimbic prefrontal cortex, amygdala, hippocampus, lateral temporal cortex, and brainstem. Several of these areas have been implicated in the development of social behavior in nonhuman primates (Machado & Bachevalier, 2003), and therefore it is likely that the dysfunction in metabolism in these brain regions may be related to some of abnormal social behaviors observed in institutionalized children. In addition, similar to the effects of institutionalization on social development, decreased metabolic activity in these regions appears to have had a long-term effect, given that most children had spent at least 2 years in their adoptive homes at the time of testing. These results suggest that early institutionalization leads to decreased metabolic function in several areas of the brain known to be linked to social behavior, and that recovery of function in these brain regions may be less reversible with intervention. It should be noted that only 10 institutionalized children were used in this study, and that all of these children had behavioral problems, which also may have led to the observed decreased metabolic function. Additional studies should be conducted on a larger, more normative sample, with a better-matched control group, before definitive conclusions can be drawn.

In a separate group of institutionalized Romanian children who had normal intelligence and few behavior problems, Chugani's group (Eluvathingal et al., 2006) used diffusion tensor imaging (DTI) to investigate the connectivity of white matter of pathways of the limbic system, because this system was highly implicated in their previous PET study. DTI measures were collected between the ages of 7 and 11 years in both institutionalized children who were adopted between the ages of 17 and 60 months and never-institutionalized age-matched control children. Tracts investigated included the uncinate fasciculus, stria terminalis, fornix, and cingulum. An overall tendency toward decreased connectivity was found in all tracts investigated among the previously institutionalized children, compared to the never-institutionalized controls. However, the only area in the brain where the institutionalized children showed a statistically significant decrease in white matter connectivity compared to control children was in the uncinate fasciculus. These tracts run from the temporal convolutions and cortical nuclei of the amygdala to the frontal lobes, medial orbital cortex, and subcallosal area (Ebeling & von Cramon, 1992). It is not clear what may have led to decreased connectivity in the uncinate fasciculus in the children who suffered early socioemotional deprivation, but the authors suggested that it may have been due to less myelination or fewer fibers, or perhaps there was disorganization of the tract (Eluvathingal et al., 2006). These results may explain the results found for EEG coherence. Since this tract appears to show the greatest dysfunction, then perhaps it may also show the greatest recovery and increase in myelination after foster care placement, thus leading to decreased EEG coherence between temporal and frontal regions. Overall, these results suggest that early institutionalization may lead to decreased brain connectivity, with the largest decrease found in the connectivity between areas of the brain (i.e., amygdala and orbital frontal cortex) that have been shown to be involved in the development of social behavior in nonhuman primates (Machado & Bachevalier, 2003).

Neurobiology of Social Behavior and Attachment Development: Evidence from Animal Studies

It has been suggested that the effects of early institutionalization on development of social behavior are primarily due to abnormal development of the limbic system and its connections to the temporal and frontal lobes caused by socioemotional deprivation (Joseph, 1999; Schore, 2001). Indeed, findings from Chugani's group (Chugani et al., 2001; Eluvathingal et al., 2006) as well as the BEIP (Marshall et al., 2008) provide some support for this claim. It was the limbic region that displayed the most significant decreases in glucose metabolism (Chugani et al., 2001), as well the least connectivity with temporal and frontal lobes (Eluvathingal et al., 2006), among children who were previously institutionalized. In addition, significant decreases in EEG coherence among children who received foster care intervention was only found between the frontal and temporal regions, suggesting the possibility that foster care may be specifically affecting the connectivity maturation between the underlying limbic system and frontal lobes (Marshall et al., 2008). Moreover, ERP attenuation during face processing (Nelson, Zeanah, & Fox, 2007; Parker et al., 2005a, 2005b) may be the result of limbic system dysfunction, given that facial emotion processing has been linked to areas of the limbic system, such as the amygdala (Adolphs et al., 1994, 1999; Thomas et al., 2001). Although these results are consistent in suggesting that the limbic system and its connectivity to the frontal and temporal regions may be mediating the patterns of dysfunctional social behavior observed in institutionalized children, these suggestions are only speculative. However, nonhuman primate and rodent research has investigated the role of the amygdala in the development of attachment and affiliative behavior, and this research offers indirect support for these suggestions.

Early studies in rhesus macaques have shown that removal of the amygdala during infancy can have severe effects on the expression of later social behavior during both the juvenile period and adulthood (Thompson, Bergland, & Towfighi, 1977; Thompson, Schwartzbaum, & Harlow, 1968; Thompson & Towfighi, 1976). More recent studies, using techniques that produce more selective damage to the amygdala while sparing surrounding tissue (Bauman, Lavenex, Mason, Capitanio, & Amaral, 2004b; Bauman, Toscano, Mason, Lavenex, & Amaral, 2006; Prather et al., 2001), have found that amygdala lesions made during the first weeks of life do not appear to affect the development of basic social behaviors. However, the timing and expression of such behaviors are significantly different from those in monkeys that did not receive amygdala lesions. For example, compared to control animals, monkeys with neonatal amygdala lesions displayed significant increases in fear-related behaviors (e.g., fear grimaces, screams) during social interactions with familiar conspecifics (Bauman et al., 2004b; Prather et al., 2001). In addition, despite increased social fear expression, monkeys with amygdala lesions still displayed significant increases in approach and affiliative behaviors toward a conspecific, compared to sham control animals and hippocampus-lesioned animals (Bauman et al., 2004b). Interestingly, more persistent and stable increases in affiliative and approach behaviors were observed when amygdala-lesioned animals were paired with a novel conspecific versus a familiar conspecific. These unusual behaviors displayed by monkeys with neonatal amygdala lesions are similar to those displayed by institutionalized children, who also display increased social withdrawal to peers (Goldfarb, 1945; Hoksbergen, Van Dijkum, & Stoutjesdijk, 2002; Roy, Rutter, & Pickles, 2000) and caregivers (Smyke et al., 2002; Zeanah et al., 2005), but increased friendliness toward strangers (O'Connor et al., 2000, 2003; Tizard & Hodges, 1978; Tizard & Rees, 1975; Zeanah et al., 2002, 2005).

Moreover, amygdala-lesioned neonatal monkeys have been shown to exhibit less aggression and to have lower social status than either control or hippocampus-lesioned animals (Bauman et al., 2006). During competition for a preferred food, monkeys with neonatal amygdala lesions were significantly more likely to be the last in obtaining the preferred food, and less likely to display contact aggression, but more likely to display increased fear grimaces. It is likely that increased social fear response may be related to the observed decrease in aggression and dominance. Bauman et al. (2006) suggested that the amygdala-lesioned monkeys were likely to be ranked lower in social status than the others because of an abnormal balance between their expression of fear and aggressive behaviors (Bauman et al., 2006). Social status in monkeys relies heavily on the ability to recruit allies and form coalitions with fellow group members (Thierry, 1990). Therefore, an individual who has a heightened sense of fear during social encounters will approach others less, and in turn will have less opportunity to go out and recruit allies, leading to a decreased likelihood of ever obtaining high social rank. In addition, having less opportunity to interact with other monkeys may lead to a decreased understanding of when and what social expressions are appropriate during different interactions. Similar to these amygdala-lesioned monkeys, institutionalized children are less successful in peer relationships (Hodges & Tizard, 1989; Roy et al., 2004) and behave in a socially inappropriate manner (O'Connor et al., 2000, 2003). However, institutionalized children have been reported by parents to display heightened levels of aggression (Smyke et al., 2002), which is the opposite pattern of behavior from that shown by amygdala-lesioned monkeys.

The studies described above elicit parallels between the dysfunctional social behaviors observed in monkeys with neonatal amygdala lesions and in institutionalized children. These interesting similarities clearly suggest that the institutional care environment affects the development of the amygdala and its connectivity with other regions of the brain. What aspects of the institutional care environment may lead to such effects on the amygdala? One strong possibility is the lack of a stable and comforting caregiver with whom a child can form an attachment bond (Bowlby, 1951; Harlow & Zimmermann, 1959; Joseph, 1996; Suomi, 1990). Therefore, examining the effects of amygdala development on attachment may help to explain whether early deprivation does indeed affect the development of the amygdala and its circuitry.

Only recently have the effects of neonatal lesioning of the amygdala been directly examined in relation to the caregiver–infant attachment bond (Bauman, Lavenex, Mason, Capitanio, & Amaral, 2004a). These researchers found that, when compared with hippocampus-lesioned and sham-lesioned control monkeys, monkeys with neonatal amygdala lesions displayed significantly increased contact time with their mothers, suggesting the development of an attachment relationship. However, when monkeys were separated from their mothers at 6 months of age and subsequently allowed to choose between their own mothers and a familiar adult female, amygdala-lesioned monkeys did not preferentially seek proximity to their mothers, nor did they produce distress vocalizations—behaviors that were present in controls and hippocampus-lesioned animals. Amygdala-lesioned monkeys were able to distinguish between their mothers and the other female, because they showed a significant increase in approaches toward their mothers over the other female, but still spent equal amounts of time in proximity to each. These results are similar to the attachment patterns exhibited by institutionalized children, who are able to form attachment relationships with their adoptive caregivers (Chisholm, 1998), but still display indiscriminate friendliness toward other adults (O'Connor et al., 2000, 2003; Rutter et al., 2001, 2007; Zeanah et al., 2002, 2005).

Bauman and colleagues' results suggest that the amygdala plays some role in facilitating an attachment bond between caregiver and infant. However, it does not address the question of whether disruption of attachment can lead to disruption in amygdala development. A recent study by Sabatini et al. (2007) addressed this issue. They found that infant monkeys separated from their mothers from either 1 week or 1 month of age until 3 months of age displayed a significant decrease in expression of the guanylate cyclase 1-alpha-3 (GUCY1A3) gene in the amygdala, compared to maternally reared monkeys. This gene has been shown to code for receptors that bind nitric oxide, which in turn stimulates the production of cyclic guanosine monophosphate (Ahern, Klyachko, & Jackson, 2002) and is thought to modulate neural projection outgrowth (He, Yu, & Baas, 2002; Hess, Patterson, Smith, & Skene, 1993; Hindley et al., 1997; Rialas et al., 2000). The decrease in the expression of this gene strongly correlated with the age at which an infant monkey was separated from its mother. Specifically, monkeys separated at 1 week of age showed the lowest expression, nonseparated monkeys showed the highest expression, and the monkeys separated at 1 month of age showed an intermediate level of expression. At 1 week of age, GUCY1A3 expression is already at adult levels, suggesting that the observed decrease was most likely due to gene down-regulation after severe deprivation. When social behavior was analyzed, levels of expression of GUCY1A3 in the amygdala were found to be negatively correlated with levels of self-comforting and positively correlated with increased expression of typical social behavior, suggesting that down-regulation of GUCY1A3 expression in the amygdala induced by early social deprivation is predictive of later social behavior development.

Although it is not known from these results how early deprivation induced changes in the expression of this particular gene in the amygdala, Sabatini et al. (2007) suggested that the changes may have arisen from alterations in the development of the connectivity of the amygdala with other regions of the brain. The amygdala and its connections are established by the first week of life in rhesus monkeys (Machado & Bachevalier, 2003); however, significant postnatal reshaping and myelination continue until 3 months of age (Webster, Ungerleider, & Bachevalier, 1991). Therefore, development of normal amygdala connectivity may be dependent upon the stimulation of a responsive and comforting caregiver. Sabatini et al. also alluded to a possible dose-dependent effect of deprivation on amygdala GUCY1A3 expression, in which longer periods of deprivation led to greater decreases in expression.

Recent research in rodents has also indicated that the amygdala plays a critical role in the development of infant attachment (Moriceau & Sullivan, 2004a, 2006; Moriceau, Wilson, Levine, & Sullivan, 2006). In a neurobiological model of attachment, Sullivan and colleagues suggest that a sensitive period for attachment development exists, in which underlying neural circuitry optimizes the likelihood of an infant's forming an attachment with a caregiver (Moriceau & Sullivan, 2005; Sullivan, 2003). In the rat, this attachment is dependent upon learning to associate olfactory information (typically the mother's) with tactile care behaviors, such as nursing and grooming. This long-lasting association develops during the first 10 days of life. Increased preference for the mother's odor is dependent upon increased release of norepinephrine (NE) from a hyperfunctioning locus coeruleus (LC), therefore leading to increased proximity-seeking behaviors toward the mother's odor (Moriceau & Sullivan, 2004b). In addition, increased proximity seeking is facilitated through a hypofunctioning amygdala, which leads to decreased capacity for developing an aversion to the mother's odor, since mother rats occasionally harm their pups (Hofer & Sullivan, 2001; Sullivan, 2003). Dramatic decreases in LC NE release, as well as increases in amygdala function, correspond to the end of the sensitive period (for a review, see Moriceau & Sullivan, 2005).

Support for this model comes from several experiments in which Sullivan and her colleagues conditioned rat pups to prefer a novel odor that was paired with either a rewarding stimulus, such as tactile stroking (Sullivan, Hofer, & Brake, 1986), or with a punishing stimulus, such as a shock (Moriceau & Sullivan, 2004a; Sullivan, Landers, Yeaman, & Wilson, 2000) or an abusive mother (Roth & Sullivan, 2005). This odor preference could only be established before postnatal day 10, after which there was no preference for the odor paired with the stroking, as well as no aversion to the odor paired with the shock (Moriceau & Sullivan, 2005; Sullivan et al., 2000).

Coincidentally, this sensitive period of odor learning parallels the rat's stress-hyporesponsive period, in which circulating corticosterone is down-regulated (Sapolsky & Meaney, 1986); this suggests that development of HPA function may also play a critical role in facilitating attachment in rats. Corticosterone has been found in several studies to influence amygdala activation during the sensitive period, because injections of exogenous corticosterone were shown to evoke a fear response to a shock-induced odor or odor of a predator in sensitive-period pups (Moriceau & Sullivan, 2004a; Moriceau et al., 2006; Takahashi, 1994; Takahashi & Rubin, 1993). In addition, shock-induced odor preference was reinstated in post-sensitive-period rat pups by depleting corticosterone following adrenalectomy (Moriceau & Sullivan, 2004a; Moriceau et al., 2006). The facilitation of amygdala activity by corticosterone also appeared to be modulated by maternal presence, in that post-sensitive-period rats continued to display a shock-induced preference if the shock was presented in the presence of an anesthetized mother (Moriceau & Sullivan, 2006). Taken together, these results suggest that corticosterone-facilitated amygdala hypofunction present during the first 10 days of life in the rat helps ensure the development of infant–caregiver attachment, and that the mother acts to modulate corticosterone levels after presentations of a stressor once the sensitive period is over, to continue facilitation of such an attachment.

How Can Animal Models Inform Us about Development of Social Behavior in Institutionalized Children?

Sullivan's model of attachment primarily focuses on the preferential learning of an odor associated with maternal caregiving, leading to attachment. Given that rat pups have limited visual and auditory input during their first week of life, this association between maternal odors and tactile stimulation through nursing and grooming is a primary mechanism by which rat pups come to identify the mother. The formation of human infant attachment is not limited to learning through only somatosensory and olfactory stimulation, but also includes visual and auditory stimulation. Moreover, all these sensory experiences do play a role in initial human social learning. For example, shortly after birth, the human infant displays preferences for the mother's voice (DeCasper & Fifer, 1980), scent (MacFarlane, 1975; Porter, Makin, Davis, & Christensen, 1992), and face (Bushnell, Sai, & Mullin, 1989). As in rat pups, it is likely that these preferences develop through pairings of such stimuli with both maternal touch and nursing.

In addition, flexibility in learning of an odor preference observed during the sensitive period (postnatal days 1–10) in rat pups is similar to the flexibility in establishing a caregiver preference observed during the first 6–9 months after birth in human infants. As described above, rats show increased likelihood of developing preferences during this sensitive period for nonmaternal odors, even if those odors are paired with a fear-evoking stimulus (e.g., shock) for post-sensitive-period rat pups (Moriceau & Sullivan, 2005). The offset of such a sensitive period corresponds to the development of increased activation in the amygdala, as well as in the stress response system (Moriceau & Sullivan, 2004a, 2006; Moriceau et al., 2006). Similarly, from birth to approximately 6–9 months of age, human infants have a preference for nonspecific social stimulation and a lack of a fear response to strangers (Cassidy & Shaver, 1999). The period of decreased fear during the first 6–9 months of life may facilitate plasticity in forming attachments with adult caregivers.

We know that by 6 months of age, infants display differential neural responses to emotional faces (Leppanen, Moulson, Vogel-Farley, & Nelson, 2007; Nelson & de Haan, 1996). Discrimination of such emotional faces has been shown to involve amygdala activation in both adults and children (Adolphs et al., 1994, 1999; Thomas et al., 2001). In addition, human infants may have a stress hyporesponse period sometime during the first year of life (Gunnar, Fisher, & The Early Experience Stress and Prevention Network, 2006). These studies suggest that human infants display down-regulation of cortisol and amygdala activity during the first 6 months of life, and that these neural systems may play a role in human infant–caregiver attachment development similar to that in the rat. In normal development of the human infant–caregiver attachment, there is coordination of these systems during the second half of the first year; this period coincides with the development of selective attachment relationships, in which infants begin to display such behaviors as proximity seeking toward specific caregivers and fear of strangers (Cassidy & Shaver, 1999).

However, in the institutional environment, developing infants come into contact with multiple caregivers who usually do not form close relationships with these children. How might this situation affect the developing neural attachment system? Given that selective attachments are facilitated through extended interactions with specific caregivers, an environment characterized by limited interaction with caregivers and/or by constantly changing caregivers may confuse infants about whom they should fear and whom they should love. The lack of distinction between safe and unsafe caregivers may arise in the absence of preferred attachments, even though it may be potentially adaptive for the institutional environment (Rutter et al., 2004). Nevertheless, it may derive from or lead to the development of connectivity within the brain differing from that of children with a few specific caregivers.

The neurobiology of these events is obviously complex, but it may plausibly involve both the HPA axis and amygdala activation. Institutionalized children display a dampened cortisol response (particularly in morning cortisol levels) while they remain in the institution (Carlson & Earls, 1997), as well as shortly after adoption (Bruce et al., 2000). This may reflect a prolonged period of cortisol down-regulation. Given that corticosterone induction modulates amygdala activation during the sensitive period in rats, this cortisol down-regulation in human infants may lead both to a prolonged period of amygdala hypoactivation and to prolongation of the period during which specific caregiver attachment develops.

During this time, reshaping and myelination of connections between the amygdala and other brain regions, such as the orbitofrontal cortex, occurs regardless of whether attachment occurs. Because there is some evidence from nonhuman primates suggesting that the formation of such connections is dependent upon stimulation from a responsive caregiver (Sabatini et al., 2007), absence of such stimulation in the institutional environment could lead either to decreased myelination or perhaps to the formation of aberrant connections to other regions of the brain. In addition, proper myelination and synaptogenesis require a properly functioning amygdala to help guide the connections (Bachevalier & Loveland, 2006; Machado & Bachevalier, 2003). Therefore, if amygdala hypoactivation occurs to prolong the period of development of specific attachments, then there is increased likelihood that connections in the brain associated with attachment will be disorganized. The longer that these connections continue to be activated in such a disorganized pattern, the less likely the infant's brain is to develop a caregiver-specific activation pattern, leading to attachment behavior that is directed toward a few select caregivers. In addition, once the infant is removed from the environment of ever-changing caregivers to one in which there are only a few caregivers, the hypoactivation period for cortisol as well as amygdala function ceases, but abnormal connectivity still remains because the sensitive period has ended.

There is some evidence of recovery of the neuroendocrine response after removal from the institution (Bruce et al., 2000), but there is also evidence of ongoing dysfunction (Gunnar et al., 2001). This prolonged effect of hypoactivation on brain connectivity may explain why both abnormal patterns of connectivity within the limbic system (Chugani et al., 2001; Eluvathingal et al., 2006; Marshall et al., 2008) and disinhibited attachment behaviors such as indiscriminate behavior (O'Connor et al., 2000, 2003; Tizard & Hodges, 1978; Tizard & Rees, 1975; Zeanah et al., 2002, 2005) are apparent long after a child has been removed from the institutional environment.

Conclusions

Early institutionalization has profound and negative effects on the development of social behavior, particularly those related to attachment. Paralleling these effects on social behavior, early institutionalization may also have an impact on neurobiological mechanisms involving the limbic system in general and the amygdala in particular. Findings from animal research further suggest that a sensitive period exists for the development of infant–caregiver attachments, which rely heavily on amygdala activation. This period is also one involving functional changes in the stress HPA system. This neurobiological reorganization may likewise be occurring in human infants over the first year of life and may affect the development of normal attachment behavior and its underlying neural circuits.

References

Adolphs, R., Tranel, D., Damasio, H., & Damasio, A. (1994). Impaired recognition of emotion in facial expressions following bilateral damage to the human amygdala. *Nature, 372*(6507), 669–672.

Adolphs, R., Tranel, D., Hamann, S., Young, A., Calder, A., & Phelps, E. (1999). Recognition of facial emotion in nine individuals with bilateral amygdala damage. *Neuropsychologia, 37*(10), 111–117.

Ahern, G. P., Klyachko, V. A., & Jackson, M. B. (2002). cGMP and S-nitrosylation: Two routes for modulation of neuronal excitability by NO. *Trends in Neurosciences, 25*(10), 510–517.

Bachevalier, J., & Loveland, K. A. (2006). The orbitofrontal–amygdala circuit and self-regulation of social-emotional behavior in autism. *Neuroscience and Biobehavioral Reviews, 30*(1), 97–117.

Bauman, M. D., Lavenex, P., Mason, W. A., Capitanio, J. P., & Amaral, D. G. (2004a). The development of mother–infant interactions after neonatal amygdala lesions in rhesus monkeys. *Journal of Neuroscience, 24*(3), 711–721.

Bauman, M. D., Lavenex, P., Mason, W. A., Capitanio, J. P., & Amaral, D. G. (2004b). The development of social behavior following neonatal amygdala lesions in rhesus monkeys. *Journal of Cognitive Neuroscience, 16*(8), 1388–1411.

Bauman, M. D., Toscano, J. E., Mason, W. A., Lavenex, P., & Amaral, D. G. (2006). The expression of social dominance following neonatal lesions of the amygdala or hippocampus in rhesus monkeys (*Macaca mulatta*). *Behavioral Neuroscience, 120*(4), 749–760.

Bowlby, J. (1951). *Maternal care and mental health.* Geneva: World Health Organization.

Bruce, J., Kroupina, M., Parker, S. W., & Gunnar, M. (2000). *The relationship between cortisol patterns, growth retardation, and developmental delays in postinstitutionalized children.* Paper presented at the International Conference on Infant Studies, Brighton, UK.

Bushnell, I. W. R., Sai, F., & Mullin, J. T. (1989). Neonatal recognition of the mother's face. *British Journal of Developmental Psychology, 7*(1), 3–15.

Carlson, M., & Earls, F. (1997). Psychological and neuroendocrinological sequelae of early social deprivation in institutionalized children in Romania. *Annals of the New York Academy of Sciences, 807*(1), 419–439.

Carver, L. J., Dawson, G., Panagiotides, H., Meltzoff, A. N., McPartland, J., Gray, J., et al. (2003). Age-related differences in neural correlates of face recognition during the toddler and preschool years. *Developmental Psychobiology, 42*(2), 148–159.

Cassidy, J., & Shaver, P. R. (Eds.). (1999). *Handbook of attachment: Theory, research, and clinical applications.* New York: Guilford Press.

Chisholm, K. (1998). A three year follow-up of attachment and indiscriminate friendliness in children adopted from Romanian orphanages. *Child Development, 69*(4), 1092–1106.

Chugani, H. T., Behen, M. E., Muzik, O., Juhasz, C., Nagy, F., & Chugani, D. C. (2001). Local brain functional activity following early deprivation: A study of postinstitutionalized Romania orphans. *NeuroImage, 14*(6), 1290–1301.

Dawson, G., Carver, L., Meltzoff, A. N., Panagiotides, H., McPartland, J., & Webb, S. (2002). Neural correlates of face and object recognition in young children with autism spectrum disorder, developmental delay, and typical development. *Child Development, 73*(3), 700–717.

Dawson, G., Klinger, L. G., Panagiotides, H., Lewy, A., & Castelloe, P. (1995). Subgroups of autistic children based on social behavior display distinct patterns of brain activity. *Journal of Abnormal Child Psychology, 23*(5), 569–583.

de Haan, M., & Nelson, C. A. (1997). Recognition of the mother's face by six-month-old infants: A neurobehavioral study. *Child Development, 68*(2), 187–210.

DeCasper, A. J., & Fifer, W. P. (1980). Of human bonding: Newborns prefer their mothers' voices. *Science, 208*(4448), 1174–1176.

Ebeling, U., & von Cramon, D. (1992). Topography of the uncinate fasciculus and adjacent temporal fiber tracts. *Acta Neurochirurgica, 115*(3–4), 143–148.

Eluvathingal, T. J., Chugani, H. T., Behen, M. E., Juhasz, C., Muzik, O., Maqbool, M., et al. (2006). Abnormal brain connectivity in children after early severe socioemotional deprivation: A diffusion tensor imaging study. *Pediatrics, 117*(6), 2093–2100.

Fries, A. B. W., & Pollak, S. D. (2004). Emotion understanding in postinstitutionalized Eastern European children. *Development and Psychopathology, 16*(2), 355–369.

Goldfarb, W. (1945). Effects of psychological deprivation in infancy and subsequent stimulation. *American Journal of Psychiatry, 102,* 18–33.

Gunnar, M. R., Fisher, P. A., & The Early Experience Stress and Prevention Network. (2006). Bringing basic research on early experience and stress neurobiology to bear on preventive interventions for neglected and maltreated children. *Development and Psychopathology, 18,* 651–677.

Gunnar, M. R., Morison, S. J., Chisholm, K., & Schuder, M. (2001). Salivary cortisol levels in children adopted from Romanian orphanages. *Development and Psychopathology, 13*(3), 611–628.

Harlow, H. F., & Zimmermann, R. R. (1959). Affectional responses in the infant monkey: Orphaned baby monkeys develop strong and persistent attachment to inanimate surrogate mothers. *Science, 130*(3373), 421–432.

Harmony, T., Marosi, E., Diaz de Leon, A. E., Becker, J., & Fernandez, T. (1990). Effect of sex, psychosocial disadvantages and biological risk factors on EEG maturations. *Electroencephalography and Clinical Neurophysiology, 75*(6), 482–491.

He, Y., Yu, W., & Baas, P. W. (2002). Microtubule reconfiguration during axonal retraction induced by nitric oxide. *Journal of Neuroscience, 22*(14), 5982–5991.

Hess, D. T., Patterson, S. I., Smith, D. S., & Skene, J. H. (1993). Neuronal growth cone collapse and inhibition of protein fatty acylation by nitric oxide. *Nature, 366*(6455), 562–565.

Hindley, S., Juurlink, B. H., Gysbers, J. W., Middlemiss, P. J., Herman, M. A., & Rathbone, M. P. (1997). Nitric oxide donors enhance neurotrophin-induced neurite outgrowth through a cGMP-dependent mechanism. *Journal of Neuroscience Research, 47*(4), 427–439.

Hodges, J., & Tizard, B. (1989). Social and family relationships of ex-institutional adolescents. *Journal of Child Psychology and Psychiatry, 30*(1), 77–97.

Hofer, M. A., & Sullivan, R. M. (2001). Toward a neurobiology of attachment. In C. A. Nelson & M. Luciana (Eds.), *Handbook of developmental cognitive neuroscience* (pp. 599–616). Cambridge, MA: MIT Press.

Hoksbergen, R., Van Dijkum, C., & Stoutjesdijk, F. (2002). Experiences of Dutch families who parent an adopted Romanian child. *Journal of Developmental and Behavioral Pediatrics, 23*(6), 403–409.

Johnson, D. E. (2000). The impact of orphanage rearing on growth and development. In C. A. Nelson (Ed.), *Minnesota Symposia on Child Psychology: Vol. 31. The effects of adversity on neurobehavioral development* (pp. 113–162). Mahwah, NJ: Erlbaum.

Johnson, D. E. (2002). Adoption and the effects on children's development. *Early Human Development, 68*(1), 39–54.

Joseph, R. (1996). The limbic system: Emotion, laterality, and unconscious mind. *Psychoanalytic Review, 79*(3), 405–456.

Joseph, R. (1999). Environmental influences on neural plasticity, the limbic system, emotional development and attachment: A review. *Child Psychiatry and Human Development, 29*(3), 189–208.

Kaler, S. R., & Freeman, B. J. (1994). Analysis of environmental deprivation: Cognitive and social development in Romanian orphans. *Journal of Child Psychology and Psychiatry, 35*(4), 769–781.

Knudsen, E. I. (2004). Sensitive periods in the development of the brain and behavior. *Journal of Cognitive Neuroscience, 16*(8), 1412–1425.

Kreppner, J., O'Connor, T. G., Dunn, J., Andersen-Wood, L., & English and Romanian Adoptees (ERA) Study Team. (1999). The pretend and social role play of children exposed to early severe deprivation. *British Journal of Developmental Psychology, 17*(3), 319–332.

Kreppner, J., O'Connor, T. G., & Rutter, M. (2001). Can inattention/overactivity be an institutional deprivation syndrome? *Journal of Abnormal Child Psychology, 29*(6), 513–528.

Leppanen, J. M., Moulson, M. C., Vogel-Farley, V. K., & Nelson, C. A. (2007). An ERP study of emotional face processing in the adult and infant brain. *Child Development, 78*(1), 232–245.

MacFarlane, A. (1975). Olfaction in the development of social preferences in the human neonate. *CIBA Foundation Symposium, 33*, 103–117.

Machado, C. J., & Bachevalier, J. (2003). Non-human primate models of childhood psychopathology: The promise and the limitations. *Journal of Child Psychology and Psychiatry, 44*(1), 64–87.

Marshall, P. J., Fox, N. A., & Bucharest Early Intervention Project (BEIP) Core Group. (2004). A comparison of the electroencephalogram between institutionalized and community children in Romania. *Journal of Cognitive Neuroscience, 16*(8), 1327–1338.

Marshall, P. J., Reeb, B. C., Fox, N. A., & Bucharest Early Intervention Project (BEIP) Core Group. (2008). Effects of early intervention on EEG power and coherence in previously institutionalized children in Romania. *Development and Psychopathology, 20*, 861–880.

Moriceau, S., & Sullivan, R. M. (2004a). Corticosterone influences on mammalian neonatal sensitive period learning. *Behavioral Neuroscience, 118*(2), 274–281.

Moriceau, S., & Sullivan, R. M. (2004b). Unique neural circuitry for neonatal olfactory learning. *Journal of Neuroscience, 24*(5), 1182–1189.

Moriceau, S., & Sullivan, R. M. (2005). Neurobiology of infant attachment. *Developmental Psychobiology, 47*(3), 230–242.

Moriceau, S., & Sullivan, R. M. (2006). Maternal presence serves as a switch between learning fear and attraction in infancy. *Nature Neuroscience, 9*(8), 1004–1006.

Moriceau, S., Wilson, D. A., Levine, S., & Sullivan, R. M. (2006). Dual circuitry for odor-shock conditioning during infancy: Corticosterone switches between fear and attraction via amygdala. *Journal of Neuroscience, 26*(25), 6737–6748.

Moulson, M. C., Fox, N. A., Zeanah, C. H., & Nelson, C. A. (in press). Early adverse experiences and the neurobiology of facial emotion processing. *Developmental Psychology.*

Nelson, C. A., & de Haan, M. (1996). Neural correlates of infants' visual responsiveness to facial expressions of emotion. *Developmental Psychobiology, 29*(7), 577–595.

Nelson, C. A., Zeanah, C. H., & Fox, N. A. (2007). The effects of early deprivation on brain–behavioral development: The Bucharest Early Intervention Project. In D. Romer & E. Walker (Eds.), *Adolescent psychopathology and the developing brain: Integrating brain and prevention science* (pp. 197–217). New York: Oxford University Press.

O'Connor, T. G., Bredenkamp, D., Rutter, M., & English and Romanian Adoptees (ERA) Study Team. (1999). Attachment disturbances and disorders in children exposed to early severe deprivation. *Infant Mental Health Journal, 20*(1), 10–29.

O'Connor, T. G., Marvin, R. S., Rutter, M., Olrick, J. T., Britner, P. A., & English and Romanian Adoptees (ERA) Study Team. (2003). Child–parent attachment following early institutional deprivation. *Development and Psychopathology, 15*(1), 19–38.

O'Connor, T. G., Rutter, M., & English and Romanian Adoptees (ERA) Study Team. (2000). Attachment disorder behavior following early severe deprivation: Extension and longitudinal follow-up. *Journal of the American Academy of Child and Adolescent Psychiatry, 39*(6), 703–712.

Parker, S. W., Nelson, C. A., & Bucharest Early Intervention Project (BEIP) Core Group. (2005a). An event-related potential study of the impact of institutional rearing on face recognition. *Development and Psychopathology, 17*(3), 621–639.

Parker, S. W., Nelson, C. A., & Bucharest Early Intervention Project (BEIP) Core Group. (2005b). The impact of early institutional rearing on the ability to discriminate facial expressions of emotion: An event-related potential study. *Child Development, 76*(1), 54–72.

Porter, R. H., Makin, J. W., Davis, L. B., & Christensen, K. M. (1992). Breast-fed infants respond to olfactory cues from their own mother and unfamiliar lactating females. *Infant Behavior and Development, 15*(1), 85–93.

Prather, M. D., Lavenex, P., Mauldin-Jourdain, M. L., Mason, W. A., Capitanio, J. P., Mendoza, S. P., et al. (2001). Increased social fear and decreased fear of objects in monkeys with neonatal amygdala lesions. *Neuroscience, 106*(4), 653–658.

Provence, S., & Lipton, R. C. (1962). *Infants in institutions: A comparison of their development with family-reared infants during the first year of life.* New York: International Universities Press.

Rialas, C. M., Nomizu, M., Patterson, M., Kleinman, H. K., Weston, C. A., & Weeks, B. S. (2000). Nitric oxide mediates laminin-induced neurite outgrowth in PC12 cells. *Experimental Cell Research, 260*(2), 268–276.

Roth, T. L., & Sullivan, R. M. (2005). Memory of early maltreatment: Neonatal behavioral and neural correlates of maternal maltreatment within the context of classical conditioning. *Biological Psychiatry, 57*(8), 823–831.

Roy, P., Rutter, M., & Pickles, A. (2000). Institutional care: Risk from family background or pattern of rearing? *Journal of Child Psychology and Psychiatry, 41*(2), 139–149.

Roy, P., Rutter, M., & Pickles, A. (2004). Institutional care: Associations between overactivity and a lack of selectivity in attachment relationships. *Journal of Child Psychology and Psychiatry, 45*(4), 866–873.

Rutter, M. (2006). The psychological effects of early institutional rearing. In P. J. Marshall & N. A. Fox (Eds.), *The development of social engagement* (pp. 355–391). New York: Oxford University Press.

Rutter, M., Andersen-Wood, L., Beckett, C., Bredenkamp, D., Castle, J., Groothues, C., et al. (1999). Quasi-autistic patterns following severe early global privation. *Journal of Child Psychology and Psychiatry, 40*(4), 537–549.

Rutter, M., Colvert, E., Kreppner, J., Beckett, C., Castle, J., Groothues, C., et al. (2007). Early adolescent outcomes for institutionally-deprived and non-deprived adoptees: I. Disinhibited attachment. *Journal of Child Psychology and Psychiatry, 48*(1), 17–30.

Rutter, M., Kreppner, J., & O'Connor, T. G. (2001). Specificity and heterogeneity in children's responses to profound institutional privation. *British Journal of Psychiatry, 179*(2), 97–103.

Rutter, M., O'Connor, T. G., & English and Romanian Adoptees (ERA) Study Team. (2004). Are there biological programming effects for psychological development?: Findings from a study of Romanian adoptees. *Developmental Psychology, 40*(1), 81–94.

Sabatini, M. J., Ebert, P., Lewis, D. A., Levitt, P., Cameron, J. L., & Mirnics, K. (2007). Amygdala gene expression correlates of social behavior in monkeys experiencing maternal separation. *Journal of Neuroscience, 27*(12), 3295–3304.

Sanchez, M. M., Ladd, C. O., & Plotsky, P. M. (2001). Early adverse experience as a developmental risk factor for later psychopathology: Evidence from rodent and primate models. *Development and Psychopathology, 13*(3), 419–449.

Sapolsky, R. M., & Meaney, M. J. (1986). Maturation of the adrenocortical stress response: Neuroendocrine control mechanisms and the stress hyporesponsive period. *Brain Research Reviews, 11*(1), 65–76.

Schore, A. N. (2001). Effects of a secure attachment relationship on right brain development, affect regulation, and infant mental health. *Infant Mental Health Journal, 22*(1–2), 7–66.

Sloutsky, V. M. (1997). Institutional care and developmental outcomes of 6- and 7-year-old children: A contextual perspective. *International Journal of Behavioral Development, 20*(1), 131–151.

Smyke, A. T., Dumitrescu, A., & Zeanah, C. H. (2002). Attachment disturbances in young children: I. The continuum of caretaking casualty. *Journal of the American Academy of Child and Adolescent Psychiatry, 41*(8), 972–982.

Spitz, R. A. (1946). Anaclitic depression. *Psychoanalytic Study of the Child, 2*, 312–342.

Sullivan, R. M. (2003). Developing a sense of safety: The neurobiology of neonatal attachment. *Annals of the New York Academy of Sciences, 1008*, 122–131.

Sullivan, R. M., Hofer, M. A., & Brake, S. C. (1986). Olfactory-guided orientation in neonatal rats is enhanced by a conditioned change in behavioral state. *Developmental Psychobiology, 19*(6), 615–623.

Sullivan, R. M., Landers, M., Yeaman, B., & Wilson, D. A. (2000). Good memories of bad events in infancy: Ontogeny of conditioned fear and the amygdala. *Nature, 407*(6800), 38–39.

Suomi, S. J. (1990). The role of tactile contact in rhesus monkey social development. In K. E. Barnard & T. B. Brazelton (Eds.), *Touch: The foundation of experience* (pp. 129–164). Madison, CT: International Universities Press.

Takahashi, K. (1994). Organizing action of corticosterone on the development of behavioral inhibition in the preweanling rat. *Developmental Brain Research, 81*(1), 121–127.

Takahashi, K., & Rubin, W. W. (1993). Corticosteroid induction of threat-induced behavioral inhibition in preweanling rats. *Behavioral Neuroscience, 107*(5), 860–866.

Tarullo, A., Chatham, M., & Gunnar, M. (2007). *Electroencephalogram power in post-institutionalized children.* Paper presented at the biennial meeting of the Society for Research in Child Development, Boston.

Thatcher, R. W. (1992). Cyclic cortical reorganization during early childhood. *Brain and Cognition, 20*(1), 24–50.

Thierry, B. (1990). Feedback loop between kinship and dominance: The macaque model. *Journal of Theoretical Biology, 145*(4), 511–522.

Thomas, K. M., Drevets, W. C., Whalen, P. J., Eccard, C., Dahl, R. E., Ryan, N. D., et al. (2001). Amygdala response to facial expressions in children and adults. *Biological Psychiatry, 49*(4), 309–316.

Thompson, C. I., Bergland, R. M., & Towfighi, J. T. (1977). Social and nonsocial behaviors of adult rhesus monkeys after amygdalectomy in infancy or adulthood. *Journal of Comparative and Physiological Psychology, 91*(3), 533–548.

Thompson, C. I., Schwartzbaum, J. S., & Harlow, H. F. (1968). Development of social fear after amygdalectomy in infant rhesus macaques. *Physiology and Behavior, 4,* 249–254.

Thompson, C. I., & Towfighi, J. T. (1976). Social behavior of juvenile rhesus monkeys after amygdalectomy in infancy. *Physiology and Behavior, 17*(5), 831–836.

Tizard, B., & Hodges, J. (1978). The effect of early institutional rearing on the development of eight-year-old children. *Journal of Child Psychology and Psychiatry, 19*(2), 99–118.

Tizard, B., & Rees, J. (1975). The effect of early institutional rearing on behavioural problems and affectional relationships of four-year-old children. *Journal of Child Psychology and Psychiatry, 16*(1), 61–73.

Vorria, P., Rutter, M., Pickles, A., Wolkind, S., & Hobsbaum, A. (1998). A comparative study of Greek children in long-term residential group care and in two-parent families: I. Social, emotional, and behavioural differences. *Journal of Child Psychology and Psychiatry, 39*(2), 225–236.

Webster, M. J., Ungerleider, L. G., & Bachevalier, J. (1991). Connections of inferior temporal areas TE and TEO with medial temporal-lobe structures in infant and adult monkeys. *Journal of Neuroscience, 11*(4), 1095–1116.

Zeanah, C. H., Nelson, C. A., Fox, N. A., Smyke, A. T., Marshall, P. J., Parker, S. W., et al. (2003). Designing research to study the effects of institutionalization on brain and behavioral development: The Bucharest Early Intervention Project. *Development and Psychopathology, 15*(4), 885–907.

Zeanah, C. H., & Smyke, A. T. (2007). *Disturbances of attachment: A perspective from the extremes.* Paper presented at the biennial meeting of the Society for Research in Child Development, Boston.

Zeanah, C. H., Smyke, A. T., & Dumitrescu, A. (2002). Attachment disturbances in young children: II. Indiscriminate behavior and institutional care. *Journal of the American Academy of Child and Adolescent Psychiatry, 41*(8), 983–989.

Zeanah, C. H., Smyke, A. T., Koga, S. F., Carlson, M., & Bucharest Early Intervention Project (BEIP) Core Group. (2005). Attachment in institutionalized and community children in Romania. *Child Development, 76*(5), 1015–1028.

CHAPTER 25

Socioemotional Development Following Early Abuse and Neglect

Challenges and Insights from Translational Research

M. MAR SANCHEZ
SETH D. POLLAK

Childhood maltreatment is a significant public health problem. In addition to risk of physical injury or death, children who experience various forms of abuse and neglect are likely to develop a variety of health problems over the course of their lives. These problems appear to result from the stress and emotional distress associated with the behavioral inconsistencies of the children's caregivers. The goal of this chapter is to synthesize what is currently known about how complex sets of neural circuitry are shaped and refined over development by children's social experience. We address questions about the role of social experience in the development of brain–behavior relations by focusing on research with populations of abused and neglected children, as well as research with rodents and nonhuman primates that can be translated or applied to humans. Despite the fact that child maltreatment is notoriously difficult to investigate empirically, it provides an important forum for understanding the role of environmental stress, individual differences, and developmental factors in the ontogenesis of social behavior.

A challenge faced by the field of affective developmental neuroscience concerns the difficulty of precisely measuring the amount of exposure any individual has had to particular emotions. However, it is possible to estimate gross differences across groups. Therefore, studying populations of children who have had extreme social experiences such as abuse and neglect can not only help inform effective prevention and intervention strategies to help promote these children's optimal development; these studies can also shed light on the ways in which social experiences influence the developing brain. There are many examples of the ways in which maltreating parents provide emotionally expressive environments for their children that deviate in important ways from normal social experience (Camras, Sachs-Alter, & Ribordy, 1996). For example, abusive parents engage in fewer positive emotional interactions with their children than nonabusive parents do (Burgess & Conger, 1978), and direct more negative affect toward their children than nonmaltreating mothers do (Trickett, Aber, Carlson, & Cicchetti, 1991). Indeed, some of our own recent work reveals that abusive parents, not surprisingly, are generally angry people, and that their overall levels of trait anger predict their children's performance on emotion recognition tasks (Pollak, Messner, Kistler, & Cohn, 2009). Later in this chapter, we review current evidence supporting the view that the experience of child maltreatment plays a causal, rather than corollary, role in children's behavioral problems.

Various techniques have increased our understanding of the ways in which children's experiences of maltreatment can lead to emotional disorders. By carefully observing and modeling aspects of child abuse in nonhuman primates, we are also able to examine the neuroanatomical and neurophysiological substrates of emotional processes in ways that are not feasible with human subjects. As humans do, primate species exhibit complex socioemotional and neural development over a protracted developmental period during which the infant is dependent on parental care (Maestripieri & Carroll, 1998b). Indeed, studies of the effects of early experience in nonhuman animals have highlighted candidate processes through which adverse parental care affects the development of the neural systems believed to underlie heightened risk for mental health problems. As it does in humans, maltreatment occurs spontaneously in nonhuman primates, both in captivity and in the wild. In particular, physical abuse has been reported in different species, with rates similar to those seen in human populations (e.g., Brent, Koban, & Ramirez, 2002; Johnson, Kamilaris, Calogero, Gold, & Chrousos, 1996; Maestripieri, 1999). In the case of macaques, infant abuse involves violent behaviors that a mother directs toward her infant, such as crushing, throwing, dragging, and stepping or sitting on the infant (Troisi & D'Amato, 1984). These maternal behaviors are clearly distinguishable from species-typical aggressive behaviors, and they cause intense infant distress and sometimes serious injury or even death. In sum, nonhuman primate models of infant maltreatment constitute unique and naturalistic models of human childhood maltreatment that can help us understand the developmental trajectory of the impact of these early adverse experiences.

Conceptual Issues in the Study of Child Abuse

Many conceptual, practical, and ethical factors complicate the study of child abuse and are important in considering how to evaluate neuroscientific approaches to the phenomenon (for a full discussion, see Pollak, 2005).

One issue is that the identification of child abuse co-occurs with a host of complex risk factors that affect child, parent, and family functioning. Therefore, it is conceptually difficult to evaluate where to place the occurrence of child maltreatment, molecular genetic components, and other latent variables in the causal chain leading to mental health problems. This issue of causality is described very nicely by Caspi and Moffitt (2006).

A second issue is that, in contrast to most studies of adult (and many child) psychiatric disorders, it is extremely complicated or impossible to obtain and verify data on the specifics of children's maltreatment experiences. Moreover, even when such data are obtained, there is no scientific agreement about how the data should be categorized. Therefore, although contemporary neuroscience-based techniques can now be applied to understanding outcomes associated with maltreatment, basic issues about operational definitions of maltreatment and details about the constitution of the samples being studied require close scrutiny. As a result, it is often impossible to determine how comparable one "maltreated" sample is to another. Studies of maltreatment among nonhuman primates help address part of this problem, in that parent–child interactions can be reliably measured and quantified in a more controlled environment—though there is still uncertainty about which operational criteria constitute different forms of maltreatment in monkeys. These definitional problems are described by Dubowitz et al. (2005).

A simple, albeit valid, response to these criticisms of research on child maltreatment is that most human research on the developmental effects of stress exposure requires us to be opportunistic. We need to address the real events that naturally occur in children's lives, even if those events are terribly messy, inconvenient, or even inconsistent with the ideals of experimental design. As described above, studies of nonhuman animals are extraordinarily useful in addressing some of these issues. But animal models cannot substitute for human studies. Animal models do not always mimic human emotional disorders; brain development, structure, and function are not identical across species; there are chromosomal differences between species; and the actual behaviors exhibited by parents and the way they are received and experienced by offspring are not identical across species. At the same time, the common denominators that do apply across species, such as poor/inadequate parental nurturance, provide critical clues and allow for human models of the effects of child abuse to be biologically sound. These important commonalities allow us to translate knowledge from experimental animal models of brain–behavior relationships to human situations.

Of course, there are more detailed ways to consider the problems of experimental methods and definitional issues. We briefly review these issues before summarizing research findings.

Defining Child Maltreatment

Problems in defining maltreatment complicate attempts to study biological mechanisms. In human studies, the term "maltreatment" is often used inclusively, so that samples of maltreated children include individuals who have experienced multiple types of adverse experiences (i.e., physical abuse, sexual abuse, emotional abuse, and/or neglect). The proportions of different types of maltreatment vary in these samples; moreover, issues such as the definition of "emotional abuse," which is nearly impossible to measure and verify, often go unaddressed. Furthermore, the ways in which different types of maltreatment are determined across studies include methods as diverse as social service agency designations, clinical interviews, self-report of parents and/or children, behavioral observations, and retrospective reports. Although all extant approaches are subject to criticism, and researchers using them must deal with significant sensitivity–specificity tradeoffs, each approach also confers some advantage. For example, the use of legal and social services records and classifications has the advantage of being objective, but the disadvantage of being based on information that is collected and classified for administrative rather than scientific purposes. Furthermore, the legal threshold for what constitutes maltreatment is very broad: Children can be designated as "neglected" for reasons ranging from leaving a young child unsupervised to failing to follow societal rules (e.g., not sending a child to school). Thus the classification a child receives in the legal or social services system may provide little theoretically driven and scientifically useful information about maltreatment. Furthermore, developmentally informed research requires knowing the severity and duration, not merely the type, of experience. Although it is challenging to derive from legal and administrative records, this information matters. For example, Manly, Cicchetti, and Barnett (1994) showed that while infrequent but severe forms of maltreatment had a negative impact on child outcomes, frequent but less severe maltreating experiences were also detrimental.

Defining maltreatment in nonhuman primates has also been a difficult endeavor, especially for neglect. Like humans, nonhuman primate mothers demonstrate a broad spectrum of caregiving styles toward their infants. Some rhesus macaque (*Macaca mulatta*) mothers show poor maternal care; for example, they provide infrequent physical contact and fail to protect infants. But it has been difficult to determine how harmful these behaviors are to the infants over time. Therefore, research in this area has taken a conservative approach, limiting definitions of "neglect" in nonhuman primates to extreme behaviors, such as complete abandonment of the infant by the mother (Maestripieri & Carroll, 2000). Among macaques, infant abandonment is exhibited almost exclusively by primiparous mothers (between 1.5–15%), and, unlike in studies of human parenting, it rarely coexists with physical abuse (e.g., Maestripieri & Carroll, 1998a).

We have been examining other forms of monkey parenting that more closely approximate the experiences of neglected human children. In rhesus monkeys, maternal rejection of an infant co-occurs with physical abuse in about 70% of cases (McCormack, Sanchez, Bardi, & Maestripieri, 2006; Sanchez, 2006). Physically abusive rhesus mothers are also anxious and rejecting towards their offspring, and show poor "nurturing/protective" care. These rejecting maternal behaviors occur very early in the infants' lives, when rhesus mothers are typically providing infants with high levels of nurturance (Suomi, 2005). Further understanding of the impact of different kinds of parental care is likely to be a promising approach, and recent data suggest that high maternal rejection is in fact a stronger predictor than physical abuse for poor developmental outcomes in maltreated rhesus infants (McCormack et al., 2006; Sanchez et al., 2007). These studies are described further below.

Co-Occurring Risk Factors

Because child maltreatment cannot be experimentally manipulated, most human studies on the topic lack meaningful experimental controls. Therefore, it is often difficult to discern the extent to which observed effects can be associated specifically with maltreatment or are the concomitant effects of other negative factors, such as poverty, in the lives of maltreated children. A threat to validity in most studies with maltreated humans is the possibility that observed associations between various forms of maltreatment and psychopathology are not due to causal relationships between the two, but either reflect latent variables causing both maltreatment and psychopathology, or are the results of accumulated life stressors (beyond maltreatment) that co-occur in families where abuse has been identified. Historically, an alternative to the view that maltreatment causes behavioral problems was the idea that poor outcomes result from heritable factors that cause the maltreatment, or risk factors in the environment that co-occur with maltreatment. However, recent behavioral and molecular genetic data support the view that the experience of abuse plays a causal role in children's behavioral problems. These data suggest that genetic risk, in combination with early traumatic experiences, dramatically increases the likelihood that children will develop mental health problems (Kim-Cohen et al., 2006). Similar studies (Jaffee et al., 2005; Kendler et al., 2000), using large samples of humans, consistently support a causal relationship between child maltreatment and the development of psychopathology. As in the data from humans, physical abuse in rhesus monkeys has a high prevalence in some maternal lineages, suggesting an intergenerational transmission of vulnerability to maltreatment. Whether this transmission along the maternal line is due to experience or to genetic heritabilities is not yet definitive. However, evidence from cross-fostering studies (e.g., Maestripieri, 2005) suggests that early experience plays an important role in the sequelae and the perpetuation of maltreatment.

Still, the potential role of co-occurring environmental risk factors (as causes, mediators, or moderators) continues to complicate research in this area. Many factors in the lives of humans are extremely difficult to quantify in any single study. For example, a child's risk of exposure to community violence is influenced by family socioeconomic status and the child's gender, ethnicity, and age (Couch, Hanson, Saunders, Kilpatrick, & Resnick, 2000). In addition, various family characteristics have been associated with an increased likelihood of exposure to violence (Osofsky, Wewers, Hann, & Fick, 1993). Our intent in raising these issues early in this chapter is not to be discouraging. Indeed, these areas of scholarly inquiry have progressed in very exciting and productive ways. However, to bridge different lines of research—involving different populations, methods, and species—it is important to be mindful of the conceptual challenges in the work. The ways that these studies inform one another is powerful, but the realities and limitations of this work (especially with human populations of at-risk children) must also be considered. And although similarities between species are exciting, there is much to be learned from the ways in which findings across species or research programs do not align.

Child Abuse and Mental Health

Over 3 million children were reported to be victims of maltreatment in the United States in 2003 (Hussey, Chang, & Kotch, 2005). After neglect, physical abuse is the most common form of childhood maltreatment. Furthermore, in addition to those who are directly victims of violence, approximately 3.3 to 10 million children witness violence at home (Straus & Gelles, 1990). Abused children are at extremely high risk for mental health challenges that include conduct/aggression problems, depression, anxiety, and substance abuse; they also lag behind their peers in social skills (Chapman, Wall, & Barth, 2004; Toth, Manly, & Cicchetti, 1992) and experience high rates of physical health problems (Mulvihill, 2005).

Many maltreated children display socioemotional behavioral difficulties even before the onset of clear symptoms of psychopathology. For example, maltreated infants as young as 12 months of age show poor affect regulation (Gaensbauer, 1982). Children who have been maltreated experience problems involving recognition (Camras et al., 1996) and regulation (Main & George, 1985) of emotional states. Physically abused children exhibit elevated aggressive behaviors and social withdrawal during peer interactions (Rogosch & Cicchetti, 1994). Physically abused children display both interpersonal withdrawal and aggression (Rogosch, Cicchetti, & Aber, 1995), attribute hostility to others (Weiss, Dodge, Bates, & Pettit, 1992), and display contextually inappropriate affect and behavior (Klimes-Dougan & Kistner, 1990). Physically abused children also tend to readily assimilate and remember pictures of angry facial expressions and cues related to aggression, even when those cues are task-irrelevant (Pollak & Tolley-Schell, 2003; Rieder & Cicchetti, 1989). Moreover, there is a positive association among frequency and severity of physical abuse, hostile attribution tendencies toward others, and attentional biases in emotion perception (Price & Glad, 2003; Shackman, Shackman, & Pollak, 2007).

The developmental outcomes of maltreatment in nonhuman primates are comparable to those reported in maltreated children. Maltreated rhesus monkey infants exhibit delayed social development, behavioral signs of distress, heightened anxiety and fearfulness, and impulsive aggression (e.g., McCormack et al., 2006; Sanchez, 2006). Similar alterations have also been observed in other primate species, such as marmosets and vervet monkeys (Fairbanks & McGuire, 1998; Johnson et al., 1996).

One approach to understanding the nature of the risk posed by child maltreatment is to note the kinds of disorders that are likely to emerge in these children. However, the diversity of developmental trajectories taken by maltreated children suggests a complex story. Some reports indicate that child abuse is linked to increased levels of depression in childhood (Kaufman, 1991; Toth et al., 1992), while other reports note increased levels of other disturbances such as conduct disorder, attention-deficit/hyperactivity disorder, and oppositional defiant disorder. Still other studies have found high levels of anxiety and social withdrawal (Famularo, Kinscherff, & Fenton, 1992). Childhood maltreatment has also been linked to substance use disorders, dissociative disorders, and posttraumatic stress disorder (PTSD) (Aarons, Brown, Hough, Garland, & Wood, 2001; Cicchetti & Toth, 1995). Yet much remains to be learned beyond these associations, such as the nature of the etiological mechanisms, the processes that serve to maintain problems in these children, and the reasons why maltreated children may develop so many different kinds of disorders.

To address questions about individual differences in children's developmental outcomes, Manly, Kim, Rogosch, and Cicchetti (2001) examined how the timing of maltreatment affected children's adjustment. Severity of emotional maltreatment in the infancy–toddler period predicted externalizing behavior and aggression during the school-age years, as did victimization by physical abuse during the preschool period. Maltreatment that occurred during the school-age period contributed significant variance even after earlier maltreatment was controlled for. Chronic maltreatment was linked with more maladaptive outcomes, especially with onset during the infancy–toddler or preschool periods. A recent report revealed that individuals who were abused earlier in life demonstrated higher levels of anxiety and depression in adulthood, whereas individuals who were older at the time of the maltreatment were more likely to evince symptoms associated with aggression and substance abuse (Kaplow & Widom, 2007). Replication and further exploration of these findings will be helpful in linking models of developmental neurobiology to the development of psychopathology. These findings underscore the importance of adopting a multifaceted approach to examining child maltreatment, and they emphasize the importance of utilizing a developmental approach and assessing multiple dimensions of such environmental threats as maltreatment to understand patterns of adaptation and maladaptation among children exposed to adversity. Because it is clear that child maltreatment is associated with elevated risk for many different kinds of mental health problems, we highlight the kinds of difficulties that have been most frequently reported, in an effort to focus on candidate developmental mechanisms affected by early experience.

Anxiety and Depression

One area of concern involves the high prevalence of anxiety and depression among formerly abused children. Of course, depression is a heavily studied phenomenon, and many different developmental factors ranging from problems with peers to stressful lives are related to depression (see, e.g., Raver, 2003). Traumatic experiences such as maltreatment may leave children hyperresponsive to emotional stimuli, particularly those that signal threat or danger, with marked shifts in arousal levels (Cummings, Vogel, Cummings, & El-Sheikh, 1989; Eisenberg et al., 1997). Early harsh caregiving experiences also undermine a child's capacity to reflect upon the affective state of both self and others (Fonagy, Target, & Gergely, 2000).

Central findings in this area concern the interactions of early adverse environments with genetic risk factors. A very consistent body of evidence for these early environment × gene interactions involves a neurotransmitter transporter gene called 5-HTT that fine-tunes transmission of serotonin (5-HT) by reuptaking it from the synaptic cleft. The gene comes in two common allelic variants: the long allele and the short allele, which confer higher and lower 5-HT reuptake efficiency on 5-HTT, respectively. Animal studies have shown that in stressful conditions, those with two long alleles cope better. Mice with one or two copies of the short allele show more fearful reactions to stresses such as loud sounds. In addition, monkeys with the short allele that are raised in stressful conditions have impaired 5-HT transmission. Caspi et al. (2003) studied 847 individuals who had undergone a variety of assessments over more than two decades, starting at the age of 3. The negative effects of maltreating experiences were stronger among people with one short allele, and stronger still for those with two short alleles. For people with two short alleles, the probability of a major depressive episode rose to more than double the risk for the subjects with two long alleles who had similar levels of life stress. More specifically, childhood abuse predicted depression after the age of 18 only in people carrying at least one short allele. Among the 11% who had experienced severe maltreatment, the subjects with two short alleles ran a 63% risk of a major depressive episode. The participants with two long alleles averaged a 30% risk, regardless of whether they had been abused as children or not. In this study, the possibility that a short allele could somehow predispose a person to experiencing maltreatment was essentially ruled out; there was no significant difference among the three genotype groups in the number of adverse experiences reported.

Aggression and Antisocial Behavior

A second area of mental health concern involves to high levels of antisocial and aggressive symptoms observed among abused children (Parker, Rubin, Price, & DeRosier, 1995). On average, children who have been abused are more aggressive toward peers than are their nonmaltreated counterparts (Dodge, Bates, & Pettit, 1990; Kaufman & Cicchetti, 1989). Many youth with child maltreatment histories develop serious conduct problems, including arrests for violent offenses (Widom & Brzustowicz, 2006). Type of abuse experience may play a role in the development of delinquency (Jonson-Reid & Barth, 2000).

Recent findings suggest that one explanation for variability in outcomes among maltreated males relates to a gene–environment interaction involving a functional polymorphism in the promoter region of the monoamine oxidase A (MAOA) gene. Specifically, maltreated boys with the MAOA genotype conferring low levels of the MAOA enzyme developed conduct disorder, antisocial personality, and violent criminality in adulthood more often than maltreated boys with a high-activity MAOA genotype did. MAOA selectively degrades 5-HT, norephinephrine, and dopamine following reuptake from the synaptic cleft, and therefore plays a key role in regulating behavior (Caspi et al., 2002; Sabol, Hu, & Hamer, 1998; Shih, Chen, & Ridd, 1999).

Stress Disorders

Child abuse also appears to increase the comorbidity of PTSD with both depression and substance abuse (Kilpatrick et al., 2003). PTSD may not be a normative response to traumatic violence. Instead, it may represent a disordered stress response related to preexisting biological and psychological vulnerabilities (Yehuda & McFarlane, 1995). Characteristics of the social environment, such as social support and family cohesion, seem to play a role in how children respond to violence (Kliewer, Lepore, Oskin, & Johnson, 1998). Genetic mechanisms may make some individuals more susceptible to stressful environments. No specific genes have been linked conclusively to PTSD. The one study that did identify a particular gene, the gene for the dopamine D2 receptor (Comings, Muhleman, & Gysin, 1996), was not replicated (Gelenter et al., 1999).

Neural Mechanisms of Risk

Emotion and Attentional Processing

The overarching model guiding the work of Pollak and colleagues has been to focus on the interactions between the plasticity of general perceptual and attentional systems, threat and stress regulatory processes, and learning mechanisms (Pollak, 2003, 2005). The work of Sanchez and her colleagues (see Maestripieri, 1999; Sanchez, 2006) with nonhuman primates has provided a rough guide for this research on abused children. Although one must always be cautious in translating basic findings across species, basic neuroscience research converges with the human studies in suggesting that higher-order functions such as selective control of attention to threat, combined with alterations in stress reactivity, may account for maltreated children's reactions to social stimuli (Pollak, 2008). Rather than using their attentional resources to attenuate emotional reactivity, physically abused children appear to overly attend to threatening cues, perhaps at the expense of other contextually relevant information (Dodge, Pettit, Bates, & Valente, 1995; Pollak & Tolley-Schell, 2003). Converging empirical support also comes from the laboratories of De Bellis (2005) and Heim (Heim, Ehlert, & Hellhammer, 2000; Heim, Plotsky, & Nemeroff, 2004).

To examine the effects of different kinds of experiences, Pollak, Cicchetti, Hornung, and Reed (2000) contrasted the emotion recognition skills of abused and neglected children. Maltreatment subtypes were classified hierarchically, such that none of the children had documented sexual abuse; neglected children did not have records indicating physical abuse; but physically abused children might also have experienced neglect. Physically abused children had experienced abuse by commission (they were injured by a parent). Neglected children experienced abuse by omission (lack of care and responsiveness from parents). The neglected children had difficulty differentiating facial expressions of emotion, whereas the physically abused children performed well, especially when differentiating angry facial expressions. These data suggest that specific kinds of experiences, rather than simply the presence of stress or heterogeneous forms of maltreatment, have differential effects. Two psychophysiological studies of children's ability to allocate attention to emotional cues revealed that while nonmaltreated children and adults responded uniformly when attending to happy, fearful, and angry faces, physically abused children displayed relative increases in brain electrical activity only when actively searching for angry faces. Abused children performed identically to controls when attending to other emotional expressions, suggesting that attentional processes directed toward anger distinguish abused children's emotion processing (Pollak, Cicchetti, Klorman, & Brumaghim, 1997; Pollak, Klorman, Brumaghim, & Cicchetti, 2001).

Most of the neuroscience-oriented research to date has not involved samples of children with relatively distinct types of maltreatment. In these cases, we refer to these samples as "maltreated" if the type of experiences represented in the sample is broad. Otherwise, we refer to samples as "abused" (for physically abused), "neglected," or "isolate-reared." We do not include studies of sexual abuse in this chapter. Consistent with observations of socially deprived rhesus monkeys, neglected children have difficulties in differentiating between and responding to expressions of emotion and formulating selective attachments to caregivers (Wismer Fries & Pollak, 2004; Wismer Fries, Zigler, Kurian, Jacoris, & Pollak, 2005). (Note, however, that socially deprived monkeys, in addition to lack of maternal caregiving, also experience sensorimotor deprivation; this is not usually the case in human abuse and neglect.)

These social and emotional difficulties may reflect neuropsychological difficulties in maltreated children that reflect alterations in brain maturation (Prasad, Kramer, & Ewing-Cobbs, 2005). Indeed, impaired cognitive functioning in socially deprived monkeys is associated with decreased white matter volume in parietal and prefrontal cortices, as well as alterations in the development of neuropeptide receptors that underlie fearful and anxious behaviors (Sanchez, Hearn, Do, Rilling, & Herndon, 1998; Sanchez, Smith, & Winslow, 2003). A recent brain imaging study of children with maltreatment and PTSD revealed decreases in regions such as the prefrontal cortex and right temporal lobe volumes, and increases in hippocampal white matter volume, in comparison to sociodemographically matched controls; these effects were particularly strong among abused boys (Tupler & De Bellis, 2006). It is not yet clear whether these brain differences reflect vulnerability to, or effects of, maltreatment.

Pollak and Kistler (2002) also sought to examine whether maltreatment experience alters children's sensory thresholds in ways that might undermine effective regulation of emotion. Categorical perception occurs when perceptual mechanisms enhance differences between categories at the expense of perception of incremental changes within a category. Perceiving via categories is adaptive, because it allows an observer to efficiently assess changes between categories that are environmentally important, while ignoring subtle changes that are not important. Children performed a task that required them to distinguish faces that had been morphed to produce a continuum on which each face differed in signal intensity. Abused children had atypical perceptual preferences that influenced how they categorized angry, but not other, facial expressions. These findings are consistent with the view that infants need to adjust or tune their preexisting perceptual mechanisms to process salient aspects of their environments. To further examine whether children exposed to high levels of threat are perceptually sensitive to threat cues, Pollak and Sinha (2002) examined whether these children could readily relate visual cues to representations of emotions. To do so, they developed a technique that could capture the sequential and content-based dynamics of emotion recognition. As predicted, physically abused children accurately identified facial displays of anger on the basis of less sensory input than did controls.

Pollak and Tolley-Schell (2003) also explored a second hypothesis: that the acquired salience of anger or threat-related signals undermines abused children's attentional control. Using a selective attention paradigm, they found that abused children demonstrated relative increases in brain electrical activity when they were required to disengage their attention from angry, but not happy, faces. Physically abused children also oriented rapidly to spatial locations primed by anger. Because abused children did not differ from controls on other types of trials, these findings provided additional support for the hypothesis that physically abused children have a specific problem involving flexible processing of anger, rather than general information-processing deficits. More recently, Shackman et al. (2007) found that abused children allocated more automatic resources when attempting to inhibit attention to their mothers' angry voices (Shackman et al., 2007). Importantly, these differences were correlated with both the magnitude of abuse endured by these children and their degree of anxiety symptoms (Shackman et al., 2007).

A critical question concerns how these perceptual differences in abused children might influence affective processes further along the path of information processing. In a recent study, Perlman, Kalish, and Pollak (2008) focused on children's abstract knowledge about emotions. They found that abused children differed from controls in their appraisals of the links between events and emotions. Abused preschool-age children, unlike controls, saw anger and sadness as possible emotional outcomes following from positive situations. These findings suggest that among the consequences of learning about emotions within a physically abusive context are non-normative intuitions about the causes of emotions. Such a cognitive difference would affect the thinking that guides children's social behavior. Another recent study, also with preschool-age children, examined children's control and regulation of attention when confronted with emotional stimuli (Pollak, Vardi, Bechner, & Curtin, 2005). In this study, multiple physiological measures were obtained while children overheard two adults engaged in a hostile argument. Once the angry exchange began, abused children maintained a state of anticipatory monitoring of the environment which continued even after the angry exchange ended. In contrast, control children were initially more aroused by the introduction of anger (suggesting some habituation to anger on the part of abused children); controls showed better regulation once they had assessed the background anger. These data suggest that physically abused children develop greater sensitivity to expressions of anger as a form of adaptation to an environment where threat signals may predict the occurrence of abuse. Although adaptive in an abusive context, such processes lead to complex information-processing atypicalities that compromise children's regulatory capacities.

The Hypothalamic–Pituitary–Adrenal Axis

One biological system implicated in children's regulation is the hypothalamic–pituitary–adrenal (HPA) axis, a neuroendocrine system particularly vulnerable to the effects of mother–infant disruption (Gunnar, 2000; Sanchez, Ladd, & Plotsky, 2001). The mammalian stress neuroendocrine response involves the HPA and sympathetic–adrenomedullary (SAM) systems. These systems, interrelated at many levels, are coordinated in the central nervous system in part by the action of corticotrophin-releasing factor (CRF [or CRH]) in the hypothalamus and in extrahypothalamic nuclei, such as the central nucleus of the amygdala and the bed nucleus of the stria terminalis (Rosen & Schulkin, 1998). The HPA and SAM systems are functional prior to birth, but undergo maturational processes in the transition to extrauterine life and postnatally.

Challenges to social relationships can be potent stimuli of the HPA axis and sources of stress (Sapolsky, 1998; Selye, 1976). These factors frequently characterize situations of child abuse. Of particular concern is that the HPA axis is still maturing when children are subjected to maltreatment, raising questions about the biological consequences of maltreatment for the developmental organization of a child's stress response system (Bremner & Vermetten, 2001). The HPA axis is a major system that mediates neuroendocrine responses to stress, resulting in the release of glucocorticoids (GCs; cortisol in humans and nonhuman primates) from the adrenal cortex (see Herman et al., 2003). Superimposed upon its circadian pattern of activity, stress activates stressor-specific pathways that converge in the hypothalamus, where this information is integrated in the paraventricular nucleus by parvocellular neurons expressing CRF. CRF is released from nerve endings in the median eminence in response to metabolic, psychological, or physical threats and stimulates the release of adrenocorticotropic hormone (ACTH) from the anterior pituitary. ACTH in turn stimulates the release of GCs (cortisol in primates) from the adrenal cortex.

GCs are highly catabolic steroid hormones that affect multiple bodily functions associated with the kinds of problems observed in abused children, including energy mobilization, immune and reproductive functions, and cognition. Alterations in the normal pattern of cortisol secretion (either higher or lower than normal) have been associated with both psychiatric and somatic illnesses (McEwen, 1998; Yehuda, Halligan, & Bierer, 2002).

In rodents, there is compelling evidence that the quality of maternal care (high vs. low) has a dramatic impact on the development of neuroendocrine and neurobiological systems regulating stress physiology and adaptive behavior in the offspring. For example, the HPA axis is immature in young rodents. This is also true of other afferent and efferent pathways of threat detection and response systems that connect with the prefrontal cortex. There is substantial evidence that the development of these systems is particularly plastic and open to modification by experience during early life (e.g., Levine, 1994; Suomi, 1997; Wismer Fries, Shirtcliff, & Pollak, 2008). In particular, we know that low-quality parental care or repeated maternal separations result in offspring that become more anxious and stress-reactive adults (e.g., Meaney & Szyf, 2005; Sanchez et al., 2001). We even know very sophisticated details of the molecular mechanisms mediating some of these effects, including how maternal care affects methylation of the glucocorticoid receptor gene, and therefore its expression levels and HPA axis function (Meaney & Szyf, 2005). Recent rodent studies also indicate that interventions in the postinfancy period may help ameliorate some, but not all, of the impacts of early inadequate parental care (Francis, Diorio, Plotsky, & Meaney, 2002). Yet the validity of rodent studies for modeling the effects of early adverse experiences on stress physiology in primates is questionable. The problems include the critical differences between rodent and primate HPA axis development, and the possibility that different developmental consequences in each species are mediated by different biological mechanisms. These differences would explain the inconsistent evidence of alterations reported in studies of rat and primate HPA axis functioning (Gunnar & Vazquez, 2006).

Some primate researchers have argued that variation in parental care results in long-term changes in the neuroendocrine and neurobiological development of the offspring (e.g., Hinde, 1974). Yet little is known about the underlying mechanisms and time course of these purported effects. One clue about potential mechanisms comes from similarities between studies of nonhuman primates and rodents that pertain to effects of early adverse caregiving on the development of the frontal system and other neocortical regions involved in emotion and attention regulation. For example, Sanchez et al. (1998) studied rhesus monkeys that were socially deprived between 2 and 12 months of age. These monkeys exhibited "executive function" deficits that had also been noted in earlier studies (e.g., Harlow, Harlow, & Suomi, 1971). Sanchez et al.'s magnetic resonance imaging (MRI) studies revealed that the animals' performance on executive function tasks was correlated with decreased white matter in parietal and prefrontal cortices. When studied 2 years later, these monkeys exhibited increased density of CRF1 receptors in the prefrontal cortex and amygdala, presumably mediating the increased fearfulness and anxiety detected in these animals (Sanchez et al., 2003). In a separate study, Mathew et al. (2003) reported neuropathological alterations in the prefrontal cortex (anterior cingulate) of adult macaques with early adverse experience. The anterior cingulate is a critical region for effortful regulation of attention and negative emotionality (Posner & Rothbart, 1998),

Are these findings relevant to children's emotion-related behavior? In humans, extreme alterations in early caregiving (such as parental loss, maltreatment, or maternal depression) have an impact on stress responses in human adulthood (Gunnar & Vazquez, 2006). There is also some evidence that unresponsive or insensitive parenting in humans is associated with larger cortisol responses to stress in toddlers (Gunnar, Brodersen, Nachmias, Buss, & Rigatuso, 1996) and enhanced fearfulness in infants, which in turn is associated with more right frontal electroencephalographic (EEG) asymmetry (Hane & Fox, 2006). But much research is still needed to understand how early caregiving regulates the development of the HPA axis and other systems involved in emotion regulation in primates.

Research by Sanchez and colleagues indicates that infant maltreatment in nonhuman primates has effects on HPA axis function that are consistent with chronic stress. Specifically, elevated cortisol levels were detected during the infants' first month of life, the period of most intense physical abuse by the mothers. This was followed by low cortisol levels, particularly in the early morning hours (McCormack et al., 2003)—a finding similar to what has been noted in institutionalized children and children in foster care (see Gunnar & Fisher, 2006). Pharmacological studies performed to analyze pituitary and adrenal function in the animals revealed blunted ACTH responses to CRF administration later in life. These findings reflect a down-regulation of CRF receptors in the pituitary; they are consistent with similar effects of negative/punitive parenting reported in another nonhuman primate, the common marmoset (*Callithrix jacchus*) (Johnson et al., 1996), and with alterations detected in girls with history of childhood abuse (De Bellis et al., 1994). This down-regulation of pituitary CRF receptors could be explained by central CRF overactivity due to sustained emotional and/or physical stress at the early ages in these different species.

Dehydroepiandrosterone and Testosterone

Children exposed to maltreatment frequently exhibit a blunted diurnal rhythm of cortisol (see Gunnar & Vazquez, 2001; Yehuda et al., 2002). The diurnal rhythm of cortisol is established early in development (Lewis & Ramsay, 1995), so it is possible that testosterone and the adrenal androgen dehydroepiandrosterone (DHEA), more so than cortisol, may emerge as important biomarkers of abuse in adolescence and as important factors in psychopathology (Parker, 1999; Salek, Bigos, & Kroboth, 2002).

DHEA enhances learning, memory, and immunocompetence; protects neurons against the toxic effects of cortisol; and reduces anxiety and depression (Wolf & Kirschbaum, 1999). In older adults, DHEA plays a protective role, counteracting negative effects of stress-related hormones like cortisol. In keeping with this view, a recent stem cell study suggests that

DHEA could play a major role in moderating the genesis of new brain cells (Suzuki, Wright, Marwah, Lardy, & Svendsen, 2004). Furthermore, it is possible that a combination of hyper-arousal of cortisol and a blunted rhythm of DHEA may amplify the impact of adverse care on the developing brain and place adolescents at risk for anxiety and depression (Goodyer, Park, Netherton, & Herbert, 2001; Michael, Jenaway, Paykel, & Herbert, 2000).

The hypothalamic–pituitary–gonadal axis, indexed through testosterone in males, inter-acts with the HPA in relation to chronic stress (Dallman et al., 2002; Viau, 2002) and has been linked to individual differences in children's problem behaviors (Granger et al., 2003). Testosterone is an anabolic steroid that exaggerates aggression, competition, dominance, and risk-taking behavior in males (Monaghan & Glickman, 1992), and is suppressed by social stress or high cortisol levels (Dallman et al., 2002), particularly during puberty (Almeida, Anselmo-Franci, Rosa e Silva, & Carvalho, 1998).

Testosterone and DHEA may be particularly important during the pubertal transi-tion, when these androgens change dramatically and behavior problems are likely to emerge (Angold & Worthman, 1993). DHEA and testosterone activity may be affected due to their sensitivity to stress (Angold, 2003; Granger et al., 2003; Mazur & Booth, 1998). Therefore, Pollak et al. (Wismer Fries et al., 2008) have recently begun examining these hormone sys-tems in abused children. Although they are not normally viewed as a stress-reactive hormones like cortisol, we believe that the evidence indicating the stress sensitivity of these hormones and their regulatory role on behaviors of interest supports the inclusion of testosterone and DHEA in studies of maltreated children.

Arginine Vasopressin and Oxytocin

In addition to CRF, other neuroactive peptides, such as arginine vasopressin (AVP) and oxy-tocin (OT), participate in the neuroendocrine, emotional, and autonomic responses to stress (e.g., Nemeroff & Vale, 2005; Sanchez et al., 2001). AVP and OT are neuropeptides that, in addition to their classical roles as neurohypophysial hormones (AVP as an antidiuretic hormone and OT in parturition, lactation, and reproductive behaviors), have emerged as important regulators of stress responses and critical mediators of affiliative behaviors and social recognition/memory (Whitaker-Azmitia, 2005). In the case of OT, animal and human studies have shown that this neuropeptide plays a critical role in mediating affiliative behav-iors (maternal behavior, attachment, and social bonding), and reduces anxiety and HPA axis responses to stress. Early adverse experiences cause persistent decreases in OT neural circuits of animals (Winslow, Noble, Lyons, Sterk, & Insel, 2003). These findings have been recently confirmed in humans as well, as demonstrated by evidence that women with histories of childhood maltreatment had lower cerebrospinal fluid (CSF) levels of OT than controls had (Heim et al., 2006). In addition, CSF OT levels in those women were negatively correlated with severity of maltreatment. This reduced OT activity could have a detrimental effect on affiliative behaviors and stress vulnerability of women with early adverse experiences.

The effects of maltreatment experiences on OT neural circuits have been further con-firmed in human studies in Pollak's lab, as demonstrated by evidence that children who experienced severe early neglect showed lower levels of salivary OT reactivity as compared with controls (Wismer Fries et al., 2005). Such findings are especially relevant, given that severely neglected children tend to have difficulties forming discriminating attachments. For example, a recent meta-analysis suggests that even many months after adoption, internation-ally adopted children and their parents have not caught up with typically reared children and their patents in attachment security (van IJzendoorn & Juffer, 2006). These children may treat strangers with visible displays of affection, while they are unable to establish a sense of security with or to feel protected by their own parents (O'Connor & Zeanah, 2003).

Serotonin

One neural system of relevance to abused children is the amygdala circuitry, implicated in the evaluation of stimulus salience and, therefore, in threat responses. Hariri et al. (2002) used functional MRI (fMRI) to directly explore the neural basis of the apparent relationship between a common allelic variant in the human 5-HTT gene and emotional behavior in adults. Subjects performed a simple perceptual processing task involving the matching of fearful and angry human facial expressions. This task has been effective at consistently engaging the amygdala (Meyer-Lindenberg et al., 2005; Tessitore et al., 2005; Wang, Vijay-raghavan, & Goldman-Rakic, 2004). Consistent with their hypothesis, subjects carrying the less functional 5-HTT short allele exhibited increased amygdala activity in comparison to subjects homozygous for the long allele. This finding suggests that the increased anxiety and fearfulness may reflect the hyperresponsiveness of the amygdala to relevant environmental stimuli (Bertolino et al., 2005; Heinz et al., 2005). Studies (in human and nonhuman animals) suggest that amygdala activation due to chronic stress is associated with changes in HPA axis function (Lopez, Vazguez, & Olson, 2004). A critical question in the early experience and stress literature is whether infancy experiences have organizing effects on the limbic HPA and threat systems, or whether adverse outcomes associated with early maltreatment reflect adversity that continues throughout childhood.

Brain 5-HT systems are involved in the control of mood, sleep, aggression, and locomotor activity—core functions showing dysregulation in children and adolescents with histories of childhood maltreatment. In fact, reduced serotonergic function has been reported in maltreated children (Kaufman et al., 1998). Alterations in brain 5-HT neurotransmission also contribute to different forms of psychopathology, including anxiety and mood disorders (Manji, Drevets, & Charney, 2001), and their related pathophysiological states, such as HPA axis dysregulation.

The brain 5-HT systems play an important role in the regulation of emotionality and stress physiology, and their development is sensitive to alterations in the early environment. In fact, previous studies have demonstrated that early adverse experiences (maternal deprivation and peer rearing) have a negative impact on brain 5-HT function, as reflected by lower CSF levels of the 5-HT metabolite 5-hydroxyindoleacetic acid (5-HIAA) than in controls (e.g., Higley, Suomi, & Linnoila, 1996). Consistent with those reports, studies by Sanchez and colleagues have demonstrated that infant maltreatment also affects the development of brain 5-HT systems in rhesus monkeys. This was reflected by reduced levels of 5-HT and 5-HIAA in CSF of maltreated monkeys, which were in turn correlated with increased anxiety in the maltreated animals (Maestripieri et al., 2006b; Sanchez et al., 2007). Based on the high comorbidity of physical abuse and high maternal rejection in animals included in the "maltreatment group," Sanchez and colleagues looked more closely at the data to determine what best predicted this finding. It was actually the high levels of maternal rejection received by the animals, and not levels of physical abuse, that were strongly associated with the observed reduction in brain 5-HT function.

The reduced CSF levels of 5-HIAA were highly stable across the first 3 years of life and were also associated in females with differences in maternal behavior with the first offspring

(Maestripieri et al., 2006a). Thus females that were physically abused by their mothers as infants and became abusive mothers themselves had lower 5-HIAA CSF levels than abused females that did not perpetuate abuse with their own offspring had. Altogether, these findings indicate the important role of early maternal care in both proper behavioral and neurobiological development of primates. They also open new questions to deepen our understanding of the biological mechanisms underlying developmental psychopathology. Consistent findings are emerging in studies of abused children. For example, maltreated children with the 5-HTT short allele and little social support had high levels of depression; however, maltreated children with the same genotype and similar levels of maltreatment, but with access to social support from other adults, showed minimal depressive symptoms (Kaufman et al., 2004). These findings not only are consistent with research in adults showing that 5-HTT allelic variation moderates the development of depression after stress, but suggest that negative outcomes may be modified by environmental factors that confer risk for psychological disorders.

Immune System

In addition to activation of the HPA axis, stress is associated with activation of innate immune responses, including the release of proinflammatory cytokines and activation of proinflammatory cytokine-signaling cascades (Pace et al., 2006; Raison, Capuron, & Miller, 2006). Relevant to the impact of early life stress on 5-HT metabolism, activation of the p38 mitogen-activated protein kinase (MAPK)-signaling cascade by cytokines including interleukin-1 (IL-1) and tumor necrosis factor-alpha (TNF-alpha) increases the expression and activity of 5-HTT in the brain (Zhu, Carneriro, Dostmann, Hewlett, & Blakely, 2005). This in turn increases 5-HT reuptake, resulting in reduced levels of 5-HT available at the synapse. Based on this effect of proinflammatory cytokines on brain 5-HT systems, a study by Sanchez et al. (2007) recently examined the relationship between activation (phosphorylation) of p38 MAPK in peripheral blood monocytes and central 5-HT function in maltreated juvenile rhesus monkeys. The data showed activation of inflammatory signaling pathways in the maltreated macaques (as reflected by an increased percentage of monocytes staining positive for p38 MAPK). In addition, the activation of inflammatory signaling pathways was associated with levels of maternal rejection received early in life, and with decreased CSF concentrations of 5-HIAA. That is, the higher the maternal rejection, the higher the inflammatory markers; and the higher the inflammatory markers, the lower the CSF 5-HIAA concentrations.

These data provide the first evidence of an *in vivo* relationship between activation of p38 MAPK-signaling pathways in monocytes and reduced brain 5-HT function/metabolism in an animal model of infant maltreatment. By increasing 5-HTT expression/activity, activation of p38 MAPK-signaling pathways would be expected to decrease synaptic availability of 5-HT and to reduce 5-HT metabolites, as was found in this study. Of note, proinflammatory cytokines, including IL-1 and TNF-alpha, are also capable of influencing the activity of the enzyme indolamine 2,3-dioxygenase, which metabolizes tryptophan to kynurenine and quinolinic acid, thereby shunting tryptophan from the synthesis of 5-HT (Raison et al., 2006). Thus, in addition to directly influencing the expression of 5-HTT, proinflammatory cytokines may influence 5-HT metabolism by altering the availability of tryptophan, the primary precursor of 5-HT. Taken together, the data suggest that increased activity in p38 MAPK pathways as a function of infant maltreatment may represent a novel mechanism by which early life stress can be translated into risk for illness. Moreover, p38 MAPK pathways may serve as a unique translational target for reversing the impact of early life stress on relevant pathophysiological endpoints, including anxiety and depression. Similarly, adults who retrospectively recall maltreatment show sustained effects on immunity, such as altered B- and cytotoxic C-cell numbers and inflammatory markers (e.g., C-reactive protein, a pattern consistent with psychological states of physiological arousal and increased autonomic activity) (Danese, Pariante, Caspi, Taylor, & Poulton, 2007). These effects also appear in children (Dorshorst, Shirtcliff, Coe, & Pollak, 2007).

Summary

Research with nonhuman primate models of infant maltreatment is crucial if we are to fully understand the causes, consequences, and underlying biological mechanisms of similar experiences in humans. Longitudinal studies performed under well-controlled experimental conditions will be essential to characterize the developmental time course of biobehavioral and neurobiological alterations. Translational challenges involved in neuroscience-based approaches to understanding the effects of child abuse include the fact that most rodent and nonhuman primate studies have focused on the effects of maternal separation or isolation rearing, which more closely approximate neglect in humans than other forms of maltreatment, such as physical abuse. In addition, most clinical studies of maltreatment involve heterogeneous samples of children with different combinations of experiences. We and our colleagues are trying to address these issues by using the types of paradigms needed to conduct sophisticated and parallel neuroscience studies with children and nonhuman primates, and doing so in ways that can build bridges between the neuroscience and preventive intervention communities of researchers. By conducting preclinical studies of at-risk human children on the one hand, and harnessing the neurophysiological precision available from nonhuman primate studies on the other hand, we hope to clarify issues such as the modulatory role of the prefrontal cortex and infralimbic regions in reactivity to threat, as well as other ways in which experience-dependent fine-tuning of attention, learning, emotion, and memory systems (Black, Jones, Nelson, & Greenough, 1998) affects emotion regulation. The development of this circuitry could certainly be influenced by early maltreatment through overactivity of endocrine systems such as the HPA axis, and may be moderated by heritable characteristics of the child. The study of altered emotion-regulating processes associated with child abuse, together with biological approaches to excavate mechanisms conferring developmental risk, will synthesize key areas in which we desperately need more information to generate new solutions to mental health problems in children and adults. These include, first a focus on the neural circuitry and neurobiology of the brain's regulation of emotion, with an emphasis on understanding adaptations and sequelae of chronic social stress exposure on affective neural circuits. Second, we must focus on the development of these circuits—specifically, the processes underlying periods of rapid neurobiological change in humans during which the brain may be particularly sensitive to contextual or environmental influences. Third, we must focus on defining and specifying ways in which the environment creates long-term effects on brain and behavior, including potential corrective experiences that might foster recovery of competencies and promote health. Each of these foci holds tremendous promise for advancement of knowledge and application to the improvement of public health.

Acknowledgments

We acknowledge support for research discussed in this chapter by the following grants from the National Institute of Mental Health (NIMH), the National Institute of Child Health and Human Development (NICHD), and the National Center for Research Resources (NCRR) of the National Institutes of Health: Nos. MH61285 and MH68858 (to Seth D. Pollak), and Nos. MH65046 and HD055255 (to M. Mar Sanchez). Infrastructure support was provided by Grant Nos. P30-HD03352 (M. Seltzer, Center Director) and RR-00165 (Yerkes National Primate Research Center Base Grant). We also wish to thank members of the NIMH-funded Early Experience, Stress and Prevention Science Network (Grant No. R21 MH65046, Megan R. Gunnar, Principal Investigator) for stimulating discussions on this topic that influenced the views presented in this review, and Mary Schlaak for help in preparation of this chapter.

References

Aarons, G. A., Brown, S. A., Hough, R. L., Garland, A. F., & Wood, P. A. (2001). Prevalence of adolescent substance use disorders across five sectors of care. *Journal of the American Academy of Child and Adolescent Psychiatry, 40*(4), 419–426.

Almeida, S. A., Anselmo-Franci, J. A., Rosa e Silva, A. A., & Carvalho, T. L. (1998). Chronic intermittent immobilization of male rats throughout sexual development: A stress protocol. *Experimental Physiology, 83*(5), 701–704.

Angold, A. (2003). Adolescent depression, cortisol and DHEA. *Psychological Medicine, 33*(4), 573–581.

Angold, A., & Worthman, C. W. (1993). Puberty onset of gender differences in rates of depression: A developmental, epidemiologic and neuroendocrine perspective. *Journal of Affective Disorders, 29*(2–3), 145–158.

Black, J. E., Jones, T. A., Nelson, C. A., & Greenough, W. T. (1998). Neuronal plasticity and the developing brain. In N. E. Alessi, J. T. Coyle, S. I. Harrison, & S. Eth (Eds.), *Handbook of child and adolescent psychiatric science and treatment* (pp. 31–53). New York: Wiley.

Bertolino, A., Arciero, G., Rubino, V., Latorre, V., De Candia, M., Mazzola, V., et al. (2005). Variation of human amygdala response during threatening stimuli as a function of 5'HTTLPR genotype and personality style. *Biological Psychiatry, 57*(12), 1517–1525.

Bremner, J. D., & Vermetten, E. (2001). Stress and development: Behavioral and biological consequences. *Development and Psychopathology, 13*(3), 473–489.

Brent, L., Koban, T., & Ramirez, S. (2002). Abnormal, abusive and stress-related behaviors in baboon mothers. *Biological Psychiatry, 52,* 1047–1056.

Burgess, R., & Conger, R. (1978). Family interaction in abusive, neglectful and normal families. *Child Development, 49,* 1163–1173.

Camras, L. A., Sachs-Alter, E., & Ribordy, S. (1996). Emotion understanding in maltreated children: Recognition of facial expressions and integration with other emotion cues. In M. Lewis & M. Sullivan (Eds.), *Emotional development in atypical children* (pp. 203–225). Hillsdale, NJ: Erlbaum.

Caspi, A., McClay, J., Moffitt, T. E., Mill, J., Martin, J., Craig, I. W., et al. (2002). Role of genotype in the cycle of violence in maltreated children. *Science, 297,* 851–854.

Caspi, A., & Moffitt, T. (2006). Gene–environment interactions in psychiatry: Joining forces with neuroscience. *Nature Reviews Neuroscience, 7,* 583–590.

Caspi, A., Sugden, K., Moffitt, T. E., Taylor, A., Craig, I. W., Harrington, H., et al. (2003). Influence of life stress on depression: Moderation by a polymorphism in the 5-HTT gene. *Science, 301*(5631), 386–389.

Chapman, M. V., Wall, A., & Barth, R. P. (2004). Children's voices: The perceptions of children in foster care. *American Journal of Orthopsychiatry, 74*(3), 293–304.

Cicchetti, D., & Toth, S. L. (1995). A developmental psychopathology perspective on child abuse and neglect. *Journal of the American Academy of Child and Adolescent Psychiatry, 34*(5), 541–565.

Comings, D. E., Muhleman, D., & Gysin, R. (1996). Dopamine D-sub-2 receptor (DRD2) gene and susceptibility to posttraumatic stress disorder: A study and replication. *Biological Psychiatry, 40*(5), 368–372.

Couch, J. L., Hanson, R. F., Saunders, B. E., Kilpatrick, D. G., & Resnick, H. S. (2000). Income, race/ethnicity, and exposure to violence in youth: Results from the National Survey of Adolescents. *Journal of Community Psychology, 28*, 1262–1276.

Cummings, E. M., Vogel, D., Cummings, J. S., & El-Sheikh, M. (1989). Children's responses to expression of anger between adults. *Child Development, 60*(6), 1392–1404.

Dallman, M. F., Viau, V. G., Bhatnagar, S., Gomez, F., Laugero, K., & Bell, M. E. (2002). Corticotropin-releasing factor, corticosteroids, stress, and sugar: Energy balance, the brain, and behavior. In D. Pfaff, A. Arnold, A. M. Etgen, S. Fahrbach, & R. T. Rubin (Eds.), *Hormones, brain, and behavior* (Vol. 1, pp. 571–632). San Diego, CA: Academic Press.

Danese, A., Pariante, C. M., Caspi, A., Taylor, A., & Poulton, R. (2007). Childhood maltreatment predicts adult inflammation in a life-course study. *Proceedings of the National Academy of Sciences USA, 104*, 1319–1324.

De Bellis, M. D. (2005). The psychobiology of neglect. *Child Maltreatment, 10*, 150–172.

De Bellis, M. D., Chrousos, G. P., Dorn, L. D., Burke, L., Helmers, K., Kling, M. A., et al. (1994). Hypothalamic–pituitary–adrenal axis dysregulation in sexually abused girls. *Journal of Clinical Endocrinology and Metabolism, 78*, 249–255.

Dodge, K. A., Bates, J. E., & Pettit, G. S. (1990). Mechanisms in the cycle of violence. *Science, 250*(4988), 1678–1683.

Dodge, K. A., Pettit, G. S., Bates, J. E., & Valente, E. (1995). Social information-processing patterns partially mediate the effect of early physical abuse on later conduct problems. *Journal of Abnormal Psychology, 104*(4), 632–643.

Dorshorst, J. J., Shirtcliff, E. A., Coe, C. L., & Pollak, S. D. (2007). *Reduced containment of herpes simplex virus type 1 after child maltreatment and institutionalization.* Poster presented at the meeting of the International Society for Psychoneuroendocrinology, Madison, WI.

Dubowitz, H., Newton, R., Litrownik, A., Lewis, T., Briggs, E., Thompson, R., et al. (2005). Examination of a conceptual model of child neglect. *Child Maltreatment, 10*, 173–189.

Eisenberg, N., Guthrie, I. K., Fabes, R. A., Reiser, M., Murphy, B. C., Holgren, R., et al. (1997). The relations of regulation and emotionality to resiliency and competent social functioning in elementary school children. *Child Development, 68*(2), 295–311.

Fairbanks, L. A., & McGuire, M. T. (1998). Long-term effects of early mothering behavior on responsiveness to the environment in vervet monkeys. *Developmental Psychobiology, 21*, 711–724.

Famularo, R., Kinscherff, R., & Fenton, T. (1992). Psychiatric diagnoses of maltreated children: Preliminary findings. *Journal of the American Academy of Child and Adolescent Psychiatry, 31*(5), 863–867.

Fonagy, P., Target, M., & Gergely, G. (2000). Attachment and borderline personality disorder: A theory and some evidence. *Psychiatric Clinics of North America, 23*(1), 103–122.

Francis, D. D., Diorio, J., Plotsky, P. M., & Meaney, M. J. (2002). Environmental enrichment reverses the effects of maternal separation on stress reactivity. *Journal of Neuroscience, 22*(18), 7840–7843.

Gaensbauer, T. J. (1982). Regulation of emotional expression in infants from two contrasting caretaking environments. *Journal of the American Academy of Child Psychiatry, 21*(2), 163–170.

Gelenter, J., Southwick, S., Goodson, S., Morgan, A., Nagy, L., & Charney, D. S. (1999). No association between D2 dopamine receptor (DRD2) 'A' system alleles, or 9DRD2 haplotypes, and posttraumatic stress disorder. *Biological Psychiatry, 45*, 620–625.

Goodyer, I. M., Park, R. J., Netherton, C. M., & Herbert, J. (2001). Possible role of cortisol and dehy-droepiandrosterone in human development and psychopathology. *British Journal of Psychiatry, 179*, 243–249.

Granger, D. A., Shirtcliff, E. A., Zahn-Waxler, C., Usher, B., Klimes-Dougan, B., & Hastings, P. (2003). Salivary testosterone diurnal variation and psychopathology in adolescent males and females: Individual differences and developmental effects. *Development and Psychopathology, 15*(2), 431–449.

Gunnar, M. R. (2000). Early adversity and the development of stress reactivity and regulation. In C. A. Nelson (Ed.), *Minnesota Symposia on Child Psychology: Vol. 31. The effects of early adversity on neurobehavioral development* (pp. 163–200). Mahwah, NJ: Erlbaum.

Gunnar, M. R., Brodersen, L., Nachmias, M., Buss, M., & Rigatuso, R. (1996). Stress reactivity and attachment security. *Developmental Psychobiology, 29*, 10–36.

Gunnar, M. R., & Fisher, P. A. (2006). Early Experience, Stress, and Prevention Network: Bringing basic research on early experience and stress neurobiology to bear on preventive interventions for neglected and maltreated children. *Developmental Psychopathology, 18*(3), 651–677.

Gunnar, M. R., & Vazquez, D. M. (2001). Low cortisol and a flattening of expected daytime rhythm: Potential indices of risk in human development. *Development and Psychopathology, 13*(3), 515–538.

Gunnar, M. R., & Vazquez, D. M. (2006). Stress neurobiology and developmental psychopathology. In D. Cicchetti & D. Cohen (Eds.), *Developmental psychopathology: Vol. 2. Developmental neuro-science* (2nd ed., pp. 533–577). Hoboken, NJ: Wiley.

Hane, A. A., & Fox, N. A. (2006). Natural variations in maternal caregiving in human infants. *Psycho-logical Science, 17*(6), 550–556.

Hariri, A. R., Mattay, V. S., Tessitore, A., Kolachana, B., Fera, F., Goldman, D., et al. (2002). Sero-tonin transporter genetic variation and the response of the human amygdala. *Science, 297*, 400–403.

Harlow, H. F., Harlow, M. K., & Suomi, S. J. (1971). From thought to therapy: Lessons from a primate laboratory. *American Scientist, 59*, 538–549.

Heim, C., Ehlert, U., & Hellhammer, D. H. (2000). The potential role of hypocortisolism in the pathophysiology of stress-related bodily disorders. *Psychoneuroendocrinology, 25*, 1–35.

Heim, C., Newport, D. J., Mletzko, T., Miller, A. H., Young, L. J., & Nemeroff, C. B. (2006). Decreased cerebrospinal fluid oxytocin concentrations associated with childhood maltreatment in adult women. *Developmental Psychobiology Abstract, 48*, 603–630.

Heim, C., Plotsky, P. M., & Nemeroff, C. B. (2004). Importance of studying the contributions of early adverse experience to neurobiological findings in depression. *Neuropsychopharmacology, 29*(4), 641–648.

Heinz, A., Braus, D. F., Smolka, M. N., Wrase, J., Puls, I., Hermann, D., et al. (2005). Amygdala–prefrontal coupling depends on a genetic variation of the serotonin transporter. *Nature Neurosci-ence, 8*(1), 20–21.

Herman, J. P., Figueiredo, H., Mueller, N. K., Ulrich-Lai, Y., Ostrander, M., Choi, D., et al. (2003). Central mechanisms of stress integration: Hierarchical circuitry controlling hypothalamo–pituitary–adrenocortical responsiveness. *Frontiers in Neuroendocrinology, 24*, 151–180.

Higley, J. D., Suomi, S. J., & Linnoila, M. (1996). A nonhuman primate model of type II alcoholism?: Part 2. Diminished social competence and excessive aggression correlates with low cerebro-spinal fluid 5-hydroxyindoleacetic acid concentrations. *Alcoholism: Clinical and Experimental Research, 20*, 643–650.

Hinde, R. A. (1974). Mother–infant relations in rhesus monkeys. In N. F. White (Ed.), *Ethology and psychiatry* (pp. 29–46). Toronto: University of Toronto Press.

Hussey, J. M., Chang, J., & Kotch, J. B. (2005). Child maltreatment in the United States: Prevalence, risk factors, and adolescent health consequences. *Pediatrics, 118*, 933–942.

Jaffee, S. R., Caspi, A., Moffitt, T. E., Dodge, K. A., Rutter, M., Taylor, A., et al. (2005). Nature vs. nurture: Genetic vulnerabilities interact with physical maltreatment to promote conduct problems. *Development and Psychopathology, 17*(1), 67–84.

Johnson, E. O., Kamilaris, T. C., Calogero, A. E., Gold, P. W., & Chrousos, G. P. (1996). Effects of early parenting on growth and development in a small primate. *Pediatric Research, 39,* 999–1005.

Jonson-Reid, M., & Barth, R. P. (2000). From maltreatment report to juvenile incarceration: The role of child welfare services. *Child Abuse and Neglect, 24*(4), 505–520.

Kaplow, J. B., & Widom, C. S. (2007). Age of onset of child maltreatment predicts long-term mental health outcomes. *Journal of Abnormal Psychology, 116*(1), 176–187.

Kaufman, J. (1991). Depressive disorders in maltreated children. *Journal of the American Academy of Child and Adolescent Psychiatry, 30*(2), 257–265.

Kaufman, J., Birmaher, B., Perel, J., Dahl, R. E., Stull, S., Brent, D., et al. (1998). Serotonergic functioning in depressed abused children: Clinical and familial correlates. *Biological Psychiatry, 44,* 973–981.

Kaufman, J., & Cicchetti, D. (1989). Effects of maltreatment on school-age children's socioemotional development: Assessments in a day-camp setting. *Developmental Psychology, 25*(4), 516–524.

Kaufman, J., Yang, B.-Z., Douglas-Palumberi, H., Houshyar, S., Lipschitz, D., Krystal, J. H., et al. (2004). Social supports and serotonin transporter gene moderate depression in maltreated children. *Proceedings of the National Academy of Sciences USA, 101,* 17316–17321.

Kendler, K. S., Bulik, C. M., Silberg, J., Hettema, J. M., Myers, J., & Prescott, C. A. (2000). Childhood sexual abuse and adult psychiatric and substance use disorders in women. *Archives of General Psychiatry, 57,* 953–959.

Kilpatrick, D. G., Ruggiero, K. J., Acierno, R., Saunders, B. E., Resnick, H. S., & Best, C. L. (2003). Violence and risk of PTSD, major depression, substance abuse/dependence, and comorbidity: Results from the National Survey of Adolescents. *Journal of Consulting and Clinical Psychology, 71*(4), 692–700.

Kim-Cohen, J., Caspi, A., Taylor, A., Williams, B., Newcombe, R., Craig, I. W., et al. (2006). MAOA, early adversity, and gene–environment interaction predicting children's mental health: New evidence and a meta-analysis. *Molecular Psychiatry, 11,* 903–913.

Kliewer, W., Lepore, S., Oskin, D., & Johnson, P. (1998). The role of social and cognitive processes in children's adjustment to community violence. *Journal of Consulting and Clinical Psychology, 66,* 199–209.

Klimes-Dougan, B., & Kistner, J. (1990). Physically abused preschoolers' responses to peers' distress. *Developmental Psychology, 26*(4), 599–602.

Levine, S. (1994). The ontogeny of the hypothalamic–pituitary–adrenal axis: The influence of maternal factors. *Annals of the New York Academy of Sciences, 746,* 275–288.

Lewis, M., & Ramsay, D. S. (1995). Developmental change in infants' responses to stress. *Child Development, 66*(3), 657–670.

Lopez, N. L., Vazquez, D. M., & Olson, S. L. (2004). An integrative approach to the neurophysiological substrates of social withdrawal and aggression. *Development and Psychopathology, 16*(1), 69–93.

Maestripieri, D. (1999). The biology of human parenting: Insights from nonhuman primates. *Neuroscience and Biobehavioral Reviews, 23,* 411–422.

Maestripieri, D. (2005). Early experience affects the intergenerational transmission of infant abuse in rhesus monkeys. *Proceedings of the National Academy of Sciences USA, 102,* 9726–9729.

Maestripieri, D., & Carroll, K. A. (1998a). Risk factors for infant abuse and neglect in group-living rhesus monkeys. *Psychological Science, 9,* 143–145.

Maestripieri, D., & Carroll, K. A. (1998b). Child abuse and neglect: Usefulness of the animal data. *Psychological Bulletin, 123,* 211–223.

Maestripieri, D., & Carroll, K. A. (2000). Causes and consequences of infant abuse and neglect in monkeys. *Aggression and Violent Behavior, 5,* 245–254.

Maestripieri, D., Lindell, S. G., Higley, J. D., Newman, T. K., McCormack, K. M., & Sanchez, M. M.

(2006a). Early maternal rejection affects the development of monoaminergic systems and adult parenting in rhesus macaques. *Behavioral Neuroscience, 120,* 1017–1024.

Maestripieri, D., McCormack, K. M., Higley, J. D., Lindell, S. G., & Sanchez, M. M. (2006b). Influence of parenting style on the offspring's behavior and CSF monoamine metabolites in cross-fostered and non-cross-fostered rhesus macaques. *Behavioural Brain Research, 175,* 90–95.

Main, M., & George, C. (1985). Responses of abused and disadvantaged toddlers to distress in agemates: A study in the day care setting. *Developmental Psychology, 21*(3), 407–412.

Manji, H. K., Drevets, W. C., & Charney, D. S. (2001). The cellular neurobiology of depression. *Nature Medicine, 7,* 541–547.

Manly, J. T., Cicchetti, D., & Barnett, D. (1994). The impact of subtype, frequency, chronicity, and severity of child maltreatment on social competence and behavior problems. *Developmental Psychopathology, 6,* 121–143.

Manly, J. T., Kim, J. E., Rogosch, F. A., & Cicchetti, D. (2001). Dimensions of child maltreatment and children's adjustment: Contributions of developmental timing and subtype. *Development and Psychopathology, 13*(4), 759–782.

Mathew, S. J., Shungu, D. C., Mao, X., Smith, E. L. P., Perera, G. M., Kegeles, L. S., et al. (2003). A magnetic resonance spectroscopic imaging study of adult nonhuman primates exposed to early-life stressors. *Biological Psychiatry, 54,* 727–735.

Mazur, A., & Booth, A. (1998). Testosterone and dominance in men. *Behavioral and Brain Sciences, 21*(3), 353–363 (discussion, 363–397).

McCormack, K. M., Grand, A., LaPrairie, J., Fulks, R., Graff, A., Maestripieri, D., et al. (2003). Behavioral and neuroendocrine outcomes of infant maltreatment in rhesus monkeys: The first four months. *Society for Neuroscience Abstracts,* Abstract Viewer No. 641.14.

McCormack, K. M., Sanchez, M. M., Bardi, M., & Maestripieri, D. (2006). Maternal care patterns and behavioral development of rhesus macaque abused infants in the first 6 months of life. *Developmental Psychobiology, 48,* 537–550.

McEwen, B. S. (1998). Protective and damaging effects of stress mediators. *New England Journal of Medicine, 338,* 171–179.

Meaney, M. J., & Szyf, M. (2005). Maternal care as a model for experience-dependent chromatin plasticity. *Trends in Neurosciences, 28,* 456–463.

Meyer-Lindenberg, A., Hariri, A. R., Munoz, K. E., Mervis, C. B., Mattay, V. S., Morris, C. A., et al. (2005). Neural correlates of genetically abnormal social cognition in Williams syndrome. *Nature Neuroscience, 8*(8), 991–993.

Michael, A., Jenaway, A., Paykel, E. S., & Herbert, J. (2000). Altered salivary dehydroepiandrosterone levels in major depression. *Biological Psychiatry, 48,* 989–995.

Monaghan, E., & Glickman, S. (1992). Hormones and aggressive behavior. In J. Becker, M. Breedlove, & D. Crews (Eds.), *Behavioral endocrinology* (pp. 261–286). Cambridge, MA: MIT Press.

Mulvihill, D. (2005). The health impact of childhood trauma: An interdisciplinary review. *Issues in Comprehensive Pediatric Nursing, 28,* 115–136.

Nemeroff, C. B., & Vale, W. W. (2005). The neurobiology of depression: Inroads to treatment and new drug discovery. *Journal of Clinical Psychiatry, 66*(Suppl. 7), 5–13.

O'Connor, T. G., & Zeanah, C. H. (2003). Attachment disorders: Assessment strategies and treatment approaches. *Attachment and Human Development, 5*(3), 223–253.

Osofsky, J. D., Wewers, S., Hann, D. M., & Fick, A. C. (1993). Chronic community violence: What is happening to our children? *Psychiatry, 56,* 36–45.

Pace, T. W. W., Mletzko, T. C., Alagbe, O., Musselman, D. L., Nemeroff, C. B., Miller, A. H., et al. (2006). Increased stress-induced plasma IL-6 levels and mononuclear cell NF-kB activation in patients with major depression and increased early life stress. *American Journal of Psychiatry, 163,* 1630–1633.

Parker, C. R., Jr. (1999). Dehydroepiandrosterone and dehydroepiandrosterone sulfate production in the human adrenal during development and aging. *Steroids, 64*(9), 640–647.

Parker, J. G., Rubin, K. H., Price, J. M., & DeRosier, M. E. (1995). Peer relationships, child develop-
    ment, and adjustment: A developmental psychopathology perspective. In D. Cicchetti & D. Cohen
    (Eds.), *Developmental psychopathology: Vol. 2. Risk, disorder, and adaptation* (pp. 96–161).
    New York: Wiley.
Perlman, S. B., Kalish, C. W., & Pollak, S. D. (2008). The role of maltreatment experience in children's
    understanding of the antecedents of emotion. *Cognition and Emotion, 22*(4), 651–670.
Pollak, S. D. (2003). Experience-dependent affective learning and risk for psychopathology in chil-
    dren. *Annals of the New York Academy of Sciences, 1008,* 102–111.
Pollak, S. D. (2005). Early adversity and mechanisms of plasticity: Integrating affective neuroscience
    with developmental approaches to psychopathology. *Development and Psychopathology, 17*(3),
    735–752.
Pollak, S. D. (2008). Mechanisms linking early experience and the emergence of emotions: Illustra-
    tions from the study of maltreated children. *Current Directions in Psychological Science, 17,*
    370–375.
Pollak, S. D., Cicchetti, D., Hornung, K., & Reed, A. (2000). Recognizing emotion in faces: Develop-
    mental effects of child abuse and neglect. *Developmental Psychology, 36*(5), 679–688.
Pollak, S. D., Cicchetti, D., Klorman, R., & Brumaghim, J. (1997). Cognitive brain event-related
    potentials and emotion processing in maltreated children. *Child Development, 68,* 773–787.
Pollak, S. D., & Kistler, D. J. (2002). Early experience is associated with the development of categori-
    cal representations for facial expressions of emotion. *Proceedings of the National Academy of
    Sciences USA, 99*(13), 9072–9076.
Pollak, S. D., Klorman, R., Brumaghim, J., & Cicchetti, D. (2001). P3b reflects maltreated children's
    reactions to facial displays of emotion. *Psychophysiology, 38,* 267–274.
Pollak, S. D., Messner, M., Kistler, D. J., & Cohn, J. F. (2009). Development of perceptual expertise in
    emotion regulation. *Cognition, 110,* 242–247.
Pollak, S. D., & Sinha, P. (2002). Effects of early experience on children's recognition of facial displays
    of emotion. *Developmental Psychology, 38,* 784–791.
Pollak, S. D., & Tolley-Schell, S. A. (2003). Selective attention to facial emotion in physically abused
    children. *Journal of Abnormal Psychology, 112*(3), 323–338.
Pollak, S. D., Vardi, S., Bechner, A. M. P., & Curtin, J. J. (2005). Physically abused children's regula-
    tion of attention in response to hostility. *Child Development, 76*(5), 968–977.
Posner, M. I., & Rothbart, M. K. (1998). Attention, self-regulation and consciousness. *Philosophical
    Transactions of the Royal Society of London, Series B, Biological Sciences, 353,* 1915–1927.
Prasad, M. R., Kramer, L. A., & Ewing-Cobbs, L. (2005). Cognitive and neuroimaging findings in
    physically abused preschoolers. *Archives of Disease in Childhood, 90,* 82–85.
Price, J. M., & Glad, K. (2003). Hostile attributional tendencies in maltreated children. *Journal of
    Abnormal Child Psychology, 31*(3), 329–343.
Raison, C. L., Capuron, C., & Miller, A. H. (2006). Cytokines sing the blues: Inflammation and the
    pathogenesis of depression. *Trends in Immunology, 227,* 24–31.
Raver, C. C. (2003). Does work pay psychologically as well as economically?: The role of maternal
    employment in predicting depressive symptoms and parenting among low-income families.
    Child Development, 74(6), 1720–1736.
Rieder, C., & Cicchetti, D. (1989). Organizational perspective on cognitive control functioning and
    cognitive–affective balance in maltreated children. *Developmental Psychology, 25,* 382–393.
Rogosch, F. A., & Cicchetti, D. (1994). Illustrating the interface of family and peer relations through
    the study of child maltreatment. *Social Development, 3*(3), 291–308.
Rogosch, F. A., Cicchetti, D., & Aber, J. L. (1995). The role of child maltreatment in early deviations
    in cognitive and affective processing abilities and later peer relationship problems. *Development
    and Psychopathology, 7*(4), 591–609.
Rosen, J. B., & Schulkin, J. (1998). From normal fear to pathological anxiety. *Psychological Review,
    105*(2), 325–350.

Sabol, S., Hu, S., & Hamer, D. (1998). A functional polymorphism in the monoamine oxidase A gene promoter. *Human Genetics, 103*, 273–279.

Salek, F. S., Bigos, K. L., & Kroboth, P. D. (2002). The influence of hormones and pharmaceutical agents on DHEA and DHEA-S concentrations: A review of clinical studies. *Journal of Clinical Pharmacology, 42*(3), 247–266.

Sanchez, M. M. (2006). The impact of early adverse care on HPA axis development: Nonhuman primate models. *Hormones and Behavior, 50*, 623–631.

Sanchez, M. M., Alagbe, O., Felger, J. C., Zhang, J., Graff, A. E., Grand, A. P., et al. (2007). Activated p38 MAPK is associated with decreased CSF 5-HIAA and increased maternal rejection during infancy in young adult rhesus monkeys. *Molecular Psychiatry, 12*(10), 895–897.

Sanchez, M. M., Hearn, E. F., Do, D., Rilling, J. K., & Herndon, J. G. (1998). Differential rearing affects corpus callosum size and cognitive function of rhesus monkeys. *Brain Research, 812*, 38–49.

Sanchez, M. M., Ladd, C. O., & Plotsky, P. M. (2001). Early adverse experience as a developmental risk factor for later psychopathology: Evidence from rodent and primate models. *Developmental Psychopathology, 13*, 419–449.

Sanchez, M. M., Smith, L. G., & Winslow, J. T. (2003). Alterations in corticotropin-releasing factor (CRF) and vasopressin systems in rhesus monkeys with early adverse experience. *Society for Neuroscience Abstracts*, Abstract Viewer No. 756.15.

Sapolsky, R. M. (1997). *The trouble with testosterone: And other essays on the biology of the human predicament*. New York: Simon & Schuster.

Selye, H. (1976). *The stress of life* (2nd ed.). New York: McGraw-Hill.

Shackman, J. E., Shackman, A. J., & Pollak, S. D. (2007). Physical abuse amplifies attention to threat and increases anxiety in children. *Emotion, 7*(4), 838–852.

Shih, J., Chen, K., & Ridd, M. (1999). Monoamine oxidase: From genes to behavior. *Annual Review Neuroscience, 22*, 197–217.

Straus, M. A., & Gelles, R. J. (1990). *Physical violence in the American families: Risk factors and adaptations in violence in 8,145 families*. New Brunswick, NJ: Transaction.

Suomi, S. J. (1997). Early determinants of behavior: Evidence from primate studies. *British Medical Bulletin, 53*, 170–184.

Suomi, S. J. (2005). Mother–infant attachment, peer relationships, and the development of social networks in rhesus monkeys. *Human Development, 48*, 67–79.

Suzuki, M., Wright, L. S., Marwah, P., Lardy, H. A., & Svendsen, C. N. (2004). Mitotic and neurogenic effects of dehydroepiandrosterone (DHEA) on human neural stem cell cultures derived from the fetal cortex. *Proceedings of the National Academy of Sciences USA, 101*(9), 3202–3207.

Tessitore, A., Hariri, A. R., Fera, F., Smith, W. G., Das, S., Weinberger, D. R., et al. (2005). Functional changes in the activity of brain regions underlying emotion processing in the elderly. *Psychiatry Research, 139*(1), 9–18.

Toth, S. L., Manly, J. T., & Cicchetti, D. (1992). Child maltreatment and vulnerability to depression. *Development and Psychopathology, 4*(1), 97–112.

Trickett, P. K., Aber, J. L., Carlson, V., & Cicchetti, D. (1991). Relationship of socioeconomic status to the etiology and developmental sequelae of physical child abuse. *Developmental Psychology, 27*(1), 148–158.

Troisi, A., & D'Amato, F. R. (1984). Ambivalence in monkey mothering: Infant abuse combined with maternal possessiveness. *Journal of Nervous and Mental Disease, 172*, 105–108.

Tupler, L. A., & De Bellis, M. D. (2006). Segmented hippocampal volume in children and adolescents with posttraumatic stress disorder. *Biological Psychiatry, 59*, 523–529.

van IJzendoorn, H. W., & Juffer, F. (2006). The Emanuel Miller Memorial Lecture 2006: Adoption as intervention: Meta-analytic evidence for massive catch-up and plasticity in physical, socioemotional, and cognitive development. *Journal of Child Psychology and Psychiatry, 27*, 1228–1245.

Viau, V. (2002). Functional cross-talk between the hypothalamic–pituitary–gonadal and adrenal axes. *Journal of Neuroendocrinology, 14*(6), 506–513.

Wang, M., Vijayraghavan, S., & Goldman-Rakic, P. S. (2004). Selective D2 receptor actions on the functional circuitry of working memory. *Science, 303*(5659), 853–856.

Weiss, B., Dodge, K. A., Bates, J. E., & Pettit, G. S. (1992). Some consequences of early harsh discipline: Child aggression and a maladaptive social information processing style. *Child Development, 63*(6), 1321–1335.

Whitaker-Azmitia, P. M. (2005). Behavioral and cellular consequences of increasing serotonergic activity during brain development: A role in autism? *International Journal of Developmental Neuroscience, 23*(1), 75–83.

Widom, C. S., & Brzustowicz, L. (2006). MAOA and the "cycle of violence": Childhood abuse and neglect, MAOA genotype, and risk for violent and antisocial behavior. *Biological Psychiatry, 60*, 684–689.

Winslow, J. T., Noble, P. L., Lyons, C. K., Sterk, S. M., & Insel, T. R. (2003). Rearing effects on cerebrospinal fluid oxytocin concentration and social buffering in rhesus monkeys. *Neuropsychopharmacology, 5*, 910–918.

Wismer Fries, A. B., & Pollak, S. D. (2004). Emotion understanding in post-institutionalized Eastern European children. *Development and Psychopathology, 16*, 1355–1369.

Wismer Fries, A. B., Shirtcliff, E. A., & Pollak, S. D. (2008). Neuroendocrine dysregulation following early social deprivation in children. *Developmental Psychobiology, 50*(6), 588–599.

Wismer Fries, A. B., Zigler, T., Kurian, J., Jacoris, S., & Pollak, S. D. (2005). Early experience in humans is associated with changes in neuropeptides critical for regulating social behaviour. *Proceedings of the National Academy of Sciences USA, 102*, 17237–17240.

Wolf, O. T., & Kirschbaum, C. (1999). Actions of dehydroepiandrosterone and its sulfate in the central nervous system: Effects on cognition and emotion in animals and humans. *Brain Research: Brain Research Reviews, 30*(3), 264–288.

Yehuda, R., Halligan, S. L., & Bierer, L. M. (2002). Cortisol levels in adult offspring of Holocaust survivors: Relation to PTSD symptom severity in the parent and child. *Psychoneuroendocrinology, 27*(1–2), 171–180.

Yehuda, R., & McFarlane, A. C. (1995). Conflict between current knowledge about posttraumatic stress disorder and its original conceptual basis. *American Journal of Psychiatry, 152*(12), 1705–1713.

Zhu, C. B., Carneriro, A. M., Dostmann, W. R., Hewlett, W. A., & Blakely, R. D. (2005). p38 MAPK activation elevates serotonin transport activity via trafficking-independent protein phosphatase 2A-dependent process. *Journal of Biological Chemistry, 280*, 15649–15658.

Author Index

521

Subject Index

"f" following a page number indicates a figure; "t" following a page number indicates a table.

Abuse
  addiction and, 363
  in adulthood, 360, 363
  defining, 499–500
  mental health and, 501–504
  neural mechanisms of risk and, 504–512
  psychopathology and, 421–422
  reward system and, 358–359
  socioemotional development following, 497–513
  studying, 498–501
Accessory basal nuclei, 48
Acetylocholine. *see also* Neurotransmitters
  studying brain structure and, 20–23
  sympathetic nervous system and, 24
Active intermodal mapping (AIM)
  imitation and, 212–214, 213f
  overview, 210
  sharing body experiences and, 215, 216f
Addiction, 360–363. *see also* Substance use
Adenine (A), 460–461
Adolescence
  abuse and, 359
  addiction and, 361
  brain development during, 164–167, 165f
  decision making and, 378–392, 380f, 388f

depression and, 399–413, 400f, 405f
  emotion perception and, 116
  mentalizing and, 159–170
  reward system and, 329–333, 332f, 334–335
  social-cognitive development during, 167–170, 168f
Adrenocorticotropic hormone (ACTH)
  child maltreatment and, 507, 508
  reward system and, 348
Adult Attachment Interview (AAI)
  maternal responsiveness and, 230
  maternal sensitivity and, 237
Adult personality, 308–309
Adult social bonds. *see also* Bonding
  abuse and, 360
  developmental manipulations of OT and AVP and, 254–256
  hormones and, 256
  neurochemical activity and, 251–252
  reward system and, 342–363
Adult-directed speech (ADS), 184–187, 185f
Affect, relationship of with temperament, 308–309
Affect discrimination, 113–114. *see also* Emotion perception
Affect vulnerability, temperament and, 305–317
Affective node, adolescence development and, 170

Affiliative behaviors
  oxytocin and, 509–510
  temporopolar cortex and, 47–48
Age groups, as an index of psychosocial development, 389–391
Agency
  empathy and, 144–145
  imitation and, 143
  self–other awareness and, 148–150
Aggression
  abuse and, 359
  aggression and, 420–421
  child maltreatment and, 501–502, 503–504, 509
  empathy and perspective taking and, 153
  origins of, 421–422
  testosterone and, 270–271, 509
Agnosia, 73–74
Agonists, 22–23
Alcohol use. *see* Substance use
Amygdala
  adolescence and, 170, 405f, 412
  anterior cingulate cortex and, 48
  attraction and, 267–268, 267f, 274–275
  autism spectrum disorder and, 442–443, 444–445
  caregiving and, 233, 234f, 235, 239f, 354–355
  child maltreatment and, 508, 510

543